MODEL THEORY

MODEL THEORY

THIRD EDITION

C. C. CHANG
University of California, Los Angeles

H. JEROME KEISLER
University of Wisconsin, Madison

DOVER PUBLICATIONS, INC.
MINEOLA, NEW YORK

Copyright

Bibliographical Note

This Dover edition, first published in 2012, is an unabridged republication of the third edition of the work, originally published as Volume 73 in the series "Studies in Logic and the Foundations of Mathematics" by North-Holland Press, Amsterdam and New York, in 1990. [first edition: 1973]. A new Preface to the Dover Edition has been specially prepared for this edition by H. Jerome Keisler.

Library of Congress Cataloging-in-Publication Data

Chang, Chen Chung, 1927–
Model theory / C.C. Chang, H. Jerome Keisler. — Dover ed.
p. cm.
Originally published: 3rd ed. Amsterdam ; New York : North-Holland Press, 1990, in series: Studies in logic and the foundations of mathematics ; v. 73.
Includes bibliographical references and index.
ISBN-13: 978-0-486-48821-9
ISBN-10: 0-486-48821-7
1. Model theory. I. Keisler, H. Jerome. II. Title.

QA9.7.C45 2012
511.3' 4—dc23

2011041607

Manufactured in the United States by RR Donnelley
48821704 2015
www.doverpublications.com

Preface to the Dover Edition

Nearly four decades have gone by since the first edition of *Model Theory* was published in 1973. The subject of model theory has seen tremendous advances, and has grown from a single tightly focused field to a collection of thriving areas of research with varying objectives. During that time, generations of researchers have learned model theory from this book. It has been widely recognized as the classic text in the field, and as the place to learn the basic prerequisites that are needed for all parts of model theory.

This volume is a reprinting of the Third Edition of *Model Theory* (1990). It contains extensive additions to the First Edition. We are pleased that this book will again be available in print and accessible to students.

Here is a brief overview of some developments in model theory that took place after the publication of the Third Edition. Some recent references are given in footnotes.

In first order model theory, the material presented in Section 7.1 has blossomed into the spectacularly successful program called classification theory. It has revealed many structural properties of stable theories, and has led to striking applications in algebraic geometry and group theory.[1] This program has been extended beyond stable theories to simple theories and to theories without the independence property.[2]

Another highly successful program in first order model theory is the study of o-minimal structures, which was inspired by Tarski's proof of quantifier elimination for the real numbers.[3]

Much of first order model theory has been extended to model theory for infinitary logics and logics with cardinality quantifiers.[4]

---

[1]See Pillay, "Geometric Stability Theory" (Clarendon 1996), and Bouscaren, "Model Theory and Algebraic Geometry" (Springer 1999).

[2]See Wagner, "Simple Theories" (Kluwer 2000), and Adler, "Introduction to Theories without the Independence Property", to appear in Archive for Mathematical Logic.

[3]See van den Dries, "Tame Topology and o-minimal Structures" (Cambridge 1998).

[4]See Baldwin, "Categoricity" (AMS 2009).

Computable model theory continues to be an active area of research which combines methods from first order model theory and computability theory.[5]

In recent years, continuous model theory has emerged as an area of research that applies many valued logic to structures in analysis in the same way that first order logic is applied to structures in algebra.[6]

Nonstandard analysis and the ultrapower construction are two closely related model-theoretic methods that continue to be applied to analysis and topology.[7]

Finite model theory is a flourishing field of research with applications to computer science and complexity theory. Infinitary logics play an important role in this area.[8]

The model theory of Peano arithmetic has important applications in proof theory and reverse mathematics.[9]

Models of set theory are central to the study of large cardinals. [10]

First order model theory has provided a template for the development of model theories for a wide variety of other logics, including modal logic, intuitionistic logic, temporal logic, logics with non-classical quantifiers, and logics arising in theoretical computer science.

The Open Problem list in Appendix B has inspired a great deal of research over the years. In the Third Edition we discussed the state of these problems as of 1989. In this edition, we have added a further update at the end of Appendix B.

University of Wisconsin, Madison, October 2011.

H. J. KEISLER

---

[5]See Ersov et.al., "Handbook of Recursive Mathematics, Vol. 1 Recursive Model Theory" (Elsevier 1998).

[6]For an introduction see the article "Model Theory of Metric Structures" by Ben Yaacov et.al., in Chatzidakis et.al., "Model Theory with Applications to Algebra and Analysis vol. 2" (Cambridge 2008).

[7]See Väth, "Nonstandard Analysis" (Birkhauser 2007), Arkeryd et.al., "Nonstandard Analysis: Theory and Applications" (Kluwer 1996), Diener et.al., "Nonstandard Analysis in Practice" (Springer 1995), and Bergelson et.al., "Ultrafilters across Mathematics" (AMS 2008).

[8]See Libkin, "Elements of Finite Model Theory" (Springer 2004).

[9]See Kaye, "Models of Peano Arithmetic" (Oxford 1991), and Simpson, "Subsystems of Second Order Arithmetic" (Cambridge 2010).

[10]See Kanamori, "The Higher Infinite" (Springer 2008).

PREFACE

Model theory is the branch of mathematical logic which deals with the connection between a formal language and its interpretations, or models. In this book we shall present the model theory of first order predicate logic, which is the simplest language that has applications to the main body of mathematics. Most of the techniques in model theory were originally developed and are still best explained in terms of first order logic.

The early pioneers in the development of model theory were Löwenheim (1915), Skolem (1920), Gödel (1930), Tarski (1931), and Malcev (1936). The subject became a separate branch of mathematical logic with the work of Henkin, Robinson, and Tarski in the late 1940's and early 1950's. Since that time it has been an active area of research.

Looking over the subject as it stands today, we feel that it can best be analyzed on the basis of a few general methods of constructing models. While the methods in their pure form are quite simple, they can be iterated and combined in a great variety of ways to yield practically all the deeper results of the theory. For this reason we have organized the book on the following plan. As a rule, we introduce a method in the first section of a chapter and then give some applications of it in the remaining sections. The basic methods of constructing models are: Constants (Section 2.1), Elementary chains (Section 3.1), Skolem functions (Section 3.3), Indiscernibles (Section 3.3), Ultraproducts (Section 4.1), and Special models (Section 5.1). In the last two chapters, 6 and 7, we present some more advanced topics which combine several of these methods. We believe that this book covers most of first-order model theory and many of its applications to algebra and set theory.

Up to now no book of this sort has been written. This has made it difficult for students and outsiders to learn about large areas of the subject. It has been necessary for them to chase down an almost unlimited number of widely scattered articles, some of which are hard to read. We do not claim to have compiled all the results in first-order model theory, but we have tried to include the important results which are indispensable for further work in this area. In addition we have included some of the more recent results which are stimulating present and probably future research. In this category are the Keisler-Shelah isomorphism theorem, the Morley categoricity theorem, the work of Ax-Kochen and Ershov in field theory, and the results of Rowbottom, Gaifman and Silver on large cardinals and the constructible universe.

First-order model theory is a prerequisite for the other types of model theory and such applications as nonstandard analysis. Other logics whose model theories have been investigated are infinitary logic, logic with additional quantifiers, many-valued logic, many sorted logic, intuitionistic logic, modal logic, second-order logic. In recent years model theory for infinitary logic has made rapid progress. Model theory for second-order logic is largely beyond present methods but has a great deal of potential importance. We hope that the availability of this book will contribute to future research in all kinds of model theory and to the discovery of more applications.

This book grew out of a number of graduate courses in model theory that we have taught at UCLA and Wisconsin. The idea of writing a textbook of this sort arose in 1963 as we were completing our earlier monograph, *Continuous Model Theory*. Some lecture notes by Keisler in 1963–64 were tried out from time to time and the present form of the book gradually evolved from them. The actual writing of the book began in early 1965. In the intervening period as the book took shape it was tested in classes, expanded in scope, and almost completely rewritten during the logic year 1967–68 at UCLA. Major changes were again made in 1971–72.

We owe a debt to the many mathematicians whose work forms the subject of this book. A tribute is due to Alfred Tarski who was the motivating and influencing force in the shaping of the theory. On a more personal level, we both received our Ph. D. degrees under his direction at the University of California, Berkeley. Space does not permit us to list the names of all the colleagues and students who at various times have read or used our manuscripts and have made many constructive suggestions and criticisms.

For the amusement of all those who gave us help, we dedicate our book to all model theorists who have never dedicated a book to themselves.

We have been supported during the writing by the Departments of Mathematics at the University of California, Los Angeles, and the University of Wisconsin, Madison, by the National Science Foundation under several research grants, and by a Fullbright grant to Chang in 1966–67 and a Sloan Fellowship to Keisler in 1966–67, 1968–69.

Invaluable assistance in the proof reading and preparation of the manuscript was rendered by Jerry Gold. Perry Smith has spent many hours helping us with the page proofs. We are grateful to Sister Kathleen Sullivan for preparing the index. We wish to thank Mrs. Gerry Formanack for her excellent typing of the manuscript.

University of California, Los Angeles — C. C. CHANG
University of Wisconsin, Madison — H. J. KEISLER

April 1973

PREFACE TO THE SECOND EDITION

The field of model theory has developed rapidly since the publication of the first edition of this book in 1973. There is an up-to-date survey of the subject and extensive references to the literature in the *Handbook of Mathematical Logic*.

Only minor changes have been made from the first edition of this book. We have added a few pages at the end of Appendix B discussing the current state of the list of open problems. A few of the problems have been completely solved, and partial results have been obtained on several others.

Throughout the book, errors, misprints, and ambiguities have been corrected. We are grateful to the many colleagues who have pointed out errors and offered suggestions. We especially wish to thank S. C. Kleene, who suggested over one hundred corrections after teaching a course from the book this spring.

University of California, Los Angeles — C. C. CHANG
University of Wisconsin, Madison — H. J. KEISLER

September 1976

PREFACE TO THE THIRD EDITION

It has been thirteen years since the Second Edition of this book was written, and as one would expect, the subject of model theory has changed radically. Model theory is now dominated by new areas which were in their infancy in 1976 and have blossomed into thriving fields in their own right. Among these fields are classification (or stability) theory, nonstandard analysis, model-theoretic algebra, recursive model theory, abstract model theory, and model theories for a host of nonfirst order logics. Model-theoretic methods have also had a major impact on set theory, recursion theory, and proof theory.

In spite of the changes in the field, this book still serves well as a beginning graduate textbook and reference work. Classical first order model theory as developed here remains a prerequisite for all of the newer branches of model theory, and many newer books have relied on this book for the necessary background.

In preparing this Third Edition, we have been careful to preserve the usefulness of the book as a first textbook in model theory. We have made no attempt to cover the whole field, but have added new topics which now belong in a first graduate course. Four new sections have been added. These sections have been placed at the end of the original chapters to minimize changes in the numbering of results. Throughout the book, new exercises have been added, usually at the end of the original exercise lists. We made a number of updates, improvements, and corrections in the main text, have updated the appendix on the current status of the open problems, and have added a list of additional references.

The new Section 2.4 introduces recursively saturated models, which have led to the simplification of many arguments in model theory by replacing large saturated or special models by countable models. As an illustration, we have replaced the proof of the Vaught two-cardinal theorem in Section 3.2 by a simpler proof using recursively saturated models.

The new Section 2.5 presents Lindström's celebrated characterization of first order logic. This result has gained importance as the launching point for the subject of abstract model theory.

Because of the growing importance of model-theoretic algebra, our treatment of model completeness which had been in Section 3.1 was greatly expanded and moved to the new Section 3.5.

The new Section 4.4 on nonstandard universes was added to provide an interface which is needed to apply results from model theory to nonstandard analysis.

We wish to thank our many colleagues and students who have given us invaluable help and encouragement on this textbook. We have been supported by several National Science Foundation grants and the Vilas Trust Fund at the University of Wisconsin.

Madison, Wisconsin, 1989 H.J. KEISLER

HOW TO USE THIS BOOK AS A TEXT

This book is written at a level appropriate to first year graduate students in mathematics. The only prerequisite is some exposure to elementary logic including the notion of a formal proof. It would be helpful if the student has had undergraduate-level courses in set theory and modern algebra. All the set theory needed for the book is presented in the Appendix which the student can use to fill in any gaps in his knowledge. The first four chapters proceed at a leisurely pace. The last three chapters proceed more rapidly and require more sophistication on the part of the student.

There is ample material for a full-year graduate course in Model Theory, and there is enough flexibility so that a variety of shorter courses can be made up. Chapter 1 contains introductory material and from Chapter 2 on there is at least one interesting theorem in every section.

The core of the subject which must be in any model theory course is composed of Sections 1.1, 1.3, 1.4, 2.1, 3.1, 4.1. The sections which are next in priority are 1.2, 2.2, 3.3, 4.3, 5.1, 6.1.

To help the instructor to make up a course, we give below a table showing the dependence of the sections in the first five chapters. This table applies only to the text itself and not to the exercises, which may depend on any earlier section.

1.4:	1.3	5.1:	3.1
1.5:	1.4	5.2:	5.1
2.1:	1.4	5.3:	5.1
2.2:	2.1	5.4:	5.1
2.3:	2.2	5.5:	5.1
2.4:	2.3	6.1:	4.3, 5.1
2.5:	2.4	6.2:	4.1
3.1:	2.1	6.3:	5.5, 6.2
3.2:	2.4, 3.1	6.4:	4.2, 4.4
3.3:	2.2, 3.1	6.5:	6.4
3.4:	3.2, 3.3	7.1:	2.3, 3.2, 3.3, 5.1
3.5:	2.4, 3.1	7.3:	3.2, 3.3, 4.2
4.1:	3.1	7.4:	7.3
4.2:	4.1		
4.3:	4.1		
4.4:	4.3		

Any hereditary set in the above partial ordering of sections can be used as a course. A very short course could consist of the core Sections 1.1, 1.3, 1.4, 2.1, 3.1, and 4.1. A one quarter course might consist of the above core plus Sections 2.2, 2.3, 2.4, 3.2, and 3.3 or 3.5; this would give a fairly complete picture of countable models. An alternative one quarter course which emphasizes ultraproducts and saturated models would add to the core the Sections 4.3, 5.1, and either 4.2, 4.4, or 6.1. All of Chapters 1 through 4 plus Sections 5.1, 6.1, and 7.1 would make an appropriate one semester course.

The exercises range from extremely easy to impossibly difficult. Exercises of more than routine difficulty are indicated by a single star; a few of the more difficult ones have double stars. Quite often improvements of the basic theorems proved in the text are put in the exercises. Some exercises are major theorems in their own right and we have included them to broaden the coverage. In order to gain an understanding of the field the student should try to do at least a third of the exercises.

At the end of the book we have included a list of unsolved problems in classical model theory. We feel that the solution of any of them would be a substantial contribution and worthy of publication. Not all of the problems originated with us.

We have collected all the historical remarks on the results in the text, the exercises, and the open problems in a separate section entitled Historical

Notes. In all probability there will be some omissions and errors for which we apologize in advance. In many cases students can find suggestions for further study in these notes.

Two final remarks on typography. The word 'iff' is used in all definitions that require it and is to mean 'if and only if'. The end of each proof is indicated by the symbol ⊣, which is meant to suggest the reverse of the common yield sign of first-order logic.

A MAPPING FROM THE SECOND TO THE THIRD EDITION

Here is a table of results and exercises in the Second Edition which have new numbers in the Third Edition. Most of these changes are in the old part of Section 3.1 on model completeness, which has been replaced by the new Section 3.5, and in Section 3.2, which has been simplified using recursively saturated models. E stands for "Exercise".

OLD NUMBER	NEW NUMBER
3.1.7	3.5.1
3.1.8	3.5.3
3.1.9	3.5.11
3.1.10	3.1.7
3.1.11	3.1.8
3.1.12	3.5.8
3.1.13	3.1.9
3.1.14	3.5.10
3.1.15	3.1.10
3.1.16	3.1.11
3.2.7	2.4.3
3.2.8	2.4.2
3.2.9	3.2.8
3.2.10	E3.2.11
3.2.11	3.2.7
3.2.12 – 3.2.16	3.2.9 – 3.2.13

E2.1.25	E2.1.20
E2.1.16 – E2.1.19	E2.1.21 – E2.1.24
E3.1.2	E3.5.2
E3.1.11	E3.5.6
E3.1.12	3.5.9
E3.1.14	E3.5.7
E3.1.16	after 3.5.18
E3.1.17	3.5.18
E3.1.18	E3.5.9
E3.2.5	2.4.3
E3.2.6	E2.4.6
E3.2.7	E2.4.8
E3.2.11	E2.4.7

CONTENTS

Chapter 1. Introduction 1
1.1. What is model theory? 1
1.2. Model theory for sentential logic 4
1.3. Languages, models and satisfaction 18
1.4. Theories and examples of theories 36
1.5. Elimination of quantifiers 49

Chapter 2. Models constructed from constants 61
2.1. Completeness and compactness 61
2.2. Refinements of the method. Omitting types and interpolation theorems 77
2.3. Countable models of complete theories 96
2.4. Recursively saturated models 109
2.5. Lindström's characterization of first order logic . . . 127

Chapter 3. Further model-theoretic constructions 136
3.1. Elementary extensions and elementary chains . . . 136
3.2. Applications of elementary chains 147
3.3. Skolem functions and indiscernibles 163
3.4. Some examples 178
3.5. Model completeness 186

Chapter 4. Ultraproducts 211
4.1. The fundamental theorem 211
4.2. Measurable cardinals 227
4.3. Regular ultrapowers 248
4.4. Nonstandard universes 262

Chapter 5. Saturated and special models 292
5.1. Saturated and special models 292
5.2. Preservation theorems 306
5.3. Applications of special models to the theory of definability . 323
5.4. Applications to field theory 342
5.5. Application to Boolean algebras 372

Chapter 6. More about ultraproducts and generalizations . . . 384
6.1. Ultraproducts which are saturated 384
6.2. Direct products, reduced products, and Horn sentences . 405
6.3. Direct products, reduced products, and Horn sentences (continued) 420
6.4. Limit ultrapowers and complete extensions . . . 447
6.5. Iterated ultrapowers 463

Chapter 7. Selected topics 482
7.1. Categoricity in power 482
7.2. An extension of Ramsey's theorem and applications; some two-cardinal theorems 509
7.3. Models of large cardinality 534
7.4. Large cardinals and the constructible universe . . . 558

Appendix A. Set theory 579

Appendix B. Open problems in classical model theory 597

Historical notes 603

References 623

Additional references 634

Index of definitions 641

Index of symbols 649

CHAPTER 1

INTRODUCTION

1.1. What is model theory?

Model theory is the branch of mathematical logic which deals with the relation between a formal language and its interpretations, or models. We shall concentrate on the model theory of first-order predicate logic, which may be called 'classical model theory'.

Let us now take a short introductory tour of model theory. We begin with the models which are structures of the kind which arise in mathematics. For example, the cyclic group of order 5, the field of rational numbers, and the partially-ordered structure consisting of all sets of integers ordered by inclusion, are models of the kind we consider. At this point we could, if we wish, study our models at once without bringing the formal language into the picture. We then would be in the area known as universal algebra, which deals with homomorphisms, substructures, free structures, direct products, and the like. The line between universal algebra and model theory is sometimes fuzzy; our own usage is explained by the equation

$$\text{universal algebra} + \text{logic} = \text{model theory}.$$

To arrive at model theory, we set up our formal language, the first-order logic with identity. We specify a list of symbols and then give precise rules by which sentences can be built up from the symbols. The reason for setting up a formal language is that we wish to use the sentences to say things about the models. This is accomplished by giving a basic *truth definition*, which specifies for each pair consisting of a sentence and a model one of the truth values *true* or *false*. The truth definition is the bridge connecting the formal language with its interpretation by means of models. If the truth value 'true' goes with the sentence φ and model $\mathfrak{A}$, we say that

φ is *true in* $\mathfrak{A}$ and also that $\mathfrak{A}$ is a *model of* φ. Otherwise we say that φ is *false in* $\mathfrak{A}$ and that $\mathfrak{A}$ is not a model of φ. Moreover, we say that $\mathfrak{A}$ is a *model of a set* Σ of sentences iff $\mathfrak{A}$ is a model of each sentence in the set Σ.

What kinds of theorems are proved in model theory? We can already give a few examples. Perhaps the earliest theorem in model theory is Löwenheim's theorem (Löwenheim, 1915): If a sentence has an infinite model, then it has a countable model. Another classical result is the compactness theorem, due to Gödel (1930) and Malcev (1936): if each finite subset of a set Σ of sentences has a model, then the whole set Σ has a model. As a third example, we may state a more recent result, due to Morley (1965). Let us say that a set Σ of sentences is *categorical in power* α iff there is, up to isomorphism, exactly one model of Σ of power α. Morley's theorem states that, if Σ is categorical in one uncountable power, then Σ is categorical in every uncountable power.

These theorems are typical results of model theory. They say something negative about the 'power of expression' of first-order predicate logic. Thus Löwenheim's theorem shows that no consistent sentence can imply that a model is uncountable. Morley's theorem shows that first-order predicate logic cannot, as far as categoricity is concerned, tell the difference between one uncountable power and another. And the compactness theorem has been used to show that many interesting properties of models cannot be expressed by a set of first-order sentences – for instance, there is no set of sentences whose models are precisely all the finite models.

The three theorems we have stated also say something positive about the existence of models having certain properties. Indeed, in almost all of the deeper theorems in model theory the key to the proof is to construct the right kind of a model. For instance, look again at Löwenheim's theorem. To prove that theorem, we must begin with an uncountable model of a given sentence and construct from it a countable model of the sentence. Likewise, to prove the compactness theorem we must construct a single model in which each sentence of Σ is true. Even Morley's theorem depends vitally on the construction of a model. To prove it we begin with the assumption that Σ has two different models of one uncountable power and construct two different models of every other uncountable power.

There are a small number of extremely important ways in which models have been constructed. For example, for various purposes they can be constructed from individual constants, from functions, from Skolem terms, or from unions of chains. These constructions give the subject of model theory unity. To a large extent, we have organized this book according to these ways of constructing models.

Another point which gives model theory unity is the distinction between *syntax* and *semantics*. Syntax refers to the purely formal structure of the language – for instance, the length of a sentence and the collection of symbols occurring in a sentence, are syntactical properties. Semantics refers to the interpretation, or meaning, of the formal language – the truth or falsity of a sentence in a model is a semantical property. As we shall soon see, much of model theory deals with the interplay of syntactical and semantical ideas.

We now turn to a brief historical sketch. The mathematical world was forced to observe that a theory may have more than one model in the 19th century, when Bolyai and Lobachevsky developed non-Euclidean geometry, and Riemann constructed a model in which the parallel postulate was false but all the other axioms were true. Later in the 19th century, Frege formally developed the predicate logic, and Cantor developed the intuitive set theory in which our models live.

Model theory is a young subject. It was not clearly visible as a separate area of research in mathematics until the early 1950's. However, its historical roots go back to the older subjects of logic, universal algebra, and set theory – and some of the early work, such as Löwenheim's theorem, is now classified as model theory. Other important early developments which contributed to the theory are: the extension of Löwenheim's theorem by Skolem (1920) and Tarski; the completeness theorem of Gödel (1930) and its generalization by Malcev (1936); the characterization of definable sets of real numbers, the rigorous definition of the truth of a sentence in a model, and the study of relational systems by Tarski (1931, 1933, 1935a); the construction of a nonstandard model of number theory by Skolem (1934); and the study of equational classes initiated by Birkhoff (1935). Model theory owes a great deal to general methods which were originally developed for special purposes in older branches of mathematics. We shall come across many instances of this in our book; to mention just one, the important notion of a saturated model (Chapter 5) goes back to the η_α-structures in the theory of simple order, due to Hausdorff (1914). The subject grew rapidly after 1950, stimulated by the papers of Henkin (1949), Tarski (1950), and Robinson (1950). The phrase 'theory of models' is due to Tarski (1954). Today the literature in the subject is quite extensive. There is a rather complete bibliography in Addison, Henkin and Tarski (1965). In recent years, the theory of models has been applied to obtain significant results in other fields, notably set theory, algebra and analysis. However, until now only a tiny part of the potential strength of model theory has been used in such applications. It will be interesting to see what happens when (and if) the full strength is used.

1.2. Model theory for sentential logic

In our introduction, Section 1.1, we gave a general idea of the flavor of model theory, but we were not yet ready to give many details. We shall now come down to earth and give a rigorous treatment of model theory for a very simple formal language, sentential logic (also known as propositional calculus). We shall quickly develop this 'toy' model theory along lines parallel to the much deeper model theory for predicate logic. The basic ideas are the decision procedure via truth tables, due to Post (1921), and Lindenbaum's theorem with the compactness theorem which follows. This section will give a preview of what lies ahead in our book.

We are assuming (see Preface) that the reader is already thoroughly familiar with sentential, and even predicate, logic. Thus we shall feel free to proceed at a fairly rapid pace. Nevertheless, we shall start from scratch, in order to show what sentential logic looks like when it is developed in the spirit of model theory.

Classical sentential logic is designed to study a set $\mathscr{S}$ of simple statements, and the compound statements built up from them. At the most intuitive level, an intended interpretation of these statements is a 'possible world', in which each statement is either true or false. We wish to replace these intuitive interpretations by a collection of precise mathematical objects which we may use as our models. The first thing which comes to mind is a function F which associates with each simple statement S one of the truth values 'true' or 'false'. Stripping away the inessentials, we shall instead take a model to be a subset A of $\mathscr{S}$; the idea is that $S \in A$ indicates that the simple statement S is true, and $S \notin A$ indicates that the simple statement S is false.

1.2.1. By a *model A for $\mathscr{S}$* we simply mean a subset A of $\mathscr{S}$.

Thus the set of all models has the power $2^{|\mathscr{S}|}$. Several relations and operations between models come to mind; for example, $A \subset B$, $\mathscr{S} - A$, and the intersection $\bigcap_{i \in I} A_i$ of a set $\{A_i : i \in I\}$ of models. Two distinguished models are the empty set $\emptyset$ and the set $\mathscr{S}$ itself.

We now set up the sentential logic as a formal language. The symbols of our language are as follows:

connectives $\wedge$ (and), $\neg$ (not);

parentheses), (;

a nonempty set $\mathscr{S}$ of sentence symbols.

Intuitively, the sentence symbols stand for simple statements, and the connectives $\wedge$, $\neg$ stand for the words used to combine simple statements into compound statements. Formally, the *sentences* of $\mathscr{S}$ are defined as follows:

1.2.2.

(i). Every sentence symbol S is a sentence.

(ii). If φ is a sentence then $(\neg\varphi)$ is a sentence.

(iii). If φ, ψ are sentences, then $(\varphi \wedge \psi)$ is a sentence.

(iv). A finite sequence of symbols is a sentence only if it can be shown to be a sentence by a finite number of applications of (i)-(iii).

Our definition of sentence of $\mathscr{S}$ may be restated as a recursive definition based on the length of a finite sequence of symbols:

A single symbol is a sentence iff it is a sentential symbol; a sequence φ of symbols of length $n > 1$ is a sentence iff there are sentences ψ and θ of length less than n such that φ is either $(\neg\psi)$ or $(\psi \wedge \theta)$.

Alternatively, our definition may be restated in set-theoretical terms:

The set of all sentences of $\mathscr{S}$ is the least set Σ of finite sequences of symbols of $\mathscr{S}$ such that each sentence symbol S belongs to Σ and, whenever ψ, θ are in Σ, then $(\neg\psi)$, $(\psi \wedge \theta)$ belong to Σ.

No matter how we may think of sentences, the important thing is that *properties of sentences can only be established through an induction based on* 1.2.2. More precisely, to show that every sentence φ has a given property P, we must establish three things: (1) Every sentence symbol S has the property P; (2) if φ is $(\neg\psi)$ and ψ has the property P, then φ has the property P; (3) if φ is $(\psi \wedge \theta)$ and ψ, θ have the property P, then φ has the property P. (The reader may check his understanding of this point by proving through induction that every sentence φ has the same number of right parentheses as it has left parentheses.)

How many sentences of $\mathscr{S}$ are there? This depends on the number of sentence symbols $S \in \mathscr{S}$. Each sentence is a finite sequence of symbols. If the set $\mathscr{S}$ is finite or countable, then there are countably many sentences of $\mathscr{S}$. Of course, not every finite sequence of symbols is a sentence; for instance, $(S_0 \wedge (\neg S_5))$ is a sentence, but $\wedge \wedge) S_3$ and $S_0 \wedge \neg S_5$ are not. If the set $\mathscr{S}$ of sentence symbols has uncountable cardinal α, then the set of sentences of $\mathscr{S}$ also has power α.

Let us pause briefly to explain the role of the Greek letters φ, ψ, Σ, etc. In the above paragraphs we have used the lower case Greek letters $\varphi, \psi, \theta, \ldots$ as names for arbitrary finite sequences of symbols of $\mathscr{S}$. These letters were needed in order to write down the definition of a sentence. From now on, we shall be much more interested in sentences than in arbitrary finite sequences of symbols. We shall hereafter use the lower case Greek letters $\varphi, \psi, \theta, \ldots$ as names for arbitrary sentences of $\mathscr{S}$. The situation is similar to elementary arithmetic, where we study natural numbers 0, 1, 2, 3, ..., but much of the time we write down letters like $m, n, x, y, \ldots$ as names for arbitrary natural numbers. Just as in arithmetic where we write things like $m = x+y$, we shall now write, for example, $\varphi = (\psi \wedge \theta)$ to express the fact that φ and $(\psi \wedge \theta)$ are the same sentence. In the above paragraphs we also used capital Greek letters $\Sigma, \Gamma, \ldots$ as names for arbitrary sets of finite sequences of symbols of $\mathscr{S}$; hereafter we shall use the capital Greek letters as names for arbitrary sets of sentences of $\mathscr{S}$. The symbols $\varphi, \psi, \theta, \ldots, \Sigma, \Gamma, \ldots$ are *not* in our list of formal symbols of our language – they are merely informal symbols which we use to talk more easily about $\mathscr{S}$.

We shall introduce abbreviations to our language in the usual way, in order to make sentences more readable. The symbols $\vee$ (or), $\rightarrow$ (implies), and $\leftrightarrow$ (if and only if) are abbreviations defined as follows:

$$
\begin{aligned}
&(\varphi \vee \psi) \quad \text{for} \quad (\neg((\neg \varphi) \wedge (\neg \psi))),\\
&(\varphi \rightarrow \psi) \quad \text{for} \quad ((\neg \varphi) \vee \psi),\\
&(\varphi \leftrightarrow \psi) \quad \text{for} \quad ((\varphi \rightarrow \psi) \wedge (\psi \rightarrow \varphi)).
\end{aligned}
$$

Of course, $\vee$, $\rightarrow$ and $\leftrightarrow$ could just as well have been included in our list of symbols as three more connectives. However, there are certain advantages to keeping our list of symbols short. For instance, 1.2.2 and proofs by induction based on it are shorter this way. At the other extreme, we could have managed with only a single connective, whose English translation is 'neither ... nor ...'. We did not do this because 'neither ... nor ...' is a rather unnatural connective.

Another abbreviation which we shall adopt is to leave out unnecessary parentheses. For instance, we shall never bother to write outer parentheses in a sentence – thus $\neg S$ is our abbreviation for the sentence $(\neg S)$. We shall follow the commonly accepted usage in dropping other parentheses. Thus $\neg$ is considered more binding than $\wedge$ and $\vee$, which in turn are more binding than $\rightarrow$ and $\leftrightarrow$. For instance, $\neg\varphi \vee \psi \rightarrow \theta \wedge \varphi$ means $((\neg\varphi)\vee\psi) \rightarrow (\theta \wedge \varphi)$.

Hereafter we shall use the single symbol $\mathscr{S}$ to denote both the set of sentence symbols and the language built on these symbols. There is no fear of confusion in this double usage since the language is determined uniquely, modulo the connectives, by the sentence symbols.

We are now ready to build a bridge between the language $\mathscr{S}$ and its models, with the definition of the truth of a sentence in a model. We shall express the fact that a sentence φ is true in a model A succinctly by the special notation

$$A \models \varphi.$$

The relation $A \models \varphi$ is defined as follows:

1.2.3.

(i). If φ is a sentence symbol S, then $A \models \varphi$ holds if and only if $S \in A$.

(ii). If φ is $\psi \wedge \theta$, then $A \models \varphi$ if and only if both $A \models \psi$ and $A \models \theta$.

(iii). If φ is $\neg\psi$, then $A \models \varphi$ iff it is not the case that $A \models \psi$.

When $A \models \varphi$, we say that φ is *true in* A, or that φ *holds in* A, or that A is a *model of* φ. When it is not the case that $A \models \varphi$, we say that φ is *false in* A, or that φ *fails in* A. The above definition of the relation $A \models \varphi$ is an example of a recursive definition based on 1.2.2. The proof that the definition is unambiguous for each sentence φ is, of course, a proof by induction based on 1.2.2.

An especially important kind of sentence is a *valid sentence*. A sentence φ is called *valid*, in symbols $\models \varphi$, iff φ holds in all models for $\mathscr{S}$, that is, iff $A \models \varphi$ for all A. Some notions closely related to validity are mentioned in the exercises.

At first glance, it seems that for $\mathscr{S}$ infinite we have to examine uncountably many different infinite models A in order to find out whether a sentence φ is valid. This is because validity is a semantical notion, defined in terms of models. However, as the reader surely knows, there is a simple and uniform test by which we can find out in only finitely many steps whether or not a given sentence φ is valid.

This decision procedure for validity is based on a syntactical notion, the notion of a tautology. Let φ be a sentence such that all the sentence symbols which occur in φ are among the $n+1$ symbols $S_0, S_1, \ldots, S_n$. Let $a_0, a_1, \ldots, a_n$ be a sequence made up of the two letters t, f. We shall call such a sequence an *assignment*.

1.2.4. The *value* of a sentence φ for the assignment $a_0, \ldots, a_n$ is defined recursively as follows:

(i). If φ is the sentence symbol S_m, $m \leqslant n$, then the value of φ is a_m.
(ii). If φ is $\neg\psi$, then the value of φ is the opposite of the value of ψ.
(iii). If φ is $\psi \wedge \theta$, then the value of φ is t if the values of ψ and θ are both t, and otherwise the value of φ is f.

Note how similar Definitions 1.2.3 and 1.2.4 are. The only essential difference is that 1.2.3 involves an infinite model A, while 1.2.4 involves only a finite assignment $a_0, \ldots, a_n$.

1.2.5. Let φ be a sentence and let $S_0, \ldots, S_n$ be all the sentence symbols occurring in φ. φ is said to be a *tautology*, in symbols $\vdash \varphi$, iff φ has the value t for every assignment $a_0, \ldots, a_n$.

We shall use both of the symbols $\models$, $\vdash$ in many ways throughout this book. To keep things straight, remember this: $\models$ is used for semantical ideas, and $\vdash$ is used for syntactical ideas.

The value of a sentence φ for an assignment $a_0, \ldots, a_n$ may be very easily computed. We first find the values of the sentence symbols occurring in φ and then work our way through the smaller sentences used in building up the sentence φ. A table showing the value of φ for each possible assignment $a_0, \ldots, a_n$ is called a *truth table* of φ. We shall assume that truth tables are already quite familiar to the reader, and that he knows how to construct a truth table of a sentence. Truth tables provide a simple and purely mechanical procedure to determine whether a sentence φ is a tautology – simply write down the truth table for φ and check to see whether φ has the value t for every assignment.

Proposition 1.2.6. *Suppose that all the sentence symbols occurring in* φ *are among* $S_0, S_1, \ldots, S_n$. *Then the value of* φ *for an assignment* $a_0, a_1, \ldots, a_n, \ldots, a_{n+m}$ *is the same as the value of* φ *for the assignment* $a_0, a_1, \ldots, a_n$.

We now prove the first of a series of theorems which state that a certain syntactical condition is equivalent to a semantical condition.

Theorem 1.2.7 (Completeness Theorem). $\vdash \varphi$ *if and only if* $\models \varphi$; *in words, a sentence is a tautology if and only if it is valid.*

PROOF. Let φ be a sentence and let all the sentence symbols in φ be among $S_0, \ldots, S_n$. Consider an arbitrary model A. For $m = 0, 1, \ldots, n$, put $a_m = \mathrm{t}$ if $S_m \in A$, and $a_m = \mathrm{f}$ if $S_m \notin A$. This gives us an assignment $a_0, a_1, \ldots, a_n$. We claim:

(1) $A \models \varphi$ if and only if the value of φ for the assignment $a_0, a_1, \ldots, a_n$ is t. This can be readily proved by induction. It is immediate if φ is a sentence symbol S_m. Assuming that (1) holds for $\varphi = \psi$ and for $\varphi = \theta$, we see at once that (1) holds for $\varphi = \neg\psi$ and $\varphi = \psi \wedge \theta$.

Now let $S_0, \ldots, S_n$ be all the sentence symbols occurring in φ. If φ is a tautology, then by (1), φ is valid. Since every assignment $a_0, a_1, \ldots, a_n$ can be obtained from some model A, it follows from (1) that, if φ is valid, then φ is a tautology. ⊣

Our decision procedure for $\vdash \varphi$ now can be used to decide whether φ is valid. Several times we shall have an occasion to use the fact that a particular sentence is a tautology, or is valid. We shall never take the trouble actually to give the proof that a sentence of $\mathscr{S}$ is valid, because the proof is always the same – we simply look at the truth table.

Let us now introduce the notion of a formal deduction in our logic $\mathscr{S}$. The *Rule of Detachment* (or *Modus Ponens*) states:

$$\text{From } \psi \text{ and } \psi \to \varphi \text{ infer } \varphi.$$

We say that φ is *inferred from* ψ, θ *by detachment* iff θ is the sentence $\psi \to \varphi$.

Now consider a finite or infinite set Σ of $\mathscr{S}$.

A sentence φ is *deducible from* Σ, in symbols $\Sigma \vdash \varphi$, iff there is a finite sequence $\psi_0, \psi_1, \ldots, \psi_n$ of sentences such that $\varphi = \psi_n$ and each sentence ψ_m is either a tautology, belongs to Σ, or is inferred from two earlier sentences of the sequence by detachment. The sequence $\psi_0, \psi_1, \ldots, \psi_n$ is called a *deduction of* φ from Σ. Note that φ is deducible from the empty set of sentences if and only if φ is a tautology.

We shall say that Σ is *inconsistent* iff we have $\Sigma \vdash \varphi$ for all sentences φ. Otherwise, we say that Σ is *consistent*. Finally, we say that Σ is *maximal consistent* iff Σ is consistent, but the only consistent set of sentences which includes Σ is Σ itself. The proposition below contains facts which can be found in most elementary logic texts.

PROPOSITION 1.2.8.

(i). *If Σ is consistent and Γ is the set of all sentences deducible from Σ, then Γ is consistent.*

(ii). *If Σ is maximal consistent and $\Sigma \vdash \varphi$, then $\varphi \in \Sigma$.*

(iii). *Σ is inconsistent if and only if $\Sigma \vdash S \wedge \neg S$ (for any $S \in \mathscr{S}$).*

(iv) (Deduction Theorem). *If $\Sigma \cup \{\psi\} \vdash \varphi$, then $\Sigma \vdash \psi \rightarrow \varphi$.*

LEMMA 1.2.9 (Lindenbaum's Theorem). *Any consistent set Σ of sentences can be enlarged to a maximal consistent set Γ of sentences.*

PROOF. Let us arrange all the sentences of $\mathscr{S}$ in a list, $\varphi_0, \varphi_1, \varphi_2, \ldots, \varphi_\alpha, \ldots$. The order in which we list them is immaterial, as long as the list associates in a one–one fashion an ordinal number with each sentence. We shall form an increasing chain

$$\Sigma = \Sigma_0 \subset \Sigma_1 \subset \Sigma_2 \subset \ldots \subset \Sigma_\alpha \subset \ldots$$

of consistent sets of sentences. If $\Sigma \cup \{\varphi_0\}$ is consistent, define $\Sigma_1 = \Sigma \cup \{\varphi_0\}$. Otherwise define $\Sigma_1 = \Sigma$. At the αth stage, we define $\Sigma_{\alpha+1} = \Sigma_\alpha \cup \{\varphi_\alpha\}$ if $\Sigma_\alpha \cup \{\varphi_\alpha\}$ is consistent, and otherwise define $\Sigma_{\alpha+1} = \Sigma_\alpha$. At limit ordinals α take unions, $\Sigma_\alpha = \bigcup_{\beta<\alpha} \Sigma_\beta$. Now let Γ be the union of all the sets Σ_α.

We claim that Γ is consistent. Suppose not. Then there is a deduction $\psi_0, \psi_1, \ldots, \psi_p$ of the sentence $S \wedge \neg S$ from Γ (see Proposition 1.2.8). Let $\theta_1, \ldots, \theta_q$ be all the sentences in Γ which are used in this deduction. We may choose α so that all of $\theta_1, \ldots, \theta_q$ belong to Σ_α. But this means that Σ_α is inconsistent (again see Proposition 1.2.8), which is a contradiction.

Having shown that Γ is consistent, we next claim that Γ is maximal consistent. For suppose Δ is consistent and $\Gamma \subset \Delta$. Let $\varphi_\alpha \in \Delta$. Then $\Sigma_\alpha \cup \{\varphi_\alpha\}$ is consistent, and hence $\Sigma_{\alpha+1} = \Sigma_\alpha \cup \{\varphi_\alpha\}$. Thus $\varphi_\alpha \in \Gamma$, and hence $\Delta = \Gamma$. ⊣

LEMMA 1.2.10. *Suppose Γ is a maximal consistent set of sentences in $\mathscr{S}$. Then*:

(i). *For each sentence φ, exactly one of the sentences $\varphi, \neg \varphi$ belongs to Γ.*

(ii). *For each pair of sentences φ, ψ, $\varphi \wedge \psi$ belongs to Γ if and only if both φ and ψ belong to Γ.*

We leave the proof as an exercise.

Now consider a set Σ of sentences of $\mathscr{S}$. We shall say that A is a *model of* Σ, $A \models \Sigma$, iff every sentence $\varphi \in \Sigma$ is true in A. Σ is said to be *satisfiable* iff it has at least one model. We now prove the most important theorem of sentential logic, which is a criterion for a set Σ to be satisfiable.

Theorem 1.2.11 (Extended Completeness Theorem). *A set Σ of sentences of $\mathscr{S}$ is consistent if and only if Σ is satisfiable.*

Proof. Assume first that Σ is satisfiable, and let $A \models \Sigma$. We show that every sentence deducible from Σ holds in A. Let $\psi_0, \psi_1, \ldots, \psi_n$ be a deduction of ψ_n from Σ. Let $m \leqslant n$. If $\psi_m \in \Sigma$ or if ψ_m is a tautology, then ψ_m holds in A. If ψ_m is inferred from two sentences $\psi_p, \psi_p \rightarrow \psi_m$ which hold in A, then ψ_m must clearly hold in A. It follows by induction on m that each of the sentences $\psi_0, \psi_1, \ldots, \psi_n$ holds in A. Since $S \wedge \neg S$ does not hold in A, it is not deducible from Σ, so Σ is consistent.

Now assume that Σ is consistent. By Lindenbaum's theorem we enlarge Σ to a maximal consistent set Γ.

We now construct a model of Σ. Let A be the set of all sentence symbols $S \in \mathscr{S}$ such that $S \in \Gamma$. We show by induction that, for each sentence φ,

(1) $$\varphi \in \Gamma \text{ if and only if } A \models \varphi.$$

By definition, (1) holds when φ is a sentence symbol S_n. Lemma 1.2.10(i) guarantees that, if (1) holds when $\varphi = \psi$, then (1) holds when $\varphi = \neg \psi$. Lemma 1.2.10(ii) guarantees that, if (1) holds when $\varphi = \psi$ and when $\varphi = \theta$, then (1) holds when $\varphi = \psi \wedge \theta$. From (1) it follows that $A \models \Gamma$, and, since $\Sigma \subset \Gamma$, $A \models \Sigma$. ⊣

We can obtain a purely semantical corollary. Σ is said to be *finitely satisfiable* iff every finite subset of Σ is satisfiable.

Corollary 1.2.12 (Compactness Theorem). *If Σ is finitely satisfiable, then Σ is satisfiable.*

Proof. Suppose Σ is not satisfiable. Then by the extended completeness theorem Σ is inconsistent. Hence, $\Sigma \vdash S \wedge \neg S$. In the deduction of the sentence $S \wedge \neg S$ from Σ only a finite set Σ_0 of sentences of Σ is used. It follows that $\Sigma_0 \vdash S \wedge \neg S$, so Σ_0 is inconsistent. Then Σ_0 is not satisfiable, so Σ is not finitely satisfiable. ⊣

Note that the converse of the compactness theorem is trivially true, i.e., every satisfiable set of sentences is finitely satisfiable.

We say that φ is a *consequence* of Σ, in symbols $\Sigma \models \varphi$, iff every model of Σ is a model of φ. The reader is asked to prove Exercises 1.2.3–1.2.6 as well as the following:

COROLLARY 1.2.13

(i). *$\Sigma \vdash \varphi$ if and only if $\Sigma \models \varphi$.*

(ii). *If $\Sigma \models \varphi$, then there is a finite subset Σ_0 of Σ such that $\Sigma_0 \models \varphi$.*

We shall conclude our model theory for sentential logic with a few applications of the compactness theorem. In these applications, the true spirit of model theory will appear, but at a very rudimentary level. Since we shall often wish to combine a finite set of sentences into a single sentence, we shall use expressions like

$$\varphi_1 \wedge \varphi_2 \wedge \ldots \wedge \varphi_n$$

and

$$\varphi_1 \vee \varphi_2 \vee \ldots \vee \varphi_n.$$

In these expressions the parentheses are assumed, for the sake of definiteness, to be associated to the right; for instance,

$$\varphi_1 \wedge \varphi_2 \wedge \varphi_3 = \varphi_1 \wedge (\varphi_2 \wedge \varphi_3).$$

First we introduce a bit more terminology. A set Γ of sentences is called a *theory*. A theory is said to be *closed* iff every consequence of Γ belongs to Γ. A set Δ of sentences is said to be a *set of axioms for* a theory Γ iff Γ and Δ have the same consequences. A theory is called *finitely axiomatizable* iff it has a finite set of axioms. Since we may form the conjunction of a finite set of axioms, a finitely axiomatizable theory actually always has a single axiom. The set $\bar{\Gamma}$ of all consequences of Γ is the unique closed theory which has Γ as a set of axioms.

PROPOSITION 1.2.14. *Δ is a set of axioms for a theory Γ if and only if Δ has exactly the same models as Γ.*

COROLLARY 1.2.15. *Let Γ_1 and Γ_2 be two theories such that the set of all models of Γ_2 is the complement of the set of all models of Γ_1. Then Γ_1 and Γ_2 are both finitely axiomatizable.*

PROOF. The set $\Gamma_1 \cup \Gamma_2$ is not satisfiable, so it is not finitely satisfiable. Thus we may choose finite sets $\Delta_1 \subset \Gamma_1$, $\Delta_2 \subset \Gamma_2$ such that $\Delta_1 \cup \Delta_2$ is not satisfiable. If $A \models \Delta_1$, then A is not a model of Γ_2, and consequently $A \models \Gamma_1$. It follows by Proposition 1.2.14 that Δ_1 is a finite set of axioms for Γ_1. Similarly Δ_2 is a finite set of axioms for Γ_2. ⊣

The next group of theorems shows connections between mathematical operations on models and syntactical properties of sentences. The first result of this group concerns positive sentences. A sentence φ is said to be *positive* iff φ is built up from sentence symbols using only the two connectives $\wedge$, $\vee$. For example, $(S_0 \wedge (S_2 \vee S_3)) \vee S_{16}$ is positive, while $\neg S_4$ and $S_3 \leftrightarrow S_3$ are not positive. A set Σ of sentences is called *increasing* iff $A \models \Sigma$ and $A \subset B$ implies $B \models \Sigma$.

THEOREM 1.2.16.

(i). *$A \subset B$ if and only if every positive sentence which holds in A holds in B.*

(ii). *A consistent theory Γ is increasing if and only if Γ has a set of positive axioms.*

(iii). *A sentence φ is increasing if and only if either φ is equivalent to a positive sentence, φ is valid, or $\neg \varphi$ is valid.*

PROOF. (i). The fact that, if $A \subset B$, then every positive sentence which holds in A holds in B, is proved by induction. First, every sentence symbol which holds in A holds in B, because of 1.2.3(i) and $A \subset B$. Using 1.2.3(ii) and Exercise 1.2.2, it can be checked that, if the condition 'if φ holds in A, then φ holds in B' is true when $\varphi = \psi$, and when $\varphi = \theta$, then it is also true when $\varphi = \psi \wedge \theta$ and when $\varphi = \psi \vee \theta$. Hence that condition is true for every positive sentence φ.

Suppose that every positive sentence which holds in A holds in B. In particular, for each $S \in \mathscr{S}$, if $A \models S$, then $B \models S$. Thus, if $S \in A$, then $S \in B$, so $A \subset B$. This proves (i).

(ii). Now let Γ be a consistent increasing theory. Let Δ be the set of all positive consequences of Γ. Suppose $B \models \Delta$. Let Σ be the set of all sentences $\neg \varphi$ such that φ is positive and $B \models \neg \varphi$. Let $\neg \varphi_1, \ldots, \neg \varphi_n \in \Sigma$. Then the sentence $\varphi_1 \vee \ldots \vee \varphi_n$ is a positive sentence which fails in B. Hence $\varphi_1 \vee \ldots \vee \varphi_n$ does not belong to Δ and is not a consequence of Γ. Thus the set $\Gamma \cup \{\neg \varphi_1, \ldots, \neg \varphi_n\}$ is satisfiable, and the set $\Gamma \cup \Sigma$ is finitely satisfiable. By the compactness theorem, $\Gamma \cup \Sigma$ has a model, say A. Now for every positive sentence φ which fails in B, $\neg \varphi \in \Sigma$, so φ fails in A. Thus every positive sentence holding in A holds in B, and by (i), $A \subset B$. Since $A \models \Gamma$ and Γ is increasing, we have $B \models \Gamma$. We conclude that every model of Δ is a model of Γ. But $\Delta \subset \bar{\Gamma}$, and therefore Δ is a set of positive axioms for Γ.

Conversely, if Γ has a set of positive axioms, then it follows from (i) that Γ is increasing.

(iii). Let φ be an increasing sentence. We may assume further that φ is satisfiable. If Γ is the set of all consequences of φ, then by (ii) Γ has a positive set Δ of axioms. Now $\varphi \in \Gamma$, so $\Delta \models \varphi$, and by Corollary 1.2.13 there is a finite subset $\{\psi_1, \dots, \psi_n\}$ of Δ such that $\{\psi_1, \dots, \psi_n\} \models \varphi$. If $n = 0$, then φ is valid. Let $n > 0$. Each ψ_m is in Δ and thus in Γ, so each ψ_m is a consequence of φ. It follows that φ is equivalent to the positive sentence $\psi_1 \wedge \dots \wedge \psi_n$.

Conversely, it follows from (i) that every positive sentence is increasing. Obviously, every valid sentence and every refutable sentence are also increasing. $\dashv$

A completely trivial fact which is analogous to part (i) of the above theorem is: $A = B$ if and only if every sentence which holds in A holds in B. We shall see later on in this book that the situation is very different in predicate logic, where a maximal consistent theory ordinarily does not even come close to characterizing a single model. This is one thing which makes model theory for predicate logic so much more interesting and difficult than model theory for sentential logic.

We now turn to another kind of sentence. By a *conditional sentence* we mean a sentence $\varphi_1 \wedge \dots \wedge \varphi_n$, where each φ_i is of one of the following three kinds:

(1) S,

(2) $\neg S_1 \vee \neg S_2 \vee \dots \vee \neg S_p$,

(3) $\neg S_1 \vee \neg S_2 \vee \dots \vee \neg S_p \vee T$.

A set Σ of sentences is said to be *preserved under finite intersections* iff $A \models \Sigma$ and $B \models \Sigma$ implies $A \cap B \models \Sigma$. Σ is said to be *preserved under arbitrary intersections* iff for every nonempty set $\{A_i : i \in I\}$ of models of Σ the intersection $\bigcap_{i \in I} A_i$ is also a model of Σ.

LEMMA 1.2.17. *A theory Γ is preserved under finite intersections if and only if Γ is preserved under arbitrary intersections.*

PROOF. Let Γ be preserved under finite intersections, let $\{A_i : i \in I\}$ be a nonempty set of models of Γ, and let $B = \bigcap_{i \in I} A_i$. Let Σ be the set of all sentences of the form S or $\neg S$ which hold in B. We show that $\Gamma \cup \Sigma$ is satisfiable. Let Σ_0 be an arbitrary finite subset of Σ, and let the negative sentences in Σ_0 be $\neg S_1, \dots, \neg S_p$. If $p = 0$, all the sentences in Σ_0 are positive, and each of the models A_i is a model of Σ_0, because $B \subset A_i$. Let $p > 0$ and choose models $A_{i_1}, \ldots, A_{i_p}$ from among the A_i such that $S_1 \notin A_{i_1}, \ldots, S_p \notin A_{i_p}$. Then $A = A_{i_1} \cap \ldots \cap A_{i_p}$ is a model of Σ_0; since Γ is preserved under finite intersections, A is also a model of Γ. We have shown that $\Gamma \cup \Sigma$ is finitely satisfiable. By the compactness theorem, $\Gamma \cup \Sigma$ has a model. But the only model of Σ is B, so B is a model of Γ. ⊣

In view of the above lemma, we may as well simply say from now on that Γ is *preserved under intersections*, since it makes no difference whether we say finite or arbitrary intersections.

THEOREM 1.2.18.

(i). *A theory Γ is preserved under intersections if and only if Γ has a set of conditional axioms.*

(ii). *A sentence φ is preserved under intersections if and only if φ is equivalent to a conditional sentence.*

PROOF. (i). We leave to the reader the proof that every conditional sentence (and hence every set of conditional sentences) is preserved under intersections.

Conversely, let Γ be preserved under intersections. Consider the set Δ of all conditional consequences of Γ. It suffices to show that every model of Δ is a model of Γ. Let B be an arbitrary model of Δ. For each $T \in \mathscr{S} - B$, let Σ_T be the set of all sentences of the form

$$S_1 \wedge \ldots \wedge S_p \wedge \neg T$$

which hold in B. We also let the sentence $\neg T$ itself be in Σ_T. We first note that the conjunction of finitely many sentences in Σ_T is again equivalent to a sentence in Σ_T. Consider a sentence $\varphi \in \Sigma_T$. Then $\neg \varphi$ is clearly equivalent to a conditional sentence ψ either of the form S or of the form

$$\neg S_1 \vee \ldots \vee \neg S_p \vee T.$$

But ψ fails in B, so ψ does not belong to Δ. This means that ψ, and hence $\neg \varphi$, is not a consequence of Γ, and it follows that $\Gamma \cup \{\varphi\}$ is satisfiable. Since Σ_T is, up to equivalence, closed under finite conjunction, we see that $\Gamma \cup \Sigma_T$ is finitely satisfiable. Applying the Compactness Theorem, we may choose a model A_T of $\Gamma \cup \Sigma_T$.

For each $T \in \mathscr{S} - B$, we have $T \notin A_T$ and $B \subset A_T$. Thus, if $\mathscr{S} - B$ is not empty, then

$$B = \bigcap_{T \notin B} A_T.$$

Since each A_T is a model of Γ and Γ is closed under intersections, we have $B \models \Gamma$. In the remaining case $B = \mathscr{S}$, we let Σ be the set of all sentences of the form

$$S_1 \wedge \ldots \wedge S_p.$$

Arguing as before, we find that $\Gamma \cup \Sigma$ is finitely satisfiable and thus has a model. But B is the only model of Σ, so again B is a model of Γ.

We have now shown that every model of Δ is a model of Γ, and it follows that Δ is a set of conditional axioms for Γ.

(ii). This follows from (i) by an argument similar to the last part of the proof of Theorem 1.2.16. ⊣

We conclude with a table which summarizes the semantical and syntactical notions that we have shown to be equivalent (some of these are done in the exercises).

TABLE 1.2.1

Syntax	Semantics
φ is a tautology, $\vdash \varphi$	φ is valid, $\models \varphi$
Σ is consistent	Σ is satisfiable
φ is inconsistent	φ is not satisfiable
φ is deducible from Σ, $\Sigma \vdash \varphi$	φ is a consequence of Σ, $\Sigma \models \varphi$
φ is equivalent to a positive sentence	φ is increasing, and not valid or refutable
φ is equivalent to a conditional sentence	φ is preserved under intersections

EXERCISES

1.2.1. Let A be a model such that $S, T \in A$ and $U, V \in \mathscr{S} - A$. Which of the following sentences are true in A?

$$U, S, T \wedge U, \neg\neg\neg U, S \rightarrow V, S \wedge (S \vee U \leftrightarrow (V \rightarrow T)).$$

1.2.2. Show that, if $\varphi = \psi \vee \theta$, then $A \models \varphi$ if and only if $A \models \psi$ or $A \models \theta$ or both. Concoct similar rules for $A \models \psi \rightarrow \theta$ and $A \models \psi \leftrightarrow \theta$.

1.2.3. A sentence φ is *satisfiable* iff it has at least one model. Show that φ is satisfiable if and only if $\neg \varphi$ is not valid.

1.2.4. A sentence φ is a *consequence of* another sentence ψ, in symbols $\psi \models \varphi$, iff every model of ψ is a model of φ. Show that $\psi \models \varphi$ if and only if $\models \psi \rightarrow \varphi$.

1.2.5. Two sentences φ and ψ are (*semantically*) *equivalent* iff they have exactly the same models. Show that φ and ψ are equivalent if and only if each one is a consequence of the other, and also if and only if $\models \varphi \leftrightarrow \psi$.

1.2.6. Prove that if φ is satisfiable and $\mathscr{S}$ is countable, then the set of all models of φ has the cardinal number of the continuum.

1.2.7* (Interpolation Theorem). Assume that $\varphi \models \psi$. Show that either (i) φ is refutable, (ii) ψ is valid, or (iii) there exists a sentence θ such that $\varphi \models \theta$, $\theta \models \psi$, and every sentence symbol which occurs in θ also occurs in both φ and ψ.

1.2.8. Prove Proposition 1.2.6.

1.2.9*

(i). For every finite set K of models, there is a set Σ of sentences such that K is the set of all models of Σ.

(ii). Give an example of a set Σ of sentences such that the set of all models of Σ is countably infinite.

(iii). Give an example of a countable set of models which cannot be represented as the set of all models of some set of sentences.

In (ii) and (iii), assume that $\mathscr{S}$ is countable.

1.2.10. If $\Sigma \vdash \varphi$ for all $\varphi \in \Gamma$ and if $\Sigma \cup \Gamma \vdash \theta$, then $\Sigma \vdash \theta$.

1.2.11. Prove that the set of all non-models of Σ is empty or of power $2^{|\mathscr{S}|}$.

1.2.12. Show that no positive sentence is valid and no positive sentence is refutable.

1.2.13. A theory Γ is said to be *complete* iff for every sentence φ, exactly one of $\Gamma \models \varphi$, $\Gamma \models \neg\varphi$ holds. For any set Σ of sentences, the following are equivalent:

(i). The set of consequences of Σ is maximal consistent.

(ii). Σ is a complete theory.

(iii). Σ has exactly one model.

(iv). There is a model A such that for all φ, $\Sigma \models \varphi$ iff $A \models \varphi$.

1.2.14. Let Γ be a consistent theory and let B be a model for $\mathscr{S}$. Prove that

B is a model of the set of all positive consequences of Γ if and only if there is a model A of Γ such that $A \subset B$.

1.2.15. Show that every conditional sentence is preserved under intersections.

1.2.16. State and prove the analogue of Exercise 1.2.14 for intersections and conditional sentences.

1.2.17*. Formulate and prove a result like Theorem 1.2.18 for unions of sets of models.

1.2.18. A set Σ of sentences is said to be *independent* iff, for each $\sigma \in \Sigma$, σ is not a consequence of $\Sigma - \{\sigma\}$. Prove that if $\mathscr{S}$ is countable, then every theory Γ in $\mathscr{S}$ has an independent set of axioms.

[*Hint*: Show that Γ has a set of axioms $\Sigma = \{\sigma_1, \sigma_2, \sigma_3, \ldots\}$ such that, for each n, $\vdash \sigma_{n+1} \to \sigma_n$ but not $\vdash \sigma_n \to \sigma_{n+1}$. Then consider the set $\{\sigma_1, \sigma_1 \to \sigma_2, \sigma_2 \to \sigma_3, \ldots\}$.]

1.2.19**. Prove without any restriction on the cardinality of $\mathscr{S}$ that every theory in $\mathscr{S}$ has an independent set of axioms. (The case where $|\mathscr{S}| = \omega_1$ is very much easier than the general case, but still a challenge.)

1.3. Languages, models and satisfaction

We begin here the development of first-order languages in a way parallel to the treatment of sentential logic in Section 1.2. First, we shall define the notions of a first-order predicate language $\mathscr{L}$ and of a model for $\mathscr{L}$. We introduce some basic relations between models – reductions and expansions, isomorphisms, submodels and extensions. We shall then develop the syntax of the language $\mathscr{L}$, defining the sets of terms, formulas and sentences, and presenting the axioms and rules of inference. Finally, we give the key definition of a sentence being true in a model for the language $\mathscr{L}$. The precise formulation of this definition is much more of a challenge in first-order logic than it was for sentential logic. At the end of this section, we state the completeness and compactness theorems (Theorems 1.3.20–1.3.22), but the proofs of these theorems are deferred until the next chapter.

We first establish a uniform notation and set of conventions for such languages and their models. A *language* $\mathscr{L}$ is a collection of symbols. These symbols are separated into three groups, *relation symbols*, *function symbols* and (*individual*) *constant symbols*. The relation and function symbols of $\mathscr{L}$

will be denoted by capital Latin letters P, F, with subscripts. Lower case Latin letters c, with subscripts, range over the constant symbols of $\mathscr{L}$. If $\mathscr{L}$ is a finite set, we may display the symbols of $\mathscr{L}$ as follows:

$$\mathscr{L} = \{P_0, \dots, P_n, F_0, \dots, F_m, c_0, \dots, c_q\}.$$

Each relation symbol P of $\mathscr{L}$ is assumed to be an n-placed relation for some integer $n \geqslant 1$, depending on P. Similarly, each function symbol F of $\mathscr{L}$ is an m-placed function symbol, where $m \geqslant 1$ and m depends on F. Note that we do not allow 0-placed relation or function symbols. When dealing with several languages at the same time, we use the letters $\mathscr{L}$, $\mathscr{L}'$, $\mathscr{L}''$, etc. If the symbols of the language are quite standard, as for example $+$ for addition, $\leqslant$ for an order relation, etc., we shall simply write

$$\mathscr{L} = \{\leqslant\}, \quad \mathscr{L} = \{\leqslant, +, \cdot, 0\}, \quad \mathscr{L} = \{+, \cdot, -, 0, 1\}, \quad \text{etc.},$$

for such languages. The number of places of the various kinds of symbols is understood to follow the standard usage. The *power*, or *cardinal* of the language $\mathscr{L}$, denoted by $||\mathscr{L}||$, is defined as

$$||\mathscr{L}|| = \omega \cup |\mathscr{L}|.$$

We say that a language $\mathscr{L}$ is countable or uncountable depending on whether $||\mathscr{L}||$ is countable or uncountable.

We occasionally pass from a given language $\mathscr{L}$ to another language $\mathscr{L}'$ which has all the symbols of $\mathscr{L}$ plus some additional symbols. In such cases we use the notation $\mathscr{L} \subset \mathscr{L}'$ and say that the language $\mathscr{L}'$ is an *expansion* of $\mathscr{L}$, and that $\mathscr{L}$ is a *reduction* of $\mathscr{L}'$. In the special case where all the symbols in $\mathscr{L}'$ but not in $\mathscr{L}$ are constant symbols, $\mathscr{L}'$ is said to be a *simple expansion* of $\mathscr{L}$. Since $\mathscr{L}$ and $\mathscr{L}'$ are just sets of symbols, the expansion $\mathscr{L}'$ may be written as $\mathscr{L}' = \mathscr{L} \cup X$, where X is the set of new symbols.

Turning now to the models for a given language $\mathscr{L}$, we first point out that the situation here is more complicated than for the sentential logic $\mathscr{S}$ in Section 1.2. There, each $S \in \mathscr{S}$ could take on at most two values, true or false. Thus the set of intended interpretations for $\mathscr{S}$ has rather simple properties, as the reader discovered. This time, each n-placed relation symbol has as its intended interpretations all n-placed relations among the objects, each m-placed function symbol has as its intended interpretations all m-placed functions from objects to objects, and, finally, each constant symbol has as intended interpretations fixed or constant objects. Therefore, a 'possible world', or model for $\mathscr{L}$ consists, first of all, of a *universe* A, a nonempty set. In this universe, each n-placed P corresponds to an n-placed *relation* $R \subset A^n$ on A, each m-placed F corresponds to an m-placed *function* $G : A^m \to A$ on A, and each constant symbol c corresponds to a *constant* $x \in A$. This correspondence is given by an *interpretation* function $\mathscr{I}$ mapping the symbols of $\mathscr{L}$ to appropriate relations, functions and constants in A. A *model for* $\mathscr{L}$ is a pair $\langle A, \mathscr{I}\rangle$. We use Gothic letters to range over models. Thus we write $\mathfrak{A} = \langle A, \mathscr{I}\rangle$, $\mathfrak{B} = \langle B, \mathscr{J}\rangle$, $\mathfrak{C} = \langle C, \mathscr{K}\rangle$, etc., with appropriate subscripts and superscripts. We shall try to be quite consistent in this respect, so that the universes of the models $\mathfrak{B}'$, $\mathfrak{B}''$, $\mathfrak{B}_i$, $\mathfrak{B}_j$, etc., are precisely the sets B', B'', B_i, B_j, etc. The relations, functions and constants of $\mathfrak{A}$ are, respectively, the images under $\mathscr{I}$ of the relation symbols, function symbols and constant symbols of $\mathscr{L}$.

Note that in a given universe A there are many different permissible interpretations of the symbols of $\mathscr{L}$. Suppose $\mathfrak{A} = \langle A, \mathscr{I}\rangle$, $\mathfrak{A}' = \langle A', \mathscr{I}'\rangle$ are models for $\mathscr{L}$ and R, R' are relations of $\mathfrak{A}$, $\mathfrak{A}'$, respectively. We say that R' is the *corresponding relation to* R if they are the interpretations of the same relation symbol in $\mathscr{L}$, i.e.

$$\mathscr{I}(P) = R \quad \text{and} \quad \mathscr{I}'(P) = R' \quad \text{for some } P \in \mathscr{L}.$$

We introduce similar conventions as regards the functions and constants. When

$$\mathscr{L} = \{P_0, \ldots, P_n, F_0, \ldots, F_m, c_0, \ldots, c_q\},$$

we write the models for $\mathscr{L}$ in displayed form as

$$\mathfrak{A} = \langle A, R_0, \ldots, R_n, G_0, \ldots, G_m, x_0, \ldots, x_q\rangle.$$

When the symbols of $\mathscr{L}$ are familiar, we shall agree to use, for instance,

$$\mathfrak{A} = \langle A, \leqslant, +, \cdot\rangle$$

for models of the language $\mathscr{L} = \{\leqslant, +, \cdot\}$. We may resort to

$$\mathfrak{A} = \langle A, \leqslant_{\mathfrak{A}}, +_{\mathfrak{A}}, \cdot_{\mathfrak{A}}\rangle, \qquad \mathfrak{B} = \langle B, \leqslant_{\mathfrak{B}}, +_{\mathfrak{B}}, \cdot_{\mathfrak{B}}\rangle, \text{ etc.,}$$

if the context of the discussion requires it.

If we start with a model $\mathfrak{A}$ for the language $\mathscr{L}$ we can always expand it to a model for the language $\mathscr{L}' = \mathscr{L} \cup X$ by giving appropriate interpretations for the symbols in X. If $\mathscr{I}'$ is any interpretation for the symbols of X in $\mathfrak{A}$, and X is disjoint from $\mathscr{L}$, then $\mathfrak{A}' = \langle A, \mathscr{I} \cup \mathscr{I}'\rangle$ is a model for $\mathscr{L}'$. In this case we say that $\mathfrak{A}'$ is an *expansion* of $\mathfrak{A}$ to $\mathscr{L}'$, and $\mathfrak{A}$ is the *reduct* of $\mathfrak{A}'$ to $\mathscr{L}$. Sometimes we use the shorter notation $(\mathfrak{A}, \mathscr{I}')$ for $\mathfrak{A}'$. Clearly, there are many ways a model $\mathfrak{A}$ for $\mathscr{L}$ can be expanded to a model

$\mathfrak{A}'$ for $\mathscr{L}'$. On the other hand, given a model $\mathfrak{A}'$ for $\mathscr{L}'$, it has only one reduction $\mathfrak{A}$ to $\mathscr{L}$. Namely, we form $\mathfrak{A}$ by restricting the interpretation function $\mathscr{I}'$ on $\mathscr{L} \cup X$ to $\mathscr{L}$. The processes of expansion and reduction do not change the universe of the model.

The *cardinal*, or *power*, of the model $\mathfrak{A}$ is the cardinal $|A|$. $\mathfrak{A}$ is said to be finite, countable or uncountable if $|A|$ is finite, countable or uncountable. Note that on a finite universe A, while there can be only finitely many different relations, functions and constants, the number of different interpretation functions $\mathscr{I}$ can be very large and depends on $|\mathscr{L}|$.

We next introduce some simple but basic notions and operations on models. The reader should go through the exercises at the end of this section in order to be familiar with them.

Two models $\mathfrak{A}$ and $\mathfrak{A}'$ for $\mathscr{L}$ are *isomorphic* iff there is a 1-1 function f mapping A onto A' satisfying:

(i). For each n-placed relation R of $\mathfrak{A}$ and the corresponding relation R' of $\mathfrak{A}'$,

$$R(x_1 \dots x_n) \quad \text{if and only if} \quad R'(f(x_1) \dots f(x_n))$$

for all $x_1, \dots, x_n$ in A.

(ii). For each m-placed function G of $\mathfrak{A}$ and the corresponding function G' of $\mathfrak{A}'$,

$$f(G(x_1 \dots x_m)) = G'(f(x_1) \dots f(x_m)),$$

for all $x_1, \dots, x_m$ in A.

(iii). For each constant x of $\mathfrak{A}$ and the corresponding constant x' of $\mathfrak{A}'$,

$$f(x) = x'.$$

A function f that satisfies the above is called *an isomorphism of* $\mathfrak{A}$ *onto* $\mathfrak{A}'$, or *an isomorphism between* $\mathfrak{A}$ *and* $\mathfrak{A}'$. We use the notation $f : \mathfrak{A} \cong \mathfrak{A}'$ to denote that f is an isomorphism of $\mathfrak{A}$ onto $\mathfrak{A}'$, and we use $\mathfrak{A} \cong \mathfrak{A}'$ for $\mathfrak{A}$ is isomorphic to $\mathfrak{A}'$. For convenience we use $\cong$ to denote the *isomorphism relation* between models for $\mathscr{L}$. It is quite clear that $\cong$ is an equivalence relation. Furthermore, it preserves powers, that is, if $\mathfrak{A} \cong \mathfrak{B}$, then $|A| = |B|$. Indeed, unless we wish to consider the particular structure of each element of A or B, for all practical purposes $\mathfrak{A}$ and $\mathfrak{B}$ are the same if they are isomorphic.

A model $\mathfrak{A}'$ is called a *submodel* of $\mathfrak{A}$ if $A' \subset A$ and:

(i). Each n-placed relation R' of $\mathfrak{A}'$ is the restriction to A' of the corresponding relation R of $\mathfrak{A}$, i.e., $R' = R \cap (A')^n$.

(ii). Each m-placed function G' of $\mathfrak{A}'$ is the restriction to A' of the corresponding function G of $\mathfrak{A}$, i.e., $G' = G|(A')^m$.

(iii). Each constant of $\mathfrak{A}'$ is the corresponding constant of $\mathfrak{A}$.

We use $\mathfrak{A}' \subset \mathfrak{A}$ to denote that $\mathfrak{A}'$ is a submodel of $\mathfrak{A}$, and the symbol $\subset$ for the submodel relation between models for $\mathscr{L}$. The reader should show that $\subset$ is a partial-order relation and that, if $\mathfrak{A} \subset \mathfrak{B}$, then $|A| \leqslant |B|$. We say that $\mathfrak{B}$ is an *extension* of $\mathfrak{A}$ if $\mathfrak{A}$ is a submodel of $\mathfrak{B}$.

Combining the above two notions, we say that $\mathfrak{A}$ is *isomorphically embedded in* $\mathfrak{B}$ if there is a model $\mathfrak{C}$ and an isomorphism f such that $f : \mathfrak{A} \cong \mathfrak{C}$ and $\mathfrak{C} \subset \mathfrak{B}$. In this case we call the function f an *isomorphic embedding of* $\mathfrak{A}$ *in* $\mathfrak{B}$. If $\mathfrak{A}$ is isomorphically embedded in $\mathfrak{B}$, then $\mathfrak{B}$ is isomorphic to an extension of $\mathfrak{A}$.

To formalize a language $\mathscr{L}$, we need the following *logical symbols* (see the corresponding development for $\mathscr{S}$ in Section 1.2.):

parentheses	), (;
variables	$v_0, v_1, \ldots, v_n, \ldots$;
connectives	$\wedge$ (and), $\neg$ (not);
quantifier	$\forall$ (for all);

and one binary relation symbol $\equiv$ (identity).

We assume, of course, that no symbol in $\mathscr{L}$ occurs in the above list. Certain strings of symbols from the above list and from $\mathscr{L}$ are called *terms*. They are defined as follows:

1.3.1.

(i). A variable is a term.

(ii). A constant symbol is a term.

(iii). If F is an m-placed function symbol and $t_1, \ldots, t_m$ are terms, then $F(t_1 \ldots t_m)$ is a term.

(iv). A string of symbols is a term only if it can be shown to be a term by a finite number of applications of (i)–(iii).

The *atomic formulas* of $\mathscr{L}$ are strings of the form given below:

1.3.2.

(i). $t_1 \equiv t_2$ is an atomic formula, where t_1 and t_2 are terms of $\mathscr{L}$.

(ii). If P is an n-placed relation symbol and $t_1, \ldots, t_n$ are terms, then $P(t_1 \ldots t_n)$ is an atomic formula.

Finally, the *formulas* of $\mathscr{L}$ are defined as follows:

1.3.3.

(i). An atomic formula is a formula.

(ii). If φ and ψ are formulas, then $(\varphi \wedge \psi)$ and $(\neg \varphi)$ are formulas.

(iii). If v is a variable and φ is a formula, then $(\forall v)\varphi$ is a formula.

(iv). A sequence of symbols is a formula only if it can be shown to be a formula by a finite number of applications of (i)–(iii).

Just as in the case of $\mathscr{S}$, we may put definitions 1.3.1 and 1.3.3 in a set-theoretical setting. Namely, the set of terms of $\mathscr{L}$ is the least set T such that

T contains all constant symbols and all variables v_n, $n = 0, 1, 2, \ldots$, and, whenever F is an m-placed function symbol and $t_1, \ldots, t_m \in T$, then $F(t_1 \ldots t_m) \in T$.

Similarly, the set of formulas of $\mathscr{L}$ is the least set Φ such that

every atomic formula belongs to Φ and, whenever $\varphi, \psi \in \Phi$ and v is a variable, then $(\varphi \wedge \psi)$, $(\neg \varphi)$, $(\forall v)\varphi$ all belong to Φ.

Note that we have tacitly used the letters t (with subscripts) to range over terms, v to range over variables, and φ, ψ to range over formulas. Again, we emphasize that *properties of terms and formulas of* $\mathscr{L}$ *can only be established by an induction based on definitions* 1.3.1 *and* 1.3.3.

We can now introduce the abbreviations $\vee$, $\rightarrow$, $\leftrightarrow$ as in Section 1.2. Furthermore, we adopt all the conventions introduced earlier. The new symbol $\exists$ (there exists) is introduced as an abbreviation defined as

$$(\exists v)\varphi \quad \text{for } \neg (\forall v) \neg \varphi.$$

Some new conventions are the following:

$$\begin{aligned} \varphi_1 \wedge \varphi_2 \wedge \ldots \wedge \varphi_n &\quad \text{for} \quad (\varphi_1 \wedge (\varphi_2 \wedge \ldots \wedge \varphi_n)); \\ \varphi_1 \vee \varphi_2 \vee \ldots \vee \varphi_n &\quad \text{for} \quad (\varphi_1 \vee (\varphi_2 \vee \ldots \vee \varphi_n)); \\ (\forall x_1 x_2 \ldots x_n)\varphi &\quad \text{for} \quad (\forall x_1)(\forall x_2) \ldots (\forall x_n)\varphi; \\ (\exists x_1 x_2 \ldots x_n)\varphi &\quad \text{for} \quad (\exists x_1)(\exists x_2) \ldots (\exists x_n)\varphi. \end{aligned}$$

At this point we assume that the reader has enough experience in first-order predicate logic to continue the development on his own. In particular, we leave it to him to decide on the notions of *subformulas*, of *free* and *bound* occurrences of a variable in a formula, and to give a proper definition (based on definitions 1.3.1 and 1.3.3) of *substitution* of a term for a variable in a formula.

We now come to an extremely important convention of notation. To make sure that the reader does not miss it, we enclose it in a box:

We use $t(v_0 \dots v_n)$ to denote a term t whose variables form a subset of $\{v_0, \dots, v_n\}$. Similarly, we use $\varphi(v_0 \dots v_n)$ to denote a formula φ whose free variables form a subset of $\{v_0, \dots, v_n\}$.

Note that we do not require that all of the variables $v_0, \dots, v_n$ be free variables of $\varphi(v_0 \dots v_n)$. In fact, $\varphi(v_0 \dots v_n)$ could even have no free variables. Also, we make no restriction on the bound variables. For example, each of the following formulas is of the form $\varphi(v_0 v_1 v_2)$:

$$(\forall v_1)(\exists v_3)R(v_0 v_1 v_3), \qquad R(v_0 v_1 v_2), \qquad S(v_0 v_2), \qquad (\forall v_4)S(v_4 v_4).$$

A *sentence* is a formula with no free variables.

Note that even if $\mathscr{L}$ has no symbols, there are still formulas of $\mathscr{L}$. These formulas are built up entirely from the identity symbol $\equiv$ and the other logical symbols listed. Such formulas are called *identity formulas* and they *occur in every language*. The following proposition is simple but important.

PROPOSITION 1.3.4. *The cardinal of the set of all formulas of $\mathscr{L}$ is $||\mathscr{L}||$.*

To make all the above syntactical notions into a *formal system* we need *logical axioms* and *rules of inference*. The logical axioms for $\mathscr{L}$ are divided into three groups.

1.3.5. *Sentential Axioms*: Every formula φ of $\mathscr{L}$ which can be obtained from a tautology ψ of $\mathscr{S}$ by (simultaneously and uniformly) substituting formulas of $\mathscr{L}$ for the sentence symbols of ψ is a logical axiom for $\mathscr{L}$. From now on we call such a formula φ a *tautology* of $\mathscr{L}$.

1.3.6. *Quantifier Axioms*:

(i). If φ, ψ are formulas of $\mathscr{L}$ and v is a variable not free in φ, then the formula

$$(\forall v)(\varphi \to \psi) \to (\varphi \to (\forall v)\psi)$$

is a logical axiom.

(ii). If φ, ψ are formulas and ψ is obtained from φ by freely substituting each free occurrence of v in φ by the term t (i.e., no variable x in t shall occur bound in ψ at the place where it is introduced), then the formula

$$(\forall v)\varphi \to \psi$$

is a logical axiom.

1.3.7. *Identity Axioms*: Suppose x, y are variables, $t(v_0 \dots v_n)$ is a term and $\varphi(v_0 \dots v_n)$ is an atomic formula. Then the formulas

$$x \equiv x,$$
$$x \equiv y \to t(v_0 \dots v_{i-1} x v_{i+1} \dots v_n) \equiv t(v_0 \dots v_{i-1} y v_{i+1} \dots v_n),$$
$$x \equiv y \to (\varphi(v_0 \dots v_{i-1} x v_{i+1} \dots v_n) \to \varphi(v_0 \dots v_{i-1} y v_{i+1} \dots v_n)),$$

are logical axioms.

There are two rules of inference.

1.3.8. *Rule of Detachment* (or *Modus Ponens*): From φ and $\varphi \to \psi$ infer ψ.

1.3.9. *Rule of Generalization*: From φ infer $(\forall x)\varphi$.

Given the axioms and the rules of inference, we assume that the resulting notions of *proof*, *length of proof*, *theorem* are already familiar to the reader. As we are dealing with the usual first-order logic with identity, we shall assume as known and make free use of all of the basic theorems and metatheorems of such formal systems.

Following standard usage, $\vdash \varphi$ means that φ is a theorem of $\mathscr{L}$. If Σ is a set of sentences of $\mathscr{L}$, then $\Sigma \vdash \varphi$ means that there is a proof of φ from the logical axioms and Σ. If $\Sigma = \{\sigma_1, \dots, \sigma_n\}$ is finite, we write $\sigma_1 \dots \sigma_n \vdash \varphi$. As the logical axioms are always assumed, we say that *there is a proof of φ from Σ*, or *φ is deducible from Σ*, whenever $\Sigma \vdash \varphi$. *Σ is inconsistent* iff every formula of $\mathscr{L}$ can be deduced from Σ. Otherwise Σ is *consistent*. A sentence σ is consistent iff $\{\sigma\}$ is. Σ is *maximal consistent* (*in $\mathscr{L}$*) iff Σ is consistent and no set of sentences (of $\mathscr{L}$) properly containing Σ is consistent. We list in the proposition below some useful, though simple, properties of consistent and maximal consistent sets of sentences. (Many of these properties are found also in Proposition 1.2.8.)

PROPOSITION 1.3.10.

(i). *Σ is consistent if and only if every finite subset of Σ is consistent.*

(ii). *Let σ be a sentence. $\Sigma \cup \{\sigma\}$ is inconsistent if and only if $\Sigma \vdash \neg\sigma$. Whence $\Sigma \cup \{\sigma\}$ is consistent if and only if $\neg\sigma$ is not deducible from Σ.*

(iii). *If Σ is maximal consistent, then for any sentences σ, τ*

$$\Sigma \vdash \sigma \textit{ if and only if } \sigma \in \Sigma;$$
$$\sigma \notin \Sigma \textit{ if and only if } \neg\sigma \in \Sigma;$$
$$\sigma \wedge \tau \in \Sigma \textit{ if and only if } \sigma \textit{ and } \tau \textit{ belong to } \Sigma.$$

(iv) (Deduction Theorem). *$\Sigma \cup \{\sigma\} \vdash \tau$ if and only if $\Sigma \vdash \sigma \to \tau$.* (*Here σ is a sentence, although τ need not be one.*)

The next proposition duplicates Lemma 1.2.9. There is no change in the proof.

PROPOSITION 1.3.11 (Lindenbaum's Theorem). *Any consistent set of sentences of $\mathscr{L}$ can be extended to a maximal consistent set of sentences of $\mathscr{L}$.*

We now come to the key definition of this section. In fact, the following definition of satisfaction is the cornerstone of model theory. We first give the motivation for the definition in a few remarks. If we compare the models of Section 1.2 and the models discussed here, we see that with the former we were only concerned with whether a statement is true or false in it, while here the situation is more complicated because the sentences of $\mathscr{L}$ say something about the individual elements of the model. The whole question of the (first-order) truths or falsities of a possible world (i.e., model) is just not a simple problem. For instance, there is no way to decide whether a given sentence of $\mathscr{L} = \{+, \cdot, S, 0\}$ is true or false in the standard model $\langle N, +, \cdot, S, 0\rangle$ of arithmetic (where S is the successor function). Whereas we have already seen in Section 1.2 that there is such a decision procedure for every model for $\mathscr{S}$ and for every sentence of $\mathscr{S}$. To define the notion

the sentence σ is true in the model $\mathfrak{A}$,

we have first to break up σ into smaller parts and to examine each part. If σ is $\neg\varphi$ or if σ is $\varphi \wedge \psi$, then we see that the truth or falsity of σ in $\mathfrak{A}$ follows once we know the truth or falsity of φ and ψ in $\mathfrak{A}$. If, on the other hand, σ is $(\forall x)\varphi$, then the same method for deciding the truth of σ breaks down as φ may not be a sentence and it would be meaningless to ask if φ is true or false in $\mathfrak{A}$.

The free variable x in φ is supposed to range over the elements of A. For each particular a in A it is meaningful to ask whether

the formula φ is true in $\mathfrak{A}$ if φ is talking about a.

If for each a in A the answer to this question is yes, then we can say that σ is true in $\mathfrak{A}$. If there exists an a in A so that the answer is no, then we say that σ is false in $\mathfrak{A}$. But in order to answer the above question, even for a fixed element of A, we shall run into the same difficulty if φ happens to be $(\forall y)\psi$. Then we are led naturally to ask whether

ψ is true in $\mathfrak{A}$ if ψ is talking about a pair of elements a and b in A.

It takes but a very small step before we see that the crucial question is the following:

Given a formula $\varphi(v_0 \dots v_p)$ and a sequence $x_0, \dots, x_p$ in A, what does it mean to say that φ is true in $\mathfrak{A}$ if the variables $v_0, \ldots, v_p$ are taken to be $x_0, \ldots, x_p$?

Our plan is to give an answer to this question first for every atomic formula $\psi(v_0 \ldots v_p)$ and all elements $x_0, \ldots, x_p$. Then, by an inductive procedure based on our inductive definition of formula (1.3.1–1.3.3), we shall give an answer for all formulas $\varphi(v_0 \ldots v_p)$ and elements $x_0, \ldots, x_p$.

There is still one difficulty with our plan: if all the free variables of a formula φ are among $v_0, \ldots, v_p$, it does not follow that all the free variables of every subformula of φ are among $v_0, \ldots, v_p$. For a quantifier makes a free variable bound. This will cause trouble in the induction part of our plan. To overcome this difficulty we observe that the following is true. If all the variables, free or bound, of a formula φ are among $v_0, \ldots, v_q$, then all the variables of every subformula of φ are also among $v_0, \ldots, v_q$. So we shall modify our plan thus: First, we answer the question for all atomic formulas $\psi(v_0 \ldots v_q)$ and all elements $x_0, \ldots, x_q$. Then by an inductive procedure we answer the question for all formulas φ such that all its *free and bound* variables are among $v_0, \ldots, v_q$, and all elements $x_0, \ldots, x_q$. Finally, we *prove* that the answer to the question for a formula $\varphi(v_0 \ldots v_p)$ and elements $x_0, \ldots, x_q$, $p \leqslant q$, depends only on the elements $x_0, \ldots, x_p$ corresponding to *free* variables of φ, so that the values of $x_{p+1}, \ldots, x_q$ are irrelevant.

We are now ready for the formal definition. The crucial notion to be defined is the following: Let φ be any formula of $\mathscr{L}$, all of whose free and bound variables are among $v_0, \ldots, v_q$, and let $x_0, \ldots, x_q$ be any sequence of elements of A. We define the predicate

1.3.12. *φ is satisfied by the sequence $x_0, \ldots, x_q$ in $\mathfrak{A}$*, or *$x_0, \ldots, x_q$ satisfies φ in $\mathfrak{A}$.*

The definition proceeds in three stages (compare with 1.3.1–1.3.3).

Let $\mathfrak{A}$ be a fixed model for $\mathscr{L}$.

1.3.13. The *value of a term $t(v_0 \ldots v_q)$ at $x_0, \ldots, x_q$* is defined as follows (we let $t[x_0 \ldots x_q]$ denote this value):

(i). If $t = v_i$, then $t[x_0 \ldots x_q] = x_i$.

(ii). If t is a constant symbol c, then $t[x_0 \ldots x_q]$ is the interpretation of c in $\mathfrak{A}$.

(iii). If $t = F(t_1 \ldots t_m)$, where F is an m-placed function symbol, then

$$t[x_0 \ldots x_q] = G(t_1[x_0 \ldots x_q] \ldots t_m[x_0 \ldots x_q]),$$

where G is the interpretation of F in $\mathfrak{A}$.

1.3.14.

(i). Suppose $\varphi(v_0 \dots v_q)$ is the atomic formula $t_1 \equiv t_2$, where $t_1(v_0 \dots v_q)$ and $t_2(v_0 \dots v_q)$ are terms. Then $x_0, \dots, x_q$ *satisfies* φ if and only if

$$t_1[x_0 \dots x_q] = t_2[x_0 \dots x_q].$$

(ii). Suppose $\varphi(v_0 \dots v_q)$ is the atomic formula $P(t_1 \dots t_n)$, where P is an n-placed relation symbol and $t_1(v_0 \dots v_q), \dots, t_n(v_0 \dots v_q)$ are terms. Then $x_0, \dots, x_q$ *satisfies* φ if and only if

$$R(t_1[x_0 \dots x_q] \dots t_n[x_0 \dots x_q]),$$

where R is the interpretation of P in $\mathfrak{A}$.

For brevity, we write

$\mathfrak{A} \models \varphi[x_0 \dots x_q]$ for: $x_0, \dots, x_q$ satisfies φ in $\mathfrak{A}$.

Thus 1.3.14 can also be formulated as:

(i). $\mathfrak{A} \models (t_1 \equiv t_2)[x_0 \dots x_q]$ if and only if $t_1[x_0 \dots x_q] = t_2[x_0 \dots x_q]$.

(ii). $\mathfrak{A} \models P(t_1 \dots t_n)[x_0 \dots x_q]$ if and only if $R(t_1[x_0 \dots x_q] \dots t_n[x_0 \dots x_q])$.

1.3.15. Suppose that φ is a formula of $\mathscr{L}$ and all free and bound variables of φ are among $v_0, \dots, v_q$.

(i). If φ is $\theta_1 \wedge \theta_2$, then

$\mathfrak{A} \models \varphi[x_0 \dots x_q]$ if and only if both $\mathfrak{A} \models \theta_1[x_0 \dots x_q]$ and $\mathfrak{A} \models \theta_2[x_0 \dots x_q]$.

(ii). If φ is $\neg\,\theta$, then

$$\mathfrak{A} \models \varphi[x_0 \dots x_q] \text{ if and only if not } \mathfrak{A} \models \theta[x_0 \dots x_q].$$

(iii). If φ is $(\forall v_i)\psi$, where $i \leqslant q$, then

$\mathfrak{A} \models \varphi[x_0 \dots x_q]$ if and only if for every $x \in A$, $\mathfrak{A} \models \psi[x_0 \dots x_{i-1} x x_{i+1} \dots x_q]$.

Our definition of 1.3.12 is now completed. As simple exercises, the reader should check that the abbreviations $\vee$, $\rightarrow$, $\leftrightarrow$, $\exists$ have their usual meanings. In particular, if φ is $(\exists v_i)\psi$, where $i \leqslant q$, then

$\mathfrak{A} \models \varphi[x_0 \dots x_q]$ if and only if there exists $x \in A$ such that

$$\mathfrak{A} \models \psi[x_0 \dots x_{i-1} x x_{i+1} \dots x_q].$$

More important, the reader should realize that we can formulate a precise definition of $t[x_0 \dots x_q]$ and $\mathfrak{A} \models \varphi[x_0 \dots x_q]$ in set theory, based upon 1.3.13–1.3.15.

Having finished our definition, our first task is to prove the proposition that the relation

$$\mathfrak{A} \models \varphi(v_0 \dots v_p)[x_0 \dots x_q]$$

depends only on $x_0, \dots, x_p$, where $p < q$. This is the last part of the plan we have outlined.

PROPOSITION 1.3.16

(i). *Let $t(v_0 \dots v_p)$ be a term and let $x_0, \dots, x_q$ and $y_0, \dots, y_r$ be two sequences such that $p \leqslant q$, $p \leqslant r$, and $x_i = y_i$ whenever v_i is a variable of t. Then*

$$t[x_0 \dots x_q] = t[y_0 \dots y_r].$$

(ii). *Let φ be a formula all of whose free and bound variables are among $v_0, \dots, v_p$, and let $x_0, \dots, x_q$ and $y_0, \dots, y_r$ be two sequences such that $p \leqslant q, p \leqslant r$, and $x_i = y_i$ whenever v_i is a free variable of φ. Then*

$$\mathfrak{A} \models \varphi[x_0 \dots x_q] \textit{ if and only if } \mathfrak{A} \models \varphi[y_0 \dots y_r].$$

REMARK. Proposition 1.3.16 shows that the value of a term t at $x_0, \dots, x_q$ and whether a formula φ is satisfied or not by a sequence $x_0, \dots, x_q$ *depend only* on those values of x_i for which v_i is a free variable, and are *independent* of the other values of the sequence as well as the length of the sequence. The length q of the sequence must be high enough to cover all the free and bound variables of t and φ in order for the expressions $t[x_0 \dots x_q]$, $\mathfrak{A} \models \varphi[x_0 \dots x_q]$ to be defined at all. We can now immediately infer that if σ is a sentence, then $\mathfrak{A} \models \sigma[x_0 \dots x_q]$ is entirely independent of the sequence $x_0, \dots, x_q$. The importance of the above proposition is that it allows us to make the following definition.

1.3.17. Let $\varphi(v_0 \dots v_p)$ be a formula all of whose free and bound variables are among $v_0, \dots, v_q$, $p \leqslant q$. Let $x_0, \dots, x_p$ be a sequence of elements of A. We say that φ *is satisfied in* $\mathfrak{A}$ *by* $x_0, \dots, x_p$,

$$\mathfrak{A} \models \varphi[x_0 \dots x_p],$$

if and only if φ is satisfied in $\mathfrak{A}$ by $x_0, \dots, x_p, \dots, x_q$ for some (or, equivalently, every) $x_{p+1}, \dots, x_q$.

Let φ be a sentence all of whose bound variables are among $v_0, \dots, v_q$. We say that $\mathfrak{A}$ *satisfies* φ, in symbols $\mathfrak{A} \models \varphi$, iff φ is satisfied in $\mathfrak{A}$ by some (or, equivalently, every) sequence $x_0, \dots, x_q$.

The proof of Proposition 1.3.16 is straightforward but tedious. We shall sketch it here as a first example of an inductive proof on the 'complexity' of formulas. We shall often omit similar easy inductive proofs in the future.

PROOF OF PROPOSITION 1.3.16

(i). If $t(v_0 \dots v_p)$ is a variable v_i, then

$$t[x_0 \dots x_q] = x_i = y_i = t[y_0 \dots y_r].$$

If $t(v_0 \dots v_p)$ is a constant symbol c, and x is the interpretation of c in $\mathfrak{A}$, then

$$t[x_0 \dots x_q] = x = t[y_0 \dots y_r].$$

Suppose $t(v_0 \dots v_p)$ is $F(t_1 \dots t_m)$, where F is an m-placed function symbol and the proposition holds for each of the terms $t_1, \dots, t_m$. This means that

$$t_i[x_0 \dots x_q] = t_i[y_0 \dots y_r], \qquad (i = 1, \dots, m).$$

Therefore, if G is the interpretation of F in $\mathfrak{A}$,

$$\begin{aligned} t[x_0 \dots x_q] &= G(t_1[x_0 \dots x_q] \dots t_m[x_0 \dots x_q]) \\ &= G(t_1[y_0 \dots y_r] \dots t_m[y_0 \dots y_r]) = t[y_0 \dots y_r]. \end{aligned}$$

This verifies (i) for all terms t.

(ii). If φ is an atomic formula $t_1 \equiv t_2$, then using (i) we see that

$$\begin{aligned} t_1[x_0 \dots x_q] &= t_1[y_0 \dots y_r], \\ t_2[x_0 \dots x_q] &= t_2[y_0 \dots y_r]. \end{aligned}$$

Therefore the following are equivalent:

$$\begin{gathered} \mathfrak{A} \models \varphi[x_0 \dots x_q], \\ t_1[x_0 \dots x_q] = t_2[x_0 \dots x_q], \\ t_1[y_0 \dots y_r] = t_2[y_0 \dots y_r], \\ \mathfrak{A} \models \varphi[y_0 \dots y_r]. \end{gathered}$$

Let φ be an atomic formula $P(t_1 \dots t_n)$, where P is an n-placed predicate symbol and $t_1, \dots, t_n$ are terms. Then, using (i), we see that the following are equivalent (where R is the interpretation of P in $\mathfrak{A}$):

$$\begin{gathered} \mathfrak{A} \models \varphi[x_0 \dots x_q], \\ R(t_1[x_0 \dots x_q] \dots t_n[x_0 \dots x_q]), \\ R(t_1[y_0 \dots y_r] \dots t_n[y_0 \dots y_r]), \\ \mathfrak{A} \models \varphi[y_0 \dots y_r]. \end{gathered}$$

Suppose now that ψ, θ are formulas, all of whose free and bound variables are among $v_0, \dots, v_p$, which satisfy part (ii) of the proposition.

If φ is $\psi \wedge \theta$, the following are equivalent:

$$\begin{gathered}\mathfrak{A} \models \varphi[x_0 \dots x_q],\\ \mathfrak{A} \models \psi[x_0 \dots x_q] \text{ and } \mathfrak{A} \models \theta[x_0 \dots x_q],\\ \mathfrak{A} \models \psi[y_0 \dots y_r] \text{ and } \mathfrak{A} \models \theta[y_0 \dots y_r],\\ \mathfrak{A} \models \varphi[y_0 \dots y_r].\end{gathered}$$

If φ is $\neg\psi$, then the following are equivalent.

$$\begin{gathered}\mathfrak{A} \models \varphi[x_0 \dots x_q],\\ \text{not } \mathfrak{A} \models \psi[x_0 \dots x_q],\\ \text{not } \mathfrak{A} \models \psi[y_0 \dots y_r],\\ \mathfrak{A} \models \varphi[y_0 \dots y_r].\end{gathered}$$

Finally, let φ be $(\forall v_i)\psi$, where $i \leqslant p$. Then the following are equivalent:

$$\begin{gathered}\mathfrak{A} \models \varphi[x_0 \dots x_q],\\ \text{for all } x \in A,\ \mathfrak{A} \models \psi[x_0 \dots x_{i-1} x x_{i+1} \dots x_q],\\ \text{for all } y \in A,\ \mathfrak{A} \models \psi[y_0 \dots y_{i-1} y y_{i+1} \dots y_r],\\ \mathfrak{A} \models \varphi[y_0 \dots y_r].\end{gathered}$$

In this last part of the proof we used the fact that the free variables of ψ are just the free variables of φ and, perhaps, v_i. Our proof is now complete. ⊣

We shall state one more elementary proposition which deals with the behavior of the satisfaction relation under the substitution of variables by terms. We omit the proof, which is another tedious but straightforward induction.

PROPOSITION 1.3.18. *Let $\varphi(v_0 \dots v_p)$ be a formula and let $t_0(v_0 \dots v_p), \dots, t_p(v_0 \dots v_p)$ be terms. Suppose that no variable occurring in any of the terms $t_0, \dots, t_p$ occurs bound in φ. Let $x_0, \dots, x_p$ be a sequence of elements of A and let $\varphi(t_0 \dots t_p)$ be the formula obtained from φ by substituting t_i for v_i $(i = 0, \dots, p)$. Then*

$$\mathfrak{A} \models \varphi(t_0 \dots t_p)[x_0 \dots x_p] \textit{ if and only if } \mathfrak{A} \models \varphi[t_0[x_0 \dots x_p] \dots t_p[x_0 \dots x_p]].$$

This proposition is especially useful in the simple case that the terms $t_0, \dots, t_p$ are constant symbols $c_0, \dots, c_p$ whose interpretations in $\mathfrak{A}$ are $a_0, \dots, a_p$. In that case, $\varphi(c_0 \dots c_p)$ is a sentence, and the proposition shows that

$$\mathfrak{A} \models \varphi(c_0 \dots c_p) \text{ if and only if } \mathfrak{A} \models \varphi[a_0 \dots a_p].$$

Thus a sentence formed by replacing a tuple variables by constant symbols is true in a model if and only if the tuple of interpretations of the constant symbols satisfies the formula in the model.

We have now completed the project started several paragraphs back. Namely, we say that a sentence

$$\sigma \textit{ is true in } \mathfrak{A}$$

iff

$$\mathfrak{A} \models \sigma[x_0 \ldots x_q] \text{ for some (or for every) sequence } x_0, \ldots, x_q \text{ of } A.$$

We use the special notation $\mathfrak{A} \models \sigma$ to denote that σ is true in $\mathfrak{A}$. This last phrase is equivalent to each of the following phrases:

σ *holds in* $\mathfrak{A}$;
$\mathfrak{A}$ *satisfies* σ;
σ *is satisfied in* $\mathfrak{A}$;
$\mathfrak{A}$ *is a model of* σ.

When it is not the case that σ holds in $\mathfrak{A}$, we say that σ *is false in* $\mathfrak{A}$, or that σ *fails in* $\mathfrak{A}$, or $\mathfrak{A}$ *is a model of* $\neg\sigma$. Given a set Σ of sentences, we say that $\mathfrak{A}$ *is a model of* Σ iff $\mathfrak{A}$ is a model of each σ in Σ; it is convenient to use the notation $\mathfrak{A} \models \Sigma$ for this notion. A sentence σ that holds in every model for $\mathscr{L}$ is called *valid*. A sentence, or a set of sentences, is *satisfiable* iff it has at least one model. $\models \sigma$ denotes that σ is a valid sentence.

A sentence φ is a *consequence of another sentence* σ, in symbols $\sigma \models \varphi$, iff every model of σ is a model of φ. A sentence φ is a *consequence of a set of sentences* Σ, in symbols $\Sigma \models \varphi$, iff every model of Σ is a model of φ. It follows that

$$\Sigma \cup \{\sigma\} \models \varphi \text{ if and only if } \Sigma \models \sigma \rightarrow \varphi.$$

Two models $\mathfrak{A}$ and $\mathfrak{B}$ for $\mathscr{L}$ are *elementarily equivalent* iff every sentence that is true in $\mathfrak{A}$ is true in $\mathfrak{B}$, and vice versa. We express this relationship between models by $\equiv$. It is easy to see that $\equiv$ is indeed an equivalence relation. The symbol we have chosen to denote elementary equivalence is exactly the same as the identity symbol for the language $\mathscr{L}$. However, no confusion can ever arise because one is a relation between models for $\mathscr{L}$ and the other is a relation between terms of $\mathscr{L}$. If the context is clear, *equivalent* shall mean elementarily equivalent.

PROPOSITION 1.3.19. *If* $\mathfrak{A} \cong \mathfrak{B}$, *then* $\mathfrak{A} \equiv \mathfrak{B}$. *In case* $\mathfrak{A}$ *is finite, then the converse is also true.*

We conclude this section by stating a number of important results without proofs, but whose proofs will be given in the next chapter.

THEOREM 1.3.20 (Gödel's Completeness Theorem). *Given any sentence σ, σ is a theorem of $\mathscr{L}$ if and only if σ is valid.*

THEOREM 1.3.21 (Extended Completeness Theorem). *Let Σ be any set of sentences. Then Σ is consistent if and only if Σ has a model.*

THEOREM 1.3.22 (Compactness Theorem). *A set of sentences Σ has a model if and only if every finite subset of Σ has a model.*

As in Section 1.2, we conclude with a table of equivalent notions.

TABLE 1.3.1

Syntax	Semantics
φ is a theorem, $\vdash \varphi$	φ is valid, $\models \varphi$
Σ is consistent	Σ has a model
φ is deducible from Σ, $\Sigma \vdash \varphi$	φ is a consequence of Σ, $\Sigma \models \varphi$

EXERCISES

1.3.1. Prove that the isomorphism relation $\cong$ is an equivalence relation. Let α be any cardinal. Show that there are at most $2^{\alpha \cup ||\mathscr{L}||}$ nonisomorphic models for $\mathscr{L}$ of power α.

1.3.2. Let $\mathfrak{A} \subsetapprox \mathfrak{B}$ mean that $\mathfrak{A}$ is isomorphically embedded in $\mathfrak{B}$. Show that the relation $\subsetapprox$ is reflexive and transitive but not antisymmetric. Let N be the set of all natural numbers 0, 1, 2, ... Decide if the following are true or false:

$$\langle N, \leqslant, +, 0\rangle \subsetapprox \langle N, \leqslant, \cdot, 1\rangle,$$
$$\langle N, \leqslant, \cdot, 1\rangle \subsetapprox \langle N, \leqslant, +, 0\rangle,$$
$$\langle N-\{0\}, \leqslant, \cdot, 1\rangle \subsetapprox \langle N, \leqslant, +, 0\rangle,$$
$$\langle N-\{0\}, \cdot, 1\rangle \subsetapprox \langle N, +, 0\rangle,$$
$$\langle N-\{0\}, \cdot\rangle \subsetapprox \langle N, +\rangle.$$

Take $\leqslant$, $+$, and $\cdot$ as the usual ordering and operations on N.

1.3.3. Let $\varphi(v_0 \dots v_n)$ be a formula of $\mathscr{L}$, and $\mathfrak{A}$ be a model for $\mathscr{L}$. Prove that:

(i). The satisfaction relation $\mathfrak{A} \models \varphi[x_0 \dots x_n]$ has a precise definition in Zermelo-Fraenkel set theory.

(ii). If $\mathfrak{A}'$ is an expansion of $\mathfrak{A}$ and $x_0, \dots, x_n \in A$, then

$$\mathfrak{A} \models \varphi[x_0 \dots x_n] \quad \text{if and only if} \quad \mathfrak{A}' \models \varphi[x_0 \dots x_n].$$

1.3.4. Prove Proposition 1.3.19. Also construct a counterexample if $\mathfrak{A}$ is not finite.

1.3.5. A sentence φ is *universal* iff it is in prenex form and all of its quantifiers are universal, i.e., $\forall$. Prove that if φ is universal and $\mathfrak{A} \subset \mathfrak{B}$ and $\mathfrak{B} \models \varphi$, then $\mathfrak{A} \models \varphi$. A sentence is *existential* iff it is in prenex form and all of its quantifiers are existential, i.e., $\exists$. Prove that if φ is existential and $\mathfrak{A} \subset \mathfrak{B}$ and $\mathfrak{A} \models \varphi$, then $\mathfrak{B} \models \varphi$. Thus universal sentences are preserved under submodels and existential sentences are preserved under extensions.

1.3.6. There are at most $2^{||\mathscr{L}||}$ nonequivalent models for $\mathscr{L}$.

1.3.7. Let $\mathfrak{A}$ and $\mathfrak{B}$ be equivalent models for $\mathscr{L}$. Suppose that every element of A is a constant of A and the same is true for $\mathfrak{B}$. Then show that $\mathfrak{A} \cong \mathfrak{B}$. If the hypothesis is assumed only for $\mathfrak{A}$, show that $\mathfrak{A}$ is isomorphically embedded in $\mathfrak{B}$.

1.3.8*. Find a necessary and sufficient condition on $\mathscr{L}$ so that there will be *exactly* $2^{||\mathscr{L}||}$ nonequivalent models. Do the same for exactly 2^{α} non-isomorphic models of power α, for each infinite cardinal α.

1.3.9. Let $\mathfrak{A}$ be a model for $\mathscr{L}$ and let X be a nonempty subset of $\mathfrak{A}$. Let

$$B = \bigcap \{C : \mathfrak{C} \subset \mathfrak{A} \quad \text{and} \quad X \subset C\}.$$

Then there is a submodel $\mathfrak{B} \subset \mathfrak{A}$ with universe B. $\mathfrak{B}$ is called the *submodel generated* by X.

1.3.10. Let $\mathfrak{A}$, X be as above and let $\mathfrak{B}$ be the submodel generated by X. Then

$$B = \{t[x_1 \ldots x_n] : t \quad \text{is a term of } \mathscr{L} \text{ and} \quad x_1, \ldots, x_n \in X\}.$$

Moreover,

$$|X| \leqslant |B| \leqslant |X| \cup ||\mathscr{L}||.$$

1.3.11. Suppose $X \subset A$ and X generates the whole model $\mathfrak{A}$. Let f be a one–one map on X into another model $\mathfrak{B}$. Then there is at most one isomorphic embedding g on $\mathfrak{A}$ into $\mathfrak{B}$ such that $f \subset g$.

1.3.12. Suppose $\mathscr{L}$ has no function or constant symbols. Then for every model $\mathfrak{A}$ for $\mathscr{L}$ and every nonempty subset $X \subset A$, $\mathfrak{A}$ has a submodel with universe X.

Suppose $\mathscr{L}$ has no function symbols. Then for every model $\mathfrak{A}$ for $\mathscr{L}$ and every nonempty $X \subset A$ which contains all the constants of $\mathfrak{A}$, $\mathfrak{A}$ has a submodel with universe X.

1.3.13. Verify all the claims in Table 1.3.1, assuming the extended completeness theorem.

1.3.14*. An element $a \in A$ of a model $\mathfrak{A}$ is said to be *definable* (in $\mathfrak{A}$) iff there is a formula $\varphi(x)$ of $\mathscr{L}$ such that a is the only element in A satisfying φ. For each $n \in \omega$, find a model $\mathfrak{A}_n$ for $\mathscr{L}$ and a language with only a finite number of symbols, which has exactly n undefinable elements. For $n = 0$ or $n > 1$, the examples are easy to find. For $n = 1$, it is much harder.

1.3.15*. Let $\mathscr{L}$ have only a finite number of relation and constant symbols, but no function symbols. Define the relations $\equiv_n$ on models for $\mathscr{L}$ as follows by induction:

$\mathfrak{A} \equiv_0 \mathfrak{B}$ iff the submodels of $\mathfrak{A}$ and $\mathfrak{B}$ generated by the constant elements are isomorphic, or else $\mathscr{L}$ has no constant symbols.

$\mathfrak{A} \equiv_{n+1} \mathfrak{B}$ iff for every $a \in A$ there exists $b \in B$ such that $(\mathfrak{A}, a) \equiv_n (\mathfrak{B}, b)$, and for every $b \in B$ there exists $a \in A$ such that $(\mathfrak{A}, a) \equiv_n (\mathfrak{B}, b)$.

Note that the definition of $\equiv_{n+1}$ depends on the fact that $\equiv_n$ has already been defined for all languages of the form $\mathscr{L} \cup \{c_1, \ldots, c_m\}$. Prove that $\mathfrak{A} \equiv \mathfrak{B}$ if and only if for all n, $\mathfrak{A} \equiv_n \mathfrak{B}$.

[*Hint*: Show that for each n, there are finitely many $\equiv_n$ classes, and each $\equiv_n$ class is the class of all models of a sentence of $\mathscr{L}$.]

1.3.16. Let $\mathscr{L}$ be as in the previous exercise. Prove that if $\mathfrak{A} \equiv_n \mathfrak{B}$ then for every sentence φ of $\mathscr{L}$ in prenex form with at most n quantifiers,

$$\mathfrak{A} \models \varphi \quad \text{if and only if} \quad \mathfrak{B} \models \varphi.$$

1.3.17*. Let $\mathscr{L}$ be as in Exercise 1.3.15. Let K be a class of models for $\mathscr{L}$. Prove that the following are equivalent:

(i). There is a sentence φ of $\mathscr{L}$ such that K is the class of all models for $\mathscr{L}$ satisfying φ.

(ii). For some $n \in \omega$, K is closed under $\equiv_n$, i.e., if $\mathfrak{A} \in K$ and $\mathfrak{A} \equiv_n \mathfrak{B}$, then $\mathfrak{B} \in K$.

[*Hint*: For each n there are only finitely many nonequivalent prenex sentences with at most n quantifiers.]

1.3.18. Show that Exercise 1.3.15 is false if either $\mathscr{L}$ has function symbols, or $\mathscr{L}$ has infinitely many constant or relation symbols. However, it is still true that if for all n, $\mathfrak{A} \equiv_n \mathfrak{B}$, then $\mathfrak{A} \equiv \mathfrak{B}$.

1.3.19. Show that two models $\mathfrak{A}$, $\mathfrak{B}$ for $\mathscr{L}$ are equivalent if and only if for all finite $\mathscr{L}_0 \subset \mathscr{L}$ the reducts of $\mathfrak{A}$ and $\mathfrak{B}$ to $\mathscr{L}_0$ are equivalent. This exercise shows that a version of Exercise 1.3.15 can be given if $\mathscr{L}$ has only relation and constant symbols.

1.3.20*. As applications of Exercise 1.3.15, show that the following pairs of models $\mathfrak{A}$, $\mathfrak{B}$ are equivalent:

(i). $\mathfrak{A} = \langle A \rangle$, $\mathfrak{B} = \langle B \rangle$, where A and B are infinite.

(ii). $\mathfrak{A} = \langle A, \leqslant \rangle$, $\mathfrak{B} = \langle B, \leqslant \rangle$, where A and B are densely ordered by $\leqslant$ with no endpoints.

(iii). $\mathfrak{A} = \langle \omega, \leqslant \rangle$, $\mathfrak{B} = \langle \omega + \omega^* + \omega, \leqslant \rangle$, where $\omega + \omega^* + \omega$ is the order type of the natural numbers followed by a copy of the integers.

(iv). $\mathfrak{A} = \langle \omega^\omega, \leqslant \rangle$, $\mathfrak{B} = \langle \omega_1, \leqslant \rangle$, where ω^ω is the ordinal exponentiation of ω to the power ω.

(v). $\mathfrak{A} = \langle S_\omega(X), \subset \rangle$, $\mathfrak{B} = \langle S_\omega(Y), \subset \rangle$, where X and Y are infinite sets, $S_\omega(X)$, $S_\omega(Y)$ are the sets of finite subsets of X and Y, respectively, and $\subset$ is the inclusion relation.

1.3.21. Let T be a set of universal sentences. Assume $T \models \forall x \exists y P(x, y)$. Prove that there exist terms $t_1(x), \ldots, t_n(x)$ such that

$$T \models \forall x \bigvee_{m=1}^{n} P(x, t_m(x)).$$

1.3.22*. Let $\mathfrak{A}$ be a countable model for a countable language. Prove that if the simple expansion $(\mathfrak{A}, b)$ has more than one automorphism for each finite sequence b of elements of A, then $\mathfrak{A}$ has 2^ω automorphisms.

1.4. Theories and examples of theories

A (*first-order*) *theory* T of $\mathscr{L}$ is a collection of sentences of $\mathscr{L}$. T is said to be *closed* iff it is closed under the $\models$ relation. In view of Table 1.3.1, this is the same as requiring that T be closed under $\vdash$. Since theories are sets of sentences of $\mathscr{L}$, we may apply the expressions

a model of a theory,
consistent theory,
satisfiable theory,

as introduced in Section 1.3.

A theory T is called *complete* (in $\mathscr{L}$) iff its set of consequences is maximal consistent. If T is a theory of $\mathscr{L}$ and $\mathscr{L} \subset \mathscr{L}'$, $\mathscr{L} \neq \mathscr{L}'$, then T is not a closed theory of $\mathscr{L}'$. On the other hand, it is easy to see that if $\mathscr{L}' \subset \mathscr{L}$, then the *restriction of a closed theory T to $\mathscr{L}'$*, in symbols $T|\mathscr{L}'$, is always a closed theory of $\mathscr{L}'$. T is a *subtheory of T'* iff $T \subset T'$. If T is a subtheory of T', then T' is an *extension* of T.

A *set of axioms of a theory T* is a set of sentences with the same consequences as T. Clearly, T is a set of axioms of T, and the empty set is a set of axioms of T if and only if T is a set of valid sentences of $\mathscr{L}$. Every set of sentences Σ is a set of axioms for the closed theory $T = \{\varphi : \Sigma \models \varphi\}$. A theory T is *finitely axiomatizable* iff it has a finite set of axioms.

The most convenient and standard way of giving a theory T is by listing a finite or infinite set of axioms for it. Another way to give a theory is as follows: Let $\mathfrak{A}$ be a model for $\mathscr{L}$; then the *theory of* $\mathfrak{A}$ is the set of all sentences which hold in $\mathfrak{A}$. The theory of any model $\mathfrak{A}$ is obviously a complete theory.

Historically, the importance of theories stems from the following two facts. Once the axioms of a theory are given, then by using the relation $\vdash$ we can find out, in a syntactical manner, all the consequences of T. On the other hand, by using the satisfaction relation, we can also study all the models of T.

By the extended completeness theorem, these two approaches give basically the same results about consequences of T. However, owing to the fact that models of T also have non-first-order properties, such as isomorphism, submodels, extensions, plus many others, the second approach leads to the field now known as model theory.

We shall give in the rest of this section some examples of theories and their models to show the intimate connections that model theory has with other branches of mathematics. In each example we describe a closed theory by a set of axioms. Some classical results will be stated without proof.

1.4.1. Let $\mathscr{L}$ consist of the single 2-placed relation symbol $\leqslant$. Using the usual notation for $\leqslant$, we write $x \leqslant y$ for $\leqslant(xy)$. The theory of *partial order* has three axioms:

(1) $(\forall xyz)(x \leqslant y \wedge y \leqslant z \rightarrow x \leqslant z)$,
(2) $(\forall xy)(x \leqslant y \wedge y \leqslant x \rightarrow x \equiv y)$,
(3) $(\forall x)(x \leqslant x)$.

They are, respectively, the transitive, antisymmetric, and reflexive properties of partial orders. Any model $\langle A, \leqslant \rangle$ of this theory consists of a nonempty set A and a partial order relation $\leqslant$ on A. If we add the comparability axiom

(4) $(\forall xy)(x \leqslant y \vee y \leqslant x)$,

we obtain the theory of *simple order* (also called *linear order*). A model $\langle A, \leqslant \rangle$ for this theory is a simply-ordered set. Adding two more axioms (writing $x \not\equiv y$ for $\neg (x \equiv y)$):

(5) $(\forall xy)(x \leqslant y \wedge x \not\equiv y \rightarrow (\exists z)(x \leqslant z \wedge z \not\equiv x \wedge z \leqslant y \wedge z \not\equiv y))$,
(6) $(\exists xy)(x \not\equiv y)$,

we then have the theory of *dense* (*simple*) *order*. The rationals with the usual $\leqslant$ is an example of a model of this theory. The theory of dense order has no finite models. If we wish to consider only dense orders *without endpoints*, we add the axioms

(7) $(\forall x)(\exists y)(x \leqslant y \wedge x \not\equiv y)$,
(8) $(\forall x)(\exists y)(y \leqslant x \wedge x \not\equiv y)$.

PROPOSITION 1.4.2. *Any two countable models of the theory of dense order without endpoints are isomorphic.*

EXAMPLE 1.4.3. Let $\mathscr{L} = \{+, \cdot, {}^{-}, 0, 1\}$, where $+, \cdot$ are 2-placed function symbols, $^{-}$ is a 1-placed function symbol, and 0 and 1 are constant symbols. The theory of *Boolean algebras* has the following axioms (where we shall assume that the following formulas all have their free variables universally quantified in front).

Associativity of $+$ and $\cdot$:
$x+(y+z) \equiv (x+y)+z$, $\quad x \cdot (y \cdot z) \equiv (x \cdot y) \cdot z$.

Commutativity of $+$ and $\cdot$:
$x+y \equiv y+x$, $\quad x \cdot y \equiv y \cdot x$.

Idempotent laws:
$x+x \equiv x$, $\quad x \cdot x \equiv x$.

Distributive laws:
$x+(y \cdot z) \equiv (x+y) \cdot (x+z)$, $\quad x \cdot (y+z) \equiv x \cdot y + x \cdot z$.

Absorption laws:
$x+(x \cdot y) \equiv x$, $\quad x \cdot (x+y) \equiv x$.

De Morgan laws:
$\overline{x+y} \equiv \bar{x} \cdot \bar{y}$, $\quad \overline{x \cdot y} \equiv \bar{x} + \bar{y}$.

Laws of zero and one:
$x+0 \equiv x$, $\quad x \cdot 0 \equiv 0$,
$x+1 \equiv 1$, $\quad x \cdot 1 \equiv x$,
$0 \not\equiv 1$,
$x+\bar{x} \equiv 1$, $\quad x \cdot \bar{x} \equiv 0$.

Law of double negation:

$\bar{\bar{x}} \equiv x.$

A model $\mathfrak{A} = \langle A, +, \cdot, ^{-}, 0, 1\rangle$ of this theory is called a Boolean algebra. (Strictly speaking, we should write $+_{\mathfrak{A}}, \cdot_{\mathfrak{A}}, -_{\mathfrak{A}}, 0_{\mathfrak{A}}, 1_{\mathfrak{A}}$ in the above model. But following our convention we shall drop the subscripts.) A partial order $\leqslant$ can be defined on A by: $x \leqslant y$ if and only if $x+y = y$. It can be shown that $\leqslant$ has a largest element, namely 1, a smallest element, namely 0, and, given any two elements $x, y \in A$, the l.u.b. (least upper bound) of x and y is $x+y$, and the g.l.b. (greatest lower bound) of x and y is $x \cdot y$.

A *field of sets* S is a collection of subsets of a nonempty set X such that both the empty set $\emptyset$ and the set X are in S and S is closed under $\cup$, $\cap$ and $^{-}$ with respect to X. It is easy to see that if S is a field of sets, then

$$\langle S, \cup, \cap, ^{-}, \emptyset, X\rangle$$

is a Boolean algebra. Conversely, we have:

PROPOSITION 1.4.4 (Representation Theorem for Boolean algebras). *Every Boolean algebra is isomorphic to a field of sets.*

An *atom* of a Boolean algebra is an element $x \neq 0$ such that there is no element y which lies properly between 0 and x, i.e., not $0 \leqslant y \leqslant x$, $0 \neq y$, $y \neq x$. A Boolean algebra is *atomic* iff every nonzero element x includes an atom. A Boolean algebra is *atomless* iff it has no atoms. There are Boolean algebras which are neither atomic nor atomless. Adding the axiom (writing $x \leqslant y$ for $x+y \equiv y$)

$(\forall x)\ (0 \not\equiv x \to (\exists y)\ (y \leqslant x \wedge 0 \not\equiv y \wedge (\forall z)\ (z \leqslant y \to z \equiv 0 \vee z \equiv y)))$

gives us the theory of *atomic Boolean algebras*; while adding the axiom

$\neg (\exists y)\ (0 \not\equiv y \wedge (\forall z)\ (z \leqslant y \to z \equiv 0 \vee z \equiv y))$

gives us the theory of *atomless Boolean algebras*.

PROPOSITION 1.4.5. *Any two countable atomless Boolean algebras are isomorphic.*

Some other relevant facts about Boolean algebras can be found in the exercises.

EXAMPLE 1.4.6. Let $\mathscr{L} = \{+, 0\}$, where $+$ is a 2-placed function symbol and 0 is a constant symbol. The theory of *groups* has the following axioms:

(1) $x+(y+z) \equiv (x+y)+z$ (associativity),

(2) $x+0 \equiv x, 0+x \equiv x$ (identity),

(3) $(\exists y)\ (x+y \equiv 0 \wedge y+x \equiv 0)$ (existence of inverse).

A model $\langle G, +, 0\rangle$ of this theory is a *group*. We obtain the theory of *Abelian groups* when we add the axiom

(4) $x+y \equiv y+x$ (commutativity).

The *order* of an element x of a group is the least n such that $x+x+\ \dots\ +x$ (n times) $= 0$. If no such n exists, the order of x is infinity. For a fixed $n \geqslant 1$, we can write down the abbreviation nx for the expression

$$x+(x+(\dots (x+x)\dots)),\ n \text{ times}.$$

Suppose p is a prime. The theory of *Abelian groups with all elements of order p* has the extra axiom

(5_p) $px \equiv 0$.

PROPOSITION 1.4.7. *Any two models of the theory of Abelian groups with all elements of order p of the same power are isomorphic.*

To obtain the theory of *Abelian groups with all elements of order* ∞ (*torsion-free*) we need an infinite list of axioms: for each $n \geqslant 1$, we add the axiom

(6_n) $x \not\equiv 0 \rightarrow nx \not\equiv 0$.

This theory is our first example of a nonfinitely axiomatizable theory. If we add a further infinite list of axioms, one for each $n \geqslant 1$,

(7_n) $(\exists y)\,(ny \equiv x)$,

we have the theory of *divisible torsion-free Abelian groups.*

PROPOSITION 1.4.8. *Any two uncountable divisible torsion-free Abelian groups of the same power are isomorphic. There are countably many such groups which are countable and not isomorphic.*

EXAMPLE 1.4.9. Let $\mathscr{L} = \{+, \cdot, 0, 1\}$, where $+$ and $\cdot$ are 2-placed function symbols and 0, 1 are constant symbols. The theory of *commutative rings* (*with unit*) has the axioms (1)–(4) listed above plus the axioms (8)–(11) given below:

(8) $1 \cdot x \equiv x \wedge x \cdot 1 \equiv x$ (1 is a unit),

(9) $x \cdot (y \cdot z) \equiv (x \cdot y) \cdot z$ (associativity of $\cdot$),

(10) $x \cdot y \equiv y \cdot x$ (commutativity of $\cdot$),

(11) $x \cdot (y+z) \equiv (x \cdot y)+(x \cdot z)$ (distributivity of $\cdot$ over $+$).

Adding one more axiom

(12) $x \cdot y \equiv 0 \rightarrow x \equiv 0 \vee y \equiv 0$ (no zero divisors),

gives us the theory of *integral domains*. Adding the two axioms

(13) $0 \not\equiv 1$,

(14) $x \not\equiv 0 \rightarrow (\exists y)(y \cdot x \equiv 1)$ (existence of multiplicative inverse),

gives the important theory of *fields*. For a fixed prime p, if we add the axiom

(15_p) $p1 \equiv 0$,

we have the theory of *fields of characteristic p*. On the other hand, if we add for all primes p the negation of (15_p), namely, all the axioms

(16) $p1 \not\equiv 0$, with p a prime,

we have the theory of *fields of characteristic zero*, each field has a unique characteristic, either prime or zero. We now introduce the abbreviation x^n for the expression

$$x \cdot (x \cdot (x \ldots x) \ldots), \ n \text{ times}.$$

The infinite list of axioms, one for each $n \geqslant 1$,

(17_n) $(\exists y)(x_n \cdot y^n + x_{n-1} \cdot y^{n-1} + \ldots + x_1 \cdot y + x_0 \equiv 0) \vee x_n \equiv 0$,

when added to the theory of fields, gives us the theory of *algebraically closed fields.*

PROPOSITION 1.4.10. *Any two uncountable algebraically closed fields of the same characteristic and power are isomorphic.*

Each axiom (17_n) says that every polynomial of degree n has a root. The theory of *real closed fields* has as axioms all the axioms for fields plus the axiom

(18) $(\forall x)(\exists y)(y^2 \equiv x \vee y^2 + x \equiv 0)$,

and two infinite lists of axioms. One is the infinite list (17_n) for all odd n, and the other is the infinite list that says that 0 is not a sum of nontrivial squares:

(18_n) $x_0^2 + x_1^2 + \ldots + x_n^2 \equiv 0 \rightarrow x_0 \equiv 0 \wedge x_1 \equiv 0 \wedge \ldots \wedge x_n \equiv 0$.

The theory of *ordered fields* is formulated in the language $\mathscr{L} = \{\leqslant, +, \cdot, 0, 1\}$. It has all the field axioms, the linear order axioms, and the additional axioms

$$x \leqslant y \rightarrow x + z \leqslant y + z,$$
$$x \leqslant y \wedge 0 \leqslant z \rightarrow x \cdot z \leqslant y \cdot z.$$

The ordered fields of rational numbers and of real numbers are examples.

Of the examples of theories we have discussed so far, the following are complete: dense order without endpoints, atomless Boolean algebras, infinite Abelian groups with all elements of order p, torsion-free divisible Abelian groups, algebraically closed fields of a given characteristic, and real closed fields. The various propositions show that each of these complete theories, except the last one, enjoys the unusual property that in some (sometimes all) infinite powers all models of the given theory of that power are isomorphic.

EXAMPLE 1.4.11. Let $\mathscr{L} = \{+, \cdot, S, 0\}$, where $+$, $\cdot$ are 2-placed function symbols, S is a 1-placed function symbol (called the successor function), and 0 is a constant symbol. *Number theory* (or *Peano arithmetic*) has the following list of axioms:

(1) $0 \not\equiv Sx$ (0 has no predecessor),
(2) $Sx \equiv Sy \rightarrow x \equiv y$ (S is one–one),
(3) $x+0 \equiv x$,
(4) $x+Sy \equiv S(x+y)$,
(5) $x \cdot 0 \equiv 0$,
(6) $x \cdot Sy \equiv (x \cdot y)+x$,

and, finally, for each formula $\varphi(v_0 \dots v_n)$ of $\mathscr{L}$, where v_0 does not occur bound in φ, the axiom

$$(7_\varphi)\quad \varphi(0v_1 \dots v_n) \wedge (\forall v_0)(\varphi(v_0 v_1 \dots v_n) \rightarrow \varphi(Sv_0 v_1 \dots v_n)) \rightarrow (\forall v_0)\varphi(v_0 \dots v_n).$$

Axioms (3) and (4) are the usual recursive definition of $+$ in terms of 0 and S, and axioms (5) and (6) are the recursive definition of $\cdot$ in terms of 0, S and $+$. The whole list of axioms (7_φ), one for each φ, is called the *axiom schema of induction.*

The *standard model* of number theory is $\langle \omega, +, \cdot, S, 0 \rangle$, where S is the successor function and $+, \cdot, 0$ have their usual meaning. All other (non-isomorphic) models are called *nonstandard. Complete number theory* (or *complete arithmetic*) is the set of all sentences φ of $\mathscr{L}$ that hold in the standard model.

There are several deep results about number theory:

Gödel's (1931) incompleteness theorem states that number theory is not complete; therefore, complete number theory is a proper extension of number theory.

No finite extension (that is, by adding a finite number of new axioms) of number theory is complete; therefore complete number theory is not finitely axiomatizable over number theory, whence it is certainly not finitely axiomatizable.

Number theory itself is not finitely axiomatizable. This was proved by Ryll–Nardzewski (1952) by the use of nonstandard models. The existence of nonstandard models of complete number theory was first shown by Skolem (1934).

We mention a number of interesting subtheories of number theory. For instance, if the induction schema (7_φ) is replaced by the single axiom

(8) $(\forall x)(x \not\equiv 0 \rightarrow (\exists y)(x \equiv Sy))$,

we obtain a finitely axiomatizable subtheory of number theory (the theory Q of Tarski, Mostowski and Robinson, 1953) which is incomplete, and no finite extension of it is complete.

In the language $\mathscr{L}' = \{S, 0\}$ obtained by leaving out the symbols $+$ and $\cdot$, the subtheory of number theory given by axioms (1), (2) and the schema (7_φ), restricted of course to formulas of $\mathscr{L}'$, is complete. However, it is still not finitely axiomatizable, as can be shown by using the compactness theorem.

In the language $\mathscr{L}'' = \{+, S, 0\}$, the axioms (1)–(4) and the schema (7_φ), again restricted to formulas of $\mathscr{L}''$, give the *additive number theory* (or *Pressburger arithmetic*). This theory is not finitely axiomatizable, but it is complete (Presburger, 1929); the completeness of the theory $\mathscr{L}'$ in the previous paragraph follows from the proof given by Presburger.

EXAMPLE 1.4.12. We shall now discuss some examples of set theories.

There are two quite different reasons to include a discussion of set theories in a book on model theory. The first reason is that, if we wish to be completely precise, we should formulate our whole treatment of model theory within an appropriate system of axiomatic set theory. Actually, we are taking the more practical approach of formulating things in an informal set theory, but it is still important that, *in principle*, we could do it all in an axiomatic set theory. We have left for the Appendix an outline of the informal set theory we are using. The other reason for discussing set theories is that they are among the most interesting and important examples of theories. The second reason is the one which concerns us at this time. The theory of models is particularly well suited to the study of models of set theory. In the Appendix we have listed the axioms for four of the most familiar set theories: Zermelo, Zermelo–Fraenkel, Bernays, and Bernays–Morse. The first two of them are formulated in the language $\mathscr{L} = \{\epsilon\}$, while the other two are formulated in the language $\mathscr{L}' = \{\epsilon, V\}$, where ϵ is a binary relation symbol and V is a unary relation symbol. Zermelo set theory is a subtheory of Zermelo–Fraenkel, and Bernays set theory is a subtheory of Bernays–Morse.

The deepest results in set theory use constructions of models. However, these constructions are often of a special nature, for models of set theory only, and are therefore outside the scope of this book. For instance, the notion of constructible sets was used by Gödel (1939) to show that if Bernays set theory is consistent, then it remains consistent if we add to it the axiom of choice and the generalized continuum hypothesis; in other words, if Bernays set theory has a model, then it has a model in which the axiom of choice and the generalized continuum hypothesis are true. The same proofs and results are also well known to hold for Zermelo–Fraenkel set theory. Cohen's forcing construction has been used by Cohen and others to obtain a remarkable series of additional consistency results (see Cohen, 1963). For example, if Bernays (or Zermelo–Fraenkel) set theory has a model, then it has a model in which the axiom of choice is false, and another model in which the axiom of choice is true but the generalized continuum hypothesis is false.

In the rest of our discussion let us use the abbreviation ZF for 'Zermelo–Fraenkel set theory'. Whether or not we can prove that ZF is consistent depends on just how much we are assuming in our intuitive set theory. If our intuitive set theory is just a replica of ZF, then we cannot prove the consistency of ZF, even if we allow the use of the axiom of choice. Similarly, for any of the other set theories T we have introduced in the Appendix, we cannot prove the consistency of T if our intuitive set theory is a replica of T. These assertions follow from the Gödel incompleteness theorem. On the other hand, in Bernays–Morse set theory we can prove the consistency of Bernays set theory and of ZF. In ZF we can prove the consistency of Zermelo set theory. If we assume the existence of an inaccessible cardinal, then we can prove that Bernays–Morse set theory as well as ZF are consistent. Bernays set theory and ZF are very close to each other, and we can prove that one is consistent if and only if the other is. We shall leave the last three results above for exercises.

Neither Zermelo set theory, nor ZF, nor Bernays–Morse set theory is finitely axiomatizable (assuming that they are consistent). But, surprisingly, Bernays set theory is finitely axiomatizable (Bernays, 1937). With its finite axiomatization it is sometimes called Bernays–Gödel set theory. Each of the four set theories in our discussion, like number theory, has the following property: if the theory is consistent, then it is not complete, and no finite extension of it is complete. This is another consequence of the Gödel incompleteness theorem.

There is no completely satisfactory notion of a 'standard' model of set theory. The closest thing to it is the notion of a *natural model*. Natural models, roughly, are models of the form $\langle M, \epsilon \rangle$, where M is a set of sets formed by starting with the empty set and repeating the operations of union and power set, while ϵ is the ϵ-relation restricted to M. More precisely, we define for each ordinal α the set $R(\alpha)$ by

$$R(0) = 0,$$

$$R(\alpha+1) = S(R(\alpha)),$$

and

$$R(\alpha) = \bigcup_{\beta<\alpha} R(\beta) \quad \text{if } \alpha \text{ is a limit ordinal.}$$

Then a *natural model* of ZF (or of Zermelo set theory) is a model of the form $\langle R(\alpha), \epsilon\rangle$. A natural model of Bernays set theory is a model of the form $\langle R(\alpha+1), \epsilon, R(\alpha)\rangle$.

None of our set theories has any countable natural models. For this reason, a somewhat weaker notion of 'standard' model is also important. A model $\langle M, \epsilon\rangle$ is said to be a *transitive model* iff ϵ is the ϵ-relation restricted to M and every element of an element of M is an element of M. For models of the language $\mathscr{L}' = \{\epsilon, V\}$ we make a similar definition. The countable transitive models are the most important models for Cohen's forcing construction.

Since number theory has just one standard model and is not complete, it has consistent extensions which have no standard models. If ZF has any transitive model at all, then it has many nonequivalent transitive models. Nevertheless, if ZF is consistent, then it has consistent extensions which have no transitive models at all. Moreover, in ZF plus the axiom of choice, we cannot prove the following: if ZF has a model, then ZF has a transitive model.

EXERCISES

1.4.1. Is there a theory of well order in the first-order language $\{\leqslant\}$?

1.4.2. Find two dense orders without endpoints of the same power which are not isomorphic.

1.4.3. Every finite Boolean algebra is atomic. If it has n atoms, then it has exactly 2^n elements. Any two finite Boolean algebras with the same number of elements are isomorphic.

1.4.4. Every finite subset of a Boolean algebra generates a finite sub-Boolean algebra.

1.4.5. Find two atomless nonisomorphic Boolean algebras of the same power.

1.4.6. Prove the following weak form of the representation theorem: Every atomic Boolean algebra is isomorphic to a field of sets.

1.4.7. Prove that the theory of *infinite models* whose axioms are the infinite list of sentences σ_n, where each sentence σ_n says that there are at least n distinct elements, is not finitely axiomatizable.

1.4.8. Prove that there is no theory T such that $\mathfrak{A}$ is a model of T if and only if $\mathfrak{A}$ is finite.

1.4.9. Let $\mathfrak{A} = \langle A, +, \cdot, ^-, 0, 1\rangle$ be a Boolean algebra. A (proper) *filter* on $\mathfrak{A}$ is a subset $D \subset A$ such that $D \neq \emptyset$, $D \neq A$, and whenever $x, y \in D$ and $x \leqslant z$, then $x \cdot y \in D$ and $z \in D$. A subset E of A is said to have the *finite intersection property* iff for all $x_1, \ldots, x_n \in E$ and all n, $x_1 \cdot x_2 \cdot \ldots \cdot x_n \neq 0$. Prove that every subset E with the finite intersection property generates a filter D in the following sense:

$$x \in D \quad \text{iff} \quad x \geqslant y_1 \cdot \ldots \cdot y_n \quad \text{for some} \quad y_1, \ldots, y_n \in E.$$

A filter D on $\mathfrak{A}$ is said to be *principal* iff for some element $a \neq 0$ in A,

$$x \in D \quad \text{iff} \quad a \leqslant x.$$

D is an *ultrafilter* on $\mathfrak{A}$ iff no proper extension of D is a filter on $\mathfrak{A}$. Prove that the only principal ultrafilters on $\mathfrak{A}$ are generated by the atoms of $\mathfrak{A}$, i.e., D is a principal ultrafilter iff for some atom $a \in A$,

$$x \in D \quad \text{iff} \quad a \leqslant x.$$

Thus, if $\mathfrak{A}$ is atomless, then all ultrafilters on $\mathfrak{A}$ are nonprincipal. Prove the following:

(i). Every nonzero element of A belongs to an ultrafilter, and more generally, every filter on $\mathfrak{A}$ can be extended to an ultrafilter.

(ii). If D is an ultrafilter on $\mathfrak{A}$, then

$$x+y \in D \quad \text{iff either} \quad x \in D \quad \text{or} \quad y \in D, \qquad x \in D \quad \text{iff} \quad \bar{x} \notin D.$$

(iii). Let X be the set of all ultrafilters on $\mathfrak{A}$. For each $a \in A$, define

$$h_a = \{D \in X : a \in D\}.$$

Show that h is an isomorphism of $\mathfrak{A}$ onto the field of sets

$$\langle \{h_a : a \in A\}, \cup, \cap, ^-, \emptyset, X\rangle.$$

This gives a proof of Proposition 1.4.4.

1.4.10. Let $\mathscr{L}$ be a first-order language and consider the equivalence relation $\vdash \varphi \leftrightarrow \psi$ on the formulas of $\mathscr{L}$. Let

$$(\varphi) = \{\psi : \vdash \varphi \leftrightarrow \psi\},$$
$$B_{\mathscr{L}} = \{(\varphi) : \varphi \text{ a formula of } \mathscr{L}\},$$
$$0_{\mathscr{L}} = (\varphi \wedge \neg \varphi),\ 1_{\mathscr{L}} = (\varphi \vee \neg \varphi).$$

Define

$$(\varphi)+(\psi) = (\varphi \vee \psi), \qquad (\varphi)\cdot(\psi) = (\varphi \wedge \psi), \qquad \overline{(\varphi)} = (\neg \varphi).$$

Then $\mathfrak{B}_{\mathscr{L}} = \langle B_{\mathscr{L}}, +, \cdot, ^{-}, 0_{\mathscr{L}}, 1_{\mathscr{L}}\rangle$ is a Boolean algebra and it is known as the *Lindenbaum algebra* of $\mathscr{L}$. We shall drop the subscript $\mathscr{L}$ if it is understood. $\mathfrak{B}$ has several important subalgebras. For each $n \in \omega$, we can define

$$B_n = \{(\varphi) : \varphi \text{ has at most the variables } v_0, v_1, \ldots, v_{n-1} \text{ free}\}.$$

Then each B_n determines a sub-Boolean algebra $\mathfrak{B}_n$ of $\mathfrak{B}$. In particular, $\mathfrak{B}_0$ is the *Lindenbaum algebra of all sentences* of $\mathscr{L}$. Prove that the following are equivalent:

(i). Lindenbaum's theorem for $\mathscr{L}$, Proposition 1.3.11.

(ii). Every filter on the Lindenbaum algebra $\mathfrak{B}_0$ can be extended to an ultrafilter.

1.4.11. Let T be any theory of $\mathscr{L}$ and define

$$D_T = \{(\varphi) : T \models \varphi\}.$$

Prove that:

(i). T is consistent iff D_T is a filter on $\mathfrak{B}_0$.

(ii). T is consistent and finitely axiomatizable iff D_T is a principal filter on $\mathfrak{B}_0$.

(iii). T is complete iff D_T is an ultrafilter in $\mathfrak{B}_0$.

(iv). T is complete and finitely axiomatizable iff D_T is a principal ultrafilter on $\mathfrak{B}_0$.

The converse of the above four equivalences also holds in the following sense. Let D be a subset of B_0 and define

$$T_D = \{\varphi : \varphi \text{ is a sentence and } (\varphi) \in D\}.$$

Then (i)–(iv) hold with T replaced by T_D and D_T by D. If T is complete, then, of course, the quotient algebra $\mathfrak{B}_0/D_T$ is the two-element algebra. In general, there is a one-to-one correspondence between complete closed extensions of T and ultrafilters in $\mathfrak{B}_0/D_T$. Show, without using the completeness theorem, that the above results still hold when the notions of closed theory, finitely axiomatizable and complete are replaced by their syntactical analogues.

1.4.12. Let $\mathscr{S}$ be the language of Section 1.2. By considering the equivalence relation $\vdash \varphi \leftrightarrow \psi$ on sentences of $\mathscr{S}$, we can define the exact analogue of $\mathfrak{B}_{\mathscr{L}}$, namely

$$\mathfrak{B}_{\mathscr{S}} = \langle B_{\mathscr{S}}, +, \cdot, ^{-}, 0_{\mathscr{S}}, 1_{\mathscr{S}} \rangle.$$

The following shows that there is a close relation between completeness theorems and representations of Boolean algebras. Prove that the following are equivalent:

(i). The completeness theorem for $\mathscr{S}$, Theorem 1.2.7.

(ii). The Lindenbaum algebra $\mathfrak{B}_{\mathscr{S}}$ is isomorphic to a field of sets.

There is an analogue of this result for $\mathscr{L}$ and $\mathfrak{B}_{\mathscr{L}}$ which we shall discuss in the exercises for Section 2.1.

1.4.13. Show that in the language $\{S, 0\}$ the theories given by the two sets of axioms (see Example 1.4.11)

(A) axioms (1), (2), and schema (7_φ),

(B) axioms (1), (2), (8), and the schema

(9_n) $Sv_0 \not\equiv v_1 \vee Sv_1 \not\equiv v_2 \vee \dots \vee Sv_n \not\equiv v_0$

are equivalent.

1.4.14. Show that if $\omega < \alpha$ and α is a limit ordinal, then $\langle R(\alpha), \epsilon \rangle$ is a model of Zermelo set theory. Hence in ZF we can prove that Zermelo set theory is consistent.

1.4.15. Let θ be an (uncountable) inaccessible cardinal. Show that $\langle R(\theta), \epsilon \rangle$ is a model of ZF, and that $\langle R(\theta+1), \epsilon, R(\theta) \rangle$ is a model of Bernays–Morse set theory.

1.4.16. Which axioms of ZF are true in the model $\langle R(\omega), \epsilon \rangle$? And in the model $\langle R(\omega+\omega), \epsilon \rangle$?

1.4.17*

(i). Let $\langle A, E, U \rangle$ be an arbitrary model of Bernays set theory. Prove that $\langle U, E \cap (U \times U) \rangle$ is a model of ZF. Hence if Bernays set theory is consistent, so is ZF.

(ii). Let $\mathfrak{A} = \langle A, E \rangle$ be an arbitrary model of ZF. We may assume that no subset of A belongs to A. Let us say that a subset $X \subset A$ is *definable* in $\mathfrak{A}$ iff there is a formula $\varphi(v_n \dots v_n)$ and elements $x_1, \dots, x_n \in A$ such that

$$X = \{x \in A : \mathfrak{A} \models \varphi[xx_1 \dots x_n]\}.$$

Let B be the set of all definable subsets X of A such that there is no $y \in A$ with $X = \{x \in A : xEy\}$, and let E' be the set of all pairs $x \in A$, $X \in B$ such that $x \in X$. Prove that $\langle A \cup B, E \cup E', A \rangle$ is a model of Bernays set theory. Hence, if ZF is consistent, so is Bernays set theory.

1.4.18*

(i). Let $\langle A, \epsilon \rangle$ be any model of the axiom of extensionality, where ϵ is the ϵ-relation restricted to A. Prove that $\langle A, \epsilon \rangle$ is isomorphic to a transitive model, called a *transitive realization* of $\langle A, \epsilon \rangle$.

(ii). Show that any two transitive models of the axiom of extensionality which are isomorphic are equal.

1.4.19. Prove that there exists a complete theory T which has arbitrarily large natural models. T may be taken to be an extension of Zermelo set theory.

1.4.20. Let $\mathfrak{A}$ and $\mathfrak{B}$ be infinite simple orderings. Prove that $\mathfrak{A}$ and $\mathfrak{B}$ satisfy exactly the same universal sentences.

1.5. Elimination of quantifiers

Each model $\mathfrak{A}$ of a theory T gives rise to a complete theory, namely the set of all sentences holding in $\mathfrak{A}$, which is an extension of T. For this reason, it is important to know something about the complete extensions of a theory. In a few fortunate cases, it is possible to give a simple description of all the complete extensions of a theory by using the method of elimination of quantifiers.

This method applies only to very special theories. Moreover, each time the method is applied to a new theory we must start from scratch in the proofs, because there are few opportunities to use general theorems about models. On the other hand, the method is extremely valuable when we want to beat a particular theory into the ground. When it can be carried out, the method of elimination of quantifiers gives a tremendous amount of information about a theory. For instance, it tells us about the behavior of all formulas, as well as all sentences, relative to the theory. Usually it also gives a uniform way of deciding whether or not a sentence belongs to the theory; in other words, it gives a proof that the theory is decidable.

The question of the decidability of a theory lies outside the scope of this book, since it is not usually considered model theory. However, it is a very important question, and in fact the most striking applications of the elimination of quantifiers are to show that certain theories are decidable. The method is also valuable as a source of examples of thoroughly understood theories, which are useful for testing conjectures and for illustrating results. The method may be thought of as a direct attack on a theory. Later on, especially in §3.5, we shall learn of several more indirect attacks on theories, which work more often but give less information in particular cases.

Before describing the method, we need some more notation. In Section 1.3, we introduced the notion of a sentence φ being a consequence of a set Σ of sentences, in symbols $\Sigma \models \varphi$. What meaning shall we give to $\Sigma \models \varphi$ if φ is a formula? We shall say that a formula $\varphi(v_0 \ldots v_n)$ is a *consequence of* Σ, symbolically $\Sigma \models \varphi$, iff for every model $\mathfrak{A}$ of Σ and every sequence $a_0, \ldots, a_n \in A$, $a_0, \ldots, a_n$ satisfies φ. It follows that the *formula* $\varphi(v_0 \ldots v_n)$ is a consequence of Σ if and only if the *sentence* $(\forall v_0 \ldots v_n)\varphi(v_0 \ldots v_n)$ is a consequence of Σ. We say that two formulas φ, ψ are *Σ-equivalent* iff $\Sigma \models \varphi \leftrightarrow \psi$.

In general, the method of elimination of quantifiers is as follows: First, depending on the theory T, we pick out an appropriate set of formulas, called *basic formulas*. By a *Boolean combination* of basic formulas we mean a formula obtained from basic formulas by repeated application of the connectives $\neg$, $\wedge$. The main result to be proved is that *every formula is T-equivalent to a Boolean combination of basic formulas.* The key step in the proof is the step where we 'eliminate quantifiers'. In fact, we may state at once a simple but general lemma which shows why the name 'elimination of quantifiers' is given to the method (the name is due to Tarski, 1935).

LEMMA 1.5.1. *Let T be a theory and let Σ be a set of formulas, called basic formulas. In order to show that every formula is T-equivalent to a Boolean combination of basic formulas, it is sufficient to show the following*:

(i). *Every atomic formula is T-equivalent to a Boolean combination of basic formulas.*

(ii). *If θ is a Boolean combination of basic formulas, then $(\exists v_m)\theta$ is T-equivalent to a Boolean combination of basic formulas.*

PROOF. Let Ψ be the set of all formulas which are T-equivalent to a Boolean combination of basic formulas. We show by induction that every formula φ belongs to Ψ. If φ is an atomic formula, then $\varphi \in \Psi$ by (i). If φ is $\neg \psi$ and $\psi \in \Psi$, it is obvious that $\varphi \in \Psi$. Similarly, if φ is $\psi_1 \wedge \psi_2$ and $\psi_1, \psi_2 \in \Psi$, then $\varphi \in \Psi$. If φ is $(\exists v_m)\psi$ and $\psi \in \Psi$, then ψ is T-equivalent to a Boolean combination θ of basic formulas. Moreover, φ is T-equivalent to $(\exists v_m)\theta$. By (ii), $(\exists v_m)\theta \in \Psi$, so $\varphi \in \Psi$. ⊣

We shall illustrate the method with two simple examples. Our first example is the theory of dense simple order without endpoints (Example 1.4.1). Let us temporarily (in this section only) call this theory Δ. As we mentioned in Section 1.4, the theory Δ is complete. The method of elimination of quantifiers is one of several ways which we shall come across for proving that theories are complete. The completeness of Δ will follow from our results below. The elimination of quantifiers was applied to the theory Δ very early, by Langford (1927).

As basic formulas we shall take the atomic formulas

$$v_m \equiv v_n, \qquad v_m \leqslant v_n.$$

The Boolean combinations of atomic formulas are precisely the formulas which have no quantifiers. In any language, formulas which have no quantifiers are called *open formulas.* We wish to prove that every formula φ is Δ-equivalent to an open formula ψ. As we carry out our arguments, we shall also keep track of which variables occur in the open formula which is Δ-equivalent to a given formula. This will be useful for applications. Before we can eliminate any quantifiers, we must take a close look at the open formulas. For convenience, we use the abbreviation

$$v_m < v_n \quad \text{for} \quad v_m \leqslant v_n \wedge \neg\, v_m \equiv v_n.$$

Let us consider $n+1$ variables $v_0, \ldots, v_n$, $n > 0$. By an *arrangement* of the variables $v_0, \ldots, v_n$ we mean a finite conjunction of the form

$$\theta_0 \wedge \theta_1 \wedge \ldots \wedge \theta_{n-1},$$

where $u_0, \ldots, u_n$ is a renumbering of $v_0, \ldots, v_n$ and each formula θ_i is either $u_i < u_{i+1}$, or else $u_i \equiv u_{i+1}$. The lemma below allows us to put every open formula into a 'normal form' built up from arrangements of the variables.

LEMMA 1.5.2. *Every open formula $\varphi(v_0 \ldots v_n)$ is Δ-equivalent either to one of the formulas $v_0 < v_0$, $v_0 \equiv v_0$, or else to the disjunction of finitely many arrangements of the variables $v_0, \ldots, v_n$.*

PROOF. First, we consider the case $n = 0$. In this case, the open formula $\varphi(v_0)$ is built up from the atomic formulas $v_0 \leqslant v_0$, $v_0 \equiv v_0$. Since $\Delta \models v_0 \leqslant v_0$ and $\Delta \models v_0 \equiv v_0$, we must have either $\Delta \models \varphi$ and $\Delta \models \varphi \leftrightarrow v_0 \equiv v_0$, or else $\Delta \models \neg\, \varphi$ and $\Delta \models \varphi \leftrightarrow v_0 < v_0$.

Let us now make three observations about arrangements (we assume that $n > 0$):

(1). There are only finitely many different arrangements of the variables $v_0, \ldots, v_n$.

(2). For each simply-ordered structure $\mathfrak{A}$, each sequence $a_0, \ldots, a_n$ satisfies some arrangement of $v_0, \ldots, v_n$.

(3). Let $\varphi(v_0 \dots v_n)$ be an open formula and let ψ be an arrangement of $v_0, \dots, v_n$. Then one or both of the formulas $\psi \to \varphi$, $\psi \to \neg\varphi$, is a consequence of the theory of simple order.

(1) should be obvious, while (2) follows easily from the fact that in a simply-ordered structure, exactly one of the relations $a < b$, $a = b$, $b < a$ holds between two elements a, b. (3) is proved by induction on the length of the open formula φ, and is left to the reader.

Now let $\varphi(v_0 \dots v_n)$ be an open formula. If $\Delta \models \neg\varphi$, then φ is Δ-equivalent to the formula $v_0 < v_0$. Assume the other possibility, that it is not the case that $\Delta \models \neg\varphi$. Consider any model $\mathfrak{A}$ of Δ and sequence $a_0, \dots, a_n$ which satisfies φ in $\mathfrak{A}$. By (2), $a_0, \dots, a_n$ also satisfies some arrangement ψ of $v_0, \dots, v_n$ in $\mathfrak{A}$. Thus we cannot have $\Delta \models \psi \to \neg\varphi$, and, by (3), we must have $\Delta \models \psi \to \varphi$. Form the disjunction θ of all arrangements ψ of $v_0, \dots, v_n$ for which $\Delta \models \psi \to \varphi$. θ is the disjunction of at least one, but only finitely many formulas, in view of (1). It follows from our remarks above that $\Delta \models \varphi \to \theta$, and from the definition of θ we see that $\Delta \models \theta \to \varphi$. So φ and θ are Δ-equivalent, and our proof is complete. ⊣

We observe that actually the above lemma is true for the theory of simple order as well as for the theory Δ. The reader may check this by going carefully through the proof, noticing that the only axioms of Δ which we actually made use of are the axioms of simple order. In the next theorem, however, we need all of the axioms of Δ.

THEOREM 1.5.3. *Every formula φ is Δ-equivalent to an open formula ψ. Moreover, if all the free variables of φ are among $v_0, \dots, v_n$, $n \geqslant 0$, then ψ can be chosen so that all its variables are among $v_0, \dots, v_n$.*

PROOF. We first prove that every formula φ is Δ-equivalent to an open formula ψ. By Lemma 1.5.1, it suffices to prove that for every open formula $\psi(v_0 \dots v_n)$, the formula $(\exists v_m)\psi$ is Δ-equivalent to an open formula. If $m > n$, then v_m does not occur at all in ψ, so $(\exists v_m)\psi$ is Δ-equivalent to ψ. We may thus assume that $m \leqslant n$. By renaming the variables we can even make $m = n$.

Using Lemma 1.5.2, we may suppose that ψ is either $v_0 < v_0$, $v_0 \equiv v_0$, or a disjunction of finitely many arrangements of $v_0, \dots, v_n$. If ψ is either $v_0 < v_0$ or $v_0 \equiv v_0$, then obviously $(\exists v_n)\psi$ is Δ-equivalent to ψ. In the remaining case, let

$$\psi = \theta_0 \vee \dots \vee \theta_p,$$

where each θ_i is an arrangement of $v_0, \ldots, v_n$. Then

$$\Delta \models (\exists v_n)\psi \leftrightarrow (\exists v_n)\theta_0 \vee \ldots \vee (\exists v_n)\theta_p.$$

We may eliminate the quantifier $(\exists v_n)$ in the following way: If $n = 1$, the only possibilities for the formulas $(\exists v_1)\theta_i$ are

$$(\exists v_1)v_0 < v_1, \qquad (\exists v_1)v_0 \equiv v_1, \qquad (\exists v_1)v_1 < v_0.$$

Each of these is a consequence of Δ, and it follows that $(\exists v_1)\psi$ is a consequence of Δ and is Δ-equivalent to $v_0 \equiv v_0$.

Let $n > 1$. Then, from each arrangement θ_i of $v_0, \ldots, v_n$, we may form in a natural way an arrangement θ_i^* of $v_0, \ldots, v_{n-1}$ obtained by leaving out v_n. It is easy to see that

$$\Delta \models (\exists v_n)\theta_i \leftrightarrow \theta_i^*, \qquad i = 0, \ldots, p,$$

and hence

$$\Delta \models (\exists v_n)\psi \leftrightarrow \theta_0^* \vee \ldots \vee \theta_p^*.$$

We have shown in each case that $(\exists v_m)\psi$ is Δ-equivalent to an open formula.

We now prove the second clause of the theorem. Our proof given above actually shows that if $\psi(v_0 \ldots v_n)$ is an open formula, $n > 0$, then $(\exists v_n)\psi$ is Δ-equivalent to an open formula of the form $\theta(v_0 \ldots v_{n-1})$. Let $\varphi(v_0 \ldots v_n)$ be an arbitrary formula, $n \geqslant 0$. Then φ is Δ-equivalent to some open formula $\psi(v_0 \ldots v_n \ldots v_{n+m})$. But φ is also Δ-equivalent to $(\exists v_{n+1}) \ldots (\exists v_{n+m})\varphi$, and hence to $(\exists v_{n+1}) \ldots (\exists v_{n+m})\psi$. The latter formula is Δ-equivalent to an open formula of the form $\theta(v_0 \ldots v_n)$, and thus φ is Δ-equivalent to θ. Our proof is complete. ⊣

The proof of the theorem also gives a decision procedure for the theory Δ. Very briefly, the decision procedure is as follows. We are given an arbitrary sentence φ and we wish to determine whether φ belongs to the theory Δ. Our first step is to put φ into prenex normal form, say (after renumbering variables),

$$(Q_0 v_0)(Q_1 v_1) \ldots (Q_n v_n)\psi,$$

where $Q_0, \ldots, Q_n$ are quantifier symbols $\exists$ or $\forall$, and ψ is open. We may assume further that Q_n is $\exists$, for otherwise we may work with $\neg \varphi$. Next, we put ψ into one of the forms $v_0 < v_0$, $v_0 \equiv v_0$, or a disjunction of finitely many arrangements of $v_0, \ldots, v_n$. Then we eliminate the quantifier $(\exists v_n)$, that is, we replace $(\exists v_n)\psi$ by a Δ-equivalent open formula $\theta(v_0 \ldots v_{n-1})$ by the process explained in the proof. After that, we repeat the process until all the variables except v_0 are eliminated. When we finish, we can tell at once whether the resulting sentence $(Q_0 v_0)\theta(v_0)$ belongs to Δ. Of course, the decision procedure can be streamlined very much if it is really going to be used.

We now obtain another consequence of the theorem.

COROLLARY 1.5.4. *The theory of dense simple order without endpoints is complete.*

PROOF. Let φ be an arbitrary sentence. By Theorem 1.5.3, φ is Δ-equivalent to an open formula $\psi(v_0)$. But for any open formula $\psi(v_0)$, we have either $\Delta \models \psi$ or $\Delta \models \neg\psi$. Hence either $\Delta \models \varphi$ or $\Delta \models \neg\varphi$, and Δ is complete. ⊣

Note that Corollary 1.5.4 is only concerned with sentences, but to prove the corollary via Theorem 1.5.3 we had to use an induction concerned with arbitrary formulas. This happens time and again in model theory, because the notion of a sentence is defined using the recursive definition of a formula. Theorem 1.5.3 also tells us something about the theories formed by adding new constant symbols to the language and taking Δ as a set of axioms. We leave this application of the theorem as an exercise.

Theorem 1.5.3 can be improved a little by taking for our basic formulas only the formulas $v_m \leqslant v_n$.

COROLLARY 1.5.5. *Every formula $\rho(v_0 \ldots v_n)$ is Δ-equivalent to a Boolean combination of formulas of the form $v_m \leqslant v_p$, where $m = 0, \ldots, n$ and $p = 0, \ldots, n$.*

PROOF. In view of Theorem 1.5.3, it is enough to observe that

$$\Delta \models v_m \equiv v_p \leftrightarrow v_m \leqslant v_p \wedge v_p \leqslant v_m. \dashv$$

We now take up our second example of the elimination of quantifiers. We shall obtain a full description of all complete closed theories in the pure identity language (see Section 1.3), which has no predicate, function or constant symbols at all. In other words, we shall describe all complete closed extensions of the theory with the empty set of axioms in the pure identity language.

As in the case of dense simple order, we begin with a lemma about arrangements. What should we mean by an arrangement this time? An arrangement of $v_0, \ldots, v_n$ will be a formula which tells which variables are equal to each other and which are unequal. To be precise, we let e be an equivalence relation over the set $\{0, 1, ..., n\}$ of indices of the variables $v_0, ..., v_n$. We define the *arrangement of* $v_0, ..., v_n$ *given by* e to be the conjunction of all the formulas

$$v_i \equiv v_j, \quad iej; \quad \text{and} \quad \neg\, v_i \equiv v_j, \quad \text{not} \quad iej.$$

LEMMA 1.5.6. *Every open formula* $\varphi(v_0 \ldots v_n)$ *is either inconsistent or is equivalent to a disjunction of finitely many arrangements of* $v_0, ..., v_n$.

The proof is very similar to that of Lemma 1.5.2, so we leave it as an exercise.

We now must decide on our set of basic formulas. It should be clear that the atomic formulas are not enough. For instance, the sentence $(\forall v_0 v_1)(v_0 \equiv v_1)$ cannot be expressed by an open formula.

For our basic formulas we take all atomic formulas

$$v_m \equiv v_n,$$

together with the sentences σ_n which state that 'there are more than n distinct elements'. Formally, σ_n, $n > 0$, may be written

$$(\forall v_1 \ldots v_n)(\exists v_0)(\neg\, v_0 \equiv v_1 \wedge \ldots \wedge \neg\, v_0 \equiv v_n).$$

For good measure, we shall define σ_0 to be a valid sentence, say $(\exists v_0)(v_0 \equiv v_0)$.

THEOREM 1.5.7. *Every formula* φ *in the pure identity language is equivalent to a Boolean combination* ψ *of basic formulas. Moreover, if all the free variables of* φ *are among* $v_0, ..., v_n$, *then* ψ *may be chosen so that all its free variables are among* $v_0, ..., v_n$. *In particular, if* φ *is a sentence, then so is* ψ.

PROOF. We first show that every formula is equivalent to a Boolean combination of basic formulas. Let $\psi(v_0 \ldots v_n)$ be an arbitrary Boolean combination of basic formulas. By Lemma 1.5.1, it suffices to prove that $(\exists v_m)\psi$ is equivalent to a Boolean combination of basic formulas. First, we note that ψ is equivalent to a formula of the form

$$(\psi_0 \wedge \theta_0) \vee \ldots \vee (\psi_p \wedge \theta_p)$$

where each ψ_i is an open formula and each θ_i is a Boolean combination of the sentences $\sigma_0, \sigma_1, \sigma_2, \ldots$. Still better, using Lemma 1.5.6, we may make each ψ_i be either the inconsistent sentence $\neg\sigma_0$ or else a disjunction of finitely many arrangements of $v_0, \ldots, v_n$.

As in the previous theorem, we may assume without loss of generality that $m = n$. In the case $n = 0$, the only arrangement of v_0 is the valid formula $v_0 \equiv v_0$, so each ψ_i is either valid, in which case it may be replaced by σ_0, or else it is the inconsistent formula $\neg\sigma_0$. Thus ψ is equivalent to a Boolean combination of the sentences $\sigma_0, \sigma_1, \ldots$, and so is $(\exists v_n)\psi$.

Assume that $n > 0$. For each arrangement ψ_i, $i \leqslant p$, form ψ_i^* by deleting all the equations and inequalities in which v_n occurs. Then ψ_i^* is an arrangement of the remaining variables $v_0, \ldots, v_{n-1}$. (If ψ_i happens to be $\neg\sigma_0$, we simply let ψ_i^* also be $\neg\sigma_0$.) Note that $(\exists v_n)\psi_i$ is *not*, in general, equivalent to ψ_i^*. (Why?) However, if e_i is the equivalence relation from which the arrangement ψ_i comes, and r_i is the number of equivalence classes in e_i, then we easily see that $(\exists v_n)\psi_i$ is equivalent to $\sigma_{r_i-1} \wedge \psi_i^*$. Also $(\exists v_n)\psi$ is equivalent to the formula

$$(\theta_0 \wedge (\exists v_n)\psi_0) \vee \ldots \vee (\theta_p \wedge (\exists v_n)\psi_p).$$

It follows that $(\exists v_n)\psi$ is equivalent to

$$(\theta_0 \wedge \sigma_{r_0-1} \wedge \psi_0^*) \vee \ldots \vee (\theta_p \wedge \sigma_{r_p-1} \wedge \psi_p^*).$$

This is indeed a Boolean combination of basic formulas, and the first part of the theorem is proved.

Now, using exactly the same trick as we used at the end of the proof of Theorem 1.5.3, we can obtain the full statement of the theorem – that each formula $\varphi(v_0 \ldots v_n)$ is equivalent to a Boolean combination $\psi(v_0 \ldots v_n)$ of basic formulas, and, if φ is a sentence, then so is ψ. ⊣

We are now ready to describe clearly all the closed theories in the pure identity language. It is easy to see that for every finite set N of positive natural numbers, there is a sentence $\sigma(N)$ whose models are precisely those $\mathfrak{A}$ such that $|A| \in N$. The reader should check that for each N, $\sigma(N)$ is a pure identity sentence, and, in fact, is a Boolean combination of $\sigma_0, \sigma_1, \ldots$. We now can conclude that, up to equivalence, the sentences $\sigma(N)$ and their negations are the only pure identity sentences.

COROLLARY 1.5.8. *For every pure identity sentence φ, there is a finite set N of positive natural numbers such that φ is equivalent either to $\sigma(N)$ or to $\neg\sigma(N)$.*

We now take up the theories. It is also easy to see that for each finite or infinite set N of positive natural numbers, there is a closed theory $\Delta(N)$ whose models are precisely those $\mathfrak{A}$ such that either $|A| \in N$ or $\mathfrak{A}$ is infinite. Again, the reader should check that each theory $\Delta(N)$ has a set of pure identity sentences for axioms. To make our notation more complete, we may as well write $\Sigma(N)$ for the closed theory which has the single axiom $\sigma(N)$, where N is finite. The next corollary shows that the $\Delta(N)$ and $\Sigma(N)$ are the only closed theories in the pure identity language.

COROLLARY 1.5.9.

(i). *The finitely axiomatizable closed theories in the pure identity language are precisely the theories $\Sigma(N)$, where N is finite, and $\Delta(N)$, where $\omega - N$ is finite.*

(ii). *The nonfinitely axiomatizable closed theories in the pure identity language are precisely the theories $\Delta(N)$, where $\omega - N$ is infinite.*

PROOF. (i). The theories $\Sigma(M)$, M finite, and $\Delta(N)$, $\omega - N$ finite, are finitely axiomatizable. Indeed, $\Sigma(M)$ has the single axiom $\sigma(M)$, and $\Delta(N)$ the single axiom $\neg\sigma(\omega - N)$. By Corollary 1.5.8, any finitely axiomatizable theory has a single axiom of the form $\sigma(N)$, or else $\neg\sigma(N)$, N finite. This proves (i).

(ii). Now let T be an arbitrary closed theory in the pure identity language. Let N be the set of all positive natural numbers such that T has a model of power in N. Then for each finite model $\mathfrak{A}$, $\mathfrak{A}$ is a model of T if and only if $|A| \in N$. If one of the sentences $\sigma(M)$, M finite, belongs to T, then all models of T are finite and $N \subset M$, and thus $T = \Sigma(N)$.

Assume now that T is not of the form $\Sigma(N)$. It follows that every sentence $\varphi \in T$ is equivalent to a sentence of the form $\neg\sigma(M)$. Let N' be the union of all sets M such that $T \models \neg\sigma(M)$. Then clearly $\mathfrak{A}$ is a model of T if and only if $\mathfrak{A}$ is infinite or $|A| \in \omega - N'$. Therefore $T = \Delta(\omega - N')$. Note also that $\omega - N' = N$, so T is $\Delta(N)$. Finally, if T is not finitely axiomatizable, then by (i) the set $\omega - N$ is infinite. This proves (ii). ⊣

Note that by our last corollary, there is no theory at all whose models consist precisely of all finite models. Likewise, if N is an arbitrary infinite set of positive natural numbers, there is no theory whose models consist of all $\mathfrak{A}$ such that $|A| \in N$. In other words, in the pure identity language, *any theory which has arbitrarily large finite models has an infinite model.* We shall see later that this is true in every other first-order language as well.

There are several very important theories which have been analyzed using the elimination of quantifiers. For example, additive number theory (Presburger, 1929), the theory of Abelian groups (Szmielew, 1955), the theory of Boolean algebras (Tarski, 1949), the theory of all well-ordered models (Mostowski and Tarski, 1949), and the theories of real closed fields and of algebraically closed fields (Tarski, 1948). As might be guessed from our two simple examples, the elimination of quantifiers becomes quite difficult in some of the more substantial cases mentioned above. In each of those cases, the method gives a decision procedure for the theory, as well as a useful classification of all formulas and all complete extensions of the theory.

Most of the interesting theories which arise in mathematics are undecidable (e.g., number theory, set theory, groups, fields, partial order), and the method of elimination of quantifiers does not work for these theories.

Exercises

1.5.1. Let $\mathscr{L}(n)$ be the language $\{\leqslant, c_0, \ldots, c_{n-1}\}$ obtained from the language $\{\leqslant\}$ by adding n constant symbols.

(i). Show that the set Δ of sentences is not complete in the language $\mathscr{L}(n)$, for $n > 1$. Show that all the complete extensions are finitely axiomatizable.

(ii)*. Describe all the complete extensions of Δ in the language $\mathscr{L}(\omega)$.

1.5.2*. Let Γ be the theory of dense simple order. Prove that Γ has exactly four complete closed extensions, which come from one of the four additional axioms:

there are no endpoints;
there is a left endpoint but no right endpoint;
there is a right endpoint but no left endpoint;
there are a right and a left endpoint.

Hint: As a set of basic formulas, take the set of all atomic formulas together with the formulas which state:

v_m is a left endpoint;
v_m is a right endpoint;
there is a left endpoint;
there is a right endpoint.

Modify the proof of Theorem 1.5.3 to show that every formula is Γ-equivalent to a Boolean combination of basic formulas.]

1.5.3*. Show by elimination of quantifiers that the theory of atomless Boolean algebras is complete.

1.5.4. Which are the complete theories in the pure identity language? State a simple criterion for two models $\mathfrak{A}$, $\mathfrak{B}$ of that language to be equivalent.

1.5.5. Describe all the complete theories in the language which has n constant symbols but no relation or function symbols. Do the same for the language with ω constant symbols.

1.5.6. Outline a decision procedure for deciding whether a given pure identity sentence is valid.

1.5.7*. Analyze the following theories using the method of elimination of quantifiers:

(i). The theory with no axioms in the language with one 1-placed relation symbol and no other symbols (Behmann, 1922).

(ii). The theory of Abelian groups with all elements being of order 3.

(iii). The theory of divisible Abelian groups.

(iv). Monadic first-order logic.

1.5.8*. This and the succeeding exercises are very long and tedious, if complete proofs are written out. The theory of one *successor function* has the axioms (1), (2) and $(7)_\varphi$ from Example 1.4.11 in the language $\mathscr{L} = \{S, 0\}$. Prove that this theory is complete by elimination of quantifiers.

1.5.9*. Prove that additive number theory (from Example 1.4.11) in the language $\mathscr{L} = \{+, S, 0\}$ is complete by elimination of quantifiers.

1.5.10*. The theory of *one equivalence relation* in the language $\mathscr{L} = \{E\}$ has the following axioms:

xEx;

$xEy \wedge yEz \rightarrow xEz$;

$xEy \rightarrow yEx$.

Give a decision procedure for this theory by the method of elimination of quantifiers.

1.5.11**. Give a decision procedure for the theory of Abelian groups (Example 1.4.6) by elimination of quantifiers. Use this to describe all complete extensions of this theory.

CHAPTER 2

MODELS CONSTRUCTED FROM CONSTANTS

2.1. Completeness and compactness

In this section, we prove the basic completeness theorem first proved by Gödel (1930). The proof we give is due to Henkin (1949) and it applies to situations somewhat more general than Gödel's original proof. This extension was already noted by Malcev (1936).

The result we prove is that every consistent set of sentences T in a language $\mathscr{L}$ has a model or, in other words, is satisfiable. The proof proceeds in two stages. We shall first show that T can be extended to another consistent set of sentences $\bar{T}$ in an expanded language $\bar{\mathscr{L}}$, having certain desirable features. Then we show that every T having these desirable features has a model. It will make no difference which of the two steps we prove first.

DEFINITION. Let T be a set of sentences of $\mathscr{L}$ and let C be a set of constant symbols of $\mathscr{L}$. (C might be a proper subset of the set of all constant symbols of $\mathscr{L}$.) We say that C is a *set of witnesses* for T in $\mathscr{L}$ iff for every formula φ of $\mathscr{L}$ with at most one free variable, say x, there is a constant $c \in C$ such that

$$T \vdash (\exists x)\varphi \rightarrow \varphi(c).$$

We say that T *has witnesses in* $\mathscr{L}$ iff T has some set C of witnesses in $\mathscr{L}$.

The meaning and usage of $\varphi(c)$ should be quite clear here and in all succeeding places in this chapter: $\varphi(c)$ is obtained from φ by replacing simultaneously all free occurrences of x in φ by the constant c. We shall be careful to use $\varphi(c)$ only when it has been made clear from the context which variable x is to be replaced by c. Otherwise the notation $\varphi(c)$ would be ambiguous. For example, if φ is a formula with the free variables x, y, we have to indicate whether $\varphi(c)$ is obtained from φ by replacing x by c or by replacing y by c. An alternative notation which is completely unambiguous is to write $\varphi(c/x)$ for the formula obtained by replacing all free occurrences of x in φ by c. However, we prefer to use $\varphi(c)$ and rely on the context for clarity rather than use the more cluttered notation $\varphi(c/x)$.

LEMMA 2.1.1. *Let T be a consistent set of sentences of $\mathscr{L}$. Let C be a set of new constant symbols of power $|C| = ||\mathscr{L}||$, and let $\overline{\mathscr{L}} = \mathscr{L} \cup C$ be the simple expansion of $\mathscr{L}$ formed by adding C. Then T can be extended to a consistent set of sentences $\overline{T}$ in $\overline{\mathscr{L}}$ which has C as a set of witnesses in $\overline{\mathscr{L}}$.*

PROOF. Let $\alpha = ||\mathscr{L}||$. For each $\beta < \alpha$, let c_β be a constant symbol which does not occur in $\mathscr{L}$ and such that $c_\beta \neq c_\gamma$ if $\beta < \gamma < \alpha$. Let $C = \{c_\beta : \beta < \alpha\}$, $\overline{\mathscr{L}} = \mathscr{L} \cup C$. Clearly $||\overline{\mathscr{L}}|| = \alpha$, so we may arrange all formulas of $\overline{\mathscr{L}}$ with at most one free variable in a sequence φ_ξ, $\xi < \alpha$. We now define an increasing sequence of sets of sentences of $\overline{\mathscr{L}}$:

$$T = T_0 \subset T_1 \subset \ldots \subset T_\xi \subset \ldots, \qquad \xi < \alpha,$$

and a sequence d_ξ, $\xi < \alpha$, of constants from C such that:

(i). each T_ξ is consistent in $\overline{\mathscr{L}}$;

(ii). if $\xi = \zeta + 1$, then $T_\xi = T_\zeta \cup \{(\exists x_\zeta)\varphi_\zeta \to \varphi_\zeta(d_\zeta)\}$; x_ζ is the free variable in φ_ζ if it has one, otherwise $x_\zeta = v_0$;

(iii). if ξ is a limit ordinal different from 0, then $T_\xi = \bigcup_{\zeta<\xi} T_\zeta$.

Suppose that T_ζ has been defined. Note that the number of sentences in T_ζ which are not sentences of $\mathscr{L}$ is smaller than α, i.e., the cardinal of the set of such sentences is less than α. Furthermore, each such sentence contains at most a finite number of constants from C. Therefore, let d_ζ be the first element of C which has not yet occurred in T_ζ. For instance, $d_0 = c_0$. We show that

$$T_{\zeta+1} = T_\zeta \cup \{(\exists x_\zeta)\varphi_\zeta \to \varphi_\zeta(d_\zeta)\}$$

is consistent. If this were not the case, then

$$T_\zeta \vdash \neg((\exists x_\zeta)\varphi_\zeta \to \varphi_\zeta(d_\zeta)).$$

By propositional logic,

$$T_\zeta \vdash (\exists x_\zeta)\varphi_\zeta \wedge \neg \varphi_\zeta(d_\zeta).$$

As d_ζ does not occur in T_ζ, we have by predicate logic,

$$T_\zeta \vdash (\forall x_\zeta)((\exists x_\zeta)\varphi_\zeta \wedge \neg \varphi_\zeta(x_\zeta)),$$

$$T_\zeta \vdash (\exists x_\zeta)\varphi_\zeta \wedge \neg (\exists x_\zeta)\varphi_\zeta,$$

which contradicts the consistency of T_ζ. If ξ is a nonzero limit ordinal, and each member of the increasing chain T_ζ, $\zeta < \xi$, is consistent, then obviously $T_\xi = \bigcup_{\zeta<\xi} T_\zeta$ is consistent. This completes the induction.

Now we let $\overline{T} = \bigcup_{\xi<\alpha} T_\xi$. It is evident that $\overline{T}$ is consistent in $\overline{\mathscr{L}}$ and $\overline{T}$ is an extension of T. Suppose that φ is a formula of $\overline{\mathscr{L}}$ with at most the variable x free. Then we may suppose that $\varphi = \varphi_\xi$ and $x = x_\xi$ for some $\xi < \alpha$. Whence the sentence

$$(\exists x_\xi)\varphi_\xi \rightarrow \varphi_\xi(d_\xi)$$

belongs to $T_{\xi+1}$ and so to $\overline{T}$. ⊣

The idea of the next lemma is just as simple, but its proof is more involved and tedious.

LEMMA 2.1.2. *Let T be a consistent set of sentences and C be a set of witnesses for T in $\mathscr{L}$. Then T has a model $\mathfrak{A}$ such that every element of $\mathfrak{A}$ is an interpretation of a constant $c \in C$.*

PROOF. First, note that if a set of sentences T has a set C of witnesses in $\mathscr{L}$, then C is also a set of witnesses for every extension of T. Second, if an extension of T has a model $\mathfrak{A}$, then $\mathfrak{A}$ is also a model of T. So we may as well assume that T is maximal consistent in $\mathscr{L}$.

For two constants $c, d \in C$, define

$$c \sim d \quad \text{iff} \quad c \equiv d \in T.$$

Because T is maximal consistent, we see that for $c, d, e \in C$,

$$c \sim c;$$
$$\text{if} \quad c \sim d \quad \text{and} \quad d \sim e, \quad \text{then} \quad c \sim e;$$
$$\text{if} \quad c \sim d \quad \text{then} \quad d \sim c.$$

So $\sim$ is an equivalence relation on C. For each $c \in C$, let

$$\tilde{c} = \{d \in C : d \sim c\}$$

be the equivalence class of c. We propose to construct a model $\mathfrak{A}$ whose set of elements A is the set of all these equivalence classes $\tilde{c}$, for $c \in C$; so we define

(1) $A = \{\tilde{c} : c \in C\}$.

We now define the relations, constants, and functions of $\mathfrak{A}$.

(i). For each n-placed relation symbol P in $\mathscr{L}$, we define an n-placed relation R' on the set C by:
for all $c_1, ..., c_n \in C$,

(2) $R'(c_1 \dots c_n)$ iff $P(c_1 \dots c_n) \in T$.

By our axioms of identity, we have

$$\vdash P(c_1 \dots c_n) \wedge c_1 \equiv d_1 \wedge \dots \wedge c_n \equiv d_n \rightarrow P(d_1 \dots d_n).$$

So $\sim$ is what is called a *congruence relation* for the relation R' on C. It follows that we may define a relation R on A by

(3) $R(\tilde{c}_1 \dots \tilde{c}_n)$ iff $P(c_1 \dots c_n) \in T$.

By (2), the definition (3) is independent of the representatives of the equivalence classes $\tilde{c}_1, ..., \tilde{c}_n$. This relation R is the interpretation of the symbol P in $\mathfrak{A}$.

(ii). Now consider a constant symbol d of $\mathscr{L}$. From predicate logic, we have

$$\vdash (\exists v_0)(d \equiv v_0).$$

So $(\exists v_0)(d \equiv v_0) \in T$, and, because T has witnesses, there is a constant $c \in C$ such that

$$(d \equiv c) \in T.$$

The constant c may not be unique, but its equivalence class is unique because, using our axioms of identity,

$$\vdash (d \equiv c \wedge d \equiv c' \rightarrow c \equiv c').$$

The constant d is interpreted in the model $\mathfrak{A}$ by the (uniquely determined) element $\tilde{c}$ of A. In particular, if $d \in C$, then d is interpreted by its own equivalence class $\tilde{d}$ in $\mathfrak{A}$, because $(d \equiv d) \in T$.

(iii). We handle the function symbols in a similar way. Let F be any m-placed function symbol of $\mathscr{L}$, and let $c_1, ..., c_m \in C$. As before, we have

$$(\exists v_0)(F(c_1 \dots c_m) \equiv v_0) \in T,$$

and because T has witnesses, there is a constant $c \in C$ such that

$$(F(c_1 \dots c_m) \equiv c) \in T.$$

Once more, we have a slight difficulty because c may not be unique, and we use our axioms of identity to obtain:

$$\vdash (F(c_1 \dots c_m) \equiv c \wedge c_1 \equiv d_1 \wedge \dots \wedge c_m \equiv d_m \wedge c \equiv d) \rightarrow F(d_1 \dots d_m) \equiv d.$$

This shows that a function G can be defined on the set A of equivalence classes by the rule

(4) $G(\tilde{c}_1 \ldots \tilde{c}_m) = \tilde{c}$ iff $(F(c_1 \ldots c_m) \equiv c) \in T$.

We leave the detailed steps of (4) to the reader. We interpret the function symbol F by the function G in the model $\mathfrak{A}$.

We have now specified the universe set and the interpretation of each symbol of $\mathscr{L}$ in $\mathfrak{A}$, so we have completed the definition of the model $\mathfrak{A}$. We have pointed out that the interpretation of each constant $c \in C$ in $\mathfrak{A}$ is its equivalence class $\tilde{c}$, and it follows that every element $\tilde{c} \in A$ is the interpretation of some constant $c \in C$.

We proceed to prove that $\mathfrak{A}$ is a model of T. First of all, using (4) as the first step of an induction, we easily show that

(5) for every term t of $\mathscr{L}$ with no free variables and for every constant $c \in C$,

$$\mathfrak{A} \models t \equiv c \quad \text{if and only if} \quad (t \equiv c) \in T.$$

Using the fact that C is a set of witnesses for T, we obtain from (5):

(6) for any two terms t_1, t_2 of $\mathscr{L}$ with no free variables,

$$\mathfrak{A} \models t_1 \equiv t_2 \quad \text{if and only if} \quad (t_1 \equiv t_2) \in T,$$

(7) for any atomic formula $P(t_1 \ldots t_n)$ of $\mathscr{L}$ containing no free variables,

$$\mathfrak{A} \models P(t_1 \ldots t_n) \quad \text{if and only if} \quad P(t_1 \ldots t_n) \in T.$$

Combining (6) and (7) will form a basis for proving:

(8) for any sentence φ of $\mathscr{L}$,

$$\mathfrak{A} \models \varphi \quad \text{if and only if} \quad \varphi \in T.$$

(8) has an unusual proof in that it is proved by induction on the length of the sentences of $\mathscr{L}$. The reader will see that the reason why this could be done is because T is maximal consistent and has witnesses in $\mathscr{L}$. Without a great deal of trouble, we have for sentences φ, ψ of $\mathscr{L}$

$$\mathfrak{A} \models \neg \varphi \quad \text{if and only if} \quad (\neg \varphi) \in T,$$

and

$$\mathfrak{A} \models \varphi \wedge \psi \quad \text{if and only if} \quad (\varphi \wedge \psi) \in T.$$

Suppose $\varphi = (\exists x)\psi$. If $\mathfrak{A} \models \varphi$, then for some $\tilde{c} \in A$, $\mathfrak{A} \models \psi[\tilde{c}]$. This means that $\mathfrak{A} \models \psi(c)$, where $\psi(c)$ is obtained from ψ by replacing all free occurrences of x by c. So $\psi(c) \in T$ and because

$$\vdash \psi(c) \rightarrow (\exists x)\psi,$$

we have $\varphi \in T$. On the other hand, if $\varphi \in T$, then because T has witnesses, there exists a constant $c \in C$ such that

$$T \vdash (\exists x)\psi \rightarrow \psi(c).$$

As T is maximal, $\psi(c) \in T$, so $\mathfrak{A} \models \psi(c)$. This gives $\mathfrak{A} \models \psi[\tilde{c}]$ and $\mathfrak{A} \models \varphi$. This shows that $\mathfrak{A}$ is a model of T. ⊣

Note that a converse of Lemma 2.1.2 is very easily proved, and, in fact:

LEMMA 2.1.3. *Let C be a set of constant symbols of $\mathscr{L}$, and let T be a set of sentences of $\mathscr{L}$. If T has a model $\mathfrak{A}$ such that every element of $\mathfrak{A}$ is an interpretation of some constant $c \in C$, then T can be extended to a consistent $\bar{T}$ in $\mathscr{L}$ for which C is a set of witnesses.*

For the proof of Lemma 2.1.3, simply let $\bar{T}$ be the set of all sentences of $\mathscr{L}$ true in $\mathfrak{A}$.

The model $\mathfrak{A}$ constructed from the constants $c \in C$ of $\mathscr{L}$ by taking suitable equivalence classes is said to be *built up from the set C of constants of $\mathscr{L}$*. Since every $a \in A$ is the interpretation of some $c \in C$, we see immediately that $|A| \leqslant |C|$. We now supply the proofs of three theorems from Chapter 1.

THEOREM 1.3.21 (Extended Completeness Theorem). *Let Σ be a set of sentences of $\mathscr{L}$. Then Σ is consistent if and only if Σ has a model.*

PROOF. The consistency of Σ if Σ has a model is a straightforward argument. So assume Σ is consistent. By Lemma 2.1.1, we consider extensions $\bar{\Sigma}$ of Σ and $\bar{\mathscr{L}}$ of $\mathscr{L}$ ($||\bar{\mathscr{L}}|| = ||\mathscr{L}||$), so that $\bar{\Sigma}$ has witnesses in $\bar{\mathscr{L}}$. By Lemma 2.1.2, let $\mathfrak{A}$ be a model of $\bar{\Sigma}$. $\mathfrak{A}$ is a model for the expanded language $\bar{\mathscr{L}}$, so let $\mathfrak{B}$ be the model for $\mathscr{L}$ which is the reduct of $\mathfrak{A}$ to $\mathscr{L}$. Because sentences in Σ do not involve the constants of $\bar{\mathscr{L}}$ not in $\mathscr{L}$, we see that $\mathfrak{B}$ is a model of Σ. ⊣

COROLLARY 2.1.4 (Downward Löwenheim–Skolem–Tarski Theorem). *Every consistent theory T in $\mathscr{L}$ has a model of power at most $||\mathscr{L}||$.*

PROOF. In the proof above we may choose $\mathfrak{A}$ so that every element is a constant, and we have $|B| = |A| \leqslant ||\bar{\mathscr{L}}|| = ||\mathscr{L}||$. ⊣

Corollary 2.1.4 gives the original theorem of Löwenheim (1915): If a sentence has a model, then it has a countable (finite or infinite) model.

THEOREM 1.3.20 (Gödel's Completeness Theorem). *A sentence of $\mathscr{L}$ is a theorem of $\mathscr{L}$ if and only if it is valid.*

PROOF. We need only concern ourselves with one direction of the theorem. If a sentence σ is not a theorem of $\mathscr{L}$, then $\{\neg\sigma\}$ is consistent in $\mathscr{L}$. By Theorem 1.3.21, $\{\neg\sigma\}$ will have a model in which σ cannot hold. Hence σ is not valid. ⊣

THEOREM 1.3.22 (Compactness Theorem). *A set of sentences Σ has a model if and only if every finite subset of Σ has a model.*

PROOF. If every finite subset of Σ has a model, then every finite subset of Σ is consistent. So Σ is consistent and Theorem 1.3.21 shows that Σ has a model. ⊣

We conclude this section with a representative list of applications or consequences of the completeness and compactness theorems. Some additional exercises can be found at the end.

COROLLARY 2.1.5. *If a theory T has arbitrarily large finite models, then it has an infinite model.*

PROOF. Let T be a theory in $\mathscr{L}$ with arbitrarily large finite models. Consider the expansion $\mathscr{L}' = \mathscr{L} \cup \{c_n : n \in \omega\}$, where c_n is a list of distinct constant symbols not in $\mathscr{L}$. Consider the set Σ of $\mathscr{L}'$ defined by

$$\Sigma = T \cup \{\neg(c_n \equiv c_m) : n < m < \omega\}.$$

Any finite subset Σ' of Σ will involve at most the constants $c_0, \ldots, c_m$, say, for some m. Let $\mathfrak{A}$ be a model of T with at least $m+1$ elements, and let $a_0, \ldots, a_m$ be a list of $m+1$ distinct elements of $\mathfrak{A}$. We can verify easily that the model $(\mathfrak{A}, a_0, \ldots, a_m)$ for the finite expansion $\mathscr{L}'' = \mathscr{L} \cup \{c_0, \ldots, c_m\}$ of $\mathscr{L}$ is a model of Σ'. So, by Theorem 1.3.22, Σ has a model. The reduction of this model to $\mathscr{L}$ gives a model of T which is clearly infinite. ⊣

COROLLARY 2.1.6 (Upward Löwenheim–Skolem–Tarski Theorem). *If T has infinite models, then it has infinite models of any given power $\alpha \geqslant \|\mathscr{L}\|$.*

PROOF. The proof is similar to that of Corollary 2.1.5. Let c_ξ, $\xi < \alpha$, be a list of distinct constant symbols not in $\mathscr{L}$, and consider the set of sentences

$\Sigma = T \cup \{\neg(c_\xi \equiv c_\eta) : \xi < \eta < \alpha\}$. Every finite subset Σ' of Σ will involve at most a finite number of the constants c_ξ. Hence any infinite model of T can be expanded to a model of Σ'. By Theorem 1.3.22, Σ has a model $\mathfrak{A}$ and by Corollary 2.1.4, this model is of power at most

$$||\mathscr{L} \cup \{c_\xi : \xi < \alpha\}|| = \alpha.$$

On the other hand, the interpretations of the constants c_ξ in $\mathfrak{A}$ must give distinct elements of A. So $\alpha \leqslant |A| \leqslant \alpha$ and $|A| = \alpha$. ⊣

A result first published by Skolem (1934) is the following:

Corollary 2.1.7. *There exist nonstandard models of complete number theory.*

Proof. Recall from 1.4.11 that complete number theory is the set of all sentences holding in the standard model $\langle \omega, +, \cdot, S, 0 \rangle$ of number theory. Since this theory has an infinite model, it has models of all infinite powers. A noncountable model of complete number theory clearly cannot be standard. ⊣

A simple but powerful device in model theory is the *method of diagrams*. Let $\mathfrak{A}$ be a model for $\mathscr{L}$. We expand the language $\mathscr{L}$ to a new language

$$\mathscr{L}_A = \mathscr{L} \cup \{c_a : a \in A\}$$

by adding a new constant symbol c_a for each element $a \in A$. It is understood that if $a \neq b$, then c_a and c_b are different symbols. We may then expand $\mathfrak{A}$ to the model

$$\mathfrak{A}_A = (\mathfrak{A}, a)_{a \in A}$$

for $\mathscr{L}_A$ by interpreting each new constant c_a by the element a. The *diagram of* $\mathfrak{A}$, denoted by $\Delta_{\mathfrak{A}}$, is the set of all atomic sentences and negations of atomic sentences of $\mathscr{L}_A$ which hold in the model $\mathfrak{A}_A$.

If X is a subset of A, then we let $\mathscr{L}_X$ be the language $\mathscr{L} \cup \{c_a : a \in X\}$ and $\mathfrak{A}_X = (\mathfrak{A}, a)_{a \in X}$ be the obvious expansion of $\mathfrak{A}$ to $\mathscr{L}_X$. If f is a mapping from X into the set of elements B of a model $\mathfrak{B}$ for $\mathscr{L}$, then $(\mathfrak{B}, fa)_{a \in X}$ is the expansion of $\mathfrak{B}$ to a model for $\mathscr{L}_X$ formed by interpreting each c_a by fa.

The method of adding new constant symbols for elements of a model is used again and again in model theory. The following proposition illustrates the usefulness of diagrams.

Proposition 2.1.8. *Let $\mathfrak{A}$ and $\mathfrak{B}$ be models for $\mathscr{L}$ and let $f : A \to B$. Then the following are equivalent:*

(a). *f is an isomorphic embedding of $\mathfrak{A}$ into $\mathfrak{B}$.*
(b). *There is an extension $\mathfrak{C} \supset \mathfrak{A}$ and an isomorphism $g : \mathfrak{C} \cong \mathfrak{B}$ such that $g \supset f$.*
(c). *$(\mathfrak{B}, fa)_{a \in A}$ is a model of the diagram of $\mathfrak{A}$.*

Proof. The implication from (b) to (a) is trivial. If (a) holds, one can extend the set A to a set C and extend the function f to a one to one function g from C onto B. Then define the relations of $\mathfrak{C}$ by the rule

$$\mathfrak{C} \models R[c_1 \ldots c_n] \text{ iff } \mathfrak{B} \models R[gc_1 \ldots gc_n],$$

and similarly for functions. This will make (b) hold for $\mathfrak{C}$ and g.

To prove the equivalence of (a) and (c), use the fact that by Proposition 1.3.18, for each formula $\varphi(x_1 \ldots x_n)$ and all $a_1, \ldots, a_n$ in A,

$$\mathfrak{A} \models \varphi[a_1 \ldots a_n] \text{ if and only if } \mathfrak{A}_A \models \varphi(a_1 \ldots a_n)$$

and

$$\mathfrak{B} \models \varphi[fa_1 \ldots fa_n] \text{ if and only if } (\mathfrak{B}, fa)_{a \in A} \models \varphi(a_1 \ldots a_n). \dashv$$

Proposition 2.1.8 shows that the following three conditions are equivalent:
(a′) *$\mathfrak{A}$ is isomorphically embeddable in $\mathfrak{B}$.*
(b′) *$\mathfrak{B}$ is isomorphic to an extension of $\mathfrak{A}$.*
(c′) *$\mathfrak{B}$ can be expanded to a model of the diagram of $\mathfrak{A}$.*

In the special case that $A \subset B$ and f is the identity mapping from A into B, Proposition 2.1.8 shows that $\mathfrak{A}$ is a submodel of $\mathfrak{B}$ if and only if $\mathfrak{B}_A$ is a model of the diagram of $\mathfrak{A}$.

Corollary 2.1.9. *Suppose that $\mathscr{L}$ has no function or constant symbols. Let T be a theory in $\mathscr{L}$ and $\mathfrak{A}$ be a model for $\mathscr{L}$. Then $\mathfrak{A}$ is isomorphically embedded in some model of T if and only if every finite submodel of $\mathfrak{A}$ is isomorphically embedded in some model of T.*

Proof. We skip the easy direction and suppose that every finite submodel of $\mathfrak{A}$ is isomorphically embedded in some model of T. We show that the set $\Sigma = T \cup \Delta_{\mathfrak{A}}$ is consistent. Every finite subset Σ' of Σ contains at most a finite number of the new constants, say $c_{a_1}, \ldots, c_{a_m}$. Because the language $\mathscr{L}$ has no function or constant symbols, the finite set $A' = \{a_1, \ldots, a_m\}$ generates a finite submodel $\mathfrak{A}'$ of $\mathfrak{A}$. Let $\mathfrak{B}'$ be a model of T in which $\mathfrak{A}'$ is isomorphically embedded. We see without difficulty that $\Sigma' \subset T \cup \Delta_{\mathfrak{A}'}$. So, by Proposition 2.1.8, $\mathfrak{B}'$ can be expanded to a model of Σ', and hence Σ' has a model. By compactness, Σ has a model $\mathfrak{B}$. By Proposition 2.1.8 again, the reduct of $\mathfrak{B}$ to $\mathscr{L}$ gives a model of T in which $\mathfrak{A}$ is isomorphically embedded. $\dashv$

We next consider two applications from the theory of fields (see 1.4.9).

COROLLARY 2.1.10. *Let T be a theory in the language $\mathscr{L} = \{+, \cdot, 0, 1\}$' which has as models fields with arbitrary high finite characteristics. Then T has a model which is a field of characteristic* 0.

PROOF. Let T' be the theory of fields and consider the set

$$\Sigma = T \cup T' \cup \{p1 \not\equiv 0 : \text{all primes } p\}.$$

Recall from Chapter 1 that $p1$ is our abbreviation for the term $1 + \dots + 1$, p times, of the language $\mathscr{L}$. A finite subset Σ' of Σ will involve a highest prime, say p. Let $\mathfrak{A}$ be a model of T which is a field, so $\mathfrak{A}$ is also a model of T', and such that the characteristic of $\mathfrak{A}$ is higher than p. Then $\mathfrak{A}$ is a model of Σ', whence by compactness, Σ has a model. This model is a model of T, is a field, and has characteristic 0. ⊣

COROLLARY 2.1.11. *There exist non-Archimedean ordered fields elementarily equivalent to the ordered field of real numbers.*

PROOF. An ordered field $\langle F, +, \cdot, 0, 1, \leqslant \rangle$ is Archimedean iff for any two positive elements a, b in F there is an n such that $na \geqslant b$. This is not expressible in first-order logic.

Let T be the set of all sentences of $\mathscr{L} = \{+, \cdot, 0, 1, \leqslant\}$ holding in the ordered field of reals. Let c be a constant symbol different from 0 and 1. Let

$$\Sigma = T \cup \{n1 \leqslant c : n \in \omega\}.$$

For every finite subset Σ' of Σ, there is an expansion of the reals to a model of Σ'. By compactness, Σ has a model in which c has an interpretation b. In this model, both 1 and b are positive; yet no finite multiple of 1 can exceed b. ⊣

Corollary 2.1.11 is the very beginning of a branch of model theory called *nonstandard analysis*. The model theory of nonstandard analysis will be developed in Section 4.4.

Consider $\Delta_{\mathfrak{A}}$, the diagram of $\mathfrak{A}$ introduced earlier. We see that Proposition 2.1.8 gives an intimate connection between models of $\Delta_{\mathfrak{A}}$ and models in which $\mathfrak{A}$ can be isomorphically embedded. By the *positive diagram of* $\mathfrak{A}$ we mean the subset of $\Delta_{\mathfrak{A}}$ which consists only of atomic sentences (no negations of atomic sentences). We shall see that positive diagrams are associated with the following notion of homomorphic embedding.

Given models $\mathfrak{A}$ and $\mathfrak{A}'$ for $\mathscr{L}$, $\mathfrak{A}$ is *homomorphic to* $\mathfrak{A}'$ iff there is a function f mapping A onto A' satisfying the following:

(i). For each n-placed relation R of $\mathfrak{A}$ and the corresponding relation R' of $\mathfrak{A}'$, and all elements $x_1, \ldots, x_n$ of A,

$$\text{if}\quad R(x_1 \ldots x_n), \quad\text{then}\quad R'(f(x_1) \ldots f(x_n)).$$

(ii). For each m-placed function G of $\mathfrak{A}$ and the corresponding G' of $\mathfrak{A}'$, and for all $x_1, \ldots, x_m$ of A,

$$f(G(x_1 \ldots x_m)) = G'(f(x_1) \ldots f(x_m)).$$

(iii). For each constant x of $\mathfrak{A}$, $f(x)$ is the corresponding constant of $\mathfrak{A}'$.

A function f satisfying the above is called a *homomorphism of* $\mathfrak{A}$ *onto* $\mathfrak{A}'$. We write $\mathfrak{A} \simeq_f \mathfrak{A}'$ to indicate that f is such a homomorphism; if it is not necessary to indicate f, we write $\mathfrak{A} \simeq \mathfrak{A}'$ for $\mathfrak{A}$ is homomorphic to $\mathfrak{A}'$. In this case we also say $\mathfrak{A}'$ is *a homomorphic image of* $\mathfrak{A}$. $\mathfrak{A}$ is *homomorphically embedded in* $\mathfrak{A}'$ iff $\mathfrak{A}$ is homomorphic to some submodel of $\mathfrak{A}'$. See Exercise 2.1.3 for some elementary properties of these notions. The next proposition corresponds to Proposition 2.1.8.

PROPOSITION 2.1.12. *Let* $\mathfrak{A}$, $\mathfrak{B}$ *be models for* $\mathscr{L}$. *Then* $\mathfrak{A}$ *is homomorphically embedded in* $\mathfrak{B}$ *if and only if some expansion of* $\mathfrak{B}$ *is a model of the positive diagram of* $\mathfrak{A}$.

COROLLARY 2.1.13. *Every partial order on a set* X *can be extended to a simple order on* X.

PROOF. Suppose that $\leqslant$ partially orders X. Let $\mathfrak{A} = \langle X, \leqslant \rangle$. Let $\{c_x : x \in X\}$ be distinct constants for $x \in X$ and let Δ be the positive diagram of $\mathfrak{A}$. Let

$$\Sigma = \Delta \cup \{c_x \not\equiv c_y : x \neq y \text{ in } X\} \cup \{\sigma\},$$

where σ is the sentence which expresses that $\leqslant$ is a simple order (see 1.4.1). Let Σ' be a finite subset of Σ involving, say, the elements $x_1, \ldots, x_n$ and the corresponding constants. We need the following fact:

(1) Every partial order $\leqslant$ on $\{x_1, \ldots, x_n\}$ can be extended to a simple order $\leqslant'$ on $\{x_1, \ldots, x_n\}$ so that $\leqslant$ is preserved, i.e., if $x_i \leqslant x_j$, then $x_i \leqslant' x_j$.

The proof of (1) is not difficult and proceeds by induction on n. Assuming (1), we see that $\langle \{x_1, \ldots, x_n\}, \leqslant' \rangle$ is a model of Σ'. By compactness, Σ has a simply ordered model $\langle Y, \leqslant' \rangle$, in which there is an element y_x corresponding to each constant c_x. Clearly the set $\{y_x : x \in X\}$ is simply ordered by $\leqslant'$. If $x \leqslant z$, then $y_x \leqslant' y_z$, and if $x \neq z$, then $y_x \neq y_z$. Using the inverse of the 1–1 function $y : x \to y_x$, we can induce a simple order on X which extends $\leqslant$. ⊣

EXERCISES

2.1.1. Show that there are also *countable* nonstandard models of complete number theory.

2.1.2. Prove the representation theorem for Boolean algebras (Proposition 1.4.4) by the method of diagrams.

[*Hint*: (a). Every atomic Boolean algebra is isomorphic to a field of sets. (b). Every finite subset of a Boolean algebra generates a finite, therefore atomic, Boolean algebra. (c). If $\mathfrak{A}$ is isomorphically embedded in a field of sets, then $\mathfrak{A}$ is isomorphic to a field of sets.]

2.1.3. Prove the following. The homomorphism relation $\simeq$ is reflexive and transitive. It is not symmetric nor antisymmetric. If $\mathfrak{A} \simeq \mathfrak{B}$, then $|A| \geqslant |B|$. A sentence σ is called *positive* iff it is built up from atomic formulas using only $\wedge$, $\vee$, $\exists$, $\forall$. If $\mathfrak{A} \simeq \mathfrak{B}$, σ is a positive sentence, and $\mathfrak{A} \models \sigma$, then $\mathfrak{B} \models \sigma$. Compare this with Exercise 1.3.5.

2.1.4. Prove the assertion (1) in Corollary 2.1.13.

2.1.5. Show that every ordered field is equivalent to a non-Archimedean ordered field.

2.1.6. Show that every group which has elements of arbitrarily large finite order is equivalent to a group which has an element of infinite order.

2.1.7. Show that every model of ZF is equivalent to a (countable) model $\langle A, E\rangle$ which has an infinite sequence

$$\dots Ex_2 Ex_1 Ex_0 .$$

Therefore every model of ZF is equivalent to a countable model which is not isomorphic to a transitive model.

2.1.8. Let $\mathfrak{A} = \langle A, \leqslant, \dots\rangle$ be an infinite model such that $\leqslant$ well orders A. Show that there is a model $\mathfrak{A}' = \langle A', \leqslant', \dots\rangle$ equivalent to $\mathfrak{A}$ such that $\leqslant'$ is not a well ordering.

2.1.9. Show that every infinite model $\mathfrak{A}$ for a language $\mathscr{L}$ has an equivalent model $\mathfrak{B}$ of power $||\mathscr{L}||$ such that not every element of B is a constant of $\mathfrak{B}$.

2.1.10. Let $\mathscr{L}$ have no function or constant symbols. Let T be a theory in $\mathscr{L}$ and $\mathfrak{A}$ be a model for $\mathscr{L}$. Then $\mathfrak{A}$ is homomorphically embedded in some model of T if and only if every finite submodel of $\mathfrak{A}$ is homomorphically embedded in some model of T. (This is a homomorphism version of Corollary 2.1.9.)

2.1.11. Let $\mathfrak{A}$ be an arbitrary infinite model and let $\alpha \geqslant ||\mathscr{L}||$. Then there is a model $\mathfrak{B}$ equivalent to $\mathfrak{A}$ such that for every formula $\varphi(x)$ with one free variable, if $\varphi(x)$ is satisfied by infinitely many different elements of $\mathfrak{B}$, then $\varphi(x)$ is satisfied by α different elements of $\mathfrak{B}$.

2.1.12. A model $\mathfrak{L}$ is said to be *finitely generated* iff there is a finite set $X \subset B$ which generates $\mathfrak{L}$ (see Exercise 1.3.9). Let T be a theory in $\mathscr{L}$ and let $\mathfrak{A}$ be a model for $\mathscr{L}$. Then $\mathfrak{A}$ is isomorphically embedded in some model of T if and only if every finitely generated submodel of $\mathfrak{A}$ is isomorphically embedded in some model of T. (Compare with Corollary 2.1.9.)

2.1.13

(i). If T_1 and T_2 are two theories such that $T_1 \cup T_2$ has no models, then there is a sentence φ such that $T_1 \models \varphi$ and $T_2 \models \neg \varphi$.

(ii). If T_1 and T_2 are two theories such that for all $\mathfrak{A}$, $\mathfrak{A}$ is a model of T_1 iff $\mathfrak{A}$ is not a model of T_2, then T_1 and T_2 are finitely axiomatizable.

2.1.14. Let $T_1 \subset T_2 \subset T_3 \subset \ldots$ be a strictly increasing sequence of closed theories in $\mathscr{L}$. Show that their union $T = \bigcup_{n<\omega} T_n$ is a consistent closed theory in $\mathscr{L}$ and it is not finitely axiomatizable.

2.1.15. Let T_n, $n \in \omega$, be a strictly increasing sequence of closed theories in a finite language $\mathscr{L}$. Prove that $\bigcup_n T_n$ has an infinite model.

2.1.16. Let T be a finitely axiomatizable theory with only a countable number of complete extensions in a language $\mathscr{L}$. Prove that T has a finitely axiomatizable complete extension in $\mathscr{L}$.

2.1.17. Prove that every complete theory T in a countable language has a model $\mathfrak{A}$ of power $\leqslant 2^\omega$ such that for every $\mathfrak{B} \models T$ and every $S \subseteq B$ there is an $R \subseteq A$ such that $(\mathfrak{B}, S)$ is elementarily equivalent to $(\mathfrak{A}, R)$.

2.1.18*. Let Δ be the theory of dense linear order without endpoints. Prove the following lemma (a), and then use (a) and the Löwenheim–Skolem–Tarski Theorem to give a simpler proof of Theorem 1.5.3 on the elimination of quantifiers for the theory Δ.

(a). Let $\mathfrak{A}$ and $\mathfrak{B}$ be countable models of $\Delta, a_1, \ldots, a_n \in A$, and $b_1, \ldots, b_n \in B$. If $a_1, \ldots, a_n$ and $b_1, \ldots, b_n$ satisfy the same arrangement, then $(\mathfrak{A}, a_1 \ldots a_n) \cong (\mathfrak{B}, b_1 \ldots b_n)$.

2.1.19*. Let $\mathscr{L} = \emptyset$ be the language of pure identity theory. Prove the following lemmas (a) and (b), and then use (a), (b), and the Löwenheim–Skolem–Tarski Theorem to give a simpler proof of Theorem 1.5.7 on the elimination of quantifiers for pure identity theory.
(a). Let $\mathfrak{A}$ and $\mathfrak{B}$ be models for $\mathscr{L}$ of the same cardinality, $a_1, \ldots, a_n \in A$, and $b_1, \ldots, b_n \in B$. If $a_1, \ldots, a_n$ and $b_1, \ldots, b_n$ satisfy the same arrangement, then $(\mathfrak{A}, a_1 \ldots a_n) \cong (\mathfrak{B}, b_1 \ldots b_n)$.
(b). Let $\varphi(x_1 \ldots x_n)$ be a formula of $\mathscr{L}$ and $\theta(x_1 \ldots x_n)$ be an arrangement. Let S be the set of all cardinals α such that $\varphi \wedge \theta$ is satisfiable in a model of cardinality α. Then either S or its complement is a finite set of finite cardinals.

2.1.20. This and some of the following exercises are designed to show an alternative method of proving the extended completeness theorem for countable languages. A generalization of this method to noncountable languages is also given later.

Let $\mathscr{L}$ be a countable language, and let T be a consistent set of sentences closed under $\vdash$. We aim to prove that T has a model. The starting point of our discussion is the countable Lindenbaum algebra $\mathfrak{B}_{\mathscr{L}}$. We have already seen in Exercise 1.4.11 that T corresponds to a filter in $\mathfrak{B}_0$, the Lindenbaum algebra of sentences of $\mathscr{L}$. It is also easy to show that the set

$$\Phi = \{\varphi : T \vdash \varphi \quad \text{and} \quad \varphi \text{ is a formula of } \mathscr{L}\}$$

is a filter in $\mathfrak{B}$. For simplicity, we shall now operate in the quotient algebra $\mathfrak{B}/\Phi$. In other words, the equivalence classes of this new algebra are given by sets of formulas

$$(\psi) = \{\varphi : T \vdash \varphi \leftrightarrow \psi\},$$

with its unit element given by the set Φ, and its zero element given by

$$\{\varphi : T \vdash \neg \varphi\}.$$

We denote this quotient algebra by $\mathfrak{B}_T$ and call it the *Lindenbaum algebra of T*. $\mathfrak{B}_T$ is obviously a countable Boolean algebra.

2.1.21. Let $\mathfrak{A}$ be any Boolean algebra and let Y be a subset of A. The *sum* of Y, or the l.u.b. of Y, is defined to be the unique $y \in A$ such that

$x \leqslant y$ for all $x \in Y$ (i.e. y is an *upper bound* for Y), and

if $z \in A$ is any upper bound for Y, then $y \leqslant z$.

We denote the sum of Y if it exists by $\bigvee Y$, or if the elements of Y are indexed by I, $\bigvee_{i \in I} y_i$. In an entirely similar manner, we can define the *product* of Y, or the g.l.b. of Y, and denote it by $\bigwedge Y$ or $\bigwedge_{i\in I} y_i$. Sums and products of arbitrary $Y \subset A$ do not necessarily exist. When they do exist, they satisfy the following identities (assume that $\bigvee_{i\in I} y_i$ exists):

$$(\bigvee_{i\in I} y_i)+x = \bigvee_{i\in I} (y_i+x),$$

$$(\bigvee_{i\in I} y_i)\cdot x = \bigvee_{i\in I} (y_i\cdot x),$$

$$\overline{\bigvee_{i\in I} y_i} = \bigwedge_{i\in I} \bar{y}_i.$$

These identities imply, of course, that the sums and products on the right-hand side also exist. We leave the duals involving $\bigwedge$ to the reader.

Let φ be any formula of $\mathscr{L}$. Let $\varphi(k/p)$ be the formula obtained from φ by first replacing all bound occurrences of v_p in φ by v_j, the first variable in the sequence $v_0, v_1, \ldots$, not occurring in φ, and then replacing all free occurrences of v_k by v_p. Show that in the Boolean algebra $\mathfrak{B}_T$,

$$\bigvee_{p\in\omega} (\varphi(k/p)) = ((\exists v_k)\varphi),$$

$$\bigwedge_{p\in\omega} (\varphi(k/p)) = ((\forall v_k)\varphi).$$

Thus sums and products of certain sets of substitution instances of a single formula φ always exist and correspond to existential and universal quantification of φ. Note that the number of such sums (and products) in $\mathfrak{B}_T$ is countable.

2.1.22. An ultrafilter D on $\mathfrak{A}$ is said to *preserve the sum* $\bigvee_{i\in I} y_i$ iff

$$\bigvee_{i\in I} y_i \in D \quad \text{if and only if some} \quad y_i \in D.$$

Similarly, *D preserves the product* $\bigwedge_{i\in I} y_i$ iff

$$\bigwedge_{i\in I} y_i \in D \quad \text{if and only if all} \quad y_i \in D.$$

Prove the following: Given a countable sequence of products $\bigwedge X_0, \bigwedge X_1, \ldots, \bigwedge X_n, \ldots$ of $\mathfrak{A}$. Then there exists an ultrafilter D on $\mathfrak{A}$ which preserves each product.

[*Hint*: Pick a sequence $x_n \in X_n$ such that no finite product of elements of the form $\bigwedge X_n + \bar{x}_n$ is equal to zero. Now consider any ultrafilter D which has as elements all $\bigwedge X_n + \bar{x}_n$.]

There is also a corresponding result about countable sequences of sums.

2.1.23. Let D be any ultrafilter on $\mathfrak{B}_T$ which preserves all the products of Exercise 2.1.21. We shall now construct a model of T from the variables

$v_0, v_1, \dots$ of $\mathscr{L}$. Since the procedure is quite similar to that of Lemma 2.1.2, we ask the reader to fill in all the details.

First define equivalence by

$$v_i \sim v_j \quad \text{iff} \quad (v_i \equiv v_j) \in D.$$

The equivalence classes are denoted by $\tilde{v}_i$.

Let c be a constant symbol of $\mathscr{L}$. Since $\vdash (\exists v_0)(c \equiv v_0)$, and D preserves sums, we see that for some i, $(c \equiv v_i) \in D$. Let the interpretation of c be the class $\tilde{v}_i$.

Let t be any term of $\mathscr{L}$ (this includes the cases of function symbols and constant symbols) and v_p be a variable not occurring in t. Then $\vdash (\exists v_p)(t \equiv v_p)$. Since D preserves sums, for some j, $(t \equiv v_j) \in D$. Let the interpretation of the term t (defined on equivalence classes $\tilde{v}_i$) be $\tilde{v}_j$.

Finally, let P be a relation symbol of $\mathscr{L}$. We define the relation R by

$$R(\tilde{v}_{i_1} \dots \tilde{v}_{i_n}) \quad \text{iff} \quad (P(v_{i_1} \dots v_{i_n})) \in D.$$

In this way, we have defined in an unambiguous manner a model $\mathfrak{A}$ for $\mathscr{L}$ with universe the set of equivalence classes $\tilde{v}_i$. Now prove by induction on the formulas $\varphi(v_0 \dots v_n)$ of $\mathscr{L}$ that

$$\mathfrak{A} \models \varphi[\tilde{v}_0 \dots \tilde{v}_n] \quad \text{iff} \quad (\varphi(v_0 \dots v_n)) \in D.$$

To pass through the cases of $\forall$ or $\exists$, we again need the fact that D preserves sums. Since $\varphi \in T$ implies that $(\varphi) \in D$, this shows that $\mathfrak{A}$ is a model of T.

2.1.24. If the language $\mathscr{L}$ is uncountable, then the number of sums and products corresponding to $\exists$ and $\forall$ in $\mathfrak{B}_T$ is also uncountable. Even though, in general, Exercise 2.1.17 fails for uncountable sequences of products in an arbitrary Boolean algebra, there is, nevertheless, a version of it which holds for the algebra $\mathfrak{B}_T$. This is because every formula φ contains only a finite number of symbols. The generalization of Exercise 2.1.17 is as follows (the proof is straightforward):

Let $\mathfrak{A}$ be a Boolean algebra, α be an infinite cardinal, and $\bigwedge X_\beta$, $\beta < \alpha$, be a sequence of products of $\mathfrak{A}$. Suppose that for all $\beta < \alpha$ and all filters E on $\mathfrak{A}$ generated by fewer than α elements,

$$\text{whenever} \quad X_\beta \subset E, \quad \text{then} \quad \bigwedge X_\beta \in E.$$

Then there exists an ultrafilter D on $\mathfrak{A}$ which preserves each product $\bigwedge X_\beta$.

Using this result, a generalization of the proof in Exercise 2.1.18 can be given for noncountable languages $\mathscr{L}$. (A technical detail should be mentioned: Before proceeding with the proof we must first expand $\mathscr{L}$ to a language $\overline{\mathscr{L}}$ with $||\mathscr{L}||$ new constant symbols. This is apparently necessary, see the proof of Lemma 2.1.1.)

2.2. Refinements of the method. Omitting types and interpolation theorems

In this section, we shall give two refinements of the method used in Section 2.1 to construct countable models with additional properties. The first refinement will lead us to the omitting types theorem. At the moment, the possible ramifications of this technique to noncountable languages and models are not yet fully understood. We shall mention only a couple of results for noncountable languages.

The starting point of our discussion is the notion of a set Σ of formulas of $\mathscr{L}$ *in the (free) variables* $x_1, \ldots, x_n$. Here we are using $x_1, x_2, \ldots$ as names for arbitrary free variables of $\mathscr{L}$. We could just as well use $v_{m_1}, v_{m_2}, \ldots$, but we abhor double subscripts. The following is a precise definition: Σ is a set of formulas of $\mathscr{L}$ in the (free) variables $x_1, \ldots, x_n$ (symbolically, $\Sigma = \Sigma(x_1 \ldots x_n)$) iff $x_1, \ldots, x_n$ are distinct individual variables and every formula σ in Σ contains at most the variables $x_1, \ldots, x_n$ free. We now introduce the convention $\sigma = \sigma(x_1 \ldots x_n)$, as we did for $\varphi = \varphi(v_0 \ldots v_n)$. If $\sigma = \sigma(x_1 \ldots x_n)$, then the notation

$$\mathfrak{A} \models \sigma[a_1 \ldots a_n]$$

means that the sequence $a_1, \ldots, a_n$ of A satisfies σ in $\mathfrak{A}$ (see the section on satisfaction). It is useful also to introduce the notation

$$\mathfrak{A} \models \Sigma[a_1 \ldots a_n]$$

to mean that for every $\sigma \in \Sigma$, $a_1, \ldots, a_n$ satisfies σ in $\mathfrak{A}$; in this case we say that $a_1, \ldots, a_n$ *satisfies*, or *realizes*, Σ *in* $\mathfrak{A}$. If $c_1, \ldots, c_n$ is a sequence of constant symbols, then $\sigma(c_1 \ldots c_n)$ denotes the sentence formed by simultaneously replacing each free occurrence of x_i, $1 \leqslant i \leqslant n$, in σ by the corresponding c_i. Sometimes we shall replace just *some* of the variables by constants. If $m \leqslant n$, the notation $\sigma(c_1 \ldots c_m x_{m+1} \ldots x_n)$ is self-explanatory.

For reasons explained in Section 2.1 (before Lemma 2.1.1), we must be careful to use the above notation only in a context where the list of variables $x_1, \ldots, x_n$ is given. A completely unambiguous notation can be introduced, but at great cost in readability. For example, we could use the notation

$$\mathfrak{A} \models \sigma[a_1/x_1 \ldots a_n/x_n] \quad \text{for} \quad \mathfrak{A} \models \sigma[a_1 \ldots a_n],$$

$$\sigma(c_1/x_1 \dots c_m/x_m x_{m+1} \dots x_n) \quad \text{for} \quad \sigma(c_1 \dots c_m x_{m+1} \dots x_n).$$

Let Σ be a set of formulas in the variables $x_1, \dots, x_n$, and let $\mathfrak{A}$ be a model for $\mathscr{L}$. We say that $\mathfrak{A}$ *realizes* Σ iff some n-tuple of elements of A satisfies Σ in $\mathfrak{A}$. We say that $\mathfrak{A}$ *omits* Σ iff $\mathfrak{A}$ does not realize Σ. The phrase *Σ is satisfiable in $\mathfrak{A}$* has exactly the same meaning as *$\mathfrak{A}$ realizes Σ*. Σ is *consistent* iff Σ is satisfiable in some model.

EXAMPLE 2.2.1. Let T be Peano arithmetic and let $\Sigma(x)$ be the set

$$\{0 \not\equiv x, S0 \not\equiv x, SS0 \not\equiv x, \dots\}.$$

An element is said to be *nonstandard* iff it realizes $\Sigma(x)$. The standard model of T omits $\Sigma(x)$, while all the nonstandard models realize $\Sigma(x)$.

EXAMPLE 2.2.2. Let T be the theory of ordered fields and let $\Sigma(x)$ be the set

$$\{1 \leqslant x, 1+1 \leqslant x, 1+1+1 \leqslant x, \dots\}.$$

An element is said to be *positive infinite* iff it realizes $\Sigma(x)$. An ordered field omits $\Sigma(x)$ if and only if it is Archimedean. The ordered fields of rationals and reals omit $\Sigma(x)$. Non-Archimedean ordered fields were constructed in the last section using the compactness theorem.

EXAMPLE 2.2.3. Let T be the theory of Abelian groups and let $\Sigma(x)$ be the set

$$\{x \not\equiv 0, 2x \not\equiv 0, 3x \not\equiv 0, \dots\}.$$

Elements which realize $\Sigma(x)$ are said to be of *infinite order*. An Abelian group which omits $\Sigma(x)$ is said to be a *torsion group*. Thus in a torsion group, every element has a finite multiple which is zero.

EXAMPLE 2.2.4. Here is an example of a set of formulas with infinitely many variables. Let T be the theory of partial order and let Σ be the set

$$\{x_1 < x_0, x_2 < x_1, x_3 < x_2, \dots\}.$$

A model $\mathfrak{A}$ of T omits Σ iff $\mathfrak{A}$ is a *well founded* partial ordering. A linear ordering $\mathfrak{A}$ omits Σ iff it is a well ordering.

EXAMPLE 2.2.5. By a *type* $\Gamma(x_1 \dots x_n)$ in the variables $x_1, \dots, x_n$ we mean a maximal consistent set of formulas of $\mathscr{L}$ in these variables. Given any model $\mathfrak{A}$ and n-tuple $a_1, \dots, a_n \in A$, the set $\Gamma(x_1 \dots x_n)$ of all formulas $\gamma(x_1 \dots x_n)$ satisfied by $a_1, \dots, a_n$ is a type, and, in fact, is the unique type realized by $a_1, \dots, a_n$. It is called the *type of* $a_1, \dots, a_n$ in $\mathfrak{A}$.

EXAMPLE 2.2.6. Let $\mathfrak{A}$ be the ordered field of real numbers. Then any two distinct elements $a, b \in A$ have different types. For if $a < b$, there is a rational number r with $a < r < b$; hence a satisfies $x < r$, while b does not. Thus $\mathfrak{A}$ realizes 2^ω different types in one variable.

The next proposition answers the question: When is a set of formulas realized by some model of a theory T? Its proof is a simple application of the compactness theorem.

PROPOSITION 2.2.7. *Let T be a theory and let $\Sigma = \Sigma(x_1 \dots x_n)$. The following are equivalent*:

(i). *T has a model which realizes Σ.*

(ii). *Every finite subset of Σ is realized in some model of T.*

(iii). *$T \cup \{(\exists x_1 \dots x_n)(\sigma_1 \wedge \dots \wedge \sigma_m) : m < \omega, \sigma_1, \dots, \sigma_m \in \Sigma\}$ is consistent.*

We shall say that a formula $\sigma(x_1 \dots x_n)$ is *consistent with* a theory T iff there is a model $\mathfrak{A}$ of T which realizes $\{\sigma\}$, and we say that $\Sigma(x_1 \dots x_n)$ is *consistent with* T iff T has a model which realizes Σ. Thus (i)–(iii) above are all equivalent to the statement that Σ is consistent with T.

We now take up the question: When is a set Σ of formulas in $x_1, \dots, x_n$ *omitted* in some model of a theory T? This is a more difficult question, and we need more than the compactness theorem to answer it. The key theorem of this section, Theorem 2.2.9, gives a necessary and sufficient condition for T to have a model which omits Σ. The ω-completeness theorem 2.2.13 is one of a long list of consequences of it. We shall use Theorem 2.2.9 in the next section, and again later on. If Σ is a finite set of formulas, then there is no problem in determining whether Σ can be omitted, because the sentence

$$\varphi = (\exists x_1 \dots x_n)(\sigma_1 \wedge \dots \wedge \sigma_m),$$

where $\Sigma = \{\sigma_1, \dots, \sigma_m\}$, and its negation $\neg \varphi$ express, respectively, that Σ is realized or omitted. Thus the interesting case is where Σ is infinite.

Let us first take another look at Lemma 2.1.2. So far, we have only used the property that every element of $\mathfrak{A}$ is the interpretation of a constant $c \in C$ in a simple way, to show that $|A| \leqslant |C|$. In this section, we shall make much more use of that property of $\mathfrak{A}$.

The central idea in dealing with our problem is the notion of a theory locally realizing a set of formulas.

Let $\Sigma = \Sigma(x_1 \dots x_n)$ be a set of formulas of $\mathscr{L}$. A theory T in $\mathscr{L}$ is said to *locally realize* Σ iff there is a formula $\varphi(x_1 \dots x_n)$ in $\mathscr{L}$ such that:

(i). φ is consistent with T.

(ii). For all $\sigma \in \Sigma$, $T \models \varphi \to \sigma$.

That is, every n-tuple in a model of T which satisfies φ realizes Σ.

We say that T *locally omits* Σ iff T does not locally realize Σ. Thus T locally omits Σ if and only if for every formula $\varphi(x_1 \dots x_n)$ which is consistent with T, there exists $\sigma \in \Sigma$ such that $\varphi \wedge \neg \sigma$ is consistent with T.

For complete theories we have a simple proposition:

PROPOSITION 2.2.8. *Let T be a complete theory in $\mathscr{L}$, and let $\Sigma = \Sigma(x_1 \dots x_n)$ be a set of formulas of $\mathscr{L}$. If T has a model which omits Σ, then T locally omits Σ.*

PROOF. The proposition may be restated as follows: If T locally realizes Σ, then every model of T realizes Σ. Suppose T locally realizes Σ and let $\varphi(x_1 \dots x_n)$ be a formula consistent with T such that $T \models \varphi \to \sigma$, $\sigma \in \Sigma$.

Let $\mathfrak{A}$ be a model of T. Since T is complete, $T \models (\exists x_1 \dots x_n)\varphi$. So some n-tuple $a_1, \dots, a_n$ satisfies φ in $\mathfrak{A}$. Then $a_1, \dots, a_n$ satisfies each $\sigma \in \Sigma$, and hence realizes Σ in $\mathfrak{A}$. ⊣

The omitting types theorem is a converse of the above proposition. It holds, in fact, for arbitrary consistent theories in a countable language.

THEOREM 2.2.9 (Omitting Types Theorem). *Let T be a consistent theory in a countable language $\mathscr{L}$, and let $\Sigma(x_1 \dots x_n)$ be a set of formulas. If T locally omits Σ, then T has a countable model which omits Σ.*

PROOF. To simplify notation, let $\Sigma(x)$ be a set of formulas in one variable x. Suppose T locally omits $\Sigma(x)$. Let $C = \{c_0, c_1, \dots\}$ be a countable set of new constant symbols not already in $\mathscr{L}$ and let $\mathscr{L}' = \mathscr{L} \cup C$. Then $\mathscr{L}'$ is countable. Arrange all the sentences of $\mathscr{L}'$ in a list $\varphi_0, \varphi_1, \varphi_2, \dots$. We shall construct an increasing sequence of consistent theories

$$T = T_0 \subset T_1 \subset \dots \subset T_m \subset \dots$$

such that:

(1). Each T_m is a consistent theory of $\mathscr{L}'$ which is a finite extension of T.

(2). Either $\varphi_m \in T_{m+1}$ or $(\neg \varphi_m) \in T_{m+1}$.

(3). If $\varphi_m = (\exists x)\psi(x)$ and $\varphi_m \in T_{m+1}$, then $\psi(c_p) \in T_{m+1}$, where c_p is the first constant not occurring in T_m or φ_m.

(4). There is a formula $\sigma(x) \in \Sigma(x)$ such that $(\neg \sigma(c_m)) \in T_{m+1}$.

Assuming we already have the theory T_m, we construct T_{m+1} as follows: Let $T_m = T \cup \{\theta_1, \ldots, \theta_r\}$, $r > 0$, and let $\theta = \theta_1 \wedge \ldots \wedge \theta_r$. Let $c_0, \ldots, c_n$ contain all the constants from C occurring in θ. Form the formula $\theta(x_m)$ of $\mathscr{L}$ by replacing each constant c_i by x_i (renaming bound variables if necessary), and prefixing by $\exists x_i$, $i \not\equiv m$. Then $\theta(x_m)$ is consistent with T. Therefore, for some $\sigma(x) \in \Sigma(x)$, $\theta(x_m) \wedge \neg\, \sigma(x_m)$ is consistent with T. Put the sentence $\neg\, \sigma(c_m)$ into T_{m+1}. This makes (4) hold.

If φ_m is consistent with $T_m \cup \{\neg\, \sigma(c_m)\}$, put φ_m into T_{m+1}. Otherwise put $(\neg\, \varphi_m)$ into T_{m+1}. This takes care of (2). If $\varphi_m = (\exists x)\psi(x)$ is consistent with $T_m \cup \{\neg\, \sigma(c_m)\}$, put $\psi(c_p)$ into T_{m+1}. This takes care of (3). The theory T_{m+1} is a consistent finite extension of T_m. Thus (1)-(4) hold for T_{m+1}.

Let $T_\omega = \bigcup_{n<\omega} T_n$. From (1) and (2) we see that T_ω is a maximal consistent theory in $\mathscr{L}'$. Let $\mathfrak{B}' = (\mathfrak{B}, b_0, b_1, \ldots)$ be a countable model of T_ω, and let $\mathfrak{A}' = (\mathfrak{A}, b_0, b_1, \ldots)$ be the submodel of $\mathfrak{B}'$ generated by the constants $b_0, b_1, \ldots$. We then see from (3) that

$$A = \{b_0, b_1, \ldots\}.$$

Moreover, using (3) and the completeness of T_ω, we can show by induction on the complexity of a sentence φ in $\mathscr{L}'$ that

$$\mathfrak{A}' \models \varphi, \qquad \mathfrak{B}' \models \varphi, \qquad T_\omega \models \varphi$$

are all equivalent. Thus $\mathfrak{A}'$ is a model of T_ω and hence $\mathfrak{A}$ is a model of T. Finally, condition (4) ensures that $\mathfrak{A}$ omits Σ. ⊣

When T is a complete theory, we see that locally omitting $\Sigma(x_1 \ldots x_n)$ is a necessary and sufficient condition for T to have a model omitting Σ. Here is a necessary and sufficient condition which works in general.

COROLLARY 2.2.10. *Let $\mathscr{L}$ be countable. A theory T has a (countable) model omitting $\Sigma(x_1 \ldots x_n)$ if and only if some complete extension of T locally omits $\Sigma(x_1 \ldots x_n)$.*

EXAMPLE 2.2.11. Consider the language $\mathscr{L} = \{+, \cdot, S, 0\}$. We abbreviate $1 = S0$, $2 = SS0$, $3 = SSS0$, ... By an *ω-model* we mean a model $\mathfrak{A}$ in which

$$A = \{0, 1, 2, 3, \ldots\} \quad ,$$

that is, $\mathfrak{A}$ omits the set $\{x \not\equiv 0, x \not\equiv 1, x \not\equiv 2, \ldots\}$. A theory T in $\mathscr{L}$ is said to be *ω-consistent* iff there is no formula $\varphi(x)$ of $\mathscr{L}$ such that

$$T \models \varphi(0), \qquad T \models \varphi(1), \qquad T \models \varphi(2), \ldots$$

and

$$T \models (\exists x) \neg \varphi(x).$$

T is said to be *ω-complete* iff for every formula $\varphi(x)$ of $\mathscr{L}$ we have

$$T \models \varphi(0),\ T \models \varphi(1),\ T \models \varphi(2), \ldots \quad \text{implies} \quad T \models (\forall x)\varphi(x).$$

It follows from the omitting types theorem that:

Proposition 2.2.12. *Let T be a consistent theory in $\mathscr{L}$.*

(i). *If T is ω-complete, then T has an ω-model.*

(ii). *If T has an ω-model, then T is ω-consistent.*

Proof. (i). We show that T locally omits the set $\Sigma(x) = \{x \not\equiv 0, x \not\equiv 1, \ldots\}$. Suppose $\theta(x)$ is consistent with T. Then $T \models (\forall x) \neg \theta(x)$ fails. By ω-completeness, there is an n such that not $T \models \neg \theta(n)$. Hence $\theta(n)$ is consistent with T, so $\theta(x) \wedge \neg x \not\equiv n$ is consistent with T. Thus T locally omits $\Sigma(x)$.

(ii). Trivial. ⊣

The *ω-rule* is the following infinite rule of proof: From $\varphi(0), \varphi(1), \varphi(2), \ldots$, infer $(\forall x)\varphi(x)$, where $\varphi(x)$ is any formula of $\mathscr{L}$. *ω-logic* is formed by adding the ω-rule to the axioms and rules of inference of the first-order logic $\mathscr{L}$ and allowing infinitely long proofs. We have the following completeness theorem for ω-logic.

Proposition 2.2.13 (ω-Completeness Theorem). *A theory T in $\mathscr{L}$ is consistent in ω-logic if and only if T has an ω-model.*

Proof. Let T' be the set of all sentences of $\mathscr{L}$ provable from T in ω-logic. Then T is consistent in ω-logic if and only if T' is consistent in $\mathscr{L}$. Moreover, T' is ω-complete. Therefore T' has an ω-model if and only if T' is consistent. ⊣

The formulation of ω-logic above is aimed at studying the standard model of arithmetic. A useful generalization, which we shall call generalized ω-logic, is aimed at studying ordinary models for first order logic enriched by a symbol for the set of natural numbers.

Example 2.2.11′. Let $\mathscr{L}'$ be a countable language which has among its symbols a special unary relation symbol N and special constant symbols

$0, 1, 2, \ldots$. By an *ω-model* for $\mathscr{L}'$ we mean a model $\mathfrak{A}$ for $\mathscr{L}'$ in which N is interpreted by the set ω of natural numbers, and $0, 1, 2, \ldots$ are interpreted by themselves. In an ω-model, ω is a subset of the universe A, but we allow A to contain elements outside of ω or even to be uncountable.

Let T_N be the special set of sentences

$$T_N = \{N(m) : m < \omega\} \cup \{\neg m \equiv n : m < n < \omega\}$$

which state that the natural numbers are distinct and belong to N. T_N holds in every ω-model for $\mathscr{L}'$. A theory T in $\mathscr{L}'$ is said to be *ω-consistent* iff there is no formula $\varphi(x)$ of $\mathscr{L}'$ such that

$$T_N \cup T \models \varphi(0),\ T_N \cup T \models \varphi(1),\ T_N \cup T \models \varphi(2), \ldots$$

and

$$T_N \cup T \models (\exists x)(N(x) \wedge \neg\varphi(x)).$$

T is said to be *ω-complete* iff for every formula $\varphi(x)$ of $\mathscr{L}'$ we have

$$T_N \cup T \models \varphi(0),\ T_N \cup T \models \varphi(1),\ T_N \cup T \models \varphi(2), \ldots$$

implies

$$T_N \cup T \models (\forall x)(N(x) \rightarrow \varphi(x)).$$

The *ω-rule* for $\mathscr{L}'$ is the infinite rule: From $\varphi(0)$, $\varphi(1)$, $\varphi(2), \ldots$, infer $(\forall x)(N(x) \rightarrow \varphi(x))$. By *generalized ω-logic* we mean first order logic for the language $\mathscr{L}'$ with T_N added as an additional set of logical axioms and the ω-rule added as an additional rule of proof.

Propositions 2.2.12 and 2.2.13 take the following form for generalized ω-logic.

PROPOSITION 2.2.12′. *Let T be a theory in $\mathscr{L}'$ such that $T_N \cup T$ is consistent.*

(i). *If T is ω-complete, then T has an ω-model.*

(ii). *If T has an ω-model, then T is ω-consistent.*

PROPOSITION 2.2.13′. *A theory T in $\mathscr{L}'$ is consistent in generalized ω-logic if and only if T has an ω-model.*

The following example shows that the omitting types theorem fails for sets of formulas with infinitely many free variables.

Example 2.2.14. Let T be the theory of dense linear order without endpoints. Thus T is complete. Let $\Sigma(x_0 x_1 x_2 \ldots)$ be the set

$$\{x_1 < x_0, x_2 < x_1, x_3 < x_2, \ldots\}.$$

As we observed before, a model $\mathfrak{A}$ omits Σ if and only if $\mathfrak{A}$ is a well ordering. But T has no well ordered models, so no model of T omits Σ. However, T does locally omit Σ, because if $\varphi(x_0 x_1 \ldots x_n)$ is consistent with T, then $\varphi \wedge \neg\, x_{n+2} < x_{n+1}$ is consistent with T.

The omitting types theorem can be generalized to the case of countably many sets of formulas.

Theorem 2.2.15 (Extended Omitting Types Theorem). *Let T be a consistent theory in a countable language $\mathscr{L}$, and for each $r < \omega$ let $\Sigma_r(x_1 \ldots x_{n_r})$ be a set of formulas in n_r variables. If T locally omits each Σ_r, then T has a countable model which omits each Σ_r.*

Proof. Similar to the proof of the omitting types theorem. The only difference is that for each r the n_r-tuples of new constants are arranged in a list:

$$s_r^r, s_{r+1}^r, s_{r+2}^r, \ldots.$$

The theories T_m are built up so that for each $r = 0, 1, \ldots, m$, there is a formula $\sigma \in \Sigma_r$ such that $(\neg\, \sigma(s_m^r)) \in T_{m+1}$. ⊣

Here is a first application of the extended omitting types theorem. It uses the notion of an *elementary extension* which plays an important role in the rest of this book.

2.2.16. $\mathfrak{B}$ is said to be an *elementary extension* of $\mathfrak{A}$, $\mathfrak{A} \prec \mathfrak{B}$, iff

(i). $\mathfrak{B}$ is an extension of $\mathfrak{A}$, $\mathfrak{A} \subset \mathfrak{B}$.

(ii). For any formula $\varphi(x_1 \ldots x_n)$ of $\mathscr{L}$ and any $a_1, \ldots, a_n \in A$, $a_1, \ldots, a_n$ satisfies φ in $\mathfrak{A}$ if and only if it satisfies φ in $\mathfrak{B}$.

When $\mathfrak{B}$ is an elementary extension of $\mathfrak{A}$ we also say that $\mathfrak{A}$ is an *elementary submodel* of $\mathfrak{B}$.

A mapping $f\colon A \to B$ is said to be an *elementary embedding* of $\mathfrak{A}$ into $\mathfrak{B}$, in symbols $f\colon \mathfrak{A} \prec \mathfrak{B}$, iff for all formulas $\varphi(x_1 \ldots x_n)$ of $\mathscr{L}$ and n-tuples $a_1, \ldots, a_n \in A$, we have

$$\mathfrak{A} \models \varphi[a_1 \ldots a_n] \text{ if and only if } \mathfrak{B} \models \varphi[fa_1 \ldots fa_n].$$

An elementary embedding of $\mathfrak{A}$ into $\mathfrak{B}$ is thus the same thing as an isomorphism of $\mathfrak{A}$ onto an elementary submodel of $\mathfrak{B}$.

The following analogue of Proposition 2.1.8 is often useful.

PROPOSITION 2.2.17. *Let $\mathfrak{A}$ and $\mathfrak{B}$ be models for $\mathscr{L}$ and let $f: A \to B$. Then the following are equivalent:*

(a). *f is an elementary embedding of $\mathfrak{A}$ into $\mathfrak{B}$.*

(b). *There is an elementary extension $\mathfrak{C} > \mathfrak{A}$ and an isomorphism g: $\mathfrak{C} \cong \mathfrak{B}$ such that $g \supset f$.*

(c). *$(\mathfrak{B}, fa)_{a \in A}$ is a model of the elementary diagram of $\mathfrak{A}$.*

Proposition 2.2.17 shows that the following three conditions are equivalent:

(a′) *$\mathfrak{A}$ is elementarily embeddable in $\mathfrak{B}$.*

(b′) *$\mathfrak{B}$ is isomorphic to an elementary extension of $\mathfrak{A}$.*

(c′) *$\mathfrak{B}$ can be expanded to a model of the elementary diagram of $\mathfrak{A}$.*

In the special case that $A \subset B$ and f is the indentity mapping from A into B, Proposition 2.2.17 shows that $\mathfrak{A}$ is an elementary submodel of $\mathfrak{B}$ if and only if $\mathfrak{B}_A$ is a model of the elementary diagram of $\mathfrak{A}$.

Let us now consider the theory ZF, Zermelo–Fraenkel set theory. A model $\mathfrak{B} = \langle B, F \rangle$ of ZF is said to be an *end extension* of a model $\mathfrak{A} = \langle A, E \rangle$ of ZF iff $\mathfrak{B}$ is a proper extension of $\mathfrak{A}$ and no member of A gets a new element, that is,

$$\text{if } a \in A \text{ and } b \in B, \text{ then } bFa \text{ implies } b \in A.$$

THEOREM 2.2.18. *Every countable model $\mathfrak{A} = \langle A, E \rangle$ of* ZF *has an end elementary extension.*

PROOF. Let $\mathscr{L}$ be the language with the symbol ϵ, a constant symbol $\bar{a}$ for each $a \in A$, and a new constant symbol c. Let T be the theory with the axioms

$$\mathrm{Th}((\mathfrak{A}, a)_{a \in A}),$$

$$c \notin \bar{a}, \quad \text{where} \quad a \in A.$$

T is consistent because every finite subset of T has a model of the form $(\mathfrak{A}, a, c)_{a \in A}$. For each $a \in A$, let $\Sigma_a(x)$ be the set of formulas

$$\Sigma_a(x) = \{x \epsilon \bar{a}\} \cup \{x \not\equiv \bar{b} : bEa\}.$$

It suffices to show that T locally omits each set $\Sigma_a(x)$. For then T has a model $(\mathfrak{B}, a, c)_{a \in A}$ which omits each $\Sigma_a(x)$. We may also assume that $A \subset B$. $\mathfrak{B}$ is an elementary extension of $\mathfrak{A}$ because $\mathrm{Th}((\mathfrak{A}, a)_{a \in A}) \subset T$, whence $(\mathfrak{A}, a)_{a \in A} \equiv (\mathfrak{B}, a)_{a \in A}$. $\mathfrak{B}$ is a proper extension because $c \in B \setminus A$. Finally, $\mathfrak{B}$ is an end extension because it omits each $\Sigma_a(x)$.

To see that T locally omits each $\Sigma_a(x)$, we note that a formula $\varphi(x, c)$ of $\mathscr{L}$ is consistent with T if and only if

$$(\mathfrak{A}, a)_{a \in A} \models (\forall y)(\exists z)(\exists x)[z \notin y \wedge \varphi(x, z)].$$

Suppose $\varphi(x, c)$ is consistent with T, but $\varphi(x, c) \wedge \neg\, x \in \bar{a}$ is not. Then $\varphi(x, c) \wedge x \in \bar{a}$ is consistent with T. Using the axiom of replacement in ZF, we see in turn that the following sentences hold in $(\mathfrak{A}, a)_{a \in A}$:

$$(\forall y)(\exists z)(\exists x)[z \notin y \wedge \varphi(x, z) \wedge x \in \bar{a}]$$

$$(\exists x)(\forall y)(\exists z)[z \notin y \wedge \varphi(x, z) \wedge x \in \bar{a}].$$

Then for some $b \in A$, $\varphi(\bar{b}, c) \wedge \bar{b} \in \bar{a}$ is consistent with T, whence $\varphi(x, c) \wedge x \equiv \bar{b}$ is consistent with T. Thus T locally omits $\Sigma_a(x)$. ⊣

The omitting types theorem as it stands is false for uncountable languages. For example, let T be the theory with the axioms

$$c_\alpha \not\equiv c_\beta, \quad \alpha < \beta < \omega_1$$

in the language $\mathscr{L}$ with constants

$$\{c_\alpha : \alpha < \omega_1\} \cup \{d_n : n < \omega\}.$$

Let $\Gamma(x)$ be the set of formulas

$$\Gamma(x) = \{x \not\equiv d_n : n < \omega\}.$$

Then T locally omits $\Gamma(x)$. However no model of T omits $\Gamma(x)$ because every model of T is uncountable but each model which omits $\Gamma(x)$ is countable.

A more complicated counterexample where the theory T is complete has been given by Fuhrken (1962).

However, the omitting types theorem can be generalized to uncountable languages if we define the notion of 'locally omits' in the proper way. Let T be a theory and $\Sigma(x_1 \dots x_n)$ a set of formulas in a language $\mathscr{L}$ of power α. We say that T *α-realizes* Σ iff there is a set $\Phi(x_1 \dots x_n)$ of fewer than α formulas of $\mathscr{L}$ such that:

(i). Φ is consistent with T,

(ii). $T \cup \Phi(x_1 \dots x_n) \models \Sigma(x_1 \dots x_n)$,

that is, in any model $\mathfrak{A}$ of T, any n-tuple which realizes Φ realizes Σ. T is said to *α-omit* $\Sigma(x_1 \dots x_n)$ iff T does not α-realize $\Sigma(x_1 \dots x_n)$. Note that if Σ has power less than α, then T α-realizes Σ trivially. Thus only sets of formulas of power α can ever be α-omitted.

THEOREM 2.2.19 (α-Omitting Types Theorem). *Let T be a consistent theory in a language $\mathscr{L}$ of power α and let $\Sigma(x_1 \dots x_n)$ be a set of formulas of $\mathscr{L}$. If T α-omits Σ, then T has a model of power $\leqslant \alpha$ which omits Σ.*

The proof is like the proof of the omitting types theorem. An important problem is to find a useful sufficient condition for a theory in an uncountable language to have a model which omits a countable set of formulas. The α-omitting types theorem is of no help here since a countable set of formulas is never α-omitted when $\alpha > \omega$.

We now turn to the interpolation theorems of Craig and Lyndon.

THEOREM 2.2.20 (Craig Interpolation Theorem). *Let φ, ψ be sentences such that $\varphi \models \psi$. Then there exists a sentence θ such that:*

(i). *$\varphi \models \theta$ and $\theta \models \psi$.*

(ii). *Every relation, function or constant symbol (excluding identity) which occurs in θ also occurs in both φ and ψ.*

The sentence θ will be called a *Craig interpolant* of φ, ψ. The identity symbol is allowed to occur in θ. The following example shows why this is necessary.

EXAMPLE 2.2.21. In each of the following, φ and ψ are sentences such that the identity symbol occurs in at most one of them, and $\varphi \models \psi$; however, φ, ψ have no Craig interpolant in which the identity symbol does not occur:

(i). φ is $(\exists x)(P(x) \wedge \neg P(x))$, ψ is $(\exists x)Q(x)$;

(ii). φ is $(\exists x)Q(x)$, ψ is $(\exists x)(P(x) \vee \neg P(x))$;

(iii). φ is $(\forall xy)(x \equiv y)$, ψ is $(\forall xy)(P(x) \leftrightarrow P(y))$.

We shall see in an exercise, however, that in the Craig interpolation theorem, if the identity symbol occurs in neither φ nor ψ, and if not $\models \neg\varphi$ and not $\models \psi$, then φ and ψ have a Craig interpolant in which the identity symbol does not occur.

PROOF OF THEOREM 2.2.20. We assume that there is no Craig interpolant θ of φ and ψ, and prove that it is not the case that $\varphi \models \psi$. To do this we construct a model of $\varphi \wedge \neg\psi$. We may assume without loss of generality that $\mathscr{L}$ is the language of all symbols which occur in either φ or ψ or both. Let $\mathscr{L}_1$ be the language of all symbols of φ, $\mathscr{L}_2$ the language of all symbols of ψ, and $\mathscr{L}_0$ the language of all symbols occurring in both φ and ψ. Thus

$$\mathscr{L}_1 \cap \mathscr{L}_2 = \mathscr{L}_0, \qquad \mathscr{L}_1 \cup \mathscr{L}_2 = \mathscr{L}.$$

Form an expansion $\mathscr{L}'$ of $\mathscr{L}$ by adding a countable set C of new constant symbols and let

$$\mathscr{L}_0' = \mathscr{L}_0 \cup C, \qquad \mathscr{L}_1' = \mathscr{L}_1 \cup C, \qquad \mathscr{L}_2' = \mathscr{L}_2 \cup C.$$

The proof will resemble the proofs of the completeness and omitting types theorems, but the notion of a consistent theory will be replaced by the more general notion of an inseparable pair of theories.

Consider a pair of theories T in $\mathscr{L}_1'$ and U in $\mathscr{L}_2'$. A sentence θ of $\mathscr{L}_0'$ is said to *separate* T and U iff

$$T \models \theta \quad \text{and} \quad U \models \neg\theta.$$

T and U are said to be *inseparable* iff no sentence θ of $\mathscr{L}_0'$ separates them. To begin with, we see that

(1) $\{\varphi\}$ and $\{\neg\psi\}$ are inseparable.

For, if $\theta(c_1 \dots c_n)$ separates $\{\varphi\}$ and $\{\neg\psi\}$ and $u_1 \dots, u_n$ are variables not occurring in $\theta(c_1 \dots c_n)$, then $(\forall u_1 \dots u_n)\theta(u_1 \dots u_n)$ is a Craig interpolant of φ and ψ, contrary to our assumption.

Now let

$$\varphi_0, \varphi_1, \varphi_2, \dots, \psi_0, \psi_1, \psi_2, \dots$$

be enumerations of all sentences of $\mathscr{L}_1'$ and of $\mathscr{L}_2'$, respectively. We shall construct two increasing sequences of theories,

$$\{\varphi\} = T_0 \subset T_1 \subset T_2 \subset \dots,$$
$$\{\neg\psi\} = U_0 \subset U_1 \subset U_2 \subset \dots$$

in $\mathscr{L}_1'$ and $\mathscr{L}_2'$, respectively, such that:

(2). T_m and U_m are inseparable finite sets of sentences.

(3). If $T_m \cup \{\varphi_m\}$ and U_m are inseparable, then $\varphi_m \in T_{m+1}$.
If T_{m+1} and $U_m \cup \{\psi_m\}$ are inseparable, then $\psi_m \in U_{m+1}$.

(4). If $\varphi_m = (\exists x)\sigma(x)$ and $\varphi_m \in T_{m+1}$, then $\sigma(c) \in T_{m+1}$ for some $c \in C$.
If $\psi_m = (\exists x)\delta(x)$ and $\psi_m \in U_{m+1}$, then $\delta(d) \in U_{m+1}$ for some $d \in C$.

Given T_m and U_m, the theories T_{m+1} and then U_{m+1} are constructed in the obvious way. For (4), use constants c and d which do not occur in T_m, U_m, φ_m or ψ_m. Then inseparability will be preserved. Let

$$T_\omega = \bigcup_{m<\omega} T_m, \qquad U_\omega = \bigcup_{m<\omega} U_m.$$

Then T_ω and U_ω are inseparable. It follows that T_ω and U_ω are each consistent. We must show that $T_\omega \cup U_\omega$ is consistent. We show first that:

(5). T_ω is a maximal consistent theory in $\mathscr{L}'_1$, and U_ω is a maximal consistent theory in $\mathscr{L}'_2$.

To show this, suppose $\varphi_m \notin T_\omega$ and $(\neg \varphi_m) \notin T_\omega$. Since $T_m \cup \{\varphi_m\}$ is separable from U_m, there exists $\theta \in \mathscr{L}'_0$ such that

$$T_\omega \models \varphi_m \rightarrow \theta, \qquad U_\omega \models \neg \theta.$$

We see by the same argument that there exists $\theta' \in \mathscr{L}'_0$ such that

$$T_\omega \models \neg \varphi_m \rightarrow \theta', \qquad U_\omega \models \neg \theta'.$$

But then

$$T_\omega \models \theta \vee \theta', \qquad U_\omega \models \neg (\theta \vee \theta'),$$

contradicting the inseparability of T_ω and U_ω. This shows that T_ω is maximal consistent in $\mathscr{L}'_1$. The maximality of U_ω is similar.

Our next observation is:

(6). $T_\omega \cap U_\omega$ is a maximal consistent theory in $\mathscr{L}'_0$.

To prove (6), let σ be a sentence of $\mathscr{L}'_0$. By (5), either $\sigma \in T_\omega$ or $(\neg \sigma) \in T_\omega$, and either $\sigma \in U_\omega$ or $(\neg \sigma) \in U_\omega$. By inseparability, we cannot have $\sigma \in T_\omega$ and $(\neg \sigma) \in U_\omega$, or vice versa. Therefore either $T_\omega \cap U_\omega \models \sigma$ or $T_\omega \cap U_\omega \models \neg \sigma$.

We are now ready to construct a model. Let $\mathfrak{B}'_1 = (\mathfrak{B}_1, b_0, b_1, \ldots)$ be a model of T_ω. Using (4) and (5), we see that the submodel $\mathfrak{A}'_1 = (\mathfrak{A}_1, b_0, b_1, \ldots)$ with universe $A_1 = \{b_0, b_1, \ldots\}$ is also a model of T_ω. Similarly, U_ω has a model $\mathfrak{A}'_2 = (\mathfrak{A}_2, d_0, d_1, \ldots)$ with universe $A_2 = \{d_0, d_1, \ldots\}$. By (6), the $\mathscr{L}'_0$ reducts of $\mathfrak{A}'_1$ and $\mathfrak{A}'_2$ are isomorphic, with b_n corresponding to d_n. We may therefore take $b_n = d_n$ for each n, whence $\mathfrak{A}_1$ and $\mathfrak{A}_2$ have the same $\mathscr{L}_0$ reduct. Let $\mathfrak{A}$ be the model for $\mathscr{L}$ with $\mathscr{L}_1$ reduct $\mathfrak{A}_1$ and $\mathscr{L}_2$ reduct $\mathfrak{A}_2$. Since $\varphi \in T_\omega$ and $(\neg \psi) \in U_\omega$, $\mathfrak{A}$ is a model of $\varphi \wedge \neg \psi$. ⊣

We give two applications of the Craig interpolation theorem. The first application deals with ways of defining a relation. Let P and P' be two new n-placed relation symbols, not in the language $\mathscr{L}$. Let $\Sigma(P)$ be a set of sentences of the language $\mathscr{L} \cup \{P\}$, and let $\Sigma(P')$ be the corresponding set of sentences of $\mathscr{L} \cup \{P'\}$ formed by replacing P everywhere by P'. We say that $\Sigma(P)$ *defines P implicitly* iff

$$\Sigma(P) \cup \Sigma(P') \models (\forall x_1 \dots x_n)[P(x_1 \dots x_n) \leftrightarrow P'(x_1 \dots x_n)].$$

Equivalently, if $(\mathfrak{A}, R)$ and $(\mathfrak{A}, R')$ are models of $\Sigma(P)$, then $R = R'$. $\Sigma(P)$ is said to *define P explicitly* iff there exists a formula $\varphi(x_1 \dots x_n)$ of $\mathscr{L}$ such that

$$\Sigma(P) \models (\forall x_1 \dots x_n)[P(x_1 \dots x_n) \leftrightarrow \varphi(x_1 \dots x_n)].$$

It is obvious that, if $\Sigma(P)$ defines P explicitly, then $\Sigma(P)$ defines P implicitly. Thus, to show that $\Sigma(P)$ does not define P explicitly, it suffices to find two models $(\mathfrak{A}, R)$ and $(\mathfrak{A}, R')$ of $\Sigma(P)$, with the same reduct $\mathfrak{A}$ to $\mathscr{L}$, such that $R \neq R'$. This is a useful classical method known as Padoa's method. We now prove the converse of Padoa's method.

THEOREM 2.2.22 (Beth's Theorem). *$\Sigma(P)$ defines P implicitly if and only if $\Sigma(P)$ defines P explicitly.*

PROOF. We prove only the 'hard' direction. Suppose that $\Sigma(P)$ defines P implicitly. Add new constants $c_1, \dots, c_n$ to $\mathscr{L}$. Then

$$\Sigma(P) \cup \Sigma(P') \models P(c_1 \dots c_n) \to P'(c_1 \dots c_n).$$

By the compactness theorem, there exist finite subsets $\Delta \subset \Sigma(P)$, $\Delta' \subset \Sigma(P')$ such that

$$\Delta \cup \Delta' \models P(c_1 \dots c_n) \to P'(c_1 \dots c_n).$$

Let $\psi(P)$ be the conjunction of all $\sigma(P) \in \Sigma(P)$ such that either $\sigma(P) \in \Delta$ or $\sigma(P') \in \Delta'$. Then

$$\psi(P) \wedge \psi(P') \models P(c_1 \dots c_n) \to P'(c_1 \dots c_n).$$

Rearranging to get all symbols P on one side and all symbols P' on the other,

$$\psi(P) \wedge P(c_1 \dots c_n) \models \psi(P') \to P'(c_1 \dots c_n).$$

Then, by the Craig interpolation theorem, there is a sentence $\theta(c_1 \dots c_n)$ of $\mathscr{L} \cup \{c_1 \dots c_n\}$ such that

(1) $$\psi(P) \wedge P(c_1 \dots c_n) \models \theta(c_1 \dots c_n),$$

(2) $$\theta(c_1 \dots c_n) \models \psi(P') \rightarrow P'(c_1 \dots c_n).$$

But any model $(\mathfrak{A}, R')$ for $\mathscr{L} \cup \{P', c_1, \dots, c_n\}$ is also a model for $\mathscr{L} \cup \{P, c_1, \dots, c_n\}$ when we interpret P by R'. Thus (2) implies

(3) $$\theta(c_1 \dots c_n) \models \psi(P) \rightarrow P(c_1 \dots c_n).$$

Now (1) and (3) yield

(4) $$\psi(P) \models P(c_1 \dots c_n) \leftrightarrow \theta(c_1 \dots c_n).$$

Since $c_1, \dots, c_n$ do not occur in $\psi(P)$ (which is built from $\Sigma(P)$), we have

$$\psi(P) \models \forall x_1 \dots x_n[P(x_1 \dots x_n) \leftrightarrow \theta(x_1 \dots x_n)],$$

where $x_1, \dots, x_n$ are variables not occurring in $\theta(c_1 \dots c_n)$. Therefore

$$\Sigma(P) \models \forall x_1 \dots x_n[P(x_1 \dots x_n) \leftrightarrow \theta(x_1 \dots x_n)]. \dashv$$

Theorem 2.2.23 (Robinson Consistency Theorem). *Let $\mathscr{L}_1$ and $\mathscr{L}_2$ be two languages and let $\mathscr{L} = \mathscr{L}_1 \cap \mathscr{L}_2$. Suppose T is a complete theory in $\mathscr{L}$, and $T_1 \supset T$, $T_2 \supset T$ are consistent theories in $\mathscr{L}_1$, $\mathscr{L}_2$, respectively. Then $T_1 \cup T_2$ is consistent in the language $\mathscr{L}_1 \cup \mathscr{L}_2$.*

Proof. Suppose $T_1 \cup T_2$ is inconsistent. Then there exist finite subsets $\Sigma_1 \subset T_1$, $\Sigma_2 \subset T_2$ such that $\Sigma_1 \cup \Sigma_2$ is inconsistent. Let σ_1 be the conjunction of Σ_1 and σ_2 the conjunction of Σ_2. It follows that $\sigma_1 \models \neg \sigma_2$. By the Craig interpolation theorem, there is a sentence θ such that $\sigma_1 \models \theta$, $\theta \models \neg \sigma_2$, and every relation, function or constant symbol occurring in θ occurs in both σ_1 and σ_2. Consequently, θ is a sentence of $\mathscr{L}_1 \cap \mathscr{L}_2 = \mathscr{L}$. Now returning to T_1 and T_2, we find that $T_1 \models \theta$. Since T_1 is consistent, $T_1 \not\models \neg \theta$, so $T \not\models \neg \theta$. Moreover, $T_2 \models \neg \theta$, and, by the consistency of T_2, $T_2 \not\models \theta$; so $T \not\models \theta$. But this contradicts the hypothesis that T is a complete theory in $\mathscr{L}$. $\dashv$

The Lyndon interpolation theorem is an improvement of the Craig interpolation theorem, but it holds only for languages which have no function or constant symbols. In order to state it, we need the notions of a positive and a negative occurrence of a symbol in a formula.

In the following discussion we shall consider only formulas which are built up using the connectives $\wedge$, $\vee$, $\neg$, and the quantifiers $\forall$, $\exists$. We do not allow the connectives $\rightarrow$, $\leftrightarrow$.

[Strictly speaking, the language $\mathscr{L}$ was defined in Section 1.2 so that the only connectives are $\wedge$ and $\neg$, and the only quantifier is $\forall$. The other connectives and $\exists$ were introduced as abbreviations. Thus we now wish to avoid using the abbreviations $\rightarrow$, $\leftrightarrow$.]

We now shall consider more closely the ways in which a symbol can occur in a sentence. Let s be a symbol of $\mathscr{L}$, and let φ be a sentence of $\mathscr{L}$. Then s is said to occur *positively* in φ iff s has an occurrence in φ which is within the scope of an even number of negation symbols. The symbol s occurs *negatively* in φ iff s has an occurrence in φ which is within the scope of an odd number of negation symbols. Remembering that s may have several different occurrences in φ, we see that there are four possibilities:

s does not occur in φ;
s occurs positively in φ;
s occurs negatively in φ;
s occurs both positively and negatively in φ.

The reason we do not want to use the abbreviations $\rightarrow$ and $\leftrightarrow$ is that they contain 'hidden' negation symbols. For example, the sentence $P(c) \rightarrow Q(c)$ is an abbreviation of $\neg (P(c) \wedge \neg Q(c))$, so P occurs negatively but not positively in it, and the constant c occurs both positively and negatively in it.

On the other hand, the abbreviations

$$\varphi \vee \psi = \neg (\neg \varphi \wedge \neg \psi), \qquad (\exists x)\varphi = \neg (\forall x) \neg \varphi$$

will not cause any trouble in deciding whether a symbol s occurs positively or negatively, because they introduce exactly two 'hidden' negation symbols about φ, ψ, and two is an even number.

THEOREM 2.2.24 (Lyndon Interpolation Theorem). *Let φ, ψ be sentences of $\mathscr{L}$ such that $\varphi \models \psi$. Then there is a sentence θ of $\mathscr{L}$ such that*:

(i). *$\varphi \models \theta$ and $\theta \models \psi$.*

(ii). *Every relation symbol* (*excluding equality*) *which occurs positively in θ occurs positively in both φ and ψ.*

(iii). *Same as* (ii) *for 'negatively'.*

The following simple example shows that we cannot find an interpolant θ which satisfies (ii) and (iii) for constant symbols:

$$(\exists x)(x \equiv c \wedge \neg R(x)) \models \neg R(c).$$

Note that c is positive on the left, negative on the right, but must occur in any interpolant.

Proof of Theorem 2.2.24. The proof is obtained by making only a very few changes in the proof of the Craig interpolation theorem. We begin by assuming that there is no sentence θ such that (i)–(iii) hold, and prove that $\varphi \wedge \neg \psi$ has a model. Form the expansion $\mathscr{L}' = \mathscr{L} \cup C$ as before.

A formula is said to be in *negation normal form* (nnf) iff it is built up from atomic formulas and their negations using $\wedge$, $\vee$, $\exists$, $\forall$. Every formula is equivalent to an nnf formula. We assume that φ and ψ are nnf formulas. Let σ^* denote the nnf of $\neg \sigma$.

This time, the notion of an inseparable pair of theories is defined as follows. Let Φ be the set of all nnf sentences σ of $\mathscr{L}'$ such that every relation symbol which occurs positively (or negatively) in σ also occurs positively (negatively) in φ. The set Ψ is defined similarly with respect to ψ. Let $\Psi^* = \{\sigma^* : \sigma \in \Psi\}$. Two theories $T \subset \Phi$ and $U \subset \Psi^*$ are said to be *inseparable* iff there is no sentence $\theta \in \Phi \cap \Psi$ such that $T \models \theta$ and $U \models \neg \theta$. Using this notion we can apply the construction given in the proof of the Craig interpolation theorem to obtain a model of $\varphi \wedge \psi^*$.

This time we enumerate the sets of sentences Φ and Ψ instead of the languages $\mathscr{L}_1'$ and $\mathscr{L}_2'$, and then construct T_n and U_n as before. Some changes are needed in the rest of the proof because the sets Φ and Ψ are not necessarily closed under negation. Instead of proving that T_ω is maximal consistent, show that if $\sigma \vee \theta \in T_\omega$ then either $\sigma \in T_\omega$ or $\theta \in T_\omega$, and similarly for U_ω. Then show that T_ω and U_ω have the same equations and inequalities, and that the set Δ for all atomic and negated atomic sentences in $T_\omega \cup U_\omega$ is consistent. Finally, let $\mathfrak{A}$ be a model of Δ whose universe is the set of all constants, and prove by induction on complexity of formulas that $\mathfrak{A}$ is a model of both T_ω and U_ω and therefore a model of $\varphi \wedge \psi^*$. ⊣

A suggestion for further reading: The book "Building Models by Games" by Hodges [1985] gives an interesting treatment of a wide variety of applications of the Henkin construction in model theory.

Exercises

2.2.1. Let T be a complete theory in a countable language, and let $\Gamma_1(x_1), \Gamma_2(x_2), \Gamma_3(x_3), \ldots$ be a countable set of sets of formulas such that each $\Gamma_n(x_n)$ is consistent with T. Prove that T has a countable model which realizes each set $\Gamma_n(x_n)$.

2.2.2. Let T be a complete theory. Show that T has a model $\mathfrak{A}$ such that every set of formulas $\Gamma(x_1, x_2, \ldots)$ which is consistent with T is realized in $\mathfrak{A}$.

2.2.3. Let $\mathfrak{A} = \langle A, \leqslant, +, \cdot, 0, 1 \rangle$ be an ordered field. An element $a \in A$ is said to be finite iff there is an $n < \omega$ such that $-n \leqslant a \leqslant n$. Suppose that for any formula $\varphi(x)$, if $\mathfrak{A} \models (\exists x)\varphi(x)$, then there is a finite $a \in A$ such that $\mathfrak{A} \models \varphi[a]$. Show that $\mathfrak{A}$ is elementarily equivalent to an Archimedean-ordered field.

2.2.4. Let T be a theory in a countable $\mathscr{L}$ and let $\Sigma(x)$ and $\Delta(y)$ be two sets of formulas of $\mathscr{L}$ which are consistent with T. Suppose that for every formula $\varphi(x, y)$ of $\mathscr{L}$ there exists $\sigma(x) \in \Sigma(x)$ such that for all $\delta_1(y), \ldots, \delta_n(y) \in \Delta(y)$: if $\{\varphi, \delta_1, \ldots, \delta_n\}$ is consistent with T, then $\{\varphi, \delta_1, \ldots, \delta_n, \neg\sigma\}$ is consistent with T. Prove that T has a model realizing $\Delta(y)$ and omitting $\Sigma(x)$.

2.2.5. Let T be a complete theory in a countable language $\mathscr{L}$. Suppose that for each $n < \omega$, T has a model $\mathfrak{A}_n$ omitting the set of formulas $\Sigma_n(x)$. Prove that T has a model $\mathfrak{A}$ which omits each $\Sigma_n(x)$.

2.2.6. Let $\mathscr{L}$ be a countable language and let $\mathscr{L}' = \mathscr{L} \cup \{P_0, P_1, \ldots\}$ be a countable expansion of $\mathscr{L}$. Let T' be a maximal consistent theory in $\mathscr{L}'$ and $\Gamma(x)$ a set of formulas of $\mathscr{L}$. Suppose that for each n, the restriction of T' to $\mathscr{L} \cup \{P_0, P_1, \ldots, P_n\}$ has a model which omits $\Gamma(x)$. Prove that T' has a model omitting $\Gamma(x)$.

2.2.7. Prove that there is an ordinal $\alpha < \omega_1$ such that every formula φ of ω-logic which has a proof has a proof of length less than α.

2.2.8. Show that the compactness theorem fails for ω-logic.

2.2.9. Show that the Löwenheim–Skolem–Tarski theorem fails for models of T which omit Σ.

2.2.10*. A model $\mathfrak{B}$ of Peano arithmetic is said to be an *end extension* of $\mathfrak{A}$ iff $\mathfrak{B}$ is a proper extension of $\mathfrak{A}$ and, for all $b \in B$ and $a \in A$, if $b < a$, then $b \in A$. Prove that every countable model of Peano arithmetic has an end elementary extension.

2.2.11*. Prove the following *Restricted Omitting Types Theorem*. Let $\mathscr{L}$ be a countable language and let T be a consistent $\forall\exists$ theory in $\mathscr{L}$, that is, a theory whose axioms are sentences of the form

$$(\forall y_1 \dots y_p)(\exists z_1 \dots z_q)\varphi$$

where φ has no quantifiers. For each $n < \omega$, let $\Sigma_n(x_1 \dots x_k)$ be a set of universal formulas of $\mathscr{L}$. Suppose that for each n and each existential formula $\theta(x_1 \dots x_k)$ consistent with T, there is a formula $\sigma(x_1 \dots x_k) \in \Sigma_n(x_1 \dots x_k)$ such that $\theta \wedge \neg\sigma$ is consistent with T. Prove that T has a countable model which omits each $\Sigma_n(x_1 \dots x_k)$.

[*Hint*: The proof is similar to that of the Extended Omitting Types Theorem.]

2.2.12*. Deduce the Craig interpolation theorem from the Robinson consistency theorem.

2.2.13. Let Σ, Γ be sets of sentences of $\mathscr{L}$ such that $\Sigma \cup \Gamma$ is inconsistent. Then there exists a sentence θ of $\mathscr{L}$ such that:

(i). $\Sigma \models \theta$ and $\Gamma \models \neg\theta$.

(ii). Every relation, function or constant symbol which occurs in θ occurs in some member of Σ and in some member of Γ.

2.2.14

(i). Show that the Robinson consistency theorem fails if T is not assumed to be complete.

(ii). Show that the Robinson consistency theorem holds if the hypothesis that T is complete is replaced by the hypothesis that T is consistent, and for $i = 1, 2$, T contains every consequence of T_i in $\mathscr{L}$.

2.2.15. Prove the Craig interpolation theorem for formulas $\varphi(x_1 \dots x_n)$, $\psi(x_1 \dots x_n)$. It can be deduced easily from the Craig interpolation theorem for sentences.

2.2.16. Assume $\mathscr{L}$ has no function or constant symbols. Suppose that a set of sentences $\Sigma(P)$ of $\mathscr{L} \cup \{P\}$ defines P implicitly. Then there is a formula $\varphi(x_1 \dots x_n)$ of $\mathscr{L}$ such that:

(i). $\Sigma(P) \vdash P(x_1 \dots x_n) \leftrightarrow \varphi(x_1 \dots x_n)$.

(ii). Any symbol of $\mathscr{L}$ which occurs in φ occurs both positively and negatively in $\Sigma(P)$.

2.2.17. Let $\mathscr{L}'$ be an expansion of the language $\mathscr{L}$ and let P be an n-placed relation symbol in $\mathscr{L}' \setminus \mathscr{L}$. Let T be a theory in $\mathscr{L}'$. Suppose that for any model $\mathfrak{A}$ for $\mathscr{L}$ and any two expansions $\mathfrak{A}'$, $\mathfrak{A}''$ of $\mathfrak{A}$ to models of T, the relations of $\mathfrak{A}'$ and $\mathfrak{A}''$ corresponding to P are the same. Prove that there exists a formula $\theta(x_1 \dots x_n)$ of $\mathscr{L}$ such that

$$T \vdash P(x_1 \ldots x_n) \leftrightarrow \theta(x_1 \ldots x_n).$$

2.2.18. Let $\mathscr{L}'$ be an expansion of $\mathscr{L}$ and let T' be a theory in $\mathscr{L}'$. Suppose that each model $\mathfrak{A}$ for $\mathscr{L}$ has at most one expansion to a model T'. Prove that there is a theory T in $\mathscr{L}$ such that the models of T are exactly the reducts of the models of T' to $\mathscr{L}$.

2.2.19*. Show that the Lyndon interpolation theorem remains true when we add the conclusion:

(iv). If φ is a universal sentence, then so is θ.

Alternatively, it holds when we add:

(iv′). If ψ is an existential sentence, then so is θ.

However, the theorem becomes false if we add both the extra conclusions (iv), (iv′) at the same time.

2.2.20. Show that the Craig and Lyndon interpolation theorems hold with the following additional conclusion:

(iv). If not $\models \neg\varphi$, not $\models \psi$, and the identity symbol occurs in neither φ nor ψ, then the identity symbol does not occur in θ.

2.2.21*. Show that there is a model $\mathfrak{A}$ of Peano arithmetic which has an infinite element x such that no $y < x$ realizes the same complete type as x in $\mathfrak{A}$.

2.2.22*. Show that Peano arithmetic has two models $\mathfrak{A}$ and $\mathfrak{B}$ such that $\langle A, + \rangle \cong \langle B, + \rangle$ but not $\mathfrak{A} \cong \mathfrak{B}$. [*Hint*: Use Beth's Theorem.]

2.2.23*. Let T be a complete theory in a countable language and let $\Gamma(x)$ be a type over T which is consistent with T and locally omitted by T. Prove that T has a model in which infinitely many elements realize $\Gamma(x)$.

2.2.24*. Let S be a set of fewer than 2^ω types $\Gamma(x)$ which are maximal consistent with T and locally omitted by T. Prove that T has a countable model which simultaneously omits each $\Gamma(x) \in S$. [*Hint*: Represent the Henkin construction by a binary tree.]

2.3. Countable models of complete theories

In this section, we assume that $\mathscr{L}$ is a countable language. We shall embark on a thorough study of countable models of a complete theory. This study will give insight into what can be expected in general. Our study will center on two kinds of countable models, the atomic models, which are 'small', and the countably saturated models, which are 'large'. We begin with the atomic models.

Consider a complete theory T in $\mathscr{L}$. A formula $\varphi(x_1 \dots x_n)$ is said to be *complete* (*in* T) iff for every formula $\psi(x_1 \dots x_n)$ exactly one of

$$T \models \varphi \rightarrow \psi, \qquad T \models \varphi \rightarrow \neg \psi$$

holds. A formula $\theta(x_1 \dots x_n)$ is said to be *completable* (*in* T) iff there is a complete formula $\varphi(x_1 \dots x_n)$ with $T \models \varphi \rightarrow \theta$. If $\theta(x_1 \dots x_n)$ is not completable it is said to be *incompletable*.

A theory T is said to be *atomic* iff every formula of $\mathscr{L}$ which is consistent with T is completable in T. A model $\mathfrak{A}$ is said to be an *atomic model* iff every n-tuple $a_1, \dots, a_n \in A$ satisfies a complete formula in $\mathrm{Th}(\mathfrak{A})$.

In this and the next chapter we shall frequently pause to illustrate our definitions with examples. We shall sometimes make assertions about the examples without proofs. These proofs usually involve a combination of standard algebraic results and the theorems in the first three chapters of this book. In Section 3.4, we return to the examples and supply proofs.

2.3.1. EXAMPLES

(1). Let T be a complete theory and let $c_0, c_1, c_2, \dots$ be constant symbols of $\mathscr{L}$. Then any formula of $\mathscr{L}$ of the form

$$x_0 \equiv c_0 \wedge x_1 \equiv c_1 \wedge \dots \wedge x_n \equiv c_n$$

is complete in T. If $\mathfrak{A}$ is a model of T such that every element of A is a constant, then $\mathfrak{A}$ is an atomic model.

(2). The standard model of number theory is an atomic model.

(3). Let T be the theory of real closed ordered fields. The ordered field of real algebraic numbers is the unique atomic model of T. For example, the ordered field of real numbers is not atomic.

(4). Every finite model is atomic.

(5). Every model of pure identity theory is atomic. This gives an example of uncountable atomic models.

(6). Every dense linear ordering without endpoints is atomic.

(7). The following theory T is a complete theory which has no completable formulas and no atomic models. The language $\mathscr{L}$ has unary relation symbols $P_0(x), P_1(x), \dots$. The axioms of T are all sentences of the form

$$(\exists x)(P_{i_1}(x) \wedge \dots \wedge P_{i_m}(x) \wedge \neg P_{j_1}(x) \wedge \dots \wedge \neg P_{j_n}(x)),$$

where the $i_1, \dots, i_m, j_1, \dots, j_n$ are all distinct.

Our first theorem about atomic models is an application of the extended omitting types theorem.

THEOREM 2.3.2 (Existence Theorem for Atomic Models). *Let T be a complete theory. Then T has a countable atomic model if and only if T is atomic.*

PROOF. First assume that T has an atomic model $\mathfrak{A}$. Let $\varphi(x_1 \dots x_n)$ be consistent with T. Then, since T is complete,

$$T \models (\exists x_1 \dots x_n)\varphi(x_1 \dots x_n).$$

Let $a_1, \dots, a_n \in A$ satisfy φ, and let $\psi(x_1 \dots x_n)$ be a complete formula satisfied by $a_1, \dots, a_n$. Then we cannot have $T \models \psi \rightarrow \neg\varphi$, so we must have $T \models \psi \rightarrow \varphi$. Hence φ is completable and T is atomic.

Now assume T is atomic. For each $n < \omega$, let $\Gamma_n(x_1 \dots x_n)$ be the set of all negations of complete formulas $\psi(x_1 \dots x_n)$ in T. Then every formula $\varphi(x_1 \dots x_n)$ which is consistent with T is completable, and hence $\varphi \wedge \neg\gamma$ is consistent with T for some $\gamma \in \Gamma_n$. Therefore T locally omits each set $\Gamma_n(x_1 \dots x_n)$. By the extended omitting types theorem, T has a countable model $\mathfrak{A}$ which omits each Γ_n. Then each $a_1, \dots, a_n \in A$ satisfies a complete formula, whence $\mathfrak{A}$ is an atomic model. ⊣

Returning to our examples, we see that complete number theory and the theory of real closed ordered fields are atomic, because they have atomic models.

THEOREM 2.3.3 (Uniqueness Theorem for Atomic Models). *If $\mathfrak{A}$ and $\mathfrak{B}$ are countable atomic models and $\mathfrak{A} \equiv \mathfrak{B}$, then $\mathfrak{A} \cong \mathfrak{B}$.*

PROOF. If $\mathfrak{A}$ or $\mathfrak{B}$ is finite, then $\mathfrak{A} \cong \mathfrak{B}$ is trivial. Let $\mathfrak{A}$ and $\mathfrak{B}$ be infinite and well-order the sets A and B with order type ω. The proof will be our first example of a *back and forth construction.* We shall see many other proofs of this type later.

Let a_0 be the first element of A and let $\varphi_0(x_0)$ be a complete formula satisfied by a_0 in $\mathfrak{A}$. Since $\mathfrak{A} \models (\exists x_0)\varphi_0(x_0)$, $\mathfrak{B} \models (\exists x_0)\varphi_0(x_0)$. Thus we may choose $b_0 \in B$, which satisfies $\varphi_0(x_0)$. Now let b_1 be the first element of $B \setminus \{b_0\}$, and let $\varphi_1(x_0 x_1)$ be a complete formula satisfied by b_0, b_1 in $\mathfrak{B}$. Then both $\mathfrak{A}$ and $\mathfrak{B}$ satisfy

$$\forall x_0(\varphi_0(x_0) \rightarrow (\exists x_1)\varphi_1(x_0 x_1)),$$

because φ_0 is complete. Therefore there exists $a_1 \in A$ such that a_0, a_1 satisfy $\varphi_1(x_0 x_1)$. Next, let a_2 be the first element of $A \setminus \{a_0, a_1\}$, and so on. Going back and forth ω times, we obtain sequences

$$a_0, a_1, a_2, \dots, b_0, b_1, b_2, \dots.$$

By going back and forth we used up all of A and B, so

$$A = \{a_0, a_1, \ldots\}, \qquad B = \{b_0, b_1, \ldots\}.$$

Moreover, for each n the n-tuples $a_0, \ldots, a_{n-1}$ and $b_0, \ldots, b_{n-1}$ satisfy the same complete formula. It follows that the mapping $a_m \to b_m$ is an isomorphism of $\mathfrak{A}$ onto $\mathfrak{B}$. ⊣

Our third result on atomic models shows that they should be thought of as 'small' models of T. First, we need to define the notion of a prime model.

$\mathfrak{A}$ is said to be a *prime model* iff $\mathfrak{A}$ is elementarily embedded in every model of $\mathrm{Th}(\mathfrak{A})$. $\mathfrak{A}$ is said to be *countably prime* iff $\mathfrak{A}$ is elementarily embedded in every countable model of $\mathrm{Th}(\mathfrak{A})$.

THEOREM 2.3.4. *The following are equivalent:*

(i). $\mathfrak{A}$ *is a countable atomic model.*

(ii). $\mathfrak{A}$ *is a prime model.*

(iii). $\mathfrak{A}$ *is a countably prime model.*

PROOF. First assume that $\mathfrak{A}$ is a countable atomic model and let $T = \mathrm{Th}(\mathfrak{A})$. The proof that $\mathfrak{A}$ is prime is one-half of the 'back and forth' construction. Let $A = \{a_0, a_1, a_2, \ldots\}$ and let $\mathfrak{B}$ be any model of T. Let $\varphi_0(x_0)$ be a complete formula satisfied by a_0. Then $T \models (\exists x_0)\varphi_0$, so we may choose $b_0 \in B$ which satisfies $\varphi_0(x_0)$. Now let $\varphi_1(x_0 x_1)$ be a complete formula satisfied by a_0, a_1. Then $T \models \varphi_0(x_0) \to (\exists x_1)\varphi_1(x_0 x_1)$. Choose $b_1 \in B$ so that b_0, b_1 satisfies φ_1, and so forth. The function $a_m \to b_m$ is an elementary embedding of $\mathfrak{A}$ into $\mathfrak{B}$.

Now assume $\mathfrak{A}$ is prime. Then $\mathfrak{A}$ is elementarily embedded in every countable model of T, so $\mathfrak{A}$ is countably prime.

Assume $\mathfrak{A}$ is countably prime. Let $a_1, \ldots, a_n \in A$ and let $\Gamma(x_1 \ldots x_n)$ be the set of all formulas $\gamma(x_1 \ldots x_n)$ of $\mathscr{L}$ satisfied by $a_1, \ldots, a_n$. For any countable model $\mathfrak{B}$ of T, we have some elementary embedding $f : \mathfrak{A} \prec \mathfrak{B}$, whence $fa_1, \ldots, fa_n$ satisfies $\Gamma(x_1 \ldots x_n)$ in $\mathfrak{B}$. Thus Γ is realized in every countable model of T. By the omitting types theorem, Γ is locally realized by T. Thus there is a formula $\varphi(x_1 \ldots x_n)$ consistent with T such that $T \models \varphi \to \gamma$ for all $\gamma \in \Gamma$. But, for each formula $\psi(x_1 \ldots x_n)$, either $\psi \in \Gamma$ or $(\neg \psi) \in \Gamma$. Thus φ is complete in T. We cannot have $T \models \varphi \to \neg \varphi$, so $\varphi \in \Gamma$. Therefore $\varphi(x_1 \ldots x_n)$ is a complete formula satisfied by $a_1, \ldots, a_n$ in $\mathfrak{A}$, and $\mathfrak{A}$ is atomic. ⊣

We now turn to the study of 'large' countable models. Given a model $\mathfrak{A}$ and a subset $Y \subset A$, the expanded model $(\mathfrak{A}, a)_{a \in Y}$ will be denoted by $\mathfrak{A}_Y$, and its language by $\mathscr{L}_Y$.

A model $\mathfrak{A}$ is said to be *ω-saturated* iff for every finite set $Y \subset A$, every set of formulas $\Gamma(x)$ of $\mathscr{L}_Y$ consistent with $\mathrm{Th}(\mathfrak{A}_Y)$ is realized in $\mathfrak{A}_Y$. A model is said to be *countably saturated* iff it is countable and ω-saturated.

To gain some intuition, we shall list some examples of countably saturated models. Note that if $\mathfrak{A}$ is ω-saturated, then so is $\mathfrak{A}_Y$ for every finite subset $Y \subset A$.

2.3.5. EXAMPLES

(1). Every countable infinite model of pure identity theory is countably saturated.

(2). The ordering of the rational numbers is countably saturated.

(3). Let T be a theory in the language with only the constant symbols $c_0, c_1, \ldots$, and axioms $c_i \not\equiv c_j$, $i < j < \omega$. There are countably many countable models of T up to isomorphism; for each $\alpha \leqslant \omega$, there is a model with exactly α elements which are not constants. The model with zero nonconstants is the atomic model. The model with ω nonconstants is the countably saturated model.

(4). Let T be the theory of algebraically closed fields of characteristic zero. Again there are countably many countable models; for each $\alpha \leqslant \omega$, there is a model of transcendence degree α over the rationals. The model of degree zero, i.e. the field of algebraic numbers, is the atomic model of T. The model of transcendence degree ω is the countably saturated model.

(5). Every finite model is countably saturated.

We need some additional notation for sets of formulas. Remember that a type in the variables $x_1, \ldots, x_n$ is a maximal consistent set $\Gamma(x_1 \ldots x_n)$ of formulas. The set T' of sentences which belong to Γ is a maximal consistent theory; we call T' the *theory of* Γ. If $T \subseteq \Gamma$, Γ is called a *type of* T. Given a model $\mathfrak{A}$ of T and an n-tuple $a_1, \ldots, a_n \in A$, the set of all formulas $\gamma(x_1 \ldots x_n)$ of $\mathscr{L}$ satisfied by $a_1, \ldots, a_n$ is a type of T, called the *type of* $a_1, \ldots, a_n$. By a *type of* $\mathfrak{A}$ we mean a type of $\mathrm{Th}(\mathfrak{A})$.

Consider a set of formulas $\Sigma(x_1 \ldots x_n)$ of $\mathscr{L}$. A formula $\varphi(x_1 \ldots x_n)$ is said to be a *consequence* of Σ, in symbols $\Sigma \models \varphi$, iff for every model $\mathfrak{A}$ and every n-tuple $a_1, \ldots, a_n \in A$, if $a_1, \ldots, a_n$ satisfies Σ, then it satisfies φ. That is,

$$\mathfrak{A} \models \Sigma[a_1 \ldots a_n] \quad \text{implies} \quad \mathfrak{A} \models \varphi[a_1 \ldots a_n].$$

We let $\Sigma(c_1 \dots c_n)$ denote the set of all consequences in $\mathscr{L} \cup \{c_1, \dots, c_n\}$ of the set

$$\{\sigma(c_1 \dots c_n) : \sigma(x_1 \dots x_n) \in \Sigma\}.$$

The notation $\Sigma(c_1 \dots c_m x_{m+1} \dots x_n)$ is defined in a similar way.

Let $\mathscr{L}' = \mathscr{L} \cup \{c_1, \dots, c_m\}$ be a finite simple expansion of $\mathscr{L}$. There is a natural one-to-one correspondence between the types $\Sigma(x_1 \dots x_n)$ of $\mathscr{L}$ and the types $\Gamma(x_{m+1} \dots x_n)$ of $\mathscr{L}'$. If $\Sigma(x_1 \dots x_n)$ is a type of $\mathscr{L}$, then

$$\Sigma' = \Sigma(c_1 \dots c_m x_{m+1} \dots x_n)$$

is a type of $\mathscr{L}'$. On the other hand, if $\Gamma(x_{m+1} \dots x_n)$ is a type of $\mathscr{L}'$, then

$$\Sigma(x_1 \dots x_n) = \{\sigma(x_1 \dots x_n) : \sigma(c_1 \dots c_m x_{m+1} \dots x_n) \in \Gamma\}$$

is the unique type of $\mathscr{L}$ such that $\Sigma' = \Gamma$. (We leave the verification of this as an exercise.)

One might wonder why we used only sets of formulas in one free variable in the definition of an ω-saturated model. At first sight, it may appear that we would obtain a stronger notion by considering sets of formulas with finitely many free variables. The next proposition shows that we do not obtain a stronger notion in this way.

PROPOSITION 2.3.6. *Let $\mathfrak{A}$ be an ω-saturated model. Then for each finite $Y \subset A$, each set of formulas $\Gamma(x_1 \dots x_n)$ of $\mathscr{L}_Y$ consistent with* $\mathrm{Th}(\mathfrak{A}_Y)$ *is realized in $\mathfrak{A}_Y$.*

PROOF. We argue by induction on n. The result holds for $n = 1$ by definition. Assume the result for $n-1$ and let $\Gamma(x_1 \dots x_n)$ be consistent with $\mathrm{Th}(\mathfrak{A}_Y)$. We may assume that Γ is closed under finite conjunctions. Let

$$\Gamma'(x_1 \dots x_{n-1}) = \{(\exists x_n)\gamma(x_1 \dots x_n) : \gamma \in \Gamma\}.$$

Then Γ' is consistent with $\mathrm{Th}(\mathfrak{A}_Y)$. By inductive hypothesis, there is an $(n-1)$-tuple $a_1, \dots, a_{n-1}$ realizing Γ' in $\mathfrak{A}_Y$. Let $Y' = Y \cup \{a_1, \dots, a_{n-1}\}$. Then Y' is still finite. Moreover, the set $\Gamma(c_1 \dots c_{n-1} x_n)$ is consistent with $\mathrm{Th}(\mathfrak{A}_{Y'})$ because for each $\gamma_1, \dots, \gamma_m \in \Gamma$, $(\exists x_n)(\gamma_1 \wedge \dots \wedge \gamma_m) \in \Gamma'$. Since $\mathfrak{A}$ is ω-saturated, there exists $a_n \in A$ realizing $\Gamma(c_1 \dots c_{n-1} x_n)$ in $\mathfrak{A}_{Y'}$. Then $a_1, \dots, a_n$ realizes Γ in $\mathfrak{A}_Y$. ⊣

Our three theorems below on countably saturated models will closely parallel our three theorems for atomic models. We shall prove an existence theorem, a uniqueness theorem, and a theorem showing that countably saturated models are 'large'.

THEOREM 2.3.7 (Existence Theorem for Countably Saturated Models). *Let T be a complete theory. Then T has a countably saturated model if and only if for each $n < \omega$, T has only countably many types in n variables.*

PROOF. Suppose first that T has a countably saturated model $\mathfrak{A}$. By Proposition 2.3.6, every type of T in n variables is realized in $\mathfrak{A}$. But no n-tuple can realize two different types in n variables. Therefore T has only countably many types.

Now suppose that for each n, T has only countably many types in n variables. Add a countable set $C = \{c_1, c_2, \ldots\}$ of new constant symbols to $\mathscr{L}$, forming $\mathscr{L}'$. For each finite subset

$$Y = \{d_1, \ldots, d_n\} \subset C,$$

the types $\Gamma(x)$ of T in $\mathscr{L}_Y$ are in one-to-one correspondence with the types $\Sigma(x_1 \ldots x_n x)$ of T in $\mathscr{L}$. Therefore T has only countably many types $\Gamma(x)$ in $\mathscr{L}_Y$. Also, there are only countably many finite subsets $Y \subset C$. Let

$$\Gamma_1(x), \Gamma_2(x), \ldots$$

be an enumeration of all types of T in all expansions $\mathscr{L}_Y$, Y a finite subset of C. Let

$$\varphi_1, \varphi_2, \ldots$$

be an enumeration of all sentences of $\mathscr{L}'$. We form an increasing sequence

$$T = T_0 \subset T_1 \subset T_2 \subset \ldots$$

of theories of $\mathscr{L}'$ such that for each $m < \omega$:

(1). T_m is a consistent theory which contains only finitely many constants from C.

(2). Either $\varphi_m \in T_{m+1}$ or $(\neg \varphi_m) \in T_{m+1}$.

(3). If $\varphi_m = (\exists x)\psi(x)$ is in T_{m+1}, then $\psi(c) \in T_{m+1}$ for some $c \in C$.

(4). If $\Gamma_m(x)$ is consistent with T_{m+1}, then $\Gamma_m(d) \subset T_{m+1}$ for some $d \in C$.

The construction of T_m is straightforward. The union $T_\omega = \bigcup_{n<\omega} T_n$ is a maximal consistent theory in $\mathscr{L}'$. Using (3) we see that T_ω has a model $\mathfrak{A}' = (\mathfrak{A}, a_1, a_2, \ldots)$ such that $A = \{a_1, a_2, \ldots\}$. Thus $\mathfrak{A}$ is a countable model of T.

It remains to prove that $\mathfrak{A}$ is ω-saturated. Let $Y \subset A$ be finite and let $\Sigma(x)$ be consistent with $\mathrm{Th}(\mathfrak{A}_Y)$. Extend $\Sigma(x)$ to a type $\Gamma(x)$ in $\mathrm{Th}(\mathfrak{A}_Y)$.

For some m, $\Gamma(x) = \Gamma_m(x)$. $\Gamma_m(x)$ is consistent with T_ω and hence with T_{m+1}. Then by (4), $\Gamma_m(c_i) \subset T_{m+1}$ for some $c_i \in C$, and it follows that a_i realizes $\Gamma(x)$ in $\mathfrak{A}_Y$. ⊣

COROLLARY 2.3.8. *If T is a complete theory with only countably many nonisomorphic countable models, then T has a countably saturated model.*

PROOF. Each type of T is realized in some countable model of T, and each countable model realizes only countably many types. Therefore T has countably many types. ⊣

THEOREM 2.3.9 (Uniqueness Theorem for Countably Saturated Models). *If $\mathfrak{A}$ and $\mathfrak{B}$ are countably saturated models and $\mathfrak{A} \equiv \mathfrak{B}$, then $\mathfrak{A}$ is isomorphic to $\mathfrak{B}$.*

PROOF. The proof uses a back and forth construction which closely parallels the proof of the uniqueness theorem for atomic models. The only difference is that instead of working with complete formulas we work with types. Using countable saturation of $\mathfrak{A}$ and $\mathfrak{B}$, we obtain two sequences

$$a_0, a_1, \ldots, \qquad b_0, b_1, \ldots,$$

such that

$$A = \{a_0, a_1, \ldots\}, \qquad B = \{b_0, b_1, \ldots\},$$

and, for each n, a_n realizes the same type in $(\mathfrak{A}, a_0, \ldots, a_{n-1})$ as b_n realizes in $(\mathfrak{B}, b_0, \ldots, b_{n-1})$. Then

$$(\mathfrak{A}, a_0, a_1, \ldots) \equiv (\mathfrak{B}, b_0, b_1, \ldots),$$

whence $\mathfrak{A} \cong \mathfrak{B}$ by the mapping $a_n \to b_n$. ⊣

The 'dual' of a prime model is a countably universal model. A model $\mathfrak{A}$ is said to be *countably universal* iff $\mathfrak{A}$ is countable and every countable model $\mathfrak{B} \equiv \mathfrak{A}$ is elementarily embedded in $\mathfrak{A}$. The next theorem shows that countably saturated models are 'large'.

THEOREM 2.3.10. *Every countably saturated model is countably universal.*

PROOF. Let $\mathfrak{B}$ be a countable model and $\mathfrak{A}$ a countably saturated model, $\mathfrak{A} \equiv \mathfrak{B}$. Let $B = \{b_0, b_1, \ldots\}$. Using one half of the back and forth construction and the saturation of $\mathfrak{A}$, we obtain a sequence $a_0, a_1, a_2, \ldots$ in A such that

$$(\mathfrak{B}, b_0, b_1, \ldots) \equiv (\mathfrak{A}, a_0, a_1, \ldots).$$

Then the mapping $b_n \to a_n$ is an elementary embedding of $\mathfrak{B}$ into $\mathfrak{A}$. ⊣

For a related necessary and sufficient condition for countable saturation see Exercise 2.3.12. Example 2.3.12 shows that the converse of Theorem 2.3.10 fails.

EXAMPLE 2.3.11. Let T be the theory with infinitely many unary relations $P_0(x), P_1(x), \ldots$, and a double sequence of constants c_{ij}, $i, j < \omega$. The axioms are

$$\begin{array}{ll} (\forall x) \neg (P_i(x) \wedge P_j(x)), & i < j < \omega, \\ P_i(c_{ij}), & i < \omega, \\ c_{ij} \not\equiv c_{ik}, & j < k < \omega. \end{array}$$

It turns out that T is a complete theory. T has 2^{ω} nonisomorphic countable models, because for each n the relation $P_n(x)$ may or may not contain any nonconstants. However, T has a countably saturated model $\mathfrak{A}$. $\mathfrak{A}$ is the model in which each $P_n(x)$ contains ω nonconstants and the complement of all the $P_n(x)$ also has power ω.

EXAMPLE 2.3.12. Let T be the theory of linear orderings in which every element has an immediate predecessor and successor. It can be shown that T is a complete theory, and the models of T are exactly the orderings obtained by taking a linear ordering $\langle A, \leqslant \rangle$ and replacing each element $a \in A$ by a copy of the ordering of the integers $\langle Z, \leqslant \rangle$. T has 2^{ω} nonisomorphic countable models. The model $\langle Z, \leqslant \rangle$ is the atomic model of T. The model $\langle B, \leqslant \rangle$ formed by replacing each rational number by a copy of $\langle Z, \leqslant \rangle$ is the countably saturated model of T. By adding one more copy of $\langle Z, \leqslant \rangle$ to the end of $\langle B, \leqslant \rangle$, we obtain a model of T which is countably universal but not countably saturated.

We conclude this section with three applications of our basic results on atomic and saturated models. We recall that a theory T is said to be *ω-categorical* iff all models of T of power ω are isomorphic.

THEOREM 2.3.13 (Characterization of ω-Categorical Theories). *Let T be a complete theory. Then the following are equivalent*:

(a). *T is ω-categorical.*

(d). *For each $n < \omega$, T has only finitely many types in $x_1, \ldots, x_n$.*

PROOF. The reader is advised to sit down before beginning this proof. We shall prove the equivalence of (a) and (d) by proving a chain of implications

$$(a) \rightarrow (b) \rightarrow (c) \rightarrow (d) \rightarrow (e) \rightarrow (f) \rightarrow (a).$$

Each of the six equivalent conditions is interesting in its own right.

Assuming (a), T is ω-categorical, we prove:

(b). *T has a model $\mathfrak{A}$ which is both countably saturated and atomic.*

Let $\mathfrak{A}$ be the unique countable model of T. Then $\mathfrak{A}$ is countably prime, so $\mathfrak{A}$ is atomic. Since T has only one (hence countably many) countable models, it has a countably saturated model. Hence $\mathfrak{A}$ is countably saturated.

Now, assuming (b), we prove:

(c). *For each $n < \omega$, each type $\Gamma(x_1 \dots x_n)$ of T contains a complete formula.*

Since $\mathfrak{A}$ is ω-saturated, the type Γ is realized in $\mathfrak{A}$ by some n-tuple $a_1, \dots, a_n$. Since $\mathfrak{A}$ is atomic, $a_1, \dots, a_n$ satisfies a complete formula $\gamma(x_1 \dots x_n)$. We cannot have $(\neg \gamma) \in \Gamma$, so γ belongs to Γ.

Assuming (c), we next prove:

(d). *For each $n < \omega$, T has only finitely many types in $x_1, \dots, x_n$.*

To prove (d), let $\Sigma(x_1 \dots x_n)$ be the set of all negations of complete formulas $\varphi(x_1 \dots x_n)$ in T. Then Σ cannot be extended to a type in $x_1, \dots, x_n$, so Σ is inconsistent with T. Therefore some finite subset

$$\{\neg \varphi_1, \dots, \neg \varphi_m\} \subset \Sigma$$

is inconsistent with T. Hence

$$T \models \neg (\neg \varphi_1 \wedge \dots \wedge \neg \varphi_m),$$

whence

$$T \models \varphi_1 \vee \dots \vee \varphi_m .$$

For each $i \leqslant m$, the set $\Gamma_i(x_1 \dots x_n)$ of all consequences of $T \cup \{\varphi_i\}$ is a type of T. But in every model of T, every n-tuple satisfies one of the φ_i, hence realizes one of the Γ_i. Therefore $\Gamma_1, \Gamma_2, \dots, \Gamma_m$ are the only types of T in $x_1, \dots, x_n$.

Now we assume (d), and prove:

(e). *For each $n < \omega$, there are only finitely many formulas $\varphi(x_1 \dots x_n)$ up to equivalence with respect to T.*

Given a formula $\varphi(x_1 \dots x_n)$, let φ^* be the set of all types $\Gamma(x_1 \dots x_n)$ of T which contain φ. Then $\varphi^* = \psi^*$ implies $T \models \varphi \leftrightarrow \psi$. But there are only finitely many types of T in $x_1, \dots, x_n$, say m. Hence there are only 2^m sets of types and therefore at most 2^m formulas up to equivalence in T.

From (e) we prove:

(f). *All models of T are atomic.*

To see this, let $\mathfrak{A}$ be a model of T and let $a_1, \dots, a_n \in A$. Let $\varphi_1(x_1 \dots x_n), \dots, \varphi_r(x_1 \dots x_n)$ be a finite list of all the formulas satisfied by $a_1, \dots, a_n$, up to equivalence in T. Then $\varphi_1 \wedge \dots \wedge \varphi_r$ is a complete formula in T which is satisfied by $a_1, \dots, a_n$ in $\mathfrak{A}$. Hence $\mathfrak{A}$ is atomic.

Finally, assuming (f), we see that any two countable models of T are atomic and elementarily equivalent, hence isomorphic. Therefore T is ω-categorical. ⊣

The next theorem can often be used to show that a theory has an atomic model.

THEOREM 2.3.14. *Any complete theory T which has a countably saturated model has a countable atomic model.*

PROOF. Assume that T has no countable atomic model. Then T is not atomic. Therefore T has a consistent incompletable formula $\varphi(x_1 \dots x_n)$. For each consistent incompletable formula $\psi(x_1 \dots x_n)$ of T, we may choose two formulas $\psi_0(x_1 \dots x_n)$ and $\psi_1(x_1 \dots x_n)$ each consistent with T such that

(1) $$T \models \psi_0 \rightarrow \psi, \qquad T \models \psi_1 \rightarrow \psi, \qquad T \models \neg(\psi_0 \wedge \psi_1).$$

ψ_0 and ψ_1 are again incompletable. In this way we obtain a tree of incompletable formulas

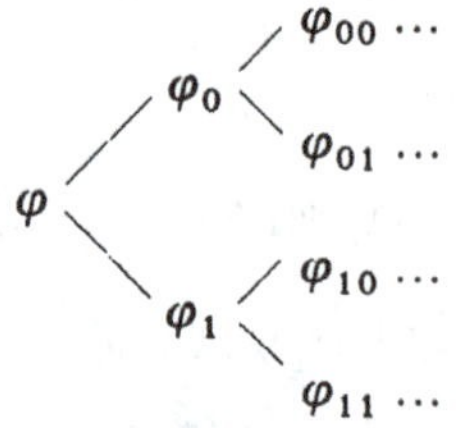

Each infinite sequence $s_0, s_1, s_2, \dots$ of zeros and ones gives a branch $\Gamma_s = \{\varphi, \varphi_{s_0}, \varphi_{s_0 s_1}, \varphi_{s_0 s_1 s_2}, \dots\}$ of the tree. There are 2^ω branches. By (1), each branch $\Gamma_s(x_1 \dots x_n)$ is a set of formulas which is consistent with T, and any two branches are inconsistent with each other. Extending each branch Γ_s to a type of T, we obtain 2^ω different types. Therefore T does not have a countably saturated model. ⊣

The converse of the above theorem is false. For example, we have already seen that the theory of real closed ordered fields has a countable atomic model. But this theory has 2^ω types and therefore has no countably saturated model. Another example of a theory with a countable atomic but no countably saturated model is complete number theory.

It is worth repeating here some of our examples of ω-categorical theories: atomless Boolean algebras; the four complete theories of dense simple order; the theory of infinite pure identity models; the theory of infinite

Abelian groups with all elements of order p (p prime); the theory of an equivalence relation with infinitely many equivalence classes and each class infinite.

We conclude this section with a surprising result of Vaught.

THEOREM 2.3.15. *No complete theory T has exactly two nonisomorphic countable models.*

PROOF. Assume T has exactly two nonisomorphic countable models. Our previous results show that T has a countably saturated model $\mathfrak{B}$ and a countable atomic model $\mathfrak{A}$ and that these two models cannot be isomorphic. Since $\mathfrak{B}$ is not atomic, it has an n-tuple $b_1, \ldots, b_n$ which does not satisfy a complete formula. Our plan is to obtain a countable atomic model $(\mathfrak{C}, c_1 \ldots c_n)$ of the complete theory $T' = \mathrm{Th}((\mathfrak{B}, b_1 \ldots b_n))$ and show that the reduct $\mathfrak{C}$ is neither ω-saturated nor atomic. Thus T will have at least three nonisomorphic countable models $\mathfrak{A}, \mathfrak{B}, \mathfrak{C}$.

Since $\mathfrak{B}$ is countably saturated, $(\mathfrak{B}, b_1 \ldots b_n)$ is countably saturated. The theory T' thus has a countably saturated model, and therefore has a countable atomic model $(\mathfrak{C}, c_1 \ldots c_n)$. The reduct $\mathfrak{C}$ is a model of T. $\mathfrak{C}$ is not atomic because the n-tuple $c_1, \ldots, c_n$ does not satisfy a complete formula. It remains to be shown that $\mathfrak{C}$ is not ω-saturated. Because T is not ω-categorical, it has infinitely many nonequivalent formulas. Therefore T' has infinitely many nonequivalent formulas. Hence no model of T' is both atomic and ω-saturated. In particular, since $(\mathfrak{C}, c_1 \ldots c_n)$ is atomic, it cannot be ω-saturated. It follows that $\mathfrak{C}$ is not ω-saturated. ⊣

EXERCISES

2.3.1. Show that if $\varphi(x_1 \ldots x_n)$ is a complete formula in T with respect to $x_1, \ldots, x_n$, then $(\exists x_n)\varphi(x_1 \ldots x_{n-1} x_n)$ is a complete formula in T with respect to $x_1, \ldots, x_{n-1}$.

2.3.2. Show that for any model $\mathfrak{A}$ the simple expansion $(\mathfrak{A}, a)_{a \in A}$ is an atomic model.

2.3.3. Let $c_1, \ldots, c_m$ be new constant symbols. Prove that for each $n \geqslant m$ the map

$$\Sigma(x_1 \ldots x_n) \to \Sigma(c_1 \ldots c_m x_{m+1} \ldots x_n)$$

is a one-to-one mapping from the types in $\mathscr{L}$ in n variables onto the types in $\mathscr{L} \cup \{c_1, \ldots, c_m\}$ in $n-m$ variables.

2.3.4. Suppose $\mathfrak{A} \equiv \mathfrak{B}$. Show that an n-tuple $a_1, \ldots, a_n \in A$ and an n-tuple $b_1, \ldots, b_n \in B$ realize the same type if and only if $(\mathfrak{A}, a_1 \ldots a_n) \equiv (\mathfrak{B}, b_1 \ldots b_n)$.

2.3.5*. Prove that a model $\mathfrak{A}$ is atomic if and only if for every finite subset Y of A, every element $a \in A$ satisfies a complete formula $\varphi(x)$ in $\mathrm{Th}(\mathfrak{A}_Y)$. Use this to show that if $\mathfrak{A}$ is atomic and $Y \subset A$ is finite, then $\mathfrak{A}_Y$ is atomic.

2.3.6. Prove that if $\mathfrak{A}$ is elementarily embedded in $\mathfrak{B}$, then every type $\Gamma(x_1 \ldots x_n)$ which is realized in $\mathfrak{A}$ is realized in $\mathfrak{B}$.

2.3.7. Let $\Sigma(x_1 \ldots x_n)$ be a type of a complete theory T. Prove that Σ is realized in every model of T if and only if Σ contains a complete formula.

2.3.8. Prove that if a complete theory T has fewer than 2^ω types, then T has an atomic model.

2.3.9*. Prove that complete number theory has no countably saturated model.

2.3.10*. Prove that no ordered field is countably saturated.

2.3.11. Prove that every complete theory which has a countably universal model has a countably saturated model.

2.3.12. Let $\mathfrak{A}$ be a countable model. Prove that $\mathfrak{A}$ is countably saturated if and only if for every finite subset Y of $\mathfrak{A}$, $\mathfrak{A}_Y$ is countably universal.

2.3.13. Show that every reduct of a countably saturated model to a sublanguage of $\mathscr{L}$ is countably saturated.

2.3.14*. Let T be a complete theory and let $\mathfrak{B}_T$ be the Lindenbaum algebra of T as defined in Exercises 1.4.10 and 2.1.15. For $n < \omega$, let $\mathfrak{B}_{n,T}$ be the Boolean subalgebra of $\mathfrak{B}_T$ determined by the formulas $\varphi(v_0 \cdots v_{n-1})$. Prove that:

(a). $\varphi(v_0 \cdots v_{n-1})$ is consistent with T if and only if $(\varphi) \neq 0$ in $\mathfrak{B}_{n,T}$.

(b). $\varphi(v_0 \cdots v_{n-1})$ is a complete formula in T if and only if (φ) is an atom of $\mathfrak{B}_{n,T}$.

(c). T is an atomic theory if and only if each $\mathfrak{B}_{n,T}$ is an atomic Boolean algebra.

(d). T is ω-categorical if and only if $\mathfrak{B}_{n,T}$ is a finite Boolean algebra for each $n < \omega$.

(e). T has a countably saturated model if and only if each $\mathfrak{B}_{n,T}$ has only countably many ultrafilters. [*Hint*: Show that types $\Sigma(v_0 \cdots v_{n-1})$ of T correspond to ultrafilters in $\mathfrak{B}_{n,T}$.]

2.3.15* (Ehrenfeucht). Let $\mathscr{L} = \{\leqslant, c_0, c_1, \ldots\}$ and let T be the theory of $\mathscr{L}$ which states that $\leqslant$ is a dense simple order without endpoints and that $c_n < c_{n+1}$, $n < \omega$. T is easily seen to be complete. There are three kinds of countable models of T. If we identify the elements of the countable model with the set of all rationals, then one of the following three cases occurs:

$$\lim_{n \to \infty} c_n = \infty;$$

$$\lim_{n \to \infty} c_n < \infty \text{ and is a rational;}$$

$$\lim_{n \to \infty} c_n < \infty \text{ and is an irrational.}$$

Determine which of these three models is countably saturated? Countable atomic? And neither?

2.3.16* (Ehrenfeucht). Modify the above example to obtain an example of a complete theory T with exactly n nonisomorphic countable models, $n \geqslant 3$. [*Hint*: Add $n-2$ 1-placed relation symbols to $\mathscr{L}$.]

2.3.17**. Let T be a theory in a countable language $\mathscr{L}$. Prove that if T has more than ω_1 nonisomorphic countable models, then T has continuum many nonisomorphic countable models. This result disappears if the continuum hypothesis holds. It is an open problem whether the hypothesis of the result can be weakened to: T has uncountably many nonisomorphic countable models (assuming that the continuum hypothesis fails).

2.3.18*. Let $\mathfrak{A}$ be a countably saturated model for an uncountable language $\mathscr{L}$. Prove that there is a countable sublanguage $\mathscr{L}' \subset \mathscr{L}$ such that for each formula φ of $\mathscr{L}$ there is a formula ψ of $\mathscr{L}'$ such that $\mathfrak{A} \models \varphi \leftrightarrow \psi$.

2.3.19*. Let $\mathfrak{A}$ and $\mathfrak{B}$ be ω-saturated models for a countable language. Show that the direct product $\mathfrak{A} \times \mathfrak{B}$ is ω-saturated.

2.3.20*. In a countable language, let T be a complete theory which is not ω-categorical. Let $\Gamma_1, \ldots, \Gamma_n$ be consistent types over T. Show that T has a countable model $\mathfrak{A}$ which realizes $\Gamma_1, \ldots, \Gamma_n$ but is not ω-saturated.

2.4. Recursively saturated models

A recursively saturated model is, roughly speaking, a model which is saturated for recursive sets of formulas. The proofs of a number of early results in model theory were simplified by using the method of recursively saturated models. In this section we shall introduce the recursively saturated models, develop their basic properties, and give some illustrations of how they are used.

There are a number of results in model theory which would be quite easy if every complete theory had a countably saturated model. But countably saturated models do not exist for complete theories with uncountably many types. However, countable recursively saturated models are often good enough, and they always exist for a complete theory in a countable language.

To prepare for the definition we must first explain what is meant by a recursive set of formulas, or more generally a recursive set of expressions (where an expression is a finite sequence of symbols of $\mathscr{L}$). Intuitively, a set of expressions is recursive if there is an algorithm which, given any expression φ, will produce the answer "yes" if φ belongs to the set and the answer "no" if φ does not belong to the set. A set of expressions is recursively enumerable if there is an algorithm which, given any expression φ, will produce the answer "yes" if φ belongs to the set and will never end if φ does not belong to the set. At the beginning of a course in Recursion Theory several equivalent mathematical definitions of recursive set are presented. The principle that these mathematical definitions are equivalent to the above intuitive notion of a recursive set is called Church's Thesis.

In this course we will only need to know two things about recursive sets: (1) There are only countably many recursive sets of formulas. (2) In a recursive language, the set of all formulas described by a "finite scheme" is recursive. The intuitive description of recursive sets above will be enough to understand our treatment of recursively saturated models. However, for the sake of completeness we shall also give a precise definition of recursive set here. We shall choose a form of the definition which is particularly easy to apply to sets of formulas. We restrict our attention in this section to the case where the language $\mathscr{L}$ is countable. We may then take each symbol of $\mathscr{L}$ to be an element of the set $R(\omega)$ of sets of finite rank. Then each finite sequence of symbols of $\mathscr{L}$ also belongs to $R(\omega)$. We begin with the notion of a recursive subset of $R(\omega)$.

DEFINITION. By a Δ_0 *formula*, or bounded quantifier formula, we mean a formula in the language with only the $\in$ symbol and equality which is built from atomic formulas using only logical connectives and the relativized quantifiers $(\forall x \in y)$ and $(\exists x \in y)$. By a Σ_1 *formula* we mean a formula which is built from Δ_0 formulas using the positive connectives $\wedge$, $\vee$, bounded quantifiers $(\forall x \in y)$, $(\exists x \in y)$, and existential quantifiers $(\exists x)$. A subset S of $R(\omega)$ is *recursively enumerable*, or r.e., if it is definable in the model $\langle R(\omega), \epsilon\rangle$ by a Σ_1 formula $\varphi(x)$, that is,

$$S = \{a \in R(\omega) : \langle R(\omega), \epsilon\rangle \models \varphi[a]\}.$$

A subset S of $R(\omega)$ is *recursive* if both S and $R(\omega)\backslash S$ are r.e.

The language $\mathscr{L}$ is said to be *recursive* if the sets of symbols of $\mathscr{L}$ and functions giving the number of places of symbols of $\mathscr{L}$ are recursive subsets of $R(\omega)$. That is, each of the sets

$\{\langle n, v_n\rangle : n \in \omega\}$,
$\{c : c \text{ is a constant symbol of } \mathscr{L}\}$,
$\{P : P \text{ is a relation symbol of } \mathscr{L}\}$,
$\{f : f \text{ is a function symbol of } \mathscr{L}\}$,
$\{\langle P, n\rangle : P \text{ is a relation symbol of } \mathscr{L} \text{ with } n \text{ places}\}$,
$\{\langle f, n\rangle : f \text{ is a function symbol of } \mathscr{L} \text{ with } n \text{ places}\}$

is a recursive subset of $R(\omega)$. For convenience we shall also include as part of the definition of a recursive language that each symbol of $\mathscr{L}$ is a natural number and that there are infinitely many natural numbers which are not symbols of $\mathscr{L}$.

We shall restrict our attention in this section to recursive languages $\mathscr{L}$. Since $R(\omega)$ is countable, every recursive language is countable and has countably many recursive sets of formulas. We can freely expand recursive languages by adding new symbols. Since any finite set is recursive, any expansion of a recursive language by finitely many new symbols is again a recursive language. Moreover, any expansion of a recursive language by a recursive set of new constants is again a recursive language, provided that there are still infinitely many natural numbers which are not used as symbols.

If $\mathscr{L}$ is recursive, then the set of all formulas of $\mathscr{L}$ is recursive, because there is an algorithm which decides whether an element of $R(\omega)$ is a formula of $\mathscr{L}$. The set of all formulas $\varphi(x_1, \ldots, x_n)$ of $\mathscr{L}$ with at most $x_1, \ldots, x_n$ free, and the set of all sentences of $\mathscr{L}$, are also recursive. We shall come across various other examples of recursive sets of formulas in this section. As a by-product of the Gödel completeness theorem, the set of all proofs in $\mathscr{L}$ is recursive and the set of all valid sentences in $\mathscr{L}$ is r.e.

We now give the key definition in this section.

Definition. Let $\mathscr{L}$ be a recursive language. A model $\mathfrak{A}$ for $\mathscr{L}$ is *recursively saturated* if for every finite set $\{c_1, \ldots, c_n\}$ of new constant symbols, every recursive set $\Gamma(x)$ of formulas of $\mathscr{L}(c_1, \ldots, c_n)$, and every n-tuple $a_1, \ldots, a_n$ of elements of A, if $\Gamma(x)$ is finitely satisfiable in $(\mathfrak{A}, a_1, \ldots, a_n)$ then $\Gamma(x)$ is realized in $(\mathfrak{A}, a_1, \ldots, a_n)$.

Examples 1. Every ω-saturated model is recursively saturated.

2. Every recursively saturated model of complete arithmetic has an infinite element, since the recursive set of formulas

$$\{n < x : n \in \omega\}$$

is finitely satisfiable.

3. Every recursively saturated real closed ordered field has a positive infinitesimal element, since the recursive set of formulas

$$\{0 < x \wedge x < 1/n : n \in \omega\}$$

is finitely satisfiable.

It is immediate that any reduct of a recursively saturated model is recursively saturated, because any recursive set of formulas in the smaller language is also a recursive set of formulas in the larger language. Also, every expansion of a recursively saturated model formed by adding finitely many new constants is again recursively saturated.

Our first theorem is that countable recursively saturated models exist.

Theorem 2.4.1 (Existence Theorem for Recursively Saturated Models). *Let $\mathscr{L}$ be a recursive language and let T be a complete theory in $\mathscr{L}$ whose models are infinite. Then T has a countable recursively saturated model.*

Proof. Before starting, let us emphasize that the theory T is not necessarily a recursive set of sentences. The proof is similar to the proof of Theorem 2.3.7, the Existence Theorem for Countably Saturated Models. However, instead of working with complete types we work with recursive sets of formulas which need not be complete.

Add a recursive set $C = \{c_1, c_2, \ldots\}$ of new constant symbols to $\mathscr{L}$ so as to form a recursive language $\mathscr{L}'$. Notice that if x_i are distinct varibles and d_i are distinct constants in C for $i = 1, \ldots, n$, then a set of formulas $\Gamma(x_1, \ldots, x_{n-1}, x_n)$ of $\mathscr{L}$ is recursive if and only if $\Gamma(d_1, \ldots, d_{n-1}, x_n)$ is recursive. Let

$$\Gamma_1(x), \Gamma_2(x), \ldots$$

be an enumeration of all recursive sets of formulas of $\mathscr{L}'$ with only x free in which only finitely many constants from C occur, and let

$$\varphi_1, \varphi_2, \ldots$$

be an enumeration of all sentences of $\mathscr{L}'$. By carrying out the construction given in the proof of Theorem 2.3.7, we now obtain a countable recursively saturated model of T. ⊣

COROLLARY 2.4.2. *Let $\mathscr{L}$ be a recursive language. Then every countable model $\mathfrak{A}$ for $\mathscr{L}$ has a countable recursively saturated elementary extension.*

PROOF. Let $\mathscr{L}'$ be a recursive expansion of $\mathscr{L}$ with a countable set C of new constant symbols. Then $\mathfrak{A}$ has an expansion $\mathfrak{A}'$ to $\mathscr{L}'$ in which every element of A is an interpretation of a constant. By Theorem 2.4.1, the elementary diagram T' of $\mathfrak{A}'$ has a countable recursively saturated model $\mathfrak{B}'$. The reduct $\mathfrak{B}$ of $\mathfrak{B}'$ to $\mathscr{L}$ is a countable recursively saturated elementary extension of $\mathfrak{A}$. ⊣

We shall see in a later chapter that the above corollary has an analogue for uncountable models. In general, countable recursively saturated models are not unique in a complete theory. Some counterexamples are indicated in the exercises. The back and forth method is not able to prove uniqueness for recursively saturated models, but it is still quite powerful, as the next few results show.

DEFINITION. A model $\mathfrak{A}$ is said to be *ω-homogeneous* if for any pair of tuples $a_1, \ldots, a_n$ and $b_1, \ldots, b_n$ of elements of A such that

$$(\mathfrak{A}, a_1, \ldots, a_n) \equiv (\mathfrak{A}, b_1, \ldots, b_n)$$

and any $c \in A$ there exists $d \in A$ such that

$$(\mathfrak{A}, a_1, \ldots, a_n, c) \equiv (\mathfrak{A}, b_1, \ldots, b_n, d).$$

A countable ω-homogeneous model is said to be *countably homogeneous*.

The methods of the preceding section can be used to show that every ω-saturated model and every atomic model is ω-homogeneous. We leave the result for atomic models as an exercise. Our next proposition is that recursively saturated models are also ω-homogeneous. This is somewhat surprising because the type of an n-tuple of elements of a model $\mathfrak{A}$ is in general not a recursive set of formulas. The trick is that the property that two n-tuples have the same type can be expressed by a recursive set of formulas, even though the type of each n-tuple is not recursive.

PROPOSITION 2.4.3. *Let $\mathscr{L}$ be a recursive language. Then every recursively saturated model for $\mathscr{L}$ is ω-homogeneous.*

PROOF. Let $\mathfrak{A}$ be a recursively saturated model for $\mathscr{L}$ and let $a_1, \ldots, a_n$ and $b_1, \ldots, b_n$ be two n-tuples in A such that

$$(\mathfrak{A}, a_1, \ldots, a_n) \equiv (\mathfrak{A}, b_1, \ldots, b_n).$$

Let $c \in A$. Choose distinct new constant symbols corresponding to $a_1, \ldots, a_n, b_1, \ldots, b_n$, and c, and form the finite expansion $\mathscr{L}'$ of $\mathscr{L}$. Let $\Gamma(x)$ be the set of all formulas of $\mathscr{L}'$ of the form

$$\varphi(a_1, \ldots, a_n, c) \leftrightarrow \varphi(b_1, \ldots, b_n, x).$$

Then $\Gamma(x)$ is a recursive set of formulas of $\mathscr{L}'$ which is finitely satisfiable in the model

$$\mathfrak{A}' = (\mathfrak{A}, a_1, \ldots, a_n, b_1, \ldots, b_n, c).$$

By recursive saturation, $\Gamma(x)$ is realized in $\mathfrak{A}'$, and this shows that $\mathfrak{A}$ is ω-homogeneous. ⊣

If $\mathfrak{A}$ is a countably homogeneous model, then a back and forth construction shows that whenever $(\mathfrak{A}, a_1, \ldots, a_n) \equiv (\mathfrak{A}, b_1, \ldots, b_n)$, we have $(\mathfrak{A}, a_1, \ldots, a_n) \cong (\mathfrak{A}, b_1, \ldots, b_n)$. (This will follow from Exercise 2.4.5 and Proposition 2.4.4 below.) The back and forth construction can be captured in a more general context with the notion of a partial isomorphism.

DEFINITION. Let $\mathfrak{A}$ and $\mathfrak{B}$ be models for a language $\mathscr{L}$. A *partial isomorphism* $I : \mathfrak{A} \cong_{\mathrm{p}} \mathfrak{B}$ between $\mathfrak{A}$ and $\mathfrak{B}$ is a relation I on the set of pairs of finite sequences $\langle a_1, \ldots, a_n \rangle$, $\langle b_1, \ldots, b_n \rangle$ of elements of A and B of the same length such that:

(i). $\emptyset \; I \; \emptyset$;

(ii). If $\langle a_1, \ldots, a_n \rangle \; I \; \langle b_1, \ldots, b_n \rangle$ then $(\mathfrak{A}, a_1, \ldots, a_n)$ and $(\mathfrak{B}, b_1, \ldots, b_n)$ satisfy the same atomic sentences of $\mathscr{L}(c_1, \ldots, c_n)$;

(iii). If $\langle a_1, \ldots, a_n \rangle \; I \; \langle b_1, \ldots, b_n \rangle$ then for all $c \in A$ there exists $d \in B$ such that $\langle a_1, \ldots, a_n, c \rangle \; I \; \langle b_1, \ldots, b_n, d \rangle$, and vice versa.

Condition (iii) is called the *back and forth* condition. Thus $\mathfrak{A}$ is ω-homogeneous if and only if the relation

$$(\mathfrak{A}, a_1, \ldots, a_n) \equiv (\mathfrak{A}, b_1, \ldots, b_n)$$

is a partial isomorphism from $\mathfrak{A}$ to $\mathfrak{A}$. We shall now consider partial isomorphisms between two different structures.

PROPOSITION 2.4.4. (i) *Any two finite or countable partially isomorphic models are isomorphic.*

(ii) *Any two partially isomorphic models are elementarily equivalent.*

PROOF. (i) By a routine back and forth construction.

(ii) Let $I : \mathfrak{A} \cong \mathfrak{B}$. Show by induction on the complexity of formulas $\varphi(x_1, \ldots, x_n)$ that if $\langle a_1, \ldots, a_n \rangle \; I \; \langle b_1, \ldots, b_n \rangle$ then

$$\mathfrak{A} \models \varphi[a_1, \ldots, a_n] \text{ iff } \mathfrak{B} \models \varphi[b_1, \ldots, b_n].$$

Then taking $n = 0$ we obtain $\mathfrak{A} \equiv \mathfrak{B}$. ⊣

We shall next prove a partial isomorphism theorem for recursively saturated models which is analogous to the uniqueness theorem for countably saturated models. In order to obtain a partial isomorphism between two different recursively saturated models, the models must be recursively saturated "together". To make this precise we introduce the notion of a model pair. To avoid complications we consider only languages which have no function symbols. In applications, the function symbols can be replaced by relations in the usual way.

DEFINITION. Let $\mathfrak{A}$ and $\mathfrak{B}$ be two models for the recursive language $\mathscr{L}$ which has no function symbols. The *model pair* $(\mathfrak{A}, \mathfrak{B})$ is the model for the language $\mathscr{L}^{\mathfrak{A}} \cup \mathscr{L}^{\mathfrak{B}}$ defined as follows. $\mathscr{L}^{\mathfrak{A}}$ is a recursive language obtained by replacing each relation symbol R of $\mathscr{L}$ by a new symbol $R^{\mathfrak{A}}$ with the same number of places, replacing each constant symbol c of $\mathscr{L}$ by a new symbol $c^{\mathfrak{A}}$, and adding one new unary relation symbol A. Identify each constant $c^{\mathfrak{A}}$ and relation $R^{\mathfrak{A}}$ with its interpretation in the model $\mathfrak{A}$. $\mathscr{L}^{\mathfrak{B}}$ is formed in a similar way. Then $(\mathfrak{A}, \mathfrak{B})$ is the model

$$(\mathfrak{A}, \mathfrak{B}) = \langle A \cup B, A, B, R^{\mathfrak{A}}, R^{\mathfrak{B}}, c^{\mathfrak{A}}, c^{\mathfrak{B}} \rangle_{R, c \in \mathscr{L}}.$$

It is easy to see that if $(\mathfrak{A}, \mathfrak{B})$ is a recursively saturated model pair, then both $\mathfrak{A}$ and $\mathfrak{B}$ are recursively saturated models for $\mathscr{L}$. However, it frequently happens that each of $\mathfrak{A}$ and $\mathfrak{B}$ is recursively saturated but the model pair $(\mathfrak{A}, \mathfrak{B})$ is not recursively saturated. For an example see Exercise 2.4.17.

In order to extend the notion of a model pair to languages with function symbols in the natural way, a function symbol of $\mathscr{L}$ should be interpreted by a pair of partial functions in $(\mathfrak{A}, \mathfrak{B})$, which are relations but not functions in our treatment. For example, if F is a unary function symbol of $\mathscr{L}$, then $F^{\mathfrak{A}}$ should be interpreted by a partial function with domain A. For this reason we may as well replace function symbols by relation symbols in the original language $\mathscr{L}$.

THEOREM 2.4.5. *Let $\mathscr{L}$ be a recursive language, and let $\mathfrak{A}$ and $\mathfrak{B}$ be elementarily equivalent models for $\mathscr{L}$ such that the model pair $(\mathfrak{A}, \mathfrak{B})$ is recursively saturated. Then $\mathfrak{A}$ is partially isomorphic to $\mathfrak{B}$. In fact, the relation*

$$(\mathfrak{A}, a_1, \ldots, a_n) \equiv (\mathfrak{B}, b_1, \ldots, b_n)$$

is a partial isomorphism.

PROOF. Let I be the elementary equivalence relation between n-tuples from $\mathfrak{A}$ and $\mathfrak{B}$. We wish to show that I is a partial isomorphism. Since $\mathfrak{A} \equiv \mathfrak{B}$, the empty sequences are in the relation I. It is immediate that any pair related by I satisfies the same atomic formulas. We must verify the back and forth condition for the relation I.

For each formula $\varphi(x_1, \ldots, x_n)$ of $\mathscr{L}$, define the formula $\varphi^{\mathfrak{A}}(x_1, \ldots, x_n)$ of $\mathscr{L}^{\mathfrak{A}}$ inductively as follows. For an atomic formula φ of $\mathscr{L}$, $\varphi^{\mathfrak{A}}$ is obtained by replacing each relation or constant symbol s of $\mathscr{L}$ by $s^{\mathfrak{A}}$. The logical connectives are passed over with no change, and the quantifiers are relativized with the rules.

$$[(\forall x)\varphi]^{\mathfrak{A}} = (\forall x)[A(x) \to \varphi^{\mathfrak{A}}],$$

$$[(\exists x)\varphi]^{\mathfrak{A}} = (\exists x)[A(x) \wedge \varphi^{\mathfrak{A}}].$$

The formula $\varphi^{\mathfrak{B}}$ is defined analogously.

We see by induction that for any formula $\varphi(x_1, \ldots, x_n)$ of $\mathscr{L}$ and any n-tuple $a_1, \ldots, a_n$ in A,

$$\mathfrak{A} \models \varphi[a_1, \ldots, a_n] \text{ iff } (\mathfrak{A}, \mathfrak{B}) \models \varphi^{\mathfrak{A}}[a_1, \ldots, a_n],$$

and similarly for $\mathfrak{B}$. It follows that whenever $a_1, \ldots, a_n$ in A and $b_1, \ldots, b_n$ in B are such that

$$(\mathfrak{A}, a_1, \ldots, a_n) \equiv (\mathfrak{B}, b_1, \ldots, b_n)$$

and $d \in A$, the set $\Gamma(x)$ consisting of $B(x)$ and all formulas of

$$\mathscr{L}^{\mathfrak{A}} \cup \mathscr{L}^{\mathfrak{B}}(a_1, \ldots, a_n, d, b_1, \ldots, b_n)$$

of the form

$$\varphi^{\mathfrak{A}}(a_1, \ldots, a_n, d) \leftrightarrow \varphi^{\mathfrak{B}}(b_1, \ldots, b_n, x)$$

is finitely satisfiable in the model pair $(\mathfrak{A}, \mathfrak{B})$. Moreover, the set $\Gamma(x)$ of formulas is recursive. It follows by recursive saturation of the model pair that there is an element $e \in B$ which realizes $\Gamma(x)$ in $(\mathfrak{A}, \mathfrak{B})$. Then

$$(\mathfrak{A}, a_1, \ldots, a_n, d) \equiv (\mathfrak{B}, b_1, \ldots, b_n, e).$$

Thus I satisfies the back and forth condition, and the proof is complete. ⊣

As a corollary we obtain a useful criterion for a theory to be complete.

Corollary 2.4.6. *Let $\mathscr{L}$ be a recursive language. A theory T in $\mathscr{L}$ is complete if and only if for any recursively saturated model pair $(\mathfrak{A}, \mathfrak{B})$ of models of T, $\mathfrak{A}$ is partially isomorphic to $\mathfrak{B}$.*

Proof. If T is complete, then the models in the pair are partially isomorphic by Theorem 2.4.5. Suppose that T is not complete. Let $\mathfrak{A}$ and $\mathfrak{B}$ be models of T which are not elementarily equivalent. Form the model pair $(\mathfrak{A}, \mathfrak{B})$. By Theorem 2.4.1 there is a countable recursively saturated model $\mathfrak{C}$ elementarily equivalent to $(\mathfrak{A}, \mathfrak{B})$. Using the relativized formulas $\varphi^{\mathfrak{A}}$ and $\varphi^{\mathfrak{B}}$ from the preceding proof, we see that $\mathfrak{C}$ is a model pair $(\mathfrak{A}', \mathfrak{B}')$ where $\mathfrak{A}' \equiv \mathfrak{A}$ and $\mathfrak{B}' \equiv \mathfrak{B}$. Then $\mathfrak{A}'$ and $\mathfrak{B}'$ are models of T but are not elementarily equivalent and hence not partially isomorphic. ⊣

There are several methods in model theory for showing that particular theories are complete. The method based on recursively saturated models and parital isomorphisms is quite powerful and easy to use. As an illustration of the method we obtain a complete set of axioms for the theory of the ordered group of integers under addition.

Example 2.4.7. The following theory T is the complete theory of the ordered group of integers under addition, $\langle Z, +, -, 0, 1, \leqslant \rangle$. The constant 1 is not necessary but is included for convenience.

The Abelian group axioms with $+, -, 0$.

The axioms for linear order.

$$x \leqslant y \rightarrow x + z \leqslant y + z.$$

1 is the least element greater than 0.

For each integer $k > 1$, the axiom that each x has a remainder modulo k,

$$(\forall x)[k | x \vee k | x - 1 \vee \ldots \vee k | x - (k - 1)]$$

where $k | x$ means $(\exists y)[ky = x]$.

PROOF. Let $(\mathfrak{A}, \mathfrak{B})$ be a recursively saturated pair of models of T. Let I be the relation such that

$$\langle a_1, \ldots, a_n \rangle \; I \; \langle b_1, \ldots, b_n \rangle$$

if and only if for each linear term $P(x_1, \ldots, x_n)$ with integer coefficients, and each integer $k > 1$,

(a) $\mathfrak{A} \models P(a_1, \ldots, a_n) > 0$ iff $\mathfrak{B} \models P(b_1, \ldots, b_n) > 0$

and

(b) $\mathfrak{A} \models k | P(a_1, \ldots, a_n)$ iff $\mathfrak{B} \models k | P(b_1, \ldots, b_n)$.

We shall show that I is a partial isomorphism from $\mathfrak{A}$ to $\mathfrak{B}$. It is trivial that $\emptyset \; I \; \emptyset$. Since each term in $\mathscr{L}$ is equal to a linear term with integer coefficients, $\langle a_1, \ldots, a_n \rangle \; I \; \langle b_1, \ldots, b_n \rangle$ implies that $\langle a_1, \ldots, a_n \rangle$ and $\langle b_1, \ldots, b_n \rangle$ satisfy the same atomic formulas.

It remains to prove that I has the back and forth property. Let $\langle a_1, \ldots, a_n \rangle \; I \; \langle b_1, \ldots, b_n \rangle$ and let $c \in A$. We must show that the recursive set $\Gamma(x)$ of all formulas of the form

$$[P(a_1, \ldots, a_n, c) > 0]^{\mathfrak{A}} \leftrightarrow [P(b_1, \ldots, b_n, x) > 0]^{\mathfrak{B}},$$

$$[k | P(a_1, \ldots, a_n, c)]^{\mathfrak{A}} \leftrightarrow [k | P(b_1, \ldots, b_n, x)]^{\mathfrak{B}},$$

$$B(x)$$

is finitely satisfiable in $(\mathfrak{A}, \mathfrak{B})$. Using the axioms of T to simplify terms we see that it suffices to show that for all linear terms $P(x_1, \ldots, x_n)$ and $Q(x_1, \ldots, x_n)$ and natural numbers $k > 1, l, m$ such that

$$\mathfrak{A} \models P(a_1, \ldots, a_n) < mc < Q(a_1, \ldots, a_n) \text{ and } c \equiv l \pmod{k}$$

there exists $x \in B$ such that

$$\mathfrak{B} \models P(b_1, \ldots, b_n) < mx < Q(b_1, \ldots, b_n) \text{ and } x \equiv l \pmod{k}.$$

If there are only finitely many elements between $P(a_1, \ldots, a_n)$ and $Q(a_1, \ldots, a_n)$ in $\mathfrak{A}$ then mc is equal to a linear term in $\langle a_1, \ldots, a_n \rangle$ and mx may be taken to be the same linear term in $\langle b_1, \ldots, b_n \rangle$. Otherwise, since $\langle a_1, \ldots, a_n \rangle \, I \, \langle b_1, \ldots, b_n \rangle$, there are infinitely many elements between $P(b_1, \ldots, b_n)$ and $Q(b_1, \ldots, b_n)$ in $\mathfrak{B}$. By the axioms of T, any infinite interval contains elements mx such that $x \equiv l \pmod{k}$. Therefore $\Gamma(x)$ is finitely satisfiable in $(\mathfrak{A}, \mathfrak{B})$. By recursive saturation, $\Gamma(x)$ is realized in $(\mathfrak{A}, \mathfrak{B})$, and thus the relation I is a partial isomorphism from $\mathfrak{A}$ to $\mathfrak{B}$. ⊣

As another application of recursive saturation we give an easy direct proof of the Robinson Consistency Theorem (Theorem 2.2.23). We saw in an exercise in Section 2.2 that the Craig Interpolation Theorem follows quickly from the Robinson Consistency Theorem, and in fact the Robinson Consistency Theorem is only needed for finite languages. On the other hand, in Section 2.2 it was shown that the full Robinson Consistency Theorem is a corollary of the Craig Interpolation Theorem. Thus we only need a direct proof of the Robinson Consistency Theorem for finite languages. In this section it is more natural to prove it for recursive languages.

2.4.8 Robinson Consistency Theorem (restated). *Let $\mathscr{L}_1$ and $\mathscr{L}_2$ be two recursive languages and let $\mathscr{L} = \mathscr{L}_1 \cap \mathscr{L}_2$. Suppose T is a complete theory in $\mathscr{L}$ and $T_1 \supset T$, $T_2 \supset T$ are consistent theories in $\mathscr{L}_1$, $\mathscr{L}_2$ respectively. Then $T_1 \cup T_2$ is consistent in the language $\mathscr{L}_1 \cup \mathscr{L}_2$.*

Proof. By replacing constant and function symbols by relation symbols in the usual way, we may assume that $\mathscr{L}_1$ and $\mathscr{L}_2$ have only relation symbols. Let $\mathfrak{A}$ be a model of T_1 and $\mathfrak{B}$ be a model of T_2. Form the model pair (in the natural extended sense for models of two languages)

$$(\mathfrak{A}, \mathfrak{B}) = \langle A \cup B, A, B, R^{\mathfrak{A}}, S^{\mathfrak{B}} : R \in \mathscr{L}_1, S \in \mathscr{L}_2 \rangle .$$

The language of $(\mathfrak{A}, \mathfrak{B})$ is the recursive language $\mathscr{L}_1^{\mathfrak{A}} \cup \mathscr{L}_2^{\mathfrak{B}}$. Let $(\mathfrak{A}', \mathfrak{B}')$ be a countable recursively saturated model which is elementarily equivalent to $(\mathfrak{A}, \mathfrak{B})$, whence $\mathfrak{A}'$ is a model of T_1 and $\mathfrak{B}'$ is a model on T_2. Let $(\mathfrak{A}_0, \mathfrak{B}_0)$ be the reduct of $(\mathfrak{A}', \mathfrak{B}')$ to the smaller language $\mathscr{L}^{\mathfrak{A}} \cup \mathscr{L}^{\mathfrak{B}}$. Then $\mathfrak{A}_0$ and $\mathfrak{B}_0$ are models of T and hence are elementarily equivalent models for $\mathscr{L}$, and $(\mathfrak{A}_0, \mathfrak{B}_0)$ is a recursively saturated model pair in the original sense. Therefore $\mathfrak{A}_0$ and $\mathfrak{B}_0$ are partially isomorphic, and since they are countable they are isomorphic by some isomorphism f. We may now expand $\mathfrak{B}'$ to a model $\mathfrak{B}''$ for $\mathscr{L}_1 \cup \mathscr{L}_2$ by interpreting each relation symbol $R \in \mathscr{L}_1$ by the f-image of its interpretation in $\mathfrak{A}'$. Then $\mathfrak{B}''$ is a model of $T_1 \cup T_2$ as required. ⊣

Recursively saturated models have a number of other applications, some of which are given in the exercises. We conclude this section with a general result about recursively saturated models, which shows that every countable recursively saturated model is "recursively saturated with respect to relations" as well as with respect to variables. We first need a useful lemma concerning the set of consequences of a recursive set of formulas.

LEMMA 2.4.9. *Let $\mathscr{L}$ and $\mathscr{L}'$ be recursive languages with $\mathscr{L} \subset \mathscr{L}'$ and let $\Gamma(x)$ be a recursive set of formulas of $\mathscr{L}'$. Then there is a recursive set $\Sigma(x)$ of formulas of $\mathscr{L}$ such that $\Sigma(x)$ and $\Gamma(x)$ have exactly the same set of consequences with at most x free in the smaller language $\mathscr{L}$.*

PROOF. Let $d_1, d_2, \ldots$ be a recursive list of all deductions from $\Gamma(x)$ of formulas of $\mathscr{L}$ in x, and let $\varphi_n(x)$ be the conjunction of n copies of the formula proved by d_n. Let $\Sigma(x)$ be the set

$$\Sigma(x) = \{\varphi_n(x) : n \in \omega\}.$$

$\Sigma(x)$ is clearly a set of formulas of $\mathscr{L}$ which has the same consequences in $\mathscr{L}$ as $\Gamma(x)$. Moreover, $\Sigma(x)$ is recursive because one can decide whether a formula $\psi(x)$ belongs to $\Sigma(x)$ by looking only at the deductions d_m where m is at most the number of $\wedge$ symbols in $\psi(x)$ plus 2. ⊣

THEOREM 2.4.10. *Let $\mathscr{L}$ and $\mathscr{L}'$ be recursive languages with $\mathscr{L} \subset \mathscr{L}'$ and let $\mathfrak{A}$ be a countable recursively saturated model for $\mathscr{L}$. Then any recursive set Γ of sentences of $\mathscr{L}'$ which is consistent with the complete theory of $\mathfrak{A}$ is satisfied in some expansion of $\mathfrak{A}$.*

We remark that if $\mathscr{L}'$ consists of $\mathscr{L}$ plus a recursive set of constant symbols, then the above theorem holds even for uncountable recursively saturated models $\mathfrak{A}$. To see this, let $\mathscr{L}_n$ be $\mathscr{L}$ plus the first n constant symbols and by Lemma 2.4.9 let Γ_n be a recursive theory in $\mathscr{L}_n$ with the same consequences in $\mathscr{L}_n$ as Γ. Because $\mathfrak{A}$ is recursively saturated there is an expansion of $\mathfrak{A}$ by adding one constant which is a model of Γ_1. Since any expansion of $\mathfrak{A}$ by finitely many constants is again recursively saturated, the process may be repeated to obtain an expansion of $\mathfrak{A}$ by countably many constants which is a model of Γ.

The proof of the general case of Theorem 2.4.10 is more difficult and requires the hypothesis that $\mathfrak{A}$ is countable.

Proof of Theorem 2.4.10. Let $\mathscr{L}'_A$ be an expansion of $\mathscr{L}'$ which has a new constant symbol for each element of A, and let $\theta_0, \theta_1, \ldots$ be a list of all sentences of $\mathscr{L}'_A$. Let T be the elementary diagram of $\mathfrak{A}$. Then $\Gamma \cup T$ is consistent. Form a sequence of sentences $\psi_0, \psi_1, \ldots$ of $\mathscr{L}'_A$ such that for each n:

(1). $\models \psi_{n+1} \rightarrow \psi_n$.
(2). ψ_n is consistent with $\Gamma \cup T$.
(3). If θ_n is consistent with $\Gamma \cup T \cup \{\psi_n\}$ then $\models \psi_{n+1} \rightarrow \theta_n$.
(4). If θ_n is of the form $(\exists x)\theta(x)$ and is consistent with $\Gamma \cup T \cup \{\psi_n\}$ then there is an $a \in A$ such that $\models \psi_{n+1} \rightarrow \theta(a)$.

Recursive saturation is needed to show that the sequence ψ_n can be chosen to satisfy property (4). Suppose $\psi_0, \ldots, \psi_n$ have been chosen to satisfy (1)–(4) and that θ_n satisfies the hypothesis of (4). Let Y be the finite set of constants from A which occur in ψ_n, and θ_n. Then

$$\Gamma'(x) = \Gamma \cup \{\psi_n\} \cup \{\theta(x)\}$$

is a recursive set of formulas of $\mathscr{L}'_Y$ which is consistent with T. By Lemma 2.4.9 there is a recursive set $\Sigma(x)$ of formulas of $\mathscr{L}_Y$ which has exactly the same set of consequences as $\Gamma'(x)$ in the language $\mathscr{L}_Y$. It follows that $\Sigma(x)$ is consistent with T, and thus is finitely satisfiable in $\mathfrak{A}_Y$. By recursive saturation there exists $a \in A$ which realizes $\Sigma(x)$ in $\mathfrak{A}_Y$.

We claim that $\Gamma'(a)$ is consistent with T. To prove this claim, suppose that $\Gamma'(a) \models \varphi(a)$ where $\varphi(x)$ is a formula of $\mathscr{L}_A$. Existentially quantifying the elements of $A - (Y \cup \{a\})$, we may take $\varphi(a)$ to be a sentence of $\mathscr{L}_Y(a)$ and $\varphi(x)$ to be a formula of $\mathscr{L}_Y$. Then $\Gamma'(x) \models \varphi(x)$, and hence $\Sigma(x) \models \varphi(x)$. Since a realizes $\Sigma(x)$ in $\mathfrak{A}_Y$, we have $\mathfrak{A}_Y \models \varphi[a]$. Therefore $\varphi(a)$ belongs to T and the claim is proved.

In view of the claim, we may take ψ_{n+1} to be $\psi_n \wedge \theta(a)$ in case (4), whence (1)–(4) will hold for $n+1$ as required. The sequence of sentences ψ_n, $n < \omega$, may then be defined by recursion. It follows from (1)–(4) that $\mathfrak{A}$ has an expansion $\mathfrak{A}'$ to $\mathscr{L}'$ such that $\mathfrak{A}'_A$ is a model of $\Gamma \cup T \cup \{\psi_n : n < \omega\}$, and in particular such that $\mathfrak{A}'$ is a model of Γ. ⊣

As an example of the preceding theorem, let T be the theory of the ordered group of integers under addition from Example 2.4.7 and let Γ be the set of sentences with an additional function symbol for multiplication consisting of the commutative ring axioms and Peano's axioms for the nonnegative elements. Then every countable recursively saturated model of T can be expanded to a model of $T \cup \Gamma$.

The results in this section can be readily extended to the case of an arbitrary countable language $\mathscr{L}$ by modifying the notion of a recursively saturated model. A set S is said to be recursive *relative to* $\mathscr{L}$ if there is an algorithm which decides whether or not an arbitrary input belongs to S but makes use of an "oracle" which will always correctly answer questions of the form "is φ a formula of $\mathscr{L}$?". Everything goes through with only minor changes when the notion of recursive saturation is replaced by recursive saturation relative to $\mathscr{L}$.

EXERCISES

2.4.1. Prove that a complete theory in a recursive language which has continuum many complete types has continuum many nonisomorphic countable recursively saturated models.

2.4.2. Let T be the complete theory of models with countably many distinct constants and no functions or relations. Prove that all recursively saturated countable models of T are isomorphic.

2.4.3. Let T be the theory of divisible torsion free Abelian groups. Prove that all recursively saturated countable models of T are isomorphic.

2.4.4. Let $\mathfrak{A}$ be a countable model for a recursive language such that for each finite sequence $\langle a_1, \ldots, a_n \rangle$ in A, every element of A realizes a recursive type in $(\mathfrak{A}, a_1, \ldots, a_n)$. Prove that $\mathfrak{A}$ is elementarily embeddable in every recursively saturated model of Th($\mathfrak{A}$).

2.4.5. Let T be the complete theory of the model $\langle \omega, \leqslant \rangle$. Prove that a model of T is countably homogeneous if and only if it is isomorphic to one of the following three models: the atomic model $\langle \omega, \leqslant \rangle$, the countably saturated model formed by adding a countable dense set of copies of $\langle Z, \leqslant \rangle$ to the end of $\langle \omega, \leqslant \rangle$, and the model formed by adding one copy of $\langle Z, \leqslant \rangle$ to the end of $\langle \omega, \leqslant \rangle$.

2.4.6. Show that every atomic model is ω-homogeneous.

2.4.7. If T is not ω-categorical, then T has at least two nonisomorphic countably homogeneous models.

2.4.8. Show that if $\mathfrak{A}$ is countably homogeneous and every type in finitely many variables in $\mathrm{Th}(\mathfrak{A})$ is realized in $\mathfrak{A}$, then $\mathfrak{A}$ is countably saturated.

2.4.9. Let $\mathfrak{A}$ and $\mathfrak{B}$ be ω-homogeneous models which realize exactly the same types in finitely many variables. Prove that $\mathfrak{A}$ and $\mathfrak{B}$ are partially isomorphic.

2.4.10*. Let $\mathfrak{A} = \langle A, \leqslant, \ldots \rangle$ be an ω-homogeneous model for a countable language such that $\leqslant$ well orders A. Prove that A has cardinality at most 2^ω.

2.4.11. A model $\mathfrak{A}$ is called almost ω-homogeneous if there is a finite sequence $a_1, \ldots, a_n$ of elements of A such that $(\mathfrak{A}, a_1, \ldots, a_n)$ is ω-homogeneous. Prove that if every countable model of theory T is almost ω-homogeneous then every model of T is almost ω-homogeneous.

2.4.12*. Let T be a complete theory in a countable language and let $\Gamma(x)$ be a type such that:

(i). T has a model which omits Γ,

(ii). for every complete type $\Sigma(y_1, \ldots, y_n)$ over T, either

$$\Sigma(y_1, \ldots, y_n) \supset \Gamma(y_i)$$

for some i, or T has a model which realizes Σ and omits Γ.

Prove that T has an ω-homogeneous model which omits Γ.

2.4.13. Let $\mathfrak{A}$ be a countably infinite recursively saturated model for a recursive language. Prove that $\mathfrak{A}$ has an automorphism which is not the identity function.

2.4.14. Let $\mathfrak{A}$ be a recursively saturated model for a recursive language, and let

$$\{\varphi_m(x_1, \ldots, x_n, y_1, \ldots, y_n) : m < \omega\}$$

be an r.e. set of formulas. Suppose that for each $m < \omega$,

$$\mathfrak{A} \models \varphi_{m+1} \rightarrow \varphi_m$$

and

$$\mathfrak{A} \models \forall x_1 \exists y_1 \forall x_2 \exists y_2 \ldots \forall x_n \exists y_n \varphi_m.$$

Prove that

$$\mathfrak{A} \models \forall x_1 \exists y_1 \forall x_2 \exists y_2 \ldots \forall x_n \exists y_n \bigwedge_{m<\omega} \varphi_m.$$

2.4.15. In an ordered field F, two elements x, y are said to realize the same cut over the rationals if for every rational number q we have $q < x$ iff $q < y$. Let F and G be two ordered fields such that $F \subset G$ and the model $\langle G, F, 0, 1, +, \cdot, \leqslant \rangle$ is recursively saturated. Prove that every element of G realizes the same cut as some element of F over the rationals.

2.4.16. Let $(\mathfrak{A}, \mathfrak{B})$ be a recursively saturated model pair. Prove that $\mathfrak{A}$ and $\mathfrak{B}$ are recursively saturated.

2.4.17. Let T be a complete theory which has uncountably many complete types in one variable. Prove that T has countable recursively saturated models $\mathfrak{A}$ and $\mathfrak{B}$ such that the model pair $(\mathfrak{A}, \mathfrak{B})$ is not recursively saturated. [*Hint*: Use Theorem 2.4.5.]

2.4.18. In a recursive language, let $\mathfrak{A}$ and $\mathfrak{B}$ be models such that $\mathfrak{B} \subset \mathfrak{A}$ and the expanded model $(\mathfrak{A}, B)$ is recursively saturated. Prove that the model pair $(\mathfrak{A}, \mathfrak{B})$ is recursively saturated.

2.4.19*. Let T be the theory in a language with countably many unary relations such that each nontrivial finite Boolean combination of relations is consistent. Show that every model of T is ω-homogeneous, and that T has countably homogeneous models which are not recursively saturated.

2.4.20. Let I be a partial isomorphism from $\mathfrak{A}$ to $\mathfrak{B}$. Show that whenever $\langle a_1, \ldots, a_n \rangle \, I \, \langle b_1, \ldots, b_n \rangle$, the expanded models $(\mathfrak{A}, a_1, \ldots, a_n)$ and $(\mathfrak{B}, b_1, \ldots, b_n)$ are partially isomorphic.

2.4.21. Let $\mathscr{L}$ be a recursive language with only relation symbols and let $(\mathfrak{A}, \mathfrak{B})$ be a countable recursively saturated model pair for $\mathscr{L}$. Prove that every existential sentence which holds in $\mathfrak{A}$ holds in $\mathfrak{B}$ if and only if $\mathfrak{A}$ is isomorphically embeddable in $\mathfrak{B}$.

2.4.22. Let $\mathscr{L}$ be a recursive language with only relation symbols and let $(\mathfrak{A}, \mathfrak{B})$ be a countable recursively saturated model pair for $\mathscr{L}$. Prove that every positive sentence which holds in $\mathfrak{A}$ holds in $\mathfrak{B}$ if and only if there is a homomorphism of $\mathfrak{A}$ onto $\mathfrak{B}$.

2.4.23. Let $\mathscr{L}$ be a recursive language with only relation symbols, let $(\mathfrak{A}, \mathfrak{B})$ be a countable recursively saturated model pair for $\mathscr{L}$, and let U be a unary relation symbol of $\mathscr{L}$. The *relativization* φ^U of a formula φ of $\mathscr{L}$ to U is defined recursively as follows.

If φ is atomic, then $\varphi^U = \varphi$,

$$(\varphi \wedge \psi)^U = \varphi^U \wedge \psi^U, (\varphi \vee \psi)^U = \varphi^U \vee \psi^U, (\neg\varphi)^U = \neg(\varphi^U),$$

$$(\forall x\varphi)^U = (\forall x)(U(x) \to \varphi^U), (\exists x\varphi)^U = (\exists x)(U(x) \wedge \varphi^U).$$

Prove that $\mathfrak{A}$ is isomorphic to the submodel of $\mathfrak{B}$ with universe $U^{\mathfrak{B}}$ if and only if for every sentence φ of $\mathscr{L}$, $\mathfrak{A} \models \varphi$ iff $\mathfrak{B} \models \varphi^U$.

2.4.24. Let $\mathscr{L}$ be a recursive language with only relation symbols, let $(\mathfrak{A}, \mathfrak{B})$ be a countable recursively saturated model pair for $\mathscr{L}$, and let E be a binary relation symbol of $\mathscr{L}$. A formula of $\mathscr{L}$ is said to be *essentially existential* if it is built up from atomic and negated atomic formulas using conjunctions, disjunctions, bounded quantifiers

$$(\forall x)(xEy \to \varphi), (\exists x)(xEy \wedge \varphi),$$

and existential quantifiers. $\mathfrak{B}$ is said to be an *outer extension* of $\mathfrak{C}$, and $\mathfrak{C}$ is said to be a *transitive submodel* of $\mathfrak{B}$, if $\mathfrak{B}$ is an extension of $\mathfrak{C}$ and for every $x \in C$ and $y \in B$, if $\mathfrak{B} \models y \, E \, x$ then $y \in C$.

Prove that $\mathfrak{B}$ is isomorphic to an outer extension of $\mathfrak{A}$ if and only if every essentially existential sentence which holds in $\mathfrak{A}$ holds in $\mathfrak{B}$.

2.4.25. By an *arithmetical set* we mean a subset of $R(\omega)$ which is definable by a formula of first order logic in the model $\langle R(\omega), \in \rangle$. A model $\mathfrak{A}$ for a recursive language $\mathscr{L}$ is said to be *arithmetically saturated* if for every finite sequence $\langle a_1, \ldots, a_n \rangle$ in A, every arithmetical set of formulas $\Gamma(x)$ in $\mathscr{L}(a_1, \ldots, a_n)$ which is finitely satisfiable in $(\mathfrak{A}, a_1, \ldots, a_n)$ is realized in $(\mathfrak{A}, a_1, \ldots, a_n)$. Prove that every countable model for $\mathscr{L}$ has a countable arithmetically saturated elementary extension.

2.4.26. Use the method of recursively saturated models to obtain a complete set of axioms for the set of natural numbers with the order relation.

2.4.27. Use the method of recursively saturated models to obtain a complete set of axioms for the rational numbers with the order relation and addition (the theory of divisible ordered Abelian groups).

2.4.28. Let T be the theory with the binary relation $\leqslant$ and unary function F and axioms stating that:

(i). $\leqslant$ is a dense simple ordering with no greatest or least element.

(ii). F is an automorphism of the ordering $\leqslant$.

(iii). For all x, x is strictly less than $F(x)$.

Prove using the method of recursively saturated models that T is complete.

2.4.29*. Use the method of recursively saturated models to show that the set of axioms for additive number theory described in Example 1.4.11 is a complete set of axioms for the set of natural numbers with zero, successor, and addition.

2.4.30*. Let $\mathscr{L}$ have one binary function symbol F and let T be the theory whose axioms state that F is a one to one function, every element belongs to the range of F, and

$$\neg y = \tau(x_1, \ldots, x_n, y)$$

where y occurs in the term $\tau(x_1, \ldots, x_n, y)$ and the term is not a single variable. Use the method of recursively saturated models to show that T is complete.

2.4.31*. Let T be a complete theory in a recursive language L, and suppose that T has a complete extension T' in the language $L \cup \{c_n : n < \omega\}$ such that T' has no atomic model. Then T has a model which is not recursively saturated.

2.4.32*. Let $\mathscr{L}$ be the language with only the binary relation ϵ and let $\mathscr{L}'$ be the language with the binary relations ϵ, S. Let $\langle R(\omega), \epsilon, \mathrm{Sat}\rangle$ be the model for $\mathscr{L}'$ where $\mathrm{Sat}(x, y)$ holds iff x is a formula of $\mathscr{L}$ and y is a tuple which satisfies x in $\langle R(\omega), \epsilon\rangle$. Show that there is a recursive set Γ of sentences of $\mathscr{L}'$ such that $\langle R(\omega), \epsilon, \mathrm{Sat}\rangle \models \Gamma$, and for each countable nonstandard model $\mathfrak{A} \equiv \langle R(\omega), \epsilon\rangle$, $\mathfrak{A}$ is recursively saturated if and only if $\mathfrak{A}$ can be expanded to a model of Γ.

2.4.33*. Prove that a countable nonstandard model of complete arithmetic is recursively saturated if and only if it is the arithmetic part of some model of Zermelo set theory.

2.4.34*. Show that a countable nonstandard model of additive number theory can be expanded to a model of Peano arithmetic if and only if it is recursively saturated.

2.5. Lindström's Characterization of First Order Logic

In this section we shall prove a result of Lindström which shows that first order logic is the only logic for which the Compactness Theorem and Downward Löwenheim Skolem Theorem hold. In order to state such a result we need a general notion of an abstract logic. This is the beginning of a subject called abstract model theory, which studies the relationship between various model theoretic results in arbitrary logics. There are many interesting logics which are richer than first order logic, such as logics with infinitely long formulas and logics with extra quantifiers. The theorem of Lindström shows that, even though there are richer logics, first order logic is of fundamental importance. No matter how we enrich first order logic, we must either give up one of the basic results which underlies our whole treatment of model theory, the Compactness Theorem or the Downward Löwenheim Skolem Theorem, or else go outside the notion of an abstract logic which is assumed in Lindström's theorem.

We shall not discuss logics other than first order logic here, and shall instead concentrate on results which characterize first order logic. To avoid long detours into side issues, we shall give a definition of abstract logic which is less general than is usually found in the literature, but is adequate for our present purposes.

Recall that a *language* $\mathscr{L}$ is a set of relation, constant, and function symbols. In order so avoid some complications which would obscure the main ideas, *we shall restrict our consideration in this section to languages* $\mathscr{L}$ *which have only relation and constant symbols, and no function symbols.*

DEFINITION 2.5.1. An *abstract logic* is a pair of classes $(l, \models_l)$ with the following properties, l is the class of *sentences* and $\models_l$ is the *satisfaction relation* of the logic $(l, \models_l)$.

(i) *Occurrence Property.* For each $\varphi \in l$ there is associated a finite language $\mathscr{L}_\varphi$, called the set of *symbols occurring in* φ. The relation $\mathfrak{A} \models_l \varphi$ is a relation between sentences φ of l and models $\mathfrak{A}$ for languages $\mathscr{L}$ which contain $\mathscr{L}_\varphi$. That is, if $\varphi \in l$ and $\mathfrak{A}$ is a model for $\mathscr{L}$, then the statement $\mathfrak{A} \models_l \varphi$ is either true or false if $\mathscr{L}$ contains $\mathscr{L}_\varphi$, and is undefined if $\mathscr{L}$ does not contain $\mathscr{L}_\varphi$.

(ii) *Expansion Property.* The relation $\mathfrak{A} \models_l \varphi$ depends only on the reduct of $\mathfrak{A}$ to $\mathscr{L}_\varphi$. That is, if $\mathfrak{A} \models_l \varphi$ and $\mathfrak{B}$ is an expansion of $\mathfrak{A}$ to a larger language, then $\mathfrak{B} \models_l \varphi$.

(iii) *Isomorphism Property.* The relation $\mathfrak{A} \models_l \varphi$ is preserved under isomorphism. That is, if $\mathfrak{A} \cong \mathfrak{B}$ and $\mathfrak{A} \models_l \varphi$ then $\mathfrak{B} \models_l \varphi$.

(iv) *Renaming Property.* The relation $\mathfrak{A} \models_l \varphi$ is preserved under renaming. Formally, let ρ be a bijection (one to one mapping) from a language $\mathscr{L}$ to a language $\rho\mathscr{L}$ which preserves the number of places of all symbols, and for each model $\mathfrak{A}$ for $\mathscr{L}$ let $\rho\mathfrak{A}$ be the model for $\rho\mathscr{L}$ induced in the obvious way by ρ. Then for each sentence $\varphi \in l$ with $\mathscr{L}_\varphi \subset \mathscr{L}$ there is a sentence $\rho\varphi \in l$ with $\mathscr{L}_{\rho\varphi} = \rho\mathscr{L}_\varphi$ such that for each model $\mathfrak{A}$ for $\mathscr{L}$, $\mathfrak{A} \models_l \varphi$ iff $\rho\mathfrak{A} \models_l \rho\varphi$.

(v) *Closure Property.* l contains all atomic sentences, l is closed under the usual first order connectives $\wedge, \vee, \neg$, $\models_l$ satisfies the usual rules for satisfaction for atomic formulas and first order connectives, and the set of symbols $\mathscr{L}_\varphi$ behaves as expected for atomic sentences and first order connectives.

(vi) *Quantifier Property.* l is closed under universal and existential quantifiers. That is, for each $\varphi \in l$ and each constant symbol $c \in \mathscr{L}_\varphi$ there are sentences $(\forall x_c)\varphi$ and $(\exists x_c)\varphi$ in l with the set of symbols $\mathscr{L}_\varphi - \{c\}$ such that:

$$\mathfrak{A} \models_l (\forall x_c)\varphi \text{ iff for all } a \in A,\ (\mathfrak{A}, a) \models_l \varphi;$$

$$\mathfrak{A} \models_l (\exists x_c)\varphi \text{ iff for some } a \in A,\ (\mathfrak{A}, a) \models_l \varphi.$$

(vii) *Relativization Property*. For each sentence $\varphi \in l$ and relation $R(x, b_1, \ldots, b_n)$ with $R, b_1, \ldots, b_n$ not in $\mathscr{L}_\varphi$, there is a new sentence $\varphi | R(x, b_1, \ldots, b_n)$, read φ *relativized to* $R(x, b_1, \ldots, b_n)$, which has the set of symbols $\mathscr{L}_\varphi \cup \{R, b_1, \ldots, b_n\}$ and is such that whenever $\mathfrak{B}$ is the submodel of a model $\mathfrak{A}$ for $\mathscr{L}_\varphi$ with universe

$$B = \{a \in A : R(a, b_1, \ldots, b_n)\},$$

we have

$$(\mathfrak{A}, R, b_1, \ldots, b_n) \models_l \varphi | R(x, b_1, \ldots, b_n) \text{ iff } \mathfrak{B} \models_l \varphi.$$

Two abstract logics are said to be *equivalent* if for each sentence of one of the logics there is a sentence of the other logic which has the same set of symbols and the same models. Two logics which are equivalent are alike except that the sentences are "renamed", and may be considered as the same. We shall be interested in characterizing first order logic up to equivalence.

Our notion of an abstract logic has no provision for free variables, and l should be thought of as a set of sentences rather than formulas. Notice that in the Quantifier Property (vi), a sentence in an expanded language with a new constant symbol takes the place of a formula with a free variable.

To shorten our notation, we shall use the symbol l both for the logic and for the class of sentences of the logic. We use $\models$ for the satisfaction relation for ordinary first order logic, and $\models_l$ for the satisfaction relation for an arbitrary abstract logic l. The most familiar example of an abstract logic is the ordinary first order logic, usually denoted by $l_{\omega,\omega} = (l_{\omega,\omega}, \models)$. The Relativization Property (vii) for first order logic holds where φ relativized to $R(x, b_1, \ldots, b_n)$ is the sentence formed by replacing each quantifier $(\forall x)\psi$ in φ by

$$(\forall x)[R(x, b_1, \ldots, b_n) \rightarrow \psi]$$

and $(\exists x)\psi$ in φ by

$$(\exists x)[R(x, b_1, \ldots, b_n) \wedge \psi].$$

The class of sentences of first order logic, $l_{\omega,\omega}$, is a proper class because it contains sentences for all languages $\mathscr{L}$. However, for each language $\mathscr{L}$, the class of sentences $\varphi \in l_{\omega,\omega}$ with $\mathscr{L}_\varphi \subseteq \mathscr{L}$ is a set. By the closure and quantifier properties, every sentence of first order logic belongs to every abstract logic l, and for each model $\mathfrak{A}$ and each first order sentence φ,

$\mathfrak{A} \models_l \varphi$ iff $\mathfrak{A} \models \varphi$. Many other examples of abstract logics have been studied extensively in the literature, but are outside the scope of this book. The book "Model-Theoretical Logics", Barwise and Feferman (1985), has a large collection of articles surveying abstract model theory and the model theory for a variety of logics beyond first order logic.

It should be emphasized that logics in the sense of Definition 2.5.1 deal with the same class of models as first order logic, and only the sentences and satisfaction relation may be different. This is a significant restriction which leaves a large loophole in Lindström's theorem. There are many examples of logics in a generalized sense which study models with additional structure and thus do not fit within our framework. These include modal logics, programming logics, and logics for models with topologies and measures. Sentential logic and ω-logic as described in this book are not examples of abstract logics in the sense of 2.5.1, because they also deal with different classes of models than first order logic. However, the notion of an abstract logic in Definition 2.5.1 is broad enough to explain what abstract model theory has to say about first order model theory.

By a *model* of a set T of sentences of an abstract logic l we mean a model $\mathfrak{A}$ such that $\mathfrak{A} \models_l \varphi$ for each $\varphi \in T$. Two models $\mathfrak{A}$ and $\mathfrak{B}$ for the same language are said to be *l-elementarily equivalent* if for each sentence $\varphi \in l$, $\mathfrak{A} \models_l \varphi$ if and only if $\mathfrak{B} \models_l \varphi$.

We shall now define two properties of abstract logics which correspond to the Compactness Theorem and the Downward Löwenheim Skolem Theorem.

DEFINITION. An abstract logic l is *countably compact* if for every countable set $T \subset l$, if every finite subset of T has a model then T has a model.

The *Löwenheim number* of l is the least cardinal α such that every sentence of l which has a model has a model of power at most α.

By the Compactness Theorem, first order logic is countably compact. In fact, first order logic is *fully compact*, that is, every finitely satisfiable set of sentences has a model. The Downward Löwenheim Skolem Theorem shows that first order logic has Löwenheim number ω.

If the Löwenheim number of an abstract logic l exists, it must be at least ω, because l contains every sentence of first order logic. For an arbitrary abstract logic, the Löwenheim number may not exist. However, the following simple result shows that it does exist if the logic does not have too many sentences. This simple result is of particular interest because it is one of the few natural results outside of set theory which uses the full power of the axiom scheme of replacement.

PROPOSITION 2.5.2. *Let l be an abstract logic such that for each finite language $\mathscr{L}$, the class $\{\varphi \in l : \mathscr{L}_\varphi \subseteq \mathscr{L}\}$ is a set. Then the Löwenheim number of l exists.*

PROOF. Because of the Renaming Property, we may restrict our attention to the class l' of sentences $\varphi \in l$ such that $\mathscr{L}_\varphi$ belongs to $R(\omega)$. By hypothesis, l' is a set. For each $\varphi \in l'$, let $\alpha(\varphi)$ be ω if φ has no models, and the least cardinal of a model of φ if φ has at least one model. By the axiom of replacement, there exists a cardinal

$$\alpha = \sup\{\alpha(\varphi) : \varphi \in l'\}.$$

α is the Löwenheim number of l. ⊣

We now prove a preliminary result which is an abstract version of Proposition 2.4.4.

PROPOSITION 2.5.3. *Let l be an abstract logic which has Löwenheim number ω. Then any two models which are partially isomorphic are l-elementarily equivalent.*

PROOF. Suppose $\mathfrak{A}$ and $\mathfrak{B}$ are models for a language $\mathscr{L}$ which are partially isomorphic by a relation I, but there is a sentence $\varphi \in l$ such that $\mathfrak{A} \models_l \varphi$ but $\mathfrak{B} \models_l \neg\varphi$. By taking reducts we may assume that $\mathscr{L} = \mathscr{L}_\varphi$. We may also assume that I is preserved under subsequences, that is, whenever $\langle a_1, \ldots, a_n\rangle \; I \; \langle b_1, \ldots, b_n\rangle$, any subsequence of $\langle a_1, \ldots, a_n\rangle$ is in the relation I to the corresponding subsequence of $\langle b_1, \ldots, b_n\rangle$. Let A' be the set of finite sequences of elements of A and let $F : A' \times A \to A'$ and $F' : A' \times A' \to A'$ be the functions

$$F(\langle a_1, \ldots, a_n\rangle, b) = \langle a_1, \ldots, a_n, b\rangle,$$

$$F'(\langle a_1, \ldots, a_m\rangle, \langle b_1, \ldots, b_n\rangle) = \langle a_1, \ldots, a_m, b_1, \ldots, b_n\rangle.$$

Form the expanded model $\mathfrak{A}'' = (A \cup A', \mathfrak{A}, F, F')$ where F and F' are ternary relations rather than binary functions. Define B', G, G', and $\mathfrak{B}''$ analogously. By the Isomorphism Property, we may take $\mathfrak{A}$ and $\mathfrak{B}$ so that the sets A, A', B, and B' are all disjoint from each other (with a different "empty sequence" in each of A', B'). Since $\mathfrak{A}$ is partially isomorphic to $\mathfrak{B}$, we may expand the model pair $(\mathfrak{A}'', \mathfrak{B}'')$ to a model $\mathfrak{C} = (\mathfrak{A}'', \mathfrak{B}'', I)$. By the closure, quantifier, and relativization properties, there is a sentence $\psi \in l$ in the language of the model $\mathfrak{C}$ which holds in $\mathfrak{C}$ and implies that

$$\mathfrak{A} \models_l \varphi, \mathfrak{B} \models_l \neg\varphi, \mathfrak{A} \cong_p \mathfrak{B}.$$

In particular, there is a first order sentence which holds in $\mathfrak{C}$ and implies that $\mathfrak{A} \cong_p \mathfrak{B}$. (This sentence uses the symbols for F and F'. The atomic formula condition depends on the fact that $\mathscr{L}$ is finite and I is closed under subsequences). Since l has Löwenheim number ω, ψ has a countable model $\mathfrak{C}_0$, from which we obtain models $\mathfrak{A}_0$ and $\mathfrak{B}_0$ for $\mathscr{L}$. $\mathfrak{A}_0$ and $\mathfrak{B}_0$ are countable models, and are partially isomorphic because $\mathfrak{C}_0$ is a model of ψ. By Proposition 2.4.4, $\mathfrak{A}_0$ and $\mathfrak{B}_0$ are isomorphic. But since $\mathfrak{C}$ is a model of ψ, $\mathfrak{A}_0 \models_l \varphi$ and $\mathfrak{B}_0 \models_l \neg\varphi$, contradicting the Isomorphism Property. This completes the proof. ⊣

THEOREM 2.5.4 (Lindström's Characterization of First Order Logic). *First order logic is, up to equivalence, the only abstract logic which is countably compact and has Löwenheim number ω.*

PROOF. Let l be a countably compact abstract logic with Löwenheim number ω. We must show that every sentence $\varphi \in l$ is l-equivalent to some first order sentence ψ, that is, for all $\mathfrak{A}$, $\mathfrak{A} \models_l \varphi$ iff $\mathfrak{A} \models \psi$. It is sufficient to consider models for a finite language $\mathscr{L}$. Let $\mathfrak{A}$ and $\mathfrak{B}$ be two models for $\mathscr{L}$. We shall define a "back and forth sequence" of relations I_k, $k \in \omega$, between $\mathfrak{A}$ and $\mathfrak{B}$. Let $\langle a_1, \ldots, a_n \rangle$ and $\langle b_1, \ldots, b_n \rangle$ be n-tuples from A and B respectively. $\langle a_1, \ldots, a_n \rangle \; I_0 \; \langle b_1, \ldots, b_n \rangle$ means that $\langle a_1, \ldots, a_n \rangle$ and $\langle b_1, \ldots, b_n \rangle$ satisfy the same atomic formulas of first order logic. By induction on k we define $\langle a_1, \ldots, a_n \rangle \; I_{k+1} \; \langle b_1, \ldots, b_n \rangle$ if and only if (1) For all $c \in A$ there exists $d \in B$ with $\langle a_1, \ldots, a_n, c \rangle \; I_k \; \langle b_1, \ldots, b_n, d \rangle$ and (2) vice versa. We shall let $\mathfrak{A} \equiv_k \mathfrak{B}$ mean that $\emptyset \; I_k \; \emptyset$ where $\emptyset$ is the empty sequence. Since $\mathscr{L}$ is finite and has no function symbols, for each k there is a finite set Γ_k of sentences of first order logic in $\mathscr{L}$ such that for all models $\mathfrak{A}$ and $\mathfrak{B}$ for $\mathscr{L}$, $\mathfrak{A} \equiv_k \mathfrak{B}$ if and only if $\mathfrak{A}$ and $\mathfrak{B}$ satisfy the same sentences of Γ_k.

Let $\varphi \in l$ be a sentence such that $\mathscr{L}_\varphi \subset \mathscr{L}$. It suffices to show that there is a $k \in \omega$ such that for all models $\mathfrak{A}$ and $\mathfrak{B}$ for $\mathscr{L}$, (3) $\mathfrak{A} \equiv_k \mathfrak{B}$ and $\mathfrak{A} \models_l \varphi$

implies $\mathfrak{B} \models_l \varphi$. When (3) is proved, it will follow that φ is equivalent to a boolean combination of sentences in Γ_k, and the proof will be complete.

Suppose (3) fails for all $k \in \omega$. Choose models $\mathfrak{A}_k, \mathfrak{B}_k$ such that (4) $\mathfrak{A}_k \equiv_k \mathfrak{B}_k$, $\mathfrak{A}_k \models_l \varphi$, and $\mathfrak{B}_k \models_l \neg\varphi$. We next adjust the models $\mathfrak{A}_k$ so that there is a model $\mathfrak{A}$ for $\mathscr{L}$ which has each $\mathfrak{A}_k$ as a submodel. By taking a subsequence we may assume that all the $\mathfrak{A}_k$ satisfy the same atomic sentences of $\mathscr{L}$. By the Isomorphism Property, we may also assume that all the $\mathfrak{A}_k$ have the same interpretations of the constants of $\mathscr{L}$. Then the union of the models $\mathfrak{A}_k$ is a model $\mathfrak{A}$ for $\mathscr{L}$ which has each $\mathfrak{A}_k$ as a submodel. We may also take the $\mathfrak{A}_k$ so that each set A_k is disjoint from ω. As in the preceding proof, let $\mathfrak{A}''$ be an expansion of $\mathfrak{A}$ with universe $A \cup A'$ and extra relations A', F, F' where A' is the set of finite sequences of elements of A, and the functions are given by $F(\boldsymbol{a}, b) = \boldsymbol{a}b$, $F'(\boldsymbol{a}, \boldsymbol{b}) = \boldsymbol{ab}$. We make similar assumptions for the $\mathfrak{B}_k$ and form the union $\mathfrak{B}$ and expansion $\mathfrak{B}''$. We may now form an expanded model $\mathfrak{C}$ from the sequences of $\mathfrak{A}_k$ and $\mathfrak{B}_k$ of the form

$$\mathfrak{C} = (\mathfrak{A}'', \mathfrak{B}'', R, S, \omega, \leqslant, I),$$

where R and S are relations such that for each $k \in \omega$,

$$A_k = \{a \in A : R(a, k)\}, \ B_k = \{b \in B : S(b, k)\},$$

$$A'_k = \{a \in A' : R(a, k)\}, \ B'_k = \{b \in B' : S(b, k)\},$$

and for each $k \in \omega$, $I(k, \boldsymbol{a}, \boldsymbol{b})$ holds if and only if $\boldsymbol{a} I_k \boldsymbol{b}$. As in the preceding proof, by the Closure, Quantifier, and Relativization Properties there is a sentence $\theta \in l$ which holds in $\mathfrak{C}$ and implies that $\langle \omega, \leqslant \rangle$ is a simple order with immediate successors and predecessors except for the first element, and (4) holds for all $k \in \omega$. By countable compactness, θ has a model

$$\mathfrak{C}\hat{} = (\mathfrak{A}\hat{}, \mathfrak{B}\hat{}, R\hat{}, S\hat{}, \omega\hat{}, \leqslant\hat{}, I\hat{}),$$

such that $\{\omega\hat{}, \leqslant\hat{}\rangle$ has a nonstandard element H. Then in terms of $I\hat{}$, (5) $\mathfrak{A}_H\hat{} \models_l \varphi$, $\mathfrak{B}_H\hat{} \models_l \neg\varphi$, and $\mathfrak{A}_H\hat{} \equiv_H \mathfrak{B}_H\hat{}$. It follows that the relation J between m-tuples given by

$$\langle a_1, \ldots, a_m \rangle J \langle b_1, \ldots, b_m \rangle \text{ iff } \langle a_1, \ldots, a_m \rangle I\hat{}_{H-m} \langle b_1, \ldots, b_m \rangle$$

is a partial isomorphism between $\mathfrak{A}_H\hat{}$ and $\mathfrak{B}_H\hat{}$. But then by Proposition 2.5.3 and the hypothesis that l has Löwenheim number ω, $\mathfrak{A}_H\hat{}$ and $\mathfrak{B}_H\hat{}$ are l-elementarily equivalent, contrary to (5). Therefore (3) holds and the proof is complete. ⊣

Theorem 2.5.4 shows that if we go beyond first order logic, we must give up one of three things: countable compactness, the Downward Löwenheim Skolem Theorem, or the properties of an abstract logic. There are good examples of each possibility. The logic $l(Q_0)$ obtained from first order logic by adding the quantifier "there exist infinitely many" is an abstract logic which has Löwenheim number ω but is not countably compact. Another such logic is the logic $l_{\omega_1,\omega}$ which is like first order logic except that it allows countable conjunctions and disjunctions. The logic $l(Q_1)$ obtained from first order logic by adding the quantifier "there exist uncountably many" is an abstract logic which is countably compact but has Löwenheim number ω_1. There is a logic l^{top} (e.g. see Flum and Ziegler (1980)) which is not an abstract logic in the above sense because it deals with models with topologies, but is countably compact (in fact fully compact) and has Löwenheim number ω.

There are several other results which characterize first order logic in the style of Lindström's Characterization Theorem, and use similar ideas in their proofs. Some of these results are stated as exercises.

Exercises

2.5.1. The *Hanf number* of an abstract logic l is the least cardinal α such that every sentence of l which has a model of power at least α has models of arbitrarily large power. (Thus the Upward Löwenheim Skolem Theorem shows that first order logic has Hanf number ω). Show that every abstract logic l, such that for every finite $\mathscr{L}$ the class $\{\varphi \in l : \mathscr{L}_\varphi \subset \mathscr{L}\}$ is a set, has a Hanf number. This is the upward analogue of Proposition 2.5.2.

2.5.2. Prove that first order logic is the only abstract logic which has Löwenheim number ω and Hanf number ω.

2.5.3. An abstract logic l is said to *pin down* the ordinal α if there is a sentence $\varphi \in l$ and unary and binary relation symbols U and R in $\mathscr{L}_\varphi$ such that φ has a model, and every model of φ is such that $\langle U, R\rangle$ is isomorphic to $\langle \alpha, \leqslant\rangle$. Show that if l is either countably compact or has Hanf number ω, then l does not pin down ω.

2.5.4. Prove that first order logic is the only abstract logic l such that any pair of partially isomorphic models is l-elementarily equivalent, and l does not pin down ω.

2.5.5. A sentence φ in an abstract logic l is said to be *l-valid* if $\mathfrak{A} \models_l \varphi$ for every model $\mathfrak{A}$ whose language contains $\mathscr{L}_\varphi$. Prove that if an abstract logic l pins down ω, then there is a finite $\mathscr{L}$ such that the set of l-valid sentences φ with $\mathscr{L}_\varphi \subset \mathscr{L}$ is not r.e., and in fact is not even arithmetical. *Hint*: You may use the fact that the set of first order sentences which hold in the standard model of arithmetic is not arithmetical.

2.5.6 (Lindström's Second Theorem). Prove that first order logic is the only abstract logic with Löwenheim number ω such that for each finite language $\mathscr{L}$, the set of l-valid sentences φ with $\mathscr{L}_\varphi \subset \mathscr{L}$ is r.e. (or even arithmetical).

2.5.7*. Prove that first order logic is the only abstract logic l such that l has Löwenheim number ω and the Robinson Consistency theory holds for l.

CHAPTER 3

FURTHER MODEL-THEORETIC CONSTRUCTIONS

3.1. Elementary extensions and elementary chains

Given two models $\mathfrak{A}$, $\mathfrak{B}$ for $\mathscr{L}$, we have already defined the two notions of $\mathfrak{A} \equiv \mathfrak{B}$ ($\mathfrak{A}$, $\mathfrak{B}$ are elementarily equivalent) and $\mathfrak{A} \subset \mathfrak{B}$ ($\mathfrak{A}$ is a submodel of $\mathfrak{B}$). Thus the natural combination of these two notions will lead to models which are submodels or extensions of an elementarily equivalent model. For example, the model $\langle \omega \setminus \{0\}, \leqslant \rangle$ is a submodel of $\langle \omega, \leqslant \rangle$, and, since they are isomorphic, they are also equivalent. However, the element 1 is the first element of $\omega \setminus \{0\}$, while it is the second element of ω. An elementary submodel is a far stronger notion, namely a submodel of a given model in which the elements in common shall have exactly the same first-order properties with respect to both models. Let us repeat the definition.

$\mathfrak{A}$ is an *elementary submodel* of $\mathfrak{B}$ iff $\mathfrak{A} \subset \mathfrak{B}$ and for all formulas φ of $\mathscr{L}$ in the variables $x_1, \ldots, x_n$, and all elements $a_1, \ldots, a_n$ in A, we have

$$\mathfrak{A} \models \varphi[a_1 \ldots a_n] \quad \text{iff} \quad \mathfrak{B} \models \varphi[a_1 \ldots a_n].$$

If $\mathfrak{A}$ is an elementary submodel of $\mathfrak{B}$, then $\mathfrak{B}$ is an *elementary extension* of $\mathfrak{A}$. Apart from some simple properties of elementary submodels and extensions, we shall answer in this section the following questions:

(1). How can we tell if a model $\mathfrak{A}$ is (isomorphic to) an elementary submodel of another model $\mathfrak{B}$?

(2). Are there any restrictions on the cardinalities of elementary submodels and extensions of a given model $\mathfrak{A}$?

(3). When can two or more models have a common elementary extension?

(4). Can the notion of elementary extensions be iterated into the transfinite?

We use the notation $\mathfrak{A} \prec \mathfrak{B}$ to denote $\mathfrak{A}$ is an elementary submodel of $\mathfrak{B}$. For convenience, the notation $\mathfrak{B} \succ \mathfrak{A}$ is also used to mean $\mathfrak{A} \prec \mathfrak{B}$.

Let $X \subset A$. We let $(\mathfrak{A}, a)_{a \in X}$, or $\mathfrak{A}_X$, denote the obvious expansion of $\mathfrak{A}$ to the language $\mathcal{L}_X = \mathcal{L} \cup \{c_a : a \in X\}$ with new constants c_a.

Proposition 3.1.1.

(i). *If* $\mathfrak{A} \prec \mathfrak{B}$, *then* $\mathfrak{A} \equiv \mathfrak{B}$.

(ii). $\mathfrak{A} \prec \mathfrak{A}$.

(iii). *If* $\mathfrak{A} \prec \mathfrak{B}$ *and* $\mathfrak{B} \prec \mathfrak{C}$, *then* $\mathfrak{A} \prec \mathfrak{C}$.

(iv). *If* $\mathfrak{A} \prec \mathfrak{C}$, $\mathfrak{B} \prec \mathfrak{C}$ *and* $\mathfrak{A} \subset \mathfrak{B}$, *then* $\mathfrak{A} \prec \mathfrak{B}$.

The proof is a simple exercise.

Proposition 3.1.2. $\mathfrak{A} \prec \mathfrak{B}$ *if and only if* $\mathfrak{A} \subset \mathfrak{B}$ *and for all formulas* $(\exists x)\varphi(xx_1 \ldots x_n)$ *in* $x_1, \ldots, x_n$ *and all* $a_1, \ldots, a_n$ *in* A,

if $\mathfrak{B} \models \exists x\varphi[a_1 \ldots a_n]$, *then there is an* $a \in A$ *such that* $\mathfrak{B} \models \varphi[aa_1 \ldots a_n]$.

Proof. We prove the nontrivial direction by the following induction on formulas φ: Whenever φ is in the variables $x_1, \ldots, x_n$ and $a_1, \ldots, a_n$ in A, then

$$\mathfrak{A} \models \varphi[a_1 \ldots a_n] \quad \text{iff} \quad \mathfrak{B} \models \varphi[a_1 \ldots a_n].$$

The induction is easy to carry out for atomic formulas, and for going through sentential connectives. In the crucial step of going from $\varphi(x_1 \ldots x_n)$ to $(\exists x_1)\varphi(x_2 \ldots x_n)$, we note the following: Given $a_2, \ldots, a_n \in A$, if $\mathfrak{A} \models (\exists x_1)\varphi[a_2 \ldots a_n]$, then there is an $a_1 \in A$ such that $\mathfrak{A} \models \varphi[a_1 \ldots a_n]$. By induction, $\mathfrak{B} \models \varphi[a_1 \ldots a_n]$, whence $\mathfrak{B} \models (\exists x_1)\varphi[a_2 \ldots a_n]$. On the other hand, if $\mathfrak{B} \models (\exists x_1)\varphi[a_2 \ldots a_n]$, then by hypothesis, there is an $a_1 \in A$ such that $\mathfrak{B} \models \varphi[a_1 \ldots a_n]$. So by induction, $\mathfrak{A} \models \varphi[a_1 \ldots a_n]$, and $\mathfrak{A} \models (\exists x_1)\varphi[a_2 \ldots a_n]$. ⊣

We recall that an elementary embedding of $\mathfrak{A}$ into $\mathfrak{B}$ is an isomorphism f of $\mathfrak{A}$ onto an elementary submodel of $\mathfrak{B}$, in symbols $f : \mathfrak{A} \prec \mathfrak{B}$. Recalling the use of the notation $\mathfrak{A} \cong\!\!\subset \mathfrak{B}$ for $\mathfrak{A}$ embedded in $\mathfrak{B}$, we shall use the notation $\mathfrak{A} \precsim \mathfrak{B}$ for $\mathfrak{A}$ elementarily embedded in $\mathfrak{B}$.

Let $\mathcal{L}_A = \mathcal{L} \cup \{c_a : a \in A\}$. The *elementary diagram* of $\mathfrak{A}$ is the theory $\mathrm{Th}(\mathfrak{A}_A)$ of all sentences of $\mathcal{L}_A$ which hold in the model $\mathfrak{A}_A = (\mathfrak{A}, a)_{a \in A}$. (Recall that the diagram of $\mathfrak{A}$ is the set of all atomic and negated atomic sentences of $\mathcal{L}_A$ which hold in $\mathfrak{A}_A$.)

Proposition 3.1.3. *Let* Γ_A *be the elementary diagram of* $\mathfrak{A}$. *Then* $\mathfrak{A} \precsim \mathfrak{B}$ *if and only if some expansion* $\mathfrak{B}'$ *of* $\mathfrak{B}$ *is a model of* Γ_A. *If* $\mathfrak{A} \subset \mathfrak{B}$, *then* $\mathfrak{A} \prec \mathfrak{B}$ *if and only if* $(\mathfrak{B}, a)_{a \in A} \models \Gamma_A$.

PROPOSITION 3.1.4. *Let $\mathfrak{F}$ be any nonempty set of elementarily equivalent models. Then there exists a model $\mathfrak{B}$ such that every model $\mathfrak{A} \in \mathfrak{F}$ is elementarily embedded in $\mathfrak{B}$.*

PROOF. For each $\mathfrak{A} \in \mathfrak{F}$, let Γ_A be its elementary diagram. We first make sure that if $\mathfrak{A} \neq \mathfrak{A}'$, then the sets of constants $\{c_a : a \in A\} \cap \{c_a : a \in A'\} = 0$. Let $\Delta = \bigcup_{\mathfrak{A} \in \mathfrak{F}} \Gamma_A$. We claim that Δ is a consistent set of sentences of $\bigcup_{\mathfrak{A} \in \mathfrak{F}} \mathscr{L}_A$. Suppose $\{\varphi_1, ..., \varphi_n\}$ is a finite subset of Δ. We may suppose that for $i \not\equiv j$, φ_i and φ_j come from different $\mathfrak{A}$'s. Then there are formulas $\varphi'_1, ..., \varphi'_n$ in the variables $x_1, ..., x_m$ and elements $a_{ij} \in A_i$, $1 \leqslant i \leqslant n$, $1 \leqslant j \leqslant m$, such that $\mathfrak{A}_i \in \mathfrak{F}$ and

$$\varphi_i = \varphi'_i(c_{a_{i1}} \dots c_{a_{im}}), \qquad 1 \leqslant i \leqslant n.$$

Then the sentence $(\exists x_1 \dots x_m)\varphi'_1 \wedge (\exists x_1 \dots x_m)\varphi'_2 \wedge \dots \wedge (\exists x_1 \dots x_m)\varphi'_n$ will hold in $\mathfrak{A}_1$, since $\mathfrak{A}_i \equiv \mathfrak{A}_j$, $1 \leqslant i, j \leqslant n$. This shows that $\{\varphi_1, ..., \varphi_n\}$ is consistent. By the compactness theorem, Δ has a model $\mathfrak{B}'$. Let $\mathfrak{B}$ be the reduct of $\mathfrak{B}'$ to the original language $\mathscr{L}$. Then Proposition 3.1.3 shows that each $\mathfrak{A} \in \mathfrak{F}$ is elementarily embedded in $\mathfrak{B}$. ⊣

THEOREM 3.1.5. *Every infinite model $\mathfrak{A}$ has arbitrarily large elementary extensions.*

PROOF. Let Γ be the elementary diagram of $\mathfrak{A}$. By the Löwenheim–Skolem–Tarski theorem, Γ has arbitrarily large models, and the result follows from Proposition 3.1.3. ⊣

Theorem 3.1.5 may be regarded as a strengthening of the Löwenheim–Skolem–Tarski theorem. Curiously, the following very special case of Theorem 3.1.5 is not easy to answer: Does every infinite model $\mathfrak{A}$ have a proper elementary extension of the same power? The answer is simple if $||\mathscr{L}|| \leqslant |A|$ (see Exercise 3.1.6). We shall give some more complete answers later in the book.

Our next theorem is a strengthening of the downward Löwenheim–Skolem–Tarski theorem.

THEOREM 3.1.6. *Let $\mathfrak{A}$ be a model of power α and let $||\mathscr{L}|| \leqslant \beta \leqslant \alpha$. Then $\mathfrak{A}$ has an elementary submodel of power β. Furthermore, given any set $X \subset A$ of power $\leqslant \beta$, $\mathfrak{A}$ has an elementary submodel of power β which contains X.*

Proof. We may assume that X has power β. For each formula $\varphi(xx_1 \dots x_n)$ and each n-tuple $a_1, \dots, a_n \in X$ such that $\mathfrak{A} \models (\exists x)\varphi[a_1 \dots a_n]$, choose an element $b \in A$ such that $\mathfrak{A} \models \varphi[ba_1 \dots a_n]$. Let X_1 be the set X plus all the b's so chosen. Since $|X| = \beta$ and $||\mathscr{L}|| \leqslant \beta$, X_1 has power β. Now repeat the process countably many times, forming a chain

$$X \subset X_1 \subset X_2 \subset \dots.$$

Let $B = \bigcup_{n<\omega} X_n$. B is closed under the functions of $\mathfrak{A}$. Each X_n has power β, so B has power β. Let $\mathfrak{B}$ be the submodel of $\mathfrak{A}$ with universe B. Consider a formula $\varphi(xx_1 \dots x_n)$ and an n-tuple $b_1, \dots, b_n \in B$ such that

$$\mathfrak{A} \models (\exists x)\varphi[b_1 \dots b_n].$$

For some $m < \omega$, we have $b_1, \dots, b_n \in X_m$. Then there exists $b \in X_{m+1}$ such that $\mathfrak{A} \models \varphi[bb_1 \dots b_n]$. Thus $b \in B$, and by Proposition 3.1.2 we have $\mathfrak{B} \prec \mathfrak{A}$. ⊣

As an immediate consequence of the above theorem, we see that if T has a model $\mathfrak{A}$ of power α which omits a set of formulas $\Sigma(x)$, and if $||\mathscr{L}|| \leqslant \beta < \alpha$, then T has a model $\mathfrak{B}$ of power β which omits $\Sigma(x)$. Any $\mathfrak{B} \prec \mathfrak{A}$ of power β will do.

Proposition 3.1.7 (Łoś–Vaught Test). *Suppose that a consistent theory T has only infinite models and T is α-categorical for some infinite cardinal $\alpha \geqslant ||\mathscr{L}||$. Then T is complete.*

Proof. It is sufficient to show that any two models $\mathfrak{A}$ and $\mathfrak{B}$ of T are equivalent. Since T has only infinite models, both $\mathfrak{A}$ and $\mathfrak{B}$ are infinite. Whence by the Löwenheim–Skolem-Tarski theorem (both downward and upward) there are models $\mathfrak{A}'$, $\mathfrak{B}'$ of power α such that $\mathfrak{A} \equiv \mathfrak{A}'$ and $\mathfrak{B} \equiv \mathfrak{B}'$. Since T is α-categorical, $\mathfrak{A}' \cong \mathfrak{B}'$, so $\mathfrak{A}' \equiv \mathfrak{B}'$ and $\mathfrak{A} \equiv \mathfrak{B}$. ⊣

For example, the following theories are categorical in some infinite power and have no finite models, so they are complete.

(1). The theory of dense simply ordered sets without endpoints; ω-categorical.

(2). The theory of atomless Boolean algebras; ω-categorical.

(3). The theory of algebraically closed fields of characteristic zero (or p); ω_1-categorical.

(4). The theory of infinite Abelian groups with all elements of order p; α-categorical for all α.

(5). The theory of countably many unequal constant symbols; ω_1-categorical.

(6). The theory of a one-to-one function of A onto A with no finite cycles; ω_1-categorical.

We now turn to the second main topic of this section, the elementary chain construction. A *chain of models* is an increasing sequence of models

$$\mathfrak{A}_0 \subset \mathfrak{A}_1 \subset \ldots \subset \mathfrak{A}_\beta \subset \ldots, \qquad \beta < \alpha,$$

whose length is an ordinal α. The *union of the chain* is the model $\mathfrak{A} = \bigcup_{\beta<\alpha}\mathfrak{A}_\beta$ which is defined as follows: The universe of $\mathfrak{A}$ is the set $A = \bigcup_{\beta<\alpha}A_\beta$. Each relation R of $\mathfrak{A}$ is the union of the corresponding relations of $\mathfrak{A}_\beta$, $R = \bigcup_{\beta<\alpha}R_\beta$. Similarly, each function G of $\mathfrak{A}$ is the union of the corresponding functions of $\mathfrak{A}_\beta$, $G = \bigcup_{\alpha<\beta}G_\beta$. The models $\mathfrak{A}_\beta$ and $\mathfrak{A}$ all have the same constants.

Here is a simple lemma:

LEMMA 3.1.8. *Given a chain $\mathfrak{A}_\beta$, $\beta<\alpha$, of models, $\bigcup_{\beta<\alpha}\mathfrak{A}_\beta$ is the unique model with universe $\bigcup_{\beta<\alpha}A_\beta$ which contains each $\mathfrak{A}_\beta$ as a submodel.*

When we iterate the notion of an elementary extension we arrive at the notion of an elementary chain. An *elementary chain* is a chain of models

$$\mathfrak{A}_0 \prec \mathfrak{A}_1 \prec \ldots \prec \mathfrak{A}_\beta \prec \ldots, \qquad \beta < \alpha,$$

such that $\mathfrak{A}_\gamma \prec \mathfrak{A}_\beta$ whenever $\gamma < \beta < \alpha$.

For example, if for each $n < \omega$, $\mathfrak{A}_n$ is the algebraically closed field of characteristic zero and transcendence degree n over the rationals, then

$$\mathfrak{A}_0 \prec \mathfrak{A}_1 \prec \ldots$$

is an elementary chain. The union $\mathfrak{A}_\omega$ of this chain is an algebraically closed field of characteristic zero and transcendence degree ω. In fact, we have $\mathfrak{A}_n \prec \mathfrak{A}_\omega$ for each n.

The following theorem is the analogue of 3.1.8 for elementary chains. In spite of its simple character, it is a very important construction.

THEOREM 3.1.9 (Elementary Chain Theorem). *Let $\mathfrak{A}_\xi$, $\xi<\alpha$, be an elementary chain of models. Then $\mathfrak{A}_\xi \prec \bigcup_{\xi<\alpha}\mathfrak{A}_\xi$ for all $\xi<\alpha$.*

PROOF. Let $\mathfrak{A} = \bigcup_{\xi<\alpha}\mathfrak{A}_\xi$. We prove the following assertion by induction on formulas φ: for all formulas φ in $x_1, \ldots, x_n$, all $\xi < \alpha$, and all elements $a_1, \ldots, a_n \in A_\xi$,

$$\mathfrak{A}_\xi \models \varphi[a_1 \ldots a_n] \text{ iff } \mathfrak{A} \models \varphi[a_1 \ldots a_n].$$

The proof is routine for atomic formulas. The induction steps involving sentential connectives are also quite easy. Assume that $\varphi = \exists x_1\psi$ is a formula in $x_2, \ldots, x_n$, $\xi < \alpha$, and $a_2, \ldots, a_n \in A_\xi$. If $\mathfrak{A}_\xi \models \varphi[a_2 \ldots a_n]$, then there is an $a_1 \in A_\xi$ such that $\mathfrak{A}_\xi \models \psi[a_1 \ldots a_n]$. So, by induction, $\mathfrak{A} \models \psi[a_1 \ldots a_n]$ and $\mathfrak{A} \models \varphi[a_2 \ldots a_n]$. On the other hand, if $\mathfrak{A} \models \varphi[a_2 \ldots a_n]$, then for some $\eta < \alpha$ and $a_1 \in A$ we have $a_1, \ldots, a_n \in A_\eta$ and $\mathfrak{A} \models \psi[a_1 \ldots a_n]$. Since $\mathfrak{A}_\xi$, $\xi < \alpha$, is a chain, we may assume that $\xi \leqq \eta$. As $a_1, \ldots, a_n$ all belong to A_η, by induction, $\mathfrak{A}_\eta \models \psi[a_1 \ldots a_n]$, so $\mathfrak{A}_\eta \models \varphi[a_2 \ldots a_n]$. Since $\mathfrak{A}_\xi \prec \mathfrak{A}_\eta$, we have, finally $\mathfrak{A}_\xi \models \varphi[a_2 \ldots a_n]$. ⊣

Here is an example which shows that we cannot replace $\prec$ by $\equiv$ in the elementary chain theorem. Let $\mathfrak{A}_0 = \langle\omega, \leqslant\rangle$ be the natural numbers with the usual ordering. For each n, form $\mathfrak{A}_n$ by adding n new elements to the beginning of the ordering $\langle\omega, \leqslant\rangle$. Then

$$\mathfrak{A}_0 \subset \mathfrak{A}_1 \subset \mathfrak{A}_2 \subset \ldots,$$

and for each n we have $\mathfrak{A}_n \equiv \mathfrak{A}_0$ and, in fact, $\mathfrak{A}_n \cong \mathfrak{A}_0$. However, the union $\mathfrak{A}_\omega = \bigcup_{n<\omega}\mathfrak{A}_n$ is an ordering which has no first element. Therefore $\mathfrak{A}_\omega \not\equiv \mathfrak{A}_0$.

To illustrate the value of elementary chains we shall use them to give a third proof of the Robinson consistency theorem, which was stated as follows:

> Let $\mathscr{L}_1$ and $\mathscr{L}_2$ be two languages and let $\mathscr{L} = \mathscr{L}_1 \cap \mathscr{L}_2$. Suppose T is a complete theory in $\mathscr{L}$ and $T_1 \supset T$, $T_2 \supset T$ are consistent theories in $\mathscr{L}_1$, $\mathscr{L}_2$, respectively. Then $T_1 \cup T_2$ is consistent in the language $\mathscr{L}_1 \cup \mathscr{L}_2$.

In Section 2.2 we proved the Craig interpolation theorem and then used it to obtain the Robinson consistency theorem, and in Section 2.4 we gave a second proof of the Robinson consistency theorem using recursively saturated models. We now give a third proof of the Robinson consistency theorem, which is essentially Robinson's original argument.

For the proof, let $\mathfrak{A}_0$ and $\mathfrak{B}_0$ be models of T_1 and T_2. To simplify notation in the proof, let

$$\mathfrak{A} \equiv_{\mathscr{L}} \mathfrak{B}, \qquad f : \mathfrak{A} \prec_{\mathscr{L}} \mathfrak{B}$$

mean that the $\mathscr{L}$-reducts of $\mathfrak{A}$ and $\mathfrak{B}$ are elementarily equivalent, and that f is an elementarily embedding of $\mathfrak{A}|\mathscr{L}$ into $\mathfrak{B}|\mathscr{L}$.

Since $\mathfrak{A}_0|\mathscr{L}$ and $\mathfrak{B}_0|\mathscr{L}$ are models of the complete theory T, we have $\mathfrak{A}_0 \equiv_{\mathscr{L}} \mathfrak{B}_0$. It follows that the elementary diagram of $\mathfrak{A}_0|\mathscr{L}$ is consistent with the elementary diagram of $\mathfrak{B}_0$. Therefore there are an elementary extension $\mathfrak{B}_1 \succ \mathfrak{B}_0$ and an embedding $f_1 : \mathfrak{A}_0 \prec_{\mathscr{L}} \mathfrak{B}_1$. Passing to the expanded language $\mathscr{L}_{A_0}$, we have

$$(\mathfrak{A}_0, a)_{a \in A_0} \equiv_{\mathscr{L}_{A_0}} (\mathfrak{B}_1, fa)_{a \in A_0}.$$

Repeating the construction in the other direction, we obtain an elementary extension $\mathfrak{A}_1 \succ \mathfrak{A}_0$ and an embedding

$$g_1 : (\mathfrak{B}_1, f_1 a)_{a \in A_0} \prec_{\mathscr{L}_{A_0}} (\mathfrak{A}_1, a)_{a \in A_0}.$$

Then g_1^{-1} is an extension of f_1. Iterating the construction we obtain a tower

$$\begin{array}{ccccccc} \mathfrak{A}_0 & \prec & \mathfrak{A}_1 & \prec & \mathfrak{A}_2 & \prec & \cdots \\ & \searrow^{f_1} & \uparrow g_1 & \searrow^{f_2} & \uparrow g_2 & & \\ \mathfrak{B}_0 & \prec & \mathfrak{B}_1 & \prec & \mathfrak{B}_2 & \prec & \cdots \end{array}$$

such that for each m

$$f_m \subset g_m^{-1} \subset f_{m+1}, \qquad f_m : \mathfrak{A}_{m-1} \prec_{\mathscr{L}} \mathfrak{B}_m, \qquad g_m : \mathfrak{B}_m \prec_{\mathscr{L}} \mathfrak{A}_m.$$

Let $\mathfrak{A} = \bigcup_{m<\omega} \mathfrak{A}_m$, $\mathfrak{B} = \bigcup_{m<\omega} \mathfrak{B}_m$. Then $\mathfrak{A}$ is a model of T_1 and $\mathfrak{B}$ is a model of T_2. Moreover, $\bigcup_{m<\omega} f_m$ is an isomorphism of $\mathfrak{A}|\mathscr{L}$ onto $\mathfrak{B}|\mathscr{L}$. Then $\mathfrak{B}$ is isomorphic to a model $\mathfrak{B}'$ such that $\mathfrak{A}|\mathscr{L} = \mathfrak{B}'|\mathscr{L}$. Piecing $\mathfrak{A}$ and $\mathfrak{B}'$ together, we obtain a model $\mathfrak{C}$ for $\mathscr{L}_1 \cup \mathscr{L}_2$ with $\mathfrak{C}|\mathscr{L}_1 = \mathfrak{A}$, $\mathfrak{C}|\mathscr{L}_2 = \mathfrak{B}'$. Then $\mathfrak{C}$ is a model of $T_1 \cup T_2$. ⊣

A construction which is very similar to the construction of elementary chains is that of a partial elementary chain. This notion lies between the notions of a chain of models and of an elementary chain of models. It is useful in obtaining some results in model theory (see Theorem 3.1.11 and some of the exercises for this section). In order to describe the construction precisely, let us now formally introduce the notions of $\mathbf{\Sigma}_n^0$ and $\mathbf{\Pi}_n^0$ formulas.

Let $\mathscr{L}$ be fixed for the following discussion. A formula φ of $\mathscr{L}$ is a $\mathbf{\Sigma}_0^0 = \mathbf{\Pi}_0^0$ *formula* iff φ contains no quantifiers. Proceeding inductively, a formula φ of $\mathscr{L}$ is a $\mathbf{\Sigma}_{n+1}^0$ (resp. $\mathbf{\Pi}_{n+1}^0$) *formula* iff $\varphi = (\exists x_1 \ldots x_m)\psi$ (resp. $(\forall x_1 \ldots x_m)\psi$), where ψ is a $\mathbf{\Pi}_n^0$ (resp. $\mathbf{\Sigma}_n^0$) formula. Obviously, every prenex formula is either a $\mathbf{\Sigma}_n^0$ or $\mathbf{\Pi}_n^0$ formula for some n. A $\mathbf{\Sigma}_n^0$ (resp. $\mathbf{\Pi}_n^0$)

formula that is a sentence is called a Σ_n^0 (resp. Π_n^0) *sentence*. Existential sentences are Σ_1^0 sentences; universal sentences are Π_1^0 sentences; universal–existential sentences are Π_2^0 sentences, and so forth.

$\mathfrak{B}$ is said to be a Σ_n^0 *extension* of $\mathfrak{A}$ iff for all Σ_n^0 formulas $\varphi(x_1 \dots x_m)$ and all $a_1, \dots, a_m \in A$,

$$\text{if} \quad \mathfrak{A} \models \varphi[a_1 \dots a_m], \quad \text{then} \quad \mathfrak{B} \models \varphi[a_1 \dots a_m].$$

A Σ_n^0*-chain of models* is a chain of models

$$\mathfrak{A}_0 \subset \mathfrak{A}_1 \subset \dots \subset \mathfrak{A}_\beta \subset \dots, \qquad \beta < \alpha,$$

such that for each $\beta < \gamma < \alpha$, $\mathfrak{A}_\gamma$ is a Σ_n^0 extension of $\mathfrak{A}_\beta$. Every extension of a model is a Σ_1^0 extension, whence every chain of models is a Σ_1^0-chain.

The next lemma is the analogue of the elementary chain theorem.

LEMMA 3.1.10 *Let* $\mathfrak{A}_\beta$, $\beta < \alpha$, *be a* Σ_n^0*-chain of models and let* $\mathfrak{A} = \bigcup_{\beta<\alpha} \mathfrak{A}_\beta$. *Then*

(i). $\mathfrak{A}$ *is a* Σ_n^0 *extension of each* $\mathfrak{A}_\beta$.

(ii). *Every* Π_{n+1}^0 *sentence which is true in all* $\mathfrak{A}_\beta$ *is true in* $\mathfrak{A}$.

PROOF. We argue by induction on n. The result already holds for $n = 0$. Assuming the result for $n-1$, we prove it for n. Let

$$\psi = (\exists x_1 \dots x_m)\varphi(x_1 \dots x_m y_1 \dots y_p)$$

be a Σ_n^0 formula, where φ is Π_{n-1}^0. Suppose $b_1, \dots, b_p \in A_\beta$ satisfies ψ in $\mathfrak{A}_\beta$. Then for some $a_1, \dots, a_m \in A_\beta$, we have

$$\mathfrak{A}_\beta \models \varphi[a_1 \dots a_m b_1 \dots b_p].$$

Let $Y = \{a_1, \dots, a_m, b_1, \dots, b_p\}$ and consider the Σ_n^0-chain $\mathfrak{A}_{\beta Y} \subset \mathfrak{A}_{\beta+1\,Y} \subset \dots$ with union $\mathfrak{A}_Y$. The Π_{n-1}^0 sentence $\varphi(a_1 \dots a_m b_1 \dots b_p)$ holds in every model of this chain and by inductive hypothesis also holds in $\mathfrak{A}_Y$. Therefore

$$\mathfrak{A} \models (\exists x_1 \dots x_m)\varphi[b_1 \dots b_p].$$

This proves that $\mathfrak{A}$ is a Σ_n^0 extension of $\mathfrak{A}_\beta$.

To prove (ii), consider a Π_{n+1}^0 sentence $(\forall x_1 \dots x_m)\theta$ which holds in all $\mathfrak{A}_\beta$, where θ is a Σ_n^0 formula. Let $a_1, \dots, a_m \in A$. Then for some $\beta < \alpha$, $a_1, \dots, a_m \in A_\beta$. These elements satisfy θ in $\mathfrak{A}_\beta$, and therefore they satisfy θ in $\mathfrak{A}$. Therefore $(\forall x_1 \dots x_m)\theta$ holds in $\mathfrak{A}$. ⊣

The last theorem of this section is a model theoretic proof of a theorem about predicate logic.

THEOREM 3.1.11. *The following are equivalent* ($n>0$):

(i). φ *is equivalent both to a* Σ^0_{n+1} *sentence and to a* Π^0_{n+1} *sentence.*

(ii). φ *is equivalent to a Boolean combination of* Σ^0_n *sentences.*

PROOF. We leave the simple proof that a Boolean combination of Σ^0_n sentences is equivalent to a Σ^0_{n+1} and a Π^0_{n+1} sentence to the reader. The point is that vacuous quantifiers may always be added to a sentence without disturbing its meaning.

In the other direction, assume (i). We shall first prove that given two models $\mathfrak{A}$ and $\mathfrak{B}$,

(1) if every Σ^0_n sentence holds in $\mathfrak{A}$ if and only if it holds in $\mathfrak{B}$, then $\mathfrak{A} \models \varphi$ if and only if $\mathfrak{B} \models \varphi$.

So let $\mathfrak{A}$, $\mathfrak{B}$ be models for which the hypothesis of (1) holds. We shall construct a Σ^0_n-chain of models

$$\mathfrak{A} = \mathfrak{A}_0 \subset \mathfrak{B}_0 \subset \mathfrak{A}_1 \subset \mathfrak{B}_1 \ldots \subset \mathfrak{A}_k \subset \mathfrak{B}_k \subset \ldots$$

such that

(2) $\mathfrak{A}_k \equiv \mathfrak{A}$ and $\mathfrak{B}_k \equiv \mathfrak{B}$ for all $k \in \omega$.

Suppose we have constructed the finite Σ^0_n-chain

$$\mathfrak{A}_0 \subset \mathfrak{B}_0 \subset \ldots \subset \mathfrak{A}_m \subset \mathfrak{B}_m,$$

so that (2) holds for all $k \leqslant m$. Let T be the collection of all Σ^0_n sentences of $\mathscr{L} \cup \{c_b : b \in B_m\}$ which hold in $\mathfrak{B}_m$. We can show that given any finite subset of T, we may take its conjunction, then existentially quantify out the new constants, and in this way obtain a Σ^0_n sentence ψ of $\mathscr{L}$ which holds in $\mathfrak{B}_m$. Using the hypothesis of (1), this sentence ψ holds in $\mathfrak{A}$. So $T \cup \mathrm{Th}(\mathfrak{A})$ is consistent and has a model $\mathfrak{A}_{m+1}$. It is easy to verify that

$$\mathfrak{A}_0 \subset \mathfrak{B}_1 \subset \ldots \mathfrak{B}_m \subset \mathfrak{A}_{m+1}$$

is still a Σ^0_n-chain. Exactly the same argument applied to $\mathfrak{A}_{m+1}$ will give us the next model $\mathfrak{B}_{m+1}$ in the Σ^0_n-chain. Suppose that φ holds in $\mathfrak{A}$, then φ holds in $\mathfrak{A}_k$ for all k. Since φ is Π^0_{n+1}, by Lemma 3.1.10, φ holds in $\bigcup_{k\in\omega}\mathfrak{A}_k = \bigcup_{k\in\omega}\mathfrak{B}_k$. If φ does not hold in $\mathfrak{B}$, then its negation $\neg\varphi$ is (equivalent to) a Π^0_{n+1} sentence holding in $\mathfrak{B}$. So again by Lemma 3.1.10, $\neg\varphi$ holds in $\bigcup_{k\in\omega}\mathfrak{B}_k$, which is a contradiction. So (1) is proved.

From (1) we now argue as follows: If (ii) is false, then for any finite collection of Σ^0_n sentences, say $\sigma_1, \ldots, \sigma_m$, we can find two models $\mathfrak{A}$ and $\mathfrak{B}$ such that

(3) $\mathfrak{A} \models \varphi$, $\mathfrak{B} \models \neg\varphi$, and $\mathfrak{A} \models \sigma_i$ iff $\mathfrak{B} \models \sigma_i$, $1 \leqslant i \leqslant m$.

This is seen as follows: Consider the 2^m conjunctions $\sigma = \sigma_1' \wedge \ldots \wedge \sigma_m'$, where each σ_i' is either σ_i or $\neg \sigma_i$. Some such conjunction σ must be consistent with both φ and $\neg \varphi$. For otherwise we have $\vdash \sigma \to \varphi$ or $\vdash \sigma \to \neg \varphi$ for all such σ, and propositional logic will show that φ is (equivalent to) a finite disjunction of such σ's, so (ii) would be true. Now from (3), a simple application of the compactness theorem will yield two models $\mathfrak{A}$ and $\mathfrak{B}$ such that $\mathfrak{A}, \mathfrak{B}$ are equivalent with respect to Σ_n^0 sentences, but not equivalent with respect to φ, contradicting (1). ⊣

Theorem 3.1.11 is true even when $n = 0$, provided that $\mathscr{L}$ has constant symbols.

Exercises

3.1.1. Let $\mathfrak{A} \subset \mathfrak{B}$. Suppose that for all elements $a_1, \ldots, a_n \in A$ and $b \in B$, there is an automorphism of $\mathfrak{B}$ onto $\mathfrak{B}$ which leaves fixed $a_1, \ldots, a_n$ but maps b onto some element in A. Then $\mathfrak{A} \prec \mathfrak{B}$.

3.1.2. Let α and β be infinite cardinals such that $\alpha \leqslant \beta$ and α is regular. Show that there is a model $\mathfrak{A}$ of complete arithmetic which has cardinality β and cofinality α.

3.1.3. Let $\mathfrak{A}$ be a ring (see Example 1.4.9), and let $\mathfrak{A}_X$, $\mathfrak{A}_Y$ be polynomial rings over $\mathfrak{A}$ with indeterminates $x \in X$ and $y \in Y$. Show that if $X \subset Y$ and X is infinite, then $\mathfrak{A}_X \prec \mathfrak{A}_Y$.

3.1.4. Let $(0, 1)$ denote the set of all rationals between 0 and 1, and let $[0, 1) = (0, 1) \cup \{0\}$. Let $\mathfrak{A} = \langle (0, 1), \leqslant \rangle$ and $\mathfrak{B} = \langle [0, 1), \leqslant \rangle$. Let $\mathfrak{A}'$ be the union of $\aleph_1$ disjoint copies of $\mathfrak{A}$ and $\aleph_0$ disjoint copies of $\mathfrak{B}$, and let $\mathfrak{B}'$ be the union of $\aleph_0$ disjoint copies of $\mathfrak{A}$ and $\aleph_1$ disjoint copies of $\mathfrak{B}$. (The *union of models* $\langle A_i, R_i \rangle$, $i \in I$, with binary relations R_i, is defined as the model $\langle \bigcup_{i \in I} A_i, \bigcup_{i \in I} R_i \rangle$.) Then

(i). $\mathfrak{A}'$ and $\mathfrak{B}'$ are isomorphically embedded in each other.

(ii). Neither of the models $\mathfrak{A}'$, $\mathfrak{B}'$ is elementarily embedded in the other.

(iii). $\mathfrak{A}' \equiv \mathfrak{B}'$, and hence (from Proposition 3.1.4) $\mathfrak{A}'$ and $\mathfrak{B}'$ are both elementarily embedded in some fixed model $\mathfrak{C}$.

3.1.5. Let $\mathfrak{A}, \mathfrak{B}$ be models for $\mathscr{L}$. The following are equivalent:

(i). There are a model $\mathfrak{C}$ and two elementary embeddings f of $\mathfrak{A}$ into $\mathfrak{C}$ and g of $\mathfrak{B}$ into $\mathfrak{C}$ such that

$$f \upharpoonright A \cap B = g \upharpoonright A \cap B.$$

(ii). $(\mathfrak{A}, a)_{a \in A \cap B} \equiv (\mathfrak{B}, a)_{a \in A \cap B}$.

3.1.6. If $||\mathscr{L}|| \leqslant |A|$, then $\mathfrak{A}$ has a proper elementary extension of the same power.

3.1.7*. Assume the GCH. Regardless of $||\mathscr{L}||$, show that every infinite model $\mathfrak{A}$ has an elementary extension in each power $\beta > |A|$.

3.1.8. A prenex sentence is a *universal–existential sentence*, or simply a $\forall\exists$-sentence, iff all of its universal quantifiers (if any) precede all of its existential quantifiers (if any). Clearly every universal or existential sentence (see Exercise 1.3.5) is a $\forall\exists$-sentence. Prove that if φ is a $\forall\exists$-sentence and $\mathfrak{A}_\xi \models \varphi$ for each member of the chain $\mathfrak{A}_\xi$, $\xi < \alpha$, then $\bigcup_{\xi<\alpha}\mathfrak{A}_\xi \models \varphi$. (This is a special case of Lemma 3.1.10.)

3.1.9. A binary relation D on a set I is said to be *directed* iff for all $i_1, i_2 \in I$, there is an $i_3 \in I$ such that $i_1 D i_3$ and $i_2 D i_3$. A collection of models $\mathfrak{A}_i$, $i \in I$, is *directed* iff there is a directed relation D on I such that for all $i, j \in I$,

$$\text{if} \quad iDj, \quad \text{then} \quad \mathfrak{A}_i \subset \mathfrak{A}_j.$$

Show that a meaningful definition can be given to $\bigcup_{i\in I}\mathfrak{A}_i$. Furthermore, each $\mathfrak{A}_i \subset \bigcup_{i\in I}\mathfrak{A}_i$.

Extend both Exercise 3.1.8 and the elementary chain theorem to directed families of models.

3.1.10*. A class K of models is said to be *closed under unions of well ordered chains* iff every union of a well ordered chain of members of K is again a member of K. Similarly, we can define the notion K *is closed under directed unions* (see Exercise 3.1.9). Prove that a class K of models is closed under unions of well ordered chains iff it is closed under directed unions.

[*Hint*: Use induction on the cardinality of the directed set.]

3.1.11. Let T be a complete theory in a countable language, let $\langle B, < \rangle$ be the full binary tree, and let $\mathfrak{A}_s$, $s \in B$ be a set of models of T such that:

(i). A complete type is realized in both $\mathfrak{A}_s$ and $\mathfrak{A}_t$ if and only if it is realized in $\mathfrak{A}_{s \cap t}$;

(ii). If $s < t$ then there is a complete type which is realized in $\mathfrak{A}_t$ but not in $\mathfrak{A}_s$;

(iii). If $s<t$ then $\mathfrak{A}_s < \mathfrak{A}_t$.

Prove that T has 2^ω pairwise nonisomorphic countable models.

3.1.12. Show that if T has a model of power α which omits a type $\Sigma(x_1 x_2 x_3 \ldots)$ and $||\mathscr{L}|| \leqslant \beta < \alpha$, then T has a model of power β which omits $\Sigma(x_1 x_2 x_3 \ldots)$.

3.1.13. Suppose that T has a model which omits a type $\Sigma(x_1 \ldots x_n)$, and every such model has a proper elementary extension omitting $\Sigma(x_1 \ldots x_n)$. Prove that T has models of arbitrarily large cardinality which omit $\Sigma(x_1 \ldots x_n)$.

3.1.14. Suppose $\mathscr{L}$ has the binary relation $\leqslant$. A model $\mathfrak{A} = \langle A, \leqslant, \ldots \rangle$ for $\mathscr{L}$ is said to be *well ordered* iff $\leqslant$ well orders A. Show that if T has a well ordered model of power α, and $||\mathscr{L}|| \leqslant \beta < \alpha$, then T has a well ordered model of power β.

3.1.15. Suppose that T is finitely axiomatizable and in some infinite power α, T has at most a finite number of nonisomorphic models of power α. Prove that for some $n \in \omega$, every formula φ is equivalent under T to a Σ_n^0 formula ψ.

3.1.16* (The reflection principle). Let φ be a sentence in the language of ZFC. Prove within ZFC: If φ then for every ordinal α there is an ordinal $\beta > \alpha$ such that $\langle R(\beta), \in \rangle \models \varphi$. [*Hint*: See the appendix on set theory for notation. The sentence φ has only finitely many subformulas. Take β to be a limit of a countable increasing sequence of ordinals β_n such that for every subformula $\psi(x_1 \ldots x_m, y)$ of φ and all $a_1, \ldots, a_m$ in $R(\beta_n)$, if $(\exists y)\psi(a_1 \ldots a_m, y)$ then $(\exists y \in R(\beta_{n+1}))\psi(a_1 \ldots a_m, y)$. The proof that $\langle R(\beta), \in \rangle \models \varphi$ is like the proof of Proposition 3.1.2.]

3.2. Applications of elementary chains

In this section we shall give several applications of elementary chains which illustrate their power in model theory.

We start with some examples of 'preservation theorems'. We say that a theory T is *preserved under submodels* iff any submodel of a model of T is a model of T. T is said to be *preserved under unions of chains* iff the union of any chain of models of T is a model of T. T is *preserved under homomorphisms* iff any homomorphic image of a model of T is a model of T.

TABLE 3.2.1

Theory	Preserved under		
	Submodels	$\bigcup$ chains	Homomorphisms
Partial order	yes	yes	no
Dense simple order	no	yes	no
Boolean algebras	yes	yes	yes
Atomic Boolean algebras	no	no	no
Groups	no	yes	yes
Groups with symbol for $-x$	yes	yes	yes
Commutative rings	no	yes	yes
Integral domains	no	yes	no
Fields	no	yes	no
Fields and one-element rings	no	yes	yes
Algebraically closed fields	no	yes	no
$(\exists x)(\forall y)R(x, y)$	no	no	yes
Peano arithmetic	no	no	no
ZF	no	no	no

The preservation theorems are results which characterize those theories which are preserved under submodels, union of chains, etc. There are several results of this type. We shall give three here and a few more in Chapters 5 and 6.

The status of some familiar theories with respect to preservation is shown in Table 3.2.1. It is a rather remarkable fact that these preservation phenomena can be explained just by the syntactical form of the axioms.

We shall use the following general lemma.

LEMMA 3.2.1. *Let T be a consistent theory in $\mathscr{L}$ and let Δ be a set of sentences of $\mathscr{L}$ which is closed under finite disjunctions. Then the following are equivalent*:

(i). *T has a set of axioms Γ such that $\Gamma \subset \Delta$.*

(ii). *If $\mathfrak{A}$ is a model of T and every sentence $\delta \in \Delta$ which holds in $\mathfrak{A}$ holds in $\mathfrak{B}$, then $\mathfrak{B}$ is a model of T.*

PROOF. It is obvious that (i) implies (ii). Assume (ii). Let Γ be the set of all sentences φ of $\mathscr{L}$ such that $T \models \varphi$ and $\varphi \in \Delta$. Then $T \models \Gamma$. We show that $\Gamma \models T$, whence Γ is a set of axioms of T. Let $\mathfrak{B}$ be a model of Γ. Let Σ be the set of sentences

$$\Sigma = \{\neg \delta : \mathfrak{B} \models \neg \delta, \delta \in \Delta\}.$$

We show that $\Sigma \cup T$ is consistent. T is assumed to be consistent. Suppose $T \cup \Sigma$ is inconsistent. Then there exist $\neg \delta_1, \ldots, \neg \delta_n \in \Sigma$, $n \geqslant 1$, such that

$T \models \neg(\neg\delta_1 \wedge \ldots \wedge \neg\delta_n)$. Hence

$$T \models \delta_1 \vee \ldots \vee \delta_n.$$

Since Δ is closed under $\vee$, we have $\delta_1 \vee \ldots \vee \delta_n \in \Delta$, whence $\delta_1 \vee \ldots \vee \delta_n \in \Gamma$ and

$$\mathfrak{B} \models \delta_1 \vee \ldots \vee \delta_n.$$

But this contradicts $\mathfrak{B} \models \neg\delta_1, \ldots, \mathfrak{B} \models \neg\delta_n$. Hence $\Sigma \cup T$ is consistent, and has a model $\mathfrak{A}$. Then every sentence $\delta \in \Delta$ which holds in $\mathfrak{A}$ holds in $\mathfrak{B}$, because otherwise $(\neg\delta) \in \Sigma$. ⊣

Our first and simplest preservation theorem below uses only the compactness theorem, not elementary chains.

Theorem 3.2.2. *A theory T is preserved under submodels if and only if T has a set of universal* (*i.e.*, Π_1^0) *axioms.*

Proof. It is easy to check that if T has a set of universal axioms, then T is preserved under submodels. Assume T is preserved under submodels. We apply Lemma 3.2.1, with Δ the set of all sentences equivalent to Π_1^0 sentences. Consider two models $\mathfrak{A} \models T$ and $\mathfrak{B}$ such that every universal sentence which holds in $\mathfrak{A}$ holds in $\mathfrak{B}$. Then every existential sentence true in $\mathfrak{B}$ is true in $\mathfrak{A}$. Consider the theory $T' = T \cup \Delta_{\mathfrak{B}}$ in the language $\mathscr{L}_B$, where $\Delta_{\mathfrak{B}}$ is the diagram of $\mathfrak{B}$. The theory T' is consistent, because, for any finite set

$$\{\theta_1(b_1 \ldots b_n), \ldots, \theta_m(b_1 \ldots b_n)\} \subset \Delta_{\mathfrak{B}},$$

the existential sentence

$$(\exists x_1 \ldots x_n)(\theta_1(x_1 \ldots x_n) \wedge \ldots \wedge \theta_m(x_1 \ldots x_n))$$

holds in $\mathfrak{B}$, hence in $\mathfrak{A}$, and hence is consistent with T. Let $\mathfrak{C}' = \mathfrak{C}_B$ be a model of T'. Then $\mathfrak{B}$ is a submodel of $\mathfrak{C}$, and $\mathfrak{C}$ is a model of T. Hence $\mathfrak{B}$ is a model of T. By Lemma 3.2.1, T has a set of universal axioms. ⊣

The next preservation theorem does use elementary chains.

Theorem 3.2.3. *A theory T is preserved under unions of chains if and only if T has a set of universal–existential* (i.e., Π_2^0) *axioms.*

Proof. ($\Leftarrow$). For the easy direction, let Γ be a set of Π_2^0 axioms for T and let $\mathfrak{A}_\beta$, $\beta < \alpha$, be a chain of models of T with union $\mathfrak{A} = \bigcup_{\beta<\alpha}\mathfrak{A}_\beta$. Consider a sentence

$$\gamma = (\forall x_1 \ldots x_m \exists y_1 \ldots y_n)\psi(x_1 \ldots x_m y_1 \ldots y_n)$$

in Γ, where ψ has no quantifiers. Each $\mathfrak{A}_\beta$ is a model of γ. Let $a_1, \ldots, a_m \in A$. For some $\beta < \alpha$, we have $a_1, \ldots, a_m \in A_\beta$. Then there exist $b_1, \ldots, b_n \in A_\beta$ such that

$$\mathfrak{A}_\beta \models \psi[a_1 \ldots a_m b_1 \ldots b_n].$$

It follows that

$$\mathfrak{A} \models \psi[a_1 \ldots a_m b_1 \ldots b_n],$$

whence $\mathfrak{A}$ satisfies γ. Hence $\mathfrak{A}$ is a model of Γ, and thus of T.

($\Rightarrow$). Now assume T is preserved under unions of chains. Let Δ be the set of all sentences which are logically equivalent to $\mathbf{\Pi}_2^0$ sentences. Then Δ is closed under finite disjunctions. Suppose $\mathfrak{A}$ and $\mathfrak{B}$ are models such that $\mathfrak{A} \models T$ and every $\mathbf{\Pi}_2^0$ sentence true in $\mathfrak{A}$ is true in $\mathfrak{B}$. Then every $\mathbf{\Sigma}_2^0$ sentence true in $\mathfrak{B}$ is true in $\mathfrak{A}$. We shall prove the following:

(1) There are models $\mathfrak{A}'$ and $\mathfrak{B}'$ such that $\mathfrak{B} \subset \mathfrak{A}'$, $\mathfrak{A}' \subset \mathfrak{B}'$, $\mathfrak{B} \prec \mathfrak{B}'$ and $\mathfrak{A} \equiv \mathfrak{A}'$.

To prove (1), we add a new constant c_b for each $b \in B$, forming the model $\mathfrak{B}_B$ and language $\mathscr{L}_B$. Let T_1 be the complete theory of $\mathfrak{A}$ in $\mathscr{L}$, and T_2 the set of all universal sentences of $\mathscr{L}_B$ which hold in $\mathfrak{B}_B$. Then $T_1 \cup T_2$ is consistent, because up to equivalence T_2 is closed under conjunction, and for any $\psi(c_{b_1} \ldots c_{b_n})$ in T_2, the $\mathbf{\Sigma}_2^0$ sentence $(\exists y_1 \ldots y_n)\psi$ holds in $\mathfrak{B}$, and hence in $\mathfrak{A}$. Let

$$\mathfrak{A}'_B = (\mathfrak{A}', b)_{b \in B}$$

be a model of $T_1 \cup T_2$. Then $\mathfrak{A} \equiv \mathfrak{A}'$ and $\mathfrak{B} \subset \mathfrak{A}'$. Moreover, every universal sentence of $\mathscr{L}_B$ true in $\mathfrak{B}_B$ is true in $\mathfrak{A}'_B$, whence every existential sentence true in $\mathfrak{A}'_B$ is true in $\mathfrak{B}_B$. Now expand the language further by adding a new constant c_a for each $a \in A' \setminus B$. Then the theory

$$D(\mathfrak{A}'_B) \cup \mathrm{Th}(\mathfrak{B}_B),$$

the diagram of $\mathfrak{A}'_B$ plus the elementary diagram of $\mathfrak{B}_B$, is consistent. It thus has a model $(\mathfrak{B}', a)_{a \in A'}$. Then $\mathfrak{A}' \subset \mathfrak{B}'$ and $\mathfrak{B} \prec \mathfrak{B}'$, so (1) is proved.

Iterating (1) we obtain a chain

$$\mathfrak{B} = \mathfrak{B}_0 \subset \mathfrak{A}_1 \subset \mathfrak{B}_1 \subset \mathfrak{A}_2 \subset \ldots$$

such that for each n,

$$\mathfrak{A}_n \equiv \mathfrak{A} \quad \text{and} \quad \mathfrak{B}_n \prec \mathfrak{B}_{n+1}.$$

Let $\mathfrak{A}_\omega$ be the union of this chain. Since each $\mathfrak{A}_n$ is a model of T, $\mathfrak{A}_\omega$ is a model of T. However, $\mathfrak{A}_\omega$ is also the union of the elementary chain

$$\mathfrak{B}_0 \prec \mathfrak{B}_1 \prec \mathfrak{B}_2 \prec \ldots.$$

Therefore by the elementary chain theorem, $\mathfrak{B}_0 \prec \mathfrak{A}_\omega$, so $\mathfrak{B} = \mathfrak{B}_0$ is also a model of T. By Lemma 3.2.1, we conclude that T has a $\mathbf{\Pi}_2^0$ set of axioms. ⊣

A formula φ is said to be *positive* iff it is built up from atomic formulas using only the connectives $\wedge$, $\vee$ and the quantifiers $\forall$, $\exists$.

THEOREM 3.2.4. *A consistent theory T is preserved under homomorphisms if and only if T has a set of positive axioms.*

PROOF. ( ). Let us first check the easy direction. A formula $\varphi(x_1 \ldots x_n)$ is *preserved under homomorphisms* iff for any homomorphism f of a model $\mathfrak{A}$ onto a model $\mathfrak{B}$ and all $a_1, \ldots, a_n \in A$, if $\mathfrak{A} \models \varphi[a_1 \ldots a_n]$, then $\mathfrak{B} \models \varphi[fa_1 \ldots fa_n]$. It suffices to prove that all positive formulas φ are preserved under homomorphisms. This is done by induction on the complexity of φ. It obviously holds when φ is atomic and passes over $\vee$ and $\wedge$. Assume $\varphi(xx_1 \ldots x_n)$ is preserved under homomorphisms. First, suppose $\mathfrak{A} \models (\exists x)\varphi[a_1 \ldots a_n]$. Then for some $a \in A$, we have $\mathfrak{A} \models \varphi[aa_1 \ldots a_n]$, $\mathfrak{B} \models \varphi[fafa_1 \ldots fa_n]$, and hence $\mathfrak{B} \models (\exists x)\varphi[fa_1 \ldots fa_n]$. Now assume $\mathfrak{A} \models (\forall x)\varphi[a_1 \ldots a_n]$. Let b be any element of B. Then for some $a \in A$ we have $b = fa$. Thus $\mathfrak{A} \models \varphi[aa_1 \ldots a_n]$, $\mathfrak{B} \models \varphi[bfa_1 \ldots fa_n]$, and $\mathfrak{B} \models (\forall x)\varphi[fa_1 \ldots fa_n]$. Therefore $(\exists x)\varphi$ and $(\forall x)\varphi$ are preserved under homomorphisms.

($\Rightarrow$). We now prove the hard direction. Let $\mathfrak{A}$ pos $\mathfrak{B}$ mean that every positive sentence true in $\mathfrak{A}$ is true in $\mathfrak{B}$. We show first that:

(1) If $\mathfrak{A}$ pos $\mathfrak{B}$, then there exist an elementary extension $\mathfrak{B}' \succ \mathfrak{B}$ and an embedding $f : \mathfrak{A} \to \mathfrak{B}'$ such that $(\mathfrak{A}, a)_{a \in A}$ pos $(\mathfrak{B}', fa)_{a \in A}$.

To prove (1), we add new constants c_a, $a \in A$, and d_b, $b \in B$. Let T_1 be the set of all positive sentences of $\mathscr{L} \cup \{c_a : a \in A\}$ true in $\mathfrak{A}_A$. Let T_2 be the set of all sentences of $\mathscr{L} \cup \{d_b : b \in B\}$ true in $\mathfrak{B}_B$. Then $T_1 \cup T_2$ is consistent because $\mathfrak{A}$ pos $\mathfrak{B}$. Let $(\mathfrak{B}', a', b)_{a \in A, b \in B}$ be a model of $T_1 \cup T_2$; T_2 shows that $\mathfrak{B}' \succ \mathfrak{B}$. Let f be the embedding $f(a) = a'$; T_1 shows that $(\mathfrak{A}, a)_{a \in A}$ pos $(\mathfrak{B}', fa)_{a \in A}$.

A similar argument can be used to prove the following dual of (1):

(2) If $\mathfrak{A}$ pos $\mathfrak{B}$, then there exist an elementary extension $\mathfrak{A}' \succ \mathfrak{A}$ and a mapping $g : B \to A'$ such that $(\mathfrak{A}', gb)_{b \in B}$ pos $(\mathfrak{B}, b)_{b \in B}$.

Now let $\mathfrak{A}_0$, $\mathfrak{B}_0$ be models such that $\mathfrak{A}_0 \models T$ and $\mathfrak{A}_0$ pos $\mathfrak{B}_0$. Iterating (1) and (2), we obtain a tower

$$\begin{array}{ccccccc}
\mathfrak{A}_0 & \prec & \mathfrak{A}_1 & \prec & \mathfrak{A}_2 & \prec & \cdots \\
 & \searrow^{f_0} & \uparrow g_1 & \searrow^{f_1} & \uparrow g_2 & & \\
\mathfrak{B}_0 & \prec & \mathfrak{B}_1 & \prec & \mathfrak{B}_2 & \prec & \cdots
\end{array}$$

such that

$$(\mathfrak{A}_0, a)_{a\in A_0} \text{ pos } (\mathfrak{B}_1, f_0 a)_{a\in A_0},$$

$$(\mathfrak{A}_1, a, g_1 b)_{a\in A_0, b\in B_1} \text{ pos } (\mathfrak{B}_1, f_0 a, b)_{a\in A_0, b\in B_1},$$

$$(\mathfrak{A}_1, a, g_1 b, a')_{a\in A_0, b\in B_1, a'\in A_1} \text{ pos } (\mathfrak{B}_2, f_0 a, b, f_1 a')_{a\in A_0, b\in B_1, a'\in A_1},$$

and so forth. It follows that for each n, f_n is a homomorphism of $\mathfrak{A}_n$ into $\mathfrak{B}_{n+1}$, and

$$f_n \subset f_{n+1}, \qquad g_{n+1}^{-1} \subset f_{n+1}.$$

Let $\mathfrak{A}_\omega = \bigcup_{n<\omega}\mathfrak{A}_n$, $\mathfrak{B}_\omega = \bigcup_{n<\omega}\mathfrak{B}_n$ be the unions of the elementary chains and $f_\omega = \bigcup_{n<\omega} f_n$. Then f_ω is a homomorphism of $\mathfrak{A}_\omega$ into $\mathfrak{B}_\omega$. f_ω maps A_ω onto B_ω because $g_n^{-1} \subset f_\omega$ for each n. By the elementary chain theorem, $\mathfrak{A}_\omega \succ \mathfrak{A}_0$, $\mathfrak{B}_\omega \succ \mathfrak{B}_0$, whence $\mathfrak{A}_\omega$ is a model of T. But T is preserved under homomorphisms, so $\mathfrak{B}_\omega \models T$ and hence $\mathfrak{B}_0 \models T$. Then, by Lemma 3.2.1, T has a set of positive axioms. ⊣

These three preservation theorems explain the phenomena in Table 3.2.1. The usual formulation of the axioms happens to be universal for those examples which are preserved under submodels, Π_2^0 for those which are closed under unions of chains, and positive for those which are closed under homomorphisms. Try writing down the axioms and see.

Each preservation theorem also has a version for single sentences.

COROLLARY 3.2.5. *A sentence is preserved under* (a) *submodels,* (b) *unions of chains,* (c) *homomorphisms, if and only if it is logically equivalent to a sentence which is* (a) *universal,* (b) *universal existential,* (c) *positive or logically false, respectively.*

PROOF. We prove (a) as an illustration. Every universal sentence is preserved under submodels. Suppose a sentence φ is preserved under submodels. Then the theory $T = \{\varphi\}$ has a set Γ of universal axioms. We have $\Gamma \models \varphi$, so by the compactness theorem, there are sentences $\gamma_1, \ldots, \gamma_n \in \Gamma$ such that $\gamma_1 \wedge \ldots \wedge \gamma_n \models \varphi$. There is a universal sentence γ equivalent to $\gamma_1 \wedge \ldots \wedge \gamma_n$. Then $\gamma \models \varphi$. But since $\Gamma \models \gamma$, we have $\varphi \models \gamma$, whence φ is logically equivalent to γ. ⊣

We shall now strike out in a different direction, working towards a generalization of the Löwenheim–Skolem–Tarski theorem which involves two cardinals. For the remainder of this section we shall work with countable languages.

Recall from Chapter 2 that a countably saturated model $\mathfrak{A}$ is a countable model which is ω-saturated, or, in other words, for all finite $X \subset A$, the model $(\mathfrak{A}, a)_{a\in X}$ realizes all types $\Sigma(x)$ in its theory. We also proved the important Theorem 2.3.7, one direction of which states that if a complete theory T (which is countable) has at most a countable number of types, then T has a countably saturated model. We shall give below another proof of this result using elementary chains.

PROPOSITION 3.2.6. *Let T be a complete theory in a countable language $\mathscr{L}$. Suppose that for each n, T has only countably many types in n variables. Then T has a countably saturated model.*

PROOF. Let $\mathfrak{A}$ be any countable model. An elementary extension $\mathfrak{B} \succ \mathfrak{A}$ is said to be *countably saturated over* $\mathfrak{A}$ iff $\mathfrak{B}$ is countable and for every finite subset $Y \subset A$, every type $\Gamma(x)$ in $\mathrm{Th}(\mathfrak{A}_Y)$ is realized in $\mathfrak{B}_Y$.

Now let $\mathfrak{A}_0$ be a countable model of T. We shall find an elementary extension $\mathfrak{A}_1 \succ \mathfrak{A}_0$ which is countably saturated over $\mathfrak{A}_0$.

Let $X \subset A_0$ be a finite subset of A_0 and let $\Sigma(x)$ be any type in x of the theory of $(\mathfrak{A}_0, a)_{a\in X}$. Since there are only a countable number of types of T, there are only a countable number of types $\Sigma(x)$ of the theory of $(\mathfrak{A}_0, a)_{a\in X}$. Since there are only a countable number of finite subsets X of A_0 altogether, the number of such types $\Sigma(x)$ is at most countable. Let $c_{\Sigma X}$ be a new constant symbol for each such $X \subset A$ and Σ. Let

$T_0 =$ (elementary diagram of $\mathfrak{A}_0$ in $\mathscr{L} \cup \{c_a : a \in A_0\}) \cup \{\Sigma(c_{\Sigma X}) : X$ a finite subset of A_0, and $\Sigma(x)$ a type of $(\mathfrak{A}_0, a)_{a\in X}\}$.

A standard argument will show that every finite subset of T_0 is consistent. So T_0 is consistent in the countable language $\mathscr{L} \cup \{c_a : a \in A_0\} \cup \{c_{\Sigma X}\}$. Let $\mathfrak{B}_1$ be any countable model of T_0, and let $\mathfrak{A}_1$ be the reduct of $\mathfrak{B}_1$ to $\mathscr{L}$. We may suppose that $\mathfrak{A}_1$ is an elementary extension of $\mathfrak{A}_0$. Then $\mathfrak{A}_1$ is countably saturated over $\mathfrak{A}_0$. We now repeat the procedure on the model $\mathfrak{A}_1$, to obtain a countable elementary extension $\mathfrak{A}_2$ of $\mathfrak{A}_1$ which is countably saturated over $\mathfrak{A}_1$, etc.

Consider the elementary chain

$$\mathfrak{A}_0 \prec \mathfrak{A}_1 \prec \mathfrak{A}_2 \ldots,$$

and let $\mathfrak{A} = \bigcup_{n\in\omega} \mathfrak{A}_n$. $\mathfrak{A}$ is still a countable model of T. We claim that $\mathfrak{A}$ is ω-saturated. Let X be a finite subset of $A = \bigcup_{n\in\omega} A_n$. There exists an n such that $X \subset A_n$. Let $\Sigma(x)$ be any type of the theory of $(\mathfrak{A}, a)_{a\in X}$. Then because $(\mathfrak{A}_{n,a}, a)_{a\in X} \prec (\mathfrak{A}, a)_{a\in X}$, $\Sigma(x)$ is a type of the theory of $(\mathfrak{A}_{n,a}, a)_{a\in X}$. Whence some $b \in A_{n+1}$ realizes Σ in $(\mathfrak{A}_{n+1}, a)_{a\in X}$. It follows that b realizes Σ in $(\mathfrak{A}, a)_{a\in X}$. $\dashv$

Note that the above construction will yield a countably saturated elementary extension of any countable model of T provided T has only a countable number of types. The construction breaks down in the case where T has a noncountable number of types. It turns out that practically the same construction will always yield a countable elementary extension which satisfies a somewhat weaker, but nevertheless useful, property.

Our next application is concerned with an analogue of the Löwenheim–Skolem–Tarski theorem. Recall that (for a countable $\mathscr{L}$), if a theory T has an infinite model, then T has an infinite model in each power. Thus every theory T will fail to distinguish between infinite cardinals. We can now ask the question whether a theory T will distinguish between pairs of infinite cardinals, in the following sense: Let the language $\mathscr{L}$ have a 1-placed relation symbol U. Let $\mathfrak{A}$ be a model for $\mathscr{L}$, and let V be the interpretation of U in $\mathfrak{A}$. Thus we may display $\mathfrak{A} = \langle A, V \ldots \rangle$. $\mathfrak{A}$ is said to be an (α, β)-model iff $|A| = \alpha$, $|V| = \beta$. We say that a theory T *admits the pair of cardinals* (α, β) iff T has an (α, β)-model. The question is:

If T admits (α, β), then what other pairs (α', β') must T admit?

The following proposition sums up all the simple properties that we can prove with the machinery available to us at present. First some notation: Let $\aleph_n(\alpha)$ be defined by induction on n as follows: $\aleph_0(\alpha) = \alpha$, $\aleph_{n+1}(\alpha) = [\aleph_n(\alpha)]^+$. Similarly, let $\beth_n(\alpha)$ be defined by $\beth_0(\alpha) = \alpha$, and $\beth_{n+1}(\alpha) = 2^{\beth_n(\alpha)}$.

PROPOSITION 3.2.7. *Let T be a theory in a countable language $\mathscr{L}$, and let α, β, γ range over infinite cardinals. Then*:

(i). *If T admits (α, β), then T admits (γ, β) for all γ such that $\alpha \geqslant \gamma \geqslant \beta$.*

(ii). *If T admits (α, β), then T admits all (γ, γ).*

(iii). *For each $n \in \omega$, there is a theory T such that T admits every $(\beth_n(\alpha), \alpha)$ and T does not admit any $(\beth_n(\alpha)^+, \alpha)$.*

(iv). *For each $n \in \omega$, there is a theory T such that T admits every $(\aleph_n(\alpha), \alpha)$ and T does not admit any $(\aleph_{n+1}(\alpha), \alpha)$.*

PROOF. (i). Suppose that $\mathfrak{A} = \langle A, V \ldots \rangle$ is a model of T with $|A| = \alpha$ and $|V| = \beta$. Let γ be such that $\alpha \geqslant \gamma \geqslant \beta$. Let X be any subset of A of cardinality γ and such that $V \subset X$. $\mathfrak{A}$ has an elementary submodel $\mathfrak{B} = \langle B, W \ldots \rangle$ such that $B \supset X$ and $|B| = \gamma$. Since $W = V \cap B = V$, we have $|W| = \beta$.

(ii). By (i) we already know that T admits (β, β). Say $\mathfrak{A} = \langle A, V \ldots \rangle$ is a model of T with $|A| = |V| = \beta$. Let $\mathscr{L}' = \mathscr{L} \cup \{F\}$, where F is a new 1-placed function symbol. Let T' be the theory in $\mathscr{L}'$ consisting of all the sentences of T together with a single sentence which says: F is a one-to-one mapping of the universe A onto V. Then T' is consistent and has an infinite model, namely the expansion $\mathfrak{A}' = (\mathfrak{A}, G)$, where G is any one-to-one function of A onto V. By the Löwenheim–Skolem–Tarski theorem, T' has a model in every infinite power γ. Obviously, the $\mathscr{L}$-reducts of these models of T' are (γ, γ)-models of T.

(iii). We shall illustrate the proof with the case $n = 1$. Suppose $\mathscr{L}$ has a 2-placed relation symbol E, as well as the 1-placed relation symbol U. Consider the following sentences of $\mathscr{L}$:

$$(\forall xy)(x \equiv y \leftrightarrow (\forall z)(E(zx) \leftrightarrow E(zy))),$$
$$(\forall x)(\neg U(x) \to (\forall y)(E(yx) \to U(y))).$$

Intuitively, these two sentences say that every element x not in U determines a subset U_x of U by the set of all y such that $E(yx)$, and, if $x \neq y$, then $U_x \neq U_y$. A moment's thought will show that if the interpretation of U in a model has cardinal α, then the cardinal of the model is at most 2^α. Whence the theory T given by the above two sentences will admit $(\beth_1(\alpha), \alpha)$ for all α, but will not admit any $(\beth_1(\alpha)^+, \alpha)$. An iteration of this idea will prove (iii).

(iv). We leave (iv) as an exercise for the reader. Its proof is slightly more subtle than the proof of (iii). ⊣

We shall need the following simple result about elementary chains of countably homogeneous models.

PROPOSITION 3.2.8.

(i). *The union of any countable elementary chain of countably homogeneous models is countably homogeneous.*

(ii). *A type in finitely many variables is realized in the union of an elementary chain if and only if it is realized in some element of the chain.*

(iii). *The union of any countable elementary chain of pairwise isomorphic countably homogeneous models is isomorphic to each member of the chain.*

PROOF. (i) and (ii) are straightforward and are left as an exercise. To prove (iii), let $\mathfrak{A}_\alpha$, $\alpha < \beta$ be such a chain and let $\mathfrak{A}$ be the union. Then by (i), $\mathfrak{A}$ is countably homogeneous. By (ii), since all the $\mathfrak{A}_\alpha$ are isomorphic, they realize exactly the same types as $\mathfrak{A}$. Then by Exercise 2.4.8, $\mathfrak{A} \cong \mathfrak{A}_\alpha$. ⊣

THEOREM 3.2.9. *If a countable theory T admits (α, β) with $\alpha > \beta \geqslant \omega$, then T admits (ω_1, ω).*

PROOF. We shall use the notion of a recursively saturated model introduced in Section 2.4. For simplicity we give the proof only for recursive languages $\mathscr{L}$. However, the proof can be readily extended to arbitrary countable languages by using the notion of a recursively saturated model relative to $\mathscr{L}$, as explained at the end of Section 2.4.

Let $\mathfrak{A} = \langle A, V, \dots \rangle$ be a model of T with $|A| = \alpha$, $|V| = \beta$. By Theorem 3.1.6, we find an elementary submodel $\mathfrak{B}$ of $\mathfrak{A}$ such that $B \supset V$ and $|B| = \beta$. Let T' be the theory of the model $(\mathfrak{A}, B)$. By Theorem 2.4.1, T' has a countable recursively saturated model $(\mathfrak{A}_0, B_0)$. Let $\mathfrak{B}_0$ be the submodel of $\mathfrak{A}_0$ with universe B_0. Then $\mathfrak{B}_0$ is a proper elementary submodel of $\mathfrak{A}_0$, and in both $\mathfrak{A}_0$ and $\mathfrak{B}_0$ the interpretation of the unary predicate symbol U is the same. By Exercise 2.4.16, the model pair $(\mathfrak{A}_0, \mathfrak{B}_0)$ is recursively saturated. Then by Theorem 2.4.5, $\mathfrak{A}_0$ and $\mathfrak{B}_0$ are partially isomorphic, and since they are countable they are isomorphic. Let us call a proper elementary extension $\mathfrak{C} > \mathfrak{C}'$ *neat* if $\mathfrak{C}$ is isomorphic to $\mathfrak{C}'$ and the interpretation of U is the same in $\mathfrak{C}$ as in $\mathfrak{C}'$. Then $\mathfrak{A}_0$ is a neat extension of $\mathfrak{B}_0$. Hence any model isomorphic to $\mathfrak{B}_0$ has a neat extension, and in particular any neat extension of $\mathfrak{B}_0$ has a neat extension. Since the model pair is recursively saturated, $\mathfrak{B}_0$ is recursively saturated, and by Proposition 2.4.3, $\mathfrak{B}_0$ is countably homogeneous. By Proposition 3.2.8 (ii), the union of any countable chain of neat extensions of $\mathfrak{B}_0$ is again a neat extension. Therefore there is an uncountable chain $\mathfrak{B}_\xi$, $\xi < \omega_1$ of neat elementary extensions of $\mathfrak{B}_0$. The union $\mathfrak{C}$ of this chain has power ω_1 but still has the same interpretation of U than $\mathfrak{B}_0$ has, and is thus a model of T of type (ω_1, ω) as required. ⊣

The proof above gives an extra fact which we state as a corollary.

COROLLARY 3.2.10. *Let T be a countable theory which has a pair of models*

$$\mathfrak{A} = \langle A, V, \dots \rangle, \qquad \mathfrak{B} = \langle B, V, \dots \rangle$$

such that $\mathfrak{B}$ is a proper elementary submodel of $\mathfrak{A}$, but V is infinite and is the same set in $\mathfrak{B}$ as in $\mathfrak{A}$. Then T admits (ω_1, ω).

The analogue of the downward Löwenheim–Skolem–Tarski theorem for two cardinals is the following statement, known as Chang's conjecture:

Every model of type (α, β), $\omega \leqslant \beta < \alpha$, has an elementary submodel of type (ω_1, ω).

We shall see in Chapter 7 that this statement cannot be proved in ZFC, and can be disproved if we assume the axiom of constructibility. The next theorem lies between Theorem 3.2.9 and Chang's conjecture. Its proof depends on the omitting-types theorem.

THEOREM 3.2.11. *Let* $\mathfrak{A} = \langle A, V \ldots \rangle$ *be a model such that* $\omega \leqslant |V| < |A|$. *Then there are two models* $\mathfrak{B} = \langle B, W \ldots \rangle$ *and* $\mathfrak{C} = \langle C, W \ldots \rangle$ *such that* $\mathfrak{B} \prec \mathfrak{A}$, $|B| = \omega$, $\mathfrak{B} \prec \mathfrak{C}$ *and* $|C| = \omega_1$.

(Note that the set W remains the same in $\mathfrak{B}$ and $\mathfrak{C}$.)

PROOF. By the downward Löwenheim–Skolem–Tarski theorem we may always assume that $|A| = |V|^+$, so $|A|$ is a regular infinite cardinal. Let $\leqslant$ be any well ordering of the set A of type $|A|$, and we may assume that the set $V \subset A$ under the well ordering $\leqslant$ is an initial segment of type $|V|$. Let $a_1, \ldots, a_n \in A$. We note that in the model $(\mathfrak{A}, \leqslant)$, if there are arbitrarily large $a \in A$ such that for some $b \in V$, $\mathfrak{A} \models \psi[baa_1 \ldots a_n]$ holds, then there is a fixed $b \in V$ such that there are arbitrarily large $a \in A$ such that $\mathfrak{A} \models \psi[baa_1 \ldots a_n]$. This is because $|A|$ is regular. In other words, the following sentence holds in $(\mathfrak{A}, \leqslant)$ (where $\psi(xyv_1 \ldots v_n)$ is an arbitrary formula of $\mathscr{L}$):

$$(1) \quad (\forall v_1 \ldots v_n)[(\forall z \exists yx)(z \leqslant y \wedge U(x) \wedge \psi(xyv_1 \ldots v_n)) \rightarrow (\exists x \forall z \exists y)(z \leqslant y \wedge U(x) \wedge \psi(xyv_1 \ldots v_n))].$$

We now prove the main step of the theorem:

(2) every countable model $(\mathfrak{B}_0, \leqslant_0) \equiv (\mathfrak{A}, \leqslant)$ has a countable proper elementary extension $(\mathfrak{B}_1, \leqslant_1)$ such that the interpretations of U in $\mathfrak{B}_0$ and in $\mathfrak{B}_1$ are the same.

So let $\mathfrak{B}_0 = \langle B_0, W \ldots \rangle$ be a model such that $(\mathfrak{B}_0, \leqslant_0) \equiv (\mathfrak{A}, \leqslant)$. Note that $\leqslant_0$ need not be a well ordering relation on B_0. Let $\mathscr{L}'$ be the expansion of $\mathscr{L}$ by adding $\{\leqslant, c\}$ and $\{c_b : b \in B_0\}$. Let T consist of the elementary diagram of $(\mathfrak{B}_0, \leqslant_0)$ plus the collection of all sentences $\{c_b < c : b \in B_0\}$. Let $\Sigma(x)$ be the set of formulas $\{U(x)\} \cup \{x \not\equiv c_b : b \in W\}$. We now look for a model of T which omits Σ. First, note that T is a consistent theory in a countable language, because, given any finite subset of T, it is possible to find an interpretation of c in the model $(\mathfrak{B}_0, \leqslant_0)$ which lies beyond all elements $b \in B_0$ such that c_b occurs in this finite set. We next verify that

T locally omits Σ.

Let $\mathfrak{B}_0^*$ be the expanded model $(\mathfrak{B}_0, \leqslant_0, b)_{b \in B_0}$. We first observe that a formula $\psi(c)$ of $\mathscr{L}'$ is consistent with T if and only if $\psi(y)$ is satisfied by arbitrarily large elements of $\mathfrak{B}_0^*$, that is,

$$\mathfrak{B}_0^* \models (\forall x \exists y)(x \leqslant y \wedge \psi(y)).$$

Suppose $(\exists x)\theta(xc)$ is consistent with T. Then either

(3) $\qquad (\exists x)(\theta(xc) \wedge \neg U(x))$ is consistent with T

or

(4) $\qquad (\exists x)(\theta(xc) \wedge U(x))$ is consistent with T.

If (3) holds, then we have found a formula $\sigma(x)$ in Σ such that $\theta(xc) \wedge \neg \sigma(x)$ is consistent with T. So let us assume that (4) holds. Then

$$\mathfrak{B}_0^* \models (\forall z \exists yx)(z \leqslant y \wedge U(x) \wedge \theta(xy)).$$

Using the equivalence of $(\mathfrak{B}_0, \leqslant_0)$ and $(\mathfrak{A}, \leqslant)$ and (1), we see that

$$\mathfrak{B}_0^* \models (\exists x \forall z \exists y)(z \leqslant y \wedge U(x) \wedge \theta(xy)).$$

Then, for some $b \in W$,

$$\mathfrak{B}_0^* \models (\forall z \exists y)(z \leqslant y \wedge U(c_b) \wedge \theta(c_b y)),$$

whence

$$\mathfrak{B}_0^* \models (\forall z \exists y)(z \leqslant y \wedge (\exists x)(\theta(xy) \wedge x \equiv c_b)).$$

This shows that $(\exists x)(\theta(xc) \wedge x \equiv c_b)$ is consistent with T. But $x \not\equiv c_b$ belongs to Σ, so we have proved that T locally omits Σ.

By the omitting types theorem, T has a countable model $\mathfrak{B}'$ omitting Σ. The reduct of $\mathfrak{B}'$ to the language of $(\mathfrak{B}_0, \leqslant_0)$ is a model

$$(\mathfrak{B}_0', \leqslant_0') \succ (\mathfrak{B}_0, \leqslant_0)$$

such that there is an element $b \in B_0' \setminus B_0$, and the interpretation of U in both $\mathfrak{B}_0$ and $\mathfrak{B}_0'$ is the set W. So (2) is proved. We now construct an ω_1-termed elementary chain of countable models $(\mathfrak{B}_\xi, \leqslant_\xi)$, $\xi < \omega_1$, such that each $(\mathfrak{B}_\xi, \leqslant_\xi) \equiv (\mathfrak{A}, \leqslant)$, $B_{\xi+1} \setminus B_\xi \neq 0$, and the interpretation of U in each $\mathfrak{B}_\xi$ is W. The model $\mathfrak{B} = \bigcup_{\xi < \omega_1} \mathfrak{B}_\xi$ is a model of T with $|B| = \omega_1$ and $|W| = \omega$. ⊣

Here is an example which shows that $\mathfrak{B}$ cannot be an arbitrary countable elementary submodel of $\mathfrak{A}$ in Theorem 3.2.11; that is, it cannot be strengthened to:

> If $\mathfrak{A} = \langle A, V, \ldots \rangle$ is a model of type (α, β), where $\omega \leqslant \beta < \alpha$, then every countable elementary submodel $\mathfrak{B} = \langle B, W, \ldots \rangle$ of $\mathfrak{A}$ has an elementary extension $\mathfrak{C} = \langle C, W, \ldots \rangle$ of power ω_1 with the same set W.

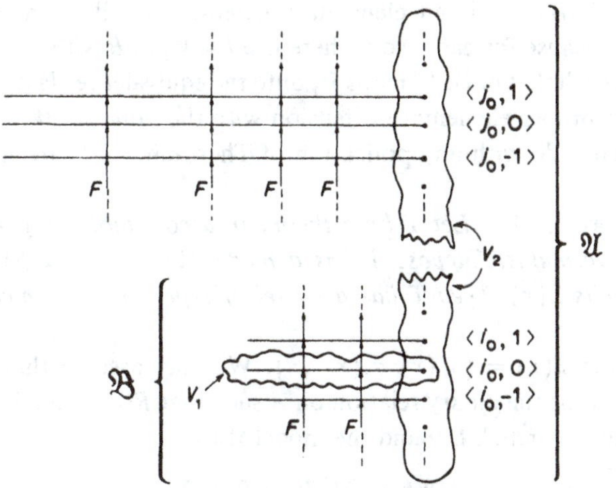

Let $\mathfrak{A} = \langle A, V, V_1, V_2, E, F\rangle$ be constructed as follows: We first draw a picture and then explain the details. Let Z be the set of all integers, I a set of power ω, and J a set of power ω_1, which is disjoint from I. Let

$$A = (I\times Z) \cup (J \times Z).$$

E is the equivalence relation over A whose classes are

$$I\times\{n\}, \qquad J\times\{n\}, \qquad n\in Z.$$

Thus E has ω countable classes and ω uncountable classes. V_1 is the equivalence class $I\times\{0\}$. Fix an element $i_0\in I$ and $j_0\in J$. V_2 is the set

$$V_2 = (\{i_0\}\times Z) \cup (\{j_0\}\times Z),$$

which contains exactly one element from each class of E. V is the countable set $V = V_1 \cup V_2$. Finally, F is the function given by

$$F(\langle i, n\rangle) = \langle i, n+1\rangle, \qquad i\in I\cup J, \qquad n\in Z.$$

F maps A one-to-one onto A, and is an automorphism of the model $\langle A, V_2, E\rangle$.

Let $\mathfrak{B}$ be the submodel of $\mathfrak{A}$ with universe $B = I\times Z$. Then $\mathfrak{B}$ is countable. Using the compactness and elementary chain theorems we can see that both $\mathfrak{A}_B$ and $\mathfrak{B}_B$ have elementary extensions $\mathfrak{A}'_B$ and $\mathfrak{B}'_B$ with ω_1 equivalence classes each of power ω_1, and $\mathfrak{A}'_B \cong \mathfrak{B}'_B$. It follows that $\mathfrak{B} \prec \mathfrak{A}$.

The set corresponding to V in $\mathfrak{B}$ is

$$W = V_1 \cup (V_2 \cap B).$$

If $\mathfrak{C} = \langle C, W, \ldots\rangle$ is an elementary extension of $\mathfrak{B}$, then we must have $\mathfrak{C} = \mathfrak{B}$, because for each $c \in C$ there is a $b \in V_2 \cap B$ with cEb, and there is an $n \in Z$ such that in $\mathfrak{B}$, F^n maps V_1 onto the equivalence class of b. Therefore $\mathfrak{B}$ has no proper elementary extension with the same set W.

We conclude with an application of Theorem 3.2.11 to omitting types.

COROLLARY 3.2.12. *Let T be a theory in a countable language and $\Sigma(x)$ a set of formulas. Suppose T has a model $\mathfrak{A}$ of type (α, β), $\omega \leqslant \beta < \alpha$, which omits $\Sigma(x)$. Then T has a model of type (ω_1, ω) which omits $\Sigma(x)$.*

PROOF. Let $\Sigma(x) = \{\sigma_0(x), \sigma_1(x), \ldots\}$. We may assume that ω is a subset of V. Let R be the binary relation on A such that $R(n, x)$ holds if and only if $n \in \omega$ and $\mathfrak{A} \models \sigma_n[x]$. Expand the model $\mathfrak{A}$ to

$$\mathfrak{A}^* = (\mathfrak{A}, R, \omega, 0, 1, 2, \ldots).$$

Then since $\mathfrak{A}$ omits $\Sigma(x)$,

$$(1) \qquad \mathfrak{A}^* \models \neg (\exists x \forall u)(\omega(u) \rightarrow R(u, x)).$$

But for each n,

$$(2) \qquad \mathfrak{A}^* \models (\forall x)(R(n, x) \leftrightarrow \sigma_n(x)).$$

Now let $\mathfrak{B}^* \prec \mathfrak{A}^*$ and $\mathfrak{C}^* \succ \mathfrak{B}^*$ with $\mathfrak{B}^*$ countable, $\mathfrak{C}^*$ of type (ω_1, ω), and U having the same interpretation in $\mathfrak{B}^*$ as in $\mathfrak{C}^*$. Then in $\mathfrak{B}^*$ and $\mathfrak{C}^*$,

$$\omega = \{0, 1, 2, \ldots\}.$$

Moreover, $\mathfrak{C}^* \equiv \mathfrak{A}^*$. It then follows from (1) and (2) that $\mathfrak{C}^*$ omits $\Sigma(x)$. Thus the reduct $\mathfrak{C}$ of $\mathfrak{C}^*$ to $\mathscr{L}$ is a model of T of type (ω_1, ω) which omits $\Sigma(x)$. ⊣

Finally, we give an example to show that the following stronger form of Corollary 3.2.12 is false.

> If $\mathfrak{A}$ is a countable model which omits $\Sigma(x)$ and $\mathfrak{A}$ has a proper elementary extension omitting $\Sigma(x)$, then $\mathfrak{A}$ has an uncountable elementary extension omitting $\Sigma(x)$.

EXAMPLE 3.2.13. Let

$$\mathfrak{A} = \langle A, V, F, a\rangle_{a \in V},$$

where V is a countable set and F is a binary function mapping $V \times V$

one-to-one onto $A \setminus V$. We represent F by a ternary relation because it is undefined outside $V \times V$. Th($\mathfrak{A}$) is ω_1-categorical. Let $\mathfrak{B}$ be the extension

$$\mathfrak{B} = \langle B, W, G, a \rangle_{a \in V}$$

of $\mathfrak{A}$, where $W = V \cup \{b_0\}$ has one extra element and G maps $W \times W$ one-to-one onto $B \setminus W$. Then $\mathfrak{B}$ is a proper elementary extension of $\mathfrak{A}$ because they have isomorphic elementary extensions of power ω_1. Both $\mathfrak{A}$ and $\mathfrak{B}$ omit the set of formulas $\Sigma(x)$ which says that x is outside U and x is the image of a pair of distinct elements which are not constants $a \in V$. Formally,

$$\Sigma(x) = \{\neg (\exists y)(F(ya) \equiv x) : a \in V\} \cup \{\neg (\exists z)(F(az) \equiv x) : a \in V\} \\ \cup \{\neg U(x)\} \cup \{(\exists y \exists z)(y \not\equiv z \wedge F(yz) \equiv x)\}.$$

However, any uncountable elementary extension of $\mathfrak{A}$ realizes $\Sigma(x)$.

This example was suggested independently by Gregory, MacKenzie and Morley.

Exercises

3.2.1. Prove that a theory T is preserved under extensions if and only if T has a set of existential axioms.

3.2.2. Prove that a consistent theory T is preserved under both homomorphisms and submodels if and only if T has a set of axioms which are both universal and positive.

3.2.3. Use the preservation theorems to prove that every theory which is preserved under submodels is preserved under unions of chains.

3.2.4*. Let T_0 be any theory. A theory T is said to be preserved under submodels relative to T_0 iff for any two models $\mathfrak{A}$, $\mathfrak{B}$ of T_0, if $\mathfrak{A} \models T$ and $\mathfrak{B} \subset \mathfrak{A}$, then $\mathfrak{B} \models T$. Prove that T is preserved under submodels relative to T_0 iff $T \cup T_0$ has a set of axioms of the form $\Gamma \cup T_0$, where Γ is a set of universal sentences. Prove similar relativized preservation theorems for unions of chains, homomorphisms and extensions.

3.2.5. Use Exercise 2.4.19 on recursively saturated models to give another proof of the preservation theorem for submodels, Theorem 3.2.2.

3.2.6. Use Exercise 2.4.20 on recursively saturated models to give another proof of the preservation theorem for homomorphic images, Theorem 3.2.4.

3.2.7. Let $\mathfrak{A}_0$ be a countable model and let $\mathfrak{B}_0$ be an elementary submodel of $\mathfrak{A}_0$. Then the model $(\mathfrak{A}_0, B_0)$ has an elementary extension $(\mathfrak{A}, B)$ such that $\mathfrak{A}$ is countably homogeneous and $\mathfrak{A} \cong \mathfrak{B}$ where $\mathfrak{B}$ is the submodel of $\mathfrak{A}$ with universe B.

3.2.8. Prove Proposition 3.2.8 (i) and (ii) on countably homogeneous models.

3.2.9. Prove that every ω-homogeneous model has a countably homogeneous elementary submodel.

3.2.10. If $\mathfrak{A}$ is countably homogeneous, $\mathfrak{B} \equiv \mathfrak{A}$, $\mathfrak{B}$ is countable, and every type $\Gamma(x_1 \dots x_n)$ realized in $\mathfrak{B}$ is realized in $\mathfrak{A}$, then $\mathfrak{B}$ is elementarily embedded in $\mathfrak{A}$.

3.2.11. Give the proof of Theorem 3.2.9 for arbitrary, not necessarily recursive, countable languages $\mathscr{L}$.

3.2.12*. Let $\mathfrak{A} = \langle \omega_1, <, R_1, R_2, \dots \rangle$ be a model in a countable language. Prove that $\mathfrak{A}$ is the union of an elementary chain $\mathfrak{A}_\alpha$, $\alpha < \omega_1$, of countable models such that each A_α is an initial segment of ω_1, i.e., a countable ordinal.

3.2.13*. Let T be a theory in a countable language. Prove that if every finite subset of T admits (ω_1, ω), then T admits (ω_1, ω).

3.2.14. Prove that every countable model of ZF has an uncountable elementary end extension. Similarly for Peano arithmetic. (See pp. 82, 91).

3.2.15. Let T be the theory with axioms stating that:

$$\leqslant \text{ is a linear ordering,}$$

$$(\forall xy)[y \leqslant x \rightarrow (\exists z)(U(z) \wedge F(x, z) = y)],$$

that is, $F(x, z)$ maps U onto the set $\{y : y \leqslant x\}$. Prove that T admits (α, β), where $\omega \leqslant \beta \leqslant \alpha$, if and only if either $\alpha = \beta$ or $\alpha = \beta^+$.

3.2.16*. Suppose that $\mathscr{L}$ has at least the 1-placed relation symbols $U_1, U_2, \dots, U_n, \dots$. Let T be a theory of $\mathscr{L}$ with a model $\mathfrak{A} = \langle A, V_1 V_2 \dots \rangle$

such that $\omega \leqslant |V_n| < |A|$ for all n. Prove that T has a model $\mathfrak{B} = \langle B, W_1 W_2 \ldots \rangle$ such that $|B| = \omega_1$ and $|W_n| = \omega$ for all n.

3.2.17*. Show that the corresponding generalization of Theorem 3.2.11 to the situation described in Exercise 3.2.16 does not hold. That is, there is no countable elementary submodel $\mathfrak{B}$ of $\mathfrak{A}$ having an elementary extension $\mathfrak{C}$ of power ω_1 such that the interpretation of each U_n is the same in $\mathfrak{B}$ and in $\mathfrak{C}$.

3.2.18*. Find a countable model $\mathfrak{A} = \langle A, V \ldots \rangle$, $V \subset A$, which has a countable proper elementary extension $\mathfrak{B} = \langle B, V \ldots \rangle$ with the same V, but the extension $\mathfrak{B}$ has no such extension. Compare this with Example 3.2.16.

3.2.19*. Let $\langle A, R \ldots \rangle \subset \langle B, S \ldots \rangle$ and R, S be binary relations. $\langle B, S \ldots \rangle$ is said to be an *end extension* of $\langle A, R \ldots \rangle$ iff $B \neq A$ and for all $a \in A$, $b \in B \setminus A$, we have aSb and not bSa. Prove that every countable model $\langle A, R \ldots \rangle$ having an elementary end extension has an elementary end extension of power ω_1. (Assume that $\mathscr{L}$ is countable. This is not the same notion as for models of ZF or Peano arithmetic.)

3.2.20*. In a recursive language, let T be a complete theory whose models are infinite. Show that T has a recursively saturated model of power ω_1 which has only countably many complete types in finitely many variables. [*Hint*: Use the method of proof of the two-cardinal Theorem 3.2.9.]

3.3. Skolem functions and indiscernibles

In this section there is less emphasis on countable languages. We shall combine the method of constructing models by Skolem functions together with another important notion, namely, that of indiscernible elements of a model. We shall first study Skolem functions. Next we prove a combinatorial result known as Ramsey's theorem. Then we apply it to obtain models generated from indiscernible elements, and finally give several applications of this notion. We shall meet this particular type of construction again later on in the book.

Given a language $\mathscr{L}$, we expand $\mathscr{L}$ to a new language $\mathscr{L}^*$ by adding new function symbols. Let F be a mapping from the set of all formulas of the form $\psi = (\exists x)\varphi$ of $\mathscr{L}$ to a list of new function symbols F_ψ. We assume that F is one-to-one and that if ψ has exactly n free variables, then F_ψ is an n-placed (Skolem) function symbol. We call the expansion $\mathscr{L} \cup \{F_\psi : \psi = (\exists x)\varphi$ a formula of $\mathscr{L}\}$ a *Skolem expansion* of $\mathscr{L}$, and we denote it by $\mathscr{L}^*$. Evidently $||\mathscr{L}^*|| = ||\mathscr{L}||$. The *Skolem theory* $\Sigma_{\mathscr{L}}$ of the language $\mathscr{L}$ in the language $\mathscr{L}^*$ has the following sentences of $\mathscr{L}^*$ as axioms:

Let $\psi = (\exists x)\varphi$ be any formula of $\mathscr{L}$ and suppose that ψ has exactly the free variables $x_1, \ldots, x_n$. Let $y_1, \ldots, y_n$ be variables not occurring in ψ. Then the sentence

$$(\forall y_1 \ldots y_n)(\psi(y_1 \ldots y_n) \to \varphi(F_\psi(y_1 \ldots y_n)y_1 \ldots y_n))$$

is an axiom of $\Sigma_{\mathscr{L}}$. (Note that $\varphi(F_\psi(y_1 \ldots y_n)y_1 \ldots y_n)$ is obtained from $\varphi(xx_1 \ldots x_n)$ by replacing all free occurrences of x in φ by the term $F_\psi(y_1 \ldots y_n)$, and all free occurrences of x_i by y_i.)

Let $\mathfrak{A}$ be a model for $\mathscr{L}$. An expansion $\mathfrak{A}^*$ of $\mathfrak{A}$ to the language $\mathscr{L}^*$ is a *Skolem expansion* of $\mathfrak{A}$ iff $\mathfrak{A}^* \models \Sigma_{\mathscr{L}}$. If T is a theory in $\mathscr{L}$, then the *Skolem expansion* of T, denoted by T^*, is the theory T^* with the set of axioms $T \cup \Sigma_{\mathscr{L}}$.

PROPOSITION 3.3.1.

(i). *Every model* $\mathfrak{A}$ *for* $\mathscr{L}$ *has a Skolem expansion* $\mathfrak{A}^*$.

(ii). *If* T *is a consistent theory in* $\mathscr{L}$, *then its Skolem expansion* T^* *is a consistent theory in* $\mathscr{L}^*$.

(iii). *Let* $\mathfrak{A}, \mathfrak{B}$ *be models for* $\mathscr{L}$, *let* $\mathfrak{B}^*$ *be a Skolem expansion of* $\mathfrak{B}$, *and let* $\mathfrak{A}^*$ *be an expansion of* $\mathfrak{A}$ *to* $\mathscr{L}^*$. *If* $\mathfrak{A}^* \subset \mathfrak{B}^*$, *then* $\mathfrak{A} \prec \mathfrak{B}$.

PROOF. (i). Let $\mathfrak{A}$ be a model for $\mathscr{L}$. Let $\psi = (\exists x)\varphi$ and suppose that ψ has exactly the free variables $x_1, \ldots, x_n$. We shall define the interpretation G_ψ of F_ψ in $\mathfrak{A}$ as follows. First well-order A. For any elements $a_1, \ldots, a_n \in A$,

if $\mathfrak{A} \models \psi[a_1 \ldots a_n]$, then let $G_\psi(a_1 \ldots a_n)$ be the first element a of A such that $\mathfrak{A} \models \varphi[aa_1 \ldots a_n]$;

and

if not $\mathfrak{A} \models \psi[a_1 \ldots a_n]$, then let $G_\psi(a_1 \ldots a_n)$ be arbitrary.

It is a simple matter to check that the expansion $\mathfrak{A}^* = (\mathfrak{A}, \{G_\psi : \psi = (\exists x)\varphi$ a formula of $\mathscr{L}\})$ is a Skolem expansion of $\mathfrak{A}$.

The proof of (ii) follows immediately from (i), and the proof of (iii) follows from Proposition 3.1.2. ⊣

Let $\mathfrak{A}^*$ be a Skolem expansion of $\mathfrak{A}$, and let $X \subset A$. The *Skolem hull of* X in $\mathfrak{A}^*$ is the smallest set Y such that $X \subset Y \subset A$, Y contains all the constants in $\mathfrak{A}$, and Y is closed under all functions in $\mathfrak{A}^*$. We let $H(X)$ denote the Skolem hull of X, and $\mathfrak{H}(X)$ the corresponding submodel of $\mathfrak{A}$ determined by the set $H(X)$.

PROPOSITION 3.3.2. *Let $\mathfrak{A}^*$ be a Skolem expansion of $\mathfrak{A}$, and let $X \subset A$. Then the Skolem hull $\mathfrak{H}(X)$ is an elementary submodel of $\mathfrak{A}$. Furthermore,* $|H(X)| \leqslant |X| \cup \|\mathscr{L}\|$.

PROOF. If we let $\mathfrak{H}(X)^*$ denote the model determined by $H(X)$ in the model $\mathfrak{A}^*$, then it is evident that $\mathfrak{H}(X)^* \subset \mathfrak{A}^*$. Since $\mathfrak{H}(X)^*$ is an expansion of $\mathfrak{H}(X)$ to $\mathscr{L}^*$, the result follows from Proposition 3.3.1(iii). ⊣

A theory T in $\mathscr{L}$ has *built-in Skolem functions* iff for every formula $\psi = (\exists x)\varphi$ with exactly the variables $x_1, \dots, x_n$ free, there is an n-placed term t_ψ of $\mathscr{L}$ such that

$$T \vdash (\forall y_1 \dots y_n)(\psi(y_1 \dots y_n) \rightarrow \varphi(t_\psi(y_1 \dots y_n)y_1 \dots y_n)).$$

(The variables $y_1, \dots, y_n$ do not occur in ψ or in t_ψ, and $\psi(y_1 \dots y_n)$, $\varphi(t_\psi(y_1 \dots y_n)y_1 \dots y_n)$ are the obvious formulas obtained from $\psi(x_1 \dots x_n)$ and $\varphi(xx_1 \dots x_n)$.)

PROPOSITION 3.3.3. *If a theory T has built-in Skolem functions, then T is model complete, i.e., whenever $\mathfrak{A}$, $\mathfrak{B}$ are two models of T and $\mathfrak{A} \subset \mathfrak{B}$, then* $\mathfrak{A} \prec \mathfrak{B}$.

PROOF. Since $\mathfrak{A}$ is a submodel of $\mathfrak{B}$, $\mathfrak{A}$ is closed under all terms t_ψ of $\mathscr{L}$. Thus the result follows from Proposition 3.1.2. ⊣

PROPOSITION 3.3.4. *Let T be a theory in $\mathscr{L}$. Then there are an expansion $\overline{\mathscr{L}}$ of $\mathscr{L}$ and an extension $\bar{T}$ of T ($\bar{T}$ a theory in $\overline{\mathscr{L}}$) such that $\bar{T}$ has built-in Skolem functions. Furthermore, every model of T has an expansion which is a model of* $\bar{T}$.

PROOF. Starting from the language $\mathscr{L} = \mathscr{L}_0$, we define an increasing sequence of expansions $\mathscr{L}_n$ by letting $\mathscr{L}_{n+1} = (\mathscr{L}_n)^*$. Note that for each n, the Skolem theory $\Sigma_{\mathscr{L}_n}$ is a set of sentences of $\mathscr{L}_{n+1}$. Let $\overline{\mathscr{L}} = \bigcup_n \mathscr{L}_n$ and let $\bar{T}$ have the set of axioms $T \cup \bigcup_n \Sigma_{\mathscr{L}_n}$. Because every formula of $\overline{\mathscr{L}}$

involves at most a finite number of symbols, we see that $\bar{T}$ has built-in Skolem functions. An induction based on Proposition 3.3.1 will prove that every model of T has an expansion which is a model of $\bar{T}$. ⊣

In order to simplify the argument for Ramsey's theorem, we shall first give a small glimpse into Section 4.1 by introducing the notion of an ultrafilter on a countable set. This notion is not strictly necessary for the argument, but it will make the ideas much clearer. The student may, if he wishes, go on ahead and read about filters and ultrafilters in Section 4.1; however, the next few paragraphs are sufficient for an understanding of the material in this section alone.

Let I be a nonempty set. A collection D of subsets of I is called an *ultrafilter over I* iff for all $X, Y \subset I$:

(a) $$X \in D \text{ and } X \subset Y \text{ implies } Y \in D;$$

(b) $$X, Y \in D \text{ implies } X \cap Y \in D;$$

(c) $$X \in D \text{ if and only if } I \setminus X \notin D.$$

D is called a *nonprincipal* ultrafilter iff, in addition to (a)–(c):

(d) $$\text{For all } i \in I, \{i\} \notin D.$$

From (a)–(d), we can draw the following immediate simple properties of nonprincipal ultrafilters D over I. From (a)–(c), it follows that $D \neq 0$, and $I \in D$ and $0 \notin D$. From (b), it follows that I must be infinite, for if $I = \{i_1, ..., i_n\}$, then by (d) and (c) we have

$$I \setminus \{i_j\} \in D \quad \text{for} \quad 1 \leqslant j \leqslant n.$$

If we now use (b) a finite number of times, we see that $0 \in D$, which is a contradiction.

A collection E of subsets of I has the *finite intersection property* iff no finite intersection of elements of E is empty. Clearly, an ultrafilter D over I has the finite intersection property. The following proposition characterizes those families with the finite intersection property which are ultrafilters over I.

PROPOSITION 3.3.5. *Let I be a nonempty set. Then the following are equivalent:*

(i). *D is an ultrafilter over I.*

(ii). *D has the finite intersection property, and is maximal with respect to this property, i.e. if E is any collection of subsets of I having the finite intersection property, then D is not a proper subset of E.*

(iii). *D is nonempty, $0 \notin D$, D satisfies (a) and (b) above, and whenever $X \cup Y \in D$, then either $X \in D$ or $Y \in D$.*

PROOF. (i) $\Rightarrow$ (ii). If D is a proper subset of E, then some $X \subset I$ will be such that $X \in E$ and $X \notin D$. It follows that $I \setminus X \in D$, so $I \setminus X \in E$, and $(I \setminus X) \cap X = 0$. This is a contradiction of the fact that E has the finite intersection property.

(ii) $\Rightarrow$ (iii). Clearly $0 \notin D$. Also $D \neq 0$, for the set $\{I\}$ has the finite intersection property. Let

$$E = \{Y \subset I: \text{ for some } \quad X \in D, X \subset Y\}.$$

E evidently includes D and has the finite intersection property. So $D = E$, whence D satisfies (a). Similarly, if we let

$$E = \{Z \subset I: \text{ for some } \quad X, Y \in D, Z = X \cap Y\},$$

then again $D = E$. So D satisfies (b).

Finally, suppose that $X \cup Y \in D$. We claim that one of the two sets $D \cup \{X\}$, $D \cup \{Y\}$ has the finite intersection property. This will show that either $X \in D$ or $Y \in D$. Suppose, on the contrary, that there are X', $Y' \in D$ such that $X' \cap X = 0$ and $Y' \cap Y = 0$. Then

$$0 = (X' \cap X) \cup (Y' \cap Y) = (X' \cup Y') \cap (X' \cup Y) \cap (X \cup Y') \cap (X \cup Y).$$

Each of the terms on the right belongs to D, and hence their intersection ($= 0$) belongs to D, which is a contradiction.

(iii) $\Rightarrow$ (i). Since $I \in D$, one and exactly one of X, $I \setminus X$ belongs to D. So (c) holds. ⊣

PROPOSITION 3.3.6. *There is a nonprincipal ultrafilter over any infinite set I.*

PROOF. We shall use Proposition 3.3.5 (ii). Let

$$E = \{I \setminus X : X \text{ is a finite subset of } I\}.$$

Then because I is infinite, E has the finite intersection property. Given any chain of families of subsets of I, E_j, $j \in J$, such that $E \subset E_j$ and each E_j has the finite intersection property, we see that $\bigcup_{j \in J} E_j$ is again a family of subsets of I with the finite intersection property. Therefore, by Zorn's Lemma, there is a maximal extension D of E. By Proposition 3.3.5 (ii), D is already an ultrafilter over I. Since $E \subset D$, no finite subset of I can belong to D, so D is nonprincipal. ⊣

Given a set X, we let $[X]^n$ denote the set of all subsets of X with exactly n elements. $S_\omega(X)$ denotes the set of all finite subsets of X. Clearly $S_\omega(X) = \bigcup_{n\in\omega}[X]^n$. Suppose that X is simply ordered (strictly) by the relation $<$. Then there is a one-to-one correspondence between elements of $[X]^n$ and strictly increasing n-termed sequences $x_1 < \ldots < x_n$ from X.

THEOREM 3.3.7 (Ramsey's Theorem). *Let I be an infinite set and let $n \in \omega$. Suppose that $[I]^n = A_0 \cup A_1$. Then there is an infinite subset $J \subset I$ such that either $[J]^n \subset A_0$ or $[J]^n \subset A_1$.*

PROOF. We first assume that I is countably infinite, because any countably infinite subset of I will satisfy the hypothesis. Let the members of I be arranged in an increasing sequence, ordered by $<$,

$$i_0 < i_1 < \ldots < i_m < \ldots.$$

We may assume that $n > 1$. Let D be any nonprincipal ultrafilter over I. Note that for all m, $\{i \in I : i_m < i\} \in D$. For each $r < n$, we shall define two subsets A_0^{n-r} and A_1^{n-r} of $[I]^{n-r}$ by induction on r as follows:

$$A_0^n = A_0 \quad \text{and} \quad A_1^n = A_1.$$

Suppose that A_0^{n-r} and A_1^{n-r} have been defined so that $[I]^{n-r} \subset A_0^{n-r} \cup A_1^{n-r}$. We let

$$A_0^{n-r-1} = \{y_1 < \ldots < y_{n-r-1} : \{i \in I : y_{n-r-1} < i \text{ and } \{y_1 \ldots y_{n-r-1}\, i\} \in A_0^{n-r}\} \in D\},$$

$$A_1^{n-r-1} = \{y_1 < \ldots < y_{n-r-1} : \{i \in I : y_{n-r-1} < i \text{ and } \{y_1 \ldots y_{n-r-1}\, i\} \in A_1^{n-r}\} \in D\}.$$

Using the properties of D, we see that

$$[I]^{n-r-1} \subset A_0^{n-r-1} \cup A_1^{n-r-1}.$$

Proceeding in this way, we finally have $I \subset A_0^1 \cup A_1^1$. The argument now splits into two symmetrical cases depending on whether $A_0^1 \in D$ or $A_1^1 \in D$. Assume that $A_0^1 \in D$. Let us define an infinite sequence

$$j_0 < j_1 < \ldots < j_m < \ldots$$

of elements of I by induction as follows. Let $j_0 \in A_0^1$. Suppose that $j_0 < \ldots < j_m$ have been defined such that

(1) for all r, $1 \leqslant r \leqslant n$, and all $y_1 < \ldots < y_r$ from $\{j_0 \ldots j_m\}$, the set $\{y_1 \ldots y_r\} \in A_0^r$.

We define j_{m+1} as follows. By (1), given $y_1 < ... < y_r$ from $\{j_0 \cdots j_m\}$, with $r < n$, the set

$$X_{y_1 \ldots y_r} = \{i \in I : y_r < i \text{ and } \{y_1 \cdots y_r i\} \in A_0^{r+1}\} \in D.$$

Since there are at most a finite number of increasing sequences of length at most $n-1$ from the set $\{j_0 \cdots j_m\}$, the number of sets $X_{y_1 \ldots y_r}$ is finite, and so their intersection Y belongs to D. Since D is nonprincipal, we may pick an element $j_{m+1} \in Y$ such that $j_m < j_{m+1}$. Now it is clear that (1) will hold with m replaced by $m+1$. Proceeding in this way we see that the infinite set $J = \{j_0 \cdots j_m \cdots\}$ can be constructed. Now it is clear that any $y_1 < \cdots < y_n$ from J will be such that $\{y_1 \cdots y_n\} \in A_0^n = A_0$, so $[J]^n \subset A_0$.

The other case when $A_1^1 \in D$ will give $[J]^n \subset A_1$. ⊣

Let $\mathfrak{A}$ be a model for $\mathscr{L}$, and let $X \subset A$ be a subset of A which carries a relation $<$ that strictly simply orders X. Note that $<$ may or may not be a relation in $\mathfrak{A}$. We say that X is a set of *elements indiscernible in* $\mathfrak{A}$ (with respect to $<$) iff for all n and all finite sequences $x_1 < \ldots < x_n$ and $y_1 < \ldots < y_n$ from X, $(\mathfrak{A}, x_1 \ldots x_n) \equiv (\mathfrak{A}, y_1 \ldots y_n)$.

If the relation $<$ is understood, then we shall not mention it explicitly. We also refer to X simply as a *set of indiscernibles in* $\mathfrak{A}$. We see that the term indiscernible means that the sequences $x_1 < \ldots < x_n$ and $y_1 < \ldots < y_n$ cannot be distinguished by any first-order formula $\varphi(v_1 \ldots v_n)$ of $\mathscr{L}$. By an *ordermorphic embedding* we shall mean an isomorphic embedding of one simply ordered structure into another. An *ordermorphism* of $\langle X, < \rangle$ is an automorphism of $\langle X, < \rangle$. This notation will help emphasize the distinction between an isomorphism of models for $\mathscr{L}$ and an ordermorphism of sets of indiscernibles in the models.

Here is a sufficient condition which can be used to obtain examples of sets of indiscernibles:

PROPOSITION 3.3.8. *Let $\langle X, < \rangle$ be a linearly ordered subset of a model $\mathfrak{A}$. Suppose that for any two increasing n-tuples $x_1 < \ldots < x_n$ and $y_1 < \ldots < y_n$ from X, there is an automorphism f of $\mathfrak{A}$ onto $\mathfrak{A}$ such that $f(x_1) = y_1, \ldots, f(x_n) = y_n$. Then X is a set of indiscernibles in $\mathfrak{A}$.*

PROOF. We have

$$f : (\mathfrak{A}, x_1 \ldots x_n) \cong (\mathfrak{A}, y_1 \ldots y_n).$$

Therefore

$$(\mathfrak{A}, x_1 \ldots x_n) \equiv (\mathfrak{A}, y_1 \ldots y_n),$$

whence X is a set of indiscernibles. ⊣

It follows, for example, that in any algebraically closed field of characteristic zero, if X is a set of algebraically independent elements, then for any linear ordering $<$ of X, X is a set of indiscernibles. Other examples are given in Exercise 3.3.13.

LEMMA 3.3.9. *Let $\mathscr{L}' = \mathscr{L} \cup \{c_n : n \in \omega\}$, where the c_n are new constants. Let T be a theory in $\mathscr{L}$ with infinite models. Then the following set T' of sentences of $\mathscr{L}'$ is consistent*:

$$T' = T \cup \{\varphi(c_{i_1} \dots c_{i_n}) \leftrightarrow \varphi(c_{j_1} \dots c_{j_n}) : \varphi(v_1 \dots v_n) \text{ is a formula of } \mathscr{L}, n \in \omega, \text{ and } i_1 < \dots < i_n, j_1 < \dots < j_n\} \cup \{\neg c_{i_1} \equiv c_{i_2} : i_1 \neq i_2\}$$

PROOF. Let $\mathfrak{A}$ be any infinite model of T, and let I be a countable infinite subset of A. Suppose that $<$ well orders I, so that

$$i_0 < i_1 < \dots < i_n < \dots$$

is a list of all the elements of I. We claim that:

(1) given any finite subset Δ of T', there is an infinite subset J_Δ of I such that for each infinite subset

$$j_0 < j_1 < \dots < j_n < \dots$$

of J_Δ, the expansion $(\mathfrak{A}, j_n)_{n\in\omega}$ satisfies Δ.

This is proved by induction on the number of sentences in Δ. So suppose (1) holds for some finite subset $\Delta \subset T'$ and let $\varphi(v_1 \dots v_m)$ be a formula of $\mathscr{L}$. We now divide $[J_\Delta]^m$ into two pieces: Let

$$A_0 = \{x_1 < \dots < x_m : x_i \in J_\Delta \text{ and } \mathfrak{A} \models \varphi[x_1 \dots x_m]\}$$

and

$$A_1 = \{x_1 < \dots < x_m : x_i \in J_\Delta \text{ and } \mathfrak{A} \models \neg\varphi[x_1 \dots x_m]\}.$$

Clearly $[J_\Delta]^m \subset A_0 \cup A_1$. By Theorem 3.3.7, there is an infinite subset $K \subset J_\Delta$ such that either $[K]^m \subset A_0$ or $[K]^m \subset A_1$. Let $k_0 < k_1 < \dots < k_n < \dots$ be an infinite subset of K. It is easy to verify that in either case the expansion $(\mathfrak{A}, k_n)_{n\in\omega}$ satisfies

$$\varphi(c_{s_1} \dots c_{s_m}) \leftrightarrow \varphi(c_{t_1} \dots c_{t_m}),$$

with $s_1 < \dots < s_m$ and $t_1 < \dots < t_m$. In addition, of course $(\mathfrak{A}, k_n)_{n\in\omega}$ still satisfies all sentences of Δ. So this proves that (1) holds whenever Δ is increased by the addition of one sentence, and the induction is complete. The consistency of T' follows immediately from (1). ⊣

THEOREM 3.3.10. *Let T be a theory in $\mathscr{L}$ with infinite models, and let $\langle X, <\rangle$ be any simply ordered set. Then there is a model $\mathfrak{A}$ of T with $X \subset A$ and such that X is a set of indiscernibles in $\mathfrak{A}$.*

PROOF. Let $\mathscr{L}' = \mathscr{L} \cup \{c_x : x \in X\}$, and let

$$T' = T \cup \{\varphi(c_{x_1} \dots c_{x_n}) \leftrightarrow \varphi(c_{y_1} \dots c_{y_n}) : \varphi(v_1 \dots v_n) \text{ a formula of } \mathscr{L}, n \in \omega, \text{ and } x_1 < \dots < x_n, y_1 < \dots < y_n \text{ from } X\} \cup \{\neg c_{x_1} \equiv c_{x_2} : x_1 \neq x_2 \text{ in } X\}.$$

Since every finite subset of X can be ordermorphically embedded in $\langle \omega, <\rangle$, we see that by Lemma 3.3.9, T' is a consistent set of sentences of $\mathscr{L}'$. Let $\mathfrak{A}'$ be any model of T', and let $\mathfrak{A}$ be the reduct of $\mathfrak{A}'$ to $\mathscr{L}$. Then $\mathfrak{A}$ is a model of T. We may, without loss of generality, identify the interpretations of c_x, $x \in X$, in $\mathfrak{A}'$with the elements x themselves. Now, by the form of T', we see that given $\varphi(v_1 \dots v_n)$ of $\mathscr{L}$ and $x_1 < \dots < x_n$ and $y_1 < \dots < y_n$ of X, we have

$$\mathfrak{A} \models \varphi[x_1 \dots x_n] \quad \text{if and only if} \quad \mathfrak{A} \models \varphi[y_1 \dots y_n].$$

So $(\mathfrak{A}, x_1 \dots x_n) \equiv (\mathfrak{A}, y_1 \dots y_n)$ and X is a set of indiscernibles in $\mathfrak{A}$. ⊣

The fact that T has models with indiscernibles of any order type is already quite remarkable. We see below that if T also has built-in Skolem functions, then this construction gives us models with even more interesting properties.

In the paragraph below, let us assume that T has built-in Skolem functions. We make the following simple observations about the models of T:

We first note by Proposition 3.3.4 that every theory can be extended to a theory with built-in Skolem functions. If $\mathfrak{A}$ is a model of T, then a Skolem expansion $\mathfrak{A}^*$ of $\mathfrak{A}$ can be obtained from $\mathfrak{A}$ by simply adding some functions which are already definable in $\mathfrak{A}$, i.e., functions which are interpretations of terms of $\mathscr{L}$ in $\mathfrak{A}$. Thus there is no essential difference between $\mathfrak{A}^*$ and $\mathfrak{A}$, and in all future discussions we shall consider them as practically the same model. Indeed, $\mathfrak{A}^*$ and $\mathfrak{A}$ are exactly the same model if the Skolem functions of $\mathfrak{A}$ are interpretations of function symbols of $\mathscr{L}$, as we know could be done. Recall also that if $X \subset A$, then the Skolem hull generated by X is the model $\mathfrak{H}(X) = \langle H(X), \dots\rangle$, where $H(X)$ is the closure of X under all terms of $\mathfrak{A}$. Also, $\mathfrak{H}(X) \prec \mathfrak{A}$. We say that a model $\mathfrak{A}$ is *generated from a set of indiscernibles* iff for some set $X \subset A$ of indiscernibles in $\mathfrak{A}$, $\mathfrak{A} = \mathfrak{H}(X)$.

The following theorem contains most of the basic and important properties of models generated from indiscernibles.

THEOREM 3.3.11. *Let X be a set of indiscernibles in a model $\mathfrak{A}$ of a theory T with built-in Skolem functions. Then*:

(a) (Subset Theorem). *If $Y \subset X$, then Y is a set of indiscernibles in $\mathfrak{H}(Y)$ with respect to the order inherited from X, and $\mathfrak{H}(Y) \prec \mathfrak{H}(X)$.*

(b) (Stretching Theorem). *Suppose that X and Y are infinite simply ordered sets. Then there exists a model $\mathfrak{B}$, in which Y is a set of indiscernibles and the sets of formulas satisfied by increasing sequences of elements from X in $\mathfrak{A}$ and from Y in $\mathfrak{B}$ are the same.*

(c) (Automorphism Theorem). *Let f be any ordermorphism of X onto X. Then f can be extended uniquely to an automorphism of $\mathfrak{H}(X)$ onto $\mathfrak{H}(X)$.*

(d) (Elementary Embedding Theorem). *Let Y be a set of indiscernibles in $\mathfrak{B}$ and such that the sets of formulas of $\mathscr{L}$ satisfied by increasing sequences of elements from X and Y are the same. Let f be a one-to-one ordermorphic embedding of X into Y; then f can be extended uniquely to an elementary embedding $\bar{f}$ of $\mathfrak{H}(X)$ into $\mathfrak{H}(Y)$. The range of $\bar{f}$ is H*(range of f).

(e) (Realizing and Omitting-Types Theorem). *Let Y, $\mathfrak{B}$ satisfy the first sentence of* (d). *Suppose also that X and Y are infinite. Then given any type $\Sigma(v_1 \dots v_n)$ of $\mathscr{L}$, $\mathfrak{H}(X)$ realizes Σ if and only if $\mathfrak{H}(Y)$ realizes Σ.*

PROOF. (a). We first note that $\mathfrak{H}(Y)$ is an elementary submodel of $\mathfrak{H}(X)$. It is quite clear that increasing sequences from Y satisfy the same formulas satisfied by increasing sequences from X. So (a) is proved.

(b). Let Σ be the set of all formulas $\varphi(v_1 \dots v_n)$ of $\mathscr{L}$ satisfied by increasing sequences $x_1 < \dots < x_n$ from X. In an expansion $\mathscr{L}' = \mathscr{L} \cup \{c_y : y \in Y\}$ of $\mathscr{L}$, let Σ' be the set of all sentences $\varphi(c_{y_1} \dots c_{y_n})$, where $\varphi \in \Sigma$ and $y_1 < \dots < y_n$ in Y. Then because X is infinite, the set Σ' is consistent. From this we can find the model $\mathfrak{B}$ with the set of indiscernibles Y.

(c) follows from (d), so we prove (d) next.

(d). Every element $y \in H(X)$ is generated from some term $t(v_1 \dots v_n)$ and some elements $x_1, \dots, x_n$ of X. We can assume that the term t and the elements x_i from X can be chosen so that t has exactly the variables $v_1, \dots, v_n$ free, $x_1 < x_2 < \dots < x_n$, and $y = t(x_1 \dots x_n)$. (Strictly speaking, we should use $t[x_1 \dots x_n]$ for the value of the term t on $x_1, \dots, x_n$. We shall dispense with the square brackets and pretend that t is a function.) We shall refer to this as a *standard representation* of y in $\mathfrak{H}(X)$.

Let $y = t(x_1 \dots x_n)$ be a standard representation of y. We define

$$\bar{f}(y) = t(f(x_1) \dots f(x_n)).$$

We first show that $\bar{f}$ is well-defined. Suppose $t'(z_1 \dots z_m)$ is another standard representation of y. Then in $\mathfrak{H}(X)$ we have $t(x_1 \dots x_n) = t'(z_1 \dots z_m)$. Let

$u_1 < \ldots < u_l$ be a listing of the set $\{x_1 \ldots x_n z_1 \ldots z_m\}$ in increasing order. Then we may express the equality $t(x_1 \ldots x_n) = t'(z_1 \ldots z_m)$ by a formula φ in terms of $u_1 \ldots u_l$, so $\mathfrak{H}(X) \models \varphi[u_1 \ldots u_l]$. By hypothesis, $\mathfrak{H}(Y) \models \varphi[f(u_1) \ldots f(u_l)]$. From this follows that $t(f(x_1) \ldots f(x_n)) = t'(f(z_1) \ldots f(z_m))$ in $\mathfrak{H}(Y)$.

Let $\varphi(v_1 \ldots v_l)$ be any formula of $\mathscr{L}$, and let $y_1, \ldots, y_l$ be such that $\mathfrak{H}(X) \models \varphi[y_1 \ldots y_l]$. Let us take standard representations of $y_1, \ldots, y_l$ given by, say, $t_1, \ldots, t_l$, together with a finite sequence of generators $x_1 < \ldots < x_n$ from X. We assume that each t_i when applied to an appropriate subsequence of $x_1 < \ldots < x_n$ gives y_i. We can now find a formula ψ containing the terms $t_1, \ldots, t_l$ and the variables $v_1, \ldots, v_n$ such that

$$\mathfrak{H}(X) \models \varphi[y_1 \ldots y_l] \quad \text{if and only if} \quad \mathfrak{H}(X) \models \psi[x_1 \ldots x_n].$$

Again, we have $\mathfrak{H}(Y) \models \psi[f(x_1) \ldots f(x_n)]$. Whence by examining the form of ψ, we see that $\mathfrak{H}(Y) \models \varphi[\bar{f}(y_1) \ldots \bar{f}(y_l)]$. So $\bar{f}$ is an isomorphism.

If $z \in H(\text{range of} f)$, then there is a standard representation $z = t(y_1 \ldots y_m)$ with $y_1, \ldots, y_m$ in the range of f. Let $x_1 < \ldots < x_m$ be in X such that $f(x_i) = y_i$. Then $\bar{f}$ maps the element $t(x_1 \ldots x_m)$ in $H(X)$ onto z. So $\bar{f}$ is onto $H(\text{range of} f)$ and the result (d) follows from (a). We leave it to the reader to verify that $\bar{f}$ is unique.

(e). Let X, Y, $\mathfrak{H}(X)$, $\mathfrak{H}(Y)$ satisfy the hypothesis of (e). Suppose that $z_1, \ldots, z_n$ in $H(X)$ realizes Σ in $\mathfrak{H}(X)$. Let $z_1, \ldots, z_n$ have standard representations in $\mathfrak{H}(X)$ and let us assume that in these representations at most the generators $x_1 < \ldots < x_m$ from X are involved. Let f be any order-preserving map of $x_1 < \ldots < x_m$ onto $y_1 < \ldots < y_m$ in Y. Then, by (d), $\mathfrak{H}(\{x_1 \ldots x_m\}) \cong \mathfrak{H}(\{y_1 \ldots y_m\})$, by the mapping $\bar{f}$. So the n-tuple of elements $\bar{f}(z_1), \ldots, \bar{f}(z_n)$ of $\mathfrak{H}(\{y_1 \ldots y_m\})$ realizes Σ. Since $\mathfrak{H}(\{y_1 \ldots y_m\}) \prec \mathfrak{H}(Y)$, we see that $\bar{f}(z_1), \ldots, \bar{f}(z_n)$ realizes Σ also in $\mathfrak{H}(Y)$. The other direction is proved analogously. ⊣

Some applications of this construction are given below and in the exercises.

COROLLARY 3.3.12. *Let $\mathscr{L}$ be a countable language and let T be a theory in $\mathscr{L}$ with infinite models. Then there is a countable collection Δ of types of $\mathscr{L}$ such that T has arbitrarily large models which realize exactly those types in Δ.*

PROOF. We first expand $\mathscr{L}$ to $\overline{\mathscr{L}}$ and extend T to a theory $\bar{T}$ in $\overline{\mathscr{L}}$ with built-in Skolem functions. $\bar{T}$ still has infinite models. Let X be a set of indiscernibles in $\mathfrak{H}(X)$, where $\mathfrak{H}(X)$ is a model of $\bar{T}$ and the ordering $<$ on X is of type ω. Then since $\overline{\mathscr{L}}$ is still countable, $\mathfrak{H}(X)$ is again countable.

So $\mathfrak{H}(X)$ realizes at most a countable number of types (of $\overline{\mathscr{L}}$). If Y is an infinite set of indiscernibles in $\mathfrak{H}(Y)$ such that increasing sequences from Y satisfy the same formulas as increasing sequences from X, then $\mathfrak{H}(Y)$ realizes exactly those types (of $\overline{\mathscr{L}}$) realized in $\mathfrak{H}(X)$. Such an $\mathfrak{H}(Y)$ exists by the stretching theorem. The reducts of $\mathfrak{H}(X)$ and $\mathfrak{H}(Y)$ to $\mathscr{L}$ are both models of T and realize exactly the same types (of $\mathscr{L}$). ⊣

COROLLARY 3.3.13. *Every infinite model has elementary extensions with arbitrarily large groups of automorphisms.*

PROOF. Let $\mathfrak{A}$ be an infinite model for $\mathscr{L}$, and let $\mathfrak{A}^*$ be an expansion of $\mathfrak{A}$ whose theory has built-in Skolem functions. Let T be the elementary diagram of $\mathfrak{A}^*$ in the language $\mathscr{L}^* \cup \{c_a : a \in A\}$. T is a theory with built-in Skolem functions, and T has an infinite model $(\mathfrak{A}^*, a)_{a\in A}$. Let X be a set of indiscernibles in $\mathfrak{H}(X)$, which is a model of T. Then the reduct of $\mathfrak{H}(X)$ to the language $\mathscr{L}$ is an elementary extension of $\mathfrak{A}$ with at least as many automorphisms as there are ordermorphisms on X. (See Exercise 3.3.7.) ⊣

COROLLARY 3.3.14. *Let $\mathscr{L}$ be a countable language and let T be a theory in $\mathscr{L}$ with infinite models. Then for every infinite cardinal α, T has a model $\mathfrak{A}$ of power α such that for every subset $B \subset A$ the expanded model $(\mathfrak{A}, b)_{b\in B}$ realizes at most $|B| \cup \omega$ types in the expanded language $\mathscr{L} \cup \{c_b : b \in B\}$.*

PROOF. Extend T to a theory $\overline{T}$ which has built-in Skolem functions in an expanded language $\overline{\mathscr{L}}$. Let $\langle X, <\rangle$ be a well ordered set of order type α, and let $\overline{\mathfrak{A}} = \mathfrak{H}(X)$ be a model of $\overline{T}$ in which X is a set of indiscernibles. Then $\overline{\mathfrak{A}}$ has power α, because $\overline{\mathscr{L}}$ is still countable. Let $B \subset A$. Choose a standard representation for each $b \in B$, and let Y be the set of all $y \in X$ which appear in one of these standard representations. Then $|Y| \leqslant |B| \cup \omega$. Let us call two sequences $x_1 < \dots < x_n$, $y_1 < \dots < y_n$ *equivalent over* Y iff for all $k \leqslant n$ and all $z \in Y$, we have $x_k \neq z$, $y_k \neq z$, and

$$x_k < z \quad \text{iff} \quad y_k < z.$$

Form the expanded language $\mathscr{L}' = \overline{\mathscr{L}} \cup \{c_z : z \in Y\}$. Whenever $x_1 < \dots < x_n$ and $y_1 < \dots < y_n$ are equivalent over Y, they satisfy the same formulas in the expanded model $(\overline{\mathfrak{A}}, z)_{z\in Y}$. It follows that for any term $t(v_1 \dots v_n)$ of $\mathscr{L}'$, the two elements $t(x_1 \dots x_n)$ and $t(y_1 \dots y_n)$ realize the same type in the model $(\overline{\mathfrak{A}}, z)_{z\in Y}$, whence also in the model $(\overline{\mathfrak{A}}, b)_{b\in B}$. Let $\mathfrak{A}$ be the reduct of $\overline{\mathfrak{A}}$ to $\mathscr{L}$. Then, whenever $x_1 < \dots < x_n$ and $y_1 < \dots < y_n$ are equivalent over Y, the elements $t(x_1 \ldots x_n)$ and $t(y_1 \ldots y_n)$ realize the same type in the model $(\mathfrak{A}, b)_{b \in B}$. However, there are at most $|Y| \cup \omega$ nonequivalent n-tuples over Y, because if we write

$$x' = \text{least } z \in Y \text{ such that } x < z,$$

or

$$x' = \infty \text{ if } Y < x,$$

then we see that $x_1 < \ldots < x_n$ and $y_1 < \ldots < y_n$ are equivalent over Y iff $x'_1 = y'_1, \ldots, x'_n = y'_n$. Moreover, every element of A is equal to some term $t(x_1 \ldots x_n)$ in the model $(\bar{\mathfrak{A}}, z)_{z \in Y}$, with $x_1 \notin Y, \ldots, x_n \notin Y$. Also there are at most $|Y| \cup \omega$ terms in the language $\mathscr{L}'$ of that model. It follows that the model $(\mathfrak{A}, b)_{b \in B}$ realizes at most $|Y| \cup \omega \leqslant |B| \cup \omega$ types. ⊣

Exercises

3.3.1. Prove that if a theory T has built-in Skolem functions and $\mathfrak{A}$ is a model of T, then the elementary diagram $\text{Th}(\mathfrak{A}_A)$ also has built-in Skolem functions.

3.3.2 (Skolem Normal Form Theorem). For every formula φ of $\mathscr{L}$ there is a universal formula ψ of $\mathscr{L}^*$ such that $\vdash \psi \rightarrow \varphi$ and $\Sigma_{\mathscr{L}} \vdash \varphi \rightarrow \psi$.

3.3.3. Let $\mathfrak{A} = \langle \omega, \leqslant \rangle$. Show that $\mathfrak{A}$ has two Skolem expansions which are not elementarily equivalent in $\mathscr{L}^*$. Thus the Skolem expansion of the complete theory $\text{Th}(\mathfrak{A})$ is no longer complete.

3.3.4. Show that if I is infinite and $[I]^n \subset A_0 \cup \ldots \cup A_r$, then there is an infinite $J \subset I$ and a k, $0 \leqslant k \leqslant r$, such that $[J]^n \subset A_k$.

3.3.5. Show that if $[I]^n \subset A_0 \cup A_1$ and D is a nonprincipal ultrafilter over I, then we cannot necessarily draw the conclusion that there is a $J \in D$ such that $[J]^n \subset A_0$ or $[J]^n \subset A_1$. ($n > 1$.)

3.3.6. Show that if $I \times I \subset A_0 \cup A_1$, then there may be no infinite $J \subset I$ such that $J \times J \subset A_0$ or $J \times J \subset A_1$.

3.3.7. Let X be a set of indiscernibles in $\mathfrak{H}(X)$. Show that there is an isomorphic embedding of the group of ordermorphisms of $\langle X, < \rangle$ into the group of automorphisms of $\mathfrak{H}(X)$. Furthermore, if the ordering $<$ on X is itself definable in $\mathfrak{H}(X)$, then this isomorphic embedding is onto. (Assume that $\mathfrak{H}(X)$ is a model for a theory with built-in Skolem functions.)

3.3.8. Generalize Corollary 3.3.12 to noncountable languages.

3.3.9*. Every infinite model $\mathfrak{A}$ has a proper elementary extension $\mathfrak{B}_0$ which has a descending chain of isomorphic elementary submodels

$$\mathfrak{B}_0 \succ \mathfrak{B}_1 \succ \dots \succ \mathfrak{B}_n \succ \dots$$

such that $\mathfrak{A} = \bigcap_{n\in\omega} \mathfrak{B}_n$.

3.3.10*. Two elements x, y are said to have the same *automorphism type* in a model $\mathfrak{A}$ iff there exists an automorphism f of $\mathfrak{A}$ such that $f(x) = y$. If x, y have the same automorphism type in $\mathfrak{A}$, then they realize the same types in the language of $\mathfrak{A}$. Show that if $\mathscr{L}$ is countable and a theory T in $\mathscr{L}$ has infinite models, then T has models $\mathfrak{A}$ of every infinite power α such that $\mathfrak{A}$ has at most ω automorphism types, and also that $\mathfrak{A}$ has 2^α distinct automorphisms.

[*Hint*: Use the Löwenheim–Skolem–Tarski Theorem to show that for each infinite cardinal α there is a dense linear ordering X of power α such that any two strictly increasing n-tuples in X are automorphic. Then take $\mathfrak{A}$ to be a Skolem hull of X.]

3.3.11*. In the above exercise, show that, for every infinite cardinal α, T has a model $\mathfrak{A}$ of power α such that for every $B \subset A$, the model $(\mathfrak{A}, b)_{b\in B}$ has at most $|B| \cup \omega$ automorphism types.

[*Hint*: Use a set of indiscernibles $\langle X, \leqslant \rangle$ where X is the set of all finite sequences of elements of $\alpha^* + \alpha$ and $\leqslant$ is the lexicographic order.]

3.3.12. Show that if $\langle X, < \rangle$ is indiscernible in $\mathfrak{A}$ and $\mathfrak{B}$ is an elementary submodel of $\mathfrak{A}$ which contains X, then $\langle X, < \rangle$ is indiscernible in $\mathfrak{B}$.

3.3.13. Use Proposition 3.3.8 and Exercise 3.3.12 to show that for each of the following examples, X is a set of indiscernibles in the model $\mathfrak{A}$:

(i). $\mathfrak{A}$ is a pure transcendental extension of a field $\mathfrak{B}$, and X is a set of algebraically independent elements over $\mathfrak{B}$.

(ii). $\mathfrak{A}$ is an algebraically closed field and X is a set of algebraically independent elements over the prime field.

(iii). $\mathfrak{A}$ is a free group and X is its set of free generators.

(iv). $\mathfrak{A}$ is the ring of polynomials over a ring $\mathfrak{B}$ in the set of variables X.

(v). $\mathfrak{A}$ is the ordered set of rationals and $\langle X, < \rangle = \mathfrak{A}$.

(vi). $\mathfrak{A}$ is a Boolean algebra and X is the set of all atoms of $\mathfrak{A}$.

(vii). $\mathfrak{A} = \langle A, E \rangle$, where E is an equivalence relation and X is one of the equivalence classes.

(viii). $\mathfrak{A}$ is as above, X is a set of nonequivalent elements, and if $x, y \in X$, then their equivalence classes are of the same power.

3.3.14*. Let $\mathfrak{A} = \langle A, < \ldots\rangle$ be an infinite model for a countable language in which $<$ is a well ordering of A. Show that for every regular cardinal $\alpha > \omega$, there is a model $\mathfrak{B} \equiv \mathfrak{A}$ such that $\mathfrak{B}$ has an increasing sequence of length α but no decreasing sequences of length α with respect to $<$.

[*Hint*: Use a set of indiscernibles of order type α.]

3.3.15**. Let $\mathfrak{A} = \langle A, < \ldots\rangle$ be an infinite model for a countable language in which $<$ is a simple ordering of A. Show that for every infinite successor cardinal α^+, there is a model $\mathfrak{B} \equiv \mathfrak{A}$ of power α^+ which has no increasing sequences or decreasing sequences of length α^+.

[*Hint*: Use a set of indiscernibles whose order type has the desired property.]

3.3.16*. Show that every model $\mathfrak{A}$ has a proper elementary extension $\mathfrak{B}$ such that there is an elementary embedding f of $\mathfrak{B}$ into $\mathfrak{B}$ with

$$A = \bigcap_{n\in\omega} f^n(B) = \{b \in B : f(b) = b\}.$$

3.3.17*. Give examples of countable decreasing elementary chains $\mathfrak{A}_0 \succ \mathfrak{A}_1 \succ \mathfrak{A}_2 \succ \ldots$ such that:

(i). $\bigcap_n A_n$ is empty.

(ii). $\mathfrak{A} = \bigcap_n \mathfrak{A}_n$ is not empty, but $\mathfrak{A}$ is not an elementary submodel of any $\mathfrak{A}_n$.

3.3.18*. Give an example, for each n, of a model $\mathfrak{A}$ and an infinite ordered set $\langle X, <\rangle$ such that $X \subset A$, any two increasing n-tuples of X realize the same types in $\mathfrak{A}$, but there exist two increasing $(n+1)$-tuples in X realizing different types in $\mathfrak{A}$.

3.3.19*. Let $\mathfrak{A}$ be a model for $\mathscr{L}$. $G(\mathfrak{A})$ shall denote the group of automorphisms of $\mathfrak{A}$. Let Σ be any consistent set of sentences of $\mathscr{L}$ and let $K(\Sigma)$ be the class of all groups H which are isomorphic to a subgroup of $G(\mathfrak{A})$ for some model $\mathfrak{A}$ of Σ. Prove that $K(\Sigma)$ is the class of all models of a set Φ of universal sentences in the language of group theory. Furthermore, show that if Σ is recursive, then Φ may be picked to be recursive.

3.3.20*. Let K be the intersection of all the classes $K(\Sigma)$, where Σ has at least one infinite model (for notation see Exercise 3.3.19). Prove that $K = K(\Sigma_0)$ when Σ_0 is the theory of simple order.

3.3.21. Let T be a complete theory with infinite models in a countable language and let κ be an infinite cardinal. Prove that T has models $\mathfrak{A}$ and $\mathfrak{B}$ of power κ where $\mathfrak{B}$ is a proper submodel of $\mathfrak{A}$ and there is an automorphism f of $\mathfrak{A}$ such that

$$\mathfrak{B} < f(\mathfrak{B}) < f(f(\mathfrak{B})) < f(f(f(\mathfrak{B}))) < \ldots$$

and

$$\mathfrak{A} = \mathfrak{B} \cup f(\mathfrak{B}) \cup f(f(\mathfrak{B})) \cup f(f(f(\mathfrak{B}))) \cup \ldots .$$

3.3.22. In a countable language, let T be a complete theory whose models are infinite. Prove that T has a family of countable models $\mathfrak{A}_S$, $S \subseteq \omega$, such that if R is a proper subset of S then $\mathfrak{A}_R$ is a proper elementary submodel of $\mathfrak{A}_S$.

3.4. Some examples

During the last two chapters we have mentioned a large number of examples and made several assertions without giving proofs. In this section we shall re-examine five of these examples in some detail, in the light of the theorems we have proved. Most of the other examples can be analyzed by methods similar to these five. One exception is the theory of real closed fields, which we shall return to in Chapter 5.

EXAMPLE 3.4.1. *The theory of dense linear order without endpoints.* Two classical models of this theory are the ordering of the reals and the ordering of the rationals. All models of T are infinite.

To prove that T is ω-categorical, we use a back and forth argument of Cantor. Let $\langle A, \leqslant \rangle$ and $\langle B, \leqslant \rangle$ be two countable dense linear orderings without endpoints. A mapping f from a finite subset $A_0 \subset A$ onto a finite subset $B_0 \subset B$ is said to be a partial isomorphism iff f is one–one and whenever $a_0 \leqslant a_1$ in $A_0, f(a_0) \leqslant f(a_1)$. It is easy to see that for any partial isomorphism f and any elements $a \in A$ and $b \in B$, there is a partial isomorphism $g \supset f$, and $a \in$ domain (g) and $b \in$ range (g). We can then construct a union of partial isomorphisms which is an isomorphism of $\langle A, \leqslant \rangle$ onto $\langle B, \leqslant \rangle$.

T is ω-categorical, T is complete and its countable model is both countably saturated and atomic. T is obviously preserved under unions of chains. T is not preserved under submodels: For example, the model $\langle \omega, \leqslant \rangle$ is a submodel of a model of T but is not dense. T is also not preserved under homomorphisms, because the one-element model is a homomorphic image of any model but is not a model of T.

Cantor's back and forth argument shows the following: if $\mathfrak{A} = \langle A, \leqslant \rangle$ is the countable model of T and $x_1 < \ldots < x_n$, $y_1 < \ldots < y_n$ in A, then $\mathfrak{A}$ has an automorphism f with $f(x_1) = y_1, \ldots, f(x_n) = y_n$. It follows that $\langle A, \leqslant \rangle$ itself is a set of indiscernibles in $\mathfrak{A}$. Moreover, every increasing n-tuple satisfies the same type $\Gamma(x_1 \ldots x_n)$ in $\mathfrak{A}$. Since $\mathfrak{A}$ is ω-saturated, every type of T is realized in $\mathfrak{A}$. Therefore in any other model $\mathfrak{B}$ of T (possibly uncountable), every increasing n-tuple also realizes $\Gamma(x_1 \ldots x_n)$. This shows that in any model $\mathfrak{B} = \langle B, \leqslant \rangle$ of T, $\langle B, \leqslant \rangle$ itself is a set of indiscernibles.

The theory T has 2^α nonisomorphic models in any uncountable power α. We can build up 2^α such models as follows. Let $\omega_1^* + \omega_1$ be the ordering consisting of a copy of ω_1 with the reverse order followed by a copy of ω_1 with the natural order, that is,

$$\ldots \xi' \ldots 3'2'1'0'0123 \ldots \xi \ldots, \qquad \xi < \omega_1.$$

Let $\eta(\omega_1^* + \omega_1)$ be the ordering formed by replacing each point in the ordering $\omega_1^* + \omega_1$ by a copy of the rationals. Let $\eta(\omega_1^* + \omega_1 + 1)$ be the ordering formed by adding one more copy of the rationals to the end of $\eta(\omega_1^* + \omega_1)$. Now for each subset $S \subset \alpha$, form the ordering $\mathfrak{A}_S$ from $\langle \alpha, \leqslant \rangle$ by replacing each $\beta \in S$ by a copy of $\eta(\omega_1^* + \omega_1)$ and each $\beta \in \alpha \setminus S$ by a copy of $\eta(\omega_1^* + \omega_1 + 1)$. If α is an uncountable cardinal, then the models $\mathfrak{A}_S$, $S \subset \alpha$, are 2^α nonisomorphic models of T of power α.

Example 3.4.2. *The theory of countably many independent unary relations.* This theory is in a language with countably many unary relation symbols $P_0(x)$, $P_1(x)$, $P_2(x)$, ... and the axioms

$$(\exists x)(P_{i_1}(x) \wedge \ldots \wedge P_{i_m}(x) \wedge \neg P_{j_1}(x) \wedge \ldots \wedge \neg P_{j_n}(x))$$

where $i_1, \ldots, i_m, j_1, \ldots, j_n$ are all distinct.

Let us first show that T is complete. All models of T are infinite. Let $\mathfrak{A}$ be any countable model of T. Using the compactness theorem, we can readily show that $\mathfrak{A}$ has an elementary extension $\mathfrak{B}$ of power 2^ω such that for each set $S \subset \omega$ there are 2^ω elements $b \in B$, with

$$\mathfrak{B} \models P_n(b) \quad \text{if} \quad n \in S,$$
$$\mathfrak{B} \models \neg P_n(b) \quad \text{if} \quad n \notin S.$$

However, up to isomorphism there is only one such model $\mathfrak{B}$. Thus any two countable models of T have isomorphic elementary extensions and hence are elementarily equivalent. We conclude that T is complete.

Let $\mathfrak{A}$ be a model of T, let $n < \omega$, and consider two elements a, b of A. Suppose a and b satisfy exactly the same atomic formulas among $P_0(x), \dots, P_n(x)$. Then the mapping which interchanges a and b is an automorphism of the reduct of $\mathfrak{A}$ to the sublanguage $\{P_0, \dots, P_n\}$. It follows that for any formula $\varphi(x)$ in which at most the symbols $P_0, \dots, P_n$ occur, we have

$$\mathfrak{A} \models \varphi[a] \quad \text{if and only if} \quad \mathfrak{A} \models \varphi[b].$$

Using the above, we see that every consistent formula $\varphi(x)$ is incompletable in T, because if all the symbols occurring in $\varphi(x)$ are among $P_0, \dots, P_n$, then any model $\mathfrak{A}$ has elements a, b with

$$\mathfrak{A} \models \varphi[a] \wedge P_{n+1}(a), \qquad \mathfrak{A} \models \varphi[b] \wedge \neg P_{n+1}(b).$$

Hence $\varphi(x)$ does not imply either $P_{n+1}(x)$ or $\neg P_{n+1}(x)$ with respect to T. Therefore T is not an atomic theory, and T has no atomic models. Consequently T has no countably saturated models.

Another consequence of the above discussion is that every formula $\varphi(x)$ is equivalent with respect to T to a finite disjunction of conjunctions of formulas $P_i(x)$ and $\neg P_j(x)$. We can readily extend this result to formulas $\varphi(x_1 \dots x_n)$ if we also allow equations and inequalities.

Two elements a, b realize the same type if and only if they satisfy exactly the same relations $P_n(x)$. So T has 2^ω types $\Gamma(x)$. Given any subset $X \subset \omega$, let $X(\mathfrak{A})$ be the set of all elements b in a model $\mathfrak{A}$ such that

$$\mathfrak{A} \models P_n(b) \quad \text{if} \quad n \in X,$$

$$\mathfrak{A} \models \neg P_n(b) \quad \text{if} \quad n \in \omega \setminus X.$$

Then for any linear ordering $\leqslant$ of $X(\mathfrak{A})$ whatsoever, $X(\mathfrak{A})$ is a set of indiscernibles in $\mathfrak{A}$.

A model $\mathfrak{A}$ for $\mathscr{L}$ is determined up to isomorphism by the cardinalities of the sets $X(\mathfrak{A})$, $X \subset \omega$. $\mathfrak{A}$ is a model of T if and only if the family of sets

$$S = \{X \subset \omega : X(\mathfrak{A}) \neq 0\}$$

is *dense* in the sense that for any two disjoint finite subsets $s, t \subset \omega$, there exists $X \in S$ such that $s \subset X$ and $t \subset \omega \setminus X$. From this we can see that there are 2^ω nonisomorphic countable models of T. Moreover, for each infinite cardinal $\alpha \leqslant 2^\omega$ there are 2^α nonisomorphic models of T of power α. And for each cardinal $\aleph_\beta > 2^\omega$ there are $|\beta|^{2^\omega}$ nonisomorphic models of T of power $\aleph_\beta$.

It is easy to find two dense sets S_1, S_2 of subsets of ω such that

$S_1 \cap S_2 = \emptyset$. From these we obtain two countable models $\mathfrak{A}_1$, $\mathfrak{A}_2$ of T such that no type $\Gamma(x)$ is realized in both $\mathfrak{A}_1$ and $\mathfrak{A}_2$. $\mathfrak{A}_1$ is formed by putting one element into $X(\mathfrak{A}_1)$ for each $X \in S_1$, and $\mathfrak{A}_2$ similarly. In fact, it is easily seen that there are 2^ω disjoint countable dense sets of subsets of ω, whence there is a family of 2^ω countable models of T such that no type $\Gamma(x)$ is realized by more than one of them.

Two n-tuples $a_1, \ldots, a_n$ and $b_1, \ldots, b_n$ realize the same type in $\mathfrak{A}$ if and only if each a_i realizes the same type as b_i and the same equations hold between the a's as between the b's. From this we can readily see that every model of T is ω-homogeneous. Given an infinite cardinal α, we can obtain a model $\mathfrak{A}$ of T of power α which realizes only countably many types of n-tuples by taking a countable dense set S and putting α elements into $X(\mathfrak{A})$ for each $X \in S$.

The theory T is preserved under extensions and hence under unions of chains. Note that all the axioms are existential. T is not preserved under submodels or homomorphisms.

EXAMPLE 3.4.3. *The theory of algebraically closed fields of characteristic zero.* A *transcendence basis* in a model $\mathfrak{A}$ of T is a maximal set $X \subset A$ with the property that no distinct n-tuple from X is a root of a nonzero polynomial in n variables over the rationals. It is a classical result of field theory that every model has a transcendence basis, all transcendence bases of $\mathfrak{A}$ have the same cardinal α(the *transcendence rank* of $\mathfrak{A}$), for each α there is a unique model of transcendence rank α, and this model has power $\alpha \cup \omega$. It follows that T is α-categorical for each uncountable cardinal α. Since all models of T are infinite, T is complete. T is also closed under unions of chains.

Moreover, T has just countably many countable models, one for each rank $\leqslant \omega$. Therefore T has a countably saturated model and an atomic model. Since the model of rank ω cannot be embedded in any model of finite rank, none of the models of finite rank is countably universal. But the countably saturated model is countably universal, so it must be the model of rank ω. The model of rank zero, the field of algebraic numbers, is a prime model of T and is thus the atomic model.

Another classical result is that if two n-tuples $a_1, \ldots, a_n$ and $b_1, \ldots, b_n$ in $\mathfrak{A}$ are roots of exactly the same polynomials over the rationals, then there is an automorphism f of $\mathfrak{A}$ with $f(a_1) = b_1, \ldots, f(a_n) = b_n$. It follows that, if X is a transcendence basis of $\mathfrak{A}$, then X is a set of indiscernibles in $\mathfrak{A}$ with respect to any simple ordering of X. It also follows that every model of T is ω-homogeneous.

EXAMPLE 3.4.4. *The theory of discrete simple orderings with an initial element.* This is the theory in the language $\mathscr{L} = \{\leqslant\}$ with the axioms:

The axioms of simple order.

$(\exists x \forall y)(x \leqslant y)$ (there is a first element).

Every element has an immediate successor.

Every element except the first has an immediate predecessor.

The model $\langle \omega, \leqslant \rangle$ is a model of T. Every other model of T can be obtained in the following way: Let $\mathfrak{A} = \langle A, \leqslant \rangle$ be any simple ordering. Form the model $\mathfrak{A}^* = \langle B, \leqslant \rangle$ of T by adding one copy of the ordering $\langle Z, \leqslant \rangle$ to the end of $\langle \omega, \leqslant \rangle$ for each element $a \in A$. More precisely,

(1) $B = \omega \cup (A \times Z)$,

(2) $\leqslant$ is the natural ordering on ω,

(3) $n \leqslant \langle a, m \rangle$ for all $n \in \omega, \langle a, m \rangle \in A \times Z$,

(4) $\langle a, m \rangle \leqslant \langle b, n \rangle$ iff $a < b$ or $a = b$ and $m \leqslant n$.

Then every model of T is isomorphic either to $\langle \omega, \leqslant \rangle$ or to some $\mathfrak{A}^*$. For example, if Q is the ordering of the rationals, then Q^* is formed by adding to the end of $\langle \omega, \leqslant \rangle$ one copy of $\langle Z, \leqslant \rangle$ for each rational. Thus T has 2^α nonisomorphic models in every infinite cardinal α.

It is convenient to work first with a new theory in the language formed by adding a symbol 0 for the first element and S for the successor function. The expanded theory $T(0, S)$ has the additional axioms

$$x \equiv 0 \leftrightarrow (\forall y)(x \leqslant y),$$

$$y = Sx \leftrightarrow x < y \wedge \neg (\exists z)(x < z \wedge z < y).$$

Every model $\langle B, \leqslant \rangle$ of T can clearly be expanded in a unique way to a model $\langle B, \leqslant, 0, S \rangle$ of $T(0, S)$. We can show that the theory $T(0, S)$ is model complete by proving that if $\mathfrak{B}$ and $\mathfrak{C}$ are models of $T(0, S)$ and $\mathfrak{C} \subset \mathfrak{B}$, then every existential sentence true in $\mathfrak{B}_C$ is true in $\mathfrak{C}_C$. Consider an existential sentence

$$(\exists x_1 \dots x_m)\varphi(x_1 \dots x_m c_1 \dots c_n)$$

true in $\mathfrak{B}_C$, where $c_1, \dots, c_n$ are elements of C. By the compactness theorem, $\mathfrak{C}$ has an elementary extension $\mathfrak{C}'$ such that before, after, and between any two copies of $\langle Z, \leqslant \rangle$ in $\mathfrak{C}$ there are infinitely many copies of $\langle Z, \leqslant \rangle$ in $\mathfrak{C}'$. Let $b_1, \dots, b_m$ satisfy φ in $\mathfrak{B}_C$. We may then choose $d_1, \dots, d_m$ in $\mathfrak{C}'$ such that the sets

$$\{b_1, \dots, b_m, c_1, \dots, c_n, 0\}, \qquad \{d_1, \dots, d_m, c_1, \dots, c_n, 0\}$$

are ordered in the same way, and the distance between two corresponding elements is either the same finite number or infinite. It follows that $d_1, \ldots, d_m$ satisfies φ in $\mathfrak{C}'_C$, so $(\exists x_1 \ldots x_m)\varphi$ holds in $\mathfrak{C}'_C$ and in $\mathfrak{C}_C$. We conclude that $T(0, S)$ is model complete.

Since $\langle \omega, \leqslant, 0, S \rangle$ is isomorphically embedded in every model of $T(0, S)$, $T(0, S)$ is a complete theory. Every countable model of $T(0, S)$ is isomorphically embedded, and hence elementarily embedded, in the model $\langle Q^*, \leqslant, 0, S \rangle$. Therefore $\langle Q^*, \leqslant, 0, S \rangle$ is a countably universal model. Since $T(0, S)$ has a countably universal model, it has a countably saturated model. We now have enough information about $T(0, S)$ to resume our study of the original theory T.

Given two models $\mathfrak{B}$, $\mathfrak{C}$ of T, the expansions $(\mathfrak{B}, 0, S)$ and $(\mathfrak{C}, 0, S)$ to $T(0, S)$ are elementarily equivalent. Therefore $\mathfrak{B}$ and $\mathfrak{C}$ are elementarily equivalent, so T is a complete theory. In fact, $T = \mathrm{Th}(\langle \omega, \leqslant \rangle)$. Given any two simple orderings $\mathfrak{A}_0 \subset \mathfrak{A}_1$, the associated models of T are such that $\mathfrak{A}_0^* \prec \mathfrak{A}_1^*$, because the operations 0 and S are the same in $\mathfrak{A}_0^*$ as in $\mathfrak{A}_1^*$.

Since the reduct of a countably saturated model is countably saturated, T has a countably saturated model. We can identify this model by a process of elimination. The model $\langle \omega, \leqslant \rangle$ is not ω-saturated because it omits the set of all 'infinite' elements. Let $\mathfrak{A} = \langle A, \leqslant \rangle$ be any countable simple ordering. If $\mathfrak{A}$ has a last element a, then $(\mathfrak{A}^*, \langle a, 0 \rangle)$ omits the set of all elements infinitely greater than $\langle a, 0 \rangle$, so $\mathfrak{A}^*$ is not ω-saturated. Similarly, if $\mathfrak{A}$ has a least element, or if $\mathfrak{A}$ is not dense, we can see that $\mathfrak{A}^*$ is not ω-saturated. The only remaining possibility is that $\mathfrak{A}$ is the ordering Q of the rationals. Therefore Q^* is the countably saturated model of T. If $\mathfrak{A}$ is a countable simple ordering which has Q as a submodel, but is not isomorphic to Q, then $Q^* \prec \mathfrak{A}^*$, and therefore $\mathfrak{A}^*$ is countably universal but not countably saturated.

T also has an atomic model. Since no other model is embedded in $\langle \omega, \leqslant \rangle$, the prime and hence atomic model of T must be $\langle \omega, \leqslant \rangle$.

T has exactly three countably homogeneous models, namely $\langle \omega, \leqslant \rangle$, Q^*, and the model 1^* formed by adding just one copy of $\langle Z, \leqslant \rangle$ to the end of $\langle \omega, \leqslant \rangle$. To see that 1^* is countably homogeneous, we note that each function f which is the identity on $\langle \omega, \leqslant \rangle$ and translates the copy of $\langle Z, \leqslant \rangle$ finitely many positions to the right or left, is an automorphism of 1^*. Suppose $a_1, \ldots, a_n$ and $b_1, \ldots, b_n$ realize the same type in 1^*. Then if a_i is in $\langle \omega, \leqslant \rangle$, we must have $a_i = b_i$. If a_j is in the copy of $\langle Z, \leqslant \rangle$, then so is b_j, say $b_j = a_j + m$. Moreover, for any other a_k in $\langle Z, \leqslant \rangle$, the distances $a_j - a_k$ and $b_j - b_k$ must be the same, so $b_k = a_k + m$. Therefore there is an automorphism of 1^* mapping each a_j onto b_j. This shows that 1^* is countably homogeneous. The proof that no other model of T is countably homogeneous is similar to the proof that no model except Q^* is countably saturated, but uses the description of types on the next page.

In any model $\mathfrak{A}^*$ of T, the set $\langle X, \leqslant \rangle$, where $X = \{\langle a, 0\rangle : a \in A\}$, is a set of indiscernibles in $\mathfrak{A}^*$. To see this, we may suppose that $\mathfrak{A}$ is a submodel of Q, so that $\mathfrak{A}^* \prec Q^*$. Given any two increasing n-tuples $x_1 < \ldots < x_n$ and $y_1 < \ldots < y_n$ from X, there is an automorphism of Q^* in which each x_i goes to y_i. Therefore

$$(Q^*, x_1 \ldots x_n) \equiv (Q^*, y_1 \ldots y_n).$$

Since $\mathfrak{A}^* \prec Q^*$, we have

$$(\mathfrak{A}^*, x_1 \ldots x_n) \equiv (\mathfrak{A}^*, y_1 \ldots y_n),$$

whence $\langle X, \leqslant \rangle$ is indiscernible in $\mathfrak{A}^*$.

Since T has a countably saturated model, it has only countably many types of n-tuples. These types can be described explicitly. Consider the following sets of formulas:

$$x_1 - x_0 = n \text{ says } x_0 \leqslant x_1 \text{ and } x_1 \text{ is the } n\text{th successor of } x_0,$$
$$x_1 - x_0 = \infty \text{ says } x_0 \leqslant x_1 \text{ and } \neg(x_1 - x_0 = n),\ n = 0, 1, 2, \ldots.$$

Then two n-tuples $x_1, \ldots, x_n$ and $y_1, \ldots, y_n$ have the same type if and only if for each i and j, x_i, x_j and y_i, y_j realize exactly the same sets of formulas of the forms

$$x_i - 0 = n, \qquad x_i - 0 = \infty, \qquad x_i - x_j = n, \qquad x_i - x_j = \infty.$$

To prove this, show that in the countably saturated model there is an automorphism mapping each x_i into y_i.

EXAMPLE 3.4.5. *Peano arithmetic.* The Gödel incompleteness theorem shows that this theory T is not complete. In fact, no finite extension $T \cup \{\varphi_1, \ldots, \varphi_n\}$ is complete. Then, using a binary tree argument, we can see that T has 2^ω complete extensions. By a classical result of Ryll-Nardzewski, T does not have a set of Π_n^0 axioms, for any n. In particular, T does not have a set of Π_2^0 axioms. Therefore, by the preservation theorem, T is not preserved under unions of chains. It follows that T is not preserved under submodels, and also that T is not model complete. T is not preserved under homomorphisms, because the one-element model does not satisfy T.

The standard model $\langle \omega, 0, S, +, \cdot \rangle$ of T is isomorphically embedded in every model of T. Using this and the fact that T is not complete, we obtain a second proof that T is not model complete.

The theory T almost has built-in Skolem functions. That is, T has an expansion $\bar{T}$ such that:

(1). $\bar{T}$ has built-in Skolem functions.

(2). Every model $\mathfrak{A}$ of T has a unique expansion $\bar{\mathfrak{A}}$ to a model of $\bar{T}$.

(3). If φ is a sentence of $\mathscr{L}$, then $T \models \varphi$ if and only if $\bar{T} \models \varphi$.

(4). For every formula $\bar{\varphi}(x_1 \dots x_n)$ of $\bar{\mathscr{L}}$ there is a formula $\varphi(x_1 \dots x_n)$ of $\mathscr{L}$ such that $\bar{T} \models \varphi \leftrightarrow \bar{\varphi}$.

The expansion $\bar{T}$ is formed by adding, for each formula $(\exists x)\varphi(xx_1 \dots x_n)$ of $\mathscr{L}$, a function symbol $F_{(\exists x)\varphi}(x_1 \dots x_n)$ and the axioms

$$(\exists x)\varphi(xx_1 \dots x_n) \to F_{(\exists x)\varphi}(x_1 \dots x_n) \equiv \text{the least } x \text{ such that } \varphi(xx_1 \dots x_n),$$
$$\neg\, \exists x\varphi(xx_1 \dots x_n) \to F_{(\exists x)\varphi}(x_1 \dots x_n) \equiv 0. \text{ So (2) will be right.}$$

By the induction scheme in Peano arithmetic, $\bar{T}$ has the properties (1)–(4). $\bar{T}$ is what is called a *definitional extension* of T. It follows from (1)–(4) that if $\mathfrak{A}$, $\mathfrak{B}$ are models of T, then $\mathfrak{A} \prec \mathfrak{B}$ if and only if $\bar{\mathfrak{A}} \subset \bar{\mathfrak{B}}$. From this we see that the intersection of any set of elementary submodels of $\mathfrak{B}$ is an elementary submodel of $\mathfrak{B}$. In particular, each model $\mathfrak{B}$ of T has a smallest elementary submodel, and for each $X \subset B$ there is a least elementary submodel of $\mathfrak{B}$ containing X.

Every complete extension T' of T is an atomic theory and thus has an atomic model. This is because if $\varphi(x_1 \dots x_n)$ is consistent with T', then the formula which states that

(5) $\varphi(x_1 \dots x_n) \wedge (\forall y_1 \dots y_n)[\varphi(y_1 \dots y_n) \to 2^{y_1}3^{y_2} \cdot \ldots \cdot p_n^{y_n} \geqslant 2^{x_1}3^{x_2} \cdot \ldots \cdot p_n^{x_n}]$

is a complete formula in T' which implies φ. The standard model $\langle \omega, 0, S, +, \cdot \rangle$ is the atomic model for its complete theory. For any model $\mathfrak{B}$ of Peano arithmetic, the model $\mathfrak{A} \prec \mathfrak{B}$ which is the intersection of all elementary submodels of $\mathfrak{B}$ must be the countable atomic model of $\text{Th}(\mathfrak{B})$. This is because $\mathfrak{A}$ has no proper elementary submodels, so no other model of $\text{Th}(\mathfrak{B})$ could be prime.

The countable atomic model $\mathfrak{A} \prec \mathfrak{B}$ can also be characterized in another way. An element $b \in B$ is *definable* in $\mathfrak{B}$ iff there is a formula $\varphi(x)$ such that b is the unique element of B which satisfies $\varphi(x)$ in $\mathfrak{B}$, i.e.,

$$\mathfrak{B} \models (\forall x)(\varphi(x) \leftrightarrow x \equiv b).$$

Then $\mathfrak{A}$ is the submodel of $\mathfrak{B}$ where A is the set of all definable elements in $\mathfrak{B}$. For every definable element in $\mathfrak{B}$ must belong to $\mathfrak{A}$, and if $a \in A$ satisfies the complete formula $\varphi(x)$, then a is definable in $\mathfrak{A}$, and hence in $\mathfrak{B}$, by the formula

$$\varphi(x) \wedge (\forall y)(\varphi(y) \to y \geqslant x).$$

Since there are only countably many formulas and every element of an atomic model of T is definable as above, we see that T has no uncountable atomic models.

Finally, we show that Peano arithmetic has no countably saturated models. For, let T' be any complete extension of T. Let $P \subset \omega$ be the set of all prime numbers. For each set $X \subset P$ of primes, let

$$\Gamma_X(x) = \{p|x : p \in X\} \cup \{\neg p|x : p \in P \setminus X\}.$$

Thus x satisfies $\Gamma_X(x)$ if and only if X is the set of all primes $p \in P$ which divide x. For each finite subset $\{\varphi_1(x), \ldots, \varphi_n(x)\} \subset \Gamma_X(x)$, we can prove in Peano arithmetic that

$$(\exists x)(\varphi_1(x) \wedge \ldots \wedge \varphi_n(x)).$$

Therefore for each complete extension T' of T, each set $\Gamma_X(x)$ is consistent with T'. But if $X_1 \neq X_2$, then $\Gamma_{X_1}(x) \cup \Gamma_{X_2}(x)$ is inconsistent. It follows that T' has 2^ω distinct types in x (at least one containing each $\Gamma_X(x)$). Therefore T' has no countably saturated model.

3.5. Model completeness

In this section we present the beginnings of a branch of model theory called model theoretic algebra. In many cases, a phenomenon from abstract algebra, especially field theory, can be generalized to a fruitful notion in model theory. The model theoretic notion may in turn lead to new ideas in other areas of algebra. This approach to model theory was strongly influenced by the work of Abraham Robinson. For example, Robinson introduced the notion of the model completion of a theory, starting with the example of the theory of algebraically closed fields as the model completion of the theory of fields. He then introduced the new theory of differentially closed fields as the model completion of the theory of differential fields. In abstract algebra, existential and quantifier free formulas play a greater role than arbitrary formulas. Model theoretic algebra, in turn, has provided a collection of results in pure model theory which concentrate on existential and open formulas. The starting point is the notion of a model complete theory.

DEFINITION. A consistent theory T is said to be *model complete* if for all models $\mathfrak{A}, \mathfrak{B}$ of T, if $\mathfrak{A} \subset \mathfrak{B}$ then $\mathfrak{A} \prec \mathfrak{B}$.

Obviously, if T is model complete then any consistent theory $U \supset T$ in the same language is model complete.

Neither one of completeness and model completeness implies the other. The theory of dense linear order with first and last elements is complete but not model complete. So are the complete theories of the models $\langle \omega, S \rangle$ and $\langle \omega, \leqslant \rangle$. For each of these theories, the natural model has an isomorphic extension which has new elements at the beginning and thus is not an elementary extension, so the theory is not model complete.

From the results on elimination of quantifiers in Section 1.4, we see that the theory of dense linear order without endpoints is model complete. This theory is also complete. In a language with two or more additional constant symbols, the theory of dense linear order is still model complete but is no longer complete. Before giving more examples, we shall give some characterizations of model completeness.

THEOREM 3.5.1. *Given a consistent theory T of $\mathscr{L}$, the following are equivalent*:

(i). *T is model complete.*

(ii). *For every model $\mathfrak{A}$ of T, the theory $T \cup \Delta_{\mathfrak{A}}$ is complete in $\mathscr{L}_{\mathfrak{A}}$, where $\Delta_{\mathfrak{A}}$ is the diagram of $\mathfrak{A}$.*

(iii). (Robinson's Test). *If $\mathfrak{A}$, $\mathfrak{B}$ are models of T and $\mathfrak{A} \subset \mathfrak{B}$, then every existential sentence which holds in $\mathfrak{B}_A$ holds in $\mathfrak{A}_A$.*

(iv). *For every existential formula $\varphi(y_1, \dots, y_n)$ there is a universal formula $\psi(y_1, \dots, y_n)$ such that $T \models \varphi \leftrightarrow \psi$.*

(v). *For every formula $\varphi(y_1, \dots, y_n)$ there is a universal formula $\psi(y_1, \dots, y_n)$ such that $T \models \varphi \leftrightarrow \psi$.*

PROOF. (i) implies (ii). Assume (i), T is model complete. Let $\mathfrak{A}$ be a model of T. Since every extension of $\mathfrak{A}$ is an elementary extension, $T \cup \Delta_{\mathfrak{A}}$ has exactly the same models as the elementary diagram $\mathrm{Th}(\mathfrak{A}_A)$, and hence is complete.

(ii) implies (iii). Assume (ii). If $\mathfrak{A}$ and $\mathfrak{B}$ are models of T with $\mathfrak{A} \subset \mathfrak{B}$, then $\mathfrak{A}_A$ and $\mathfrak{B}_A$ are both models of the complete theory $T \cup \Delta_{\mathfrak{A}}$, so every existential sentence true in $\mathfrak{B}_A$ holds in $\mathfrak{A}_A$. Thus (iii) follows.

(iii) implies (iv). Assume (iii). Let $\varphi(y_1, \dots, y_n)$ be an existential formula. It is convenient to add new constants and form the existential sentence $\varphi(c_1, \dots, c_n)$ of $\mathscr{L}' = \mathscr{L} \cup \{c_1, \dots, c_n\}$. Let Γ be the set of all universal sentences γ of $\mathscr{L}'$ such that $T \models \varphi \to \gamma$. Let $(\mathfrak{A}, b_1, \dots, b_n)$ be a model of $T \cup \Gamma$, and let $\Delta_{\mathfrak{A}}$ be the diagram of $\mathfrak{A}$. Each finite conjunction $\theta(a_1, \dots, a_n, b_1, \dots, b_n)$ of sentences of $\Delta_{\mathfrak{A}}$ is consistent with

$T \cup \{\varphi\}$, because the universal sentence $(\forall x_1, \ldots, x_n) \neg\theta(x_1, \ldots, x_n, b_1, \ldots, b_n)$ is false in $\mathfrak{A}$, thus does not belong to Γ and is not a consequence of $T \cup \{\varphi\}$. Therefore $T \cup \{\varphi\} \cup \Delta_{\mathfrak{A}}$ has a model $\mathfrak{B}_A$. Then $\mathfrak{A} \subset \mathfrak{B}$ and $\mathfrak{B}$ is a model of T. By (iii), every existential sentence true in $\mathfrak{B}_A$ is true in $\mathfrak{A}_A$. In particular, the existential sentence φ holds in $(\mathfrak{A}, b_1, \ldots, b_n)$. We conclude that every model of $T \cup \Gamma$ is a model of φ, i.e. $T \cup \Gamma \models \varphi$. By the compactness theorem, there are $\gamma_1, \ldots, \gamma_n \in \Gamma$ such that $T \models \gamma_1 \wedge \ldots \wedge \gamma_n \rightarrow \varphi$. It follows that $T \models \gamma_1 \wedge \ldots \wedge \gamma_n \leftrightarrow \varphi$. Moving quantifiers to the front and replacing $c_1, \ldots, c_n$ by $y_1, \ldots, y_n$, we obtain a universal formula ψ such that $T \models \varphi(y_1, \ldots, y_n) \leftrightarrow \psi(y_1, \ldots, y_n)$. This proves (iv).

(iv) implies (v) by induction on the complexity of formulas.

(v) implies (i). Finally we assume (v). Let $\mathfrak{A}$, $\mathfrak{B}$ be models of T with $\mathfrak{A} \subset \mathfrak{B}$. Let $a_1, \ldots, a_n \in A$ and suppose $\mathfrak{B} \models \varphi[a_1, \ldots, a_n]$. Let ψ be universal and $T \models \varphi \leftrightarrow \psi$. Then we have $\mathfrak{B} \models \psi[a_1, \ldots, a_n]$, $\mathfrak{A} \models \psi[a_1, \ldots, a_n]$, and $\mathfrak{A} \models \varphi[a_1, \ldots, a_n]$. The same argument holds for $\neg\varphi$. Therefore $\mathfrak{A} \prec \mathfrak{B}$ and T is model complete. ⊣

Robinson's Test, condition (iii) in the above proposition, is often used to show that particular theories are model complete. One of the original examples which served to motivate the notion of model completeness is the theory ACF of algebraically closed fields. The model completeness of ACF is closely related to a classical algebraic result, the Hilbert nullstellensatz. One form of the nullstellensatz is as follows. Let F and G be algebraically closed fields with $F \subset G$ and let $\Sigma(x_1, \ldots, x_n)$ be a finite set of polynomial equations with coefficients in F; if $\Sigma(x_1, \ldots, x_n)$ has a solution in G then $\Sigma(x_1, \ldots, x_n)$ has a solution in F. In field theory, any term is equal to a polynomial, and any inequality $\neg(p = q)$ can be replaced by an equation with one extra existentially quantified variable, $(\exists x)[x \cdot (p - q) = 1]$. Thus the nullstellensatz implies that for any pair of algebraically closed fields $F \subset G$, any existential sentence with constants from F which holds in G holds in F. By Robinson's Test (iii), the theory ACF is model complete. Notice that ACF is not a complete theory, because fields of different characteristic are not elementarily equivalent.

Here is a list of examples of model complete theories.

EXAMPLES 3.5.2.

(1). The complete theory of the model $\langle \omega, 0, S \rangle$.

(2). The complete theory of the ordered group of integers under addition with a constant for 1, $\langle Z, +, -, 0, 1, \leqslant \rangle$, which was considered in Example 2.4.7. (Without a constant for 1 the theory would not be model complete, because the even integers would be an isomorphic submodel which would fail to satisfy the formula stating that there is an element between 0 and 2).

(3). The theory of countably many independent unary relations.
(4). The theory of divisible torsion free Abelian groups.
(5). The theory of divisible ordered Abelian groups.
(6). The theory of algebraically closed fields.
(7). The theory of real closed fields.
(8). The theory of real closed ordered fields.

Each of these theories can be proved to be model complete by Robinson's Test. However, the methods we shall develop in this section will lead to other, easier proofs of model completeness. Some of the examples of model completeness will be left as exercises. The model completeness of the theories of real closed fields and real closed ordered fields depend on results from algebra which will be explained in detail in Section 5.4.

Sometimes a theory which is not model complete can be made model complete by adding a symbol for a constant, function, or relation which is already definable by a formula in the original theory. For example, the theory of $\langle \omega, S \rangle$ is not model complete but the theory of $\langle \omega, 0, S \rangle$ is model complete. The constant 0 is definable in $\mathrm{Th}(\langle \omega, S \rangle)$ as the unique element which is not in the range of S.

As another example of model completeness, we recall that by Proposition 3.3.3, every theory with built-in Skolem functions is model complete.

We have seen in Section 3.4 that condition (v) of Theorem 3.5.1 is false for Peano arithmetic. Therefore Peano arithmetic is not model complete. However, the definitional extension of Peano arithmetic with built in Skolem functions is model complete. For the same reasons, complete number theory, that is, the complete theory of the standard model $\langle \omega, +, \cdot, S, 0 \rangle$, is not model complete but has a model complete definitional extension with built in Skolem functions.

By examining the proof of Theorem 3.5.1 we obtain two useful corollaries.

COROLLARY 3.5.3. *Suppose T is a consistent theory in a language $\mathscr{L}$ with $\alpha \geqslant ||\mathscr{L}||$. Then T is model complete if and only if for all models $\mathfrak{A}$ and $\mathfrak{B}$*

of T such that $\mathfrak{A} \subset \mathfrak{B}$ and $\mathfrak{A}$ and $\mathfrak{B}$ are either finite or of power α, every existential sentence true in $\mathfrak{B}_A$ is true in $\mathfrak{A}_A$.

PROOF. Same as the proof of Theorem 3.5.1, but using models of power α in the step from (iii) to (iv). ⊣

COROLLARY 3.5.4. *Suppose T is a consistent theory in a recursive language $\mathscr{L}$. Then for T to be model complete it is necessary and sufficient that T have the following property*:

(∗) *For all models $\mathfrak{A}$ and $\mathfrak{B}$ of T such that $\mathfrak{A} \subset \mathfrak{B}$ and the expanded model $(\mathfrak{B}, A)$ is recursively saturated, and every tuple $c_1, \ldots, c_n$ in A, the models $(\mathfrak{A}, c_1, \ldots, c_n)$ and $(\mathfrak{B}, c_1, \ldots, c_n)$ are partially isomorphic.*

PROOF. Suppose first that T is model complete. Then for any pair of models $\mathfrak{A}$ and $\mathfrak{B}$ as in (∗), $\mathfrak{B}$ is an elementary extension of $\mathfrak{A}$, and thus for all $c_1, \ldots, c_n$ in A, $(\mathfrak{A}, c_1, \ldots, c_n) \equiv (\mathfrak{B}, c_1, \ldots, c_n)$. Since $(\mathfrak{B}, A)$ is recursively saturated, it follows from the proof of Theorem 2.4.5 that $(\mathfrak{A}, c_1, \ldots, c_n)$ and $(\mathfrak{B}, c_1, \ldots, c_n)$ are partially isomorphic.

Now assume that T has property (∗). Since $\mathscr{L}$ is recursive, it is countable. We show that T satisfies the criterion in Corollary 3.5.3. Let $\mathfrak{A}$ and $\mathfrak{B}$ be finite or countable models of T such that $\mathfrak{A} \subset \mathfrak{B}$. Then the expanded model $(\mathfrak{B}, A)$ has a recursively saturated countable elementary extension $(\mathfrak{B}', A')$. It follows from (∗) that every existential sentence (in fact every sentence) true in $\mathfrak{B}'_A$ is true in $\mathfrak{A}'_A$, and therefore every existential sentence true in $\mathfrak{B}_A$ is true in $\mathfrak{A}_A$, as required in 3.5.3. Therefore T is model complete. ⊣

This corollary can be used to show that theories are model complete by a method which is parallel to the method of showing that theories are complete with recursively saturated model pairs. It can be applied to all of the theories listed in Example 3.5.2.

EXAMPLE 3.5.5. We prove that the complete theory T of the model $\langle Z, +, -, 0, 1, \leqslant \rangle$ is model complete. We shall make use of the complete set of axioms for T given in Example 2.4.7. Let $\mathfrak{A}$, $\mathfrak{B}$ be models of T such that $\mathfrak{A} \subset \mathfrak{B}$ and the pair $(\mathfrak{B}, A)$ is recursively saturated, and let $c_1, \ldots, c_m$ be a tuple of elements of A. Let I be the relation on pairs of tuples from A and B given in Example 2.4.7, that is,

$$\langle a_1, \ldots, a_n \rangle I \langle b_1, \ldots, b_n \rangle$$

if and only if for each linear term $P(x_1, \ldots, x_n)$ with integer coefficients and each integer $k > 1$,

(a) $$\mathfrak{A} \models P(a_1, \ldots, a_n) > 0 \text{ iff } \mathfrak{B} \models P(b_1, \ldots, b_n) > 0$$

and

(b) $$\mathfrak{A} \models k | P(a_1, \ldots, a_n) \text{ iff } \mathfrak{B} \models k | P(b_1, \ldots, b_n).$$

Let J be the relation on pairs of tuples from A and B such that

$$\langle a_1, \ldots, a_n \rangle J \langle b_1, \ldots, b_n \rangle$$

if and only if

$$\langle c_1, \ldots, c_m, a_1, \ldots, a_n \rangle I \langle c_1, \ldots, c_m, b_1, \ldots, b_n \rangle.$$

We shall show that J is a partial isomorphism from $(\mathfrak{A}, c_1, \ldots, c_m)$ to $(\mathfrak{B}, c_1, \ldots, c_m)$. It will then follow from Corollary 3.5.4 that T is model complete. By the argument of Example 2.4.7, I is a partial isomorphism from $\mathfrak{A}$ to $\mathfrak{B}$. Therefore J preserves atomic sentences and satisfies the back and forth condition. It remains to show that $\emptyset J \emptyset$, that is,

$$\langle c_1, \ldots, c_m \rangle I \langle c_1, \ldots, c_m \rangle.$$

Let $P(x_1, \ldots, x_m)$ be a linear term with integer coefficients and let k be an integer greater than 1. Since $\mathfrak{A} \subset \mathfrak{B}$, we have

$$\mathfrak{A} \models P(c_1, \ldots, c_m) \text{ iff } \mathfrak{B} \models P(c_1, \ldots, c_m)$$

and

$$\mathfrak{A} \models k | P(c_1, \ldots, c_m) \text{ implies } \mathfrak{B} \models k | P(c_1, \ldots, c_m).$$

Suppose that

$$\text{not } \mathfrak{A} \models k | P(c_1, \ldots, c_m).$$

From the axioms of T, there is a positive integer $h < k$ such that

$$\mathfrak{A} \models k | P(c_1, \ldots, c_m) - h.$$

Since $\mathfrak{A} \subset \mathfrak{B}$,

$$\mathfrak{B} \models k | P(c_1, \ldots, c_m) - h.$$

It now follows from the axioms of T that

$$\text{not } \mathfrak{B} \models k | P(c_1, \ldots, c_m).$$

Therefore $\emptyset J \emptyset$, and J is a partial isomorphism as required. ⊣

We now introduce the important notion of an existentially closed model, which is suggested by the notion of an algebraically closed field and is related to Robinson's test for model completeness in Theorem 3.5.1 (iii).

DEFINITION. A model $\mathfrak{A}$ is said to be *existentially closed* in a model $\mathfrak{B}$ if $\mathfrak{A} \subset \mathfrak{B}$ and every existential sentence which holds in $\mathfrak{B}_A$ holds in $\mathfrak{A}_A$. $\mathfrak{A}$ is said to be existentially closed for T if $\mathfrak{A}$ is existentially closed in every model $\mathfrak{B}$ of T such that $\mathfrak{A} \subset \mathfrak{B}$.

By Theorem 3.5.1, a consistent theory T is model complete if and only if every model of T is existentially closed for T. Corollary 3.5.3 states that if $||\mathscr{L}|| \leq \alpha$, then T is model complete if and only if every model of T which is finite or of power α is existentially closed for T.

DEFINITION. Given a theory T, $T_\forall$ denotes the set of all universal consequences of T, and $T_{\forall\exists}$ denotes the set of all universal existential consequences of T. Two theories T and U are *cotheories* if $T_\forall = U_\forall$.

REMARK 3.5.6. Here is a list of easy facts about existentially closed models and cotheories whose proofs are left as exercises.

(1). $\mathfrak{A}$ is a model of $T_\forall$ if and only if $\mathfrak{A}$ is a submodel of some model of T.

(2). T and U are cotheories if and only if every model of T can be extended to a model of U and every model of U can be extended to a model of T.

(3). $T_\forall$ is a cotheory of T, and is the unique cotheory of T which has a set of universal axioms.

(4). If $\mathfrak{A} \subset \mathfrak{B} \subset \mathfrak{C}$ and $\mathfrak{A}$ is existentially closed in $\mathfrak{C}$, then $\mathfrak{A}$ is existentially closed in $\mathfrak{B}$.

(5). If T and U are cotheories then a model $\mathfrak{A}$ is existentially closed for T if and only if it is existentially closed for U.

(6). The union of a chain of models which are existentially closed for T is existentially closed for T.

(7). If $\mathfrak{A}$ is existentially closed in $\mathfrak{B}$, then every $\forall\exists$ sentence true in $\mathfrak{B}$ is true in $\mathfrak{A}$.

(8). Every model of $T_\forall$ which is existentially closed for T is a model of $T_{\forall\exists}$.

Let us consider again the theory ACF of algebraically closed fields.

Let F denote the theory of fields and ID the theory of integral domains with unit. It is known that every integral domain with unit can be extended to a field and every field can be extended to an algebraically closed field. Thus by (2), the theories, ACF, F, and ID are cotheories of each other. The theory ID has a universal set of axioms, and therefore ID is equivalent to $\mathrm{ACF}_\forall$ and to $F_\forall$. The theory ACF is model complete, so every model of ACF is extentially closed for ACF. By (5), every model of ACF is existentially closed for F and ID as well.

Similarly, the theories of real closed ordered fields and ordered fields are cotheories of each other, and each real closed ordered field is existentially closed for the theory of ordered fields.

The next lemma is an existence result for existentially closed models.

LEMMA 3.5.7. *Suppose T is a theory which is preserved under unions of chains. Then for each cardinal $\alpha \geqslant ||\mathscr{L}||$, every model of T of power α can be extended to a model of T of power α which is existentially closed for T.*

PROOF. Let $\mathfrak{A}$ be a model of T of power $\alpha \geqslant ||\mathscr{L}||$. Let $\{\varphi_\beta : \beta < \alpha\}$ be an enumeration of all existential sentences of $\mathscr{L}_A$. We form a chain of models of T of power α,

$$\mathfrak{A} = \mathfrak{A}_0 \subset \mathfrak{A}_1 \subset \ldots \subset \mathfrak{A}_\beta \subset \ldots,$$

such that if φ_β holds in some extension of $\mathfrak{A}_{\beta A}$ which is a model of T, then φ_β holds in $\mathfrak{A}_{\beta+1A}$. We can continue through the limit ordinals because T is preserved under unions of chains. Let $\mathfrak{A}'$ be the union of the chain of models $\mathfrak{A}_\beta$, $\beta < \alpha$. Then every existential sentence which holds in some extension of $\mathfrak{A}'_A$ which is a model of T holds in $\mathfrak{A}'_A$. Repeating the construction ω times we obtain a chain of models

$$\mathfrak{A} = \mathfrak{B}_0 \subset \mathfrak{B}_1 \subset \ldots$$

of T such that any existential sentence with constants from B_n which holds in some extension of $\mathfrak{B}_{n+1}$ which is a model of T holds in $\mathfrak{B}_{n+1}$. It follows that the union of the chain of models $\mathfrak{B}_n$, $n < \omega$, is existentially closed for T and is a model of T of power α which is an extension of $\mathfrak{A}$. ⊣

The following result sometimes gives a very easy proof that a theory is model complete.

THEOREM 3.5.8 (Lindström). *Let T be a theory in a countable language* *such that*:

(i). *All models of T are infinite.*

(ii). *T is preserved under unions of chains.*

(iii). *T is α-categorical for some infinite cardinal α.*

Then T is model complete.

PROOF. By (iii), T has infinite models. By Lemma 3.5.7, T has a model of power α which is existentially closed for T. Since T is α-categorical, every model of T of power α is existentially closed for T. Then by Corollary 3.5.3 and (i), T is model complete. ⊣

EXAMPLES 3.5.9. We have seen in Section 3.1 that each of the following theories (1)–(6) is categorical in some infinite power and has no finite models, and is therefore complete by the Los-Vaught Test. Each of these theories is also preserved under unions of chains. Therefore by Theorem 3.5.8, each of these theories is model complete.

(1). The theory of dense simply ordered sets without endpoints (which is a cotheory of the theory of simple order).

(2). The theory of atomless Boolean algebras (which is a cotheory of the theory of Boolean algebras).

(3). The theory of algebraically closed fields of characteristic zero (which is a cotheory of the theory of fields of characteristic zero). Similarly for characteristic p.

(4). The theory of infinite Abelian groups with all elements of order p.

(5). The theory of countably many unequal constant symbols.

(6). The theory of a one-to-one function of A onto A with no finite cycles.

(7). The theory of divisible torsion free Abelian groups.

Theorem 3.5.8 can also be used to give a very simple proof that the theory of algebraically closed fields is model complete. If $\mathfrak{A}$ and $\mathfrak{B}$ are algebraically closed fields and $\mathfrak{A} \subset \mathfrak{B}$, then $\mathfrak{A}$ and $\mathfrak{B}$ have the same characteristic. Since the theory of algebraically closed fields of a given characteristic is model complete by 3.5.8, $\mathfrak{A} \prec \mathfrak{B}$. Therefore the theory of algebraically closed fields is model complete.

An example of an ω_1-categorical complete theory which is not model complete is the theory of the model $\langle \omega, S \rangle$. An example of an ω-categorical complete theory which is not model complete is the theory of equivalence relations with infinitely many infinite classes and one finite class which has just one element.

Let us record a useful property of model complete theories which follows at once from the Elementary Chain Theorem and the preservation theorem for unions of chains.

PROPOSITION 3.5.10. *If T is model complete then T is preserved under unions of chains and T has a set of $\forall\exists$ axioms.*

Recall from Section 2.3 that a model $\mathfrak{A}$ of a theory T is prime iff $\mathfrak{A}$ is elementarily embeddable in every model of T. We now introduce the analogous notion for isomorphic embeddings.

DEFINITION. A model $\mathfrak{A}$ of a theory T is said to be *algebraically prime* iff $\mathfrak{A}$ is isomorphically embeddable in every model of T.

Since every elementary embedding is an isomorphic embedding, every prime model of a theory T is algebraically prime. If T is model complete, then every isomorphic embedding is an elementary embedding, so every algebraically prime model is prime. See Exercise 4.4.20 for an example of a complete theory which has algebraically prime models (in fact infinitely many) but no prime model. A theory which is not complete cannot have a prime model, but may have an algebraically prime model. Here are some examples:

EXAMPLES. The field of rational numbers is an algebraically prime model of the theory of fields of characteristic zero.

The field of integers modulo p is an algebraically prime model of the theory of fields of characteristic p.

The two-element Boolean algebra is an algebraically prime model of the theory of Boolean algebras.

The one-element group is an algebraically prime model of the theory of groups.

Every countable model of the theory of dense simple orders (with or without endpoints) is algebraically prime.

We shall now introduce two properties of a theory which say that models can be fitted together.

DEFINITION. A theory T has the *joint embedding property* if for any two models $\mathfrak{A}$, $\mathfrak{B}$ of T there is a model $\mathfrak{C}$ of T such that both $\mathfrak{A}$ and $\mathfrak{B}$ are isomorphically embeddable in $\mathfrak{C}$.

T has the *amalgamation property* if for any three models, $\mathfrak{A}$, $\mathfrak{B}$, and $\mathfrak{C}$

of T and isomorphic embeddings

$$f:\mathfrak{C}\to\mathfrak{A},\ g:\mathfrak{C}\to\mathfrak{B}$$

there is a model $\mathfrak{C}'$ of T and isomorphic embeddings

$$f':\mathfrak{A}\to\mathfrak{C}',\ g':\mathfrak{B}\to\mathfrak{C}'$$

such that the diagram commutes. The model $\mathfrak{C}'$ is said to *amalgamate* $\mathfrak{A}$ and $\mathfrak{B}$ over $\mathfrak{C}$.

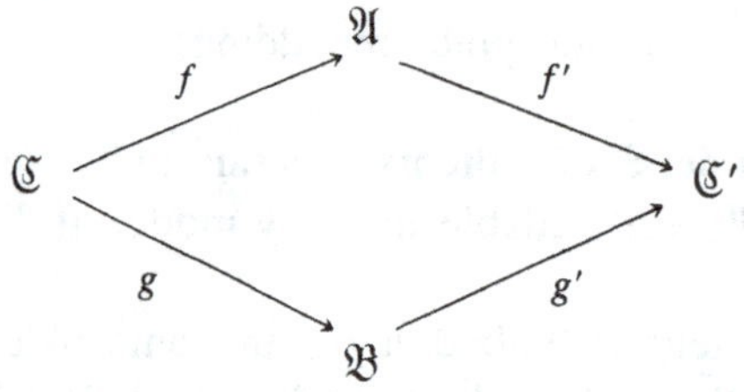

Notice that T has the amalgamation property if and only if for every model $\mathfrak{C}$ of T, the theory $T\cup\Delta_{\mathfrak{C}}$ has the joint embedding property where $\Delta_{\mathfrak{C}}$ is the diagram of $\mathfrak{C}$. By the compactness theorem, every complete theory satisfies the analogues of the joint embedding property and the amalgamation property for elementary embeddings. Therefore every complete theory has the joint embedding property.

As a special case, it follows that every model complete theory has the amalgamation property. For if T is model complete, then for each model $\mathfrak{C}$ of T, $T\cup\Delta_{\mathfrak{C}}$ is complete and thus has the joint embedding property.

Examples. Some well known theories from algebra which have the amalgamation property are the theories of fields, ordered fields, groups, Abelian groups, and Boolean algebras.

The theory of fields does not have the joint embedding property because fields of different characteristics cannot be embedded in a common field. The theory of algebraically closed fields is an example of a model complete theory which does not have the joint embedding property. The theories of fields of characteristic p, ordered fields, groups, Abelian groups, and Boolean algebras do have the joint embedding property. An example of a theory which has the joint embedding property but does not have the amalgamation property is the theory of equivalence relations such that at most one equivalence class has more than one element.

The next proposition gives two simple conditions under which a model complete theory must be complete.

Proposition 3.5.11. *Let T be a model complete theory.*

(i) (Prime Model Test). *If T has an algebraically prime model then T is complete.*

(ii) *T is complete if and only if T has the joint embedding property.*

Proof. Both parts follow at once from the fact that any isomorphic embedding between models of T is an elementary embedding. ⊣

Examples. The ordered field of real algebraic numbers is an algebraically prime model of the theory of real closed ordered fields. Since the theory of real closed ordered fields is model complete, it is complete by the Prime Model Test. Two possible converses of Proposition 3.5.1 fail. The theory of countably many unary relations is complete but has no algebraically prime model. The complete theories of each of the models $\langle \omega, S \rangle$, $\langle \omega, \leqslant \rangle$, and $\langle \omega, +, \cdot, S, 0 \rangle$ have algebraically prime models and the joint embedding property, but are not model complete. The theory of algebraically closed fields is model complete but does not have an algebraically prime model. All of the other model complete theories listed in Example 3.5.2, and all the model complete theories listed in Example 3.5.9, have algebraically prime models and hence are complete.

The theories of algebraically closed fields and real closed ordered fields were the original examples which led Robinson to introduce the notion of a model complete theory. We now take up the notions which generalize the relationship between the theory of algebraically closed fields and the theory of fields, and the parallel relationship between the theory of real closed ordered fields and the theory of ordered fields. Robinson first introduced the notion of a model completion for this purpose. More recently, it was found that the more general but simpler notion of a model companion is better.

Definition. A theory T^* is a *model companion* of T if T^* is a cotheory of T and T^* is model complete.

Examples 3.5.12. Here are some examples of model companions from algebra.

(1). The theory of atomless Boolean algebras is a model companion of the theory of Boolean algebras.

(2). The theory of dense simple order without endpoints is a model companion of the theory of simple order.

(3). The theory of divisible torsion free Abelian groups is a model companion of the theory of torsion free Abelian groups.

(4). The theory of divisible ordered Abelian groups is a model companion of the theory of ordered Abelian groups.

(5). The theory of algebraically closed fields is a model companion of the theory of fields.

(6). The theory of real closed ordered fields is a model companion of the theory of ordered fields.

(7). The theory of real closed fields is a model companion of the theory of formally real fields. (A field is formally real if no finite sum of squares of nonzero elements is equal to zero. It can be shown that a field can be expanded to an ordered field if and only if it is formally real.)

We now develop some properties of the model companion.

PROPOSITION 3.5.13. *Every theory has at most one model companion up to logical equivalence.*

PROOF. Let T^* and T^{**} be model companions of T. Then T^* and T^{**} are model complete cotheories of each other. Let $\mathfrak{A}_1$ be a model of T^*. By repeated extension, there is a chain of models

$$\mathfrak{A}_1 \subset \mathfrak{A}_2 \subset \mathfrak{A}_3 \subset \ldots$$

such that $\mathfrak{A}_n$ is a model of T^* for odd n and of T^{**} for even n. Let $\mathfrak{A}$ be the union of this chain. The $\mathfrak{A}_n$ for odd n form an elementary chain, and by the Elementary Chain Theorem $\mathfrak{A}$ is an elementary extension of $\mathfrak{A}_1$. Similarly, $\mathfrak{A}$ is an elementary extension of $\mathfrak{A}_2$. Therefore $\mathfrak{A}_1$ is a model of T^{**}. In a similar way, every model of T^{**} is a model of T^*, so T^* and T^{**} are logically equivalent. ⊣

REMARK 3.5.14. If T^* is a model companion of T then every model of T^* is existentially closed for T and is a model of $T_{\forall\exists}$.

PROOF. By 3.5.6 (4) and (7). ⊣

PROPOSITION 3.5.15. *Let T be a theory which is preserved under unions of chains and let K be the class of all models of T which are existentially closed for T. Then* (i). *If T has a model companion T^*, then K is the class of all models of T^*.*

(ii). *T has a model companion if and only if K is an elementary class (the class of all models of some theory T').*

PROOF. (i) Let T^* be the model companion of T. By 3.5.14, every model of T^* belongs to K. Now let $\mathfrak{A}$ be any model in the class K. Since $\mathfrak{A}$ is a model of T and T^* is a cotheory of T, $\mathfrak{A}$ can be extended to a model $\mathfrak{B}$ of T^*. $\mathfrak{A}$ is existentially closed for T and hence for T^*. Then $\mathfrak{A}$ is existentially closed in $\mathfrak{B}$ and hence every $\forall\exists$ sentence true in $\mathfrak{B}$ is true in $\mathfrak{A}$. By proposition 3.5.10, T^* has an $\forall\exists$ set of axioms, so $\mathfrak{A}$ is a model of T^*.

(ii) If T has a model companion then (i) shows that K is an elementary class. Suppose K is an elementary class, say the class of all models of a theory T'. Then every model of T' is a model of T. By Lemma 3.5.7, every model of T can be extended to a model of T'. Thus T and T' are cotheories. Therefore every model of T' is existentially closed for T'. By Robinson's Test, Theorem 3.5.1 (iii), T' is model complete and hence a model companion of T. ⊣

Proposition 3.5.15 can be used to show that several familiar theories from algebra which are preserved under unions of chains do not have model companions. These include the theories of groups, commutative rings, and division rings. For each of these theories, the class of all existentially closed models of the theory is not elementary. As an illustration we sketch the argument for the theory of groups.

EXAMPLE 3.5.16. *The theory of groups has no model companion.*

PROOF. It is known that for each group G and any two elements $a, b \in G$ of the same order, there is a group $H \supset G$ in which a and b are conjugate, that is, $b = c^{-1}ac$ for some $c \in H$. The proof uses methods from group theory and will not be given here. Since each group can be extended to an existentially closed group, there is an existentially closed group G which contains elements of arbitrarily large finite order. Two elements in a group which are conjugate have the same order. Therefore, for each finite n there is a pair of elements in G of order greater than n which are not conjugate in G. By compactness, G has an elementary extension G' which has a pair of elements of infinite order which are not conjugate in G'. Thus G' is not existentially closed, and the class of existentially closed groups is not elementary. By 3.5.15, the theory of groups has no model companion. ⊣

Here is a syntactical characterization of $\forall\exists$ theories which have model companions. Its proof uses the Restricted Omitting Types Theorem stated in Exercise 2.2.11.

THEOREM 3.5.17. *Let T be a consistent $\forall\exists$ theory in a countable language $\mathcal{L}$. Then T has a model companion if and only if condition* (i) *below holds.*

(i) *For every existential formula $\theta(x_1 \ldots x_n)$ there is a universal formula $\theta'(x_1 \ldots x_n)$ such that:*

(a) $$T \vdash \theta(x_1 \ldots x_n) \rightarrow \theta'(x_1 \ldots x_n).$$

(b) *For every universal formula $\varphi(x_1 \ldots x_n)$,*

if $T, \theta' \rightarrow \theta \vdash \varphi$, then $T \vdash \varphi$.

Moreover, if (i) *holds, then the theory*

$$T' = T \cup \{(\forall x_1 \ldots x_n)(\theta' \rightarrow \theta) : \theta \text{ is existential}\}$$

is a model companion of T.

PROOF. First suppose that T' is a model companion of T. Let $\theta(x_1 \ldots x_n)$ be an existential formula. T' is model complete, so there is a universal formula θ' such that $T' \vdash \theta \leftrightarrow \theta'$. We prove (a) and (b). Let $\mathfrak{A}$ be a model of T and let $\mathfrak{A} \models \theta[a_1 \ldots a_n]$. Since T' is a cotheory of T, there is an extension $\mathfrak{B} \supset \mathfrak{A}$ such that $\mathfrak{B} \models T'$. θ is existential, so $\mathfrak{B} \models \theta[a_1 \ldots a_n]$. Using the fact that $\mathfrak{B} \models T'$, we have $\mathfrak{B} \models \theta'[a_1 \ldots a_n]$. Since θ' is universal, $\mathfrak{A} \models \theta'[a_1 \ldots a_n]$. This proves (a).

Now let φ be a universal formula which is not a consequence of T. Suppose $\mathfrak{A}$ is a model of T and $\mathfrak{A} \models \neg\varphi[a_1 \ldots a_n]$. Extend $\mathfrak{A}$ to a model $\mathfrak{B}$ of T'. Then $\mathfrak{B} \models \neg\varphi[a_1 \ldots a_n]$. Since $\mathfrak{B}$ is a model of T', we have $\mathfrak{B} \models \theta' \rightarrow \theta[a_1 \ldots a_n]$. T' is a cotheory of T, so $\mathfrak{B}$ can be extended to a model $\mathfrak{C}$ of T. Both $\neg\varphi$ and $(\theta' \rightarrow \theta)$ are equivalent to existential formulas, so $\mathfrak{C} \models \neg\varphi[a_1 \ldots a_n]$ and $\mathfrak{C} \models (\theta' \rightarrow \theta)[a_1 \ldots a_n]$. This shows that φ is not a consequence of T and $(\theta' \rightarrow \theta)$, and (b) follows.

For the converse, assume (i). Let

$$T' = T \cup \{(\forall x_1 \ldots x_n)(\theta' \rightarrow \theta) : \theta \text{ is existential}\}.$$

We must prove that T' is a model companion of T. Using (a), we see that for each existential formula θ, $T' \models \theta \leftrightarrow \theta'$ and θ' is universal, so T' is model complete by Theorem 3.5.1 (iv). For each existential formula $\theta(x_1 \dots x_n)$, let Σ_θ be the single universal formula

$$\Sigma_\theta = \{\neg[\theta'(x_1 \dots x_n) \to \theta(x_1 \dots x_n)]\}.$$

Let $\mathfrak{A}$ be a countable model of T and let $T_{\mathfrak{A}}$ be the union of T and the diagram of $\mathfrak{A}$. Using (b) and the Restricted Omitting Types Theorem, we see that $T_{\mathfrak{A}}$ has a countable model $\mathfrak{B}_A$ omitting each set Σ_θ. Let $\mathfrak{B}$ be the reduct of $\mathfrak{B}_A$ to the original language $\mathscr{L}$. $\mathfrak{B} \supset \mathfrak{A}$ and by (a), $\mathfrak{B}$ is a model of T'. Therefore T' is a cotheory of T, and thus a model companion of T. ⊣

We shall now introduce the notion of a model completion of a theory.

DEFINITION. A theory T^* is a *model completion* of a theory T if T^* is a model companion of T and for every model $\mathfrak{A}$ of T with diagram $\Delta_{\mathfrak{A}}$, $T^* \cup \Delta_{\mathfrak{A}}$ is complete.

The notion of model completion may be regarded as a uniqueness condition. It adds to the notion of model companion the condition that whenever $\mathfrak{A}$ is a model of T and $\mathfrak{B}$, $\mathfrak{C}$ are two extensions of $\mathfrak{A}$ which are models of T^*, $\mathfrak{B}_A$ and $\mathfrak{C}_A$ satisfy the same sentences. The following result gives a relationship between model completions and the amalgamation property.

PROPOSITION 3.5.18. *Let T^* be a model companion of T. The following are equivalent*:

(i). *T^* is a model completion of T.*
(ii). *T has the amalgamation property.*

PROOF. Assume (i), that T^* is a model completion of T. Let $\mathfrak{A}$, $\mathfrak{B}$, and $\mathfrak{C}$ be models of T with isomorphic embeddings $f : \mathfrak{C} \to \mathfrak{A}$, $g : \mathfrak{C} \to \mathfrak{B}$. Since T^* is a cotheory of T, there are models $\mathfrak{A}'$, $\mathfrak{B}'$ of T^* with $\mathfrak{A} \subset \mathfrak{A}'$ and $\mathfrak{B} \subset \mathfrak{B}'$. Then $(\mathfrak{A}', fc)_{c \in C}$ and $(\mathfrak{B}', gc)_{c \in C}$ are both models of the complete theory $T^* \cup \Delta_{\mathfrak{C}}$. Any complete theory has the joint embedding property, so these models have a common extension $(\mathfrak{C}', hc)_{c \in C}$ which is also a model of $T^* \cup \Delta_{\mathfrak{C}}$. Because T and T^* are cotheories, $\mathfrak{C}'$ has an extension $\mathfrak{C}''$ which is a model of T. Then $\mathfrak{C}''$ amalgamates $\mathfrak{A}$ and $\mathfrak{B}$ over $\mathfrak{C}$. This shows that T has the amalgamation property, so (ii) holds.

Now assume (ii), that T has the amalgamation property. Let $\mathfrak{C}$ be a model of T and let $(\mathfrak{A}, fc)_{c\in C}$, $(\mathfrak{B}, gc)_{c\in C}$ be models of $T^* \cup \Delta_{\mathfrak{C}}$. We must show that these models are elementarily equivalent. Since T and T^* are cotheories, $\mathfrak{A}$ and $\mathfrak{B}$ can be extended to models $\mathfrak{A}'$ and $\mathfrak{B}'$ of T. By the amalgamation property of T, there is a model $\mathfrak{C}'$ of T and isomorphic embeddings $f' : \mathfrak{A}' \to \mathfrak{C}'$, $g' : \mathfrak{B}' \to \mathfrak{C}'$ which amalgamate $\mathfrak{A}'$ and $\mathfrak{B}'$ over $\mathfrak{C}$. Again using the fact that T and T^* are cotheories, $\mathfrak{C}'$ can be extended to a model $\mathfrak{C}''$ of T^*. We have the following commutative diagram.

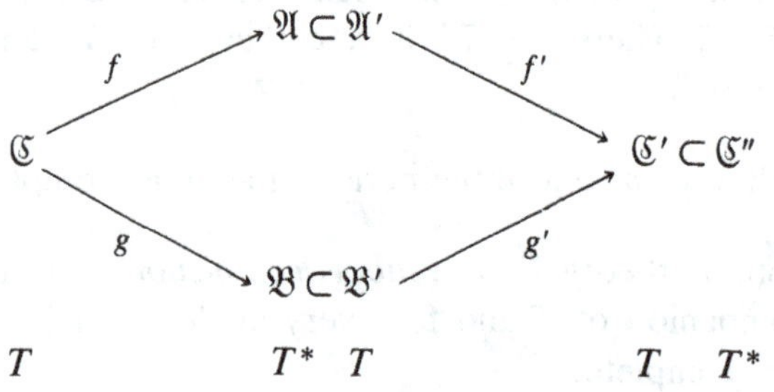

Since T^* is model complete, the embeddings of $\mathfrak{A}$ and $\mathfrak{B}$ into $\mathfrak{C}''$ are elementary. It follows that $(\mathfrak{A}, fc)_{c\in C} \equiv (\mathfrak{B}, gc)_{c\in C}$ as required. Therefore T^* is a model completion of T, and (i) holds. ⊣

We now introduce a condition on a theory which is stronger than model completeness.

Definition. A theory T is said to *admit elimination of quantifiers* if for every formula $\varphi(x_1, \dots, x_n)$ of $\mathscr{L}$ there is an open formula $\psi(x_1, \dots, x_n)$ of $\mathscr{L}$ such that

$$T \models \varphi(x_1, \dots, x_n) \leftrightarrow \psi(x_1, \dots, x_n).$$

We see from Theorem 3.5.1 (iv) that every theory which admits elimination of quantifiers is model complete.

Proposition 3.5.19. *Let T be model complete. Then the following are equivalent.*

(i). *T is a model completion of $T_\forall$.*
(ii). *$T_\forall$ has the amalgamation property.*
(iii). *T admits elimination of quantifiers.*

Proof. T is always a cotheory of $T_\forall$, so a model complete T is a model companion of $T_\forall$. Thus the equivalence of (i) and (ii) follows from 3.5.18. We shall prove that (i) is equivalent to (iii). The proof is similar to the proof in Theorem 3.5.1 that a theory T is model complete if and only if every formula is equivalent to an existential formula with respect to T.

Assume (i), that T is a model completion of $T_\forall$. Let $\varphi(x_1, \ldots, x_n)$ be any formula of $\mathscr{L}$. Let $\Sigma(x_1, \ldots, x_n)$ be the set of all open consequences of $T \cup \varphi(x_1, \ldots, x_n)$ in the variables $x_1, \ldots, x_n$. Add new constant symbols $c_1, \ldots, c_n$. Let $(\mathfrak{A}, a_1, \ldots, a_n)$ be any model of $T \cup \Sigma(c_1, \ldots, c_n)$. Let D be the diagram of the submodel of $\mathfrak{A}$ generated by $a_1, \ldots, a_n$. Then $T \cup D$ must be consistent with $\varphi(c_1, \ldots, c_n)$, because it is consistent with $T \cup \Sigma(c_1, \ldots, c_n)$. By (i), $T \cup D$ is complete, and therefore $T \cup D \models \varphi(c_1, \ldots, c_n)$. Therefore $(\mathfrak{A}, a_1, \ldots, a_n)$ is a model of $\varphi(c_1, \ldots, c_n)$. This shows that

$$T \cup \Sigma(x_1, \ldots, x_n) \models \varphi(x_1, \ldots, x_n).$$

Since Σ is closed under finite conjunctions, there is a sentence $\psi \in \Sigma$ such that

$$T \cup \psi(x_1, \ldots, x_n) \models \varphi(x_1, \ldots, x_n).$$

Then $\psi(x_1, \ldots, x_n)$ is an open formula which is equivalent to $\varphi(x_1, \ldots, x_n)$ with respect to T, and (iii) holds.

Now assume (iii), that T admits elimination of quantifiers. Let $\mathfrak{C}$ be a model of $T_\forall$ and let $(\mathfrak{A}, fc)_{c \in C}$, $(\mathfrak{B}, gc)_{c \in C}$ be models of $T \cup \Delta_\mathfrak{C}$ where $\Delta_\mathfrak{C}$ is the diagram of $\mathfrak{C}$. The models $(\mathfrak{A}, fc)_{c \in C}$ and $(\mathfrak{B}, gc)_{c \in C}$ satisfy the same open sentences. By elimination of quantifiers, each sentence of $\mathscr{L}_C$ is equivalent to an open sentence of $\mathscr{L}_C$ with respect to T. Therefore $(\mathfrak{A}, fc)_{c \in C} \equiv (\mathfrak{B}, gc)_{c \in C}$, and the theory $T \cup \Delta_\mathfrak{C}$ is complete. This proves that T is a model completion of $T_\forall$. ⊣

Condition (i) of the above proposition is sometimes called submodel completeness. It states that for every model $\mathfrak{A}$ of T and every submodel $\mathfrak{C} \subset \mathfrak{A}$, the theory $T \cup \Delta_\mathfrak{C}$ is complete. By comparison, criterion (ii) for model completeness in Theorem 3.5.1 states that for every model $\mathfrak{A}$ of T, $T \cup \Delta_\mathfrak{A}$ is complete.

In each of the theories T in Examples 3.5.2 and 3.5.9 except the theory of real closed fields and the theory of $\langle \omega, +, -, 0, 1, \leqslant \rangle$, it can be shown that T admits elimination of quantifiers by showing that $T_\forall$ has the amalgamation property. For example, if T is the theory of atomless Boolean algebras, then $T_\forall$ is the theory of Boolean algebras, which has the amalgamation property. By Proposition 3.5.17, the theory of atomless Boolean algebras admits elimination of quantifiers and is a model completion of the theory of Boolean algebras.

The theory RCF of real closed fields does not admit elimination of quantifiers. In fact, the formula $(\exists x)\ x \cdot x = y$, which says that y is nonnegative, is not equivalent to an open formula. This is because any open formula $\psi(y)$ in one variable is equivalent to a finite Boolean combination of equations of the form $p(y) = 0$ where p is a polynomial with rational coefficients, and therefore in each model $\mathfrak{A}$, one of the sets

$$\{a \in A : \mathfrak{A} \models \psi[a]\}, \{a \in A : \mathfrak{A} \models \neg\psi[a]\}$$

must be finite. Thus by Proposition 3.5.19, RCF is not a model completion of $\mathrm{RCF}_\forall$, and $\mathrm{RCF}_\forall$ does not have the amalgamation property. On the other hand, RCF does have the amalgamation property because it is model complete.

Similarly, in the theory of the model $\langle \omega, +, -, 0, 1, \leqslant \rangle$, the formula $(\exists x)\ x + x = y$, which says that y is divisible by 2, is not equivalent to an open formula.

The elimination of quantifiers for the theories of algebraically closed fields and for real closed ordered fields are important classical results of Tarski and Robinson. Macintyre, McKenna, and van den Dries (1983) obtained converses of these results. They showed that the theory of algebraically closed fields of characteristic 0 or p are the only complete extensions of the theory of infinite fields which admit elimination of quantifiers, and the theory of real closed ordered fields is the only complete extension of the theory of ordered fields which admits elimination of quantifiers. More recently, analogous questions for other areas of algebra have been studied by several authors.

The following theorem gives a syntactical characterization of universal theories which have model completions. It is a consequence of Theorem 3.5.17.

THEOREM 3.5.20. *Let T be a consistent universal theory in a countable language $\mathscr{L}$. Then T has a model completion if and only if condition* (i) *below holds.*

(i) *For every existential formula $\theta(x_1 \dots x_n)$ there is an open formula $\theta'(x_1 \dots x_n)$ such that:*

(a) $$T \vdash \theta(x_1 \dots x_n) \to \theta'(x_1 \dots x_n).$$

(b) *For every universal formula* $\varphi(x_1 \dots x_n)$,

$$\textit{if } T, \theta' \to \theta \vdash \varphi, \textit{ then } T \vdash \varphi.$$

Moreover, if (i) *holds, then the theory*

$$T' = T \cup \{(\forall x_1 \dots x_n)(\theta' \to \theta) : \theta \textit{ is existential}\}$$

is a model completion of T.

Proof. First suppose T has a model completion T'. T' admits elimination of quantifiers by 3.5.19. Therefore for each existential formula $\theta(x_1 \dots x_n)$ there is an open formula $\theta'(x_1 \dots x_n)$ such that $T' \models \theta \leftrightarrow \theta'$. As in the proof of 3.5.17, (a) and (b) hold for T'.

For the converse, assume (i) and let

$$T' = T \cup \{(\forall x_1 \dots x_n)(\theta' \to \theta) : \theta \text{ is existential}\}.$$

By Theorem 3.5.17, T' is a model companion of T. Since $T' \models T$, it follows from (a) that

$$T' \models \theta \leftrightarrow \theta'$$

for each existential formula θ, so T' admits elimination of quantifiers. Since T is universal, $T = T'_\forall$. By Proposition 3.5.19, T' is a model completion of T. ⊣

As we have seen, many theories, including some familiar theories from algebra, do not have model companions. Several constructions have been discovered which share many of the properties of the model companion but exist for every theory T.

Definition. A mapping $T \to T^*$ from theories to theories in $\mathscr{L}$ is said to be a *companion operator* if:

(1). For each theory T, T^* is a cotheory of T;
(2). if T and U are cotheories then $T^* = U^*$;
(3). every model of T^* is a model of $T_{\forall\exists}$.

Proposition 3.5.21. *Suppose* $*$ *is a companion operator. Then for every theory* T *which has a model companion,* T^* *is the model companion of* T.

PROOF. Let T' be the model companion of T. Then

$$T^* = (T')^* \supseteq (T')_{\forall\exists} = T',$$

so every model of T^* is a model of T'. Since T' is model complete, it follows that T^* is model complete, and since T^* is a cotheory of T it must be the model companion of T. ⊣

Two examples of companion operators are the finite forcing companion and the infinite forcing companion introduced by A. Robinson. The definition and properties of these and other companion operators is left to the exercises.

EXERCISES

3.5.1. Show that the following theories are not model complete:

(a) The theory of divisible torsion free Abelian groups.

(b) The theory of divisible ordered Abelian groups.

3.5.2. Show that the following theories are model complete:

(a) The theory of divisible torsion free Abelian groups.

(b) The theory of divisible ordered Abelian groups.

(c) Any complete theory in a language with only unary relation symbols and constant symbols.

(d) The complete theory of $\langle [0,1], \leqslant, 0, 1\rangle$, that is, the theory of dense simple order with constant symbols for the first and last element.

(e) The complete theory of any set with a bijection (one to one onto function).

3.5.3. Which equivalence relations have model complete theories?

3.5.4. Which reducts of $\langle \omega, +, S, \leqslant, 0, 1\rangle$ have model complete theories?

3.5.5. Prove the results stated in Remark 3.5.6.

3.5.6. Let T be a theory in a countable language $\mathscr{L}$. Suppose that for every finite or countable model $\mathfrak{A}$ of T, $T \cup \Delta_{\mathfrak{A}}$ is complete. Prove that T is model complete.

3.5.7. Show that each theory in Exercise 3.5.2 admits elimination of quantifiers.

3.5.8. Which reducts of $\langle \omega, +, S, \leqslant, 0, 1 \rangle$ have theories which admit elimination of quantifiers?

3.5.9. Let T be the theory in a language with three unary relations U, V, W and two binary relations R, S and axioms which state that:

The union of U, W, and W is the whole model;

U, V, and W are pairwise disjoint;

R is a one to one function of U onto V;

S is a one to one function of the union of U and V onto W;

the model is infinite.

Prove that T is complete and in fact α-categorical for each infinite α, and T is model complete, but T does not admit elimination of quantifiers.

3.5.10. Prove that the theory of equivalence relations with infinitely many infinite classes is the model completion of the theory of equivalence relations.

3.5.11. Prove that the theory of models $\langle A, f \rangle$ where f is a bijection of A has a model completion, namely the theory of models $\langle A, f \rangle$ where f is a bijection with infinitely many cycles of each finite size.

3.5.12. Show that the theory of groups in the language with symbols for the unit element, product, and inverse, is universal and has the amalgamation property but does not have a model companion.

3.5.13. Let T be the universal theory with the axioms

$$\forall x \, \forall y [R(x, y) \rightarrow P_n(x)], \; n = 0, 1, 2, \ldots .$$

Show that T has the amalgamation property but does not have a model companion.

3.5.14*. Prove that the theory of graphs (symmetric binary relations) has a model completion.

3.5.15*. Prove that the theory of Abelian groups has a model completion.

3.5.16*. Prove that each of the following theories has no model companion:

(a) Commutative rings.

(b) Division algebras.
(c) Peano arithmetic.
(d) Complete number theory.
(e) ZFC.

3.5.17*. Let T be a consistent theory in a countable recursive language $\mathscr{L}$. Prove that T admits elimination of quantifiers if and only if for all models $\mathfrak{A}$, $\mathfrak{B}$ of T such that the model pair $(\mathfrak{A}, \mathfrak{B})$ is recursively saturated, and every subset $C \subset A \cap B$, the simple expansions $(\mathfrak{A}, c_1 \dots c_n)$ and $(\mathfrak{B}, c_1 \dots c_n)$ are partially isomorphic for all $c_1 \dots c_n \in C$.

3.5.18*. Let T be preserved under unions of chains, and let K be the class of all models which are existentially closed for T. Prove that K is the largest class L of models such that:

(i). Every model of T can be extended to a model in L;
(ii). if $\mathfrak{A}$ and $\mathfrak{B}$ belong to L and $\mathfrak{A} \subset \mathfrak{B}$ then $\mathfrak{A}$ is existentially closed in $\mathfrak{B}$.

3.5.19*. Let $\mathfrak{A}$ and $\mathfrak{B}$ be models of T and let $\mathfrak{C}$ be a model which is existentially closed for T, and let $f : \mathfrak{C} \to \mathfrak{A}$, $g : \mathfrak{C} \to \mathfrak{B}$ be isomorphic embeddings. Then there is a model $\mathfrak{C}'$ of T and isomorphic embeddings $f' : \mathfrak{A} \to \mathfrak{C}'$, $g' : \mathfrak{B} \to \mathfrak{C}'$ such that the diagram commutes. That is, models of T can always be amalgamated if the amalgam is existentially closed for T.

3.5.20*. Suppose $\mathfrak{A}$ and $\mathfrak{B}$ are models of T which are existentially closed for T, and $\mathfrak{A} \subset \mathfrak{B}$. Then every $\forall\exists\forall$ sentence true in $\mathfrak{B}$ is true in $\mathfrak{A}$. [*Hint*: Extend $\mathfrak{B}$ to an elementary extension of $\mathfrak{A}$.]

3.5.21*. If T is complete and ω-categorical and has a model companion T^*, then T^* is ω-categorical. [*Hint*: Use the characterization of ω-categoricity from Section 2.3.]

3.5.22*. Let $\mathscr{L}$ be the language with countably many unary function symbols, and let T be the theory in $\mathscr{L}$ whose axioms state that each function is one to one, has no finite loops, and that the ranges of the functions are pairwise disjoint. Prove that T is a complete universal theory which has countably many algebraically prime models and no prime model.

3.5.23*. For each theory T, let $T^{\#}$ be the set of all $\forall\exists$ sentences φ such that $T_{\forall} \cup \{\varphi\}$ is a cotheory of T. Prove that # is a companion operator.

3.5.24*. Let T be a theory in $\mathscr{L}$, $\mathfrak{A}$ a model of $T_{\forall}$ and φ a sentence of $\mathscr{L}_A$. The *infinite forcing relation* $\mathfrak{A} \Vdash \varphi$ is defined inductively by:

If φ is atomic then $\mathfrak{A} \Vdash \varphi$ iff $\mathfrak{A} \models \varphi$;

$\mathfrak{A} \Vdash \varphi \vee \psi$ iff $\mathfrak{A} \Vdash \varphi$ or $\mathfrak{A} \Vdash \psi$;

$\mathfrak{A} \Vdash (\exists x)\varphi(x)$ iff there exists $a \in A$ such that $\mathfrak{A} \Vdash \varphi(a)$;

$\mathfrak{A} \Vdash \neg\varphi$ iff there is no model $\mathfrak{B} \supset \mathfrak{A}$ of $T_{\forall}$ such that $\mathfrak{B} \Vdash \varphi$.

Prove the following results.

(a) If $\mathfrak{A}, \mathfrak{B}$ are models of $T_{\forall}$, $\mathfrak{A} \subset \mathfrak{B}$, and $\mathfrak{B} \Vdash \varphi$ then $\mathfrak{A} \Vdash \varphi$.

(b) Every model $\mathfrak{A}$ of $T_{\forall}$ can be extended to a model $\mathfrak{B}$ of $T_{\forall}$ such that for every sentence φ of $\mathscr{L}_B$, $\mathfrak{B} \Vdash \varphi$ or $\mathfrak{B} \Vdash \neg\varphi$. Such a model is said to be *infinitely generic*.

(c) A model $\mathfrak{B}$ of $T_{\forall}$ is infinitely generic for T if and only if for every sentence φ of $\mathscr{L}_B$, $\mathfrak{B} \Vdash \varphi$ iff $\mathfrak{B} \models \varphi$.

(d) Every infinitely generic model for T is existentially closed for T.

(e) If T has a model companion T^*, then the class of models of T^* is equal to the class of all infinitely generic models for T.

(f) The *infinite forcing companion* of T is the theory of the class of infinitely generic models for T. The infinite forcing companion is a companion operator.

(g) The infinite forcing companion of T is complete if and only if T has the joint embedding property.

3.5.25*. Let T be a theory in $\mathscr{L}$ and let C be a countable set of new constant symbols outside of $\mathscr{L}$. A *condition* for T is a finite set of atomic and negated atomic sentences of $\mathscr{L}_C$ which is consistent with T. The *finite forcing relation* $p \Vdash \varphi$ between conditions for T and sentences φ of $\mathscr{L}_C$ is defined as follows.

If φ is atomic then $p \Vdash \varphi$ iff $\varphi \in p$;

$p \Vdash \varphi \vee \psi$ iff $p \Vdash \varphi$ or $p \Vdash \psi$;

$p \Vdash (\exists x)\varphi(x)$ iff $p \Vdash \varphi(c)$ for some $c \in C$;

$p \Vdash \neg\varphi$ iff there is no condition $q \supseteq p$ with $q \Vdash \varphi$.

We say that p *weakly forces* φ, $p \Vdash^{w} \varphi$, if $p \Vdash \neg\neg\varphi$. Prove the following results.

(a) If $p \Vdash \varphi$ and $q \supseteq p$ then $q \Vdash \varphi$.

(b) A *finitely generic set* is a set G of sentences of $\mathscr{L}_C$ such that each finite subset of G is a condition, and for each sentence φ of $\mathscr{L}_C$, there is a finite $p \subset G$ which forces either φ or $\neg\varphi$. Write $G \Vdash \varphi$ if $p \Vdash \varphi$ for some

$p \subset G$. Each condition is contained in a finitely generic set. $p \Vdash^w \varphi$ iff $G \Vdash \varphi$ for each finitely generic $G \supset p$.

(c) For each finitely generic set G, there is a model $\mathfrak{A}(G)$ for $\mathscr{L}_C$ such that $\{\varphi : G \Vdash \varphi\}$ is the elementary diagram of $\mathfrak{A}(G)$. The reduct of $\mathfrak{A}(G)$ to $\mathscr{L}$ is called a *finitely generic model* for T.

(d) Every finitely generic model for T is a countable model of $T_{\forall\exists}$ which is existentially closed for T.

(e) If T has a model companion T^*, then the class of all finitely generic models for T is equal to the class of all countable models of T^*.

(f) The *finite forcing companion* of T is the theory of the class of finitely generic models for T. The finite forcing companion is a companion operator.

(g) The finite forcing companion of T is complete if and only if T has the joint embedding property.

3.5.26*. This is a refinement of Exercise 2.2.11 (the Restricted Omitting Types Theorem). Let T be an $\forall\exists$ theory and $\Sigma(x)$ a set of universal formulas such that every existential formula $\theta(x)$ consistent with T, there is a $\sigma(x) \in \Sigma(x)$ such that $\theta \wedge \neg\sigma$ is consistent with T. Show that there is a finitely generic model for T which omits $\Sigma(x)$.

CHAPTER 4

ULTRAPRODUCTS

4.1. The fundamental theorem

In the preceding two chapters we have developed three basic methods of constructing models: from constants, from Skolem functions, and from elementary chains. This chapter contains a treatment of another basic method of constructing models, the ultraproduct construction. This method originated with Skolem in the 1930's, and has been used extensively since the work of Łoś in 1955. In this section we shall define the notions of an ultrafilter and ultraproduct and prove the important results which establish the connection between ultraproducts and the first-order properties of models. At the same time, we shall introduce the more general reduced product construction, which we shall meet again in Chapter 6.

We begin with the notion of an ultrafilter over a set I, and the proof of the ultrafilter theorem which concerns their existence.

Let I be a nonempty set. We recall that $S(I)$ is the set of all subsets of I. A *filter D over I* is defined to be a set $D \subset S(I)$ such that:

$$I \in D;$$

$$\text{if} \quad X, Y \in D, \quad \text{then} \quad X \cap Y \in D;$$

$$\text{if} \quad X \in D \quad \text{and} \quad X \subset Z \subset I, \quad \text{then} \quad Z \in D.$$

We observe that every filter D is a *nonempty* set since $I \in D$. Examples of filters are:

The *trivial filter* $D = \{I\}$.

The *improper filter* $D = S(I)$.

For each $Y \subset I$, the filter $D = \{X \subset I : Y \subset X\}$; this filter is called the *principal filter* generated by Y.

The *Fréchet filter* $D = \{X \in S(I) : I \setminus X \text{ is finite}\}$.

D is said to be a *proper filter* iff it is not the improper filter $S(I)$. The proposition below shows how we can make a filter beginning with an arbitrary subset of $S(I)$. First we give a definition:

Let E be a subset of $S(I)$. By the *filter generated by E* we mean the intersection D of all filters over I which include E:

$$D = \bigcap\{F : E \subset F \quad \text{and} \quad F \text{ is a filter over } I\}.$$

E is said to have the *finite intersection property* iff the intersection of any finite number of elements of E is nonempty.

PROPOSITION 4.1.1. *Let E be any subset of S(I) and let D be the filter generated by E. Then*:

(i). *D is a filter over I.*

(ii). *D is the set of all $X \in S(I)$ such that either $X = I$ or for some $Y_1, \ldots, Y_n \in E$,*

$$Y_1 \cap \ldots \cap Y_n \subset X.$$

(iii). *D is a proper filter iff E has the finite intersection property.*

PROOF. (i) is almost immediate.

To prove (ii), let D' be the set of all $X \in S(I)$ such that $X = I$ or for some $Y_1, \ldots, Y_n \in E$, $Y_1 \cap \ldots \cap Y_n \subset X$. We show that $D = D'$. We have put $I \in D'$. Let $X, X' \in D'$, and let $Y_i, Y_j' \in E$ be such that

$$Y_1 \cap \ldots \cap Y_n \subset X, \qquad Y_1' \cap \ldots \cap Y_m' \subset X'.$$

If $X \subset Z \subset I$, then

$$Y_1 \cap \ldots \cap Y_n \subset Z,$$

so $Z \in D'$. Moreover,

$$Y_1 \cap \ldots \cap Y_n \cap Y_1' \cap \ldots \cap Y_m' \subset X \cap X',$$

so $X \cap X' \in D'$. Therefore D' is a filter over I. Obviously $E \subset D'$. It follows that $D \subset D'$.

Now consider any filter F over I which includes E. Then $I \in F$. For any $Y_1, \ldots, Y_n \in E$, we have $Y_1 \cap \ldots \cap Y_n \in F$, and hence any $X \in S(I)$ which includes $Y_1 \cap \ldots \cap Y_n$ belongs to F. Thus $D' \subset F$. This shows that $D' \subset D$, whence $D = D'$.

(iii) follows easily from (ii). ⊣

We pause to give one more example of a filter which will be particularly important in our study. Let J be an infinite set, and let $I = S_\omega(J)$ be the set of all finite subsets of J. For each $j \in J$, let

$$\hat{\jmath} = \{i \in I : j \in i\}$$

be the set of all finite subsets of J which contain j. Now let

$$E = \{\hat{\jmath} : j \in J\}.$$

Then E is a subset of $S(I)$, and we may form the filter D generated by E. From Proposition 4.1.1, D is just the set of all subsets X of I such that for some $i \in I$, every $i' \in I$ which includes i belongs to X. Moreover, E has the finite intersection property, whence D is a proper filter.

We now turn to ultrafilters. D is said to be an *ultrafilter over I* iff D is a filter over I such that for all $X \in S(I)$,

$$X \in D \quad \text{if and only if} \quad (I \setminus X) \notin D.$$

If we simply say that D is an ultrafilter, we shall tacitly assume that D is an ultrafilter over the set $I = \bigcup D$.

PROPOSITION 4.1.2. *The following are equivalent*:

(i). *D is an ultrafilter over I.*

(ii). *D is a maximal proper filter over I. That is, D is a proper filter over I, and the only proper filter over I which includes D is D itself.*

PROOF. (i) $\Rightarrow$ (ii). Assume (i). Then $0 \notin D$, because $I \in D$ and $0 = I \setminus I$. Hence D is a proper filter. Let F be any proper filter over I which includes D. If $X \in F$ and $X \notin D$, then $I \setminus X \in D$, whence $I \setminus X \in F$, and

$$0 = X \cap (I \setminus X) \in F.$$

This contradicts the assumption that F is proper. Thus $F \subset D$, so $F = D$, and (ii) holds.

(ii) $\Rightarrow$ (i). Assume (ii). Consider any set $X \in S(I)$. We cannot have both $X \in D$ and $I \setminus X \in D$, because then $0 \in D$, whence every $Y \in S(I)$ is in D, and D is not proper. It suffices now to prove that, if $I \setminus X \notin D$, then $X \in D$. Suppose $I \setminus X \notin D$. Let $E = D \cup \{X\}$, and let F be the filter generated by E. Consider any $Y_1, \ldots, Y_n \in E$, and let

$$Z = Y_1 \cap \ldots \cap Y_n.$$

Since D is closed under finite intersections, we either have $Z \in D$ or $Z = Y \cap X$ for some $Y \in D$. In the first case $Z \neq 0$, because $0 \notin D$. In the second case, we also have $Z \neq 0$, for otherwise we would have $Y \cap X = 0$, $Y \subset I \setminus X$, whence $I \setminus X \in D$. Thus in any case, $Z \neq 0$. By Proposition 4.1.1, we see that $0 \notin F$. This means that F is a proper filter including D, so by (ii), $F = D$. Therefore $E \subset D$ and $X \in D$. This proves (i). ⊣

We now prove an important theorem about the existence of ultrafilters.

PROPOSITION 4.1.3 (Ultrafilter Theorem). *If $E \subset S(I)$ and E has the finite intersection property, then there exists an ultrafilter D over I such that $E \subset D$.*

PROOF. By Proposition 4.1.1, the filter F generated by E does not contain the empty set, whence F is proper. Moreover, if C is any nonempty chain of proper filters over I, then $\bigcup C$ is a proper filter over I. This follows very easily from the definition of proper filter, and is left as an exercise. Furthermore, if each $D \in C$ includes E, then $\bigcup C$ includes E. It follows by Zorn's lemma that the class $\mathscr{E}$ of all proper filters over I including E has a maximal element, say D. Thus $E \subset D$. D is a maximal proper filter over I, because if D' is a proper filter including D, then $E \subset D'$, and so D' belongs to $\mathscr{E}$ and $D' = D$. Thus, by Proposition 4.1.2, D is an ultrafilter over I. ⊣

COROLLARY 4.1.4. *Any proper filter over I can be extended to an ultrafilter over I.*

PROOF. Every proper filter has the finite intersection property (see Exercise 4.1.1). ⊣

We are now ready to introduce the reduced product and ultraproduct constructions. The latter is merely a special case of the former. We first apply the construction to sets, and then to models.

Suppose I is a nonempty set, D is a proper filter over I, and for each $i \in I$, A_i is a nonempty set. Let

$$C = \prod_{i \in I} A_i$$

be the Cartesian product of these sets. Thus C is the set of all functions f with domain I such that for each $i \in I$, $f(i) \in A_i$. For two functions $f, g \in C$, we say that f and g are *D-equivalent*, in symbols $f =_D g$, iff

$$\{i \in I : f(i) = g(i)\} \in D.$$

To continue, we need the following:

PROPOSITION 4.1.5. *The relation* $=_D$ *is an equivalence relation over the set* C.

The proof is left as an exercise. Now let f_D be the equivalence class of f:

$$f_D = \{g \in C : f =_D g\}.$$

We then define the *reduced product of* A_i *modulo* D to be the set of all equivalence classes of $=_D$. It is denoted by $\prod_D A_i$. Thus

$$\prod_D A_i = \{f_D : f \in \prod_{i \in I} A_i\}.$$

We shall call the set I the *index set* for $\prod_D A_i$. In the special case where D is an ultrafilter over I, the reduced product $\prod_D A_i$ is called an *ultraproduct*. In the case when all the sets A_i are the same, say $A_i = A$, the reduced product may be written $\prod_D A$, and it is called the *reduced power of* A *modulo* D. In particular, if D is an ultrafilter, then $\prod_D A$ is called the *ultrapower of* A *modulo* D.

Strictly speaking, our notation $\prod_D A_i$ is not completely precise. From the filter D we can recover the index set $I = \bigcup D$, so there is no need to write an I in our notation. However, there may be some ambiguity if the sets depend on more than one variable, for example $\prod_D A_{ij}$. The reduced product really depends on both the filter D and the *function* $\langle A_i : i \in I \rangle$. So in the few cases where the need arises we shall use the more complete notation

$$\prod_D \langle A_i : i \in I \rangle$$

for the reduced product $\prod_D A_i$. For example, $\prod_D A_{ij}$ is ambiguous, and could mean any one of

$$\prod_D \langle A_{ij} : i \in I \rangle, \qquad \prod_D \langle A_{ij} : j \in I \rangle, \qquad \prod_D \langle A_{ij} : \langle i, j \rangle \in I \rangle,$$

in our more complete notation.

There are a number of interesting and difficult problems of a purely set-theoretical nature concerning the existence of various kinds of ultrafilters and the cardinality of ultraproducts. We shall postpone a discussion of these problems to the next two sections and proceed immediately to the fundamental model-theoretic results on ultraproducts. For these are the results which make ultraproducts important.

We now give the definition of reduced product of models. Let I be a nonempty set, let D be a proper filter over I, and for each $i \in I$ let $\mathfrak{A}_i$ be a model for $\mathscr{L}$. We have our usual convention that in $\mathfrak{A}_i$ relation symbols P are interpreted by R_i, functions F by G_i, and constants c by a_i.

4.1.6. The *reduced product* $\prod_D \mathfrak{A}_i$ is the model for $\mathscr{L}$ described as follows:

(i). The universe set of $\prod_D \mathfrak{A}_i$ is $\prod_D A_i$.

(ii). Let P be an n-placed relation symbol of $\mathscr{L}$. The interpretation of P in $\prod_D \mathfrak{A}_i$ is the relation S such that

$$S(f_D^1 \dots f_D^n) \quad \text{if and only if} \quad \{i \in I : R_i(f^1(i) \dots f^n(i))\} \in D.$$

(iii). Let F be an n-placed function symbol of $\mathscr{L}$. Then F is interpreted in $\prod_D \mathfrak{A}_i$ by the function H given by

$$H(f_D^1 \dots f_D^n) = \langle G_i(f^1(i) \dots f^n(i)) : i \in I \rangle_D.$$

(iv). Let c be a constant of $\mathscr{L}$. Then c is interpreted by the element $b \in \prod_D A_i$, where

$$b = \langle a_i : i \in I \rangle_D.$$

To show that the above definition is consistent, we must check that $S(f_D^1 \dots f_D^n)$ and $H(f_D^1 \dots f_D^n)$ depend only on the equivalence classes $f_D^1, \dots, f_D^n$, and not on the representatives $f^1, \dots, f^n$ of these equivalence classes. We state this fact as a proposition and leave the proof as an exercise.

PROPOSITION 4.1.7. *Let I, D, $\mathfrak{A}_i$ be as in definition* 4.1.6. *Suppose that* $f^1 =_D g^1, \dots, f^n =_D g^n$. *Then*

$$\{i \in I : R_i(f^1(i) \dots f^n(i))\} \in D \quad \textit{if and only if}$$
$$\{i \in I : R_i(g^1(i) \dots g^n(i))\} \in D,$$

and

$$\langle G_i(f^1(i) \dots f^n(i)) : i \in I \rangle =_D \langle G_i(g^1(i) \dots g^n(i)) : i \in I \rangle.$$

We now prove two important theorems about ultraproducts. The first of these, which we have labeled the expansion theorem, has to do with the passage from the language $\mathscr{L}$ to an expansion $\mathscr{L}'$ of $\mathscr{L}$. It is a very simple result, but is an extremely powerful tool when combined with other facts about ultraproducts. Actually, the expansion theorem concerns arbitrary reduced products.

THEOREM 4.1.8 (Expansion Theorem). *Let $\mathscr{L}'$ be an expansion of $\mathscr{L}$. Let I be a nonempty set and for each $i \in I$ let $\mathfrak{A}_i$ be a model for $\mathscr{L}$ and $\mathfrak{B}_i$ an expansion of $\mathfrak{A}_i$ to $\mathscr{L}'$. Let D be a filter over I. Then the reduced product $\prod_D \mathfrak{B}_i$ is an expansion of the reduced product $\prod_D \mathfrak{A}_i$ to $\mathscr{L}'$.*

PROOF. For each i, the models $\mathfrak{A}_i$ and $\mathfrak{B}_i$ have the same universe set, $A_i = B_i$. Therefore the reduced products have the same universe sets, $\prod_D A_i = \prod_D B_i$. Since $\mathfrak{B}_i$ is an expansion of $\mathfrak{A}_i$, each symbol of $\mathscr{L}$ has the same interpretation in $\mathfrak{B}_i$ as in $\mathfrak{A}_i$. From definition 4.1.6, we see that the interpretation of a symbol of $\mathscr{L}$ in $\prod_D \mathfrak{A}_i$ depends only on its interpretations in the models $\mathfrak{A}_i$, and on the universe sets and the filter D. It follows that each symbol of $\mathscr{L}$ has the same interpretation in $\prod_D \mathfrak{B}_i$ as in $\prod_D \mathfrak{A}_i$, so the former is an expansion of the latter. ⊣

Our next result is the 'fundamental theorem' of ultraproducts. It holds only for ultraproducts, and not for arbitrary reduced products. From now on, we shall concentrate exclusively on ultraproducts, and shall not discuss reduced products again until Chapter 6.

THEOREM 4.1.9 (The Fundamental Theorem of Ultraproducts). *Let $\mathfrak{B}$ be the ultraproduct $\prod_D \mathfrak{A}_i$, and let I be the index set. Then:*

(i). *For any term $t(x_1 \ldots x_n)$ of $\mathscr{L}$ and elements $f_D^1, \ldots, f_D^n \in B$, we have*

$$t_{\mathfrak{B}}[f_D^1 \ldots f_D^n] = \langle t_{\mathfrak{A}_i}[f^1(i) \ldots f^n(i)] : i \in I \rangle_D .$$

(ii). *Given any formula $\varphi(x_1 \ldots x_n)$ of $\mathscr{L}$ and $f_D^1, \ldots, f_D^n \in B$, we have*

$$\mathfrak{B} \models \varphi[f_D^1 \ldots f_D^n] \quad \text{if and only if} \quad \{i \in I : \mathfrak{A}_i \models \varphi[f^1(i) \ldots f^n(i)]\} \in D.$$

(iii). *For any sentence φ of $\mathscr{L}$,*

$$\mathfrak{B} \models \varphi \quad \text{if and only if} \quad \{i \in I : \mathfrak{A}_i \models \varphi\} \in D.$$

PROOF. (iii) is an immediate consequence of (i) and (ii). The proofs of (i) and (ii) are by induction on the terms and formulas, respectively.

(i). From the definition of reduced product we see that (i) holds whenever $t(x_1 \ldots x_n)$ is a term of the form $F(x_1 \ldots x_n)$ and also whenever $t(x_1 \ldots x_n)$ is a constant symbol or variable. Suppose that

$$t(x_1 \ldots x_n) = F(t_1(x_1 \ldots x_n) \ldots t_m(x_1 \ldots x_n)),$$

where F is a function symbol of $\mathscr{L}$ and the terms $t_1, \ldots, t_m$ all satisfy (i). Then by the definition of interpretation of terms,

$$t_{\mathfrak{B}}[f_D^1 \ldots f_D^n] = H(t_{1\mathfrak{B}}[f_D^1 \ldots f_D^n] \ldots t_{m\mathfrak{B}}[f_D^1 \ldots f_D^n]),$$

where H is the interpretation of F in $\mathfrak{B}$. By (i), for $t_1, \ldots, t_m$, we have for

$k = 1, \ldots, m,$

$$t_{k\mathfrak{B}}[f_D^1 \ldots f_D^n] = g_D^k,$$

where

$$g^k = \langle t_{k\mathfrak{A}_i}[f^1(i) \ldots f^n(i)] : i \in I \rangle.$$

By 4.1.6 (iii),

$$H(g_D^1 \ldots g_D^m) = \langle G_i(g^1(i) \ldots g^m(i)) : i \in I \rangle_D.$$

Again using the definition of interpretation of terms,

$$t_{\mathfrak{A}_i}[f^1(i) \ldots f^n(i)] = G_i(g^1(i) \ldots g^m(i)).$$

Combining, we obtain

$$t_{\mathfrak{B}}[f_D^1 \ldots f_D^n] = H(g_D^1 \ldots g_D^m) = \langle t_{\mathfrak{A}_i}[f^1(i) \ldots f^n(i)] : i \in I \rangle_D.$$

Thus $t(x_1 \ldots x_n)$ satisfies (i).

(ii). The proof that (ii) holds for all atomic formulas is similar to the above proof of (i), and we shall leave the details to the reader. We only point out that (i) is used in the proof of (ii) for atomic formulas.

Suppose that $\varphi = \neg\psi(x_1 \ldots x_n)$ and (ii) holds for $\psi(x_1 \ldots x_n)$. Then the following are equivalent:

$$\mathfrak{B} \models \varphi[f_D^1 \ldots f_D^n];$$

$$\text{not} \quad \mathfrak{B} \models \psi[f_D^1 \ldots f_D^n];$$

$$\{i \in I : \mathfrak{A}_i \models \psi[f^1(i) \ldots f^n(i)]\} \notin D;$$

$$\{i \in I : \text{not } \mathfrak{A}_i \models \psi[f^1(i) \ldots f^n(i)]\} \in D;$$

$$\{i \in I : \mathfrak{A}_i \models \varphi[f^1(i) \ldots f^n(i)]\} \in D.$$

The fact that D is an ultrafilter is used to show that the third and fourth lines in the above list are equivalent. Indeed, this is the only point in the entire proof of the theorem where we need the fact that D is an ultrafilter, and not merely a proper filter.

The next step is to prove that if ψ and θ satisfy (ii), then so does $\psi \wedge \theta$. This is done by writing a string of equivalences like the one we used for $\neg\psi$. This time the crucial fact about D which we need is that

$$X \cap Y \in D \quad \text{if and only if} \quad X \in D \quad \text{and} \quad Y \in D.$$

Every filter has this property. The details of this step in the proof are straightforward. They are left to the reader.

For the last part of our induction, suppose that

$$\varphi(x_1 \dots x_n) = (\exists x_0)\psi(x_0 x_1 \dots x_n),$$

and that (ii) holds for ψ. Then the following statements are equivalent:

$$\mathfrak{B} \models \varphi[f_D^1 \dots f_D^n];$$

$$\text{there exists } f_D^0 \in B \text{ such that } \mathfrak{B} \models \psi[f_D^0 \dots f_D^n];$$

(1) there exists $f_D^0 \in B$ such that $\{i \in I : \mathfrak{A}_i \models \psi[f^0(i) \dots f^n(i)]\} \in D$.

Since $\mathfrak{A}_i \models \psi[f^0(i) \dots f^n(i)]$ implies that $\mathfrak{A}_i \models \varphi[f^1(i) \dots f^n(i)]$, the statement (1) above implies

(2) $$\{i \in I : \mathfrak{A}_i \models \varphi[f^1(i) \dots f^n(i)]\} \in D.$$

On the other hand, if (2) holds, then we may pick a function $f^0 \in \prod_{i \in I} A_i$ such that (1) holds. Hence (1) is equivalent to (2), and

$$\mathfrak{B} \models \varphi[f_D^1 \dots f_D^n]$$

is also equivalent to (2). This shows that the formula φ has the property (ii). Our induction is now completed. ⊣

Corollary 4.1.10. *Let $\prod_D \mathfrak{A}$ be an ultrapower of $\mathfrak{A}$. Then $\prod_D \mathfrak{A} \equiv \mathfrak{A}$.*

Let us now give a few applications of the fundamental theorem. Further applications can be found in the exercises. Our first application is another proof of the compactness theorem (Theorem 1.3.22).

Corollary 4.1.11 (An ultraproduct version of the compactness theorem). *Let Σ be a set of sentences of $\mathscr{L}$, let $I = S_\omega(\Sigma)$ be the set of all finite subsets of Σ, and for each $i \in I$, let $\mathfrak{A}_i$ be a model of i. Then there exists an ultrafilter D over I such that the ultraproduct $\prod_D \mathfrak{A}_i$ is a model of Σ.*

Proof. For each $\sigma \in \Sigma$, let $\hat{\sigma}$ be the set of all $i \in I$ such that $\sigma \in i$. The set

$$E = \{\hat{\sigma} : \sigma \in \Sigma\}$$

has the finite intersection property because

$$\{\sigma_1, \dots, \sigma_n\} \in \hat{\sigma}_1 \cap \dots \cap \hat{\sigma}_n.$$

It follows by the ultrafilter theorem that E can be extended to an ultrafilter D over I. Any ultrafilter D over I which includes E will do. If $i \in \hat{\sigma}$, then

$\sigma \in i$, whence $\mathfrak{A}_i \models \sigma$. Thus for each $\sigma \in \Sigma$,

$$\{i \in I : \mathfrak{A}_i \models \sigma\} \supset \hat{\sigma} \quad \text{and} \quad \hat{\sigma} \in D.$$

Therefore

$$\{i \in I : \mathfrak{A}_i \models \sigma\} \in D.$$

By the fundamental theorem, $\prod_D \mathfrak{A}_i \models \sigma$ for all $\sigma \in \Sigma$. Hence $\prod_D \mathfrak{A}_i$ is a model of Σ. ⊣

Our next application gives characterizations of the notions of 'elementary class' and 'basic elementary class' in terms of ultraproducts and elementary equivalence. This theorem is best stated in terms of classes of models. These classes will be so large that they cannot be sets (see the Appendix on set theory). A class K of models for $\mathscr{L}$ is said to be an *elementary class* iff there exists a theory T in $\mathscr{L}$ such that K is exactly the class of all models of T. K is a *basic elementary class* iff there is a sentence φ of $\mathscr{L}$ such that K is the class of all models of φ. A class K of models for $\mathscr{L}$ is said to be *closed under elementary equivalence* iff $\mathfrak{A} \in K$, $\mathfrak{A} \equiv \mathfrak{B}$ implies $\mathfrak{B} \in K$. K is said to be *closed under ultraproducts* iff every ultraproduct $\prod_D \mathfrak{A}_i$ of a family of models $\mathfrak{A}_i \in K$ belongs to K.

THEOREM 4.1.12. *Let K be an arbitrary class of models. Then:*

(i). *K is an elementary class if and only if K is closed under ultraproducts and elementary equivalence.*

(ii). *K is a basic elementary class if and only if both K and the complement of K are closed under ultraproducts and elementary equivalence.*

PROOF. (i). Every elementary class is obviously closed under elementary equivalence. The fundamental theorem shows that if $\mathfrak{A}_i \models \varphi$ for all $i \in I$, then $\prod_D \mathfrak{A}_i \models \varphi$. Thus every elementary class is also closed under ultraproducts.

Let K be any class of models which is closed under ultraproducts and elementary equivalence. Let T be the set of all sentences of $\mathscr{L}$ which hold in all $\mathfrak{A} \in K$. Then T is a theory in $\mathscr{L}$ and every $\mathfrak{A} \in K$ is a model of T. Let $\mathfrak{B}$ be any model of T. Let Σ be the set of all sentences true in $\mathfrak{B}$, and let $I = S_\omega(\Sigma)$. For each $i = \{\sigma_1, \ldots, \sigma_n\} \in I$, there exists a model $\mathfrak{A}_i \in K$ which is a model of i, for otherwise the sentence $\neg(\sigma_1 \wedge \ldots \wedge \sigma_n)$ would belong to T and yet be false in $\mathfrak{B}$. Choose a model $\mathfrak{A}_i \in K$ of i for each $i \in I$. By Corollary 4.1.11, there exists an ultraproduct $\prod_D \mathfrak{A}_i$ which is a model of Σ. Since K is closed under ultraproducts, $\prod_D \mathfrak{A}_i \in K$. But every model of Σ

is elementarily equivalent to $\mathfrak{B}$, so $\prod_D \mathfrak{A}_i \equiv \mathfrak{B}$. Therefore $\mathfrak{B} \in K$. Hence K is the class of all models of T, and so K is an elementary class.

(ii). This follows easily from (i) and the compactness theorem (see Exercise 2.1.13). We leave the details for an exercise. ⊣

Our third application shows that each model $\mathfrak{A}$ is elementarily embeddable in every ultrapower of $\mathfrak{A}$ in a natural way. We now define the *natural embedding* of $\mathfrak{A}$ into $\prod_D \mathfrak{A}$. This embedding is extremely important and will be used many times in our treatment of ultraproducts.

Let I be a nonempty set, D a proper filter over I, and $\mathfrak{A}$ a model. The *natural embedding* d of $\mathfrak{A}$ into $\prod_D \mathfrak{A}$ is the function d such that $d(a)$ is the equivalence class of the constant function with value a,

$$d(a) = \langle a : i \in I \rangle_D.$$

The range of d is denoted by $d(A)$, and the restriction of $\prod_D \mathfrak{A}$ to $d(A)$ is denoted by $d(\mathfrak{A})$.

Corollary 4.1.13. *Let $\mathfrak{A}$ be a model and D an ultrafilter. Then the natural embedding of $\mathfrak{A}$ into the ultrapower $\prod_D \mathfrak{A}$ is an elementary embedding.*

Proof. Let $\varphi(x_1 \dots x_n)$ be a formula of $\mathscr{L}$ and $a_1, \dots, a_n \in A$. Then, by the fundamental theorem, the following are equivalent:

$$\prod_D \mathfrak{A} \models \varphi[d(a_1) \dots d(a_n)];$$

$$\{i \in I : \mathfrak{A} \models \varphi[a_1 \dots a_n]\} \in D;$$

$$\mathfrak{A} \models \varphi[a_1 \dots a_n]. \dashv$$

Corollary 4.1.13 shows that d is an isomorphism of $\mathfrak{A}$ onto $d(\mathfrak{A})$ and that $d(\mathfrak{A})$ is an elementary submodel of the ultrapower $\prod_D \mathfrak{A}$. One of the topics of the next section will be to determine when d is a proper embedding.

Our last application is a very simple one which depends on both the expansion theorem and the fundamental theorem. It shows that certain second-order formulas are 'preserved under ultraproducts'. This application will be useful in the next section. We wish to consider quantifiers over relation and function symbols, as well as over individual variables.

By a Σ_1^1 formula over $\mathscr{L}$ we mean a formula ψ of the following form:

$$(*) \qquad (\exists P_1 \dots P_m F_1 \dots F_n)\varphi,$$

where the P_i and F_j are new relation and function symbols not occurring in $\mathscr{L}$, and where φ is a formula in the expanded first-order language

$$\mathscr{L}' = \mathscr{L} \cup \{P_1 \dots P_m F_1 \dots F_n\}.$$

Thus a Σ_1^1 formula is a second-order formula all of whose relation and function quantifiers occur at the beginning and are existential. Satisfaction for Σ_1^1 formulas is defined in the obvious way. If φ is a sentence, then ψ holds in a model $\mathfrak{A}$ for $\mathscr{L}$ if and only if there exists an expansion

$$\mathfrak{A}' = (\mathfrak{A}, R_1 \dots R_m G_1 \dots G_n)$$

of $\mathfrak{A}$ to $\mathscr{L}'$ such that φ holds in $\mathfrak{A}'$. If φ has a free variable x, then $\mathfrak{A} \models \psi[a]$ if and only if there exists an expansion $\mathfrak{A}'$ of $\mathfrak{A}$ to $\mathscr{L}'$ such that $\mathfrak{A}' \models \varphi[a]$.

COROLLARY 4.1.14 (Σ_1^1 formulas are preserved under ultraproducts). *Suppose* $\mathfrak{B}$ *is the ultraproduct* $\prod_D \mathfrak{A}_i$, $f_D^1, \dots, f_D^r \in B$, *and* $\psi(x_1 \dots x_r)$ *is a* Σ_1^1 *formula. If*

$$\{i \in I : \mathfrak{A}_i \models \psi[f^1(i) \dots f^r(i)]\} \in D,$$

then

$$\mathfrak{B} \models \psi[f_D^1 \dots f_D^r].$$

PROOF. Let ψ be as in (*) and let

$$X = \{i \in I : \mathfrak{A}_i \models \psi[f^1(i) \dots f^r(i)]\}.$$

For each $i \in X$, let $\mathfrak{A}_i'$ be an expansion of $\mathfrak{A}_i$ to $\mathscr{L}'$ such that

$$\mathfrak{A}_i' \models \varphi[f^1(i) \dots f^r(i)].$$

Since $X \in D$, we ave by the fundamental theorem,

$$\prod_D \mathfrak{A}_i' \models \varphi[f_D^1 \dots f_D^r].$$

By the expansion theorem, $\prod_D \mathfrak{A}_i'$ is an expansion of $\mathfrak{B} = \prod_D \mathfrak{A}_i$ to $\mathscr{L}'$. Therefore

$$\mathfrak{B} \models \psi[f_D^1 \dots f_D^r]. \dashv$$

The reader may wonder why the notation Σ_1^1 is used. It is just one instance of the currently popular way of classifying formulas. In general, a Σ_n^m formula is a formula of $(m+1)$th-order logic which is of the form

$$(\exists X_{11} \dots X_{1r_1} \forall X_{21} \dots X_{2r_2} \dots \mathrm{Q} X_{n1} \dots X_{nr_n})\varphi,$$

where φ is in mth-order logic. The subscript n stands for the number of blocks of quantifiers. Q is either $\exists$ or $\forall$ depending on whether n is odd or even. If the formula has n blocks of $(m+1)$th-order quantifiers beginning with a universal quantifier, it is called a Π_n^m formula. As we have explained in our introduction to this book, we shall almost always deal only with first-order logic. The Σ_n^0 and Π_n^0 formulas are first-order, and we have already introduced them in Section 3.1.

EXERCISES

4.1.1. Let D be a filter over I. Show that the following are equivalent:

(i). D is a proper filter.

(ii). $0 \notin D$.

(iii). D has the finite intersection property.

4.1.2.

(i). The intersection of any set of filters over I is a filter over I.

(ii). The union of any chain of proper filters over I is a proper filter over I.

4.1.3. Let D be an ultrafilter over I and let $X \in D$. Then $D \cap S(X)$ is an ultrafilter over X. Similarly for proper filters.

4.1.4. D is a principal ultrafilter over I if and only if $D = \{X \in S(I) : i \in X\}$ for some $i \in I$.

4.1.5. If X is infinite, then there exist nonprincipal ultrafilters over X.

4.1.6. A filter D is principal if and only if $\bigcap D \in D$. Every filter over a finite set is principal.

4.1.7. Let D be a proper filter over I. D is an ultrafilter if and only if for all $X, Y \in S(I)$, $X \cup Y \in D$ implies $X \in D$ or $Y \in D$.

4.1.8. Let E be a countable subset of $S(\omega)$. Then the filter generated by E cannot be a nonprincipal ultrafilter.

4.1.9. Prove Propositions 4.1.5 and 4.1.7.

4.1.10. Suppose D is the principal ultrafilter where $\{j\} \in D$. Prove that $\prod_D \mathfrak{A}_i$ is isomorphic to $\mathfrak{A}_j$.

4.1.11. Let D be a proper filter over I, let $X \in D$, and let $E = D \cap S(X)$.

Prove that

$$\prod_D \mathfrak{A}_i \cong \prod_E \mathfrak{A}_x.$$

(Cf. Exercise 4.1.3.)

4.1.12. The *direct product* of the models $\mathfrak{A}_i$, $i \in I$ is defined as follows.

The universe set is the Cartesian product $\prod_{i \in I} A_i$. A relation $S(f^1 \dots f^n)$ holds in the direct product if and only if the corresponding relation $R_i(f^1(i) \dots f^n(i))$ holds in $\mathfrak{A}_i$ for all $i \in I$. The functions H are defined by

$$H(f^1 \dots f^n) = \langle G_i(f^1(i) \dots f^n(i)) : i \in I \rangle,$$

and the constants b are defined by

$$b = \langle a_i : i \in I \rangle.$$

Prove that the direct product is isomorphic to the trivial reduced product $\prod_{\{I\}} \mathfrak{A}_i$.

4.1.13. If D, E are proper filters over I and $D \subset E$, then $\prod_E \mathfrak{A}_i$ is a homomorphic image of $\prod_D \mathfrak{A}_i$. Hence every reduced product is a homomorphic image of the direct product of the models $\mathfrak{A}_i$. (See Section 2.1 for the definitions of homomorphism and homomorphic image.)

4.1.14. Show that there exists an ultraproduct $\prod_D A_i$ of finite sets A_i which is infinite.

4.1.15. Let D be a proper filter over I. If each $\mathfrak{A}_i$ is isomorphically embedded in $\mathfrak{B}_i$, then $\prod_D \mathfrak{A}_i$ is isomorphically embedded in $\prod_D \mathfrak{B}_i$. If each $\mathfrak{A}_i$ is isomorphic to $\mathfrak{B}_i$, then $\prod_D \mathfrak{A}_i$ is isomorphic to $\prod_D \mathfrak{B}_i$. If each $\mathfrak{A}_i$ is a homomorphic image of $\mathfrak{B}_i$, then $\prod_D \mathfrak{A}_i$ is a homomorphic image of $\prod_D \mathfrak{B}_i$.

4.1.16. Let D be an ultrafilter over I. If $\mathfrak{A}_i \equiv \mathfrak{B}_i$ for all $i \in I$, then $\prod_D \mathfrak{A}_i \equiv \prod_D \mathfrak{B}_i$. If $\mathfrak{A}_i$ is elementarily embedded in $\mathfrak{B}_i$ for all $i \in I$, then $\prod_D \mathfrak{A}_i$ is elementarily embedded in $\prod_D \mathfrak{B}_i$.

4.1.17. A class K of models for $\mathscr{L}$ is said to be a *pseudo-elementary class* iff for some expansion $\mathscr{L}'$ of $\mathscr{L}$ and some elementary class K' for $\mathscr{L}'$, K is the class of all reducts of models in K' to $\mathscr{L}$. Prove that every pseudo-elementary class is closed under ultraproducts. Prove that if Γ is a set of Σ_1^1 sentences, then the class of all models of Γ is a pseudo-elementary class.

4.1.18. Let K be a class of models for $\mathscr{L}$. Let M be the class of all models $\mathfrak{A}$ such that $\mathfrak{A}$ is elementarily equivalent to an ultraproduct of members of K. Prove that M is an elementary class, and is the least elementary class which includes K.

4.1.19. Give the proof of Theorem 4.1.12 (ii).

4.1.20. The following are equivalent:

(i). K is a basic elementary class.

(ii). There exists a finitely axiomatizable theory T in $\mathscr{L}$ such that K is the class of all models of T.

(iii). Both K and its complement are elementary classes.

4.1.21. Let D be a principal ultrafilter. Prove that $d(\mathfrak{A}) = \prod_D \mathfrak{A}$, whence d is an isomorphism of $\mathfrak{A}$ onto $\prod_D \mathfrak{A}$.

4.1.22. Let D be a proper filter. Prove that d is an isomorphic embedding of $\mathfrak{A}$ into $\prod_D \mathfrak{A}$.

4.1.23. Let K and M be two classes of models. Let T_1 be the theory of K (the set of all sentences which hold in every model in K), and let T_2 be the theory of M. Prove that $T_1 \cup T_2$ is consistent if and only if some ultraproduct of members of K is elementarily equivalent to some ultraproduct of members of M.

4.1.24. Let $\mathfrak{A}_\alpha$, $\alpha < \beta$, be an elementary chain of length $\beta > 0$. Let D be an ultrafilter over β such that for each $\alpha < \beta$, the set $\{\gamma : \alpha \leqslant \gamma < \beta\}$ belongs to D. Prove that $\bigcup_{\alpha<\beta} \mathfrak{A}_\alpha$ is elementarily embedded in the ultraproduct $\prod_D \mathfrak{A}_\alpha$.

4.1.25. In the above exercise suppose that $\mathfrak{A}_\alpha$, $\alpha < \beta$, is only a chain of models, and D is only a proper filter. Prove that $\bigcup_{\alpha<\beta} \mathfrak{A}_\alpha$ is isomorphically embedded in the reduced product $\prod_D \mathfrak{A}_\alpha$.

4.1.26. Using Theorem 4.1.12, prove that none of the following theories is finitely axiomatizable:

(i) infinite models of pure identity theory;

(ii) fields of characteristic zero;

(iii) real closed fields;

(iv) algebraically closed fields;

(v) divisible Abelian groups;

(vi) torsion-free Abelian groups;

(vii) the theory of the model $\langle \omega, S \rangle$, where S is the successor function.

4.1.27. Show that none of the following classes of models is closed under elementary equivalence (use Corollary 4.1.10):

(i) the class of free groups;

(ii) the class of torsion groups;

(iii) the class of simple groups;

(iv) the class of all rings isomorphic to polynomial rings over the field of rational numbers.

4.1.28. Prove that every model $\mathfrak{A}$ can be isomorphically embedded in some ultraproduct of finite submodels of $\mathfrak{A}$. (For this problem, assume that $\mathscr{L}$ has no function or constant symbols, so we can be sure that finite submodels exist.) This gives a stronger form of Corollary 2.1.9.

4.1.29. Give an example of a Π_1^1 formula which is not preserved under ultraproducts.

4.1.30*. Let F_i, $i \in I$ be a family of fields. From the direct product $R = \prod_{i \in I} F_i$. Thus R is a ring. For each ultrafilter D over I, let

$$M_D = \{f \in R : \{i \in I : f(i) = 0\} \in D\}.$$

Prove the following:

(i). For each D, M_D is a maximal ideal in R.

(ii). For every maximal ideal M in R, there is an ultrafilter D over I such that $M = M_D$.

(iii). The ultraproduct $\prod_D F_i$ is isomorphic to the quotient field R/M.

Thus ultraproducts of fields are essentially the same thing as quotient fields of direct products of fields. Show that the same results hold for division rings (which have all the field axioms except for commutativity of multiplication).

4.1.31. Let D be an ultrafilter and let $\mathfrak{A} \times \mathfrak{B}$ be the direct product of $\mathfrak{A}$ and $\mathfrak{B}$. Prove that $\prod_D (\mathfrak{A} \times \mathfrak{B}) \cong \prod_D \mathfrak{A} \times \prod_D \mathfrak{B}$.

4.1.32. Let D be a nonprincipal ultrafilter over ω. Prove that for every infinite set A, the natural embedding $D : A \to \prod_D A$ is a proper embedding. Prove the same result under the weaker hypothesis that the index set I of D can be partitioned into countably many sets none of which belongs to D. (Such ultrafilters are called *countably incomplete* in Section 4.3.)

4.1.33. Let D be a nonprincipal ultrafilter over ω. Prove that the ultrapower $\prod_D \langle \omega, \leq \rangle$ is isomorphic to a proper initial segment of $\prod_D(\prod_D \langle \omega, \leq \rangle)$. [*Hint*: Use the fact that $\langle \omega, \leq \rangle$ is isomorphic to an initial segment of $\prod_D \langle \omega, \leq \rangle$.]

4.1.34*. Let $\mathfrak{A}$ be a model with the property that each subset U of A is a relation of $\mathfrak{A}$ and each function $f : A \to A$ is a function of $\mathfrak{A}$. Suppose $\mathfrak{A} \prec \mathfrak{B}$ and there is an element $b \in B$ such that $\mathfrak{B}$ has no proper submodels containing b. Prove that there is an ultrafilter D over A such that $\mathfrak{B} \cong \prod_D A$.

4.1.35*. Let D be an ultrafilter over a set I and let $f : I \to I$. Prove that there is a set $X \in D$ such that either $f(i) = i$ for all $i \in X$, or $f(i) \notin X$ for all $i \in X$. [*Hint*: Let $J = \{i \in I : f(i) \neq i\}$. Show that J can be partitioned into 3 sets X_1, X_2, X_3 such that for each $n \in \{1, 2, 3\}$, $f(i) \notin X_n$ for all $i \in X_n$.]

4.2. Measurable cardinals

In this section we shall study ultraproducts of a special kind. These are ultraproducts where the ultrafilter is α-complete. We shall apply these ultraproducts to the problem of the existence of α-complete ultrafilters. This problem has had a profound influence on the recent development of set theory.

Let α be an infinite cardinal. A filter D over I is said to be *α-complete* iff the intersection of any non-empty set of fewer than α elements of D belongs to D, that is,

$$E \subset D \quad \text{and} \quad |E| < \alpha \quad \text{implies} \quad \bigcap E \in D.$$

We see at once that:

PROPOSITION 4.2.1.

(i). *Every filter is ω-complete.*

(ii). *A filter D is α-complete for all α if and only if D is principal.*

(iii). *If $\alpha < \beta$, then every β-complete filter is α-complete.*

A slightly less trivial proposition is the following:

PROPOSITION 4.2.2. *Let D be a filter over a set I of power α. If D is α^+-complete, then D is principal.*

PROOF. Let E be the set of all complements of singletons which belong to D,

$$E = \{I \setminus \{i\} : I \setminus \{i\} \in D\}.$$

Since $|I| = \alpha$, $|E| \leqslant \alpha < \alpha^+$. D is α^+-complete and $E \subset D$, so $\bigcap E \in D$. On the other hand, if $X \in D$, then $\bigcap E \subset X$. For if $i \notin X$, then $X \subset I \setminus \{i\}$, whence $I \setminus \{i\} \in D$, $I \setminus \{i\} \in E$, and $i \notin \bigcap E$. Thus D is the principal filter generated by $\bigcap E$. ⊣

We shall be mainly interested in nonprincipal α-complete ultrafilters. The above proposition shows that α must be at most the power of I.

If α is a regular cardinal, then the set D of all sets $X \subset \alpha$ for which $|\alpha \setminus X| < \alpha$ is an α-complete nonprincipal filter over α. We shall see below that α-complete nonprincipal ultrafilters are much harder to come by. The following is a useful lemma.

LEMMA 4.2.3. *Let D be an ultrafilter over a set I. Then D is α-complete if and only if for every partition of I into fewer than α parts, one of the parts belongs to D.*

PROOF. Suppose D is α-complete and let $I = \bigcup_{\eta<\beta} X_\eta$ be a partition of I into β parts, where $\beta < \alpha$. Then

$$\bigcap_{\eta<\beta} (I \setminus X_\eta) = 0 \notin D.$$

Hence for some $\eta < \beta$, $I \setminus X_\eta \notin D$, whence $X_\eta \in D$.

Now suppose that for every partition of I into fewer than α parts, one part is in D. Let E be a subset of D of power less than α. Let $E = \{e_\eta : \eta < \beta\}$ be an enumeration of E, where $\beta < \alpha$. We define a function f on I into $\beta+1$ as follows. If $i \in \bigcap E$, then $f(i) = \beta$. If $i \in I \setminus \bigcap E$, then $f(i)$ is the least $\eta < \beta$ such that $i \notin e_\eta$. By our assumption, there exists $\eta < \beta+1$ such that $f^{-1}(\eta) \in D$. For each $\eta < \beta$, $f(i) = \eta$ implies $i \notin e_\eta$, so $f^{-1}(\eta) \cap e_\eta = 0$. Since $e_\eta \in D$, $f^{-1}(\eta) \notin D$. Therefore we must have $f^{-1}(\beta) \in D$. But $f^{-1}(\beta) = \bigcap E$, so $\bigcap E \in D$. Hence D is α-complete. ⊣

We now prove a result which tells when the natural embedding maps $\mathfrak{A}$ onto $\prod_D \mathfrak{A}$.

PROPOSITION 4.2.4. *Let $\mathfrak{A}$ be a model of power α and let D be an ultrafilter. Then the natural embedding d maps $\mathfrak{A}$ onto $\prod_D \mathfrak{A}$ if and only if D is α^+-complete.*

PROOF. Suppose D is α^+-complete. Let $f_D \in \prod_D A$. Then f maps I into A. Since $|A| = \alpha$, the partition $I = \bigcup\{f^{-1}(a) : a \in A\}$ partitions I into fewer than α^+ parts. By Lemma 4.2.3, there exists $a \in A$ such that $f^{-1}(a) \in D$. Then $f = {}_D\langle a : i \in I\rangle$, so $f_D = d(a)$. This shows that $d(A) = \prod_D A$.

Now suppose d maps $\mathfrak{A}$ onto $\prod_D \mathfrak{A}$. Let $I = \bigcup_{\eta<\beta} X_\eta$ be a partition of I into $\beta < \alpha^+$ parts. Since $|\beta| \leqslant \alpha = |A|$, we may renumber the sets X_η with indices from A, say

$$I = \bigcup_{a\in B} X_a,$$

where $B \subset A$. Let f be the function on I into A given by: $f(i) = a$ if and only if $i \in X_a$. Then $f_D \in \prod_D A = d(A)$, so $f_D = d(a)$ for some $a \in A$. This means that $f^{-1}(a) \in D$. But $f^{-1}(a) = X_a$, so $X_a \in D$. By Lemma 4.2.3, D is α^+-complete. ⊣

COROLLARY 4.2.5. *If either $\mathfrak{A}$ is finite or D is a principal ultrafilter, then d maps $\mathfrak{A}$ onto $\prod_D \mathfrak{A}$.*

The interesting ultrapowers are those in which d maps $\mathfrak{A}$ properly into $\prod_D \mathfrak{A}$.

A cardinal α is said to be *measurable* iff there exists a nonprincipal α-complete ultrafilter over α. Obviously, ω is measurable. We shall learn from the results below that the first uncountable measurable cardinal, if it exists at all, must be very large. We observe that if one set of power α has an α-complete nonprincipal ultrafilter, then so does every other set of power α.

LEMMA 4.2.6. *Suppose D is an α-complete ultrafilter over I and $f : I \to J$. Then the set $E = \{Y \subset J : f^{-1}(Y) \in D\}$ is an α-complete ultrafilter over J.*

The proof is left as an exercise.

PROPOSITION 4.2.7. *Let D be a nonprincipal ultrafilter. Then there is a greatest cardinal α such that D is α-complete. Moreover, α is a measurable cardinal.*

PROOF. Since D is not principal, there exists a least cardinal β such that D is not β-complete. β cannot be a limit cardinal, because if D is γ^+-complete for all $\gamma < \beta$, then D is closed under all intersections of fewer than β sets, and hence is β-complete. Therefore, β is a successor cardinal, $\beta = \alpha^+$; and α is the greatest cardinal such that D is α-complete.

D is not α^+-complete, so by Lemma 4.2.3 there is a partition $I = \bigcup_{\eta<\alpha} X_\eta$ of I into α parts such that none of the sets X_η belongs to D. Let f be the function on I onto α given by $f(i) = \eta$ if and only if $i \in X_\eta$. By Lemma 4.2.6. the set

$$E = \{Y \subset \alpha : f^{-1}(Y) \in D\}$$

is an α-complete ultrafilter over α. Moreover, E is nonprincipal. For if E is principal, then $\{\eta\} \in E$ for some $\eta < \alpha$, whence $X_\eta = f^{-1}(\{\eta\}) \in D$. Therefore α is measurable. ⊣

COROLLARY 4.2.8. *If $|A|$ is less than the first uncountable measurable cardinal (or if there is none), and if D is an ω_1-complete ultrafilter, then d maps $\mathfrak{A}$ onto $\prod_D \mathfrak{A}$.*

PROOF. If there is no measurable cardinal $> \omega$, then by Proposition 4.2.7, D is principal. If there is a least measurable cardinal $\alpha > \omega$, then by Proposition 4.2.7, D is α-complete. Since $|A| < \alpha$, D is $|A|^+$-complete. The result now follows from Proposition 4.2.4. ⊣

We now wish to prove a stronger form of the fundamental theorem for α-complete ultrafilters. To do this, we need to introduce the infinitary language $\mathscr{L}_\alpha$. The language $\mathscr{L}_\alpha$ has α individual variables instead of only countably many. The set of formulas of $\mathscr{L}_\alpha$ is obtained by adding to the rules of formation for $\mathscr{L}$ the following two additional rules, which permit infinite conjunctions and quantifiers:

4.2.9. If Φ is a set of formulas of $\mathscr{L}_\alpha$ of power $|\Phi| < \alpha$, then $\bigwedge \Phi$ is a formula of $\mathscr{L}_\alpha$.

4.2.10. If φ is a formula of $\mathscr{L}_\alpha$ and V is a set of variables of power $|V| < \alpha$, then $(\forall V)\varphi$ is a formula of $\mathscr{L}_\alpha$.

Thus $\mathscr{L}_\omega$ is just the usual logic $\mathscr{L}$.

The models for $\mathscr{L}_\alpha$ are exactly the same as the models for $\mathscr{L}$. If α is a regular cardinal, then each formula of $\mathscr{L}_\alpha$ has fewer than α symbols. It should be noted that if $\alpha > \omega$, then a formula may have infinitely many free variables. The notion of truth of a formula of $\mathscr{L}_\alpha$ in a model can be defined in a precise way by adding to the definition of truth for formulas of $\mathscr{L}$. Infinite disjunctions, $\bigvee \Phi$, and existential quantifiers, $(\exists V)\varphi$, are introduced as abbreviations in the obvious way. We shall leave the details as an exercise.

The most interesting example of a property which can be expressed in $\mathscr{L}_{\omega_1}$ is that of a well founded relation. The following sentence states that the relation $P(x, y)$ is well founded:

$$(\forall x_0 x_1 x_2 \ldots) \neg \bigwedge_{n<\omega} P(x_{n+1} x_n).$$

When added to the sentence of $\mathscr{L}$ stating that $P(xy)$ is a simple ordering, we obtain the statement that $P(xy)$ is a well ordering.

The sentence

$$\neg (\exists x) \bigwedge \{\sigma(x) : \sigma \in \Sigma\}$$

of $\mathscr{L}_{\omega_1}$ expresses the fact that the type $\Sigma(x)$ is omitted, and the negation of that sentence says that $\Sigma(x)$ is realized. The sentence

$$(\forall x)(x \equiv 0 \vee x \equiv 1 \vee \ldots)$$

is true in $\mathfrak{A}$ if and only if $\mathfrak{A}$ is an ω-model. This is also a sentence of $\mathscr{L}_{\omega_1}$.

THEOREM 4.2.11. *Let $\mathfrak{B}$ be an ultraproduct $\prod_D \mathfrak{A}_i$, where I is the index set and D is an α-complete ultrafilter. Then:*

(i). *Given any formula $\varphi(x_1 x_2 \ldots)$ of $\mathscr{L}_\alpha$ and $f_D^1, f_D^2, \ldots \in B$, we have*

$$\mathfrak{B} \models \varphi[f_D^1 f_D^2 \ldots] \quad \textit{if and only if} \quad \{i \in I : \mathfrak{A}_i \models \varphi[f^1(i) f^2(i) \ldots]\} \in D.$$

(The list $x_1, x_2, \ldots$ of variables may be infinite.)

(ii). *For any sentence φ of $\mathscr{L}_\alpha$,*

$$\mathfrak{B} \models \varphi \quad \textit{if and only if} \quad \{i \in I : \mathfrak{A}_i \models \varphi\} \in D.$$

PROOF. (ii) is a special case of (i). By the fundamental theorem, we already know that (i) holds for formulas of $\mathscr{L}$, and, in particular, for atomic formulas. The proof that for any formula ψ of $\mathscr{L}_\alpha$, if (i) holds for ψ, then (i) holds for $\neg \psi$, is exactly the same as before.

Suppose Φ is a set of formulas of $\mathscr{L}_\alpha$, $|\Phi| < \alpha$, and (i) holds for all $\varphi \in \Phi$. Then for any $f_D^1, f_D^2, \ldots$ in B, the following are equivalent:

$$\mathfrak{B} \models (\bigwedge \Phi)[f_D^1 f_D^2 \ldots];$$

$$\text{for all} \quad \varphi \in \Phi, \qquad \mathfrak{B} \models \varphi[f_D^1 f_D^2 \ldots];$$

$$\text{for all } \varphi \in \Phi, \quad \{i \in I : \mathfrak{A}_i \models \varphi[f^1(i) f^2(i) \ldots]\} \in D;$$

$$\{i \in I : \text{for all } \varphi \in \Phi, \ \mathfrak{A}_i \models \varphi[f^1(i) f^2(i) \ldots]\} \in D;$$

$$\{i \in I : \mathfrak{A}_i \models (\bigwedge \Phi)[f^1(i) f^2(i) \ldots]\} \in D.$$

The equivalence of the third and fourth lines uses the α-completeness of D. Hence $\bigwedge \Phi$ has the property (i).

Now suppose a formula $\psi(x_1 x_2 \ldots y_1 y_2 \ldots)$ of $\mathscr{L}_\alpha$ has the property (i), $\{y_1 y_2, \ldots\}$ is a set of fewer than α variables, and

$$\varphi = (\exists y_1 y_2 \ldots)\psi.$$

We shall prove that φ has the property (i). This is done by observing that the following are equivalent:

$$\mathfrak{B} \models \varphi[f_D^1 f_D^2 \ldots];$$

there exist $g_D^1, g_D^2, \ldots \in B$ such that

$$\mathfrak{B} \models \psi[f_D^1 f_D^2 \ldots g_D^1 g_D^2 \ldots];$$

there exist $g_D^1, g_D^2, \ldots \in B$ such that

$$\{i \in I : \mathfrak{A}_i \models \psi[f^1(i) f^2(i) \ldots g^1(i) g^2(i) \ldots]\} \in D;$$

$$\{i \in I : \mathfrak{A}_i \models \varphi[f^1(i) f^2(i) \ldots]\} \in D.$$

From this, it follows that (i) holds for all formulas of $\mathscr{L}_\alpha$ because the infinite universal quantifier can be expressed in the usual way in terms of the infinite existential quantifier and negation. ⊣

As an example of Theorem 4.2.11, if D is an ω_1-complete ultrafilter, then any ultraproduct of well ordered structures modulo D is well ordered, and any ultraproduct of well founded structures modulo D is well founded. We shall indicate some applications of this theorem in the exercises. One particularly important application is the following weak compactness theorem for infinite languages:

THEOREM 4.2.12 (Weak Compactness Theorem). *Let α be a measurable cardinal. If Σ is a set of sentences of $\mathscr{L}_\alpha$ such that $|\Sigma| = \alpha$ and every subset Σ_0 of Σ of power $|\Sigma_0| < \alpha$ has a model, then Σ has a model.*

PROOF. Since α is measurable, there exists an α-complete nonprincipal ultrafilter D over α. Since D contains no singletons and is α-complete, D contains no sets of power less than α. Hence for each $\gamma < \alpha$,

$$\{\beta : \gamma < \beta < \alpha\} \in D.$$

Let us enumerate Σ as $\Sigma = \{\sigma_\beta : \beta < \alpha\}$. For each $\beta < \alpha$, there is a model $\mathfrak{A}_\beta$ of the set $\{\sigma_\gamma : \gamma < \beta\}$. Let $\mathfrak{B}$ be the ultraproduct $\prod_D \mathfrak{A}_\beta$. Then for each

$\sigma_\gamma \in \Sigma$, we have

$$\{\beta < \alpha : \mathfrak{A}_\beta \models \sigma_\gamma\} \supset \{\beta : \gamma < \beta < \alpha\} \in D.$$

By Theorem 4.2.11, $\mathfrak{B}$ is a model of Σ. ⊣

A cardinal α which satisfies the conclusion of Theorem 4.2.12 above is said to be *weakly compact*. Thus every measurable cardinal is weakly compact. We shall see later that the converse does not hold, and in fact every uncountable measurable cardinal α has many weakly compact cardinals $\beta < \alpha$.

LEMMA 4.2.13. *Let α be an uncountable measurable cardinal and let D be a nonprincipal α-complete ultrafilter over α. Form the ultrapower $\mathfrak{B} = \prod_D\langle \alpha, < \rangle$. Then*:

(i). *$\mathfrak{B}$ is a well-ordered structure of order type greater than α.*

(ii). *For every $\gamma < \alpha$, $d(\gamma)$ is the γth element of $\mathfrak{B}$.*

PROOF. $\mathfrak{B}$ is well-ordered by Theorem 4.2.11. d is an isomorphic embedding, so $\mathfrak{B}$ has order type at least α. For each $\gamma < \alpha$, let $\bar{\gamma}$ be the γth element of $\mathfrak{B}$. Let $\gamma < \alpha$ and suppose (ii) holds for all $\delta < \gamma$. This means that $d(\delta) = \bar{\delta}$ for all $\delta < \gamma$. Add a constant c_δ for each $\delta < \alpha$. Thus c_δ is interpreted by δ in the model $\langle \alpha, <, \delta\rangle_{\delta<\alpha}$, and by $d(\delta)$ in $(\mathfrak{B}, d(\delta))_{\delta<\alpha}$. In the model $\langle \alpha, <, \delta\rangle_{\delta<\alpha}$, the sentence

$$(1) \qquad (\forall x)(x < c_\gamma \leftrightarrow \dot{\bigvee} \{x \equiv c_\delta : \delta < \gamma\})$$

of $\mathscr{L}_\alpha$ holds. Hence, by Theorem 4.2.11, (1) holds in $(\mathfrak{B}, d(\delta))_{\delta<\alpha}$. Thus $d(\gamma)$ is the γth element of $\mathfrak{B}$, for the set of elements $< d(\gamma)$ is exactly $\{\bar{\delta} : \delta < \gamma\}$. (ii) now follows by induction.

To prove (i) we note that D is not principal, so D is not α^+-complete, whence by Proposition 4.2.4 the natural embedding d maps α properly into B. But, by (ii), $d(\alpha)$ is an initial segment of $\mathfrak{B}$. Thus $d(\alpha)$ is a proper initial segment, and (i) follows. ⊣

THEOREM 4.2.14. *Let α be an uncountable measurable cardinal. Then*:

(i). *α is inaccessible.*

(ii). *α is the αth inaccessible cardinal.*

PROOF. Let D be a nonprincipal α-complete ultrafilter over α. Consider the model $\mathfrak{A} = \langle \alpha, <, b\rangle_{b\in\alpha}$ and form the ultrapower $\mathfrak{B} = \prod_D \mathfrak{A}$. Let $\mathfrak{B}$ have order type β, and for $\gamma < \beta$ let $\bar{\gamma}$ be the γth element of $\mathfrak{B}$. By Lemma

4.2.13, we have $\beta > \alpha$, and $d(\gamma) = \bar{\gamma}$ for all $\gamma < \alpha$. Thus each constant c_γ, $\gamma < \alpha$, is interpreted by γ in $\mathfrak{A}$ and by $d(\gamma) = \bar{\gamma}$ in $\mathfrak{B}$. We are now ready to prove (i).

(i). First, we show that α is regular. Suppose α is singular. Then for some $\gamma < \alpha$, there is a function F on γ into α such that the range of F is cofinal in $\langle \alpha, < \rangle$. Define $F(\delta) = 0$ for $\alpha > \delta \geqslant \gamma$, and form the model $(\mathfrak{A}, F)$ and the ultrapower

$$\prod_D (\mathfrak{A}, F) = (\mathfrak{B}, G).$$

For each $\delta < \alpha$, we have

$$G(\bar{\delta}) = G(d(\delta)) = d(F(\delta)) = \overline{F(\delta)} < \bar{\alpha}.$$

Hence in $(\mathfrak{B}, G)$,

(1) $$(\exists x)(\forall y)(y < c_\gamma \to F(y) < x)$$

holds, with $\bar{\alpha}$ for x. However, since the range of F is cofinal in $\langle \alpha, < \rangle$, the sentence

(2) $$(\forall x)(\exists y)(y < c_\gamma \wedge x < F(y))$$

holds in $(\mathfrak{A}, F)$, thus also in $(\mathfrak{B}, G)$. But (1) and (2) contradict each other. Therefore α is regular.

Now we show that α is a strong limit cardinal, that is, $\gamma < \alpha$ implies $2^\gamma < \alpha$. Suppose instead that for some γ,

$$\gamma < \alpha \leqslant 2^\gamma.$$

Then there exists a one–one function F on α into $S(\gamma)$. Let $R \subset \alpha \times \gamma$ be the binary relation 'representing' F, namely

$$R(\eta, \delta) \quad \text{if and only if} \quad \delta \in F(\eta).$$

Form the ultrapower

$$(\mathfrak{B}, S) = \prod_D (\mathfrak{A}, R).$$

Let $\bar{F}$ be the function on β defined by

$$\bar{F}(\eta) = \{\delta < \beta : S(\bar{\eta}, \bar{\delta})\}.$$

The function $\bar{F}$ is a one–one function on β into $S(\gamma)$, because the two sentences below hold in $(\mathfrak{A}, R)$ and hence in $(\mathfrak{B}, S)$:

$$(\forall xy)(R(xy) \to y < c_\gamma),$$

$$(\forall xy)[x \not\equiv y \to (\exists z) \neg (R(xz) \leftrightarrow R(yz))].$$

Moreover, $F(\eta) = \bar{F}(\eta)$ for all $\eta < \alpha$, because d maps $(\mathfrak{A}, F)$ isomorphically into $(\mathfrak{B}, G)$. It follows that the set $X = \bar{F}(\alpha)$ does not belong to the range of F, while $X \in S(\gamma)$. Thus the sentence

$$(\exists x)(\forall y)(R(xy) \leftrightarrow \bigvee \{y \equiv c_\delta : \delta \in X\})$$

is false in $(\mathfrak{A}, R)$, but true in $(\mathfrak{B}, S)$ with $\bar{\alpha}$ for x. This contradiction shows that α is a strong limit cardinal, and (i) is proved.

(ii). This part of the proof can best be understood by using the fact that Σ_1^1 formulas are preserved under ultraproducts (Corollary 4.1.14). Let X be the class of all inaccessible cardinals, Y the class of all ordinals which are not regular cardinals, and Z the class of all ordinals which are not strong limit cardinals. Thus $\gamma \in X$ if and only if $\gamma \notin Y \cup Z$. We wish to show that $X \cap \alpha$ is cofinal in α. Then, since α is regular, it will follow that $|X \cap \alpha| = \alpha$, whence α is the αth inaccessible cardinal and (ii) holds. Since

$$\alpha \subset X \cup Y \cup Z,$$

it suffices to prove that for each $\gamma < \alpha$,

(3) there exists δ such that $\gamma \leqslant \delta < \alpha$ and $\delta \notin Y \cup Z$.

Suppose that (3) fails for some $\gamma < \alpha$. Then for all $\delta < \alpha$,

(4) $$\delta < \gamma \quad \text{or} \quad \delta \in Y \quad \text{or} \quad \delta \in Z.$$

There is a Σ_1^1 formula $\psi_Y(u)$ such that for any model $\langle \alpha', < \rangle$ with α' an ordinal, and any $\delta < \alpha'$,

(5) $$\delta \in Y \quad \text{if and only if} \quad \langle \alpha', < \rangle \models \psi_Y[\delta].$$

$\psi_Y(u)$ is just a formalization of the statement:

'There exists $y < u$ and a function $F : y \to u$ such that the range of F is cofinal in u'.

There is also a Σ_1^1 formula $\psi_Z(u)$ such that for any ordinal α', and any $\delta < \alpha'$,

(6) $$\delta \in Z \quad \text{if and only if} \quad \langle \alpha', < \rangle \models \psi_Z[\delta].$$

$\psi_Z(u)$ is obtained by formalizing the statement:

'There exists $y < u$ and a relation $R \subset u \times y$ such that R represents a one–one function on u into $S(y)$'.

We have already explained how a function $F : u \to S(y)$ can be represented by a relation $R \subset u \times y$, and how to say that F is one–one using R. From (4)–(6), we see that the formula

(7) $$x < c_\gamma \vee \psi_Y(x) \vee \psi_Z(x)$$

is satisfied in $\langle\alpha, <\rangle$ by all $x \in \alpha$. Moving the second-order quantifiers to the front, we see that (7) is equivalent to a Σ_1^1 formula. Therefore, by Corollary 4.1.14, for any $f_D \in B$ the formula (7) is satisfied in $\langle B, <\rangle$ by f_D. Since $\langle B, <\rangle$ is isomorphic to $\langle\beta, <\rangle$, the formula (7) is satisfied in $\langle\beta, <\rangle$ by all $x \in \beta$. Putting $x = \alpha$, we have

$$\alpha < \gamma \quad \text{or} \quad \langle\beta, <\rangle \models \psi_Y[\alpha] \quad \text{or} \quad \langle\beta, <\rangle \models \psi_Z[\alpha].$$

Using (5) and (6), with $\beta = \alpha'$, we see that

$$\alpha < \gamma \quad \text{or} \quad \alpha \in Y \quad \text{or} \quad \alpha \in Z.$$

But this contradicts our assumption that $\gamma < \alpha$ and α is inaccessible. Therefore (3) holds, and our proof is complete. ⊣

Part (ii) of the above proof also works when α is an inaccessible weakly compact cardinal (see Exercise 4.2.9). The argument can be extended to show that even larger cardinals are less than α.

Our next application of measurable cardinals concerns axiomatic set theory. We prove that if there exists an uncountable measurable cardinal, then the axiom of constructibility fails. Equivalently, the axiom of constructibility implies that ω is the only measurable cardinal. The axiom of constructibility was introduced by Gödel in 1938 to prove that the generalized continuum hypothesis and the axiom of choice are consistent with ZF (or BG) provided that ZF is consistent. Gödel proved first that if ZF is consistent, then so is ZF plus the axiom of constructibility. Secondly, the axiom of constructibility implies the axiom of choice and the generalized continuum hypothesis. These classical results belong to set theory rather than model theory, and are thus outside the scope of this book. We shall present the result in such a way that a minimum knowledge of set theory is needed.

To begin with, let us review the rank function $R(\alpha)$. For each ordinal α, the set $R(\alpha)$ is defined as follows (see the Appendix):

$$R(0) = 0,$$

$$R(\alpha+1) = S(R(\alpha)),$$

and for α a limit ordinal,

$$R(\alpha) = \bigcup_{\beta<\alpha} R(\beta).$$

The sets $R(\alpha)$ are important because they provide natural models of set theory. In Exercise 1.4.15 it was shown that when θ is an uncountable inaccessible cardinal, $\langle R(\theta), \in\rangle$ is a model of Zermelo–Fraenkel set theory and $\langle R(\theta+1), \in, R(\theta)\rangle$ is a model of Bernays–Morse set theory.

We list some elementary facts which will be needed for $R(\alpha)$ in the following proposition.

PROPOSITION 4.2.15.

(i). *The set* $R(\alpha)$ *is transitive, that is, every element of an element of* $R(\alpha)$ *is an element of* $R(\alpha)$ (*see Section* 1.4).

(ii). *If* $\alpha \leqslant \beta$, *then* $R(\alpha) \subset R(\beta)$.

(iii). $R(\omega+\alpha)$ *has power* $\beth_\alpha$.

(iv). *If* α *is an ordinal, then* $\alpha \subset R(\alpha)$, *but* $\alpha \notin R(\alpha)$.

We shall also use another set-theoretic function, $H(\alpha)$, the set of all sets which are hereditarily of power less than α. For each cardinal α, we define $H(\alpha)$ by the condition:

$x \in H(\alpha)$ iff there is a transitive set y such that $x \subset y$ and $|y| < \alpha$.

Here are some basic facts about $H(\alpha)$. The theory ZF$-$P, described in the Appendix, is Zermelo–Fraenkel set theory minus the power set axiom.

PROPOSITION 4.2.16.

(i). $H(\alpha)$ *is a transitive subset of* $R(\alpha)$.

(ii). *If* $\alpha \leqslant \beta$, *then* $H(\alpha) \subset H(\beta)$.

(iii). $\alpha \subset H(\alpha)$.

(iv). *If* α *is an uncountable regular cardinal, then* $\langle H(\alpha), \in\rangle$ *is a model of* ZF$-$P.

It is not obvious from the definition that $H(\alpha)$ is even a set, but this follows from part (i) above. To prove (i), we show by induction on the rank of y that if y is transitive and $|y| < \alpha$, then $y \in R(\alpha)$. The other parts are also not hard to prove. It turns out that $H(\omega) = R(\omega)$, and, more generally, $H(\alpha) = R(\alpha)$ when α is inaccessible. However, $H(\omega_1)$ is a rather small subset of $R(\omega_1)$, in fact $H(\omega_1)$ has power 2^ω, while $R(\omega_1)$ has power $\beth_{\omega_1}$.

We now have a few things to say about well-founded models. Recall that a binary relation E is said to be *well founded* iff there are no infinite sequences decreasing with respect to E, that is, no infinite sequences $x_0, x_1, \ldots$ such that

$$x_1 E x_0, \; x_2 E x_1, \; \ldots .$$

For example, the $\in$-relation is well founded. Let $\langle B, E\rangle$ be a well founded model. We shall say that an element $b \in B$ is an *ordinal* of $\langle B, E\rangle$ iff

$$\langle B, E\rangle \models (\forall yz)[(y \in b \wedge z \in b \rightarrow y \in z \vee z \in y \vee y \equiv z) \\ \wedge (y \in b \wedge z \in y \rightarrow z \in b)].$$

It is easy to see that x is an ordinal of $\langle H(\alpha), \in\rangle$ if and only if $x \in \alpha$. We need one more preliminary result.

Proposition 4.2.17. *Suppose $\langle B, E\rangle$ is a well-founded model of* ZF$-$P. *Then the set of ordinals of $\langle B, E\rangle$ is well ordered by E.*

Proof. Let x be any ordinal of $\langle B, E\rangle$. Over the set $X = \{y \in B : yEx\}$, the relation E must be transitive. For if $uEyEz$, but not uEz, then either $u = z$ or zEu, and in either case we have an infinite sequence decreasing with respect to E (either $y, u, y, u, \ldots$, or $z, y, u, z, y, u, \ldots$). Therefore E simply orders X. Since E is well founded, it well orders X. We may now simply imitate the proof in ZF that the ordinals are well ordered by the $\in$ relation (see the Appendix) to show that the ordinals of $\langle B, E\rangle$ are well ordered by E. ⊣

Let $\langle B, E\rangle$ be a well-founded model of ZF$-$P. By the *order type* of $\langle B, E\rangle$, we mean the order type of the set of ordinals of $\langle B, E\rangle$ under the relation E. We may now state the axiom of constructibility. We state it in an unusual way so that we can get by with a minimum knowledge of set theory. However, in ZF with the axiom of choice, our formulation here can be shown to be equivalent to the more familiar statements of the axiom of constructibility.

Axiom of Constructibility. *For every regular cardinal $\alpha > \omega$, every well-founded model $\langle B, E\rangle$ of* ZF$-$P *of order type α is isomorphic to $\langle H(\alpha), \in\rangle$.*

This axiom states that $H(\alpha)$ is very narrow. We can see this more clearly in the following equivalent statement of the axiom of constructibility (which is less convenient for our intended use):

For every regular cardinal $\alpha > \omega$, there is no proper subset $M \subset H(\alpha)$ such that M is transitive, $\alpha \subset M$, and $\langle M, \in\rangle$ is a model of ZF$-$P.

Theorem 4.2.18 (Scott's Theorem).

(i). *If there exists an uncountable measurable cardinal, then the axiom of constructibility fails.*

(ii). *The axiom of constructibility implies that ω is the only measurable cardinal.*

PROOF. Let α be the first uncountable measurable cardinal. Let $\beta = |2^{2^{2^\alpha}}|^+$. Then $R(\alpha+3)$ is a transitive set of power less than β, and β is a regular cardinal. It follows that $R(\alpha+3) \in H(\beta)$ and $\langle H(\beta), \in \rangle$ is a model of ZF−P.

Let D be a nonprincipal α-complete ultrafilter over α, and form the ultrapower

$$\langle B, E \rangle = \Pi_D \langle H(\beta), \in \rangle.$$

Then $\langle B, E \rangle$ is a well founded model of ZF−P. We claim that $\langle B, E \rangle$ has order type β. The natural embedding d maps β isomorphically into the ordinals of $\langle B, E \rangle$, so $\langle B, E \rangle$ has order type at least β. Let x be any ordinal of $\langle B, E \rangle$. Then $x = f_D$ for some function f on α into β. Since $\mathrm{cf}(\beta) > \alpha$, f is not cofinal in β. Thus there exists $\gamma < \beta$ such that $f \in {}^\alpha\gamma$. It follows that, if yEx, then $y = g_D$ for some $g \in {}^\alpha\gamma$. Therefore the set $\{y : yEx\}$ has power $\leqslant \gamma^\alpha$, and since $\gamma \leqslant 2^{2^{2^\alpha}}$, $\gamma^\alpha < \beta$. This shows that every ordinal of $\langle B, E \rangle$ has fewer than β predecessors, and it follows that $\langle B, E \rangle$ has order type at most β. Our claim is established.

Let $\varphi(x)$ be the formula of set theory stating that 'x is the first uncountable measurable cardinal'. When $\varphi(x)$ is written out in detail, we see that the quantifiers in $\varphi(x)$ can be restricted to, say, $S(S(S(x)))$. Thus in the model $\langle H(\beta), \in \rangle$, an element b satisfies $\varphi(x)$ if and only if b really is the first uncountable measurable cardinal, $b = \alpha$. In the ultrapower $\langle B, E \rangle$, the unique element satisfying $\varphi(x)$ is thus $d(\alpha)$. The ordinal $d(\alpha)$ is greater than the αth ordinal of $\langle B, E \rangle$. For we see from Lemma 4.2.13 that there is a function $f \in {}^\alpha\alpha$ such that f_D has α predecessors, and clearly $f_D E d(\alpha)$. This shows that $\langle B, E \rangle$ is not isomorphic to $\langle H(\beta), \in \rangle$. For any isomorphism would have to map α to the αth ordinal $\bar{\alpha}$ of $\langle B, E \rangle$, and $\bar{\alpha}$ does not satisfy $\varphi(x)$. So the axiom of constructibility fails. ⊣

We shall now prove another theorem about ultraproducts modulo an α-complete ultrafilter. This result will not hold for weakly compact cardinals. First, we need to study normal ultrafilters. Normal ultrafilters are α-complete nonprincipal ultrafilters over α which have a valuable extra property. We only define normal ultrafilters over a set α which is an uncountable measurable cardinal. We shall not attempt to give any notion of a normal ultrafilter over an arbitrary set I.

An ultrafilter D over α is said to be *normal* iff $\alpha > \omega$ and

(1). D is nonprincipal;

(2). D is α-complete;

(3). In the ultrapower $\prod_D\langle\alpha, <\rangle$, the αth element is f_D, where f is the identity function on α, i.e., $f(\gamma) = \gamma$ for all $\gamma < \alpha$.

An alternative form of (3) is:

Proposition 4.2.19. *Let D be a nonprincipal α-complete ultrafilter over α. Then D is normal if and only if for every function $g \in {}^\alpha\alpha$ such that $\{\beta : g(\beta) < \beta\} \in D$, there exists $\gamma < \alpha$ with*

$$\{i \in \alpha : g(i) = \gamma\} \in D.$$

Briefly, every decreasing function on α is equivalent to a constant function modulo D.

We leave the proof as an exercise.

Proposition 4.2.20. *If α is a measurable cardinal and $\alpha > \omega$, then there exists a normal ultrafilter over α.*

Proof. Let E be an arbitrary α-complete nonprincipal ultrafilter over α. Form the ultrapower $\prod_E\langle\alpha, <\rangle = \mathfrak{B}$, and let f_E be the αth element of $\mathfrak{B}$. Define

$$D = \{X \subset \alpha : f^{-1}(X) \in E\}.$$

D is an α-complete ultrafilter over α because $0 \notin D$, and, for all $X \in S(\alpha)$ and $C \subset S(\alpha)$, (see Lemma 4.2.6)

$$f^{-1}(\alpha \setminus X) = \alpha \setminus f^{-1}(X),$$
$$f^{-1}(\bigcup C) = \bigcup f^{-1}(C),$$
$$f^{-1}(\bigcap C) = \bigcap f^{-1}(C).$$

D is nonprincipal because for each $\gamma < \alpha$,

$$f_E = \bar{\alpha} \neq \bar{\gamma} = d(\gamma),$$

whence

$$f^{-1}(\{\gamma\}) = \{i \in \alpha : f(i) = \gamma\} \notin E,$$

and so $\{\gamma\} \notin D$. (As before, $\bar{\gamma}$ is the γth element of $\mathfrak{B}$.)

Finally, we show that D is normal. Let $g \in {}^\alpha\alpha$ be such that

$$X = \{\beta : g(\beta) < \beta\} \in D.$$

We shall use Proposition 4.2.19. Let $h = g \circ f$. Then for all $\beta \in f^{-1}(X)$,

$$h(\beta) = g(f(\beta)) < f(\beta).$$

$f^{-1}(X) \in E$, so $h_E < f_E = \bar{\alpha}$. Thus for some $\gamma = \alpha$, $h_E = \bar{\gamma}$. By Theorem 4.2.14, $\bar{\gamma} = d(\gamma)$. So

$$\{i \in \alpha : h(i) = \gamma\} \in E.$$

But

$$\{i : h(i) = \gamma\} = \{i : g(f(i)) = \gamma\} = f^{-1}(\{j : g(j) = \gamma\}),$$

whence

$$f^{-1}(\{j : g(j) = \gamma\}) \in E.$$

Thus, by Proposition 4.2.19, D is normal. ⊣

So now we know that normal ultrafilters exist. There does not seem to be any useful counterpart of normal ultrafilters for the measurable cardinal ω because the definition depended strongly on the well ordering of the ultraproduct. Our next theorem will give an indication of how valuable normal ultrafilters are.

We now prove a result concerning ultraproducts of $R(\alpha)$'s.

Theorem 4.2.21. *Let α be an uncountable measurable cardinal and let D be a normal ultrafilter over α. Then*

$$\langle R(\alpha+1), \in\rangle \cong \prod_D \langle R(\beta+1), \in\rangle.$$

Moreover, an isomorphism π is defined as follows: For each $x \in R(\alpha+1)$,

$$\pi(x) = \langle x \cap R(\beta) : \beta < \alpha\rangle_D.$$

Proof. Let the ultraproduct be denoted by

$$\prod_D \langle R(\beta+1), \in\rangle = \langle B, E\rangle.$$

We must show that π is one–one, onto, and for all $x, y \in R(\alpha+1)$,

$$\text{(1)} \qquad x \in y \quad \text{if and only if} \quad \pi x E \pi y.$$

To see that π is one–one, suppose $x, y \in R(\alpha+1)$ and $x \neq y$. Then there is an element z which belongs to one and not the other, say $z \in x \setminus y$. Since $x \in R(\alpha+1) = S(R(\alpha))$, $z \in R(\alpha)$. α is a limit ordinal, so $z \in R(\gamma)$ for some $\gamma < \alpha$. Then, whenever $\gamma \leqslant \beta < \alpha$, we have

$$z = z \cap R(\beta) \quad \text{and} \quad z \in (x \cap R(\beta)) \setminus (y \cap R(\beta)).$$

But the set of all β, where $\gamma \leqslant \beta < \alpha$, belongs to D, since D is nonprincipal and α-complete. Hence

$$\pi z E \pi x \quad \text{and not} \quad \pi z E \pi y.$$

This shows that $\pi x \neq \pi y$, so π is one–one.

Next we verify (1). Suppose $x \in y$. Then $x \in R(\alpha)$, so $x \in R(\gamma)$ for some $\gamma < \alpha$. Whenever $\gamma \leqslant \beta < \alpha$, we have

$$x = x \cap R(\beta) \quad \text{and} \quad x \in y \cap R(\beta).$$

It follows that $\pi x E \pi y$.

We next prove that π is onto. Let $f_D \in B$. First suppose that

$$X = \{\beta \in \alpha : f(\beta) \in R(\beta)\} \in D.$$

For each β, let

$$g(\beta) = \text{least } \gamma \text{ such that } f(\beta) \in R(\gamma+1).$$

Then, whenever $\beta \in X$, we have $g(\beta) < \beta$. For if β is a limit ordinal, then $f(\beta) \in R(\gamma) \subset R(\gamma+1)$ for some $\gamma < \beta$. And if $\beta = \gamma+1$, then $g(\beta) \leqslant \gamma < \beta$. D is normal, so by Proposition 4.2.19, there exists $\gamma < \alpha$, for which

$$Y = \{\beta : g(\beta) = \gamma\} \in D.$$

The set $R(\gamma+1)$ has power $\beth_{\gamma+1}$. But, by Theorem 4.2.14, α is inaccessible, so

$$\alpha = \beth_\alpha > \beth_{\gamma+1}.$$

Let us consider the partition of α such that one of the partition classes is $\alpha \setminus Y$, and for each set $u \in R(\gamma+1)$, the set

$$\{\beta : f(\beta) = u\}$$

is a partition class. There are only $\beth_{\gamma+1}$ partition classes, and D is α-complete, so one class belongs to D. We cannot have $\alpha \setminus Y \in D$, so there is a $u \in R(\gamma+1)$ such that

$$\{\beta : f(\beta) = u\} \in D.$$

But when $\gamma < \beta$ we have $u \cap R(\beta) = u$, so

$$\{\beta : f(\beta) = u \cap R(\beta)\} \in D.$$

It follows that $f_D = \pi(u)$; hence f_D is in the range of π.

Our argument thus far shows a little bit more:

(2) If there exists $g_D \in B$ such that $h_D E g_D$, then there exists $u \in R(\alpha)$ with $h_D = \pi u$.

For we have

$$\{\beta : h(\beta) \in g(\beta)\} \in D,$$

whence

$$\{\beta : h(\beta) \in R(\beta)\} \in D,$$

and our argument gives a u. This shows the other direction of (1).

It remains to consider an arbitrary $f_D \in B$. Let

$$x = \{y \in R(\alpha) : \pi y E f_D\}.$$

Then $x \in R(\alpha+1)$. We claim that $\pi x = f_D$. Since the axiom of extensionality holds in $\langle R(\alpha+1), \in\rangle$ and in $\langle B, E\rangle$, it suffices to prove that

(3) $$h_D E f_D \quad \text{if and only if} \quad h_D E \pi x.$$

If $h_D E f_D$, then by (2) we have $h_D = \pi u$ for some $u \in R(\alpha)$. Then $u \in x$, so by (1), $h_D = \pi u E \pi x$. If $h_D E \pi x$, then we use (2) again to get a $u \in R(\alpha)$ with $h_D = \pi u$. But then $\pi u E \pi x$, so by (1), $u \in x$. By the definition of x, $h_D = \pi u E f_D$. This proves (3), so $\pi x = f_D$, and π maps $R(\alpha+1)$ onto B. ⊣

Corollary 4.2.22. *Let α be an uncountable measurable cardinal and D a normal ultrafilter over α. Then for any formula $\varphi(x_1 \dots x_n)$ and any $S_1, \dots, S_n \in R(\alpha+1)$, we have*

$$\langle R(\alpha+1), \in\rangle \models \varphi[S_1 \dots S_n]$$

if and only if

$$\{\beta \in \alpha : \langle R(\beta+1), \in\rangle \models \varphi[S_1 \cap R(\beta) \dots S_n \cap R(\beta)]\} \in D.$$

If φ is a sentence,

$$\langle R(\alpha+1), \in\rangle \models \varphi$$

if and only if

$$\{\beta \in \alpha : \langle R(\beta+1), \in\rangle \models \varphi\} \in D.$$

We conclude this section with another application showing that uncountable measurable cardinals are very large.

Theorem 4.2.23. *Let α be an uncountable measurable cardinal, and let D be a normal ultrafilter over α. Then*:

(i). *The set of all inaccessible cardinals $\beta < \alpha$ belongs to D.*

(ii). *The set of all weakly compact cardinals $\beta < \alpha$ belongs to D.*

(iii). *α is the αth weakly compact inaccessible cardinal.*

Proof. (iii) follows from (i) and (ii) because every set in D has power α.

(i). When we formalize the definition of an inaccessible cardinal in the natural way, we obtain a sentence φ such that for all ordinals β,

$\langle R(\beta+1), \in\rangle \models \varphi$ if and only if β is an inaccessible cardinal. Since α is inaccessible by Theorem 4.2.14, φ holds in $\langle R(\alpha+1), \in\rangle$; hence by Corollary 4.2.22, (i) holds.

(ii). Since weak compactness is defined in terms of the model theory of the infinitary logic $\mathscr{L}_\alpha$, it is not very easy to formalize its definition in terms of the $\in$ relation in set theory. We shall instead use a set-theoretic property which, for inaccessible cardinals, is equivalent to weak compactness. This equivalence will be left as one of the exercises. A binary relation xTy is said to be a *tree* over X iff:

(1) T is transitive, i.e. xTy and yTz implies xTz;

(2) T is well-founded;

(3) if yTx and zTx, then either yTz, zTy or $y = z$;

(4) T has a *least element* t such that tTx for all $x \in T \setminus \{t\}$.

It follows from (2) that we never have xTx, and never have $xTyTx$. Let T be a tree over X. For each $x \in X$, the set of predecessors of x is well ordered by T, because it is linearly ordered and well founded. Let $o(x)$, the *order of* x, be the order type of the set of predecessors of x. The *order of* T, $o(T)$, is defined to be the supremum of the orders of $x \in X$:

$$o(T) = \bigcup_{x \in X} o(x).$$

The reader should convince himself that the graph of a tree looks like a tree. A *branch* of T is a subset $B \subset X$ such that:

(5) if $x \in B$ and yTx, then $y \in B$;

(6) if $x, y \in B$, then either xTy, yTx or $x = y$.

Thus any branch of T is simply ordered, and hence well ordered, by T. The *order* of a branch B is just the order type of B under T. We say that a cardinal α has the *tree property* (or ramification property) iff

> for every tree T on α of order α such that, for each $\beta < \alpha$, fewer than α elements have order β, T has a branch of order α.

We leave as an exercise the fact that an inaccessible cardinal α has the tree property if and only if α is weakly compact. When we formalize the above definition of a tree and of the tree property, we obtain a sentence ψ such that, for all cardinals β, $\langle R(\beta+1), \in\rangle \models \psi$ if and only if β has the tree property. Since α is weakly compact (Theorem 4.2.12), and inaccessible, we have $\langle R(\alpha+1), \in\rangle \models \psi$. Therefore the set

$$U = \{\beta < \alpha : \langle R(\beta+1), \in\rangle \models \psi\} \in D.$$

By (i), the set of all $\beta \in U$ such that β is an inaccessible cardinal belongs to D, and (ii) follows. $\dashv$

EXERCISES

4.2.1. Let α be measurable. Prove that if $|I| \geqslant \alpha$, there exists an α-complete nonprincipal ultrafilter over I.

4.2.2. Let $\mathfrak{A}$ be a model of power $|A| \leqslant \alpha$ and suppose $||\mathscr{L}|| \leqslant \alpha$. Show that there is a sentence φ of $\mathscr{L}_{\alpha^+}$ such that for all $\mathfrak{B}$, $\mathfrak{B} \models \varphi$ if and only if $\mathfrak{B} \cong \mathfrak{A}$.

4.2.3*. Let $\mathfrak{A}$ be a model of power $|A| \leqslant \alpha$. Prove that for any $\mathfrak{B}$, $\mathfrak{A} \equiv \mathfrak{B}$ in the language $\mathscr{L}_{\alpha^+}$ if and only if $\mathfrak{A} \cong \mathfrak{B}$ (even when $\mathscr{L}$ has more than α symbols).

4.2.4. Let $\mathfrak{A} = \langle \alpha, S \rangle_{S \subset \alpha}$ and let $\mathfrak{B} = \langle B, T_S \rangle_{S \subset \alpha}$ be an elementary extension of $\mathfrak{A}$ in the sense of the language $\mathscr{L}_\alpha$. For any $b \in B$, let

$$D_b = \{S \subset \alpha : T_S(b)\}.$$

Prove that each D_b is an α-complete ultrafilter over α. Moreover, D_b is principal if and only if $b \in \alpha$.

4.2.5. A cardinal α is said to be *strongly compact* iff the following compactness theorem holds in the language $\mathscr{L}_\alpha$: if Σ is a set of sentences of $\mathscr{L}_\alpha$ and every set $\Sigma_0 \subset \Sigma$ of power $|\Sigma_0| < \alpha$ has a model, then Σ has a model. Prove that every strongly compact cardinal is measurable.

[*Hint*: Use the preceding exercise.]

4.2.6*. Model-theoretic characterizations of measurability: Prove that each of the following conditions on α are equivalent:

(i). α is a measurable cardinal.

(ii). Suppose, for each $\beta < \alpha$, Σ_β is a set of sentences of the language $\mathscr{L}_\alpha$, and for each $\gamma < \alpha$ the set $\bigcup_{\beta<\gamma}\Sigma_\beta$ has a model. Then $\bigcup_{\beta<\alpha}\Sigma_\beta$ has a model ('medium compactness').

(iii). Every model $\mathfrak{A}$ of power $|A| = \alpha$ has a proper elementary extension in the sense of the language $\mathscr{L}_\alpha$.

(iv). The model $\langle R(\alpha), \in, S \rangle_{S \subset R(\alpha)}$ has a proper elementary extension $\mathfrak{B} = \langle B, E, T_S \rangle_{S \subset R(\alpha)}$ such that for all $b \in B$ and $a \in R(\alpha)$, $b \, E \, a$ implies $b \in R(\alpha)$.

4.2.7. Show that for any Σ_1^1 formula $\psi(x_1 \dots x_n)$, the formula $(\forall x_1)\psi(x_1 \dots x_n)$ is equivalent to a Σ_1^1 formula.

[*Hint*: Replace quantified relation and function symbols by symbols with one more argument place.]

4.2.8. Suppose α is an uncountable weakly compact cardinal, and let $\mathfrak{A} = \langle \alpha, <, R_1 \dots R_n \rangle$. Prove that there exists an ordinal $\beta > \alpha$ and a model $\mathfrak{B} = \langle \beta, <, S_1 \dots S_n \rangle$ such that $\mathfrak{A} \prec \mathfrak{B}$ and every Σ_1^1 sentence which holds in $\mathfrak{A}$ holds in $\mathfrak{B}$.

4.2.9*. Prove that if $\alpha > \omega$ is an inaccessible weakly compact cardinal, then α is the αth inaccessible cardinal.

[*Hint*: Adapt the proof of Theorem 4.2.14 (ii), using the preceding two exercises.]

4.2.10. A set $C \subset \alpha$ is said to be *unbounded* iff $\bigcup C = \alpha$, and is said to be *closed* iff, for all nonempty subsets $X \subset C$, either $\bigcup X \in C$ or $\bigcup X = \alpha$. Prove that if α is regular, then the intersection of fewer than α closed and unbounded subsets of α is again a closed and unbounded subset of α.

4.2.11. Suppose $\mathrm{cf}(\alpha) > \omega$, and C is a closed and unbounded subset of α. Prove that the set

$$C' = \{\beta \in \alpha : \beta = \bigcup (C \cap \beta)\}$$

of all limit points of C is again a closed and unbounded subset of α.

4.2.12. α is said to be a *Mahlo number* iff every closed and unbounded subset of α contains an inaccessible cardinal. Prove that if α is a Mahlo number, then α is the αth inaccessible cardinal.

4.2.13*. Let $\alpha > \omega$ be an inaccessible weakly compact cardinal.
(i). Prove that α is a Mahlo number.
(ii). Prove that α is the αth Mahlo number.

4.2.14*. Let X_β, $\beta < \alpha$, be a sequence of subsets of α. By the *diagonal intersection* of the sets X_β we mean the set

$${}^{d}\bigcap_{\beta<\alpha} X_\beta = \{\gamma \in \alpha : \gamma \in \bigcap_{\beta<\gamma} X_\beta\}.$$

Prove that D is a normal ultrafilter over α if and only if D is a nonprincipal α-complete ultrafilter and D is closed under diagonal intersections, that is,

$$\text{if} \quad X_\beta \in D \quad \text{for all} \quad \beta < \alpha, \quad \text{then} \quad {}^{d}\bigcap_{\beta<\alpha} X_\beta \in D.$$

4.2.15*. Let D be a normal ultrafilter over α. Prove that every closed and unbounded subset of α belongs to D.

[*Hint*: Use Theorem 4.2.19.]

4.2.16* (Mahlo's operation). Let $X \subset \alpha$. Define $M(X)$ to be the set of all $\beta < \alpha$ such that every closed and unbounded subset of β contains an element of X. (Thus, if X is the set of all inaccessible cardinals $\beta < \alpha$, then $M(X)$ is the set of all Mahlo numbers $\beta < \alpha$.) Prove that every normal ultrafilter D over α is closed under the Mahlo operation, that is, $X \in D$ implies $M(X) \in D$.

4.2.17. Let D be a normal ultrafilter over α, let $g \in {}^{\alpha}\alpha$, and let E be the set

$$E = \{X \subset \alpha : g^{-1}(X) \in D\}.$$

Prove that if E is a normal ultrafilter over α, then $E = D$.

4.2.18. Let φ be a Σ_1^2 sentence in the language of the model $\langle R(\alpha), \in, S\rangle$. Suppose that α is a measurable cardinal and D is a normal ultrafilter over α. Prove that if

$$\{\beta \in \alpha : \langle R(\beta), \in, S \cap R(\beta)\rangle \models \varphi\} \in D,$$

then

$$\langle R(\alpha), \in, S\rangle \models \varphi.$$

4.2.19. A cardinal α is said to be Π_n^m*-indescribable* iff for every set $S \subset R(\alpha)$ and every Π_n^m sentence φ, if

$$\langle R(\alpha), \in, S\rangle \models \varphi,$$

then there exists $\beta < \alpha$ such that

$$\langle R(\beta), \in, S \cap R(\beta)\rangle \models \varphi.$$

Σ_n^m-indescribability is defined analogously.

(i). Prove that every measurable cardinal $\alpha > \omega$ is Π_1^2-indescribable.

(ii). Prove that the first measurable cardinal $\alpha > \omega$ is not Σ_1^2-indescribable.

4.2.20*. A cardinal β is said to be *indescribable* iff it is Π_n^m-indescribable for all $m, n < \omega$. Prove that if α is the first uncountable measurable cardinal, then there exists an indescribable cardinal $\beta < \alpha$.

[*Hint*: Let D be a normal ultrafilter over α. Show that in the ultrapower $\prod_D \langle R(\alpha), \in\rangle$ the formal statement 'x is indescribable' is satisfied by the element $\bar{\alpha}$.]

4.2.21*. Equivalent formulations of weak compactness. Prove that for any inaccessible cardinal α, the following are equivalent:

(i). α is weakly compact.

(ii). Every model of power α, which has at most α relations and functions, has a proper elementary extension in the sense of the language $\mathscr{L}_\alpha$.

(iii). Every model of the form $\mathfrak{A} = \langle R(\alpha), \in, S_1 \dots S_n \rangle$ has a proper elementary extension $\mathfrak{B} = \langle B, E, T_1 \dots T_n \rangle$ such that for all $b \in B$ and $a \in R(\alpha)$, bEa implies $b \in R(\alpha)$.

(iv). Either $\alpha = \omega$ or α is $\mathbf{\Pi}_1^1$-indescribable.

(v). α has the tree property (see the proof of Theorem 4.2.23).

(vi). (Lindenbaum's Theorem for $\mathscr{L}_\alpha$.) Let $|\mathscr{L}| \leqslant \alpha$. Every set Σ of sentences of $\mathscr{L}_\alpha$ of power $|\Sigma| \leqslant \alpha$, with the property that every subset of Σ of power $< \alpha$ has a model, can be extended to a maximal set with that property.

[*Hint*: The easiest order to prove the equivalences is

$$\text{(i)} \to \text{(ii)} \to \text{(iv)} \to \text{(v)} \to \text{(vi)} \to \text{(i)}, \text{ and } \text{(ii)} \to \text{(iii)} \to \text{(v)}.]$$

4.3. Regular ultrapowers

In the previous section, we investigated ultraproducts taken modulo an α-complete ultrafilter. We shall now study the opposite case – ultraproducts modulo an ultrafilter which is not ω_1-complete. This section may be read independently of Section 4.2.

We shall begin by investigating ultrafilters of various kinds, namely countably incomplete, uniform, and regular.

4.3.1. A filter D is said to be *countably incomplete* iff there exists a countable set $E \subset D$ such that $\bigcap E \notin D$. (This is the same as not ω_1-complete, in the terminology of Section 4.2.)

4.3.2. Let α be a cardinal. A proper filter D over I is said to be *α-regular* iff there exists a set $E \subset D$ of power $|E| = \alpha$ such that each $i \in I$ belongs to only finitely many $e \in E$.

PROPOSITION 4.3.3. *A filter D over I is ω-regular if and only if there exists a countable decreasing chain*

$$I = I_0 \supset I_1 \supset I_2 \supset \dots$$

of elements $I_n \in D$ such that $\bigcap_n I_n = 0$.

The above proposition is often useful in dealing with ultraproducts. The proof is left as an exercise. It is obvious that the notion of α-regular becomes stronger as α increases. In view of the next proposition, we may think of α-regularity as a stronger and stronger kind of countable incompleteness as α increases.

PROPOSITION 4.3.4. *An ultrafilter D is ω-regular if and only if D is countably incomplete. Any ω-regular filter is countably incomplete.*

PROOF. Suppose D is ω-regular. Let $E \subset D$, $|E| = \omega$, be such that each $i \in I$ belongs to only finitely many $e \in E$. Then $\bigcap E = 0$, and, since D is a proper filter, $\bigcap E \notin D$, whence D is countably incomplete.

Let D be a countably incomplete ultrafilter. Take a set $E = \{e_0, e_1, ...\} \subset D$ such that $\bigcap E \notin D$. Let

$$e'_0 = e_0 \setminus \bigcap E, \qquad e'_{n+1} = e'_n \cap e_{n+1}.$$

Then $E' = \{e'_0, e'_1, ...\} \subset D$, and $\bigcap E' = 0$. Therefore D is ω-regular. ⊣

Our next task is to establish the existence of α-regular ultrafilters.

PROPOSITION 4.3.5. *For any set I of infinite power α, there exists an α-regular ultrafilter D over I.*

PROOF. It suffices to show that some set J of power α has an α-regular ultrafilter over it. We consider the set $J = S_\omega(\alpha)$ of all finite subsets of α. For each $\beta \in \alpha$, let

$$\hat{\beta} = \{j \in J : \beta \in j\},$$

and let

$$E = \{\hat{\beta} : \beta \in \alpha\}.$$

Then E has power α. Moreover, each $j \in J$ belongs to only finitely many $\hat{\beta} \in E$, because j is finite, and $j \in \hat{\beta}$ means $\beta \in j$. It follows that any proper filter over J which includes E is α-regular. E has the finite intersection property, because

$$\{\beta_1, ..., \beta_n\} \in \hat{\beta}_1 \cap ... \cap \hat{\beta}_n.$$

Hence by the ultrafilter theorem, E can be extended to an ultrafilter D over J, whence D is an α-regular ultrafilter. ⊣

We turn now to the cardinality problem for ultraproducts. Given an ultrafilter D over I and nonempty sets A_i, $i \in I$, what is the cardinality of the ultraproduct $|\prod_D A_i|$? It is a special case of Exercise 4.1.15 that the cardinality of the ultraproduct $\prod_D A_i$ depends only on the cardinality of the sets A_i. Thus the cardinality problem is really a problem in the arithmetic of cardinals. Some cases of the problem are still unsolved and appear quite difficult. However, we shall prove some partial answers here which shed much light on the question, and we can later apply these results to model theory. We begin with a list of elementary facts.

PROPOSITION 4.3.6. *Let D be a proper filter over I. Then*:

(i). *If $|A_i| = |B_i|$ for all $i \in I$, then $|\prod_D A_i| = |\prod_D B_i|$.*
(ii). *If $|A_i| \leqslant |B_i|$ for all $i \in I$, then $|\prod_D A_i| \leqslant |\prod_D B_i|$.*
(iii). $|\prod_D A_i| \leqslant |\prod_{i \in I} A_i|$.
(iv). $|A| \leqslant |\prod_D A| \leqslant |A|^{|I|}$.

The next two propositions give some more substantial information about cardinalities of ultrapowers. The first result shows that when D is regular, then $\prod_D A$ has the largest possible power.

PROPOSITION 4.3.7. *Suppose D is an α-regular filter over a set I of power α. If A is infinite, then*

$$|\prod_D A| = |A|^\alpha.$$

PROOF. Proposition 4.3.6 (iv) gives the upper bound

$$|\prod_D A| \leqslant |A|^\alpha.$$

We now show that for any α-regular filter D, regardless of the power of A,

$$(1) \qquad |A|^\alpha \leqslant |\prod_D A|.$$

Since D is α-regular, there exists $E \subset D$ of power $|E| = \alpha$ such that each $i \in I$ belongs to only finitely many $e \in E$. Let B be the set of all finite sequences of elements of A. Since A is infinite, we have $|A| = |B|$. Thus to prove (1), it suffices to find a one–one function

$$\pi : {}^E A \to \prod_D B.$$

For each $g : E \to A$, we shall define a function $g' : I \to B$. We intend to define π by putting $\pi g = g'_D$. First, we choose a simple ordering $\leqslant$ of the set E. For each $i \in I$, define the element $g'(i) \in B$ as follows. Let

(2) $$\langle e_1 \dots e_n \rangle$$

be the finite set of all $e \in E$ such that $i \in e$, arranged in increasing order with respect to $\leqslant$. Then

$$g'(i) = \langle g(e_1) \dots g(e_n) \rangle.$$

Thus g' maps I into B. Now define $\pi g = g'_D$. Then π maps ${}^E A$ into $\prod_D B$. It remains to show that π is one–one. Suppose $g, h \in {}^E A$ and $g \neq h$. Then for some $e \in E$, $g(e) \neq h(e)$. Now, for any $i \in e$, e will occur in the finite sequence (2) of sets containing i, say $e = e_k$. Then

$$g'(i) = \langle \dots g(e_k) \dots \rangle \neq \langle \dots h(e_k) \dots \rangle = h'(i).$$

Thus $e \in D$ and $g'(i) \neq h'(i)$ for all $i \in e$, whence

$$\pi g = g'_D \neq h'_D = \pi h.$$

Therefore π is one–one, and (1) follows. ⊣

In the case when A is finite, $\prod_D A$ always has the same power as A, for the fact that A has n elements is expressible by a single sentence of $\mathscr{L}$.

Corollary 4.3.8. *If $\mathfrak{A}$ is infinite, then $\mathfrak{A}$ has ultrapowers of arbitrarily large power.*

The above corollary gives another proof of the Löwenheim–Skolem–Tarski theorem, 3.1.5 (because $\mathfrak{A}$ is elementarily embedded in its ultrapower – Corollary 4.1.13). Next we show that certain cardinals cannot be cardinals of ultrapowers, at least when D is countably incomplete.

Proposition 4.3.9. *Let D be an ω-regular filter. If A is infinite, then*

$$|\prod_D A| = |\prod_D A|^\omega.$$

Proof. Let B be the set of all finite sequences in A. Then $|B| = |A|$ because A is infinite. So it suffices to prove that

(1) $$|\prod_D A|^\omega \leqslant |\prod_D B|.$$

It suffices to find a mapping τ on a subset of $\prod_D B$ onto ${}^\omega(\prod_D A)$. To do this, it suffices to find a mapping σ on ${}^\omega({}^I A)$ into ${}^I B$ such that:

(2) if $g, h \in {}^{\omega}({}^{I}A)$ and $\sigma g =_D \sigma h$, then, for all n, $g(n) =_D h(n)$.

For then we may define the mapping τ by:

$$\text{if } \sigma g = f, \text{ then } \tau(f_D) = \langle g(0)_D g(1)_D \dots \rangle.$$

Since D is ω-regular, there is a sequence $I = I_0 \supset I_1 \supset \dots$ of sets $I_n \in D$ such that $\bigcap_n I_n = 0$ (Proposition 4.3.3). Then for each $i \in I$, there is a unique integer $n(i)$ such that

$$i \in I_{n(i)} \setminus I_{n(i)+1}.$$

For each function $g \in {}^{\omega}({}^{I}A)$, define $\sigma g \in {}^{I}B$ by setting

$$(\sigma g)(i) = \langle g(1)(i) \dots g(n(i))(i) \rangle.$$

It remains to verify (2). Suppose g, h are in ${}^{\omega}({}^{I}A)$, and $\sigma g =_D \sigma h$. Then the set

$$X = \{i \in I : (\sigma g)(i) = (\sigma h)(i)\} \in D.$$

For each $n < \omega$, we have $X \cap I_n \in D$. Whenever $i \in X \cap I_n$, $n \leqslant n(i)$, and thus $g(n)(i) = h(n)(i)$. It follows that $g(n) =_D h(n)$ for all n. Thus (2) holds and hence (1) holds. ⊣

It follows, for example, that no ultrapower of a set modulo a countably incomplete ultrafilter can have power ω, $\beth_\omega$, or even any power of cofinality ω. Other similar cardinality theorems are known, and we shall include them in our list of exercises. Before moving on to our next topic, we wish to say a few words about unsolved problems. The following questions remain completely open:

Does there exist a countably incomplete ultrafilter D such that

$$|\prod_D \omega| \text{ is singular?}$$

$$|\prod_D \omega| \text{ is inaccessible?}$$

$$|\prod_D \omega| < |\prod_D (2^\omega)|?$$

We can state a more sophisticated problem along this line if we introduce another notion. A filter D over I is said to be *uniform* iff every member of D has the same cardinality $|I|$. We then ask:

Does there exist a countably incomplete uniform ultrafilter over α which is not α-regular?

Does there exist a countably incomplete uniform ultrafilter D over α such that $|\prod_D \omega| < 2^\alpha$?

We have already answered the above two questions negatively for $\alpha = \omega$. By Proposition 4.3.6, the second question implies the first.

Although these questions originated from model theory, they really belong more to set theory, and consistency results have been found which involve large cardinals in one direction and the constructible sets or the core model of Dodd and Jensen (1981) in the other direction.

Some negative results on the first question: Prikry (1971) showed that the axiom of constructibility implies that every uniform ultrafilter over ω_1 is regular. Jensen (unpublished) improved this result to ω_n. Using the core model, Donder (1988) proved that if ZFC is consistent, then so is ZFC+ 'every uniform ultrafilter on an infinite set is regular'.

Some positive results on the second question: From Exercise 6.5.9 it follows that both questions have affirmative answers if α is a measurable cardinal. Magidor (1979) has shown that if

$$\text{ZFC} + [\text{there is a huge cardinal}]$$

is consistent, then so are each of the following:

$$\text{ZFC} + [\text{there is a uniform ultrafilter } D \text{ over } \omega_3 \text{ with } |\textstyle\prod_D \omega_1| \leqslant \omega_3],$$

$$\text{ZFC} + [\text{there is a uniform ultrafilter } D \text{ over } \omega_2 \text{ with } |\textstyle\prod_D \omega_0| \leqslant \omega_2].$$

In each case, D cannot be regular because of Proposition 4.3.7. Foreman, Magidor, and Shelah have obtained an analogous result for ω_1. Huge cardinals are defined in Kanamori, Reinhart, and Solovay (1978).

Some additional results in ZFC about cardinalities of ultraproducts can be found in Exercise 4.3.13–4.3.18 and in Shelah (1978), Chapter VI. For some relationships between cardinalities of ultrapowers and descendingly complete ultrafilters (defined in Exercise 4.3.10), see Adler and Jorgensen (1972).

Let us now study the model-theoretic properties of regular ultrapowers. We first give an application of the cardinality result, Proposition 4.3.7, to the Löwenheim–Skolem problem for two cardinals considered in Section 3.2. Let $\mathscr{L}$ have among its symbols a 1-placed relation symbol U. Let us use the term (α, β)*-model* for a model $\mathfrak{A}$ for $\mathscr{L}$ such that the universe A has power α and the interpretation of U has power β. Thus a theory T in $\mathscr{L}$ admits (α, β) iff T has an (α, β)-model.

THEOREM 4.3.10. *If a theory T admits (α, β) and $\beta \geqslant \omega$, then for all cardinals γ, T admits $(\alpha^\gamma, \beta^\gamma)$. In fact, every (α, β)-model has an elementary extension, which is an $(\alpha^\gamma, \beta^\gamma)$-model.*

PROOF. Let $\mathfrak{A}$ be an (α, β)-model of T, and let V be the interpretation of U in $\mathfrak{A}$. We may assume that γ is infinite. Let D be a γ-regular ultrafilter over a set of power γ. Now form the ultrapower $\prod_D \mathfrak{A}$. By Proposition 4.3.7,

$$(1) \qquad |\prod_D A| = \alpha^\gamma, \qquad |\prod_D V| = \beta^\gamma.$$

Let V' be the interpretation of U in $\prod_D \mathfrak{A}$. We shall show that V' has the same power as $\prod_D V$. For each $f \in {}^I V$, we are considering two different equivalence classes of f, both ordinarily denoted by f_D. We now need to distinguish them, so we shall write

$$f_D^A = \{g \in {}^I A : f =_D g\} = f_D \quad \text{in the sense of} \quad \prod_D A,$$

$$f_D^V = \{g \in {}^I V : f =_D g\} = f_D \quad \text{in the sense of} \quad \prod_D V.$$

For any $f, g \in {}^I V$, we have

$$(2) \qquad f_D^A = g_D^A \quad \text{iff} \quad f =_D g \quad \text{iff} \quad f_D^V = g_D^V.$$

Moreover, for any $f \in {}^I A$ such that $f_D^A \in V'$, we have

$$\{i \in I : f(i) \in V\} \in D.$$

Thus if we define $f' \in {}^I A$ by

$$f'(i) = f(i) \qquad \text{if } f(i) \in V,$$

$$f'(i) = \text{any element of } V \text{ otherwise},$$

we have

$$(3) \qquad f' \in {}^I V \quad \text{and} \quad f_D^A = f_D'^A.$$

From (2) and (3), it follows that the mapping π such that $\pi(f_D^V) = f_D^A$ is a one–one map of $\prod_D V$ onto V'. Therefore, by (1), $\prod_D \mathfrak{A}$ is an $(\alpha^\gamma, \beta^\gamma)$-model. ⊣

COROLLARY 4.3.11 (Narrowing a Gap). *Assume the G.C.H. Suppose $\alpha \geqslant \alpha' \geqslant \beta' \geqslant \beta \geqslant \omega$ and $\|\mathscr{L}\| \leqslant \alpha'$. Then every theory T in $\mathscr{L}$ which admits (α, β) admits (α', β').*

PROOF. By Proposition 3.2.7 (i) (generalized to the case $\|\mathscr{L}\| \leqslant \alpha'$), we may assume that $\beta < \beta'$. First take the case that β' is a successor cardinal, $\beta' = \gamma^+$. Then $\beta \leqslant \gamma$, so by the G.C.H., $\beta' = \beta^\gamma$. By Theorem 4.3.10, T admits (α^γ, β'). Then, using Proposition 3.2.7 (i), we see that T admits (α', β').

The case where β' is a limit cardinal requires an iteration of Theorem 4.3.10. We form an elementary chain $\mathfrak{A}_\gamma$, $\beta \leqslant \gamma < \beta'$, such that each $\mathfrak{A}_\gamma$ is an (α, γ)-model of T. Let $\mathfrak{A}_\beta$ be any (α, β)-model of T. Suppose $\gamma < \beta'$ and we have constructed $\mathfrak{A}_\eta$, $\beta \leqslant \eta < \gamma$. If γ is a limit ordinal, let $\mathfrak{A}_\gamma = \bigcup\{\mathfrak{A}_\eta : \beta \leqslant \eta < \gamma\}$. Then $\mathfrak{A}_\gamma$ is an (α, γ)-model of T. If $\gamma = \eta^+$, then $\gamma = \eta^\eta$, so by Theorem 4.3.10 there is an (α^η, γ)-model $\mathfrak{B} \succ \mathfrak{A}_\eta$. By Proposition 3.2.7 (i), there is an (α, γ)-model $\mathfrak{A}_\gamma$ such that $\mathfrak{A}_\eta \prec \mathfrak{A}_\gamma \prec \mathfrak{B}$. By induction, we have the desired elementary chain. Now let $\mathfrak{A}' = \bigcup\{\mathfrak{A}_\gamma : \beta \leqslant \gamma < \beta'\}$. Then $\mathfrak{A}'$ is an (α, β')-model of T. Thus T admits (α, β'); hence T admits (α', β'). ⊣

In Section 6.5 we shall eliminate the G.C.H. from Corollary 4.3.11 by using iterated ultrapowers.

A model $\mathfrak{A}$ is said to be *α-universal* iff every model $\mathfrak{B}$ of power less than α which is elementarily equivalent to $\mathfrak{A}$ is elementarily embedded in $\mathfrak{A}$, i.e.,

$$\mathfrak{B} \equiv \mathfrak{A} \quad \text{and} \quad |B| < \alpha \quad \text{implies} \quad \mathfrak{B} \precsim \mathfrak{A}.$$

THEOREM 4.3.12. *Suppose $||\mathscr{L}|| \leqslant \alpha$ and D is an α-regular ultrafilter. Then for every model $\mathfrak{A}$, the ultrapower $\prod_D \mathfrak{A}$ is α^+-universal.*

PROOF. Let $E \subset D$, where $|E| = \alpha$ and each $i \in I$ belongs to only finitely many $e \in E$. Also, let $\mathfrak{B}$ be a model such that $\mathfrak{B} \equiv \mathfrak{A}$ and $|B| \leqslant \alpha$. We are to show that $\mathfrak{B}$ is elementarily embedded in $\prod_D \mathfrak{A}$. Form the elementary diagram Γ_B of $\mathfrak{B}$ in the expanded language $\mathscr{L}_B$ which contains a new constant for each $b \in B$. It suffices to find an expansion $(\prod_D \mathfrak{A}, a_b)_{b \in B}$ of the ultrapower $\prod_D \mathfrak{A}$ which is a model of Γ_B. Since $||\mathscr{L}|| \leqslant \alpha$ and $|B| \leqslant \alpha$, we have $|\Gamma_B| \leqslant \alpha$. Thus there is a one–one function H on Γ_B into E. Now consider an element $i \in I$. Since there are only finitely many $e \in E$ such that $i \in e$, there are only finitely many sentences $\varphi \in \Gamma_B$ such that $i \in H(\varphi)$. Let

$$\{\varphi_1, \ldots, \varphi_n\} = \{\varphi \in \Gamma_B : i \in H(\varphi)\}.$$

The sentence $\varphi_1 \wedge \ldots \wedge \varphi_n$ holds in $(\mathfrak{B}, b)_{b \in B}$, and therefore is consistent with the theory of $\mathfrak{A}$. It follows that there exists an expansion

$$(\mathfrak{A}, f_b(i))_{b \in B}$$

of $\mathfrak{A}$ to $\mathscr{L}_B$ which is a model of $\varphi_1 \wedge \ldots \wedge \varphi_n$. When we choose such an expansion for each $i \in I$, we obtain functions $f_b \in {}^I A$, $b \in B$, such that for all $i \in I$ and $\varphi \in \Gamma_B$,

$$i \in H(\varphi) \quad \text{implies} \quad (\mathfrak{A}, f_b(i))_{b \in B} \models \varphi.$$

Moreover, for each $\varphi \in \Gamma_B$, $H(\varphi) \in D$, whence by the fundamental theorem,

$$\prod_D (\mathfrak{A}, f_b(i))_{b \in B} \models \varphi. \tag{1}$$

For each b, let $a_b = (f_b)_D$. Then, by the definition of ultraproduct,

$$\prod_D (\mathfrak{A}, f_b(i))_{b \in B} = (\prod_D \mathfrak{A}, a_b)_{b \in B}.$$

Thus (1) shows that $(\prod_D \mathfrak{A}, a_b)_{b \in B}$ is a model of Γ_B, so $\mathfrak{B} \precsim \prod_D \mathfrak{A}$. ⊣

COROLLARY 4.3.13 (Frayne's Theorem). *$\mathfrak{A} \equiv \mathfrak{B}$ if and only if $\mathfrak{A}$ is elementarily embedded in some ultrapower $\prod_D \mathfrak{B}$ of $\mathfrak{B}$.*

PROOF. If $\mathfrak{A} \precsim \prod_D \mathfrak{B}$, then $\mathfrak{A} \equiv \prod_D \mathfrak{B} \equiv \mathfrak{B}$. For the converse, suppose $\mathfrak{A} \equiv \mathfrak{B}$. Let α be the maximum of $||\mathscr{L}||$ and $|A|$, and let D be an α-regular ultrafilter over α. Then, by Theorem 4.3.12, $\prod_D \mathfrak{B}$ is α^+-universal, whence $\mathfrak{A} \precsim \prod_D \mathfrak{B}$. ⊣

Our next corollary will be used in Chapter 5 to prove the existence of saturated models. Let $\mathfrak{B}$ be an elementary extension of $\mathfrak{A}$ and let β be a cardinal. For each set $Y \subset A$, let $\mathscr{L}_Y$ be the expansion of $\mathscr{L}$ obtained by adding a new constant symbol for each $y \in Y$. We shall say that $\mathfrak{B}$ is *β-saturated over* $\mathfrak{A}$ iff it has the following property:

> For every set $Y \subset A$ of power $|Y| < \beta$, every set $\Sigma(x)$ of formulas of $\mathscr{L}_Y$ which is consistent with the theory of $(\mathfrak{A}, y)_{y \in Y}$ is satisfiable in $(\mathfrak{B}, y)_{y \in Y}$.

A model $\mathfrak{B}$ is said to be *β-saturated* iff $\mathfrak{B}$ is β-saturated over $\mathfrak{B}$. In the case $\beta = \omega$, this agrees with the definition of an ω-saturated model given in Section 2.3.

COROLLARY 4.3.14. *Let $\mathfrak{A}$ be a model for a language $\mathscr{L}$ of power $\leqslant \alpha$. Then the ultrapower of $\mathfrak{A}$ modulo an α-regular ultrafilter is isomorphic to an elementary extension of $\mathfrak{A}$ which is α^+-saturated over $\mathfrak{A}$.*

Proof. Let D be an α-regular ultrafilter and form the ultrapower $\prod_D \mathfrak{A}$. Let $\mathfrak{B}$ be an elementary extension of $\mathfrak{A}$ such that the natural embedding $d : \mathfrak{A} \to \prod_D \mathfrak{A}$ can be extended to an isomorphism $f : \mathfrak{B} \cong \prod_D \mathfrak{A}$. Let $Y \subset A$, $|Y| < \alpha^+$. Then $||\mathscr{L}_Y|| \leqslant \alpha$. Therefore, by Theorem 4.3.12, the ultrapower $\prod_D(\mathfrak{A}, y)_{y \in Y}$ is α^+-universal. But

$$\prod_D (\mathfrak{A}, y)_{y\in Y} = (\prod_D \mathfrak{A}, d(y))_{y\in Y} \cong (\mathfrak{B}, y)_{y\in Y},$$

so $(\mathfrak{B}, y)_{y\in Y}$ is α^+-universal. Any set $\Sigma(x)$ of formulas of $\mathscr{L}_Y$ consistent with the theory of $(\mathfrak{A}, y)_{y\in Y}$ is satisfiable in some model $(\mathfrak{C}, c_y)_{y\in Y} \equiv (\mathfrak{A}, y)_{y\in Y}$ of power $\leqslant \alpha$. Then $(\mathfrak{C}, c_y)_{y\in Y}$ is elementarily embedded in $(\mathfrak{B}, y)_{y\in Y}$, and it follows that $\Sigma(x)$ is satisfiable in $(\mathfrak{B}, y)_{y\in Y}$. ⊣

Theorem 4.3.12 is not true for arbitrary ultraproducts (instead of ultrapowers). In Section 6.1 we shall investigate ultrafilters D for which Theorem 4.3.12 does hold for ultraproducts. (We shall call them α^+-good ultrafilters.)

There are two interesting questions which remain open in connection with Theorem 4.3.12.

Is $\prod_D \mathfrak{A}$ α^{++}-universal for every α-regular ultrafilter D and every model $\mathfrak{A}$?

Does $\mathfrak{A} \equiv \mathfrak{B}$, $|A|, |B| \leqslant 2^\alpha$, D an α-regular ultrafilter over α, and $\alpha^+ = 2^\alpha$ imply that $\prod_D \mathfrak{A} \cong \prod_D \mathfrak{B}$?

For both questions, we must assume that $||\mathscr{L}|| \leqslant \alpha$, otherwise counterexamples are known. In Chapter 6 we shall see that the answer to both questions is yes when D has the stronger property of being α^+-good. Note that the first question is a strengthening of Theorem 4.3.12 where α^+-universal is improved to α^{++}-universal. Exercise 4.3.36 states a result which lends some plausibility to the second question. It shows that $\prod_D \mathfrak{A}$ and $\prod_D \mathfrak{B}$ are very much alike, and, in fact, share properties expressible in the infinitary language $\mathscr{L}_\alpha$.

Exercises

4.3.1. Prove Proposition 4.3.3.

4.3.2.

(i). Every uniform ultrafilter over an infinite set is nonprincipal.

(ii). An ultrafilter D over I is uniform if and only if D contains every set $X \subset I$ such that $|I \setminus X| < |I|$.

4.3.3. If D is an arbitrary ultrafilter and J is an element of D of smallest cardinality, then $D \cap S(J)$ is a uniform ultrafilter over J. Similarly for proper filters. Since

$$\prod_D \mathfrak{A}_i \cong \prod_{D \cap S(J)} \mathfrak{A}_j$$

(Exercise 4.1.11), this shows that in the study of ultraproducts we need only consider uniform ultrafilters.

4.3.4. Let I be an infinite set of power α. Then every α-regular filter over I is uniform.

4.3.5. Let I be an infinite set of power α. Then there does not exist an α^+-regular filter over I.

4.3.6. If D is an α-regular ultrafilter over I and $J \in D$, then $D \cap S(J)$ is an α-regular ultrafilter over J. A similar result holds for filters.

4.3.7. Let D be an ultrafilter over I and let $f : I \to J$. Let E be the set

$$E = \{X \subset J : f^{-1}(X) \in D\}.$$

Show that E is an ultrafilter over J, and, furthermore, if E is α-regular, then D is α-regular. (Cf. Lemma 4.2.6.)

4.3.8. A proper filter D over I is α-regular if and only if there exists a function $f : I \to S_\omega(\alpha)$ such that for each $\beta \in \alpha$,

$$\{i \in I : \beta \in f(i)\} \in D.$$

4.3.9*. Let I be an infinite set of power α. If $E \subset S(I)$, $|E| \leqslant \alpha$, and the filter generated by E is uniform, then E can be extended to an α-regular ultrafilter D over I.

4.3.10. An ultrafilter D is said to be *descendingly α-complete* iff for every descending chain X_γ, $\gamma < \alpha$ of sets $X_\gamma \in D$, of length α, the intersection $\bigcap_{\gamma<\alpha} X_\gamma \in D$. Prove the following:

(i). D is descendingly ω-complete iff D is countably complete.

(ii). D is descendingly α-complete iff D is descendingly $\mathrm{cf}(\alpha)$-complete.

(iii). D is α-complete iff for all $\beta < \alpha$, D is descendingly β-complete.

(iv). If D is a uniform ultrafilter over a set of power α, then D is not descendingly α-complete. However, D is descendingly β-complete whenever $\mathrm{cf}(\beta) > \alpha$.

4.3.11*. If D is a descendingly α-complete ultrafilter, then D is not α-regular.

4.3.12**. If α is regular, then every descendingly α-complete ultrafilter is descendingly α^+-complete. Thus, if $m \leqslant n < \omega$, then no uniform ultrafilter over ω_n is descendingly ω_m-complete.

4.3.13*. Suppose D is a uniform ultrafilter over a set of power α, and $|A| = \alpha$. Then $|\prod_D A| > \alpha$.

4.3.14*. If D is a countably incomplete ultrafilter, then either

$$|\prod_D A_i| < \omega \quad \text{or} \quad |\prod_D A_i| \geqslant 2^\omega.$$

Thus, in all cases, $|\prod_D A_i| \neq \omega$.

4.3.15*. If D is any ultrafilter and $|A| = |A|^\omega$, then $|\prod_D A| = |\prod_D A|^\omega$.

4.3.16*. Suppose D is an ultrafilter which is not descendingly α-complete. If $|A| = |A|^\beta$ whenever $0 < \beta < \alpha$, then $|\prod_D A| = |\prod_D A|^\alpha$. (Proposition 4.3.9 is the special case where $\alpha = \omega$.)

4.3.17*. The notion of a tree of order α was introduced during the proof of Theorem 4.2.23. A tree T of order α is said to be a *Kurepa tree* iff T has more than α branches of order α, while for each $\beta < \alpha$, T has fewer than α elements of order β. Suppose that there exists a Kurepa tree of order α^+. Then for every uniform ultrafilter D over α^+,

$$|\prod_D \alpha| > \alpha^+.$$

4.3.18*. For any ultrafilter D, and any cardinal α,

$$|\prod_D (2^\alpha)| \leqslant 2^{|\prod_D \alpha|}, \quad \text{and} \quad |\prod_D (\alpha^+)| \leqslant |\prod_D \alpha|^+.$$

If D is a uniform ultrafilter over α^+, or over 2^α, then $|\prod_D \alpha| > \alpha$.

4.3.19. Let D be a proper filter, let $\mathfrak{A} \subset \mathfrak{B}$, and form the expanded model $(\mathfrak{B}, A)$ and its reduced power $\prod_D(\mathfrak{B}, A)$. Let A' be the (1-placed) relation in $\prod_D(\mathfrak{B}, A)$ corresponding to A, and let $\mathfrak{A}'$ be the reduct to $\mathscr{L}$ of the submodel of $\prod_D(\mathfrak{B}, A)$ with universe A'. Prove that $\mathfrak{A}' \cong \prod_D \mathfrak{A}$.

4.3.20. If a theory T admits (α, β, γ), then T admits $(\alpha^\delta, \beta^\delta, \gamma^\delta)$ (provided $\gamma \geqslant \omega$). (Where 'T admits (α, β, γ)' is defined in the obvious way.)

4.3.21. A subset $X \subset B$ is said to be *definable* in the model $\mathfrak{B}$ iff there is a formula $\varphi(x_0 x_1 \ldots x_n)$ and elements $b_1, \ldots, b_n \in B$ such that

$$X = \{b_0 \in B : \mathfrak{B} \models \varphi[b_0 b_1 \ldots b_n]\}.$$

Suppose D is a countably incomplete ultrafilter. Let $\mathfrak{A}$ be any model and form the ultrapower $\prod_D \mathfrak{A}$. Show that any definable subset of $\prod_D \mathfrak{A}$ is either finite or has power $\geqslant 2^\omega$.

4.3.22. A class K of models for $\mathscr{L}$ is said to be *relatively compact* iff for every set Σ of sentences of $\mathscr{L}$, if every finite subset of Σ has a model in K, then Σ has a model in K. Let $||\mathscr{L}|| \leqslant \alpha$ and suppose D is an α-regular ultrafilter. Then any class K of models for $\mathscr{L}$ which is closed under ultraproducts modulo D is relatively compact.

4.3.23. Every finite model is α-universal.

4.3.24. Suppose $||\mathscr{L}|| \leqslant \alpha$. Let $\mathscr{L}'$ be a simple expansion of $\mathscr{L}$ formed by adding α new constants. Then a model $\mathfrak{A}$ for $\mathscr{L}$ is α^+-universal if and only if every set Σ of sentences of $\mathscr{L}'$ which is consistent with the theory of $\mathfrak{A}$ has a model $\mathfrak{A}'$ which is an expansion of $\mathfrak{A}$.

4.3.25. Let $||\mathscr{L}|| \leqslant \alpha$; then every reduct of an α^+-universal model is α^+-universal.

4.3.26. Let D be an α-regular ultrafilter. Then for any model $\mathfrak{A}$, every reduct of the ultrapower $\prod_D \mathfrak{A}$ to a language of power $\leqslant \alpha$ is α^+-universal.

4.3.27. Suppose D is a countably incomplete ultrafilter and $\mathscr{L}$ is countable. Let $\mathfrak{A}$ be a model for $\mathscr{L}$. Then every type $\Sigma(x_1 \dots x_n)$ which is realized in some model $\mathfrak{B} \equiv \mathfrak{A}$ is realized in $\prod_D \mathfrak{A}$.

4.3.28*. Let D be an α-regular ultrafilter. Then the model $\langle \alpha^+, < \rangle$ is isomorphically embedded in the ultrapower $\prod_D \langle \omega, < \rangle$.

4.3.29*. Call $\mathfrak{B}$ an *ultrapower extension* of $\mathfrak{A}$ iff $\mathfrak{A} \subset \mathfrak{B}$ and for some ultrafilter D, the natural embedding $d : \mathfrak{A} \to \prod_D \mathfrak{A}$ can be extended to an isomorphism $f : \mathfrak{B} \cong \prod_D \mathfrak{A}$. By an *ultrapower chain* over $\mathfrak{A}$ we mean a chain $\mathfrak{A} = \mathfrak{A}_0 \subset \mathfrak{A}_1 \subset \mathfrak{A}_2 \dots$ of length ω such that each $\mathfrak{A}_{n+1}$ is an ultrapower extension of $\mathfrak{A}_n$. Show that every ultrapower extension of $\mathfrak{A}$ is an elementary extension, and every ultrapower chain is an elementary chain. Prove that $\mathfrak{A} \equiv \mathfrak{B}$ if and only if there exist ultrapower chains

$$\mathfrak{A} = \mathfrak{A}_0 \subset \mathfrak{A}_1 \subset \dots, \qquad \mathfrak{B} = \mathfrak{B}_0 \subset \mathfrak{B}_1 \subset \dots$$

over $\mathfrak{A}$ and $\mathfrak{B}$ such that

$$\bigcup_{n<\omega} \mathfrak{A}_n \cong \bigcup_{n<\omega} \mathfrak{B}_n.$$

This gives an algebraic characterization of elementary equivalence.

4.3.30. Prove that K is an elementary class iff K is closed under ultraproducts, and both K and its complement are closed under isomorphisms and under unions of ultrapower chains. Give a similar characterization of basic elementary classes.

4.3.31. In Exercise 4.3.29, suppose $||\mathscr{L}|| \leqslant \beth_\alpha$, $|A| \leqslant \beth_\alpha$, and $|B| \leqslant \beth_\alpha$. Show that the ultrapower chains may be chosen so that the unions have power $\beth_{\alpha+\omega}$ (or else are finite).

4.3.32*. Prove that an ultrafilter D is α-regular if and only if for every language $\mathscr{L}$ of power $||\mathscr{L}|| \leqslant \alpha$ and every model $\mathfrak{A}$ for $\mathscr{L}$, $\prod_D \mathfrak{A}$ is α^+-universal.

4.3.33*. Give an example showing that Theorem 4.3.12 fails if the hypothesis $||\mathscr{L}|| \leqslant \alpha$ is weakened to $||\mathscr{L}|| \leqslant \alpha^+$.

4.3.34. An ultrafilter D is descendingly α-complete iff the set $d(\alpha)$ is cofinal in the ultrapower $\prod_D \langle \alpha, < \rangle$.

4.3.35*. Let $||\mathscr{L}|| \leqslant \alpha$ and let D be an α-regular ultrafilter. Let $\mathfrak{A}$, $\mathfrak{B}$ be infinite models such that $\mathfrak{A} \equiv \mathfrak{B}$ and form the ultrapowers $\prod_D \mathfrak{A}$, $\prod_D \mathfrak{B}$. Suppose $\mathfrak{A}' \prec \prod_D \mathfrak{A}$, $\mathfrak{B}' \prec \prod_D \mathfrak{B}$, and $|A'|, |B'| \leqslant \alpha$. Then there exist models $\mathfrak{A}''$, $\mathfrak{B}''$ of power α such that

$$\mathfrak{A}' \prec \mathfrak{A}'' \prec \prod_D \mathfrak{A}, \qquad \mathfrak{B}' \prec \mathfrak{B}'' \prec \prod_D \mathfrak{B}, \qquad \mathfrak{A}'' \cong \mathfrak{B}''.$$

4.3.36. Suppose D is an α-regular ultrafilter, and $\mathfrak{A} \equiv \mathfrak{B}$. Then the ultrapowers $\prod_D \mathfrak{A}$ and $\prod_D \mathfrak{B}$ are elementarily equivalent in the sense of the infinitary language $\mathscr{L}_\alpha$. (The language $\mathscr{L}_\alpha$ is described in Section 4.2.)**

4.3.37. Let D be a uniform ultrafilter over ω_1 and let T be a theory in a language of power ω_1. For each class K of models let K^* be the class of models of the form $\prod_D \mathfrak{A}_\alpha$ where each $\mathfrak{A}_\alpha \in K$. Suppose that each finite subset of T has a model in K. Prove that T has a model in the class $(K^*)^*$.

4.3.38*. Let $\mathfrak{A} = \langle A, \leqslant, \ldots \rangle$ be an infinite model for a countable language such that $\leqslant$ simply orders A. Let D be a nonprincipal ultrafilter over ω. Prove without using the continuum hypothesis that $\prod_D \mathfrak{A}$ has a set X of indiscernibles of power 2^ω.

[*Hint*: Find an X which is an infinite ultraproduct of finite subsets of A and use Exercise 4.3.13.]

4.3.39. Let $\mathfrak{A} = \langle A, E, \ldots \rangle$ be a model such that E is an equivalence relation and each equivalence class of E is countable. Prove that $\mathfrak{A}$ has an elementary extension $\mathfrak{B} = \langle B, F, \ldots \rangle$ such that every equivalence class of F has cardinality 2^ω.

4.3.40. Let D and E be ultrafilters over sets I and J. We say that $E \leqslant D$ if there is a function $f : I \rightarrow J$ such that $E = f^{-1}[D]$ in the sense of Exercise 4.3.7. (The relation $E \leqslant D$ is clearly transitive. It is called the Keisler-Rudin ordering in the literature.) Show that $D \leqslant E$ and $E \leqslant D$ if and only if there is a one to one function h of I onto J such that

$$E = \{X \subset J : h^{-1}(X) \in D\}.$$

[*Hint*: Use Exercise 4.1.35.]

4.3.41*. Prove that $E \leqslant D$ if and only if for every model $\mathfrak{A}$, $\prod_E \mathfrak{A}$ is elementarily embeddable in $\prod_D \mathfrak{A}$.

4.3.42. An nonprincipal ultrafilter D over ω is said to be *selective*, if for every partition $\omega = \bigcup_{n<\omega} X_n$ of ω into countably many disjoint sets, either some X_n belongs to D or there is a choice set $Y \in D$ such that $|Y \cap X_n| = 1$ for each n. Prove that D is selective if and only if for every nonprincipal ultrafilter E, $E \leqslant D$ implies $D \leqslant E$.

4.3.43*. Assuming the continuum hypothesis, prove that selective ultrafilters exist.

4.4. Nonstandard universes

The purpose of this section is to provide a bridge between classical model theory and Abraham Robinson's nonstandard analysis, which was first introduced in Robinson (1961). We believe that Robinson's choice of the name "nonstandard analysis" is unfortunate, and instead use the name "Robinsonian analysis" here. Robinsonian analysis has been used extensively as a method of applying model theory to other areas of mathematics and the natural sciences. There are several good books on the subject (see the Historical Notes), and we shall not go into Robinsonian analysis itself here.

Our emphasis will be on model-theoretic questions, usually involving the satisfaction relation, which arise in the foundations of Robinsonian analysis. We shall deal with the existence of models with certain properties, and semantic relationships between two models. The ultrapower is the key method of constructing models in Robinsonian analysis.

Several approaches to Robinsonian analyis have been proposed in the literature, but in practice most research in the subject uses one of two approaches: superstructures and internal set theory. The intuitive idea underlying all approaches is to start with the set-theoretic universe $\langle V, \in \rangle$ and form an elementary extension $\langle W, E\rangle$ in which all infinite sets are enlarged. This turns out to be awkward because V is not a set, and worse yet, E is not well founded and thus cannot be represented as the $\in$ relation on some collection of sets. Superstructures and internal set theory deal with this problem in different ways. In this section we shall first treat the superstructure approach and then internal set theory.

The superstructure approach works with models in a language which has a symbol $\in$ for set membership, and concentrates on formulas which have bounded quantifiers $(\forall x \in y)$ and $(\exists x \in y)$ instead of the ordinary quantifiers $\forall x$ and $\exists x$. Many model theoretic constructions have analogues for bounded quantifier formulas.

Given a set X, the superstructure over X, denoted by $V(X)$, is obtained from X by taking the power set countably many times. The precise definition is as follows. For each natural number n, we define the set $V_n(X)$ recursively by

$$V_0(X) = X, V_{n+1}(X) = V_n(X) \cup S(V_n(X)).$$

The *superstructure* $V(X)$ is the union

$$V(X) = \bigcup_{n<\omega} V_n(X).$$

We shall often work with the model $\langle V(X), \in \rangle$ where $\in$ is the set membership relation. The elements of X will play the role of individuals rather than sets. To avoid irrelevant side issues, we shall confine our attention to the case where there are no extra membership relationships in $\langle V(X), \in \rangle$ involving elements of X. The set X is said to be a *base set* if $\emptyset \notin X$ and for all $x \in X$, $x \cap V(X) = \emptyset$. It is easy to replace any set X by a base set of the same size; see Exercise 4.4.1 for one method. Notice that any subset of a base set is also a base set. From now on, it will always be understood that X is a base set.

In a superstructure $V(X)$ over a base set X, the elements of X will be called *individuals relative to* $V(X)$ and the elements of $V(X) \setminus X$ will be called *sets relative to* $V(X)$. The individuals can be characterized as those elements $x \in V(X)$ such that $x \neq \emptyset$ but $x \cap V(X) = \emptyset$.

It is clear from the definition that

$$V_0(X) \subset V_1(X) \subset \ldots \subset V_n(X) \subset \ldots, V_n(X) \subset V(X)$$

and

$$V_0(X) \in V_1(X) \in \ldots \in V_n(X) \in \ldots, V_n(X) \in V(X).$$

Moreover, if X is infinite, the cardinalities of $V_n(X)$ and $V(X)$ are given by

$$|V_n(X)| = \beth_n(|X|), |V(X)| = \beth_\omega(|X|).$$

Our first lemma shows that the collection of all sets relative to $V(X)$ at level $n+1$ is equal to the field of all subsets of $V_n(X)$.

LEMMA 4.4.1. *For each natural number n,*

$$V_{n+1}(X) = X \cup S(V_n(X))$$

and

$$V_{n+1}(X) \setminus X = S(V_n(X)).$$

PROOF. We prove the first equation by induction on n. We see at once from the definitions that the lemma holds for $n=0$. Assume that the lemma holds for $n-1$, so that $V_n(X) = X \cup S(V_{n-1}(X))$. Using the definition of $V_{n+1}(X)$ and the fact that $V_{n-1}(X) \subset V_n(X)$, we have

$$\begin{aligned} V_{n+1}(X) &= V_n(X) \cup S(V_n(X)) \\ &= X \cup S(V_{n-1}(X)) \cup S(V_n(X)) \\ &\subset X \cup S(V_n(X)) \cup S(V_n(X)) \\ &= X \cup S(V_n(X)) \\ &\subset V_{n+1}(X). \end{aligned}$$

This completes the induction.

The second equation follows because X is a base set. ⊣

COROLLARY 4.4.2. *Let X be a base set. Suppose $a \in V(X)$, and $a \in b \in V_n(X)$. Then $n > 0$ and $a \in V_{n-1}(X)$.*

PROOF. Since X is a base set and $a \in b \cap V(X)$, we cannot have $b \in X$, so $n > 0$. By Lemma 4.4.1, $b \in X \cup S(V_{n-1}(X))$. Therefore $b \in S(V_{n-1}(X))$, and hence $a \in V_{n-1}(X)$, as required. ⊣

Our next lemma shows that the collection of sets of a superstructure is closed under simple set-theoretic operations. The proof is routine and is left as an exercise.

LEMMA 4.4.3. *Let $V(X)$ be a superstructure with base set X.*

(i). *If $x_1, \dots, x_m \in V_n(X)$ then $\{x_1, \dots, x_m\} \in V_{n+1}(X) \setminus X$.*

(ii). *If $x_1, \dots, x_m \in V_n(X)$ then $\langle x_1, \dots, x_m \rangle \in V_{n+2(m-1)}(X)$.*

(iii). *If $u \in V_n(X) \setminus X$ and $v \subset u$, then $v \in V_n(X) \setminus X$.*

(iv). *If $u, v \in V_n(X) \setminus X$ then $u \times v \in V_{n+3}(X) \setminus X$.*

(v). *If $u \in V_n(X) \setminus X$ then $S(u) \in V_{n+1}(X) \setminus X$.*

It is sometimes useful to consider models $\mathfrak{A}$ for some language $\mathscr{L}$ such that the universe A is a set relative to $V(X)$. By Lemma 4.4.3, each relation or function over A will be an element of the superstructure $V(A)$.

Throughout this section, $\mathscr{L}_\in$ will be the language $\{\in\}$ with the single binary relation symbol $\in$, so that the superstructure $\langle V(X), \in \rangle$ is a model for $\mathscr{L}_\in$.

We shall use the following abbreviations in $\mathscr{L}_\in$, called *bounded quantifiers*.

$$(\forall x \in y)\ \varphi \text{ means } (\forall x)(x \in y \rightarrow \varphi),$$

$$(\exists x \in y)\ \varphi \text{ means } (\exists x)(x \in y \wedge \varphi).$$

A *bounded quantifier formula* is a formula obtained from atomic formulas using connectives and bounded quantifiers.

Formulas in the language $\mathscr{L}_\in$ become more readable when abbreviations such as the subset symbol $\subset$ or the ordered pair symbol $\langle x, y \rangle$ are used instead of expressing everything directly in terms of $\in$ and $\equiv$. The following lemma shows that we can introduce these abbreviations as bounded quantifier formulas. In the abbreviations, some of the variables must be interpreted as sets relative to $V(X)$ rather than arbitrary elements of $V(X)$. We can see the reason for this by considering the empty set. In a superstructure $V(X)$, $\emptyset$ is the only set relative to $V(X)$ which has no elements in $V(X)$, but each individual relative to $V(X)$ also has no elements in $V(X)$.

LEMMA 4.4.4. *Let $n \in \omega$. There are bounded quantifier formulas $\varphi_0, \dots, \varphi_6$ of $\mathscr{L}_\in$ such that for every superstructure $V(X)$ with base set X,* *all elements* $x_1, \ldots, x_n$ *of* $V(X)$, *and all sets* u, v, w *relative to* $V(X)$, *the following hold.*

(0). $u = \emptyset$ *iff* $\langle V(X), \in \rangle \models \varphi_0[u]$

(1). $u = \{x_1, \ldots, x_n\}$ *iff* $\langle V(X), \in \rangle \models \varphi_1[u, x_1 \ldots x_n]$

(2). $u = \langle x_1, \ldots, x_n \rangle$ *iff* $\langle V(X), \in \rangle \models \varphi_2[u, x_1 \ldots x_n]$

(3). $u \subset v$ *iff* $\langle V(X), \in \rangle \models \varphi_3[u, v]$

(4). $u = v \times w$ *iff* $\langle V(X), \in \rangle \models \varphi_4[uvw]$

(5). $u : v \to w$ *iff* $\langle V(X), \in \rangle \models \varphi_5[uvw]$

(6). $u \in V_n(X)$ *iff* $\langle V(X), \in \rangle \models \varphi_6[u]$

Proof. We give the proof for a few cases to illustrate the method. In the case $n = 2$, the required abbreviation φ_1 for $u = \{x, y\}$ is the bounded quantifier formula

$$x \in u \wedge y \in u \wedge (\forall z \in u)(z \equiv x \vee z \equiv y).$$

Using φ_1, we obtain an abbreviation φ_2 for $u = \langle x, y \rangle$:

$$(\exists v \in u)(\exists w \in u)(\varphi_1(uvw) \wedge \varphi_1(vxx) \wedge \varphi_1(wxy)).$$

The proof of (6) is by induction on n. ⊣

In Lemma 4.4.4 above, it is important that the same abbreviation formula works for every superstructure. We now consider mappings from one superstructure into another.

Let $\mathfrak{A}$ and $\mathfrak{B}$ be models for $\mathscr{L}_\in$. A mapping $f : A \to B$ is said to be a *bounded elementary embedding* of $\mathfrak{A}$ into $\mathfrak{B}$, in symbols $f : \mathfrak{A} \prec_b \mathfrak{B}$, iff f is an isomorphic embedding of $\mathfrak{A}$ into $\mathfrak{B}$ and for every bounded quantifier formula $\varphi(x_1, \ldots, x_n)$ and $a_1, \ldots, a_n$ in A,

$$\mathfrak{A} \models \varphi[a_1, \ldots, a_n] \text{ if and only if } \mathfrak{B} \models \varphi[fa_1, \ldots, fa_n].$$

Thus a bounded elementary embedding is the analogue of an elementary embedding for bounded quantifier formulas.

We say that $\mathfrak{B}$ is a *bounded elementary extension* of $\mathfrak{A}$, and $\mathfrak{A}$ is a *bounded elementary submodel* of $\mathfrak{B}$, in symbols $\mathfrak{A} \prec_b \mathfrak{B}$, if $\mathfrak{A}$ is a submodel of $\mathfrak{B}$ and the identity mapping is a bounded elementary embedding.

Definition. By a *nonstandard universe* we mean a triple

$$\langle V(X), V(Y), * \rangle$$

such that:

(a) X and Y are infinite base sets;

(b) (Transfer Principle)

$$* : \langle V(X), \in \rangle \prec_b \langle V(Y), \in \rangle$$

is a bounded elementary embedding:

(c) $$^*X = Y;$$

and

(d) for every infinite subset A of X, $\{^*a : a \in A\}$ is a proper subset of $^*A$.

The set $^*A$ is called the *star* of A.

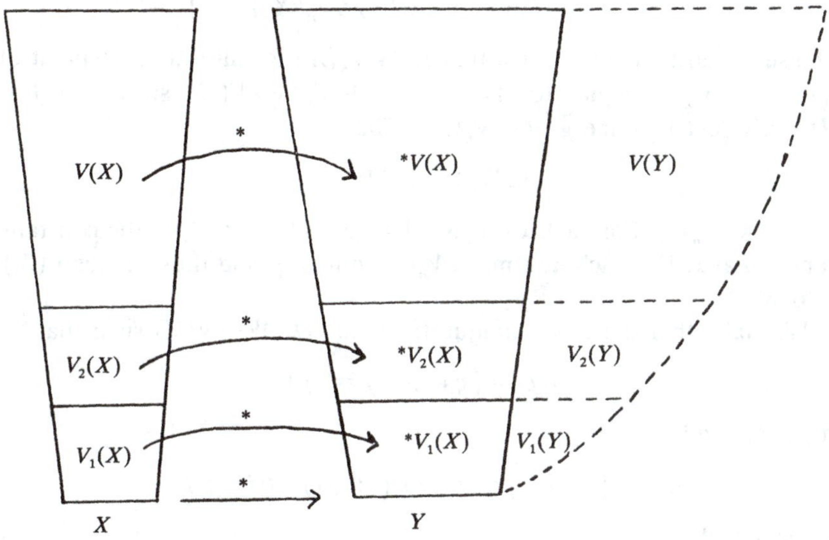

A nonstandard universe

Condition (d) in the definition of a nonstandard universe can be considerably weakened; see Exercise 4.4.28.

We shall give two proofs of the existence of nonstandard universes, first a direct construction using ultrapowers, and then an indirect construction using the compactness theorem.

THEOREM 4.4.5 (Existence of Nonstandard Universes). *Let $V(X)$ be a superstructure with an infinite base set X. Then there exists a nonstandard universe $\langle V(X), V(Y), * \rangle$.*

PROOF VIA ULTRAPOWERS. For this proof we shall introduce a higher order extension of the ultrapower construction which we call a bounded ultrapower. We shall use the ordinary ultrapower of the base set X to form the base set Y, with $*$ being the natural embedding d of X into Y, and then define the restriction of $*$ to $V_n(X)$ by recursion on n.

Let D be any countably incomplete ultrafilter over an index set I. Let Y be the ordinary ultrapower $\prod_D X$ of the set X. We shall assume hereafter that the index set I has been chosen so that Y is a base set; see Exercise 4.4.2. For each natural number n, let

$$W_n = \{f \in {}^I V(X) : \{i : f(i) \in V_n(X)\} \in D\},$$

the set of all functions f such that $f(i) \in V_n(X)$ for all i in an element of D. Thus W_n is equal to the set of all $f : I \to V(X)$ such that f is D-equivalent to some $g : I \to V_n(X)$. Then

$$W_0 \subset W_1 \subset \ldots \subset W_n \subset \ldots$$

Let $W = \bigcup_n W_n$. For each $x \in V(X)$, let $c(x) = \langle x : i \in I \rangle$ be the constant function at x. For each n, c maps $V_n(X)$ into W_n, and thus c maps $V(X)$ into W.

We claim that there is a unique function $/D : W \to V(Y)$ such that:

$$f/D = \{g \in {}^I X : f =_D g\}$$

if $f \in W_0$, and

$$f/D = \{g/D : g \in W \text{ and } \{i : g(i) \in f(i)\} \in D\}$$

if $f \in W \setminus W_0$.

Here is a proof sketch of the claim. For $f \in W_0$, f/D belongs to the ultrapower Y. One can readily show by induction that for each $n \in \omega$ there is a unique function $h_n : W_n \to V_n(Y)$ such that

$$h_n(f) = f/D$$

for $f \in W_0$, and

$$h_n(f) = \{h_n(g) : g \in W \text{ and } \{i : g(i) \in f(i)\} \in D\}$$

for $f \in W_n \setminus W_0$. Moreover, the functions h_n form an increasing chain. The union of the functions h_n is the required function $/D$. This establishes the claim.

Now define the embedding $*: V(X) \to V(Y)$ by $^*A = c(A)/D$. Then for each individual $a \in X$, $^*a = d(a)$. And for each set A relative to $V(X)$,

$$^*A = \{f/D : f \in W \text{ and } \{i : f(i) \in A\} \in D\}.$$

We call the triple $\langle V(X), V(Y), *\rangle$ defined in this way the *bounded ultrapower* of $V(X)$ modulo D.

We now show that any bounded ultrapower is a nonstandard universe. It is clear from the definition that $^*X = Y$. Let A be an infinite subset of X. Since D is countably incomplete, it follows from Proposition 4.2.4 (or Exercise 4.1.32) that the natural embedding $d : A \to \prod_D A$ is a proper embedding, and therefore $\{^*a : a \in A\}$ is a proper subset of $^*A$.

We need the following analogue of Łoś' Theorem.

SUBLEMMA 4.4.6. *For every bounded quantifier formula* $\varphi(x_1 \dots x_n)$ *and every tuple* $f_1 \dots f_n \in W$,

$$\langle V(Y), \in \rangle \models \varphi[f_1/D \dots f_n/D] \textit{ if and only if}$$
$$\{i : \langle V(X), \in \rangle \models \varphi[f_1(i) \dots f_n(i)]\} \in D.$$

The proof of this sublemma is by induction on the complexity of bounded quantifier formulas, and is left to the reader. The definition of $/D$ takes care of the case where φ is an atomic formula.

Taking $f_1, \dots, f_n$ to be constant functions $c(a_1), \dots, c(a_n)$, we conclude that

$$\langle V(Y), \in \rangle \models \varphi[^*a_1 \dots {}^*a_n] \text{ if and only if } \langle V(X), \in \rangle \models \varphi[a_1 \dots a_n],$$

that is, $*$ is a bounded elementary embedding. This completes the proof that $\langle V(X), V(Y), *\rangle$ is a nonstandard universe. ⊣

The bounded ultrapower construction can still be defined in the same way even if D is a countably complete ultrafilter. If D is a principal ultrafilter, the bounded ultrapower modulo D is trivial in the sense that $*$ is an isomorphism of $\langle V(X), \in \rangle$ onto $\langle V(Y), \in \rangle$. If D is a countably complete nonprincipal ultrafilter, so that I must be at least as large as the first uncountable measurable cardinal μ, and $|X| \geqslant \mu$, then the bounded ultrapower is not trivial. It fails to be a nonstandard universe and instead has the property that for any infinite subset A of X of power less μ, $A = {}^*A$. In the case of a countably complete ultrafilter D, the bounded ultrapower of $V(X)$ modulo D is the same as the full ultrapower of the model $\langle V(X), \in \rangle$ modulo D.

The second proof of the existence of nonstandard universes, via the compactness theorem, is more like Robinson's original proof. We shall break this proof into a series of lemmas which are useful in their own right.

DEFINITION. Let $\mathfrak{B} = \langle B, E \rangle$ be a model for $\mathscr{L}_\in$. A submodel $\mathfrak{A}$ of $\mathfrak{B}$ is said to be a *transitive submodel* if whenever $a \in A$, $b \in B$, and bEa, we have $b \in A$.

For example, Lemma 4.4.2 shows that for each n, $\langle V_n(X), \in \rangle$ is a transitive submodel of $\langle V(X), \in \rangle$.

LEMMA 4.4.7. *Let $\mathfrak{B}$ be a model for $\mathscr{L}_\in$. Every transitive submodel $\mathfrak{A}$ of $\mathfrak{B}$ is a bounded elementary submodel.*

PROOF. By a straightforward induction on the complexity of bounded quantifier formulas φ, we see that for all $a_1, \ldots, a_n$ in A, $\mathfrak{B} \models \varphi[a_1 \ldots a_n]$ if and only if $\mathfrak{A} \models \varphi[a_1 \ldots a_n]$. ⊣

A generalization of Lemma 4.4.7 to languages with extra symbols, and a much deeper converse result, are stated in Exercise 2.4.22.

We now wish to truncate a model $\mathfrak{B}$ of $\mathscr{L}_\in$ by chopping off all elements which are not at a finite "level" over the base set. To do this, we need a formal analogue of the relation $y \in V_n(X)$ in the language $\mathscr{L}_\in$. Define by recursion on n the bounded quantifier formulas $\nu_n(y, x)$ as follows:

$$\nu_0(y, x) \text{ is } y \in x;$$

$$\nu_{n+1}(y, x) \text{ is } \nu_n(y, x) \vee (\forall z \in y)\nu_n(z, x).$$

Let BASE(x) be the bounded quantifier formula

$$\neg(\exists y \in x)(\exists z \in y) z \equiv z,$$

which states that each element of x is "disjoint from the universe".

Comparing the definitions, we see that for any superstructure $V(X)$ over a base set X,

$$V_n(X) = \{a \in V(X) : \langle V(X), \in \rangle \models \nu_n[a, X]\}$$

and

$$\langle V(X), \in \rangle \models \mathrm{BASE}[X].$$

Given a model $\mathfrak{A}$ for $\mathscr{L}_\in$ and an element $X \in A$, define the *truncation* of $\mathfrak{A}$ over X to be the submodel $\mathfrak{B}$ of $\mathfrak{A}$ with universe

$$B = \{a \in A : \text{for some } n < \omega, \mathfrak{A} \models \nu_n[a, X]\}.$$

LEMMA 4.4.8. *Let X be a base set, let $\mathfrak{B}$ be a bounded elementary extension of $\langle V(X), \in \rangle$, and let $\mathfrak{A}$ be the truncation of $\mathfrak{B}$ over X. Then $\mathfrak{A}$ is its own truncation over X, and*

$$\langle V(X), \in \rangle \prec_b \mathfrak{A} \prec_b \mathfrak{B}.$$

PROOF: By the argument for Corollary 4.4.2, we see that $\mathfrak{A}$ is a transitive submodel of $\mathfrak{B}$. By Lemma 4.4.7, $\mathfrak{A} \prec_b \mathfrak{B}$, and the result follows. ⊣

We shall call a model $\mathfrak{A}$ for $\mathscr{L}_\in$ *truncated over* Y if $\mathfrak{A}$ is its own truncation over Y. For example, the model $\langle V(X), \in \rangle$ is truncated over X.

THEOREM 4.4.9 (Mostowski Collapsing Theorem). *Let $\mathfrak{A} = \langle A, E \rangle$ be a model for $\mathscr{L}_\in$ and X be an element of A such that*

(a) *$\mathfrak{A}$ is truncated over X;*

(b) $\mathfrak{A} \models \mathrm{BASE}[X]$;

(c) (Extensionality over X)

$$\mathfrak{A} \models (\forall u)(\forall v)(u \in x \vee v \in x \vee (u \equiv v \leftrightarrow (\forall z)(z \in u \leftrightarrow z \in v)))[X].$$

(d) *$Y = \{a \in A : aEX\}$ is a base set.*

Then there is a unique bounded elementary embedding

$$h : \mathfrak{A} \to \langle V(Y), \in \rangle$$

such that $h(a) = a$ for all $a \in Y$, $h(X) = Y$, and the range of h is a transitive submodel of $\langle V(Y), \in \rangle$.

PROOF. We define the restriction of h to the set

$$A_n = \{a \in A : \mathfrak{A} \models \nu_n[a, X]\}$$

by recursion on n. We have $A_0 = Y$. For $a \in A_0$ we define $h(a) = a$ as required by the theorem. Suppose $h(a)$ is defined for $a \in A_n$. Let $a \in A_{n+1} \setminus A_n$. Then whenever bEa we have $b \in A_n$. We may therefore define

$$h(a) = \{h(b) : bEa\}.$$

Since $\mathfrak{A}$ is truncated over X, $A = \bigcup_{n<\omega} A_n$, so we have defined the function h with domain A. By induction on n, h maps each set A_n into $V_n(Y)$, and thus h maps A into $V(Y)$. From the definition, the mapping h from $\mathfrak{A}$ into $\langle V(Y), \in \rangle$ preserves atomic formulas of the form $u \in v$, that is, aEb implies $h(a) \in h(b)$. Using the Extensionality property (c), one can show by induction on n that for each n, h restrictied to A_n is one to one, and hence h is one to one.

For $a, b \in A$, $h(a) \in h(b)$ implies aEb. This is because if $h(a) \in h(b)$ then $h(a) = h(c)$ for some cEb, and since h is one to one, $c = a$. It follows that h is an isomorphic embedding of $\mathfrak{A}$ into $\langle V(Y), \in \rangle$. Let C be the range of h, so that h is an isomorphism of $\mathfrak{A}$ onto a submodel $\mathfrak{C}$ of $\langle V(Y), \in \rangle$. We see from the definition of h that $h(X) = Y$ and $\mathfrak{C}$ is a transitive submodel of $\langle V(Y), \in \rangle$. By Lemma 4.4.7, $\mathfrak{C}$ is a bounded elementary submodel of $\langle V(Y), \in \rangle$, and thus h is a bounded elementary embedding of $\mathfrak{A}$ into $\langle V(Y), \in \rangle$.

To prove uniqueness, let h' be any other mapping with the required properties. Since h' is a bounded elementary embedding, for each n, $a \in A_n$ if and only if $h'(a) \in V_n(Y)$. Using this and the other properties of h', one shows by induction on n that $h'(a) = h(a)$ for all $a \in A_n$. ⊣

In Theorem 4.4.9, the mapping $h : \mathfrak{A} \to \langle V(Y), \in \rangle$ is called the *Mostowski collapse* of $\mathfrak{A}$.

Proof of Theorem 4.4.5 via Compactness Theorem. By the compactness theorem, $\langle V(X), \in \rangle$ has an elementary extension $\mathfrak{B} = \langle B, E \rangle$ such that every infinite subset $U \subset X$ is enlarged, that is, there exists $a \in B \setminus V(X)$ such that aEU. We may choose $\mathfrak{B}$ so that the set $Y = \{b \in B : bEX\}$ is a base set. Let $\mathfrak{A}$ be the truncation of $\mathfrak{B}$ over X. By Lemma 4.4.8, $\mathfrak{A}$ is a bounded elementary submodel of $\mathfrak{B}$ and a bounded elementary extension of $\langle V(X), \in \rangle$.

Since X is a base set, $\mathfrak{A} \models \mathrm{BASE}[X]$. We shall leave the proof that $\mathfrak{A}$ has the Extensionality property 4.4.9(c) as an exercise. Let $h : \mathfrak{A} \to \langle V(Y), \in \rangle$ be the unique bounded elementary embedding from

Theorem 4.4.9, and let $*: V(X) \to V(Y)$ be the restriction of h to $V(X)$. Then $*$ is a composition of two bounded elementary embeddings, and therefore $*$ is a bounded elementary embedding of $\langle V(X), \in \rangle$ into $\langle V(Y), \in \rangle$. Moreover, $^*X = Y$, and $\{^*x : x \in U\}$ is a proper subset of $^*U$ for each infinite subset U of X. Thus $\langle V(X), V(Y), * \rangle$ is a nonstandard universe. ⊣

Now that we have shown that nonstandard universes exist, we shall study their properties. Assume hereafter that $\langle V(X), V(Y), * \rangle$ is a nonstandard universe.

The condition $^*X = Y$ guarantees that $x \in X$ iff $^*x \in Y$. Thus if x is an individual relative to $V(X)$ then $^*x$ is an individual relative to $V(Y)$, and if u is a set relative to $V(X)$ then $^*u$ is a set relative to $V(Y)$.

All of the bounded quantifier formulas in Lemma 4.4.4 are preserved under $*$. For example, by 4.4.4 (3), if A and B are sets relative to $V(X)$ and $A \subset B$, then $^*A \subset {}^*B$. By 4.4.4 (5), if $f : A \to B$ then $^*f : {}^*A \to {}^*B$.

Let A be a set relative to $V(X)$, that is, $A \in V(X) \setminus X$. A is both an element of $V(X)$ and a subset of $V(X)$. It is important to distinguish between the element $^*A$ of $V(Y)$ and the subset $\{^*a : a \in A\}$ of $V(Y)$. We sometimes use the notation

$$^\sigma A = \{^*a : a \in A\}.$$

Since $x = y$ is a bounded quantifier formula, $*$ is a one to one map of A onto $^\sigma A$, and thus $^\sigma A$ has the same size as A. Since $x \in y$ is a bounded quantifier formula, $a \in A$ implies $^*a \in {}^*A$, and therefore $^\sigma A \subset {}^*A$. With this notation, condition (d) in the definition of a nonstandard universe states that for every infinite $A \subset X$, $^\sigma A$ is a proper subset of $^*A$. In particular, $^\sigma X$ is a proper subset of $Y = {}^*X$.

One can think of the mapping $*$ as an expander which blows up each infinite set A relative to $V(X)$ to a larger set $^*A$ relative to $V(Y)$.

By renaming the elements of Y, we can assume without loss of generality that X is a subset of Y and $^*x = x$ for each $x \in X$. This simplifies notation in Robinsonian analysis by reducing the number of *'s. Under this simplifying assumption, $^\sigma X = X$, and in fact $^\sigma A = A$ for every subset $A \subset X$. At higher levels, $^\sigma A$ will usually be different from A.

More precisely, given a nonstandard universe $\langle V(X), V(Y), * \rangle$, we may form the new nonstandard universe $\langle V(X), V(Z), \rho \rangle$ where $Z = X \cup (Y \setminus {}^\sigma X)$, and $\rho(a) = f(^*a)$ where f is the isomorphism $f : \langle V(Y), \in \rangle \cong \langle V(Z), \in \rangle$ such that $f(y) = y$ if $y \in Y \setminus {}^\sigma X$ and $f(^*x) = x$ if $x \in X$.

As usual, when we do this it is understood that $\langle V(X), V(Y), *\rangle$ is such that Z is a base set. By Exercises 4.4.2 and 4.4.3, in the case of a bounded ultrapower the base set X and ultrafilter index set I can always be chosen so that $X \cup \prod_D X$ is a base set, whence $Z = X \cup (\prod_D X \setminus {}^\sigma X)$ is also a base set.

The following lemma is used to compute the star of a set which is definable by a bounded quantifier formula from constants in $V(X)$.

Lemma 4.4.10. *Let $\varphi(x_1 \ldots x_n, y)$ be a bounded quantifier formula. Then for all $b_1, \ldots, b_n \in V(X)$ and all sets A relative to $V(X)$,*

$$^*\{y \in A : \langle V(X), \in\rangle \models \varphi[b_1 \ldots b_n, y]\}$$
$$= \{y \in {}^*A : \langle V(Y), \in\rangle \models \varphi[{}^*b_1 \ldots {}^*b_n, y]\}.$$

Proof. Let

$$B = \{y \in A : \langle V(X), \in\rangle \models \varphi[b_1 \ldots b_n, y]\}.$$

Then $B \in V(X)$, and

$$\langle V(X), \in\rangle \models (\forall y \in A)(y \in B \leftrightarrow \varphi[b_1 \ldots b_n, y]).$$

This is a bounded quantifier formula, so

$$\langle V(Y), \in\rangle \models (\forall y \in {}^*A)(y \in {}^*B \leftrightarrow \varphi[{}^*b_1 \ldots {}^*b_n, y]).$$

Also, ${}^*B \subset {}^*A$, and the result follows. ⊣

Combining Lemmas 4.4.4 and 4.4.10, we see that the simplest set-theoretic operations are preserved by the bounded elementary embedding $*$.

Corollary 4.4.11. *Let A, B and f be sets of a superstructure $V(X)$.*

(1). *$*$ maps $\langle S(A), \cup, \cap, \setminus\rangle$ isomorphically, into $\langle S({}^*A), \cup, \cap, \setminus\rangle$.*
(2). *If A is finite then ${}^*A = {}^\sigma A$.*
(3). ${}^*(A \times B) = {}^*A \times {}^*B$.
(4). *If $S(A) \subset B$, then ${}^*(S(A)) = S({}^*A) \cap {}^*B$.*
(5). *If $V_{n+1}(X) \subset B$, then ${}^*(V_n(X)) = V_n(Y) \cap {}^*B$.*

We now consider models whose universe sets are sets relative to $V(X)$. The next proposition shows that for such models, the bounded elementary embedding $*$ gives rise to an elementary embedding in the original sense.

Let $\mathscr{L}$ be a first order language and let $\mathfrak{A}$ be a model for $\mathscr{L}$ whose universe set A is a set relative to $V(X)$. By Lemma 4.4.3, each finitary relation over A belongs to $V(X)$, and thus the interpretation of each symbol of $\mathscr{L}$ belongs to $V(X)$. Abusing notation, we denote by $^*\mathfrak{A}$ the model for $\mathscr{L}$ with universe $^*A$ such that for each symbol P of $\mathscr{L}$ with interpretation $I(P)$ in $\mathfrak{A}$, P has interpretation $^*I(P)$ in $^*\mathfrak{A}$. (There is an abuse of notation because the interpretation function for $^*\mathfrak{A}$ has the original language $\mathscr{L}$ instead of $^*\mathscr{L}$ as its domain.)

Proposition 4.4.12. *If $\mathfrak{A}$ is a model whose universe set A is a set relative to $V(X)$, then the restriction of $*$ to A is an elementary embedding of $\mathfrak{A}$ into $^*\mathfrak{A}$.*

Proof. Since each formula contains only finitely many symbols of $\mathscr{L}$, it suffices to prove the result in the case that $\mathscr{L}$ is finite, say $\mathscr{L} = \{P_1, \dots, P_k\}$. Let R_j be the interpretation of P_j in $\mathfrak{A}$, and identify $\mathfrak{A}$ with the tuple $\langle A, R_1, \dots, R_k \rangle$. Then $\mathfrak{A}$ is a set relative to $V(X)$, and $\mathfrak{A} \in V_m(X)$ for some m. By Lemma 4.4.10,

$$^*\mathfrak{A} = \langle ^*A, ^*R_1, \dots, ^*R_k \rangle.$$

One can show by induction on the complexity of formulas that for each formula $\varphi(x_1 \dots x_n)$ of $\mathscr{L}$, there is a bounded quantifier formula $\varphi'(x_1 \dots x_n, a, v)$ of $\mathscr{L}_\in$ such that for all $b_1, \dots, b_n \in A$,

$$(1) \qquad \mathfrak{A} \models \varphi[b_1 \dots b_n] \text{ iff } \langle V(X), \in \rangle \models \varphi'[b_1 \dots b_n, \mathfrak{A}, V_m(X)],$$

and for all $b_1, \dots, b_n \in {}^*A$,

$$(2) \qquad ^*\mathfrak{A} \models \varphi[b_1, \dots, b_n] \text{ iff } \langle V(Y), \in \rangle \models \varphi'[b_1 \dots b_n, ^*\mathfrak{A}, V_m(Y)].$$

Since $*$ is a bounded elementary embedding, it follows from (1) and (2) that the restriction of $*$ to A is an ordinary elementary embedding from $\mathfrak{A}$ into $^*\mathfrak{A}$. ⊣

Proposition 4.4.12 is often applied in the simplified case where X is a subset of Y and $^*x = x$ for all $x \in X$. In this case, if the universe A of $\mathfrak{A}$ is an infinite subset of X, then $^*\mathfrak{A}$ is a proper elementary extension of $\mathfrak{A}$.

Examples. In these examples, suppose $^*x = x$ for all $x \in X$.

Let $\mathfrak{A}$ be a standard model of arithmetic with universe $A \subset X$. Then $^*\mathfrak{A}$ is a nonstandard model of complete arithmetic.

Let $\mathfrak{B}$ be isomorphic to the ordered field of real numbers with universe $B \subset X$ and a symbol for the set N of positive integers of $\mathfrak{B}$. Then $^*\mathfrak{B}$ is a proper elementary extension of $\mathfrak{B}$ and $^*N$ is a proper extension of N. The elements of B are called standard real numbers, and the elements of $^*B$ hyperreal numbers. Abraham Robinson showed that the hyperreal number field could be used to develop the calculus in a rigorous manner which mirrors the intuitive use of infinitesimals by Leibniz and his contemporaries. Infinitesimals arise as follows. A hyperreal number is infinitesimal iff its absolute value is less than any positive standard real, finite iff its absolute value is less than some positive standard real, and infinite iff it is not finite.

0 is the only standard infinitesimal. Here is a proof that positive infinite elements and positive infinitesimals exist. Each element of N has finitely many predecessors in N, and hence cannot have any new predecessors in $^*N$. Thus the new elements of $^*N$ must be greater than all elements of N in $^*\mathfrak{B}$, and hence positive infinite. This shows that the ordered field $^*\mathfrak{B}$ is nonarchimedean, that is, it has positive infinite elements. Using the ordered fields properties, it follows that the reciprocal of a positive infinite hyperreal number is positive infinitesimal. Thus positive infinitesimals exist in $^*\mathfrak{B}$.

Now consider an expansion of the standard real model $\mathfrak{B}$ in the preceding paragraph which has symbols for additional real functions f of finitely many variables. Then $^*f$ is an extension of f to $^*\mathfrak{B}$ which satisfies the same first order formulas. Robinson's approach to calculus starts with the fact that the notion of a limit in calculus can be expressed very elegantly in terms of infinitesimals. But that is another story.

We now turn to the important notion of an internal set. An element $A \in V(Y)$ is said to be *internal* iff A is an element of $^*V_n(X)$ for some n, or equivalently, A is an element of $^*B$ for some set B relative to $V(X)$. The set of all internal elements of $V(Y)$ is denoted by $^*V(X)$. Sets relative to $V(Y)$ which are not internal are called *external*.

By 4.4.11 (5), $^*(V_n(X))$ is the set of all internal elements of $V_n(Y)$,

$$^*(V_n(X)) = V_n(Y) \cap {}^*V(X).$$

An element $B \in V(Y)$ is called *standard* if $B = {}^*A$ for some $A \in V(X)$. We denote by $^\sigma V(X)$ the set of all standard sets. Notice that the mapping $*$ is an isomorphism of $\langle V(X), \in \rangle$ onto $\langle {}^\sigma V(X), \in \rangle$, and that $\langle {}^\sigma V(X), \in \rangle$ is a bounded elementary submodel of $\langle V(Y), \in \rangle$.

In practice, the next two propositions are the usual means of showing that sets are internal.

PROPOSITION 4.4.13. *For each* $A \in V(X)$, $^*A$ *is internal. That is,* ${}^{\sigma}V(X) \subset {}^*V(X)$.

PROOF. We have $A \in V_n(X)$ for some n, and therefore $^*A \in {}^*V_n(X)$. $\dashv$

PROPOSITION 4.4.14 (Internal Definition Principle). *Let* $\varphi(x_1 \ldots x_m, y)$ *be a bounded quantifier formula. If* $A_1, \ldots, A_m, B$ *are internal, then the set*

$$\{y \in B : \langle V(Y), \in \rangle \models \varphi[A_1 \ldots A_m, y]\}$$

is internal.

PROOF. Take k such that $A_1, \ldots, A_m, B \in {}^*V_k(X)$. Let $\psi(v)$ be the bounded quantifier formula

$$(\forall x_1 \ldots x_m, b \in v)(\exists u \in v)(\forall y \in v)(y \in u \leftrightarrow (y \in b \wedge \varphi(x_1 \ldots x_m, y))).$$

Then

$$\langle V(X), \in \rangle \models \psi[V_k(X)].$$

Therefore

$$\langle V(Y), \in \rangle \models \psi[{}^*V_k(X)].$$

Taking A_i for x_i and B for b, it follows that the set

$$\{y \in B : \langle V(Y), \in \rangle \models \varphi[A_1 \ldots A_m, y]\}$$

belongs to $^*V_k(X)$ and hence is internal. $\dashv$

EXAMPLES. Since $Y = {}^*X$ and $X \in V(X)$, every element of Y is internal. It follows from 4.4.14 that for each internal set A, the set of all internal subsets of A is closed under unions, intersections, and complements. Moreover, if $\mathfrak{A}$ is a model for $\mathscr{L}$ whose universe is a set relative to $V(X)$, then each relation on $^*A$ which is definable in $^*\mathfrak{A}$ by a formula of $\mathscr{L}$ with constants from $^*A$ is internal.

Notice that by Lemma 4.4.10, Proposition 4.4.14 also holds when "internal" is everywhere replaced by "standard".

COROLLARY 4.4.15. *The model* $\langle {}^*V(X), \in \rangle$ *of internal sets is a transitive submodel of* $\langle V(Y), \in \rangle$. *Moreover,*

$$* : \langle V(X), \in \rangle \prec_b \langle {}^*V(X), \in \rangle,$$

and

$$\langle {}^*V(X), \in \rangle \prec_b \langle V(Y), \in \rangle.$$

PROOF. Let $A \in V(Y)$, $A \in B$, and $B \in {}^*V(X)$. For some n, $B \in {}^*V_{n+1}(X)$, so by Lemma 4.1.2 and the Transfer Principle, $A \in {}^*V_n(X)$. Therefore $A \in {}^*V(X)$. This shows that $\langle {}^*V(X), \in \rangle$ is a transitive submodel of $\langle V(Y), \in \rangle$. The remaining two statements follow by Proposition 4.4.13 and Lemma 4.4.7. ⊣

We remark that $\langle {}^\sigma V(X), \in \rangle$ is not a transitive submodel of $\langle V(Y), \in \rangle$, because ${}^*X \in {}^\sigma V(X)$ but ${}^*X$ has elements which do not belong to ${}^\sigma X$ and hence do not belong to ${}^\sigma V(X)$.

EXAMPLE. We can now give an example of an external set. Suppose $\mathfrak{A}$ is a standard model of arithmetic whose universe A is a subset of X. Then every set in $V_1(X)$ which meets A has a least element in $\mathfrak{A}$. By the Transfer Principle, every set in ${}^*V_1(X)$ which meets ${}^*A$ has a least element in ${}^*\mathfrak{A}$. By 4.4.11 (5), every internal set in $V_1(Y)$ belongs to ${}^*V_1(X)$. The set ${}^*A \setminus A$ of infinite elements of ${}^*\mathfrak{A}$ belongs to $V_1(Y)$, meets ${}^*A$, and has no least element in ${}^*\mathfrak{A}$. Therefore ${}^*A \setminus A$ does not belong to ${}^*V_1(X)$ and must be external. It follows that A is external. By a similar argument, any nonempty subset of ${}^*A$ which has no least element in ${}^*\mathfrak{A}$ is external. See the exercises for other examples.

Instead of the model $\langle V(X), \in \rangle$, one might wish to work with an expansion $\langle V(X), \in, R_i \rangle_{i \in I}$. Given an expansion $\mathscr{L} \supset \mathscr{L}_\in$, the notions of a bounded quantifier formula and a bounded elementary embedding are defined exactly as before, except that we start with the set of atomic formulas of $\mathscr{L}$ instead of $\mathscr{L}_\in$. The next theorem shows that our results on nonstandard universes can be extended "for free" to such expansions.

We first extend the star notation to subsets of the superstructure $V(X)$ in the following way. Given any subset $U \subset V(X)$, define ${}^*U$ to be the union of the chain

$${}^*U = \bigcup_{n<\omega} {}^*(U \cap V_n(X)).$$

For each n, $U \cap V_n(X)$ belongs to $V_{n+1}(X)$ and hence to $V(X)$, so ${}^*(U \cap V_n(X))$ is an element of $V(Y)$. In the case that U is already a set relative to $V(X)$, we have $U \subset V_n(X)$ for some n, and the new and old definitions of $*U$ agree. Since any relation R over $V(X)$ is a subset of $V(X)$, the star notation also applies to relations over $V(X)$.

THEOREM 4.4.16 (Expansion Theorem). *Let* $\mathfrak{A} = \langle V(X), \in, R_i \rangle$ *be an expansion of* $\langle V(X), \in \rangle$. *Then* $\mathfrak{B} = \langle *V(X), \in, *R_i \rangle_{i \in I}$ *is the unique expansion of* $\langle *V(X), \in \rangle$ *such that* $*$ *is a bounded elementary embedding from* $\mathfrak{A}$ *into* $\mathfrak{B}$.

PROOF. Let $\mathscr{L}$ be the expanded language $\mathscr{L}_\in \cup \{P_i : i \in I\}$. For convenience suppose that $\mathscr{L}$ has only relation symbols. We first prove that $* : \mathfrak{A} \prec_b \mathfrak{B}$. Let $\mathfrak{A}_n$ be the submodel of $\mathfrak{A}$ with universe $V_n(X)$, and $\mathfrak{B}_n$ be the submodel of $\mathfrak{B}$ with universe $*V_n(V)$. The proof of Lemma 4.4.7 goes through for the expanded language $\mathscr{L}$, so that $\mathfrak{A}_n \prec_b \mathfrak{A}$ and $\mathfrak{B}_n \prec_b \mathfrak{B}$. By Lemma 4.4.3 and the definition of $*R_i$, we see that $\mathfrak{B}_n = *\mathfrak{A}_n$ for each n. By Proposition 4.4.12, $* : \mathfrak{A}_n \prec \mathfrak{B}_n$ for each n. Consider a bounded quantifier formula $\varphi(x_1 \ldots x_m)$ and $a_1, \ldots, a_m$ in $V(X)$. For sufficiently large n, $a_1, \ldots, a_m$ all belong to $V_n(X)$. Thus the following are equivalent:

$$\mathfrak{A} \models \varphi[a_1 \ldots a_m]$$

$$\mathfrak{A}_n \models \varphi[a_1 \ldots a_m]$$

$$\mathfrak{B}_n \models \varphi[*a_1 \ldots *a_m]$$

$$\mathfrak{B} \models \varphi[*a_1 \ldots *a_m],$$

proving that $* : \mathfrak{A} \prec_b \mathfrak{B}$.

It remains to prove uniqueness. Let $\mathfrak{C} = \langle *V(X), \in, T_i \rangle_{i \in I}$ be an expansion of $\langle *V(X), \in \rangle$ such that $* : \mathfrak{A} \prec_b \mathfrak{C}$. Let $i \in I$ and for each n let $S_n = R_i \cap V_n(X)$. Then $V_n(X), S_n \in V(X)$, and the equation $S_n = R_i \cap V_n(X)$ is expressible by a bounded quantifier formula $\varphi(s, v)$ in $\mathscr{L}$. Since $*$ is a bounded elementary embedding, we have

$$\mathfrak{C} \models \varphi[*S_n, *V_n(X)]$$

for each n. It follows that

$$T_i \cap *V_n(X) = *S_n = *R_i \cap *V_n(X)$$

for each n, and hence $T_i = *R_i$ and $\mathfrak{C} = \mathfrak{B}$ as required. ⊣

EXAMPLES. If FIN is the set of all finite sets relative to $V(X)$, then *FIN, the set of all *-*finite sets*, is the union

$$^*\mathrm{FIN} = \bigcup_{n<\omega} {}^*\{u \in V_n(X) \setminus X : u \text{ is finite}\}.$$

The expansion theorem 4.4.16 shows that *FIN is the collection of all sets u relative to $V(Y)$ such that for some $H \in {}^*N$, there is an internal function f mapping $\{K \in {}^*N : K < H\}$ onto u.

If we let S denote the power set function on $V(X) \setminus X$, then S regarded as a set of ordered pairs is a subset of $V(X)$, and $^*S$ is a function mapping $^*V(X) \setminus {}^*X$ into itself. Theorem 4.4.16 shows that $^*S$ extends S in the sense that for every set A relative to $V(X)$, $({}^*S)({}^*A) = {}^*(S(A))$.

The next proposition characterizes $^*S$ in another way.

PROPOSITION 4.4.17. *For each internal set B relative to $V(Y)$, $({}^*S)(B)$ is the set of all internal subsets of B.*

PROOF. Let B be internal, so that $B \in {}^*V_n(X)$ for some n. Since $^*V_n(X) = {}^*V(X) \cap V_n(Y)$, every internal subset of B belongs to $^*V_n(X)$. We have

$$\langle V(X), \in, S\rangle \models (\forall u \in V_n(X))(\forall y \in V_n(X))(y \in S(u) \leftrightarrow y \subset u).$$

By 4.4.16, this formula is also satisfied by $^*V_n(X)$ in $\langle {}^*V(X), \in, {}^*S\rangle$. Taking B for u, we have

$$(\forall y \in {}^*V_n(X))(y \in ({}^*S)(B) \leftrightarrow y \subset B).$$

Since every internal subset of B belongs to $^*V_n(X)$, $({}^*S)(B)$ is the set of all internal subsets of B. ⊣

We now prove two results which establish relationships between internal sets and ultrapowers.

PROPOSITION 4.4.18. *Suppose that $\langle V(X), V(Y), *\rangle$ is a bounded ultrapower of $V(X)$ modulo D, and let h be the Mostowski collapse of the truncation of the ordinary ultrapower of $\langle V(X), \in\rangle$ modulo D. Then*

$$^*V(X) = \mathrm{range}(/D) = \mathrm{range}(h).$$

PROOF. We prove the first equation. As in our construction of the bounded ultrapower,

$$W_n = \{f \in {}^I V(X) : \{i : f(i) \in V_n(X)\} \in D\}, W = \bigcup_n W_n.$$

For each n, we have

$$\begin{aligned} {}^*V_n(X) &= c(V_n(X))/D \\ &= \{g/D : g \in W \text{ and } \{i : g(i) \in V_n(X)\} \in D\} \\ &= \{g/D : g \in W_n\}. \end{aligned}$$

Taking unions of each side over n we have

$${}^*V(X) = \{g/D : g \in W\}. \dashv$$

THEOREM 4.4.19 (Local Ultrapower Theorem). *Let $\langle V(X), V(Y), *\rangle$ be a nonstandard universe. For every internal set $B \in {}^*V(X)$ there is a bounded ultrapower $\langle V(X), V(Z), \rho\rangle$ and a bounded elementary embedding*

$$h : ({}^\rho V(X), \in\rangle \to \langle {}^*V(X), \in\rangle$$

such that B belongs to the range of h, and ${}^*A = h({}^\rho A)$ for all $A \in V(X)$.

PROOF. Let I be a set relative to $V(X)$ such that $B \in {}^*I$. Let J be a countably infinite subset of X, choose an element $b \in {}^*J \setminus J$ and put $\beta = \langle B, b\rangle$ (this is done to make that we end up with a countably incomplete ultrafilter). Let

$$D = \{C \subset I \times J : \beta \in {}^*C\}.$$

Then D is an ultrafilter over $I \times J$. D is countably incomplete because for each $j \in J$, $I \times \{j\} \notin D$. Let $\langle V(X), V(Z), \rho\rangle$ be the bounded ultrapower of $V(X)$ modulo D, and let ${}^\rho V(X)$ be its collection of internal sets. By Proposition 4.4.18.

$${}^\rho V(X) = \{f/D : f \in W\}.$$

Since $I \times J \in V(X) \setminus X$, $W \subset V(X)$ and thus ${}^*f$ is defined and is a function with domain ${}^*(I \times J) = {}^*I \times {}^*J$ for each $f \in W$. Define $h : {}^\rho V(X) \to {}^*V(X)$ by

$$h(f/D) = ({}^*f)(\beta)$$

h is a well-defined function because if $f/D = e/D$ then

$$\{k \in I \times J : f(k) = e(k)\} \in D,$$

so

$$\beta \in {}^*\{k \in I \times J : f(k) = e(k)\}$$

and hence $({}^*f)(\beta) = ({}^*e)(\beta)$.

We have $h(g/D) = B$ where g is the function $g(i, j) = i$, and hence B is in the range of h. It remains to show that h is a bounded elementary embedding. Using Łoś' theorem (Sublemma 4.4.6) and Lemma 4.4.10, we see that for all $f_1, \ldots, f_n \in W$ and all bounded quantifier formulas $\varphi(x_1 \ldots x_n)$, the following are equivalent.

$$\langle {}^pV(X), \in\rangle \models \varphi[f_1/D \ldots f_n/D],$$
$$\{k \in I \times J : \langle V(X), \in\rangle \models \varphi[f_1(k) \ldots f_n(k)]\} \in D,$$
$$\beta \in {}^*\{k \in I \times J : \langle V(X), \in\rangle \models \varphi[f_1(k) \ldots f_n(k)]\},$$
$$\beta \in \{k \in {}^*(I \times J) : \langle {}^*V(X), \in\rangle \models \varphi[{}^*f_1(k) \ldots {}^*f_n(k)]\},$$
$$\langle {}^*V(X), \in\rangle \models \varphi[({}^*f_1)(\beta) \ldots ({}^*f_n)(\beta)],$$
$$\langle {}^*V(X), \in\rangle \models \varphi[h(f_1/D) \ldots h(f_n/D)].$$

This completes the proof. ⊣

This theorem gives us some information about cardinalities of internal sets.

PROPOSITION 4.4.20. *Each infinite internal set relative to $V(Y)$ has cardinality at least 2^ω. Every countably infinite set relative to $V(Y)$ is external.*

PROOF. Let B be an infinite internal set relative to $V(Y)$. Let $\langle V(X), V(Z), \rho\rangle$ and $h : {}^pV(X) \to {}^*V(X)$ be the bounded ultrapower and bounded elementary embedding from theorem 4.4.19, so that B is in the range of h. The bounded ultrapower is taken modulo a countably incomplete ultrafilter D over a set I. We have $B = h(C)$ for some $C \in {}^pV(X)$. Since B is infinite, C is infinite. Moreover, $C = f/D$ for some n and some $f : I \to V_{n+1}(X)$. To prove that $|C| \geq 2^\omega$, one must find 2^ω functions $g : I \to V_n(X)$ which are unequal modulo D and such that $\{i : g(i) \in f(i)\} \in D$. This follows from Exercise 4.3.14, which states that any ultraproduct modulo a countably incomplete ultrafilter which is infinite has cardinality at least 2^ω. ⊣

We now turn to saturated nonstandard universes. We shall start out by stating a bounded quantifier analogue of our previous definition of saturation. We shall see that in the case of nonstandard universes, this definition can be considerably simplified. Let $\mathscr{L}_{V(X)}$ and $\mathscr{L}_{\mathrm{INT}}$ be the extensions of $\mathscr{L}_\in$ formed by adding constants for all elements of $V(X)$ or $^*V(X)$, respectively. A constant symbol $c \in V(X)$ will be interpreted by c in $\langle V(X), \in \rangle$ and by $^*c$ in $\langle V(Y), \in \rangle$. In the definitions which follow, α is an uncountable cardinal.

We shall say that the superstructure $\langle V(X), V(Y), *\rangle$ is *α-saturated over* $V(X)$, or is an *α-enlargement*, iff for every $n < \omega$, every set $\Sigma(x)$ of fewer than α bounded quantifier formulas of $\mathscr{L}_{V(X)}$ which is finitely satisfiable in $\langle V(X), \in \rangle$ by elements of $V_n(X)$ is satisfiable in $\langle V(Y), \in \rangle$ by an element of $^*V_n(X)$.

$\langle V(X), V(Y), \in \rangle$ is *saturated over* $V(X)$, or is an *enlargement*, iff for every $n < \omega$, every set $\Sigma(x)$ of bounded quantifier formulas of $\mathscr{L}_{V(X)}$ which is finitely satisfiable in $\langle V(X), \in \rangle$ by elements of $V_n(X)$ is satisfiable in $\langle V(Y), \in \rangle$ by an element of $^*V_n(X)$.

Thus $\langle V(X), V(Y), *\rangle$ is an enlargement if and only if it is α-saturated over $V(X)$ for all cardinals α, and also if and only if it is $|V(X)|^+$-saturated over $V(X)$.

We shall say that the superstructure $\langle V(X), V(Y), *\rangle$ is *α-saturated* iff for every $n < \omega$, every set $\Sigma(x)$ of fewer than α bounded quantifier formulas of $\mathscr{L}_{\mathrm{INT}}$ which is finitely satisfiable in $\langle V(Y), \in \rangle$ by elements of $^*V_n(X)$ is satisfiable in $\langle V(Y), \in \rangle$ by an element of $^*V_n(X)$.

COROLLARY 4.4.21. *Every α-saturated nonstandard universe is an α-enlargement.*

PROOF. By Proposition 4.4.13. ⊣

Several characterizations of enlargements and α-saturated nonstandard universes are given in the problem set. The following theorem shows that enlargements exist.

THEOREM 4.4.22. *Let D be an α-regular ultrafilter. Then the bounded ultrapower of $V(X)$ modulo D is an α^+-enlargement.*

PROOF. Comparing the definitions of ultrapower and bounded ultrapower, one can check by induction on n that there is an isomorphism

$$h : \prod_D \langle V_n(X), \in \rangle \cong \langle {}^*V_n(X), \in \rangle$$

such that for each $A \in V_n(X)$, $^*A = h(d(A))$. The result now follows from Corollary 4.3.14. ⊣

In many applications of Robinsonian analysis, an ω_1-saturated nonstandard universe is needed. The next theorem will be used to show that ω_1-saturated nonstandard universes exist.

THEOREM 4.4.23 (Comprehensiveness Theorem). *Let $\langle V(X), V(Y), *\rangle$ be a bounded ultrapower. Let A and B be sets relative to $V(X)$ and let $F: A \to {}^*B$. Then there exists an internal function $G: {}^*A \to {}^*B$ such that $G({}^*a) = F(a)$ for all $a \in A$.*

PROOF. Let the bounded ultrapower be modulo the ultrafilter D over I. For each $a \in A$, choose a representative $f(a): I \to B$ such that $F(a) = f(a)/D$. Let $g: I \to {}^AB$ be the function defined by $g(i)(a) = f(a)(i)$, and let $G = g/D$. G is internal. For each $i \in I$ we have $g(i): A \to B$, so $G: {}^*A \to {}^*B$. Finally, for each $a \in A$, we have

$$G({}^*a) = \langle g(i)(a)\rangle_{i\in I}/D = \langle f(a)(i)\rangle_{i\in I}/D = f(a)/D = F(a),$$

as required. ⊣

COROLLARY 4.4.24. *Every bounded ultrapower is an ω_1-saturated nonstandard universe.*

PROOF. Let $\{\varphi_n(x): n \in N\}$ be a countable set of bounded quantifier formulas of $\mathscr{L}_{\mathrm{INT}}$ which is finitely satisfiable in $\langle V(Y), \in\rangle$ by elements of ${}^*V_k(X)$. Let B_n be the set of elements of ${}^*V_k(X)$ which satisfy φ_n. We must show that $\bigcap_n B_n \neq \emptyset$. By 4.4.14, each B_n is internal. We may assume without loss of generality that $N \subset X$. By 4.4.23, there is an internal function $G: {}^*N \to {}^*C$ such that $G(n) = B_n$ for all $n \in N$. The internal set

$$\{H \in {}^*N : \textstyle\bigcap_{K<H} G(K) \neq \emptyset\}$$

contains N and therefore contains an infinite H. It follows that

$$\textstyle\bigcap_{n\in N} G(n) = \bigcap_{n\in N} B_n \neq \emptyset. \quad ⊣$$

Exercise 4.4.29 shows that there are nonstandard universes, and even enlargements, which are not ω_1-saturated.

The analogue of Corollary 4.4.24 for ordinary ultrapowers, that any countably incomplete ultrapower is ω_1-saturated, will be proved in Section 6.1 by a different argument.

For arbitrary cardinals α, the existence of α-saturated nonstandard universes will follow from the results in the next section, 5.1.

We now take up the topic of internal set theory, which was introduced by Nelson (1977). It works within an expansion of ZFC which has a new unary predicate St(x) for "standard" and additional axioms which reflect the intuitive idea of a "rich" elementary extension of the original universe.

DEFINITION. The language of internal set theory has the binary relation symbol $\in$ and the unary relation symbol St. A formula is called *internal* iff the symbol St does not occur in it. We use the abbreviations

$$(\forall^{\mathrm{St}} x)\varphi \text{ for } (\forall x)\mathrm{St}(x) \rightarrow \varphi, \ (\exists^{\mathrm{St}} x)\varphi \text{ for } (\exists x)(\mathrm{St}(x) \wedge \varphi),$$

$$(\forall^{\mathrm{StFin}} x)\varphi \text{ for } (\forall x)(\mathrm{St}(x) \wedge x \text{ is finite}) \rightarrow \varphi.$$

Internal set theory, IST, has the following axiom schemes.

The axioms of ZFC;

Idealization: $(\forall^{\mathrm{StFin}} y)(\exists x)(\forall z \in y)\varphi \rightarrow (\exists x)(\forall^{\mathrm{St}} z)\varphi$, where φ is an internal formula in which y does not occur;

Standardization: $(\forall^{\mathrm{St}} x)(\exists^{\mathrm{St}} y)(\forall^{\mathrm{St}} z)(z \in y \leftrightarrow z \in x \wedge \Phi)$, where Φ is an arbitrary formula in which y does not occur;

Transfer: $(\forall^{\mathrm{St}} x_1)\ldots(\forall^{\mathrm{St}} x_n)((\forall^{\mathrm{St}} y)\varphi \rightarrow (\forall y)\varphi)$, where φ is an internal formula which has at most the free variables $x_1, \ldots, x_n, y$.

Here is an example of a model of IST.

THEOREM 4.4.25. *Let* $\mathfrak{A} = \langle R(\xi), \in \rangle$ *where* ξ *is a limit ordinal* $>\omega$, *and let* $\mathfrak{B} = \langle B, E \rangle$ *be an* $(\beth_\xi)^+$*-saturated elementary extension of* $\mathfrak{A}$. *Then* $\mathfrak{B}' = \langle B, E, R(\xi) \rangle$ *satisfies the Idealization, Standardization, and Transfer schemes. If* $\mathfrak{A}$ *is a model of* ZFC, *then* $\mathfrak{B}$ *is a model of* IST.

PROOF. The interpretation of St in $\mathfrak{B}'$ is $R(\xi)$. Since $\mathfrak{A} \prec \mathfrak{B}$, $\mathfrak{B}'$ satisfies the Transfer scheme. The model $\mathfrak{A}$ has power $\beth_\xi$. Finiteness can be expressed in $\mathfrak{B}'$ because ξ is a limit ordinal $>\omega$. Since $\mathfrak{B}$ is $(\beth_\xi)^+$-saturated, the idealization scheme holds in $\mathfrak{B}'$. Standardization holds in $\mathfrak{B}'$ because, for any $x \in R(\xi)$, the set

$$y = \{z \in x : \mathfrak{B}' \models \Phi[z, \ldots]\}$$

belongs to $R(\xi)$. ⊣

In the above theorem, if $\mathfrak{B}$ is only assumed to be $(\beth_\xi)^+$-saturated over $\mathfrak{A}$, then $\mathfrak{B}'$ will still satisfy the Standardization and Transfer schemes, and the restricted Idealization scheme where only x and z occur free in φ.

We shall now use Theorem 4.4.25 to show that IST is a conservative extension of ZFC, and thus the consistency of IST follows from the consistency of ZFC. The proof will depend on the existence of α-saturated elementary extensions, which will be proved in Section 5.1.

THEOREM 4.4.26. IST *is a conservative extension of ZFC. That is, any sentence of the language of* ZFC *which is provable from* IST *is already provable from* ZFC.

PROOF: Let θ be a sentence of the language of ZFC which is provable from IST. Then θ is provable from a finite subset $\mathrm{ZFC}_0 \subset \mathrm{ZFC}$ and the Idealization, Standardization, and Transfer schemes. To show that θ is provable from ZFC, we formalize the argument in the next paragraph within ZFC.

We assume the negation of θ and get a contradiction. By the reflection principle (Exercise 3.1.16), there is a limit ordinal $\xi > \omega$ such that ZFC_0 and $\neg\theta$ hold in $\mathfrak{A} = \langle R(\xi), \in \rangle$. By the existence theorem for saturated models, Theorem 5.1.4, $\mathfrak{A}$ has a $(\beth_\xi)^+$-saturated elementary extension $\mathfrak{B}$. By Theorem 4.4.25, $\mathfrak{B}' = (\mathfrak{B}, R(\xi))$ is a model of ZFC_0 and the Idealization, Standardization, and Transfer schemes. Therefore $\mathfrak{B}' \models \theta$, contradicting $\mathfrak{A} \models \neg\theta$. ⊣

We conclude with a comparison of the superstructure and internal set theory approaches. The superstructure approach works within ordinary ZFC set theory, at the price of restricting the universe to the finite $\in$-levels. It uses the interplay between the three structures $\langle V(X), \in \rangle$, $\langle {}^*V(X), \in \rangle$, and $\langle V(Y), \in \rangle$, that is, the collections of standard, internal, and external sets. The external sets often play an essential part; for example, in many applications one uses the σ-algebra generated by the algebra of internal subsets of a set.

The internal set theory approach is syntactic in character, and works within a conservative extension of ZFC. It has the expository advantage of avoiding any mention of models. There are no stars, and the new predicate $\mathrm{St}(x)$ appears instead. All the $\in$ levels of the original set-theoretic universe are kept, but the price is the loss of the external sets. Seen from the outside, it uses the interplay between the two structures $\langle \mathrm{St}, \in \rangle$ and $\langle V, \in \rangle$ of standard and internal sets, where $\langle \mathrm{St}, \in \rangle$ is the usual set-theoretic universe and $\langle V, \in \rangle$ is a non well-founded elementary extension. Particular external sets whose elements are internal can be treated as formulas involving St(x). A disadvantage is that the language cannot talk about sets of external sets, such as the σ-algebra generated by an algebra of internal sets. In practice, internal set theory has been adequate for certain areas of Robinsonian analysis (e.g. singular perturbations), but inadequate for others (e.g. probability theory, Banach spaces).

Exercises

In the exercises which follow, it is understood that $\langle V(X), V(Y), * \rangle$ is a nonstandard universe and $^*x = x$ for all $x \in X$.

4.4.1. Let α be an infinite ordinal and let X be a set such that every element of an element of X has rank α. (Ranks of sets are defined in the Appendix.) Prove that X is a base set. [*Hint*: Prove by induction on n that each element of $V_n(X)$ has rank β where either $\beta < n$ or $\alpha < \beta \leq \alpha + n + 1$.]

4.4.2. Let X be a set of rank β and let D be an ultrafilter over a set I of rank γ such that $\gamma \geq \beta + \omega$. Prove that the ultrapower $\prod_D X$ is a base set. [*Hint*: Prove that each function $f : I \to X$ has the same rank, γ if γ is a limit ordinal, and $\gamma + 2$ if γ is a successor ordinal. Then use Exercise 4.4.1.]

4.4.3. Let α and β be ordinals such that $\omega \leq \alpha$ and $\alpha + \omega \leq \beta$. Let X and Y be sets such that every element of an element of X has rank α and every element of an element of Y has rank β. Prove that $X \cup Y$ is a base set. [*Hint*: No element of $V(X \cup Y)$ has rank α or β.]

4.4.4. Prove that the composition of two bounded elementary embeddings is a bounded elementary embedding.

4.4.5 (Bounded elementary chain theorem). Let $\mathfrak{A}_\alpha$, $\alpha < \gamma$, be a chain of models for $\mathscr{L}_\in$ such that $\mathfrak{A}_\alpha \prec_b \mathfrak{A}_\beta$ whenever $\alpha < \beta < \gamma$. Prove that $\mathfrak{A}_0 \prec_b \bigcup_{\alpha<\gamma} \mathfrak{A}_\alpha$.

4.4.6. Give the proof of Lemma 4.4.3, and complete the proof of Lemma 4.4.4.

4.4.7. Let $\varphi(x_1 \dots x_m)$ be a formula of $\mathscr{L}_\epsilon$ of the form $(\exists y_1 \dots y_n)(\forall z_1 \dots z_p)\psi$ where ψ is a bounded quantifier formula. Then for all $a_1, \dots, a_m$ in A, if $\langle V(X), \in \rangle \models \varphi[a_1 \dots a_m]$ then $\langle {}^*V(X), \in \rangle \models \varphi[a_1 \dots a_m]$.

4.4.8. Prove that for every infinite set A relative to $V(X)$ (not necessarily a subset of X), ${}^\sigma A$ is a proper subset of ${}^*A$.

4.4.9. Prove that for each $A \in V(X)$ there is a subset $A_0 \subset X$ such that for all $U \subset X$, $A \in V(U)$ if and only if $A_0 \subset U$.

4.4.10. Show that for each set A relative to $V(X)$, the following are equivalent:

${}^*A = A$;

${}^\sigma A = A$;

${}^*A \in V(X)$;

$A \in V(Z)$ for some finite subset $Z \subset X$.

[*Hint*: Use the preceding exercise.]

4.4.11. Let F be an internal function which is a set relative to $V(Y)$. Prove that the domain and range of F are internal sets.

4.4.12. Given sets A and B, let EXP(A, B) be the set ${}^BA$ of all functions from B into A. Prove that for any two sets A, B relative to $V(X)$, ${}^*(\mathrm{EXP}(A, B))$ is the set of all internal functions $f : {}^*B \to {}^*A$.

4.4.13. Let F be a function $F : V(X) \to V(X)$, and suppose that for each n there exists k such that F restricted to $V_n(X)$ maps $V_n(X)$ into $V_k(X)$. Show that

$$ {}^*F = \bigcup_{n<\omega} {}^*(F \cap V_n(X)) $$

is a total function from ${}^*V(X)$ into itself.

4.4.14. The binary function EXP from pairs of sets into sets may be regarded as a subset of $V(X) \setminus X$, so that *EXP is a subset of ${}^*V(X) \setminus Y$. Prove that for any two internal sets A, B relative to $V(Y)$, $({}^*\mathrm{EXP})(A, B)$ is the set of all internal functions $f : {}^*B \to {}^*A$.

4.4.15. In the definition of a nonstandard universe, prove that the Transfer Principle (b) may be replaced by the statement that $*$ is a bounded elementary embedding of $\langle V(X), \in \rangle$ into $\langle U, \in \rangle$ for some transitive submodel $\langle U, \in \rangle$ of $\langle V(Y), \in \rangle$.

4.4.16 (Overspill Principle). Let $\mathfrak{A}$ be a model isomorphic to the ordered field of real numbers with universe $A \subset X$. Prove that any internal subset of $^*A$ which contains arbitrarily large infinitesimals contains arbitrarily small positive noninfinitesimals, and that any internal subset of $^*A$ which contains arbitrarily large finite elements contains arbitrarily small positive infinite elements.

4.4.17. Let $\mathfrak{A}$ be as in the preceding exercise. Prove that any internal subset of $^*A$ which contains all sufficiently large infinitesimals contains all sufficiently small positive noninfinitesimals, and that any internal subset of $^*A$ which contains all sufficiently large finite elements contains all sufficiently small positive infinite elements.

4.4.18. Let $\mathfrak{A}$ be as in the preceding exercises, and suppose that the nonstandard universe is ω_1-saturated. Prove that every countable set of infinitesimals has an infinitesimal upper bound.

4.4.19. Suppose that the nonstandard universe is ω_1-saturated. Prove that the union of a countable strictly increasing chain of internal subsets of $V_n(Y)$ is external.

4.4.20. Suppose that the nonstandard universe is ω_1-saturated. Let B be a subset of $V_n(Y)$ such that both B and $V_n(Y) \setminus B$ are unions of countable sets of internal sets. Prove that B is internal.

4.4.21. Suppose that the set N of natural numbers is a subset of X. Prove that $\langle V(X), V(Y), * \rangle$ is ω_1-saturated if and only if every countable sequence $\langle A_n : n \in N \rangle$ of internal sets can be extended to an internal sequence $\langle B_n : n \in {}^*N \rangle$.

4.4.22. Suppose the nonstandard universe is an α-enlargement, where $\alpha > \omega$. Prove that for every model $\mathfrak{A}$ for a language $\mathscr{L}$ whose universe set A is a subset of X, $^*\mathfrak{A}$ is α-saturated over $\mathfrak{A}$.

4.4.23. Suppose the nonstandard universe is α-saturted, where $\alpha > \omega$.

Let $\mathfrak{A}$ be a model for a language $\mathscr{L}$ such that the universe A and each relation and function of $\mathfrak{A}$ is internal. Prove that $\mathfrak{A}$ is an α-saturated model.

4.4.24. (i). A nonstandard universe $\langle V(X), V(Y), *\rangle$ is an α-enlargement if and only if each collection U of fewer than α standard sets relative to $V(Y)$, if U has the finite intersection property then U has nonempty intersection.

(ii). $\langle V(X), V(Y), *\rangle$ is α-saturated if and only if for each set U of fewer than α internal sets relative to $V(Y)$, if U has the finite intersection property, then U has nonempty intersection.

4.4.25. A binary relation R with domain B is said to be *concurrent* if for every finite subset C of B there exists y such that for all $x \in C$, $\langle x, y\rangle \in R$. Prove that the nonstandard universe is an enlargement if and only if for every concurrent relation R with domain B such that R is a set relative to $V(X)$ there exists y such that for all $x \in B$, $\langle {}^*x, y\rangle \in {}^*R$.

4.4.26. Prove that a nonstandard universe is an enlargement if and only if it is a $|V(X)|$-enlargement.

4.4.27. Prove that a nonstandard universe is an enlargement if and only if for each set A relative to $V(X)$ there is a $*$-finite set B such that ${}^\sigma A \subset B$.

4.4.28*. Prove that in the definition of a nonstandard universe, the condition (d) can be replaced by the weaker condition that ${}^\sigma A$ is a proper subset of ${}^*A$ for some infinite $A \subset X$ of cardinality less than the first measurable cardinal.

4.4.29*. Prove that for each infinite base set A, there is a nonstandard universe $\langle V(X), V(Y), *\rangle$ which is an enlargement but is not ω_1-saturated. [*Hint*: Form the union of a countable elementary chain beginning with $\langle V(X), \in\rangle$, then truncate and take the Mostowski collapse.]

4.4.30*. Let $\langle V(X), V(Y), *\rangle$ be an enlargement and let $\mathfrak{A}$ and $\mathfrak{B}$ be models for a language $\mathscr{L}$ whose universe sets A and B are sets relative to $V(X)$. Prove that if $\mathfrak{A} \equiv \mathfrak{B}$, then $\mathfrak{A}$ is elementarily embeddable in ${}^*\mathfrak{B}$.

4.4.31*. Let D be an α-regular ultrafilter and let $\langle V(X), V(Y), *\rangle$ be the bounded ultrapower of $V(X)$ modulo D. Prove that $\langle V(X), V(Y), *\rangle$ is α^+-universal in the sense that for any nonstandard universe $\langle V(X), V(Z), \rho\rangle$ with at most α internal sets there is a bounded elementary embedding

$$h : \langle {}^{\rho}V(X), \in\rangle \to \langle {}^{*}V(X), \in\rangle$$

such that $h({}^{\rho}A) = {}^{*}A$ for all $A \in V(X)$.

4.4.32*. Let U be a set of internal sets, $U \subset {}^{*}V(X)$. By the *Skolem hull* of U we mean the set

$$H(U) = \{({}^{*}F)(u) : F \in \text{FUNC and } u \in U \cap \text{Domain}({}^{*}F)\}$$

where FUNC is the set of all functions which are sets relative to $V(X)$. Prove that for any U,

$$\langle {}^{\sigma}V(X), \in\rangle \prec_{\mathrm{b}} \langle H(U), \in\rangle \prec_{\mathrm{b}} \langle {}^{*}V(X), \in\rangle,$$

and that $H(U)$ is the intersection of all sets W such that

$${}^{\sigma}V(X) \cup U \subset W \text{ and } \langle W, \in\rangle \prec_{\mathrm{b}} \langle {}^{*}V(X), \in\rangle.$$

4.4.33. Let $\langle V(X), V(Y), *\rangle$ be a bounded ultrapower modulo an ultrafilter D over a set I relative to $V(X)$. Prove that ${}^{*}V(Y)$ is the Skolem hull of the singleton $\{\text{id}/D\}$ where id is the identity function on I.

4.4.34. Show that for each base set X there exists a bounded ultrapower of $V(X)$ such that ${}^{*}V(X)$ is not the Skolem hull of a singleton. [*Hint*: Get an upper bound for the cardinality of the Skolem hull $H(U)$.]

4.4.35*. Prove that ${}^{*}V(Y)$ is the Skolem hull of a singleton if and only if $\langle V(X), V(Y), *\rangle$ is a bounded ultrapower of $V(X)$ modulo an ultrafilter D over a set I relative to $V(X)$ (renaming the elements of Y if necessary). [*Hint*: Use the proof of the Local Ultrapower Theorem.]

4.4.36*. Call a nonstandard universe *minimal* iff ${}^{*}V(X)$ and ${}^{\sigma}V(X)$ are the only sets W such that $\langle {}^{\sigma}V(X), \in\rangle \prec_{\mathrm{b}} \langle W, \in\rangle \prec_{\mathrm{b}} \langle {}^{*}V(X), \in\rangle$. Prove that a nonstandard universe is minimal if and only if it is a bounded ultrapower with respect to a selective ultrafilter over ω (renaming the elements of Y if necessary). Selective ultrafilters are defined in Exercise 4.3.42.

CHAPTER 5

SATURATED AND SPECIAL MODELS

5.1. Saturated and special models

We have already defined in Section 2.3 the notion of an ω-saturated model. To repeat the definition, a model $\mathfrak{A}$ is *ω-saturated* iff for every finite subset $X \subset A$, the expansion $(\mathfrak{A}, a)_{a\in X}$ realizes every type $\Sigma(v)$ that is consistent with its theory. So an ω-saturated model is rich in the number of types that it realizes; essentially, it realizes the maximum number of types. We also proved in Section 2.3 that any two equivalent countable ω-saturated models are isomorphic; furthermore, if a theory T has only a countable number of finite types, then T has countably saturated models. We shall develop in this section some simple properties of generalizations of ω-saturated models. The generalizations we have in mind are the notions of an α-saturated model, a saturated model, and a special model.

Let α be a cardinal. A model $\mathfrak{A}$ is *α-saturated* iff for every subset $X \subset A$ with fewer than α elements, the expansion $(\mathfrak{A}, a)_{a\in X}$ realizes every type $\Sigma(v)$ of the language $\mathscr{L} \cup \{c_a : a \in X\}$ which is consistent with the theory of $(\mathfrak{A}, a)_{a\in X}$. In working with α-saturated models, we frequently want to enumerate the elements of X in the expansion $(\mathfrak{A}, a)_{a\in X}$. Thus if ξ is an ordinal less than α and $X = \{a_\eta : \eta < \xi\}$, then we write $(\mathfrak{A}, a_\eta)_{\eta<\xi}$. There is then no ambiguity when we write $(\mathfrak{A}, a_\eta)_{\eta<\xi} \equiv (\mathfrak{B}, b_\eta)_{\eta<\xi}$.

A model $\mathfrak{A}$ is *saturated* iff it is $|A|$-saturated.

PROPOSITION 5.1.1.

(i). *If α is a limit cardinal, then $\mathfrak{A}$ is α-saturated if and only if $\mathfrak{A}$ is β-saturated for all cardinals $\beta < \alpha$.*

(ii). *$\mathfrak{A}$ is α-saturated if and only if for every ordinal $\xi < \alpha$ and every $a \in {}^{\xi}A$, the model $(\mathfrak{A}, a_\eta)_{\eta<\xi}$ realizes all types $\Sigma(x_1 \dots x_n)$ consistent with its theory.*

(iii). *$\mathfrak{A}$ is α-saturated if and only if for every ordinal $\xi < \alpha$ and every $a \in {}^{\xi}A$, the model $(\mathfrak{A}, a_\eta)_{\eta<\xi}$ is α-saturated.*

(iv). *If $\mathfrak{A}$ is α-saturated, then every reduct of $\mathfrak{A}$ is α-saturated.*

Parts (ii) *and* (iii) *require that α be infinite.*

The proof of Proposition 5.1.1 is left to the reader.

PROPOSITION 5.1.2.

(i). *If $\mathfrak{A}$ is α-saturated and infinite, then $|A| \geqslant \alpha$.*

(ii). *$\mathfrak{A}$ is finite if and only if $\mathfrak{A}$ is α-saturated for all α.*

PROOF. (i). Suppose $|A| < \alpha$. Let $a \in {}^{|A|}A$ be an enumeration of all the elements of A, and let $\Sigma(v)$ be a type such that

$$\Sigma(v) \supset \{v \not\equiv c_a : a \in A\}.$$

Σ is a type of $(\mathfrak{A}, a)_{a\in A}$ and Σ cannot be realized in $(\mathfrak{A}, a)_{a\in A}$.

(ii). If $\mathfrak{A}$ is infinite, then by (i), $\mathfrak{A}$ is not $|A|^+$-saturated. On the other hand, suppose that $\mathfrak{A}$ is finite. Let $X \subset A$, and let $\Sigma(v)$ be a type of $(\mathfrak{A}, a)_{a\in X}$. If $\Sigma(v)$ is not realized in $\mathfrak{A}$, then for every $b \in A$, there is a formula $\sigma_b \in \Sigma$ such that $(\mathfrak{A}, a)_{a\in X} \models \neg\sigma_b[b]$. This shows that the finite subset $\{\sigma_b : b \in A\}$ of Σ is not consistent in $(\mathfrak{A}, a)_{a\in X}$. ⊣

We next restate an earlier result from Chapter 4.

LEMMA 5.1.3. *Suppose $||\mathscr{L}|| \leqslant \alpha$ and $\omega \leqslant |A| \leqslant 2^\alpha$. Then there exists an elementary extension $\mathfrak{B}$ of $\mathfrak{A}$ of power 2^α such that for every $X \subset A$ of power α, $(\mathfrak{B}, a)_{a\in X}$ realizes every type $\Sigma(v)$ of $(\mathfrak{A}, a)_{a\in X}$.*

PROOF. This lemma already follows from Corollary 4.3.14. The proof there requires the ultrapower construction modulo a regular ultrafilter. While the proof given there is not difficult, the student should realize that the compactness theorem is already adequate for this construction. For this reason, we sketch an alternate proof.

Since $|A| \leqslant 2^\alpha$, the number of subsets $X \subset A$ of power α is at most 2^α. The language $\mathscr{L}_X = \mathscr{L} \cup \{c_a : a \in X\}$ still has at most α symbols. Whence the total number of types $\Sigma(v)$ of $\mathscr{L}_X$ is at most 2^α. For each $X \subset A$ of power α and each type $\Sigma(v)$ of $(\mathfrak{A}, a)_{a\in X}$, we introduce a new constant $c_{X\Sigma}$. Let

$T =$ the elementary diagram of $\mathfrak{A}$ in $\mathscr{L}_A$ plus all the sentences $\Sigma(c_{X\Sigma})$ for every $X \subset A$ of power α and every type $\Sigma(v)$ of $(\mathfrak{A}, a)_{a\in X}$.

We can show that T is consistent and has an infinite model. Since T has at most 2^α symbols, T has a model of power 2^α. The reduct of this model to $\mathscr{L}$ is the desired model $\mathfrak{B}$. ⊣

LEMMA 5.1.4 (Existence of α^+-saturated Models of Power 2^α). *Suppose* $\|\mathscr{L}\| \leq \alpha$ *and* $\omega \leq |A| \leq 2^\alpha$. *Then there is an* α^+*-saturated elementary extension of* $\mathfrak{A}$ *of power* 2^α.

PROOF. We construct an elementary chain $\mathfrak{B}_\xi$, $\xi < 2^\alpha$ of length 2^α such that

(1) each $\mathfrak{B}_\xi$ is an elementary extension of $\mathfrak{A}$ of power 2^α;

(2) for every $X \subset B_\xi$ of power α, $(\mathfrak{B}_{\xi+1}, a)_{a\in X}$ realizes every type $\Sigma(v)$ of $(\mathfrak{B}_\xi, a)_{a\in X}$.

We take $\mathfrak{B}_0$ to be the model constructed in Lemma 5.1.3. If η is a limit ordinal, $0 < \eta < 2^\alpha$, we let $\mathfrak{B}_\eta = \bigcup_{\xi<\eta}\mathfrak{B}_\xi$. If $\eta = \xi+1$, we let $\mathfrak{B}_{\xi+1}$ be the elementary extension of $\mathfrak{B}_\xi$ given by Lemma 5.1.3 with $\mathfrak{A}$ replaced by $\mathfrak{B}_\xi$. Let $\mathfrak{B} = \bigcup_{\eta<2^\alpha}\mathfrak{B}_\eta$. Clearly $\mathfrak{B}$ is an elementary extension of $\mathfrak{A}$ of power 2^α. Let $X \subset B$ be of power α, and let $\Sigma(v)$ be a type of $(\mathfrak{B}, a)_{a\in X}$. Since 2^α has cofinality greater than α, there is some $\xi < 2^\alpha$ such that $X \subset B_\xi$. Now $\mathfrak{B}_\xi$ is an elementary submodel of $\mathfrak{B}$. So $\Sigma(v)$ is also a type of $(\mathfrak{B}_\xi, a)_{a\in X}$. By our construction, some $b \in B_{\xi+1}$ realizes Σ, i.e.,

$$(\mathfrak{B}_{\xi+1}, a)_{a\in X} \models \Sigma[b].$$

Since $\mathfrak{B}_{\xi+1} \prec \mathfrak{B}$, we see that $(\mathfrak{B}, a)_{a\in X} \models \Sigma[b]$. ⊣

PROPOSITION 5.1.5.

(i). (GCH.) *Let* $\|\mathscr{L}\| \leq \alpha$. *Then every theory* T *in* $\mathscr{L}$ *having infinite models has a saturated model in each regular power* $\beta > \alpha$.

(ii). *Let* $\|\mathscr{L}\| \leq \alpha$. *Then every theory* T *in* $\mathscr{L}$ *having infinite models has a saturated model in each inaccessible power* $\gamma > \alpha$.

PROOF. (i). The case $\beta = \gamma^+$ is immediate. In the case where β is a regular limit cardinal, the proof follows from the argument given below for (ii) and the GCH.

(ii). Let $\mathfrak{A}_\alpha$ be a model of T of power α. By Lemma 5.1.4, we construct an elementary chain $\mathfrak{A}_\beta$, where β runs over all cardinals between α and γ, such that

for limit cardinals β, $\mathfrak{A}_\beta = \bigcup_{\delta<\beta}\mathfrak{A}_\delta$;
$\mathfrak{A}_{\beta^+}$ is a β^+-saturated model of power 2^β.

Then the union $\mathfrak{A}_\gamma = \bigcup_{\beta<\gamma} \mathfrak{A}_\beta$ is a γ-saturated model of T of power γ, because

$$|A_\gamma| = \bigcup_{\beta<\gamma} 2^\beta = \gamma,$$

and each subset $X \subset A_\gamma$ of power $|X| < \gamma$ is included in some A_{β^+}, where $|X| \leqslant \beta$. ⊣

In one of the exercises, the student is asked to prove that the GCH cannot be eliminated from Proposition 5.1.5(i). Thus the existence of infinite saturated models is, in general, limited to cases where we assume either the GCH or the existence of inaccessible cardinals. Of course, as we have already noted at the beginning of this section, very special theories T may have infinite (countably) saturated models. It turns out that for many purposes the notion of a saturated model can be replaced by that of a special model. For example, every saturated model is special, but not all special models are saturated. More importantly, the existence of special models does not require the GCH or inaccessible cardinals. We next take up the study of special models.

A model $\mathfrak{A}$ is said to be *special* iff $\mathfrak{A}$ is the union of an elementary chain $\mathfrak{A}_\beta$, $\beta < |A|$, where β is a cardinal and each $\mathfrak{A}_\beta$ is β^+-saturated. The chain of models $\mathfrak{A}_\beta$, $\beta < |A|$, is called a *specializing chain* of $\mathfrak{A}$. Note that β is always a cardinal smaller than $|A|$, and nothing is said about the cardinal of $\mathfrak{A}_\beta$. Of course, it follows from Proposition 5.1.2 that either $\mathfrak{A}_\beta$ is finite or $|A_\beta| \geqslant \beta^+$.

PROPOSITION 5.1.6.

(i). *Every saturated model is special.*

(ii). *Every finite model is special.*

(iii). *A model of power α^+ is saturated if and only if it is special.*

(iv). *If α is a regular limit cardinal, then a model of power α is saturated if and only if it is special.*

(v). *If $\mathfrak{A}$ is special, then every reduct of $\mathfrak{A}$ is special.*

(vi). (GCH.) *If T has an infinite model, it has a special model in each power $\alpha > ||\mathscr{L}||$.*

PROOF. (i). For each cardinal $\beta < |A|$, define $\mathfrak{A}_\beta = \mathfrak{A}$. Then, since $\mathfrak{A}$ is $|A|$-saturated, $\mathfrak{A}_\beta$ is β^+-saturated and $\bigcup_{\beta<\alpha} \mathfrak{A}_\beta = \mathfrak{A}$.

(ii). This follows from (i) and Proposition 5.1.2.

(iii). Let $\mathfrak{A}$ be a model of power α^+. If $\mathfrak{A}$ is saturated, then, by (i), $\mathfrak{A}$ is special. Suppose $\mathfrak{A}$ is special and let $\mathfrak{A}_\beta$, $\beta < \alpha^+$, be a specializing chain of $\mathfrak{A}$. This chain has a last member $\mathfrak{A}_\alpha$, which must be $\mathfrak{A}$. But $\mathfrak{A}_\alpha$ is α^+-saturated, so $\mathfrak{A}$ is.

(iv). Let $\mathfrak{A}_\beta$, $\beta < \alpha$, be a specializing chain of $\mathfrak{A}$. If $X \subset A$ and $|X| < \alpha$, then $X \subset A_\beta$ and $|X| \leqslant \beta$ for some $\beta < \alpha$, and every type $\Sigma(v)$ of $(\mathfrak{A}, a)_{a \in X}$ is realized in $\mathfrak{A}_\beta$.

(v). This follows from Proposition 5.1.1 (iv).

(vi). Apply Proposition 5.1.5 (i) to get a specializing chain. ⊣

Let α be an infinite cardinal. α^* is defined to be the cardinal sum $\bigcup_{\beta<\alpha} 2^\beta$, where β ranges over cardinals less than α and 2^β is cardinal exponentiation of 2 to the power β. We shall prove that special models exist in each power $\alpha = \alpha^*$. But first we shall state some simple properties of the *-operation.

PROPOSITION 5.1.7.

(i). $\alpha = \alpha^*$ *if and only if* $2^\beta \leqslant \alpha$ *for all* $\beta < \alpha$.

(ii). *There are arbitrarily large cardinals* α *such that* $\alpha = \alpha^*$.

(iii). $(\alpha^+) = (\alpha^+)^*$ *if and only if* $2^\alpha = \alpha^+$.

(iv). *The* GCH *is equivalent to* $\alpha = \alpha^*$ *for all infinite cardinals* α.

PROPOSITION 5.1.8 (Existence of Special Models). *Suppose that* $\alpha = \alpha^*$, $\|\mathscr{L}\| < \alpha$, *and* $\omega \leqslant |A| < \alpha$. *Then there is a special elementary extension* $\mathfrak{B}$ *of* $\mathfrak{A}$ *of power* α.

PROOF. If $\alpha = \gamma^+$, then by Proposition 5.1.7 (iii), $2^\gamma = \gamma^+$. So, by Lemma 5.1.4, $\mathfrak{A}$ has a saturated elementary extension of power $\gamma^+ = \alpha$. So let us assume that α is a limit cardinal. This means that if $\beta < \alpha$, then $\beta^+ < \alpha$ and $2^\beta \leqslant \alpha$. Let $\gamma = \|\mathscr{L}\| \cup |A|$. We construct an elementary chain of models $\mathfrak{B}_\beta$, β a cardinal less than α, as follows (we use Lemma 5.1.4 repeatedly): Let $\mathfrak{B}_\gamma$ be any γ^+-saturated elementary extension of $\mathfrak{A}$ of power 2^γ. For any cardinal $\delta < \gamma$, let $\mathfrak{B}_\delta = \mathfrak{B}_\gamma$. Let $\gamma \leqslant \beta < \alpha$. Given $\mathfrak{B}_\beta$ of power at most 2^β, let $\mathfrak{B}_{\beta^+}$ be any β^{++}-saturated elementary extension of power $2^{(\beta^+)}$. Given $\mathfrak{B}_\delta$, $\delta < \beta$, where β is a limit cardinal less than α, we first take the union $\mathfrak{B}' = \bigcup_{\delta<\beta} \mathfrak{B}_\delta$, which is an infinite model of power at most 2^β; then we let $\mathfrak{B}_\beta$ be any β^+-saturated elementary extension of $\mathfrak{B}'$ of power 2^β. Now the chain $\mathfrak{B}_\beta$, $\beta < \alpha$, is a specializing chain of the model $\bigcup_{\beta<\alpha} \mathfrak{B}_\beta$ with power α. ⊣

We ask the reader to show as an exercise that with a little more work we can extend the inequalities $\omega \leqslant |A| < \alpha$ in Proposition 5.1.8 to $\omega \leqslant |A| \leqslant \alpha$.

Now that we have established the existence of special models, we shall next consider them from the point of view of the notions of α-homogeneous models and α-universal models. Recall that we have already defined α-universal models in Section 4.3 and ω-homogeneous models in Section 3.2. We now introduce the corresponding notion of an α-homogeneous model for an infinite cardinal α. A model $\mathfrak{A}$ is *α-homogeneous* iff for all $\xi < \alpha$, $a \in {}^{\xi}A$, $b \in {}^{\xi}A$ and $c \in A$, if $(\mathfrak{A}, a_\eta)_{\eta<\xi} \equiv (\mathfrak{A}, b_\eta)_{\eta<\xi}$, then there is a $d \in A$ such that $(\mathfrak{A}, a_\eta, c)_{\eta<\xi} \equiv (\mathfrak{A}, b_\eta, d)_{\eta<\xi}$. Let us also recall that a model $\mathfrak{A}$ is *α-universal* iff for every model $\mathfrak{B} \equiv \mathfrak{A}$ of power smaller than α, $\mathfrak{B}$ is elementarily embedded in $\mathfrak{A}$. Given a class K of models, we say that $\mathfrak{A}$ is *α-universal with respect to K* iff every model $\mathfrak{B} \in K$ of power smaller than α is elementarily embedded in $\mathfrak{A}$. Finally, we say that $\mathfrak{A}$ is *homogeneous* (or *universal*) iff $\mathfrak{A}$ is $|A|$-homogeneous (or $|A|$-universal).

The next proposition is an analogue of Proposition 5.1.1.

Proposition 5.1.9. *Let α be infinite.*

(i). *$\mathfrak{A}$ is α-homogeneous (α-universal) if and only if $\mathfrak{A}$ is β-homogeneous (β-universal) for all $\beta \leqslant \alpha$.*

(ii). *If α is a limit cardinal, then $\mathfrak{A}$ is α-homogeneous (α-universal) if and only if $\mathfrak{A}$ is β-homogeneous (β-universal) for all $\beta < \alpha$.*

(iii). *If $\mathfrak{A}$ is α-saturated, then $\mathfrak{A}$ is α-homogeneous.*

(iv). *$\mathfrak{A}$ is α-homogeneous if and only if for all $\xi < \alpha$ and all $a \in {}^{\xi}A$, $(\mathfrak{A}, a_\eta)_{\eta<\xi}$ is α-homogeneous.*

The proofs are left as an exercise.

Lemma 5.1.10. *Suppose that $\mathfrak{A}$ is α-saturated, $\mathfrak{A} \equiv \mathfrak{B}$ and $b \in {}^{\alpha}B$. Then there exists an $a \in {}^{\alpha}A$ such that $(\mathfrak{A}, a_\xi)_{\xi<\alpha} \equiv (\mathfrak{B}, b_\xi)_{\xi<\alpha}$.*

Proof. We find the sequence a_ξ, $\xi < \alpha$, by transfinite recursion on $\xi < \alpha$. Suppose that for all $\eta < \xi$ we have $a_\eta \in A$ such that

$$(\mathfrak{A}, a_\eta)_{\eta<\xi} \equiv (\mathfrak{B}, b_\eta)_{\eta<\xi}.$$

Let $\Sigma(v)$ be the type of the element b_ξ in $(\mathfrak{B}, b_\eta)_{\eta<\xi}$. Clearly, Σ is a type of $(\mathfrak{A}, a_\eta)_{\eta<\xi}$; whence Σ is realized by an element, say $a_\xi \in A$. It follows that

$$(\mathfrak{A}, a_\eta)_{\eta\leqslant\xi} \equiv (\mathfrak{B}, b_\eta)_{\eta\leqslant\xi}.$$

We have found the element a_ξ and our recursion is complete. The conclusion follows. ⊣

The two-sided version of Lemma 5.1.10 is as follows:

LEMMA 5.1.11 (Back and Forth Lemma). *Suppose that* α *is infinite,* $\mathfrak{A}$ *and* $\mathfrak{B}$ *are both* α*-saturated, and* $\mathfrak{A} \equiv \mathfrak{B}$. *Let* $a \in {}^{\alpha}A$ *and* $b \in {}^{\alpha}B$. *Then there are* $\bar{a} \in {}^{\alpha}A$ *and* $\bar{b} \in {}^{\alpha}B$ *such that*

$$\text{range}\,(a) \subset \text{range}\,(\bar{a}),$$

$$\text{range}\,(b) \subset \text{range}\,(\bar{b}),$$

$$(\mathfrak{A}, \bar{a}_\xi)_{\xi<\alpha} \equiv (\mathfrak{B}, \bar{b}_\xi)_{\xi<\alpha}.$$

PROOF. We represent each ordinal $\xi < \alpha$ uniquely as a sum $\xi = \lambda + n$, where λ is a limit ordinal and $n \in \omega$. ξ is said to be *even* if n is even; otherwise ξ is *odd*. We wish to find two sequences $\bar{a} \in {}^{\alpha}A$, $\bar{b} \in {}^{\alpha}B$ such that for all $\xi < \alpha$

(a). For all $\eta < \xi$, if $\eta = \lambda + 2n$ is even, then $\bar{a}_\eta = a_{\lambda+n}$.

(b). For all $\eta < \xi$, if $\eta = \lambda + (2n+1)$ is odd, then $\bar{b}_\eta = b_{\lambda+n}$.

(c). $(\mathfrak{A}, \bar{a}_\eta)_{\eta<\xi} \equiv (\mathfrak{B}, \bar{b}_\eta)_{\eta<\xi}$.

Suppose that $\xi < \alpha$ and $\bar{a}_\eta$, $\bar{b}_\eta$, $\eta < \xi$, have been found so that (a), (b) and (c) hold for ξ. We now find $\bar{a}_\xi$ and $\bar{b}_\xi$. if $\xi = \lambda + 2n$ is even, first let $\bar{a}_\xi = a_{\lambda+n}$. Next let $\Sigma(v)$ be the type of $\bar{a}_\xi$ in $(\mathfrak{A}, \bar{a}_\eta)_{\eta<\xi}$. By (c), Σ is a type of $(\mathfrak{B}, \bar{b}_\eta)_{\eta<\xi}$. So Σ is realized by some element $\bar{b}_\xi \in B$. Clearly,

$$(\mathfrak{A}, \bar{a}_\eta)_{\eta\leqslant\xi} \equiv (\mathfrak{B}, \bar{b}_\eta)_{\eta\leqslant\xi}.$$

The case when $\xi = \omega \cdot \lambda + (2n+1)$ is odd is treated similarly; we let $\bar{b}_\xi = b_{\omega\cdot\lambda+n}$ and find $\bar{a}_\xi$. This defines $\bar{a}$ and $\bar{b}$ by transfinite recursion. The sequences $\bar{a} \in {}^{\alpha}A$ and $\bar{b} \in {}^{\alpha}B$ will satisfy the conclusions of the lemma. $\dashv$

The above two lemmas will now be our tools for the rest of this section.

THEOREM 5.1.12. α*-saturated models are* α^+*-universal.*

PROOF. Let $\mathfrak{A}$ be α-saturated and let $\mathfrak{B}$ be any equivalent model of power $\leqslant \alpha$. Let $b \in {}^{\alpha}B$ be an enumeration of all the elements of B. By Lemma 5.1.10, there is an $a \in {}^{\alpha}A$ such that

$$(\mathfrak{A}, a_\xi)_{\xi<\alpha} \equiv (\mathfrak{B}, b_\xi)_{\xi<\alpha}.$$

From this and Proposition 3.1.3, $\mathfrak{B} \precsim \mathfrak{A}$. $\dashv$

THEOREM 5.1.13 (Uniqueness of Saturated Models). *Let* $\mathfrak{A}$ *and* $\mathfrak{B}$ *be equivalent saturated models of the same power. Then* $\mathfrak{A} \cong \mathfrak{B}$.

PROOF. Let $|A| = |B| = \alpha$ and let $a \in {}^\alpha A$, $b \in {}^\alpha B$ be enumerations of A and B, respectively. By Lemma 5.1.11, there are "extensions" $\bar{a} \in {}^\alpha A$, $\bar{b} \in {}^\alpha B$ of a and b such that

$$(\mathfrak{A}, \bar{a}_\xi)_{\xi<\alpha} \equiv (\mathfrak{B}, \bar{b}_\xi)_{\xi<\alpha}.$$

Since $\bar{a}$ and $\bar{b}$ still enumerate A and B, the isomorphism of $\mathfrak{A}$ and $\mathfrak{B}$ follows. ⊣

THEOREM 5.1.14 (α-saturated models are exactly the α-homogeneous and α-universal models). *Suppose that $||\mathscr{L}|| < \alpha$. The following three conditions on $\mathfrak{A}$ are equivalent*:

(i). $\mathfrak{A}$ *is α-saturated.*

(ii). $\mathfrak{A}$ *is α-homogeneous and α^+-universal.*

(iii). $\mathfrak{A}$ *is α-homogeneous and α-universal.*

PROOF. The implication (i) ⇒ (ii) follows from Proposition 5.1.9 (iii) and Theorem 5.1.12.

(iii) trivially follows from (ii).

We now prove (iii) ⇒ (i). Assume $\mathfrak{A}$ is α-homogeneous and α-universal. Let $\xi < \alpha$, $a \in {}^\xi A$, and let $\Sigma(v)$ be a type of $(\mathfrak{A}, a_\eta)_{\eta<\xi}$. Let T be the set of sentences consisting of the theory of $(\mathfrak{A}, a_\eta)_{\eta<\xi}$ in $\mathscr{L} \cup \{c_\eta : \eta < \xi\}$ plus all sentences $\Sigma(c)$. Then T is clearly consistent in a language of power smaller than α. Let $\mathfrak{B}'$ be a model of T. We may consider $\mathfrak{B}' = (\mathfrak{B}, b_\eta, d)_{\eta<\xi}$, where $\mathfrak{B}$ is the $\mathscr{L}$-reduct of $\mathfrak{B}'$. Then

$$(1) \qquad (\mathfrak{A}, a_\eta)_{\eta<\xi} \equiv (\mathfrak{B}, b_\eta)_{\eta<\xi}.$$

Since $\mathfrak{A}$ is α-universal and $|B| < \alpha$, the model $\mathfrak{B}$ can be elementarily embedded into $\mathfrak{A}$. To avoid the introduction of new letters, let us assume that $\mathfrak{B}$ itself is an elementary submodel of $\mathfrak{A}$. So we can replace (1) by

$$(\mathfrak{A}, a_\eta)_{\eta<\xi} \equiv (\mathfrak{A}, b_\eta)_{\eta<\xi}.$$

Since $\mathfrak{A}$ is α-homogeneous, there is a $c \in A$ such that

$$(\mathfrak{A}, a_\eta, c)_{\eta<\xi} \equiv (\mathfrak{A}, b_\eta, d)_{\eta<\xi} \equiv (\mathfrak{B}, b_\eta, d)_{\eta<\xi}.$$

So c realizes the type Σ in $(\mathfrak{A}, a_\eta)_{\eta<\xi}$. ⊣

Note that the equivalence of Theorem 5.1.14 (i) and (ii) still holds if $||\mathscr{L}|| = \alpha$.

COROLLARY 5.1.15 (Saturated models are homogeneous and universal). *If $||\mathscr{L}|| < |A|$, then $\mathfrak{A}$ is saturated if and only if it is homogeneous and universal.*

This follows immediately from Theorem 5.1.14.

We now turn to special models.

THEOREM 5.1.16. *If $\mathfrak{A}$ is special, then $\mathfrak{A}$ is $|A|^+$-universal.*

PROOF. Let $\alpha = |A|$. If α is not a limit cardinal, then $\mathfrak{A}$ is saturated, so $\mathfrak{A}$ is α^+-universal. So assume α is a limit cardinal. Let $\mathfrak{A}_\beta$, $\beta < \alpha$, be a specializing chain of $\mathfrak{A}$. Let $\mathfrak{B}$ be any equivalent model of power $\leqslant \alpha$, and let $b \in {}^\alpha B$ be an enumeration of B. We now find an $a \in {}^\alpha A$ such that for all $\beta \leqslant \alpha$, $a \mid \beta \in {}^\beta A_\beta$ and

$$(1) \qquad (\mathfrak{A}_\beta, a_\xi)_{\xi<\beta} \equiv (\mathfrak{B}, b_\xi)_{\xi<\beta}.$$

Suppose we found $a \mid \delta$ for all $\delta < \beta$. If $\beta = \gamma^+$, then from (1) we have

$$(\mathfrak{A}_\beta, a_\xi)_{\xi<\gamma} \equiv (\mathfrak{B}, b_\xi)_{\xi<\gamma}.$$

Applying Lemma 5.1.10 to $(\mathfrak{A}_\beta, a_\xi)_{\xi<\gamma}$, we can find elements a_ξ, $\gamma \leqslant \xi < \beta$, in A_β such that (1) holds. If β is a limit cardinal, then obviously each a_ξ, $\xi < \beta$, is in A_β and again (1) holds. Once $a \in {}^\alpha A$ is found, then $(\mathfrak{A}, a_\xi)_{\xi<\alpha} \equiv (\mathfrak{B}, b_\xi)_{\xi<\alpha}$. Since b enumerates B, this proves the theorem. ⊣

THEOREM 5.1.17 (Uniqueness of Special Models). *If $\mathfrak{A}$ and $\mathfrak{B}$ are equivalent special models of the same power, then $\mathfrak{A} \cong \mathfrak{B}$.*

PROOF. The proof uses the back and forth technique applied to the specializing chains $\mathfrak{A}_\beta$, $\beta < \alpha$, and $\mathfrak{B}_\beta$, $\beta < \alpha$. We replace the element by element construction of Lemma 5.1.11 by a block by block construction, using Lemma 5.1.10. We leave the details to the reader. ⊣

We conclude this section with a study of homogeneous models.

LEMMA 5.1.18. *Let $\beta \leqslant \alpha$, let $\mathfrak{A}$ be α-homogeneous, and for each $\xi < \beta$, let $a^\xi \in {}^\xi A$. Suppose that for all $\xi \leqslant \eta < \beta$,*

$$(\mathfrak{A}, a^\xi_\lambda)_{\lambda<\xi} \equiv (\mathfrak{A}, a^\eta_\lambda)_{\lambda<\xi}.$$

Then there is an $a \in {}^\beta A$ such that for all $\xi < \beta$,

$$(\mathfrak{A}, a_\lambda)_{\lambda<\xi} \equiv (\mathfrak{A}, a^\xi_\lambda)_{\lambda<\xi}.$$

PROOF. Suppose that for all $\lambda < \xi < \beta$, we have constructed $a_\lambda \in A$ such that

(1) for all $\eta \leqslant \xi$, $(\mathfrak{A}, a_\lambda)_{\lambda<\eta} \equiv (\mathfrak{A}, a_\lambda^\eta)_{\lambda<\eta}$.

We wish to extend the sequence a_λ, $\lambda < \xi$, by one more element. If $\xi = \zeta+1$, then

(2) $$(\mathfrak{A}, a_\lambda)_{\lambda\leqslant\zeta} \equiv (\mathfrak{A}, a_\lambda^\xi)_{\lambda\leqslant\zeta}.$$

By hypothesis,

(3) $$(\mathfrak{A}, a_\lambda^\xi)_{\lambda\leqslant\zeta} \equiv (\mathfrak{A}, a_\lambda^{\xi+1})_{\lambda\leqslant\zeta}.$$

Putting (2) and (3) together and using the α-homogeneity of $\mathfrak{A}$, we see that there is an $a_\xi \in A$ such that

(4) $$(\mathfrak{A}, a_\lambda)_{\lambda<\xi+1} \equiv (\mathfrak{A}, a_\lambda^{\xi+1})_{\lambda<\xi+1}.$$

Suppose ξ is a limit ordinal. Let $a^{\xi+1}$ be given. We claim that

(5) $$(\mathfrak{A}, a_\lambda^{\xi+1})_{\lambda<\xi} \equiv (\mathfrak{A}, a_\lambda)_{\lambda<\xi}.$$

This is the usual situation when formulas have finite lengths. Of course, we make heavy use of the fact that ξ is a limit ordinal; whence, if (1) holds for all $\xi' < \xi$, then (1) holds also for ξ. So $(\mathfrak{A}, a_\lambda)_{\lambda<\xi} \equiv (\mathfrak{A}, a_\lambda^\xi)_{\lambda<\xi}$ and from this (5) follows. By (5) and α-homogeneity, we again can find $a_\xi \in A$ such that (4) holds. ⊣

LEMMA 5.1.19. *Suppose that $\mathfrak{A}$ is α-homogeneous and every type realized in $\mathfrak{B}$ is realized in $\mathfrak{A}$, i.e., for all $b_1, \ldots, b_n \in B$ $(n \in \omega)$, there are $a_1, \ldots, a_n \in A$ such that $(\mathfrak{A}, a_1 \ldots a_n) \equiv (B, b_1 \ldots b_n)$. Then for each $b \in {}^\alpha B$, there is an $a \in {}^\alpha A$ such that $(\mathfrak{A}, a_\xi)_{\xi<\alpha} \equiv (\mathfrak{B}, b_\xi)_{\xi<\alpha}$.*

PROOF. We prove the following by induction on infinite cardinals $\beta \leqslant \alpha$:

(1) for every $b \in {}^\beta B$, there is an $a \in {}^\beta A$ such that $(\mathfrak{B}, b_\xi)_{\xi<\beta} \equiv (\mathfrak{A}, a_\xi)_{\xi<\beta}$.

(1) is true for $\beta = \omega$. This can be seen as follows: Let $b \in {}^\omega B$. For each $n \in \omega$, let b^n be the restriction of b to n. By our hypothesis, there are $a^n \in {}^n A$ such that

$$(\mathfrak{B}, b_m)_{m<n} \equiv (\mathfrak{A}, a_m^n)_{m<n} \quad \text{for all} \quad n.$$

Now the sequences a^n, $n \in \omega$, satisfy the hypothesis of Lemma 5.1.18. Whence there is an $a \in {}^\omega A$ such that

$$(\mathfrak{A}, a_m)_{m<n} \equiv (\mathfrak{A}, a_m^n)_{m<n} \quad \text{for all} \quad n.$$

This gives $(\mathfrak{A}, a_m)_{m<\omega} \equiv (\mathfrak{B}, b_m)_{m<\omega}$. The other two cases of $\beta = \gamma^+$ or β

a limit cardinal are entirely analogous to the case $\beta = \omega$. We only need to note that if (1) holds for an infinite cardinal β, then by rearranging the terms of the sequences, we have the necessary hypothesis for Lemma 5.1.18. Namely,

for every ordinal $\xi < \beta^+$ and every $b \in {}^\xi B$, there is an $a \in {}^\xi A$ such that $(\mathfrak{B}, b_\eta)_{\eta<\xi} \equiv (\mathfrak{A}, a_\eta)_{\eta<\xi}$. ⊣

Lemma 5.1.20. *Suppose that α is infinite; both $\mathfrak{A}$ and $\mathfrak{B}$ are α-homogeneous and $\mathfrak{A}$ and $\mathfrak{B}$ realize the same types. Then for all $a \in {}^\alpha A$ and $b \in {}^\alpha B$, there are $\bar{a} \in {}^\alpha A$ and $\bar{b} \in {}^\alpha B$ such that*

$$\text{range}(a) \subset \text{range}(\bar{a}),$$
$$\text{range}(b) \subset \text{range}(\bar{b}),$$
$$(\mathfrak{A}, \bar{a}_\xi)_{\xi<\alpha} \equiv (\mathfrak{B}, \bar{b}_\xi)_{\xi<\alpha}.$$

Proof. This is proved in exactly the same manner as the proof of Lemma 5.1.11. We add, however, the new twist introduced in Lemma 5.1.19. The proper form of induction is the following: For all infinite cardinals $\beta \leqslant \alpha$, we have

for all $a \in {}^\beta A$, $b \in {}^\beta B$, there are $\bar{a} \in {}^\beta A$, $\bar{b} \in {}^\beta B$ such that range $(a) \subset \text{range}(\bar{a})$, range$(b) \subset$ range$(\bar{b})$ and $(\mathfrak{A}, \bar{a}_\xi)_{\xi<\beta} \equiv (\mathfrak{B}, \bar{b}_\xi)_{\xi<\beta}$.

There are no special difficulties involved, so we leave the proof as an exercise. ⊣

Proposition 5.1.21. *Every α-homogeneous model is α^+-universal with respect to the class K of models defined by*

$\mathfrak{B} \in K$ iff every type realized in $\mathfrak{B}$ is realized in $\mathfrak{A}$.

Proof. Let $\mathfrak{B} \in K$, $|B| \leqslant \alpha$, and let $b \in {}^\alpha B$ be an enumeration of B. By Lemma 5.1.19, there is an $a \in {}^\alpha A$ such that

$$(\mathfrak{B}, b_\xi)_{\xi<\alpha} \equiv (\mathfrak{A}, a_\xi)_{\xi<\alpha}.$$

Whence $\mathfrak{B} \precsim \mathfrak{A}$. ⊣

Theorem 5.1.22. *Suppose that $\mathfrak{A}$ and $\mathfrak{B}$ are two homogeneous models of the same power α realizing exactly the same types. Then $\mathfrak{A} \cong \mathfrak{B}$.*

Proof. Let α be infinite and let $a \in {}^\alpha A$, $b \in {}^\alpha B$ be enumerations of A and B, respectively. By Lemma 5.1.20, there are $\bar{a} \in {}^\alpha A$, $\bar{b} \in {}^\alpha B$ such that

$$(\mathfrak{A}, \bar{a}_\xi)_{\xi<\alpha} \equiv (\mathfrak{B}, \bar{b}_\xi)_{\xi<\alpha},$$

and $\bar{a}$ and $\bar{b}$ are again enumerations of A and B. It follows that $\mathfrak{A} \cong \mathfrak{B}$. If α is finite, then, since $\mathfrak{A}$ and $\mathfrak{B}$ realize the same types, they are elementarily equivalent and therefore isomorphic. ⊣

Exercises

5.1.1. The union of an elementary chain $\mathfrak{A}_\eta$, $\eta < \alpha$, of α-saturated models is again α-saturated, if α is regular.

5.1.2. Find a union of a countable elementary chain of ω_1-saturated models which is not ω_1-saturated.

5.1.3. Let $\mathfrak{A}$ be β^+-saturated, and let $X \subset A$ be such that $||\mathscr{L}|| \leqslant |X| \leqslant \beta$. Then there exists an $|X|^+$-saturated elementary submodel of $\mathfrak{A}$ of power at most $2^{|X|}$ containing the set X. Do the same exercise with β^+-saturated and $|X|^+$-saturated replaced by β^+-homogeneous and $|X|^+$-homogeneous.

5.1.4. Let $||\mathscr{L}|| = \alpha$. Find a theory T of $\mathscr{L}$ having infinite models such that T has a saturated model of power α^+ if and only if $\alpha^+ = 2^\alpha$.

5.1.5. If α is weakly inaccessible, then every special model of power α is saturated.

5.1.6. Suppose $\mathfrak{A}$ is special of power α. Then show that $\mathfrak{A}$ has a specializing chain $\mathfrak{A}_\beta$, $\beta < \alpha$, such that $|A_\beta| \leqslant 2^\beta$ whenever $||\mathscr{L}|| \leqslant \beta < \alpha$.

5.1.7. Suppose $||\mathscr{L}|| < \alpha$, $\alpha = \alpha^* = |A| = |A'|$, $\mathfrak{A}$ has a specializing chain $\mathfrak{A}_\beta$, $\beta < \alpha$, and $\mathfrak{A} \prec \mathfrak{A}'$. Prove that there are a special elementary extension $\mathfrak{B}$ of $\mathfrak{A}'$ of power α and a specializing chain $\mathfrak{B}_\beta$, $\beta < \alpha$, of $\mathfrak{B}$ such that $\mathfrak{A}_\beta \prec \mathfrak{B}_\beta$ for all $\beta < \alpha$.

5.1.8. Show that if $\mathfrak{A}$ is α-universal and $||\mathscr{L}|| < \alpha$, then every reduct of $\mathfrak{A}$ is again α-universal.

5.1.9*. Our definition of α-homogeneous models is such that in the case where α is a singular cardinal it is difficult to show that there are α-homogeneous models of power α. Suppose α is an infinite cardinal. We say that a model $\mathfrak{A}$ of power α is *weakly homogeneous* iff $\mathfrak{A}$ is the union of an elementary chain, $\mathfrak{A} = \bigcup_{\beta<\alpha} \mathfrak{A}_\beta$, such that each $\mathfrak{A}_\beta$ is β^+-homogeneous. Every special model $\mathfrak{A}$ is weakly homogeneous. If α is regular, then for models of power α weak homogeneity is equivalent to homogeneity. Prove analogues of Proposition 5.1.21 and Theorem 5.1.22 (and the associated necessary lemmas) for weakly homogeneous models of power α with $||\mathscr{L}|| < \alpha$.

5.1.10*. Let $\mathscr{L}$ be a countable language. Prove that if a theory T in $\mathscr{L}$ has at most a countable number of countably homogeneous models, then it has in each power α at most a countable number of (weakly) homogeneous models of power α.

[*Hint*: Show that every homogeneous model of T of power α realizes at most a countable number of types.]

5.1.11* (GCH). Let $||\mathscr{L}|| = \omega$ and let $\mathfrak{A}$ be a weakly homogeneous model of power β. Prove that for every cardinal γ, $\omega_1 \leqslant \gamma < \beta$, there exists an elementary submodel $\mathfrak{B}$ of $\mathfrak{A}$ such that $\mathfrak{B}$ is weakly homogeneous and $\mathfrak{B}$ has power γ.

5.1.12 (GCH). Suppose $||\mathscr{L}|| = \omega$ and $\omega < \alpha < \beta$. Let T be a theory in $\mathscr{L}$. Then the number of (nonisomorphic) weakly homogeneous models of T of power α is greater than or equal to the number of weakly homogeneous models of T of power β.

5.1.13* (GCH). Give an example of a complete theory which has three homogeneous models of power ω_0, two homogeneous models of power ω_1, and one homogeneous model of power ω_2.

5.1.14. Let $||\mathscr{L}|| = \omega$. Show that in each power α there are at most 2^{2^ω} nonisomorphic homogeneous models of power α. Find a theory which has exactly 2^{2^ω} nonisomorphic homogeneous models of each power $\alpha \geqslant 2^\omega$.

5.1.15*. Find an example of a complete theory T in a countable language such that

(i) all models of T of power $> 2^\omega$ are homogeneous;

(ii) T has a model of power 2^ω which is not homogeneous.

5.1.16* (GCH). Give an example of a complete theory which has exactly ω nonisomorphic (weakly) homogeneous models in each infinite power. Do the same for 2^ω, for 2 and for 3.

For which finite n can such an example be found? (This is an open question.)

Added in Third Edition: Kudaibergenov (1986) has announced a solution to this problem, that for each $n < \omega$ there is a theory which has n homogeneous models in each power. Earlier partial results are in Lo (1980).

5.1.17. Let $\mathscr{L}$ be countable with at least a 2-place relation symbol and α be an infinite cardinal. Show that there is an infinite α^+-saturated model $\mathfrak{A}$ for $\mathscr{L}$ such that for no $B \subset A$, $\omega \leqslant |B| \leqslant \alpha$, is the model $(\mathfrak{A}, B)$ ω_1-saturated.

[*Hint*: Consider an α^+-saturated model equivalent to $\langle R(\omega), \in \rangle$.]

5.1.18. Prove that every infinite saturated model in a countable language is isomorphic to a proper elementary submodel of itself.

5.1.19. Let $\mathfrak{A}$ be a saturated model for a countable language where $|A|$ is a regular cardinal α. Prove that $\mathfrak{A}$ is the union of an elementary chain $\mathfrak{A}_\beta$, $\beta < \alpha$, of models isomorphic to $\mathfrak{A}$.

5.1.20. Show that for each regular cardinal α, complete arithmetic has models of arbitrarily large cardinality which are α-saturated, but have cofinality α and thus are not α^+-saturated.

5.1.21. Let $V(X)$ be a superstructure (as defined in Section 4.4) such that X is infinite and $|V(X)| \leq 2^\alpha$. Prove that there is an α^+-saturated nonstandard universe $\langle V(X), V(Y), * \rangle$ which has 2^α internal sets. (Note that if in addition $\alpha^+ = 2^\alpha$, then the nonstandard universe is saturated.)

5.1.22. Let α be an inaccessible cardinal and let $V(X)$ be a superstructure such that $\omega \leq |X| < \alpha$. Prove that there is a saturated superstructure $\langle V(X), V(Y), * \rangle$ which has α internal sets.

5.1.23. Let $\langle V(X), V(Y), * \rangle$ and $\langle V(X), V(Z), \rho \rangle$ be saturated nonstandard universes with $|Y| = |Z|$. Prove that the superstructures are isomorphic, that is, there is an isomorphism

$$h : \langle V(Y), \in \rangle \cong \langle V(Z), \in \rangle$$

such that for all $A \in V(X)$, $h(*A) = {}^\rho A$.

5.1.24*. A superstructure $\langle V(X), V(Y), * \rangle$ is said to have the *α-isomorphism property* if for any two models $\mathfrak{A}$, $\mathfrak{B}$ for a language $\mathscr{L}'$ with fewer than α symbols, if $\mathfrak{A} \equiv \mathfrak{B}$ and the universes and all relations and functions of $\mathfrak{A}$ and $\mathfrak{B}$ are internal, then $\mathfrak{A} \cong \mathfrak{B}$.

Prove that for each cardinal α and each infinite base set X, there exists a nonstandard universe $\langle V(X), V(Y), * \rangle$ with the α-isomorphism property. [*Hint*: Use Exercise 4.4.28 and form a bounded elementary chain.]

5.1.25*. Suppose that $\langle V(X), V(Y), * \rangle$ is a nonstandard universe which is an enlargement and has the α-isomorphism property. Prove that it is an α-saturated nonstandard universe.

5.1.26*. Prove that every saturated nonstandard universe with α internal sets has the α-isomorphism property.

5.2. Preservation theorems

In Section 3.2 we used elementary chains to prove a group of results called preservation theorems. In Exercises 3.2.5 and 3.2.6, the student was asked to give different proofs using recursively saturated models. A more uniform method for proving preservation theorems is the use of saturated or special models. We shall give alternate proofs of the preservation theorems from Chapter 3 (submodels, unions of chains, and homomorphisms) by this method and then go on to some additional preservation theorems.

We shall frequently use the first general lemma of Section 3.2, so we repeat it here for convenience.

LEMMA 3.2.1 (repeated). *Let T be a consistent theory of $\mathscr{L}$ and let Δ be a set of sentences of T closed under finite disjunctions. Then the following are equivalent*:

(i). *T has a set of axioms Γ such that $\Gamma \subset \Delta$.*

(ii). *If $\mathfrak{A}$ is a model of T and every sentence $\delta \in \Delta$ which holds in $\mathfrak{A}$ holds in $\mathfrak{B}$, then $\mathfrak{B}$ is a model of T.*

Our first lemma is an analogue of the result that α-saturated models are α^+-universal.

LEMMA 5.2.1. *Suppose that $\mathfrak{B}$ is α-saturated, $\alpha \geqslant |A|$, and every existential sentence holding in $\mathfrak{A}$ holds in $\mathfrak{B}$. Then $\mathfrak{A}$ is isomorphically embeddable in $\mathfrak{B}$.*

PROOF. For the purposes of this (and later) proofs we introduce the temporary notation $\mathfrak{B}_1(\mathrm{E})\mathfrak{B}_2$ to mean that every existential sentence holding on $\mathfrak{B}_1$ holds on $\mathfrak{B}_2$. Let $a \in {}^{\alpha}A$ be an enumeration of A. We find the sequence b_ξ, $\xi < \alpha$, by induction on $\xi < \alpha$. Suppose that for all $\eta < \xi$, we have $b_\eta \in B$ such that

$$(\mathfrak{A}, a_\eta)_{\eta<\xi}(\mathrm{E})(\mathfrak{B}, b_\eta)_{\eta<\xi}.$$

Let $\Sigma(v)$ be the set of all existential formulas $\sigma(v)$ satisfied by a_ξ in $(\mathfrak{A}, a_\eta)_{\eta<\xi}$. By taking finite conjunctions φ of formulas σ in Σ and quantifying by $(\exists v)$, we obtain sentences which are equivalent to existential sentences. Whence all such sentences $(\exists v)\varphi$ hold in $(\mathfrak{B}, b_\eta)_{\eta<\xi}$. Thus Σ is consistent with the theory of $(\mathfrak{B}, b_\eta)_{\eta<\xi}$. We can now extend Σ to some (complete) type $\bar{\Sigma}(v)$ of $(\mathfrak{B}, b_\eta)_{\eta<\xi}$. Since $(\mathfrak{B}, b_\eta)_{\eta<\xi}$ is still α-saturated, $\bar{\Sigma}$, and hence Σ, is realized by some element $b_\xi \in B$. So we have

$$(\mathfrak{A}, a_\eta)_{\eta\leqslant\xi}(\mathrm{E})(\mathfrak{B}, b_\eta)_{\eta\leqslant\xi}.$$

So our induction is complete. Since α is a limit ordinal, we also have $(\mathfrak{A}, a_\xi)_{\xi<\alpha}(\mathrm{E})(\mathfrak{B}, b_\xi)_{\xi<\alpha}$. Since existential sentences contain all atomic and negations of atomic sentences, the mapping $a_\xi \to b_\xi$ is an isomorphic embedding of $\mathfrak{A}$ into $\mathfrak{B}$. ⊣

Recall that a sentence φ is preserved under extensions iff for all models $\mathfrak{A} \subset \mathfrak{B}$, $\mathfrak{A} \models \varphi$ implies $\mathfrak{B} \models \varphi$.

PROPOSITION 5.2.2. *The following two conditions on models $\mathfrak{A}$ and $\mathfrak{B}$ are equivalent*:

(i). *Every existential sentence holding in $\mathfrak{A}$ holds in $\mathfrak{B}$.*

(ii). *$\mathfrak{A}$ is isomorphically embeddable in some elementary extension of $\mathfrak{B}$.*

PROOF. (ii) ⇒ (i). Since existential sentences are preserved under extensions, this is trivial.

(i) ⇒ (ii). Let β be a cardinal such that $|A| \leqslant \beta$ and $||\mathscr{L}|| \leqslant \beta$. By the existence theorem for saturated models, $\mathfrak{B}$ has a β^+-saturated elementary extension $\mathfrak{B}'$, which is either finite or of power 2^β. By Lemma 5.2.1, $\mathfrak{A}$ is elementarily embeddable in $\mathfrak{B}$. ⊣

THEOREM 5.2.3. *A theory T is preserved under extensions if and only if T has a set of existential axioms.*

PROOF. We prove only the hard direction. Assume T is preserved under extensions. The disjunction of finitely many existential sentences is equivalent to an existential sentence, so we may use Lemma 3.2.1. Let $\mathfrak{A}$ be a model of T and suppose every existential sentence true in $\mathfrak{A}$ is true in $\mathfrak{B}$. Then $\mathfrak{A}$ is isomorphically embeddable in some $\mathfrak{B}' \succ \mathfrak{B}$. Hence $\mathfrak{B}'$, and therefore $\mathfrak{B}$, is a model of T. By Lemma 3.2.1, T has a set of existential axioms. ⊣

THEOREM 5.2.4. *A theory is preserved under submodels if and only if T has a set of universal axioms.*

PROOF. This is similar to that of the preceding theorem. ⊣

We now turn our attention to the universal–existential sentences, namely the Π_2^0 sentences.

PROPOSITION 5.2.5. *The following are equivalent*:

(i). *Every universal–existential sentence holding in* $\mathfrak{A}$ *also holds in* $\mathfrak{B}$.

(ii). *There are two models* $\mathfrak{B}'$ *and* $\mathfrak{A}'$ *such that* $\mathfrak{B} \subset \mathfrak{A}' \subset \mathfrak{B}'$, $\mathfrak{B} \prec \mathfrak{B}'$ *and* $\mathfrak{A} \prec \mathfrak{A}'$.

PROOF. (ii) ⇒ (i). Let φ be a universal–existential sentence holding in $\mathfrak{A}$. For simplicity, assume that $\varphi = (\forall x \exists y)\sigma$, where σ is quantifier-free. We show that $\mathfrak{B} \models \varphi$. So let $b \in B$. Thus $(\mathfrak{B}, b) \equiv (\mathfrak{B}', b)$. Since $b \in A'$ and $\mathfrak{A}' \models \varphi$, we have $\mathfrak{A}' \models (\exists y)\sigma[b]$. As existential sentences are preserved under extensions, $\mathfrak{B}' \models (\exists y)\sigma[b]$. This means $\mathfrak{B} \models (\exists y)\sigma[b]$. Whence $\mathfrak{B} \models (\forall x \exists y)\sigma$. So (i) holds.

(i) ⇒ (ii). From (i) it easily follows that

every existential–universal (Σ_2^0) sentence holding in $\mathfrak{B}$ also holds in $\mathfrak{A}$.

Let $\mathfrak{A}'$ be a $|B|$-saturated elementary extension of $\mathfrak{A}$ such that $|B| \leqslant |A'|$ and let $b \in {}^{\alpha}B$ be an enumeration of B. Using exactly the same induction as in Lemma 5.2.1 (we leave the simple details as an exercise), we can find $a \in {}^{\alpha}A'$ such that

every existential–universal sentence holding on $(\mathfrak{B}, b_\xi)_{\xi<\alpha}$ also holds on $(\mathfrak{A}', a_\xi)_{\xi<\alpha}$.

This implies that every existential sentence holding in $(\mathfrak{A}', a_\xi)_{\xi<\alpha}$ holds in $(\mathfrak{B}, b_\xi)_{\xi<\alpha}$. Now apply Proposition 5.2.2 and obtain in a straightforward manner an elementary extension $(\mathfrak{B}', b_\xi)_{\xi<\alpha}$ of $(\mathfrak{B}, b_\xi)_{\xi<\alpha}$ in which $(\mathfrak{A}', a_\xi)_{\xi<\alpha}$ is isomorphically embeddable. We can assume without loss of generality that

$$\mathfrak{B} \subset \mathfrak{A}' \subset \mathfrak{B}', \qquad \mathfrak{A} \prec \mathfrak{A}', \qquad \mathfrak{B} \prec \mathfrak{B}'. \dashv$$

We say that three models $\mathfrak{A}$, $\mathfrak{B}$, $\mathfrak{C}$ form a *1-sandwich* iff $\mathfrak{A} \subset \mathfrak{B} \subset \mathfrak{C}$ and $\mathfrak{A} \prec \mathfrak{C}$. We say that the model $\mathfrak{A}$ *is sandwiched by* $\mathfrak{B}$ iff there are elementary extensions $\mathfrak{A}' \succ \mathfrak{A}$, $\mathfrak{B}' \succ \mathfrak{B}$ such that $\mathfrak{B}$, $\mathfrak{A}'$, $\mathfrak{B}'$ form a 1-sandwich.

THEOREM 5.2.6. *The following are equivalent*:

(i). *T has a set of universal–existential axioms.*

(ii). *T is preserved under unions of chains of models.*

(iii). *T is preserved under* 1-*sandwiches, i.e., if* $\mathfrak{A} \models T$ *and* $\mathfrak{A}$ *is sandwiched by* $\mathfrak{B}$, *then* $\mathfrak{B} \models T$.

PROOF. We proceed in the direction (i) $\Rightarrow$ (ii) $\Rightarrow$ (iii) $\Rightarrow$ (i).

(i) $\Rightarrow$ (ii). This is easy.

(ii) $\Rightarrow$ (iii). Suppose that $\mathfrak{A}$ is sandwiched by $\mathfrak{B}$. We shall construct a chain of models,

(1) $\mathfrak{B}_0 \subset \mathfrak{A}_0 \subset \mathfrak{B}_1 \subset \mathfrak{A}_1 \subset \ldots \subset \mathfrak{B}_n \subset \mathfrak{A}_n \subset \ldots$

such that $\mathfrak{B}_0 = \mathfrak{B}$, each triple $\mathfrak{B}_n$, $\mathfrak{A}_n$, $\mathfrak{B}_{n+1}$ forms a 1-sandwich, and $\mathfrak{A}_0 \succ \mathfrak{A}$, and each $\mathfrak{A}_{n+1}$ is equivalent to $\mathfrak{A}$. In order to define the models $\mathfrak{A}_n$, $\mathfrak{B}_n$ inductively, we see that we only need the following:

(2) If $\mathfrak{B}_n$, $\mathfrak{A}_n$, $\mathfrak{B}_{n+1}$ form a 1-sandwich, then there are an elementarily equivalent $\mathfrak{A}_{n+1} \equiv \mathfrak{A}_n$ and an elementary extension $\mathfrak{B}_{n+2}$ of $\mathfrak{B}_{n+1}$ such that $\mathfrak{B}_{n+1}$, $\mathfrak{A}_{n+1}$, $\mathfrak{B}_{n+2}$ form a 1-sandwich.

To prove (2), let $\mathscr{L}' = \mathscr{L} \cup \{c_b : b \in B_{n+1}\} \cup \{U\}$, where U is a 1-placed relation symbol, and consider the theory T' in $\mathscr{L}'$ given by

$$\{\text{the elementary diagram of } \mathfrak{B}_{n+1}\} \cup \{\varphi^{(U)} : \mathfrak{A}_n \models \varphi\} \cup \{U(c_b) : b \in B_{n+1}\}.$$

Here $\varphi^{(U)}$ is the relativization of the sentence φ (see Exercise 5.2.20). So any model of T' will be an elementary extension $\mathfrak{B}_{n+2}$ of $\mathfrak{B}_{n+1}$ which contains a subset U that includes at least all the elements of $\mathfrak{B}_{n+1}$, and such that the submodel determined by the interpretation of U in $\mathfrak{B}_{n+2}$ is a model equivalent to $\mathfrak{A}_n$. The consistency of T' is shown as follows: Any finite subset of T' consists of three parts:

a sentence $\sigma(c_{b_1} \ldots c_{b_m})$ such that $\mathfrak{B}_{n+1} \models \sigma[b_1 \ldots b_m]$,

a sentence $\varphi^{(U)}$ such that $\mathfrak{A}_n \models \varphi$,

the sentence $U(c_{b_1}) \wedge \ldots \wedge U(c_{b_m})$.

Since $\mathfrak{B}_n \prec \mathfrak{B}_{n+1}$, there will be $d_1, \ldots, d_m \in B_n$ such that

$$\mathfrak{B}_n \models \sigma[d_1 \ldots d_m], \qquad \mathfrak{B}_{n+1} \models \sigma[d_1 \ldots d_m].$$

Now the model $(\mathfrak{B}_{n+1}, A_n, d_1 \ldots d_m)$ clearly will be a model of this finite subset of T'. So T' has a model, and (2) is proved.

Now consider the sequence of models in the chain (1). Since T holds in $\mathfrak{A}_0$, we have $\mathfrak{A}_n \models T$ for all n. Whence $\bigcup_{n \in \omega} \mathfrak{A}_n \models T$. On the other hand, the chain of models $\mathfrak{B}_n$, $n \in \omega$, is an elementary chain, so $\mathfrak{B} \prec \bigcup_{n \in \omega} \mathfrak{B}_n = \bigcup_{n \in \omega} \mathfrak{A}_n$. Whence $\mathfrak{B} \models T$.

(iii) $\Rightarrow$ (i). Assume (iii). Then (i) follows by Proposition 5.2.5 using Lemma 3.2.1. ⊣

There are natural generalizations of Theorem 5.2.6 to $\mathbf{\Sigma}_2^0$ sentences and to $\mathbf{\Pi}_n^0$ and $\mathbf{\Sigma}_n^0$ sentences with $n \geqslant 3$. For the sake of completeness, we shall give below a generalization of Theorem 5.2.6 for $\mathbf{\Pi}_{2n}^0$ sentences. The reader is asked in an exercise to provide analogous generalizations for $\mathbf{\Pi}_m^0$ sentences when m is odd and for Σ_n^0 sentences for $n \geqslant 2$. The chain of $2n+1$ models

$$\mathfrak{B}_0 \subset \mathfrak{A}_0 \subset \mathfrak{B}_1 \subset \mathfrak{A}_1 \subset \ldots \subset \mathfrak{B}_{n-1} \subset \mathfrak{A}_{n-1} \subset \mathfrak{B}_n$$

is an *n-sandwich* iff

$$\mathfrak{B}_0 \prec \mathfrak{B}_1 \prec \ldots \prec \mathfrak{B}_n, \qquad \mathfrak{A}_0 \prec \mathfrak{A}_1 \prec \ldots \prec \mathfrak{A}_{n-1}.$$

We say that the model $\mathfrak{A}$ is *n-sandwiched by the model* $\mathfrak{B}_0$ iff there are elementary chains

$$\mathfrak{A} \prec \mathfrak{A}_0 \prec \mathfrak{A}_1 \prec \ldots \prec \mathfrak{A}_{n-1}, \qquad \mathfrak{B}_0 \prec \ldots \prec \mathfrak{B}_n$$

such that the chain

$$\mathfrak{B}_0 \subset \mathfrak{A}_0 \subset \ldots \subset \mathfrak{A}_{n-1} \subset \mathfrak{B}_n$$

is an n-sandwich. The next proposition generalizes Proposition 5.2.5.

PROPOSITION 5.2.7. *Let $n \geqslant 1$. The following are equivalent:*

(i). *Every Π_{2n}^0 sentence holding in $\mathfrak{A}$ holds in $\mathfrak{B}$.*

(ii). *There is an n-sandwich*

$$\mathfrak{B}_0 \subset \mathfrak{A}_0 \subset \ldots \subset \mathfrak{A}_{n-1} \subset \mathfrak{B}_n$$

such that $\mathfrak{B}_0 = \mathfrak{B}$ and $\mathfrak{A} \prec \mathfrak{A}_0$.

PROOF. The reader is asked first to review briefly the proof of Proposition 5.2.5.

(ii) ⇒ (i). This is based on the following assertion proved by induction on k, $1 \leqslant k \leqslant n$:

> for every Π_{2k}^0 formula $\varphi(x_1 \ldots x_l)$ and all $b_1, \ldots, b_l \in B_m$, with $0 \leqslant m \leqslant n-k$, if $\mathfrak{A}_m \models \varphi[b_1 \ldots b_l]$, then $\mathfrak{B}_m \models \varphi[b_1 \ldots b_l]$.

Note that when $k = 1$, the first stage of the induction for all m, $0 \leqslant m \leqslant n-1$, is essentially given by one direction of Proposition 5.2.5. There is no special difficulty in carrying the induction from k to $k+1$. So we shall assume that the assertion is proved. When $k = n$, we see that every Π_{2n}^0 sentence holding on $\mathfrak{A}_0$ holds on $\mathfrak{B}_0$, as was to be proved.

(i) ⇒ (ii). Let us start with the models $\mathfrak{A}$ and $\mathfrak{B}_0 = \mathfrak{B}$. Let $\mathfrak{A}_0$ be any $|B_0|$-saturated elementary extension of $\mathfrak{A}$ and let $\bar{b}^0 \in {}^{\beta_0}B_0$ be an enumeration of $\mathfrak{B}_0$. So

$$\text{every } \Pi_{2n}^0 \text{ sentence holding on } \mathfrak{A} \text{ also holds on } \mathfrak{B}_0,$$

and hence

(1) every Σ^0_{2n} sentence holding on $\mathfrak{B}_0$ also holds on $\mathfrak{A}$.

By examining the proof of Proposition 5.2.5, (1) leads to

(2) there is an $a^0 \in {}^{\beta_0}A_0$ such that every Σ^0_{2n} sentence holding in $(\mathfrak{B}_0, \bar{b}^0_\xi)_{\xi<\beta_0}$ also holds in $(\mathfrak{A}_0, a^0_\xi)_{\xi<\beta_0}$.

Deleting the outermost existential quantifiers and stating the result in the contrapositive form we find that (2) yields

(3) every Σ^0_{2n-1} sentence holding in $(\mathfrak{A}_0, a^0_\xi)_{\xi<\beta_0}$ also holds in $(\mathfrak{B}_0, \bar{b}^0_\xi)_{\xi<\beta_0}$.

Now let $\mathfrak{B}_1$ be any $|A_0|$-saturated elementary extension of $\mathfrak{B}_0$ and let $\bar{a}^0 \in {}^{\alpha_0}A_0$ be an extension of a_0 which enumerates $\mathfrak{A}_0$. Then (3) gives

(4) there is a $b^1 \in {}^{\alpha_0}B_1$ such that every Σ^0_{n-1} sentence holding in $(\mathfrak{A}_0, \bar{a}^0_\xi)_{\xi<\alpha_0}$ holds in $(\mathfrak{B}_1, b^1_\xi)_{\xi<\alpha_0}$.

From (4) we have

(5) every $\Sigma^0_{2(n-1)}$ sentence holding in $\mathfrak{B}_1$ holds in $\mathfrak{A}_0$.

Furthermore, we already have the 1-sandwich

$$\mathfrak{B}_0 \subset \mathfrak{A}_0 \subset \mathfrak{B}_1, \qquad \mathfrak{A} \prec \mathfrak{A}_0, \qquad \mathfrak{B}_0 \prec \mathfrak{B}_1.$$

Clearly, using (5) in the same way we used (1), we can extend the 1-sandwich to a 2-sandwich

$$\mathfrak{B}_0 \subset \mathfrak{A}_0 \subset \mathfrak{B}_1 \subset \mathfrak{A}_1 \subset \mathfrak{B}_2, \qquad \mathfrak{A} \prec \mathfrak{A}_0 \prec \mathfrak{A}_1, \qquad \mathfrak{B}_0 \prec \mathfrak{B}_1 \prec \mathfrak{B}_2,$$

and

every $\Sigma^0_{2(n-2)}$ sentence holding in $\mathfrak{B}_2$ also holds in $\mathfrak{A}_1$.

It is now clear how an inductive proof can be given, so that we can construct the n-sandwich as required in (ii). ⊣

The next theorem generalizes Theorem 5.2.6.

THEOREM 5.2.8 (Keisler Sandwich Theorem). *Let T be a theory of $\mathscr{L}$ and let $n \geqslant 1$. The following are equivalent:*

(i). *T has a set of Π^0_{2n} axioms.*

(ii). *T is preserved under unions of Σ^0_{2n-1}-chains.*

(iii). *T is preserved under n-sandwiches, i.e., if $\mathfrak{A} \models T$ and $\mathfrak{A}$ is n-sandwiched by $\mathfrak{B}$, then $\mathfrak{B} \models T$.*

PROOF. (i)$\Rightarrow$(ii). This follows by Lemma 3.1.10.

(ii)$\Rightarrow$(iii). Suppose that $\mathfrak{A}$ is n-sandwiched by $\mathfrak{B} = \mathfrak{B}_0$ and $\mathfrak{A} \models T$. Then by an induction very much similar to the one in Proposition 5.2.7, we have

(1) for every Σ^0_{2n-1} formula $\varphi(x_1 \dots x_l)$ and all $a_1, \dots, a_l \in A_0$, if

$\mathfrak{B}_1 \models \varphi[a_1 \dots a_l]$, then $\mathfrak{A}_0 \models \varphi[a_1 \dots a_l]$.

Consider the three models $\mathfrak{B}_0 \subset \mathfrak{A}_0 \subset \mathfrak{B}_1$ such that $\mathfrak{B}_0 \prec \mathfrak{B}_1$ and (1) holds for the pair $\mathfrak{A}_0, \mathfrak{B}_1$. We shall extend this chain by adding two more models,

$$\mathfrak{B}_0 \subset \mathfrak{A}_0 \subset \mathfrak{B}_1 \subset \mathfrak{A}_1 \subset \mathfrak{B}_2,$$

such that

(2) $\mathfrak{B}_0 \prec \mathfrak{B}_1 \prec \mathfrak{B}_2$, $\mathfrak{A}_0 \equiv \mathfrak{A}_1$, and (1) holds for the pairs $\mathfrak{A}_0, \mathfrak{A}_1$ and $\mathfrak{A}_1, \mathfrak{B}_2$.

To prove (2), extending the corresponding idea in Theorem 5.2.6, let

$$\mathscr{L}' = \mathscr{L} \cup \{c_b : b \in B_1\} \cup \{U\}$$

and consider the theory T given by the following sentences of $\mathscr{L}'$:

the elementary diagram of $\mathfrak{B}_1$,

$\{\theta^{(U)} : \mathfrak{A}_0 \models \theta$ and θ a sentence of $\mathscr{L}\}$,

$\{U(c_b) : b \in B_1\}$,

$\{\psi^{(U)}(c_{a_1} \dots c_{a_l}) : a_1, \dots, a_l \in A_0;\ \psi(x_1 \dots x_l)$ a Π^0_{2n-1} formula such that $\mathfrak{A}_0 \models \psi[a_1 \dots a_l]\}$,

$\{(\forall x_1 \dots x_l)(\psi \wedge U(x_1) \wedge \dots \wedge U(x_l) \rightarrow \psi^{(U)}) : \psi(x_1 \dots x_l)$ is a Σ^0_{2n-1} formula of $\mathscr{L}\}$.

A typical finite subset of the sentences of T will consist of five parts:

(3) a sentence $\sigma(c_{b_1} \dots c_{b_m})$ such that $\mathfrak{B}_1 \models \sigma[b_1 \dots b_m]$,
a sentence $\theta^{(U)}$, where $\mathfrak{A}_0 \models \theta$,
a sentence $U(c_{b_1}) \wedge \dots \wedge U(c_{b_m})$,
a sentence $\psi_1^{(U)}(c_{a_1} \dots c_{a_l})$ such that $a_1, \dots, a_l$ is among $b_1, \dots, b_m$ and ψ_1 is a Π^0_{2n-1} formula and $\mathfrak{A}_0 \models \psi_1[a_1 \dots a_l]$,
a sentence $(\forall x_1 \dots x_l)(\psi_2 \wedge U(x_1) \wedge \dots \wedge U(x_l) \rightarrow \psi_2^{(U)})$, where ψ_2 is a Σ^0_{2n-1} formula.

By (1) we have $\mathfrak{B}_1 \models \psi_1[a_1 \dots a_l]$. Since $\mathfrak{B}_0 \prec \mathfrak{B}_1$, we can find elements $d_1, \dots, d_m \in B_0$ such that the model $(\mathfrak{B}_1, A_0, d_1 \dots d_m)$ satisfies all the sentences in (3). Whence T is consistent. Now let $\mathfrak{B}_2$ be a model of T and let A_1 be the interpretation of U in $\mathfrak{B}_2$. Then we see easily that (2) holds. Repeating this procedure, we obtain by induction a chain of models

$$\mathfrak{B}_0 \subset \mathfrak{A}_0 \subset \mathfrak{B}_1 \subset \mathfrak{A}_1 \subset \dots \subset \mathfrak{B}_n \subset \mathfrak{A}_n \subset \mathfrak{B}_{n+1} \dots$$

such that

$$\mathfrak{B}_0 \prec \mathfrak{B}_1 \prec \ldots \prec \mathfrak{B}_n \prec \ldots,$$

the models $\mathfrak{A}_n$, $n \in \omega$, form a Σ^0_{2n-1}-chain, and

$$\mathfrak{B}_0 \prec \bigcup_{n\in\omega} \mathfrak{B}_n = \bigcup_{n\in\omega} \mathfrak{A}_n.$$

Since each $\mathfrak{A}_n \equiv \mathfrak{A}$, φ holds in the union. So φ holds in $\mathfrak{B}_0$. This proves (iii).

(iii) $\Rightarrow$ (i). This follows from Proposition 5.2.7. ⊣

The sandwich theorem has an application to Peano arithmetic. It has been shown by other methods that Peano arithmetic does not have a Π^0_n set of axioms, for any n. It follows that for each n, Peano arithmetic is not preserved under n-sandwiches. That is, there is a chain of models

$$\mathfrak{B}_0 \subset \mathfrak{A}_0 \subset \ldots \subset \mathfrak{A}_{n-1} \subset \mathfrak{B}_n$$

such that

$$\mathfrak{A}_0 \prec \ldots \prec \mathfrak{A}_{n-1}, \qquad \mathfrak{B}_0 \prec \ldots \prec \mathfrak{B}_n,$$

and $\mathfrak{A}_0$ is a model of Peano arithmetic but $\mathfrak{B}_0$ is not.

It is also known that ZF has no Π^0_n set of axioms, so ZF is also not preserved under n-sandwiches.

The next few results are concerned with positive sentences and homomorphisms. Recall that positive formulas are built up from atomic formulas and $\wedge$, $\vee$, $\exists$, $\forall$. We say that a formula is *negative* iff it is built up from negations of atomic formulas and $\wedge$, $\vee$, $\exists$, $\forall$. Obviously φ is equivalent to a positive sentence if and only if $\neg\varphi$ is equivalent to a negative sentence. Recall also that a homomorphism of $\mathfrak{A}$ onto $\mathfrak{B}$ is a mapping f of A onto B which preserves all atomic formulas, i.e., if $\sigma(x_1 \ldots x_n)$ is an atomic formula and $\mathfrak{A} \models \sigma[a_1 \ldots a_n]$, then $\mathfrak{B} \models \sigma[f(a_1) \ldots f(a_n)]$.

The next two lemmas are generalizations of Lemmas 5.1.10 and 5.1.11. We adopt the temporary convention that $\mathfrak{B}_1(\mathrm{P})\mathfrak{B}_2(\mathfrak{B}_1(\mathrm{N})\mathfrak{B}_2)$ shall mean that every positive (negative) sentence holding on $\mathfrak{B}_1$ holds on $\mathfrak{B}_2$. More generally, if Φ is an arbitrary set of formulas, we write $\mathfrak{B}_1(\Phi)\mathfrak{B}_2$ for 'every sentence of Φ holding in $\mathfrak{B}_1$ holds in $\mathfrak{B}_2$'.

Lemma 5.2.9. *Let Φ be the set of formulas of $\mathscr{L}$ which are equivalent to positive (negative) formulas. Suppose that $\mathfrak{B}$ is α-saturated, $\mathfrak{A}(\Phi)\mathfrak{B}$ and $a \in {}^\alpha A$. Then there is $b \in {}^\alpha B$ such that*

$$(\mathfrak{A}, a_\xi)_{\xi<\alpha}(\Phi)(\mathfrak{B}, b_\xi)_{\xi<\alpha}.$$

Proof. We first mention that the proof only requires that Φ be a set of formulas which is closed under conjunction and existential quantification.

Whence our proof for $\Phi = \mathrm{P}$ will work equally well for $\Phi = \mathrm{N}$, or even $\Phi =$ the set of all formulas of $\mathscr{L}$ (Lemma 5.1.10).

We find the sequence b_ξ, $\xi < \alpha$, by induction. Suppose $\xi < \alpha$, and for all $\eta < \xi$, we have b_η such that

$$(\mathfrak{A}, a_\eta)_{\eta<\xi}(\Phi)(\mathfrak{B}, b_\eta)_{\eta<\xi}.$$

Let $\Sigma(v)$ be the set of all formulas σ of Φ (with respect to the language of $(\mathfrak{A}, a_\eta)_{\eta<\xi}$) in the variable v such that

$$(\mathfrak{A}, a_\eta)_{\eta<\xi} \models \sigma[a_\xi].$$

By using the basic properties of Φ, we see that

$$(\mathfrak{B}, b_\eta)_{\eta<\xi} \models (\exists v)\sigma \qquad \text{for all} \quad \sigma \in \Phi.$$

Whence Σ can be extended to a complete type of $(\mathfrak{B}, b_\eta)_{\eta<\xi}$. As $(\mathfrak{B}, b_\eta)_{\eta<\xi}$ is α-saturated, some element $b_\xi \in B$ realizes Σ, so that

$$(\mathfrak{A}, a_\eta)_{\eta\leq\xi}(\Phi)(\mathfrak{B}, b_\eta)_{\eta\leq\xi}.$$

Our induction is complete and the lemma is proved. ⊣

Lemma 5.2.10. *Suppose that α is infinite, $\mathfrak{A}$, $\mathfrak{B}$ are α-saturated, $a \in {}^\alpha A$, $b \in {}^\alpha B$, and $\mathfrak{A}(\mathrm{P})\mathfrak{B}$. Then there are $\bar{a} \in {}^\alpha A$, $\bar{b} \in {}^\alpha B$ such that*

$$\text{range}\,(a) \subset \text{range}\,(\bar{a}),$$

$$\text{range}\,(b) \subset \text{range}\,(\bar{b}),$$

$$(\mathfrak{A}, \bar{a}_\xi)_{\xi<\alpha}(\mathrm{P})(\mathfrak{B}, \bar{b}_\xi)_{\xi<\alpha}.$$

Proof. (Please first briefly review the proof of Lemma 5.1.11.) We shall find the sequences $\bar{a}$, $\bar{b}$ such that

if $\xi = \omega \cdot \lambda + 2n$ is even, then $\bar{a}_\xi = a_{\omega\cdot\lambda+n}$;
if $\xi = \omega \cdot \lambda + 2(n+1)$ is odd, then $\bar{b}_\xi = b_{\omega\cdot\lambda+n}$;
$(\mathfrak{A}, \bar{a}_\eta)_{\eta<\xi}(\mathrm{P})(\mathfrak{B}, \bar{b}_\eta)_{\eta<\xi}$.

If ξ is even, we fix $\bar{a}_\xi$ as above, and find $\bar{b}_\xi \in B$ as in Lemma 5.2.9. If ξ is odd, we fix $\bar{b}_\xi$ as above and note that

$$(\mathfrak{B}, \bar{b}_\eta)_{\eta<\xi}(\mathrm{N})(\mathfrak{A}, \bar{a}_\eta)_{\eta<\xi}.$$

Whence again we find $\bar{a}_\xi \in A$ as in Lemma 5.2.9. ⊣

Proposition 5.2.11. *Suppose that $\mathfrak{A}$ and $\mathfrak{B}$ are special models such that every positive sentence holding in $\mathfrak{A}$ also holds in $\mathfrak{B}$. Assume that either $\mathfrak{B}$ is finite or $|B| = |A|$. Then $\mathfrak{A} \simeq \mathfrak{B}$.*

PROOF. We sketch below a proof in the case where B is infinite and $|A| = |B|$. However, if B is finite, then regardless of whether A is infinite or not, the same proof will work. The point to remember is that finite models are α-saturated for all cardinals α. So Lemma 5.2.10 applies.

If $\alpha = \gamma^+$, then $\mathfrak{A}$ and $\mathfrak{B}$ are both α-saturated, of power α. Let $a \in {}^\alpha A$, $b \in {}^\alpha B$ enumerate A and B, respectively. Then there are $\bar{a} \in {}^\alpha A$, $\bar{b} \in {}^\alpha B$ which satisfy Lemma 5.2.10. Whence $\bar{a}$, $\bar{b}$ must still both enumerate A, B, and the mapping $\bar{a}_\xi \to \bar{b}_\xi$, for $\xi < \alpha$, gives a homomorphism of $\mathfrak{A}$ onto $\mathfrak{B}$.

If α is a limit cardinal, let $\mathfrak{A}_\beta$, $\beta < \alpha$, and $\mathfrak{B}_\beta$, $\beta < \alpha$, be specializing chains of $\mathfrak{A}$ and $\mathfrak{B}$, respectively. We use a back and forth argument as in the uniqueness theorem, and leave the details to the reader. ⊣

PROPOSITION 5.2.12. *The following are equivalent*:

(i). *Every positive sentence holding on $\mathfrak{A}$ also holds on $\mathfrak{B}$.*

(ii). *There are elementary extensions $\mathfrak{A} \prec \mathfrak{A}'$, $\mathfrak{B} \prec \mathfrak{B}'$ such that $\mathfrak{B}'$ is a homomorphic image of $\mathfrak{A}'$.*

PROOF. (ii) ⇒ (i). This follows since positive sentences are preserved under homomorphisms.

(i) ⇒ (ii). Note that if $\mathfrak{A}$ or $\mathfrak{B}$ is finite, then $|A| \geqslant |B|$. We find special elementary extensions $\mathfrak{A}'$ of $\mathfrak{A}$ and $\mathfrak{B}'$ of $\mathfrak{B}$ such that $|A'| \geqslant |B'|$, with equality holding if $\mathfrak{B}$ is infinite and $\mathfrak{A}'(\mathrm{P})\mathfrak{B}'$. By Proposition 5.2.11, $\mathfrak{B}'$ is a homomorphic image of $\mathfrak{A}'$. ⊣

Lyndon's homomorphism theorem now follows at once using Lemma 3.2.1. We repeat the statement here:

THEOREM 5.2.13. *A consistent theory T is preserved under homomorphisms if and only if T has a positive set of axioms.*

The next three results are not, strictly speaking, preservation theorems, nor do their proofs require the use of α-saturated or special models. Nevertheless, they give us a better insight into the general problem of elementary classes closed under certain operations and they lead to some interesting (and difficult) exercises at the end of this section. We shall state the results for single sentences and leave the corresponding results for classes to the exercises.

PROPOSITION 5.2.14. *Let φ be a sentence of $\mathscr{L}$. Suppose that whenever $\mathfrak{A}$ is a model of φ and $\mathfrak{B}_1$ and $\mathfrak{B}_2$ are submodels of $\mathfrak{A}$, which are themselves models of φ, then $\mathfrak{B}_1 \cap \mathfrak{B}_2$ is a model of φ. Then, given any nonempty collection X of models of φ which are submodels of a model of φ, their intersection $\bigcap X$ is again a model of φ.*

Proof. (The definitions of $\mathfrak{B}_1 \cap \mathfrak{B}_2$ and $\bigcap X$ are taken in the natural sense.) Let $\mathfrak{A}$ be a model of φ and let X be a nonempty collection of models of φ which are submodels of $\mathfrak{A}$. Let $\mathscr{L}' = \mathscr{L} \cup \{c_a : a \in A\} \cup \{U\}$, where U is a new 1-placed relation symbol. Let T be the theory in $\mathscr{L}'$ given by the following sentences:

the diagram of $\mathfrak{A}$ in $\mathscr{L} \cup \{c_a : a \in A\}$,
the relativized sentence $\varphi^{(U)}$,
the sentences $U(c_a)$ for all elements a in the intersection $\bigcap X$,
the sentences $\neg U(c_a)$ for all elements $a \in A$ not belonging to $\bigcap X$.

Given any finite subset T' of T, it involves at most a finite number of sentences $\neg U(c_{a_1}), \ldots, \neg U(c_{a_n})$. Suppose that for $1 \leqslant i \leqslant n$,

$$a_i \notin B_i, \quad \mathfrak{B}_i \in X.$$

Let $\mathfrak{B} = \bigcap_{1 \leqslant i \leqslant n} \mathfrak{B}_i$. By hypothesis, $\mathfrak{B}$ is a model of φ. Whence the model

$$(\mathfrak{A}, B, a)_{a \in A}$$

is a model of T', with U being interpreted by the set B. So T is consistent. Let $\mathfrak{B}$ be a model of T. Clearly $\mathfrak{A}$ is isomorphically embeddable in $\mathfrak{B}$ onto a model $\mathfrak{A}_1$, and also the interpretation of U in $\mathfrak{B}$ gives rise to another submodel $\mathfrak{A}_2$ of $\mathfrak{B}$. Both $\mathfrak{A}_1$ and $\mathfrak{A}_2$ are models of φ, whence $\mathfrak{A}_1 \cap \mathfrak{A}_2$ is a model of φ. But it is clear that $\mathfrak{A}_1 \cap \mathfrak{A}_2 \cong \bigcap X$. Hence $\bigcap X$ is a model of φ. ⊣

Proposition 5.2.15. *Let φ be a sentence of $\mathscr{L}$ satisfying the hypothesis of Proposition* 5.2.14. *Then φ is equivalent to a universal–existential sentence.*

Proof. We prove that φ is closed under unions of chains. The result then follows from Theorem 5.2.6. There is no loss of generality if we prove that φ is closed under unions of chains of type ω. (In any case, see Exercises 5.2.13 and 5.2.14.) So suppose that

$$\mathfrak{A}_0 \subset \mathfrak{A}_1 \subset \ldots \subset \mathfrak{A}_n \subset \ldots$$

is a chain of models of φ. Let $\mathfrak{A} = \bigcup_{n \in \omega} \mathfrak{A}_n$. By using the diagram of $\mathfrak{A}$ and the compactness theorem, we can easily prove that $\mathfrak{A}$ is a submodel of some model $\mathfrak{B}$ of φ. Our argument now parallels that of Proposition 5.2.12. Let T be a theory in $\mathscr{L} \cup \{c_b : b \in B\} \cup \{U\}$ given by the following sentences:

the diagram of $\mathfrak{B}$ in $\mathscr{L} \cup \{c_b : b \in B\}$,
$\varphi^{(U)}$,
$U(c_a)$ for all $a \in A$,
$\neg U(c_a)$ for all $a \in B \setminus A$.

Again T is consistent. So it has a model $\mathfrak{B}'$ of φ. Now both $\mathfrak{B}$ and the model with the interpretation of U in $\mathfrak{B}'$ as universe are models of φ. Hence their intersection is a model of φ. But this intersection is isomorphic to $\mathfrak{A}$. ⊣

PROPOSITION 5.2.16. *Suppose that φ is preserved under nonempty descending intersections, i.e., whenever*

$$\mathfrak{A}_0 \supset \mathfrak{A}_1 \supset \ldots \supset \mathfrak{A}_n \supset \ldots$$

is a sequence of model of φ, and their intersection is nonempty, then their intersection is again a model of φ. Then φ is equivalent to a universal–existential sentence.

PROOF. As in the proof of Proposition 5.2.15, we shall use Theorem 5.2.6 and show that φ is preserved under unions of chains. So let

$$\mathfrak{A}_0 \subset \ldots \subset \mathfrak{A}_n \subset \ldots$$

be a chain of models of φ. We may assume that the union $\mathfrak{A} = \bigcup_{n\in\omega} \mathfrak{A}_n$ is a submodel of some model $\mathfrak{B}_0$ of φ. Let $U_1, U_2, \ldots, U_n, \ldots$ be a sequence of new 1-placed relation symbols, and we shall construct an increasing chain of models

$$\mathfrak{B}_0 \subset \mathfrak{B}_1 \subset \ldots \subset \mathfrak{B}_n \subset \ldots,$$

where each model $\mathfrak{B}_n$ is a model for $\mathscr{L}_n = \mathscr{L} \cup \{U_i : 1 \leqslant i \leqslant n\}$ and such that the $\mathscr{L}_n$ reduct of the chain

$$\mathfrak{B}_n \subset \mathfrak{B}_{n+1} \subset \ldots$$

is an elementary chain in the sense of $\mathscr{L}_n$. Furthermore, we require that the interpretation of U_{n+1} in $\mathfrak{B}_{n+1}$, say V^{n+1}_{n+1}, satisfies the following:

$$V^{n+1}_{n+1} \subset V^{n+1}_n \text{ (the interpretation of } U_n \text{ in } \mathfrak{B}_{n+1}),$$
$$A \subset V^{n+1}_{n+1}, \qquad V^{n+1}_{n+1} \cap (B_n \setminus A) = 0,$$
$$\mathfrak{B}_{n+1} \models \varphi^{(U_{n+1})}.$$

Suppose we have the sequence $\mathfrak{B}_0 \subset \ldots \subset \mathfrak{B}_n$ satisfying the above. Then we see that in the model $\mathfrak{B}_n$,

$$A \subset V^n_n \subset V^n_{n-1} \subset \ldots \subset V^n_1,$$
$$\mathfrak{B}_n \models \varphi^{(U_i)} \quad \text{for} \quad 1 \leqslant i \leqslant n.$$

We wish to find a model $\mathfrak{B}_{n+1}$ for $\mathscr{L}_{n+1}$ such that

$$\mathfrak{B}_n \prec \text{the } \mathscr{L}_n\text{-reduct of } \mathfrak{B}_{n+1},$$

and, in $\mathfrak{B}_{n+1}$, we have

$$A \subset V_{n+1}^{n+1} \subset V_n^{n+1} \subset \ldots \subset V_1^{n+1},$$
$$V_{i+1}^{n+1} \cap (B_n \setminus A) = 0,$$
$$\mathfrak{B}_{n+1} \models \varphi^{(U_i)}, \qquad 1 \leqslant i \leqslant n+1.$$

To obtain $\mathfrak{B}_{n+1}$, let T be given by the $\mathscr{L}_n$ elementary diagram of $\mathfrak{B}_n$ plus the sentences which express (using the full strength of $\mathscr{L}_{n+1}$ plus constants for the elements of B_n)

$$A \subset U_{n+1} \subset U_n,$$
$$\varphi^{(U_{n+1})},$$
$$U_{n+1} \cap (B_n \setminus A) = 0.$$

The consistency of any finite portion of T can be shown in the model $\mathfrak{B}_n$ by an appropriate choice of A_m for U_{n+1}. So T is consistent, and we may assume that it has a model $\mathfrak{B}_{n+1}$ which extends the chain to

$$\mathfrak{B}_0 \subset \mathfrak{B}_1 \subset \ldots \subset \mathfrak{B}_{n+1}.$$

Let $\mathfrak{B} = \bigcup_{n\in\omega} \mathfrak{B}_n$ and let $V_1, V_2, \ldots$ be the interpretations of $U_1, U_2, \ldots$ in $\mathfrak{B}$. Then we see that

$$A \subset \ldots \subset V_n \subset \ldots \subset V_1,$$
$$V_{n+1} \cap (B_n \setminus A) = 0,$$
$$\mathfrak{B} \models \varphi^{(U_n)}.$$

So the submodels of $\mathfrak{B}$ with universe V_n with respect to the original language $\mathscr{L}$ is a descending chain of models of φ, whose intersection is $\mathfrak{A}$. Hence $\mathfrak{A} \models \varphi$. ⊣

The techniques involved in the above proofs can be extended to give other results; see the exercises.

Exercises

5.2.1*. Suppose that $\mathscr{L}$ has only a finite number of relation symbols and no function or constant symbols. Prove that K is the class of models of a set of universal sentences iff for all models $\mathfrak{A}$, $\mathfrak{A} \in K$ iff every finite submodel of $\mathfrak{A}$ is isomorphically embeddable in some model in K. Prove a version of this if $\mathscr{L}$ has infinitely many relation symbols. Prove another version of this for arbitrary $\mathscr{L}$.

5.2.2*. Assume $\mathscr{L}$ is as in Exercise 5.2.1. Prove that K is the class of models of a single universal sentence iff there is a number n such that for all models $\mathfrak{A}$, $\mathfrak{A} \in K$ iff every n-element submodel of $\mathfrak{A}$ is isomorphically embeddable in some model of K. Is there an analogue when $\mathscr{L}$ is arbitrary?

5.2.3. Let K be an elementary class. Show that the class $\bar{K}$ of all submodels of models of K is the class of models of some set of universal sentences.

[*Hint*: Use Exercise 5.2.1, the version for arbitrary $\mathscr{L}$.]

5.2.4*. Show by a counterexample that if K is a basic elementary class, then the class $\bar{K}$ as in Exercise 5.2.3 need not be the class of models of a single universal sentence.

5.2.5*. Find counterexamples for the analogues of Exercise 5.2.3, where $\bar{K}$ is the class of all extensions of models of K, all homomorphic images of models of K, and all unions of chains of models of K.

5.2.6. A sentence is *existential–positive* iff it is both existential and positive. Prove that a sentence φ is preserved under both homomorphisms and extensions iff either it is equivalent to an existential–positive sentence, or else $\vdash \neg \varphi$. Do the same for universal–positive and homomorphism and submodels. Are there analogues for universal–existential–positive sentences?

5.2.7. Give model-theoretic characterizations, in the manner of Theorem 5.2.8, of theories with Σ^0_{2n} axioms. By using sandwiches of even lengths, e.g., start with $\mathfrak{B}_0$ and end with $\mathfrak{A}_n$, characterize theories in Π^0_m and Σ^0_m for m odd. Extend all these results to elementary classes.

5.2.8. The notion of n-sandwiches can be given the following purely algebraic formulation. Prove that the model $\mathfrak{A}$ is n-sandwiched by $\mathfrak{B}_0$ iff there is a $(2n+1)$-chain of models

$$\mathfrak{B}_0 \subset \mathfrak{A}_0 \subset \mathfrak{B}_1 \subset \ldots \subset \mathfrak{A}_{n-1} \subset \mathfrak{B}_n$$

such that each $\mathfrak{B}_{k+1}$ is isomorphic to an ultrapower of $\mathfrak{B}_k$, $0 \leqslant k < n$, and similarly for the models $\mathfrak{A}_k$, $0 \leqslant k < n$. Then all references to syntactical notions or the notion of satisfaction are deleted.

5.2.9. By appealing to the proof of Theorem 5.2.6, show the following for an elementary class K: K is closed under unions of infinite chains of length ω iff K is closed under arbitrary directed unions. (For the definition of directed unions, see Exercise 3.1.9. Exercise 3.1.10 is also relevant.)

5.2.10. Change the hypothesis of Proposition 5.2.14 to the following: Whenever $\mathfrak{B}_1, \mathfrak{B}_2 \subset \mathfrak{A}$ are three models of φ and $\mathfrak{B}_1 \cap \mathfrak{B}_2 \neq 0$, then $\mathfrak{B}_1 \cap \mathfrak{B}_2$ is a model of φ. Then prove that the conclusion can be similarly changed to: Whenever the intersection of a nonempty family of submodels of φ is nonempty, then it is a model of φ.

5.2.11. Show that there are sentences φ which are preserved under nonempty descending intersections but which are not preserved under nonempty intersections of two submodels.

5.2.12**. Characterize the sentences φ, or sets of sentences Σ, which are preserved under nonempty intersections of two submodels. (As a starting point, see Exercise 5.2.18.)

5.2.13**. Prove that a sentence φ is preserved under nonempty descending intersections if and only if it is preserved under unions of chains and nonempty intersections of two elementary submodels.

5.2.14*. Prove that φ is preserved under (nonempty) simple infinite descending intersections if and only if φ is preserved under (nonempty) directed descending intersections.

5.2.15. Prove the following more general version of Proposition 5.2.16. Suppose that φ is closed under nonempty descending intersections. Let $\mathfrak{A}$ be a model such that

(i) some extension of $\mathfrak{A}$ is a model of φ,

(ii) for every extension $\mathfrak{B}$ of $\mathfrak{A}$ which is a model of φ and for all finite subsets $A' \subset A$ and $B' \subset B \setminus A$, there is a model $\mathfrak{C}$ of φ such that $A' \subset C \subset B$ and $C \cap B' = 0$.

Then $\mathfrak{A}$ is a model of φ.

5.2.16*. Find a sentence φ which is preserved under unions of chains but not preserved under nonempty descending intersections.

5.2.17*. Let U be a 1-placed relation symbol of $\mathscr{L}$ and T be a complete theory of $\mathscr{L}$. Prove that the following are equivalent:

(i). There is a model $\mathfrak{A} = \langle A, V \ldots\rangle$ of T such that V is a descending intersection of submodels of $\mathfrak{A}$ and of T.

(ii). There is a model $\mathfrak{A} = \langle A, V \ldots\rangle$ of T such that for every finite subset $X \subset A \setminus V$, there is a submodel $\mathfrak{B} = \langle B, V \ldots\rangle$ of T such that $B \supset V$ and $B \cap X = 0$.

5.2.18. Suppose that a sentence φ has the following form (ψ open):

$$\varphi = (\forall x_1 \ldots x_n \exists y_1 \ldots y_m)[\psi(x_1 \ldots x_n y_1 \ldots y_m)$$
$$\wedge (\forall z_1 \ldots z_m)(\psi(x_1 \ldots x_n z_1 \ldots z_m) \to \bigwedge_{1 \leq i \leq m} y_i \equiv z_i)].$$

Show that φ is preserved under intersections of submodels.

5.2.19*. A model $\mathfrak{A}$ is said to be *simple* iff every homomorphism mapping $\mathfrak{A}$ onto $\mathfrak{B}$ is an isomorphism of $\mathfrak{A}$ onto $\mathfrak{B}$. Let T be a theory of $\mathscr{L}$. Then every model of T is simple if and only if every formula φ of $\mathscr{L}$ is equivalent under T to a positive formula.

5.2.20*. Let U be a 1-placed relation in $\mathscr{L}$. Given a model $\mathfrak{A} = \langle A, U \ldots\rangle$ for $\mathscr{L}$, let $\mathfrak{A}|U$ be the least submodel of $\mathfrak{A}$ which includes U. The *relativization* φ^U of a formula φ of $\mathscr{L}$ to U is defined in the following recursive way:

If φ is atomic, then $\varphi^U = \varphi$;

$$(\varphi \wedge \psi)^U = \varphi^U \wedge \psi^U, \qquad (\varphi \vee \psi)^U = \varphi^U \vee \psi^U, \qquad (\neg \varphi)^U = \neg \varphi^U;$$
$$((\forall x)\varphi)^U = (\forall x)(U(x) \rightarrow \varphi^U);$$
$$((\exists x)\varphi)^U = (\exists x)(U(x) \wedge \varphi^U).$$

Prove that φ^U is logically equivalent to $(\varphi^U)^U$. Prove that for every theory T and sentence φ of $\mathscr{L}$, the following conditions are equivalent:

(i). There is a sentence ψ of $\mathscr{L}$ such that

$$T \vdash \varphi \leftrightarrow (\psi^U).$$

(ii). For any two models $\mathfrak{A} = \langle A, U \ldots\rangle$ and $\mathfrak{B} = \langle B, V \ldots\rangle$ of T, if $\mathfrak{A}|U \cong \mathfrak{B}|V$ and $\mathfrak{A} \models \varphi$, then $\mathfrak{B} \models \varphi$.

5.2.21*. We say that $\mathfrak{A}$ is the *union of a set* K of models iff each $\mathfrak{B} \in K$ is a submodel of $\mathfrak{A}$ and $A = \bigcup\{B : \mathfrak{B} \in K\}$. Prove that a sentence φ is preserved under unions if and only if it is logically equivalent to a sentence of the form $(\forall x \exists y_1 \ldots y_n)\psi$, where ψ is quantifier-free.

5.2.22*. By a *strong homomorphism* of $\mathfrak{A}$ onto $\mathfrak{B}$ we mean a homomorphism f of $\mathfrak{A}$ onto $\mathfrak{B}$ such that for every n-placed relation R of $\mathfrak{A}$ and the corresponding relation S of $\mathfrak{B}$, we have

$$S = \{\langle fa_1 \ldots fa_n\rangle : \langle a_1 \ldots a_n\rangle \in R\}.$$

Assume $\mathscr{L}$ contains only binary relation symbols. Prove that a sentence φ is preserved under strong homomorphisms iff φ is equivalent to a sentence of the form

$$(\exists x_1 \forall y_1 z_1 \exists x_2 \forall y_2 z_2 \ldots \exists x_m \forall y_m z_m) \bigwedge_{i=1}^{r} \bigvee_{j=1}^{s} \theta_{ij},$$

where each θ_{ij} is either atomic or of the form

$$\neg P(y_k z_k).$$

Generalize the result to arbitrary $\mathscr{L}$. Give a sentence which is closed under homomorphisms but not closed under strong homomorphisms.

5.2.23**. The *direct product* $\mathfrak{A} \times \mathfrak{B}$ of two models $\mathfrak{A}$, $\mathfrak{B}$ is defined in Exercise 4.1.12. We say that $\mathfrak{A}$ is a *direct factor* of $\mathfrak{C}$ iff there exists a model $\mathfrak{B}$ such that either $\mathfrak{A} \times \mathfrak{B} = \mathfrak{C}$ or $\mathfrak{B} \times \mathfrak{A} = \mathfrak{C}$. Give a syntactical characterization of sentences preserved under direct factors. Give an example of a sentence closed under strong homomorphisms but not under direct factors.

5.2.24*. By a *direct system* of type ω, we mean a sequence of models $\mathfrak{A}_0, \mathfrak{A}_1, \mathfrak{A}_2, \ldots$ and a sequence of mappings $f_0, f_1, f_2, \ldots$ such that each f_n is a homomorphism on $\mathfrak{A}_n$ into $\mathfrak{A}_{n+1}$. A model $\mathfrak{A}$ will be said to be a *direct limit* of the direct system iff for each n, there is a homomorphism g_n on $\mathfrak{A}_n$ into $\mathfrak{A}$ such that:

(i) for all $n < \omega$ and $a \in A_n$, $g_n(a) = g_{n+1}f_n(a)$;

(ii) A is the union of the ranges of the functions $g_0, g_1, g_2, \ldots$;

(iii) for any other model $\mathfrak{A}'$ and homomorphisms g_n' satisfying (i) and (ii) above, there is a homomorphism h on $\mathfrak{A}$ onto $\mathfrak{A}'$ such that for each n, $g_n \circ h = g_n'$.

Prove that the direct limit exists and is unique up to isomorphism. Prove that a sentence φ is preserved under direct limits iff φ is equivalent to a sentence of the form

$$\bigwedge_{i=1}^{m} (\forall x_1 \ldots x_s)(\psi_i \to (\exists y_1 \ldots y_t)\theta_i),$$

where the ψ_i and θ_i are quantifier-free positive formulas.

5.2.25*. By an *inverse system* of type ω, we mean a sequence of models $\mathfrak{A}_0, \mathfrak{A}_1, \mathfrak{A}_2, \ldots$ and mappings $f_0, f_1, f_2, \ldots$ such that each f_n is a homomorphism of a submodel of $\mathfrak{A}_{n+1}$ onto $\mathfrak{A}_n$. $\mathfrak{A}$ is said to be an *inverse limit* of the sequences iff for each n, there is a homomorphism g_n on a submodel $\mathfrak{B}_n \subset \mathfrak{A}$ onto $\mathfrak{A}_n$ such that

(i) for all n and all $b \in B_n$, $f_n g_{n+1}(b) = g_n(b)$;

(ii) $\mathfrak{A} = \bigcup_{n<\omega} \mathfrak{B}_n$;

(iii) if $\mathfrak{A}'$, g_n' also satisfy (i) and (ii) above, then there is a homomorphism h of $\mathfrak{A}'$ onto $\mathfrak{A}$ such that for each n, $h \circ g_n = g_n'$.

Show that the inverse limit of any inverse system exists and is unique. Prove that a sentence φ is preserved under inverse limits iff φ is equivalent to a sentence of the form

$$\bigwedge_{i=1}^{m} (\forall x_1 \ldots x_s)((\forall y_1 \ldots y_t)\psi_i \to \theta_i),$$

where ψ_i and θ_i are quantifier-free positive formulas.

5.2.26*. Let $\mathscr{L} = \{\epsilon\}$, where ϵ is a binary relation symbol. Given two models $\mathfrak{A} = \langle A, E\rangle$, $\mathfrak{A}' = \langle A', E'\rangle$ for $\mathscr{L}$, we say that $\mathfrak{A}'$ is an *outer extension* of $\mathfrak{A}$ iff $\mathfrak{A} \subset \mathfrak{A}'$ and for all $a \in A$ and $b \in A'$, if $bE'a$, then $b \in A$. A formula φ of $\mathscr{L}$ is called *essentially existential* iff it is obtained from atomic and negations of atomic formulas by means of

conjunction: $\theta_1 \wedge \theta_2$;
disjunction: $\theta_1 \vee \theta_2$;
restricted quantifications: $(\forall x)(x \epsilon y \to \theta)$, $(\exists x)(x \epsilon y \wedge \theta)$;
existential quantifications: $(\exists y)\varphi$.

Prove that a sentence φ of $\mathscr{L}$ is preserved under outer extensions iff φ is equivalent to an essentially existential sentence. Generalize this result to cases when $\mathscr{L}$ may contain symbols (relations and functions) other than ϵ.

5.3. Applications of special models to the theory of interpolation and definability

This section contains applications of special models to prove some more general results in interpolation and definability. We have already encountered in Section 2.2 the two interpolation theorems of Craig and Lyndon and the definability theorem of Beth. The kind of proof employed in Section 2.2 is, of course, quite elementary when compared with proofs requiring the use of special or saturated models. We shall see, however, that there are several generalizations whose proofs use special models. At the moment, the use of special models in these results (Theorem 5.3.6 and some of the exercises) appears to be essential.

By way of introducing the use of special models in interpolation and definability results, and as a refresher to the reader, we first give a new proof of the Robinson consistency theorem, from which the reader should be able to deduce the Craig interpolation theorem.

THEOREM 2.2.23 (Robinson Consistency Theorem). *Let $\mathscr{L}_1$ and $\mathscr{L}_2$ be two languages and let $\mathscr{L} = \mathscr{L}_1 \cap \mathscr{L}_2$. Suppose T is a complete theory in $\mathscr{L}$, and $T_1 \supset T$, $T_2 \supset T$ are consistent theories in $\mathscr{L}_1$ and $\mathscr{L}_2$, respectively. Then $T_1 \cup T_2$ is consistent in the language $\mathscr{L}_1 \cup \mathscr{L}_2$.*

PROOF. Suppose that $\mathfrak{A}_1 \models T_1$ and $\mathfrak{A}_2 \models T_2$. Let $\mathfrak{B}_1$ be the $\mathscr{L}$ reduct of $\mathfrak{A}_1$, and similarly let $\mathfrak{B}_2$ be the $\mathscr{L}$ reduct of $\mathfrak{A}_2$. Since $\mathfrak{B}_1 \models T$, $\mathfrak{B}_2 \models T$ and T is complete, we have $\mathfrak{B}_1 \equiv \mathfrak{B}_2$. Now let $\overline{\mathfrak{A}}_1$, $\overline{\mathfrak{A}}_2$ be special elementary extensions of $\mathfrak{A}_1$ and $\mathfrak{A}_2$ of the same (finite or infinite) power. Let $\overline{\mathfrak{B}}_1$ and $\overline{\mathfrak{B}}_2$ be the reducts of $\overline{\mathfrak{A}}_1$ and $\overline{\mathfrak{A}}_2$, respectively, to $\mathscr{L}$. Then we see that

$\mathfrak{B}_1 \prec \overline{\mathfrak{B}}_1$, and $\mathfrak{B}_2 \prec \overline{\mathfrak{B}}_2$,

$\overline{\mathfrak{B}}_1$ and $\overline{\mathfrak{B}}_2$ are equivalent special models of the same power.

Hence $\overline{\mathfrak{B}}_1 \cong \overline{\mathfrak{B}}_2$. Using this, we can find an expansion $\mathfrak{A}$ of $\overline{\mathfrak{A}}_1$ by adding interpretations of the symbols in $\mathscr{L}_2 \setminus \mathscr{L}_1$ which are images of the corresponding interpretations in $\overline{\mathfrak{A}}_2$ under the isomorphism mapping $\overline{\mathfrak{B}}_2$ onto $\overline{\mathfrak{B}}_1$. $\mathfrak{A}$ then will be a model for $\mathscr{L}_1 \cup \mathscr{L}_2$ such that its $\mathscr{L}_1$-reduct is $\overline{\mathfrak{A}}_1$ and its $\mathscr{L}_2$-reduct is isomorphic to $\overline{\mathfrak{A}}_2$. So every sentence of $T_1 \cup T_2$ holds in $\mathfrak{A}$, whence $T_1 \cup T_2$ is consistent. ⊣

Let P and P' be two new n-placed relation symbols not in $\mathscr{L}$. Let $\Sigma(P)$ be a set of sentences of $\mathscr{L} \cup \{P\}$, and let $\Sigma(P')$ be the corresponding set of sentences of $\mathscr{L} \cup \{P'\}$ formed by replacing P everywhere by P'. Recall from Chapter 2 that $\Sigma(P)$ defines P implicitly iff

$$\Sigma(P), \Sigma(P') \models (\forall x_1 \dots x_n)(P(x_1 \dots x_n) \leftrightarrow P'(x_1 \dots x_n)).$$

Phrased slightly differently, this notion is equivalent to the following: Given any model $\mathfrak{A}$ for $\mathscr{L}$, there is at most one n-placed relation R over A such that

$$(\mathfrak{A}, R) \models \Sigma(P).$$

If $\Sigma(P)$ were a finite set of sentences (any argument involving the compactness theorem will reduce $\Sigma(P)$ to this case), then the condition that there be at most one relation R can be written thus:

$$(^*) \qquad \mathfrak{A} \models (\forall PP')(\Sigma(P) \wedge \Sigma(P') \rightarrow P = P').$$

The sentence appearing to the right of $\models$ is a second-order sentence, in fact, a Π^1_1 sentence (see the discussion in Chapter 4). The interpretation of $\models$ is, of course, the usual (standard) satisfaction predicate with respect to second-order sentences. By Beth's theorem 2.2.22, (*) is true for every model for $\mathscr{L}$ iff $\Sigma(P)$ defines P explicitly, i.e., there is a formula $\varphi(x_1 \dots x_n)$ of $\mathscr{L}$ such that

$$\Sigma(P) \models (\forall x_1 \dots x_n)(P(x_1 \dots x_n) \leftrightarrow \varphi(x_1 \dots x_n)).$$

From the form of Beth's theorem, it would follow that an important model-theoretic question to ask is the following:

> Given $\Sigma(P)$ and given $\mathfrak{A}$ (a model for $\mathscr{L}$), exactly how many (or how few) relations R are there on A such that $(\mathfrak{A}, R) \models \Sigma(P)$?

This question for an arbitrary model $\mathfrak{A}$ is extremely difficult to answer, as it would already imply a certain knowledge of second-order logic, which we do not as yet have. If the question were posed for the class of all models $\mathfrak{A}$ for $\mathscr{L}$, as, for example, in Beth's theorem and in some of our later theorems in this section, very elegant answers can be found. It turns out that for special models, we can answer quite satisfactorily the question whether

$$\mathfrak{A} \models (\exists P)\bigwedge \Sigma(P),$$

as the following proposition shows.

PROPOSITION 5.3.1. *Let $\Sigma(P; x_1 \dots x_n)$ be a set of formulas of $\mathscr{L} \cup \{P\}$ in the free variables $x_1, \dots, x_n$. Then there exists a set Φ of formulas of $\mathscr{L}$ in the variables $x_1, \dots, x_n$ satisfying:*

(i). *every formula $\varphi(x_1 \dots x_n)$ in Φ contains only those symbols of $\mathscr{L}$ occurring in the set Σ;*

(ii). $\models (\forall x_1 \dots x_n)((\exists P)\bigwedge \Sigma \rightarrow \bigwedge \Phi)$;

(iii). *for every special model $\mathfrak{A}$ for $\mathscr{L}$, which is either finite or is of power $\omega < |A| = |A|^*$, $||\mathscr{L}|| < |A|$,*

$$\mathfrak{A} \models (\forall x_1 \dots x_n)((\exists P)\bigwedge \Sigma \leftrightarrow \bigwedge \Phi).$$

PROOF. We first define the set Φ as follows: Φ consists of all those formulas $\varphi(x_1 \dots x_n)$ of $\mathscr{L}$ containing only symbols occurring in Σ such that

$$\models (\forall x_1 \dots x_n)((\exists P)\bigwedge \Sigma \rightarrow \varphi).$$

Then it is clear that (i) and (ii) hold.

To prove (iii), let $\mathfrak{A}$ satisfy the hypothesis of (iii) and let $a_1, \dots, a_n \in A$ be such that

(1) $$A \models \bigwedge \Phi[a_1 \dots a_n].$$

We first show that

(2) the theory of $(\mathfrak{A}, a_1 \dots a_n) \cup \Sigma(P; c_1 \dots c_n)$ is consistent in the language $\mathscr{L}' = \mathscr{L} \cup \{P, c_1, \dots, c_n\}$.

If (2) were false, then for some $\sigma(x_1 \dots x_n)$ of $\mathscr{L}$, we have

$$\mathfrak{A} \models \sigma[a_1 \dots a_n],$$

$$\Sigma(P; c_1 \dots c_n) \models \neg\sigma(c_1 \dots c_n).$$

By the interpolation theorem, we may assume that $\neg\sigma(x_1 \dots x_n)$ contains only symbols occurring in Σ. This shows that $\neg\sigma(x_1 \dots x_n) \in \Phi$, contradicting (1). Now let $(\mathfrak{B}, R, b_1 \dots b_n)$ be a special model for $\mathscr{L}'$ which satisfies the set of sentences in (2) and which has power $|B| = |A|$. Then since

$$(\mathfrak{A}, a_1 \dots a_n) \equiv (\mathfrak{B}, b_1 \dots b_n),$$

and both are special models of the same power, we have

$$(\mathfrak{A}, a_1 \dots a_n) \cong (\mathfrak{B}, b_1 \dots b_n).$$

The isomorphism naturally carries the relation R on B onto a relation S on A such that

$$(\mathfrak{A}, S, a_1 \dots a_n) \models \Sigma(P; c_1 \dots c_n). \dashv$$

Some remarks on this simple-looking proposition are in order here. If $\mathscr{L}$ is countable and some sort of meaningful Gödel numbers are assigned to the formulas of $\mathscr{L}$, then it is easily seen that the set Φ will be recursively enumerable in Σ. If Σ is a single formula or sentence, then Φ is recursively enumerable. Proposition 5.3.1 can be improved in two directions. First of all, if Σ is a single sentence or formula, then Φ can be taken to be primitive recursive, i.e., we can give an explicit description of the set Φ, so that the conclusion still holds. Secondly, we can dispense with the condition that the power of $\mathfrak{A}$ must be a cardinal $\alpha = \alpha^*$, and, in the case where Σ is a single sentence the restriction that $|A| > ||\mathscr{L}||$ is also not necessary. Finally, the use of the Craig interpolation theorem in the proof is not essential, as we can restrict ourselves to those symbols of $\mathscr{L}$ which do occur in some formula in Σ.

We next prove a generalization of the Craig interpolation theorem. At its simplest level, the Craig interpolation theorem for formulas can be stated in the following fashion: Let $\varphi(P; x_1 \dots x_n)$ be a formula of $\mathscr{L} \cup \{P\}$ and let $\psi(Q; x_1 \dots x_n)$ be a formula of $\mathscr{L} \cup \{Q\}$. We assume that P and Q are relation symbols not already in $\mathscr{L}$. Then the following are equivalent:

(i). $\models (\forall x_1 \dots x_n)((\exists P)\varphi \to (\forall Q)\psi)$;

(ii). there is a formula $\theta(x_1 \dots x_n)$ of $\mathscr{L}$ such that

$$\models (\forall x_1 \dots x_n)[((\exists P)\varphi \to \theta) \wedge (\theta \to (\forall Q)\psi)].$$

The generalization which follows is stated in a similar way.

THEOREM 5.3.2. *Let φ and ψ be as in the above, and let $\mathsf{Q}_1, \dots, \mathsf{Q}_n$ be any string of quantifiers, $\forall$ or $\exists$. Then the following are equivalent*:

(i). $\models (\mathsf{Q}_1 x_1 \dots \mathsf{Q}_n x_n)((\exists P)\varphi \to (\forall Q)\psi)$.

(ii). *There is a formula $\theta(x_1 \dots x_n)$ of $\mathscr{L}$ such that*

$$\models (\mathsf{Q}_1 x_1 \dots \mathsf{Q}_n x_n)[((\exists P)\varphi \to \theta) \wedge (\theta \to (\forall Q)\psi)].$$

PROOF. (ii) $\Rightarrow$ (i). This is trivial and follows from logic.

(i) ⇒ (ii). By Proposition 5.3.1, let

$$\sigma_1, \sigma_2, \ldots, \sigma_m, \ldots$$

be a list of all formulas of $\mathscr{L}$ in Φ (from Proposition 5.3.1) with respect to the formula $(\exists P)\varphi$, and let

$$\tau_1, \tau_2, \ldots, \tau_m, \ldots$$

be a list of all negations of formulas of $\mathscr{L}$ in Φ (from Proposition 5.3.1) with respect to the formula $(\exists Q) \neg \psi$. We may assume by taking finite conjunctions that

$$\models \sigma_{m+1} \rightarrow \sigma_m,$$

and by taking finite disjunctions that

$$\models \tau_m \rightarrow \tau_{m+1}.$$

We already know that

(1) $$\models (\exists P)\varphi \rightarrow \bigwedge_m \sigma_m,$$

(2) $$\models \bigvee_m \tau_m \rightarrow (\forall Q)\psi.$$

We claim that

(3) there exists an m such that $\models (\mathrm{Q}_1 x_1 \ldots \mathrm{Q}_n x_n)(\sigma_m \rightarrow \tau_m)$.

First of all, (3) is certainly sufficient to prove the theorem, as by taking $\theta = \sigma_m$ or $\theta = \tau_m$, and by using (1), (2), (3) and some predicate logic, we have the conclusion (ii). (This is a good time to brush up on some of the rules of predicate logic.) So it is sufficient to prove (3). Suppose that (3) is false. Then

(4) for each m, there exists a model $\mathfrak{A}_m$ such that

$$\mathfrak{A}_m \models (\overline{\mathrm{Q}}_1 x_1 \ldots \overline{\mathrm{Q}}_n x_n)(\sigma_m \wedge \neg \tau_m),$$

where $\overline{\mathrm{Q}}_i = \exists$ if $\mathrm{Q}_i = \forall$ and $\overline{\mathrm{Q}}_i = \forall$ if $\mathrm{Q}_i = \exists$. The sentences in (4) become logically stronger as m becomes larger. Therefore, by the compactness theorem, the set of sentences

(5) $$(\overline{\mathrm{Q}}_1 x_1 \ldots \overline{\mathrm{Q}}_n x_n)(\sigma_m \wedge \neg \tau_m), \qquad m = 1, 2, \ldots,$$

is consistent and has a model. Since σ_m and τ_m contain only symbols occurring in φ or ψ, we may assume that $\mathscr{L}$ is countable. Let $\mathfrak{A}$ be a special model of the sentences in (5) of power $\beth_{\omega_1}$. It is a simple fact to verify that not only is $\mathfrak{A}$ special, but also

(6) $\mathfrak{A}$ is ω_1-saturated.

Schematically, we have

$$\mathfrak{A} \models \bigwedge_m (\overline{Q}_1 x_1 \ldots \overline{Q}_n x_n)(\sigma_m \wedge \neg \tau_m).$$

We show that, by using (6),

(7) $$\mathfrak{A} \models (\overline{Q}_1 x_1 \ldots \overline{Q}_n x_n)[\bigwedge_m (\sigma_m \wedge \neg \tau_m)].$$

This is proved by pushing the infinite conjunction $\bigwedge_m$ through the quantifiers $\overline{Q}$ one by one. If $\overline{Q}_i = \forall$, then clearly $\bigwedge_m \overline{Q}_i = \overline{Q}_i \bigwedge_m$. If $\overline{Q}_i = \exists$, then we first show:

(8) let $a_1, \ldots, a_{i-1}$ be any sequence of elements in A, and let $(\exists x_i)\eta_m(x_1 \ldots x_{i-1} x_i)$, $m = 1, 2, \ldots$, be any sequence of formulas of $\mathscr{L}$ such that $\models \eta_{m+1} \to \eta_m$. Then

$$\mathfrak{A} \models \bigwedge_m (\exists x_i)\eta_m[a_1 \ldots a_{i-1} x_i] \quad \text{iff} \quad \mathfrak{A} \models (\exists x_i) \bigwedge_m \eta_m[a_1 \ldots a_{i-1} x_i].$$

The proof of (8) is immediate, since, by (6), $(\mathfrak{A}, a_1 \ldots a_{i-1})$ is still ω_1-saturated, and we simply find an $a_i \in A$ such that

$$\mathfrak{A} \models \eta_m[a_1 \ldots a_i] \qquad \text{for all} \quad m = 1, 2, \ldots.$$

Now we see that

$$\mathfrak{A} \models (\overline{Q}_1 x_1 \ldots \overline{Q}_{i-1} x_{i-1}) \bigwedge_m (\overline{Q}_i x_i \ldots \overline{Q}_n x_n)(\sigma_m \wedge \neg \tau_m)$$

iff

$$\mathfrak{A} \models (\overline{Q}_1 x_1 \ldots \overline{Q}_i x_i) \bigwedge_m (\overline{Q}_{i+1} x_{i+1} \ldots \overline{Q}_n x_n)(\sigma_m \wedge \neg \tau_m).$$

Whence, by induction, (7) holds. Since $(\beth_{\omega_1}) = (\beth_{\omega_1})^*$, the conclusions of Proposition 5.3.1 with respect to φ and ψ both hold for $\mathfrak{A}$. So (by predicate logic again)

$$\mathfrak{A} \models (\overline{Q}_1 x_1 \ldots \overline{Q}_n x_n)(\bigwedge_m \sigma_m \wedge \neg \bigvee_m \tau_m),$$

$$\mathfrak{A} \models (\overline{Q}_1 x_1 \ldots \overline{Q}_n x_n)((\exists P)\varphi \wedge \neg (\forall Q)\psi),$$

$$\mathfrak{A} \models \neg (Q_1 x_1 \ldots Q_n x_n)((\exists P)\varphi \to (\forall Q)\psi).$$

This contradicts (i). So (3) must hold and the theorem is proved. ⊣

Further refinements of Theorem 5.3.2 can be found in the exercises.

Our next few results are generalizations of Beth's theorem in various directions. No essential use of special models is needed again until Theorem 5.3.6, although the proofs of Theorems 5.3.3 and 5.3.4, and Proposition 5.3.5 are simpler with special models.

Let $\Sigma(P)$ be a set of sentences of $\mathscr{L} \cup \{P\}$, P a new n-placed relation symbol. $\Sigma(P)$ is said to define P *explicitly up to disjunction* iff there are a finite number of formulas $\varphi_1(x_1 \dots x_n), \dots, \varphi_m(x_1 \dots x_n)$ of $\mathscr{L}$ such that

$$\Sigma(P) \models \bigvee_{1 \leqslant i \leqslant m} (\forall x_1 \dots x_n)(P(x_1 \dots x_n) \leftrightarrow \varphi_i(x_1 \dots x_n)).$$

$\Sigma(P)$ is said to define P *explicitly up to parameters* iff there is a formula $\varphi(x_1 \dots x_n y_1 \dots y_m)$ of $\mathscr{L}$ such that

$$\Sigma(P) \models (\exists y_1 \dots y_m \forall x_1 \dots x_n)(P(x_1 \dots x_n) \leftrightarrow \varphi(x_1 \dots x_n y_1 \dots y_m)).$$

Finally, $\Sigma(P)$ is said to define P *explicitly up to parameters and disjunction* iff there are formulas $\varphi_1(x_1 \dots x_n y_1 \dots y_m), \dots, \varphi_k(x_1 \dots x_n y_1 \dots y_m)$ of $\mathscr{L}$ such that

$$\Sigma(P) \models \bigvee_{1 \leqslant i \leqslant k} (\exists y_1 \dots y_m \forall x_1 \dots x_n)(P(x_1 \dots x_n) \leftrightarrow \varphi_i(x \dots x_n y_1 \dots y_m)).$$

By the compactness theorem, every result that we prove in what follows extends obviously to sets of sentences of $\mathscr{L} \cup \{P\}$, once we can prove it for single sentences of $\mathscr{L} \cup \{P\}$. It will be equally clear that instead of an n-placed relation symbol P, we can, in order to save on notation, use a 1-placed relation symbol U. The reader is asked to verify these claims as exercises. If $\varphi(U)$ is a sentence of $\mathscr{L} \cup \{U\}$, we sometimes shall simply write φ for $\varphi(U)$.

The next theorem slightly extends the range of Beth's theorem.

THEOREM 5.3.3 (Svenonius' Theorem). *Let φ be a sentence of $\mathscr{L} \cup \{U\}$, U a new 1-placed predicate. The following are equivalent:*

(i). *For every model $\mathfrak{A}$ for $\mathscr{L}$, if any two expansions $(\mathfrak{A}, X_1)$ and $(\mathfrak{A}, X_2)$ of $\mathfrak{A}$ to models of φ are isomorphic, then $X_1 = X_2$.*

(ii). *φ defines U explicitly up to disjunction.*

PROOF. (ii) $\Rightarrow$ (i). Suppose that $\theta_1(x), \dots, \theta_k(x)$ are formulas of $\mathscr{L}$ such that

$$\varphi \models \bigvee_{1 \leqslant i \leqslant k} (\forall x)(U(x) \leftrightarrow \theta_i). \tag{1}$$

Let $(\mathfrak{A}, X)$ be an expansion of $\mathfrak{A}$ which is a model of φ. This means that for some $\theta_i(x)$,

$$X = \{a \in A : \mathfrak{A} \models \theta_i[a]\}.$$

So X remains fixed under any automorphism of $\mathfrak{A}$, and, in particular, under any isomorphism of $(\mathfrak{A}, X)$ onto an $(\mathfrak{A}, X')$.

(i) ⇒ (ii). Suppose that (ii) does not hold, i.e., for no formulas $\theta_1(x), \dots, \theta_k(x)$ of $\mathscr{L}$ does (1) hold. Then the set

$$\Sigma = \{\varphi\} \cup \{\neg (\forall x)(U(x) \leftrightarrow \theta(x)) : \theta(x) \text{ a formula of } \mathscr{L}\}$$

of sentences of $\mathscr{L} \cup \{U\}$ is consistent. Let T be any complete extension of Σ in $\mathscr{L} \cup \{U\}$. Note that T does not define U explicitly, as there is no formula $\theta(x)$ of $\mathscr{L}$ such that

$$T \models (\forall x)(U(x) \leftrightarrow \theta(x)).$$

Whence, by Beth's theorem, there is a model $\mathfrak{A}$ for $\mathscr{L}$ which has two different expansions into models of T:

$$(\mathfrak{A}, X_1) \models T, \qquad (\mathfrak{A}, X_2) \models T, \qquad X_1 \neq X_2.$$

Since T is complete, we also have

$$(\mathfrak{A}, X_1) \equiv (\mathfrak{A}, X_2).$$

Consider the model $(\mathfrak{A}, X_1 X_2)$. Let $(\mathfrak{B}, Y_1 Y_2)$ be a special elementary extension of it with respect to the language $\mathscr{L} \cup \{U, U'\}$. Clearly, $(\mathfrak{B}, Y_1)$ and $(\mathfrak{B}, Y_2)$ are special, of the same power, and equivalent. So, because $(\mathfrak{A}, X_1 X_2) \models \neg (\forall x)(U(x) \leftrightarrow U'(x))$, we see that

$$(\mathfrak{B}, Y_1) \cong (\mathfrak{B}, Y_2),$$

$$(\mathfrak{B}, Y_1 Y_2) \models \neg (\forall x)(U(x) \leftrightarrow U'(x)),$$

$$Y_1 \neq Y_2.$$

This contradicts (i). ⊣

The following finite analogue of Beth's theorem is more difficult to prove. It gives one answer to the question we posed just before Proposition 5.3.1.

THEOREM 5.3.4 (Kueker Finite Definability Theorem). *Let φ be a sentence of $\mathscr{L} \cup \{U\}$, and let $n \geqslant 1$. Then the following are equivalent:*

$(\mathrm{i})_n$. *For every model $\mathfrak{A}$ for $\mathscr{L}$, there are at most n subsets $X \subset A$ such that $(\mathfrak{A}, X) \models \varphi$.*

$(\mathrm{ii})_n$. *There are formulas $\sigma(v_1 \dots v_k)$, $\theta_i(xv_1 \dots v_k)$, $1 \leqslant i \leqslant n$, of $\mathscr{L}$ such* *that*

$$\varphi \models (\exists v_1 \dots v_k)\sigma,$$

$$\varphi \models (\forall v_1 \dots v_k)[\sigma \to \bigvee_{1 \leqslant i \leqslant n} (\forall x)(U(x) \leftrightarrow \theta_i)].$$

PROOF. $(\text{ii})_n \Rightarrow (\text{i})_n$. This is easy to verify.

$(\text{i})_n \Rightarrow (\text{ii})_n$. We shall first prove an easy consequence of Proposition 5.3.1. *Let $\theta(P; y)$ be a formula of $\mathscr{L} \cup \{P\}$. Then the following are equivalent*:

(1) $$\models (\exists y \forall P)\theta.$$

(2) *There is a formula $\sigma(y)$ of $\mathscr{L}$ such that $\models (\exists y)\sigma$ and $\models (\forall y)(\sigma \to \theta)$.*

The proof of (1) from (2) is easy and we leave it to the reader.

Assume (1). Then, by Proposition 5.3.1, there is a set $\Sigma(y)$ of formulas of $\mathscr{L}$ such that

$$\mathfrak{A} \models (\forall y)((\forall P)\theta \leftrightarrow \bigvee \Sigma),$$

for every special model $\mathfrak{A}$ for $\mathscr{L}$ such that $|A| = |A|^*$ and $||\mathscr{L}|| < |A|$ if $\mathfrak{A}$ is infinite. By (1), we have

$$\mathfrak{A} \models (\exists y)\sigma \quad \text{for some} \quad \sigma \in \Sigma.$$

Since every model for $\mathscr{L}$ is equivalent to some such special model $\mathfrak{A}$, we see that

$$\models \bigvee \{(\exists y)\sigma : \sigma \in \Sigma\}.$$

Hence it follows from compactness that there are $\sigma_1, \dots, \sigma_m \in \Sigma$ such that

$$\models (\exists y)\sigma_1 \vee \dots \vee (\exists y)\sigma_m.$$

So, defining $\sigma = \sigma_1 \vee \dots \vee \sigma_m$, we find that (2) holds.

Returning now to the proof of the theorem, assume $(\text{i})_n$. We first show that there are formulas $\psi_i(U; v_1 \dots v_k)$ of $\mathscr{L} \cup \{U\}$, $1 \leqslant i \leqslant n$, such that

(3) $$\models (\exists v_1 \dots v_k \forall U U' \forall x)[(\varphi(U) \to \bigvee_{1 \leqslant i \leqslant n} \psi_i(U; v_1 \dots v_k)) \wedge \bigwedge_{1 \leqslant i \leqslant n} (\varphi(U) \wedge \psi_i(U) \wedge \varphi(U') \wedge \psi_i(U') \to (U(x) \to U'(x)))].$$

Condition (3) simply says that given any model $\mathfrak{A}$ for $\mathscr{L}$, we can find elements $a_1, \dots, a_k \in A$ such that at most one subset $X \subset A$ satisfies

$$(\mathfrak{A}, X) \models \varphi \wedge \psi_i[a_1 \dots a_k].$$

We now describe a procedure by which each ψ_i can be written down. Given $\mathfrak{A}$, we first assume that $\mathfrak{A}$ has exactly n sets $X_1, \ldots, X_n$ such that

$$(\mathfrak{A}, X_i) \models \varphi, \qquad i = 1, \ldots, n.$$

We may assume that one of the X_i, say X_n, is such that

$$X_n \not\subseteq X_i \quad \text{for} \quad i = 1, \ldots, n-1.$$

Let

$$\psi_1(U; v_1 \ldots v_{n-1}) = U(v_1) \wedge U(v_2) \wedge \ldots \wedge U(v_{n-1}),$$

and let $a_i \in X_n \setminus X_i$ for $i = 1, \ldots, n-1$. Then X_n is the only subset of A such that

$$(\mathfrak{A}, X_n) \models \varphi \wedge \psi_1[a_1 \ldots a_{n-1}].$$

Next we may assume that

$$X_{n-1} \not\subseteq X_i, \quad \text{for} \quad i = 1, \ldots, n-2.$$

We write

$$\psi_2(U; v_1 \ldots v_{2n-3}) = (\neg \psi_1) \wedge U(v_n) \wedge \ldots \wedge U(v_{2n-3}),$$

and find $a_{n-1+i} \in X_{n-1} \setminus X_i$, $i = 1, \ldots, n-2$, so that X_{n-1} is the only subset of A satisfying

$$(\mathfrak{A}, X_{n-1}) \models \varphi \wedge \psi_2[a_1 \ldots a_{2n-3}].$$

Proceeding in this fashion, we obtain ψ_i, $i = 1, \ldots, n$. The integer k is simply the total number of distinct variables required to write down ψ_n. Certainly our choice of the elements $a_1, \ldots, a_k \in A$ is such that (3) holds for $\mathfrak{A}$. In the case where $\mathfrak{A}$ has fewer than n subsets X such that $(\mathfrak{A}, X) \models \varphi$, we can still prove that (3) holds by suitably identifying some of the a_i. So (3) is proved.

Let θ be the subformula in (3) immediately following the set quantifiers. We now apply the implication (1) $\Rightarrow$ (2) to θ. (It is clear that in this application we can replace $(\exists y)$ by $(\exists v_1 \ldots v_k)$ and $(\forall P)$ by $(\forall UU')$.) So we obtain a formula $\sigma(v_1 \ldots v_k)$ of $\mathscr{L}$ such that

$$\models (\exists v_1 \ldots v_k)\sigma,$$

$$\models (\forall v_1 \ldots v_k)(\sigma \to (\varphi(U) \to \bigvee_{1 \leq i \leq n} \psi_i(U))), \tag{4}$$

$$\models (\forall v_1 \ldots v_k x)[\sigma \wedge \varphi(U) \wedge \psi_i(U) \wedge \varphi(U') \wedge \psi_i(U') \to (U(x) \to U'(x))], \quad i = 1, \ldots, n. \tag{5}$$

Using propositional logic, we can rewrite each part of (5) as an implication of two formulas, one containing U and the other containing U'. We now apply the Craig interpolation theorem to each part of (5) and obtain formulas $\theta_i(v_1 \dots v_k x)$ of $\mathscr{L}$, $i = 1, \dots, n$, such that

$$\models (\forall v_1 \dots v_k x)[(\sigma \wedge \varphi(U) \wedge \psi_i(U) \wedge U(x) \rightarrow \theta_i) \wedge (\theta_i \rightarrow (\varphi(U') \wedge \psi_i(U') \rightarrow U'(x)))], \qquad i = 1, \dots, n.$$

After identifying U and U', and lifting out φ, we obtain

$$\varphi \models (\forall v_1 \dots v_k x)[\sigma \wedge \psi_i(U) \rightarrow (U(x) \rightarrow \theta_i(x))], \qquad i = 1, \dots, n,$$

$$\varphi \models (\forall v_1 \dots v_k x)[\psi_i(U) \rightarrow (\theta_i \rightarrow U(x))], \qquad i = 1, \dots, m.$$

From the latter two statements and (4) we obtain the desired conclusion

$$\varphi \models (\forall v_1 \dots v_k)[\sigma \rightarrow \bigvee_{1 \leqslant i \leqslant n} \forall x(U(x) \leftrightarrow \theta_i)]. \dashv$$

The finite analogue of Theorem 5.3.3 is the following:

PROPOSITION 5.3.5. *Let φ be a sentence of $\mathscr{L} \cup \{U\}$ and let $n \geqslant 1$. Then the following are equivalent*:

(i). *For every model $(\mathfrak{A}, X)$ of φ, there are at most n sets Y such that $(\mathfrak{A}, X) \cong (\mathfrak{A}, Y)$.*

(ii). *There are formulas $\sigma_j(v_1 \dots v_k)$, $\theta_{ij}(xv_1 \dots v_k)$, $1 \leqslant j \leqslant m$, $1 \leqslant i \leqslant n$, of $\mathscr{L}$ such that*

$$\varphi \models \bigvee_{1 \leqslant j \leqslant m} \{(\exists v_1 \dots v_k)\sigma_j \wedge (\forall v_1 \dots v_k)(\sigma_j \rightarrow \bigvee_{1 \leqslant i \leqslant n} (\forall x)(U(x) \leftrightarrow \theta_{ij}))\}.$$

The proof is similar to the proof of Theorem 5.3.3 and we leave it as an exercise.

We next consider an infinite analogue of Beth's theorem and Theorem 5.3.3. Note that condition (v) of Theorem 5.3.6 states that φ defines U explicitly up to disjunction and with parameters. Theorem 5.3.6 provides another answer to our earlier question.

THEOREM 5.3.6 (Chang–Makkai Theorem). *Let φ be a sentence of $\mathscr{L} \cup \{U\}$. The following five conditions are equivalent*:

(i). *For every infinite model $\mathfrak{A}$ for $\mathscr{L}$,*

$$|\{X : X \subset A \quad \text{and} \quad (\mathfrak{A}, X) \models \varphi\}| < |A|^+.$$

(ii). *For every infinite model* $(\mathfrak{A}, X) \models \varphi$,

$$|\{Y : (\mathfrak{A}, X) \cong (\mathfrak{A}, Y)\}| < |A|^+.$$

(iii). *For every infinite model* $\mathfrak{A}$ *for* $\mathscr{L}$,

$$|\{X : X \subset A \quad \text{and} \quad (\mathfrak{A}, X) \models \varphi\}| < 2^{|A|}.$$

(iv). *For every infinite model* $(\mathfrak{A}, X) \models \varphi$,

$$|\{Y : (\mathfrak{A}, X) \cong (\mathfrak{A}, Y)\}| < 2^{|A|}.$$

(v). *There are a finite number of formulas* $\theta_1(xv_1 \ldots v_m), \ldots, \theta_n(xv_1 \ldots v_m)$ *of* $\mathscr{L}$ *such that*

$$\varphi \models \bigvee_{1 \leqslant i \leqslant n} (\exists v_1 \ldots v_m \forall x)(U(x) \leftrightarrow \theta_i).$$

Proof. The set of m-tuples of an infinite set A has cardinal $|A|$. So it follows that (v) implies all the other conditions (i), (ii), (iii) and (iv). Among the four conditions (i)–(iv), (iv) is the weakest, since if it fails, then so will (i)–(iii). So it is sufficient to prove that (iv) implies (v). We shall show that if (v) fails, then (iv) also fails. This will then prove the theorem.

In order to simplify the proof somewhat, we shall assume that $\mathscr{L}$ is a countable language; in fact, we may assume that $\mathscr{L}$ contains only the symbols occurring in φ. Suppose (v) does not hold. Then the set

$$\{\varphi\} \cup \{(\forall v_1 v_2 \ldots) \neg (\forall x)(U(x) \leftrightarrow \theta(xv_1 v_2 \ldots)) : \theta(xv_1 v_2 \ldots) \text{ is a formula of } \mathscr{L}\}$$

is consistent, whence it has a model $(\mathfrak{A}, X)$. In the model $\mathfrak{A}$, the set X can never be defined by any formula of $\mathscr{L}$ with any finite number of parameters from A. So A must be infinite. We can assume that the model $(\mathfrak{A}, X)$ is special and is of power $\beth_\omega$. This means that $(\mathfrak{A}, X)$ is the union of a specializing elementary chain

$$(\mathfrak{A}_0, X_0) \prec (\mathfrak{A}_1, X_1) \prec \ldots \prec (\mathfrak{A}_n, X_n) \prec \ldots$$

(we have only taken a subchain determined by the cardinals $\beth_n$, and we are using n as an index instead of $\beth_n$), where each $(\mathfrak{A}_n, X_n)$ is a $\beth_n^+$-saturated model of power $\beth_{n+1}$. We may also suppose that the set A is well-ordered in such a way that

$$A = \{a_\xi : \xi < \beth_\omega\},$$

$$A_n = \{a_\xi : \xi < \beth_{n+1}\} \quad \text{for all} \quad n < \omega.$$

The model $(\mathfrak{A}, X)$ will be shown to violate (iv), i.e.

$$|\{Y : (\mathfrak{A}, X) \cong (\mathfrak{A}, Y)\}| = 2^{|A|}.$$

This means that we have to create $2^{|A|}$ automorphisms of $\mathfrak{A}$ such that distinct automorphisms will give rise to distinct subsets of A when applied to the set X.

Let ${}^{<|A|}2$ denote the set of all functions whose domain is some ordinal $\xi < |A|$, and whose range has at most two values 0, 1. That is,

$${}^{<|A|}2 = \{f : f \in {}^{\xi}2 \quad \text{and} \quad \xi < |A|\}.$$

We shall define two functions G and H such that the following hold:
(1) Domain (G) = domain (H) = ${}^{<|A|}2$.
(2) If $f \subset g$, then $G(f) \subset G(g)$ and $H(f) \subset H(g)$.
(3) If $f \in {}^{\xi}2$ and $\xi < \beth_n$, then

$$G(f), H(f) \in {}^{\xi}A_n, \qquad (\mathfrak{A}, G(f)_\eta)_{\eta<\xi} \equiv (\mathfrak{A}, H(f)_\eta)_{\eta<\xi}.$$

(4) If $f \in {}^{\xi}2$ and $\xi = \lambda + 3m$ with λ a limit ordinal, then

$$G(f \cup \{\langle \xi 0 \rangle\}) = G(f \cup \{\langle \xi 1 \rangle\}) = G(f) \cup \{\langle \xi, a_{\lambda+m} \rangle\}.$$

(5) If $f \in {}^{\xi}2$ and $\xi = \lambda + 3m + 1$ with λ a limit ordinal, then

$$H(f \cup \{\langle \xi 0 \rangle\}) = H(f \cup \{\langle \xi 1 \rangle\}) = H(f) \cup \{\langle \xi a_{\lambda+m} \rangle\}.$$

(6) If $f \in {}^{\xi}2$, $\xi < \beth_n$, and $\xi = \lambda + 3m + 2$ with λ a limit ordinal, then there are two elements $b_0 \in X_n$ and $b_1 \notin X_n$ such that

$$G(f \cup \{\langle \xi 0 \rangle\}) = G(f) \cup \{\langle \xi b_0 \rangle\},$$
$$G(f \cup \{\langle \xi 1 \rangle\}) = G(f) \cup \{\langle \xi b_1 \rangle\},$$
$$H(f \cup \{\langle \xi 0 \rangle\}) = H(f \cup \{\langle \xi 1 \rangle\}).$$

The existence of the functions G and H is proved by induction on ξ, the domain of f. There are precisely three cases of the induction corresponding to cases (4), (5) and (6). So suppose that G and H have been defined on all functions $f \in {}^{<|A|}2$ such that domain f is less than ξ. If ξ is a limit ordinal, then the extension of G, H to $g \in {}^{\xi}2$ is obvious. So let us deal only with the cases when $\xi = \eta + 1$.

Case 1. $\xi < \beth_n$, $\xi = \lambda + 3m$, λ a limit ordinal. Let $g \in {}^{\xi}2$. Then for some $f \in {}^{\eta}2$, $g = f \cup \{\langle \xi 0 \rangle\}$ or $g = f \cup \{\langle \xi 1 \rangle\}$. In either case, define

$G(g) = G(f) \cup \{\langle \xi a_{\lambda+m} \rangle\}$. We now have to find an element $a \in A_n$ such that

$$(7) \qquad (\mathfrak{A}, G(f)_\zeta, a_{\lambda+m})_{\zeta<\eta} \equiv (\mathfrak{A}, H(f)_\zeta, a)_{\zeta<\eta}.$$

But by inductive hypothesis we have

$$(\mathfrak{A}, G(f)_\zeta)_{\zeta<\eta} \equiv (\mathfrak{A}, H(f)_\zeta)_{\zeta<\eta}.$$

Whence

$$(\mathfrak{A}_n, G(f)_\zeta)_{\zeta<\eta} \equiv (\mathfrak{A}_n, H(f)_\zeta)_{\zeta<\eta}.$$

Since $a_{\lambda+m} \in A_n$, and $\mathfrak{A}_n$ is $\beth_n^+$-saturated, we find by the usual technique an element $a \in A_n$ which satisfies (7). Then we define

$$H(g) = H(f) \cup \{\langle \xi a \rangle\}.$$

Case 2. $\xi < \beth_n$, $\xi = \lambda+3m+1$, λ a limit ordinal. Here the argument is entirely analogous to Case 1, except that we extend the definition of H first by (5), and then extend G.

Case 3. $\xi < \beth_n$, $\xi = \lambda+3m+2$, λ a limit ordinal. Let $f \in {}^\eta 2$. Consider the two possible extensions of f,

$$g_0 = f \cup \{\langle \xi 0 \rangle\} \quad \text{and} \quad g_1 = f \cup \{\langle \xi 1 \rangle\}.$$

In order to satisfy (6), we shall have to find three elements $b_0, b_1, a \in A_n$ such that

$$b_0 \in X_n, \qquad b_1 \notin X_n,$$
$$G(g_0) = G(f) \cup \{\langle \xi b_0 \rangle\},$$
$$G(g_1) = G(f) \cup \{\langle \xi b_1 \rangle\},$$
$$H(g_0) = H(g_1) = H(f) \cup \{\langle \xi a \rangle\}.$$

We shall first find b_0 and b_1 such that:

$$(8) \qquad b_0 \in X_n, \qquad b_1 \notin X_n,$$
$$(\mathfrak{A}_n, G(f)_\zeta, b_0)_{\zeta<\eta} \equiv (\mathfrak{A}_n, G(f)_\zeta, b_1)_{\zeta<\eta}.$$

If (8) is never true, then

(9) for all $b_0 \in X_n$, $b_1 \notin X_n$, there is a formula $\varphi_{b_0 b_1}(x)$ of $\mathscr{L} \cup \{c_\zeta : \zeta < \eta\}$ such that

$$(\mathfrak{A}_n, G(f)_\zeta)_{\zeta<\eta} \models \varphi_{b_0 b_1}[b_0],$$
$$(\mathfrak{A}_n, G(f)_\zeta)_{\zeta<\eta} \models \neg\, \varphi_{b_0 b_1}[b_1].$$

Fix $b_0 \in X_n$. Let $\Sigma_{b_0} = \{\varphi_{b_0 b_1} : b_1 \notin X_n\}$. Then by (9),

$$\{a \in A_n : (\mathfrak{A}_n, G(f)_\zeta)_{\zeta<\eta} \models \Sigma_{b_0}[a]\} \subset X_n. \tag{10}$$

We now see that some finite subset $\Sigma' \subset \Sigma_{b_0}$ also has the property that

$$\{a \in A_n : (\mathfrak{A}_n, G(f)_\zeta)_{\zeta<\eta} \models \Sigma'_{b_0}[a]\} \subset X_n. \tag{11}$$

For otherwise, by the $\beth_n^+$-saturatedness of $(\mathfrak{A}_n, X_n)$, we can find an element $b_1 \notin X_n$ such that

$$(\mathfrak{A}_n, G(f)_\zeta)_{\zeta<\eta} \models \Sigma_{b_0}[b_1],$$

which contradicts (10). Let σ_{b_0} be the conjunction of this finite subset Σ' from (11). Clearly,

$$(\mathfrak{A}_n, G(f)_\zeta)_{\zeta<\eta} \models \sigma_{b_0}[b_0].$$

So, letting b_0 range over X_n, and letting $\Sigma = \{\sigma_{b_0} : b_0 \in X_n\}$, we see that, by (11),

$$X_n = \{a \in A_n : (\mathfrak{A}_n, G(f)_\zeta)_{\zeta<\eta} \models \Sigma[a]\}.$$

By the same reasoning which established (11), using again the $\beth_n^+$-saturatedness of $(\mathfrak{A}_n, X_n)$, we see that some finite conjunction σ of formulas from Σ satisfies

$$X_n = \{a \in A_n : (\mathfrak{A}_n, G(f)_\zeta)_{\zeta<\eta} \models \sigma[a]\}.$$

This means that X_n is definable in $\mathfrak{A}_n$ from a formula σ of $\mathscr{L}$ with some parameters in A_n. Since $(\mathfrak{A}_n, X_n) \prec (\mathfrak{A}, X)$, this is a contradiction. So (8) is proved.

Since $(\mathfrak{A}_n, G(f)_\zeta)_{\zeta<\eta} \equiv (\mathfrak{A}_n, H(f)_\zeta)_{\zeta<\eta}$ by inductive hypothesis, we use the $\beth_n^+$-saturatedness of $\mathfrak{A}_n$ once more to find an element $a \in A_n$ such that

$$(\mathfrak{A}_n, G(f)_\zeta, b_0)_{\zeta<\eta} \equiv (\mathfrak{A}_n, H(f)_\zeta, a)_{\zeta<\eta}.$$

By (8), we also have

$$(\mathfrak{A}_n, G(f)_\zeta, b_1)_{\zeta<\eta} \equiv (\mathfrak{A}_n, H(f)_\zeta, a)_{\zeta<\eta}.$$

We have now finished the proof of Case 3.

Once G, H are defined on ${}^{<|A|}2$ satisfying (1)–(6), we can easily extend them in the natural way to functions defined on ${}^{|A|}2$, so that, given any $f \in {}^{|A|}2$,

$$(\mathfrak{A}, G(f)_\xi)_{\xi<|A|} \equiv (\mathfrak{A}, H(f)_\xi)_{\xi<|A|}.$$

Since

$$\text{range}\,(G(f)) = A = \text{range}\,(H(f)),$$

this defines an automorphism $G(f)_\xi \to H(f)_\xi$ of $\mathfrak{A}$ onto $\mathfrak{A}$. If two functions $f, g \in {}^{|A|}2$ differ at the first place on an ordinal $\xi = \lambda+3m+2$ with λ a limit ordinal, then the two automorphisms

$$G(f)_\xi \to H(f)_\xi, \qquad G(g)_\xi \to H(g)_\xi$$

will map the set X onto distinct subsets of A. This is because by (6),

$$a = H(f)_\xi = H(g)_\xi,$$

and, say $f_\xi = 0$ and $g_\xi = 1$,

$$G(f)_\xi \in X \quad \text{and} \quad G(g)_\xi \notin X.$$

So the image of X under $G(f)_\xi \to H(f)_\xi$ contains a and the image of X under $G(g)_\xi \to H(g)_\xi$ excludes a. Obviously, there are $2^{|A|}$ such functions, whence there are $2^{|A|}$ distinct automorphic images of X. So (iv) fails. Hence the theorem is proved. ⊣

Exercises

5.3.1*. Assume that the set $\Sigma(P; x_1 \dots x_n)$ of Proposition 5.3.1 is a single formula of $\mathscr{L} \cup \{P\}$. Improve Proposition 5.3.1 to the following: There is a primitive recursive set Φ of formulas of $\mathscr{L}$ (containing only those symbols in Σ) such that for all special models $\mathfrak{A}$ for $\mathscr{L}$,

$$\mathfrak{A} \models (\forall x_1 \dots x_n)((\exists P)\Sigma \leftrightarrow \bigwedge \Phi).$$

[*Hint*: Assume without loss of generality that P is a 1-placed function symbol and that $\Sigma = (\exists P \forall x)\varphi(P(x)x)$, where $\varphi(yx)$ is a formula of $\mathscr{L}$ and $\varphi(P(x)x)$ is obtained from $\varphi(yx)$ by replacing y by $P(x)$. Then intuitively Σ implies each of the sentences:

$(\forall x_1 \exists y_1)\varphi(y_1 x_1)$,

$(\forall x_1 \exists y_1 \forall x_2 \exists y_2)((x_1 \equiv x_2 \to y_1 \equiv y_2) \wedge \varphi(y_1 x_1) \wedge \varphi(y_2 x_2))$,

⋮

$(\forall x_1 \exists y_1 \forall x_2 \exists y_2 \dots \forall x_n \exists y_n)(\bigwedge_{1 \leqslant i < j \leqslant n} (x_i \equiv x_j \to y_i \equiv y_j) \wedge \bigwedge_{1 \leqslant i \leqslant n} \varphi(y_i x_i))$.

⋮

Show that if $\mathfrak{A}$ is a special model for $\mathscr{L}$, then $\mathfrak{A} \models (\exists P)\Sigma$ iff $\mathfrak{A}$ satisfies each sentence of the list.]

5.3.2*. Prove the following generalization of Theorem 5.3.2. Let

$$\varphi(P; S_1 \dots S_m; x_1 \dots x_n)$$

be a formula of $\mathscr{L} \cup \{P, S_1, \dots, S_m\}$,

$$\psi(Q; S_1 \dots S_m; x_1 \dots x_n)$$

be a formula of $\mathscr{L} \cup \{Q, S_1, \dots, S_m\}$, where $P, Q, S_1, \dots, S_m$ are new relation (or function) symbols. Let $\mathscr{Q}$ be a sequence of quantifiers on the (second-order) variables $S_1, \dots, S_m$ and the (first-order) variables $x_1, \dots, x_n$, such that any second-order quantifier in $\mathscr{Q}$ occurs universally. We make no stipulation on the order in which these quantifiers in $\mathscr{Q}$ are to occur. Then the following are equivalent:

(i). $\models \mathscr{Q}((\exists P)\varphi \rightarrow (\forall Q)\psi)$.

(ii). There is a formula $\theta(S_1 \dots S_m; x_1 \dots x_n)$ of $\mathscr{L} \cup \{S_1, \dots, S_m\}$ such that

$$\models \mathscr{Q}[((\exists P)\varphi \rightarrow \theta) \wedge (\theta \rightarrow (\forall Q)\psi)].$$

[*Hint*: Try to 'code' the relations and functions on a model $\mathfrak{A}$ by using elements of A and relations with one more place. This is where the assumption that second-order variables must be quantified universally is crucial.]

5.3.3. Deduce Beth's theorem 2.2.22 from Theorem 5.3.3 directly.

5.3.4. Verify that the proofs of Theorems 5.3.3 and 5.3.4, Proposition 5.3.5, and Theorem 5.3.6 go through using a set $\Sigma(P)$ of sentences of $\mathscr{L} \cup \{P\}$, where P is not necessarily 1-placed.

5.3.5. Give a proof of Proposition 5.3.1 without the cardinality restriction $|A| = |A|^*$ in Proposition 5.3.1 (iii).

[*Note*: This is easier than Exercise 5.3.1 above.]

5.3.6. Show by a counterexample that Theorem 5.3.4 cannot be simplified to: for $n \geqslant 1$, Theorem 5.3.4 $(\mathrm{i})_n$ is equivalent to

(ii′). There are n formulas $\theta_1(x), \dots, \theta_n(x)$ of $\mathscr{L}$ such that

$$\varphi \models \bigvee_{1 \leqslant i \leqslant n} (\forall x)(U(x) \leftrightarrow \theta_i(x)).$$

5.3.7. Prove Proposition 5.3.5.

5.3.8*. The following is a countable common generalization of Theorem 5.3.4 and Proposition 5.3.5 (compare with Theorem 5.3.6). Let φ be a sentence of $\mathscr{L} \cup \{U\}$. Then the following are equivalent:

(i). For every model $\mathfrak{A}$ for $\mathscr{L}$, there are at most a finite number of subsets X of A such that $(\mathfrak{A}, X) \models \varphi$.

(ii). For every model $(\mathfrak{A}, X)$ of φ, there are at most a finite number of $Y \subset A$ such that $(\mathfrak{A}, X) \cong (\mathfrak{A}, Y)$.

(iii). There are a number n and formulas $\sigma(v_1 \dots v_k)$, $\theta_i(xv_1 \dots v_k)$, $1 \leqslant i \leqslant n$, of $\mathscr{L}$ such that

$$\varphi \models (\exists v_1 \dots v_k)\sigma,$$

$$\varphi \models (\forall v_1 \dots v_k)[\sigma \to \bigvee_{1 \leqslant i \leqslant n} (\forall x)(U(x) \leftrightarrow \theta_i)].$$

5.3.9. Let φ be a sentence of $\mathscr{L} \cup \{U_1, \dots, U_n\}$. Then the following are equivalent:

(i). For every model $\mathfrak{A}$ for $\mathscr{L}$ there is at most one n-tuple of subsets $X_1, \dots, X_n$ of A such that

$$(\mathfrak{A}, X_1 \dots X_n) \models \varphi.$$

(ii). There are formulas $\theta_1(x), \dots, \theta_n(x)$ of $\mathscr{L}$ such that

$$\varphi \models \bigwedge_{1 \leqslant i \leqslant n} (\forall x)(U_i(x) \leftrightarrow \theta_i(x)).$$

5.3.10. Let φ be a sentence of $\mathscr{L} \cup \{U\}$. The following are equivalent:

(i). For every model $\mathfrak{A}$ for $\mathscr{L}$, the set

$$\{X : X \subset A \quad \text{and} \quad (\mathfrak{A}, X) \models \varphi\}$$

is a chain of subsets of A.

(ii). There is a formula $\varphi(xy)$ of $\mathscr{L}$ such that

$$\varphi \models (\forall xy)[(U(x) \to (\varphi(xy) \to U(y))) \wedge (\neg U(x) \to (U(y) \to \varphi(xy)))].$$

5.3.11*. Let K be an elementary class of partially ordered structures $\mathfrak{A} = \langle A, \leqslant \rangle$. A subset $X \subset A$ is called *hereditary* iff $a \in X$ and $b \leqslant a$ imply $b \in X$. Suppose that every automorphism of every model $\mathfrak{A}$ in K maps every hereditary subset X of A onto another set $Y \subset A$ which is comparable with X, i.e., $X \subset Y$ or $Y \subset X$. Then there exists an n such that in every model in K every set of pairwise incomparable (under $\leqslant$) elements has at most n elements.

5.3.12. The equivalence of conditions (1) and (2) in the proof of Theorem 5.3.4 can be generalized as follows: Let $\theta(S_1 \dots S_k; x_1 \dots x_k)$ be a formula of $\mathscr{L} \cup \{S_1, \dots, S_k\}$. Then the following are equivalent:

(i). $\models (\exists x_1 \forall S_1 \exists x_2 \forall S_2 \ldots \exists x_k \forall S_k)\theta$.

(ii). There are formulas $\sigma_i(S_1 \ldots S_{i-1}; x_1 \ldots x_i)$ of $\mathscr{L} \cup \{S_1, \ldots, S_{i-1}\}$, $i = 1, \ldots, k$, such that:

(a) $\models (\exists y_1)\sigma_1$;

(b) $\models (\forall x_1 \ldots x_i)(\sigma_i \to (\exists x_{i+1})\sigma_{i+1})$, $i = 1, \ldots, k-1$;

(c) $\models (\forall x_1 \ldots x_k)(\sigma_k \to \theta)$.

The interpolation theorem in Exercise 5.3.2 (as well as Theorem 5.3.2) can now be improved as follows: Let us suppose (without loss of generality) that (in Exercise 5.3.2) $m = n = k$ and $\mathscr{Q}$ is the sequence of quantifiers $(\exists x_1 \forall S_1 \ldots \exists x_k \forall S_k)$. Then the following is equivalent to both Exercise 5.3.2 (i) and Exercise 5.3.2 (ii):

(iii). There are formulas $\sigma_i(S_1 \ldots S_{i-1}; x_1 \ldots x_i)$ of $\mathscr{L} \cup \{S_1 \ldots S_{i-1}\}$, $i = 1, \ldots, k$, and a formula $\theta(S_1 \ldots S_k; x_1 \ldots x_k)$ of $\mathscr{L} \cup \{S_1 \ldots S_k\}$ such that (a), (b), and

$$\models (\forall x_1 \ldots x_k)[(\sigma_k \wedge (\exists P)\varphi \to \theta) \wedge (\theta \to (\forall Q)\psi)].$$

5.3.13. Prove Beth's theorem directly from Theorem 5.3.6.

5.3.14. Show that Theorem 5.3.6 is still true if conditions (i)–(iv) are modified in the following manner: delete the word 'infinite' throughout and replace $|A|^+$ by $|A|^+ \cup \omega$ and $2^{|A|}$ by $2^{|A|} \cup \omega$.

5.3.15. Use Theorem 5.3.6 to prove that every special model of $\mathscr{L}$ of power α greater than $||\mathscr{L}||$ has 2^α automorphisms. It follows that every saturated model of power $\alpha > ||\mathscr{L}||$ has 2^α automorphisms.

5.3.16*. Prove, by techniques similar to those employed in Theorem 5.3.6, that every countable model for a countable language has a continuum number of automorphisms if it has more than a countable number of automorphisms.

5.3.17*. Let $\mathfrak{A}$ be a countable model for a countable $\mathscr{L}$, and let $X \subset A$. If

$$|\{Y : Y \subset A \quad \text{and} \quad (\mathfrak{A}, Y) \cong (\mathfrak{A}, X)\}| > \omega,$$

then

$$|\{Y : Y \subset A \quad \text{and} \quad (\mathfrak{A}, Y) \cong (\mathfrak{A}, X)\}| = 2^\omega.$$

5.3.18. Using Exercise 5.3.17, show by means of a simple argument that Theorem 5.3.6 holds when 'infinite' is replaced everywhere by 'countably infinite', assuming $\mathscr{L}$ is countable.

[*Hint*: Code the subsets by elements of a model and then use the downward two-cardinal theorem 3.2.12.]

5.3.19*. Let $\varphi(P, Q)$ be a sentence of $\mathscr{L} \cup \{P, Q\}$. Show that the following two conditions are equivalent:

(i). For every sentence $\tau(P)$ of $\mathscr{L} \cup \{P\}$, if $\models \varphi \rightarrow \tau$, then $\models \tau$.

(ii). For every sentence $\tau(P)$ of $\mathscr{L} \cup \{P\}$, if $\tau(P)$ has a model, then $\tau(P) \wedge \varphi(P, Q)$ has a model.

On the other hand, show by an example that the following definability result is not true: (i) is equivalent to

(i′). Every model $\mathfrak{A}$ for $\mathscr{L} \cup \{P\}$ has an expansion $\mathfrak{A}^*$ which is a model of $\varphi(P, Q)$.

5.3.20. Give a proof of Theorem 5.3.2 using countable recursively saturated models instead of special models.

5.3.21. Prove Svenonius' Theorem 5.3.3 using countable recursively saturated models instead of special models.

5.3.22. Prove Kueker's Finite Definability Theorem 5.3.4 using countable recursively saturated models instead of special models.

5.3.23. Prove the Chang-Makkai Theorem 5.3.6 but modified by replacing "infinite model" everywhere by "countably infinite model", using countable recursively saturated models instead of special models.

5.4. Applications to field theory

Saturated or special models give us a very powerful method of proving that certain theories are complete. In this section we shall apply this method to some extensions of the theory of fields. The method depends upon the following simple consequence of the results of Section 5.1.

PROPOSITION 5.4.1. *Let $\mathscr{L}$ be a countable language and let T be a consistent theory in $\mathscr{L}$, all of whose models are infinite.*

(i). *Assume the continuum hypothesis. Then T is complete if and only if any two saturated models of T of power ω_1 are isomorphic.*

(ii). *T is complete if and only if any two special models of T of power $\beth_\omega$ are isomorphic.*

PROOF. (i). If T is complete then any two saturated models of T of the same power are elementarily equivalent, and hence isomorphic. If T is not complete, then T has two different complete extensions, say T_1 and T_2. Since all models of T are infinite, T_1 and T_2 have ω_1-saturated models $\mathfrak{A}_1$, $\mathfrak{A}_2$ of power $2^\omega = \omega_1$. Then $\mathfrak{A}_1$, $\mathfrak{A}_2$ are both models of T. But they are not elementarily equivalent, and hence not isomorphic.

The proof of (ii) is similar. ⊣

We shall use Proposition 5.4.1 (i) to show that theories are complete under the assumption of the continuum hypothesis. In each case there will be a very similar proof using Proposition 5.4.1 (ii) which shows that the theory is complete without assuming the continuum hypothesis. However, special models are more complicated to deal with, and these extra complications would tend to obscure the main ideas of the proofs. So we shall be using the continuum hypothesis only to simplify the proof, and it will always be possible to eliminate it by using special models instead of saturated models.

Saturated models are in fact generalizations of the 'η_α sets' of Hausdorff. The η_α sets, which are very special kinds of densely ordered sets, still provide us with an instructive example of saturated models. Let $\langle A, \leqslant \rangle$ be a densely ordered set without endpoints. If X, Y are subsets of A, then we write $X < Y$ iff $x < y$ for all $x \in X$, $y \in Y$ (where $x < y$ means $x \leqslant y$ and $x \neq y$). If $z \in A$ and $X \subset A$, then $X < z$ means $X < \{z\}$, and $z < X$ means $\{z\} < X$. Let α be an ordinal. We say that $\langle A, \leqslant \rangle$ is an *η_α set* iff $\langle A, \leqslant \rangle$ is a densely ordered set without endpoints and, for all $X, Y \subset A$ of power less than ω_α, if $X < Y$, then there exists $z \in A$ such that $X < z < Y$. Note, in particular, that in an η_α set, every subset of power less than ω_α has an upper bound and a lower bound, for we always have $X < \emptyset$ and $\emptyset < Y$, where $\emptyset$ is the empty set.

PROPOSITION 5.4.2. *Let $\langle A, \leqslant \rangle$ be a densely ordered set without endpoints and let α be an ordinal. Then $\langle A, \leqslant \rangle$ is ω_α-saturated if and only if it is an* η_α set.

PROOF. Suppose $\langle A, \leqslant \rangle = \mathfrak{A}$ is ω_α-saturated. Let X, Y be subsets of A of power less than ω_α such that $X < Y$, and form the expanded model $(\mathfrak{A}, x, y)_{x \in X, y \in Y} = \mathfrak{A}'$. Let c_x, c_y be the corresponding constant symbols. Then the set of formulas

(1) $$\Sigma(v) = \{c_x < v : x \in X\} \cup \{v < c_y : y \in Y\}$$

is consistent with the theory of $\mathfrak{A}'$. Since $|X| < \omega_\alpha$ and $|Y| < \omega_\alpha$, $\Sigma(v)$ is satisfied in $\mathfrak{A}'$ by some element $z \in A$. Then $X < z < Y$.

The converse depends on the following fact:

(2) If $\mathfrak{A}$, $\mathfrak{B}$ are densely ordered sets without endpoints, $\varphi(x_1 \dots x_n)$ is a formula of $\mathscr{L}$, and the expansions $(\mathfrak{A}, a_1 \dots a_n)$, $(\mathfrak{B}, b_1 \dots b_n)$ satisfy exactly the same atomic sentences, then

$$\mathfrak{A} \models \varphi[a_1 \dots a_n] \text{ iff } \mathfrak{B} \models \varphi[b_1 \dots b_n].$$

The proof of (2) is a straightforward induction on the complexity of the formula φ.

Assume that $\mathfrak{A}$ is an η_α set. Let A_0 be a subset of A of power $|A_0| < \omega_\alpha$, and let $\Gamma(v)$ be a type in the expanded model $\mathfrak{A}' = (\mathfrak{A}, a)_{a \in A_0}$. We must show that $\Gamma(v)$ is realized in $\mathfrak{A}'$. If the formula $c_a \equiv v$ belongs to $\Gamma(v)$ for some $a \in A_0$, then $\Gamma(v)$ is just the type determined by the formula $c_a \equiv v$, and a satisfies $\Gamma(v)$ in $\mathfrak{A}'$.

Suppose now that for all $a \in A_0$, the formula $c_a \not\equiv v$ belongs to $\Gamma(v)$. Let

$$X = \{a \in A_0 : (c_a < v) \in \Gamma(v)\}, \qquad Y = \{a \in A_0 : (v < c_a) \in \Gamma(v)\}.$$

Then $X \cup Y = A_0$, $X < Y$, and $|X|, |Y| < \omega_\alpha$. Hence there exists $a^* \in A$ such that $X < a^* < Y$. Let c^* be a new individual constant. Then $\Gamma(c^*)$ is consistent and therefore has a model

$$(\mathfrak{B}, b_a, b^*)_{a \in A_0} = (\mathfrak{B}', b^*),$$

where b^* is the interpretation of c^*. Hence we have

$$b_a < b^* \quad \text{for} \quad a \in X, \qquad b^* < b_a \quad \text{for} \quad a \in Y.$$

It follows that the models $(\mathfrak{A}', a^*)$ and $(\mathfrak{B}', b^*)$ satisfy exactly the same atomic sentences, and $\mathfrak{B}$ is a densely ordered set without endpoints. Therefore, by (2),

$$(\mathfrak{A}', a^*) \equiv (\mathfrak{B}', b^*).$$

Since $(\mathfrak{B}', b^*)$ is a model of $\Gamma(c^*)$, we conclude that $(\mathfrak{A}', a^*)$ is a model of $\Gamma(c^*)$, whence a^* satisfies $\Gamma(v)$. ⊣

As a first illustration of our method of proving completeness, we shall prove that the theory of real closed fields is complete. We wish to indicate where the methods of model theory are used in the proof. First, we shall state a purely algebraic lemma. While this lemma is a deep result, its proof uses standard algebraic methods which are quite outside the subject matter of this book. Accordingly, we shall omit the proof of the lemma and ask the reader either to take it for granted or to read about it elsewhere.

We recall from Section 1.4 that the theory of real closed fields has as a set of axioms the field axioms plus axioms stating that every polynomial of odd degree has a root, 0 is not a sum of nontrivial squares, and for all x, either x or $-x$ has a square root. It follows at once that every real closed field has characteristic zero, and hence is infinite.

The best known real closed field is the field of real numbers. In the theory of real closed fields, let $x \leqslant y$ be the abbreviation of

$$(\exists z)(z^2 \equiv y-x).$$

Then $\leqslant$ is a dense ordering of any real closed field, and $\leqslant$ has no greatest or least element. Thus each real closed field F can be expanded in a unique way to an *ordered* real closed field $(F, \leqslant)$. It will be sometimes more convenient to work with ordered real closed fields.

It is easily seen from the field axioms that the intersection of any family of subfields of a field is a field. If F is a field, F_0 is a subfield of F, and $x \in F$, let us denote by $F_0(x)$ the least subfield F_1 of F such that $F_0 \subset F_1$ and $x \in F_1$. By the Löwenheim–Skolem–Tarski theorem,

$$|F_0(x)| \leqslant |F_0| \cup \omega.$$

This can also be shown using the fact that $F_0(x)$ is the set of all values of rational functions of x with coefficients in F_0.

An element $x \in F$ is said to be *algebraic over* F_0 iff x is a root of a nonzero polynomial with coefficients in F_0. We shall denote by $\bar{F}_0$ the set of all elements $x \in F$ which are algebraic over F_0. Thus $F_0 \subset \bar{F}_0$. $\bar{F}_0$ is called the *relative algebraic closure* of F_0 in F. F_1 is said to be a *real closure* of F_0 iff F_1 is a real closed field and F_0 is a subfield of F_1 and every $x \in F_1$ is algebraic over F_0, i.e. $F_1 \subset \bar{F}_0$. In our discussion of fields, it will be easiest to use the same symbol F for a field and for its universe set; thus

$$F = \langle F, +, \cdot, 0, 1 \rangle.$$

If there is an ordering $\leqslant$ on F, then $(F, \leqslant)$ will denote the expansion of F to the language $\{+, \cdot, 0, 1, \leqslant\}$, while $\langle F, \leqslant \rangle$ will be the corresponding reduct which is a model for the language $\{\leqslant\}$.

LEMMA 5.4.3

(i). *Let F be a real closed field with ordering $\leqslant$, and let F_0 be the smallest* *subfield of F. Then* $(F_0, \leqslant)$ *is isomorphic to the field of rational numbers with the usual order.*

(ii). *Let F be an arbitrary field and let* F_0 *be a subfield of F. Then the relative algebraic closure* $\bar{F}_0$ *is a subfield of F, and*

$$\bar{\bar{F}}_0 = \bar{F}_0, \qquad |\bar{F}_0| \leqslant |F_0| \cup \omega.$$

Moreover, if F is a real closed field, then $\bar{F}_0$ *is a real closure of* F_0.

(iii). *Let F be a real closed field and let* F_0 *be a real closed subfield of F. Then for all* $x, y \in F_0$, *we have* $x \leqslant y$ *in the sense of* F_0 *iff* $x \leqslant y$ *in the sense of F. Also,* $F_0 = \bar{F}_0$.

(iv). *Let* $(F, \leqslant)$ *and* $(G, \leqslant)$ *be ordered real closed fields, let* F_0, G_0 *be real closed subfields, and let* $f : F_0 \cong G_0$ *be an isomorphism between them. Suppose* $x \in F \setminus F_0$, $y \in G \setminus G_0$, *and for all* $a \in F_0$,

$$a \leqslant x \text{ iff } f(a) \leqslant y.$$

Then we can extend f to an isomorphism $g : (F_0(x), \leqslant) \cong (G_0(y), \leqslant)$ *such that* $g(x) = y$.

(v). (Uniqueness of the real closure.) *Let* F_0, G_0 *be fields, let* F_1, G_1 *be real closures of* F_0 *and* G_0, *and form the ordered real closed fields* $(F_1, \leqslant)$, $(G_1, \leqslant)$. *If* $f : (F_0, \leqslant) \cong (G_0, \leqslant)$, *then f can be extended to an isomorphism* $g : (F_1, \leqslant) \cong (G_1, \leqslant)$.

The first three parts of the above lemma are elementary facts which are easy to prove, while the last two parts are substantial results in field theory. Part (v) states that every ordered field has at most one real closure. In general, it is not true that every field has at most one real closure. That is, in (v) we cannot weaken the hypothesis $f: (F_0, \leqslant) \cong (G_0, \leqslant)$ to $f: F_0 \cong G_0$.

THEOREM 5.4.4 (Tarski's Theorem). *The theory of real closed fields is complete.*

PROOF. We shall assume the continuum hypothesis and use Proposition 5.4.1 (i). As we have explained above, the continuum hypothesis can be avoided by using Proposition 5.4.1 (ii) and with a similar but more complicated proof.

Let F, G be any two saturated real closed fields of power ω_1. By Proposition 5.4.1 it suffices to prove that F and G are isomorphic. Since the order relation $\leqslant$ is definable, the ordered real closed fields $(F, \leqslant)$ and $(G, \leqslant)$ are still saturated. Then all their reducts are saturated. In particular, the ordered sets $\langle F, \leqslant \rangle$ and $\langle G, \leqslant \rangle$ are η_1 sets. We may enumerate the sets F, G with order type ω_1:

$$F = \{a_\alpha : \alpha < \omega_1\}, \qquad G = \{b_\alpha : \alpha < \omega_1\}.$$

We shall form two ascending chains F_α, G_α, and an ascending chain of mappings f_α, such that for all $\alpha < \omega_1$,
(1) F_α and G_α are countable real closed subfields of F and G;
(2) $f_\alpha : F_\alpha \cong G_\alpha$;
(3) $\{a_\gamma : \gamma < \alpha\} \subset F_\alpha$ and $\{b_\gamma : \gamma < \alpha\} \subset G_\alpha$.
Consider the minimal subfields F_0', G_0'. By Lemma 5.4.3 (i), $(F_0', \leqslant)$ and $(G_0', \leqslant)$ are isomorphic. Let $F_0 = \bar{F}_0'$, $G_0 = \bar{G}_0'$. Then by Lemma 5.4.3 (v), there exists an isomorphism $f_0 : F_0 \cong G_0$. If α is a limit ordinal, we let

$$F_\alpha = \bigcup_{\delta<\alpha} F_\delta, \qquad G_\alpha = \bigcup_{\delta<\alpha} G_\delta, \qquad f_\alpha = \bigcup_{\delta<\alpha} f_\delta.$$

Assuming (1)–(3) for all $\beta < \alpha$, it easily follows that (1)–(3) hold for α. In particular, a union of a chain of real closed fields is a real closed field, because the axioms are $\mathbf{\Pi}_2^0$.

Now assume that (1)–(3) hold for α. We use a back and forth argument. Since $\langle G, \leqslant \rangle$ is an η_1 set and F_α is countable, there exists $y \in G$ such that for all $c \in F_\alpha$,

$$c \leqslant a_\alpha \text{ iff } f_\alpha(c) \leqslant y, \qquad c = a_\alpha \text{ iff } f_\alpha(c) = y.$$

Then by Lemma 5.4.3 (iv), f_α can be extended to an isomorphism

$$f : (F_\alpha(a_\alpha), \leqslant) \cong (G_\alpha(y), \leqslant).$$

By Lemma 5.4.3 (v), f in turn can be extended to an isomorphism

$$g : \overline{F_\alpha(a_\alpha)} \cong \overline{G_\alpha(y)}.$$

Moreover, $\overline{F_\alpha(a_\alpha)}$ and $\overline{G_\alpha(y)}$ are still countable. Now using the fact that $\langle F, \leqslant \rangle$ is an η_1 set, we can find an $x \in F$ such that for all $d \in \overline{G_\alpha(y)}$,

$$d \leqslant b_\alpha \text{ iff } g^{-1}(d) \leqslant x, \qquad d = b_\alpha \text{ iff } g^{-1}(d) = x.$$

Then g can be extended to

$$h : (\overline{F_\alpha(a_\alpha)}(x), \leqslant) \cong (\overline{G_\alpha(y)}(b_\alpha), \leqslant).$$

Let

$$F_{\alpha+1} = \overline{\overline{F_\alpha(a_\alpha)}(x)}, \qquad G_{\alpha+1} = \overline{\overline{G_\alpha(y)}(b_\alpha)}.$$

Then h can be extended to a mapping

$$f_{\alpha+1} : F_{\alpha+1} \cong G_{\alpha+1}.$$

Now $f_\alpha \subset f \subset g \subset h \subset f_{\alpha+1}$, and $a_\alpha \in F_{\alpha+1}$, $b_\alpha \in G_{\alpha+1}$, so (1)–(3) hold for $\alpha+1$.

Finally, let $f_{\omega_1} = \bigcup_{\alpha<\omega_1} f_\alpha$. Then from (1)–(3) we see that $f_{\omega_1} : F \cong G$. ⊣

Note that in the above proof we did not use the full assumption that F and G are ω_1-saturated models, but only that $\langle F, \leqslant \rangle$ and $\langle G, \leqslant \rangle$ are η_1 sets. Moreover, the argument remains valid for any successor cardinal $\omega_{\alpha+1}$ instead of ω_1. Thus the proof of Theorem 5.4.4 gives the following extra information, for each ordinal α:

Any two real closed fields whose order structures are $\eta_{\alpha+1}$ sets of power $\omega_{\alpha+1}$ are isomorphic (Erdös, Gillman and Henrickson, 1955).

A similar argument can also be used to prove:

Every real closed field whose order structure is $\omega_{\alpha+1}$-saturated (i.e., an $\eta_{\alpha+1}$ set), is $\omega_{\alpha+1}$-saturated.

Tarski's original proof of Theorem 5.4.4 also showed that the theory of real closed ordered fields admits elimination of quantifiers. This can also be proved using Lemma 5.4.3 and the results of Section 3.5. See exercise 5.4.4.

Macintyre, McKenna, and van den Dries (1983) proved the following converse result: The theory of real closed ordered fields is the only complete extension of the theory of ordered fields which admits elimination of quantifiers.

A similar proof shows that the theory of algebraically closed fields of characteristic p is complete. However, this already follows from ω_1-categoricity. A more interesting case is the theory of all models (F, U), where F is an algebraically closed field and U is a subfield. Given such a model (F, U), an element x of F is said to be of *degree n over U* iff n is the least positive integer such that x is a root of a polynomial of degree n with coefficients in U. Thus x has degree 1 over U iff $x \in U$, and x has degree n over U for some $n < \omega$ iff x is algebraic over U. We say that x is *transcendental over U*, or has *infinite degree over U*, iff for all n, x does not have degree n over U. A subset X of F is said to be *transcendental* (or *algebraically independent*) *over* U iff for any $x_1, \ldots, x_n \in X$, the only polynomial with n variables and coefficients in U having $x_1, \ldots, x_n$ as a root is the zero polynomial. Thus $\{x\}$ is transcendental over U iff x is.

LEMMA 5.4.5.

(i). *Let F, G be algebraically closed fields, let U, V be subfields, and let $f : U \cong V$. Suppose $X \subset F$ is a maximal transcendental set over U, $Y \subset G$ is a maximal transcendental set over V, and $|X| = |Y|$. Then f can be extended to an isomorphism $g : F \cong G$.*

(ii). *If F is an algebraically closed field and U is a subfield of F, then any two maximal transcendental sets over U have the same cardinality.*

(iii). *Let F be an algebraically closed field and let U be a subfield of F. Then exactly one of the following statements holds:*

(a) *$F = U$.*

(b) *U is a real closed field and F is the algebraic closure of U.*

(c) *For each $n < \omega$, there is an x in F of degree $> n$ (perhaps infinite) over U.*

THEOREM 5.4.6. *Suppose F, G are algebraically closed fields, U and V are subfields, and $U \equiv V$. Assume that neither (F, U) nor (G, V) is of the type* (a) *or* (b) *in Lemma 5.4.5. Then*

$$(F, U) \equiv (G, V).$$

PROOF. We may assume that the models (F, U) and (G, V) are saturated of power ω_1, and prove that they are isomorphic. The union of a chain of transcendental sets over U is obviously transcendental over U, whence by Zorn's lemma there exists a maximal transcendental set X over U. There is a set $\Sigma(v)$ of formulas which is satisfied by x if and only if x is transcendental over U. Just take $\Sigma(v) = \{\neg \sigma_n(v) : n < \omega\}$, where $\sigma_n(v)$ says that v is of degree n over U. By Lemma 5.4.5 (iii), every finite subset of $\Sigma(v)$ is satisfiable in (F, U), so $\Sigma(v)$ is satisfiable. This shows that X is nonempty. Now form a set $\Sigma(v_1 \ldots v_n)$ of formulas saying that $v_1, \ldots, v_n$ are transcendental over U. If $x \in X$, then for each $m < \omega$, the n-tuple

$$x, x^m, x^{m^2}, \ldots, x^{m^n}$$

cannot be a root of any nonzero polynomial of degree less than m with coefficients in U. For if it were, then we would obtain a nonzero polynomial with coefficients in U and x as a root. It follows that $\Sigma(v_1 \ldots v_n)$ is finitely satisfiable in (F, U), so there exist transcendental sets of power n over U. By Lemma 5.4.5 (ii), the maximal set X must be infinite. We may form the expanded model $(F, U, x)_{x \in X}$, and in this model there is a set of formulas $\Gamma(v)$ saying that $X \cup \{v\}$ is transcendental over U. Since X is infinite, every finite subset of $\Gamma(v)$ is satisfied in $(F, U, x)_{x\in X}$ by an element of X not mentioned in the finite subset of $\Gamma(v)$. If X is countable, then by saturatedness, $\Gamma(v)$ is satisfiable in $(F, U, x)_{x\in X}$. But X is a maximal transcendental set over U, so $\Gamma(v)$ is not satisfiable. Therefore X is uncountable, and $|X| = \omega_1$.

Exactly the same argument shows that there is a maximal transcendental set Y over V of power ω_1. Since (F, U) and (G, V) are ω_1-saturated, the subfields U and V are ω_1-saturated. Hence either both are finite or both have power ω_1. In either case, since $U \equiv V$, there is an isomorphism $f: U \cong V$. By Lemma 5.4.5 (i), f may be extended to an isomorphism $g: F \cong G$. Since $f \subset g$, g maps U onto V; hence $g: (F, U) \cong (G, V)$. ⊣

As a third example, we shall show that if two fields of characteristic zero are elementarily equivalent, then their fields of formal power series are elementarily equivalent. We shall actually prove a much more general result about valued fields, but first we describe the classical case.

Consider a field H, and let $H(t)$ be a pure transcendental extension of H. We shall define a *valuation* on $H(t)$ in the following way. We set val $(0) = 0$. For each element $a \neq 0$ in H, define val $(a) = 1$. For each nonzero polynomial

$$b = a_0 + a_1 t + \dots + a_n t^n$$

with coefficients in H, let val $(b) = t^m$, where a_m is the first nonzero coefficient, i.e.

$$a_0 = 0, \dots, a_{m-1} = 0, \qquad a_m \neq 0.$$

In particular, val $(t^m) = t^m$. Finally, if b and c are two polynomials with coefficients in H, and with $c \neq 0$, define

$$\text{val}\left(\frac{b}{c}\right) = \frac{\text{val}(b)}{\text{val}(c)}.$$

Let $\mathbb{Z}$ be the set of integers and let

$$V = \{t^m : m \in \mathbb{Z}\}.$$

Then $\langle V, \cdot, 1\rangle$ is a subgroup of $\langle H(t) \setminus \{0\}, \cdot, 1\rangle$ and val maps $H(t)$ onto $V \cup \{0\}$. We introduce a simple ordering $\leqslant$ on the set $V \cup \{0\}$ by making 0 the greatest element and putting

$$t^m \leqslant t^n \quad \text{if and only if} \quad m \leqslant n.$$

Thus the structure $\langle V, \cdot, 1, \leqslant\rangle$ is isomorphic to $\langle \mathbb{Z}, +, 0, \leqslant\rangle$.

The structure

$$\langle H(t), +, \cdot, 0, 1, V, \leqslant, \text{val}\rangle$$

is an example of a valued field. We now list the axioms for the theory of valued fields. All of these axioms are easily seen to hold in $H(t)$.

5.4.7. A *valued field* (with cross section) is a model

$$F = \langle F, +, \cdot, 0, 1, V, \leqslant, \text{val}\rangle,$$

where

(a) $\langle F, +, \cdot, 0, 1\rangle$ is a field;

(b) $\langle V, \cdot, 1\rangle$ is a group with unit 1 (the *value group*);

(c) $\leqslant$ is a simple ordering of the set $V \cup \{0\}$ with greatest element 0;

(d) for all $x, y, z \in V$, $x \leqslant y$ implies $x \cdot z \leqslant y \cdot z$;

(e) val is a function from F onto $V \cup \{0\}$;

(f) for all $x, y \in F$,

$$\text{val}\,(x \cdot y) = \text{val}\,(x) \cdot \text{val}\,(y),$$

$$\text{val}\,(x+y) \geqslant \min\,\{\text{val}\,(x), \text{val}\,(y)\},$$

$$\text{val}\,(x) = 0 \quad \text{iff} \quad x = 0;$$

(g) for all $x \in V$, val $(x) = x$ ('cross section' axiom).

We shall now extend the valued field $H(t)$ to a valued field $H((t))$ which is the *completion* of $H(t)$. It is constructed by a process analogous to the construction of the real numbers from the rational numbers. One of the standard definitions of a real number is as an equivalence class of Cauchy sequences of rationals. A Cauchy sequence of rationals is a sequence $\langle x_n : n < \omega\rangle$ such that $\lim_{m,n\to\infty}|x_m - x_n| = 0$; two sequences $\langle x_n\rangle$, $\langle y_n\rangle$ are equivalent iff $\lim_{n\to\infty}|x_n - y_n| = 0$. The sums and products are defined pointwise, e.g., $\langle x_n\rangle + \langle y_n\rangle = \langle x_n + y_n\rangle$.

In a similar way, a *Cauchy sequence* in the valued field $H(t)$ is a sequence $\langle x_n : n < \omega\rangle$ of elements of $H(t)$ such that for all $y \in V$, there exists an $N < \omega$ such that

$$\text{val}\,(x_m - x_n) > y \quad \text{for all} \quad m, n \geqslant N.$$

Two Cauchy sequences $\langle x_n\rangle$, $\langle y_n\rangle$ are *equivalent* iff, for all $z \in V$, there exists an $N < \omega$ such that

$$\text{val}\,(x_n - y_n) > z \quad \text{for all} \quad n \geqslant N.$$

Define $H((t))$ as the set of all equivalence classes of Cauchy sequences in

$H(t)$. It turns out that when we define sums and products of Cauchy sequences pointwise, $H((t))$ is a field. When we identify each element $x \in H(t)$ with the equivalence class of the constant sequence $\langle x, x, \ldots\rangle$, then $H(t)$ becomes a subfield of $H((t))$. Moreover, if $\langle x_n\rangle$ is any Cauchy sequence not equivalent to the zero sequence, then the function val (x_n) of n is eventually constant, i.e., there exists $N < \omega$ and $y \in V$ such that

$$\text{val}\,(x_n) = y \quad \text{for all} \quad n \geqslant N.$$

Therefore the valuation function on $H(t)$ may be extended to a function val on $H((t))$ into $V \cup \{0\}$ by defining val $(\langle x_n\rangle)$ to be the eventual value of val (x_n). It is easy to check that the structure

$$\langle H((t)), +, \cdot, 0, 1, V, \leqslant, \text{val}\rangle$$

is a valued field. $H((t))$ is called the field of *formal power series* over H, and it is also called the *completion* of $H(t)$.

The valued field $H((t))$ also has another, much deeper property, known as Hensel's lemma. To state Hensel's lemma we need more notation. Consider a valued field F. By the *valuation ring*, we mean the subring $R(F)$ of the field $\langle F, +, \cdot, 0, 1\rangle$, whose elements are all $x \in F$ with val $(x) \geqslant 1$. It follows from axiom (f) that $R(F)$ is a ring with zero 0 and unit 1. Furthermore, the set

$$M(F) = \{x \in F : \text{val}\,(x) > 1\}$$

is a maximal ideal in $R(F)$, because if an ideal I in $R(F)$ contains an element x with val $(x) = 1$, then $x^{-1} \in R(F)$, whence $1 = x \cdot x^{-1} \in I$. Therefore the quotient ring

$$F^* = R(F)/M(F)$$

is a field. F^* is called the *residue class field* of F. For each $x \in R(F)$, let

$$x^* = x/M(F)$$

be the residue class of x in F^*. Thus $*$ is a homomorphism of $R(F)$ onto F^*. It follows that if F has prime characteristic p, then so does F^*. Moreover, if F^* has characteristic 0, so does F.

In the special case $F = H((t))$, we have $H((t))^* \cong H(t)^*$. In fact, we have the following lemma:

LEMMA 5.4.8. *H is a subfield of $R(H((t)))$ and the restriction of $*$ to H maps H isomorphically onto $H((t))^*$. Thus we may identify H with the residue class field of $H((t))$.*

We denote by $R(F)[t]$ and $F^*[t]$ the rings of all polynomials in the variable t with coefficients in $R(F)$ and F^*, respectively. For each $p(t) \in R(F)[t]$, let $p^*(t) \in F^*[t]$ be formed by replacing each coefficient a of $p(t)$ by a^*. Then Hensel's lemma is the following:

5.4.9 (Hensel's Lemma). *If $p(t) \in R(F)[t]$, $p(t)$ has leading coefficient 1, and in $F^*[t]$ we have $p^*(t) = q'(t)r'(t)$, where $q'(t)$ and $r'(t)$ are relatively prime, then there is a factorization $p(t) = q(t)r(t)$ in $R(F)[t]$ such that $q^*(t) = q'(t)$, $r^*(t) = r'(t)$.*

By a *Hensel field* we mean a valued field in which Hensel's lemma holds. A classical result is:

LEMMA 5.4.10. *For any field H, the valued field $H((t))$ of formal power series is a Hensel field.*

To return to model theory, we have:

LEMMA 5.4.11. *The class of all Hensel fields is an elementary class.*

We leave the details of the proof of Lemma 5.4.11 as an exercise. The only difficulty is to express Hensel's lemma in $\mathscr{L}$. For each degree n, Hensel's lemma for $p(t)$ of degree n can be expressed by a single sentence of $\mathscr{L}$.

Given a valued field F, the value group is denoted by

$$\text{val}\,(F) = \langle V, \cdot, 1, \leqslant \rangle.$$

We shall prove the following theorem:

THEOREM 5.4.12. *Suppose F and G are Hensel fields such that F^* has characteristic zero, $F^* \equiv G^*$, and* $\text{val}\,(F) \equiv \text{val}\,(G)$. *Then $F \equiv G$.*

A helpful reference on valued fields is the monograph of Ribenboim (1967), hereafter referred to as [R]. We shall give the proof of the above theorem modulo some results in [R] of a purely algebraic nature. All the uses of model-theoretic methods in the proof will be given here. A few more definitions are needed.

Consider a valued field F. If $\langle G, +, \cdot, 0, 1\rangle$ is a subfield of $\langle F, +, \cdot, 0, 1\rangle$, we let

$$\text{val}_F\,(G) = \{\text{val}\,(x) : x \in G \quad \text{and} \quad x \neq 0\},$$

and

$$G^{*F} = \{x^* : x \in G \quad \text{and} \quad \text{val}(x) \geqslant 1\}.$$

G is said to be a *valued subfield* of F iff G is a subfield of F and $\text{val}_F(G) \subset G$. It is easy to check that every valued subfield of F is a valued field. If G is a valued subfield of F, then $\text{val}_F(G) = \text{val}(G)$, and G^{*F} becomes G^* when we identify $x/M(F)$ with $x/M(G)$. For any $X \subset F$, there is a least-valued subfield of F which includes X. Let us recall that if G is a subfield of F, then the *relative algebraic closure* $\bar{G}$ of G in F is the set of all $x \in F$ algebraic over G. We know from Lemma 5.4.3 (ii) that G is a subfield of F. We say that G is *relatively algebraically closed in F* iff $G = \bar{G}$.

If G is a valued subfield of F, the value group $\text{val}(G)$ is said to be *closed under roots* iff for any $y \in \text{val}(F)$ and any positive integer n, if $y^n \in \text{val}(G)$, then $y \in \text{val}(G)$. That is, any nth root of an element of $\text{val}(G)$ in $\text{val}(F)$ belongs to $\text{val}(G)$. Clearly, if G is a relatively algebraically closed valued subfield of F, then $\text{val}(G)$ is closed under roots.

We shall now introduce the analogue of the real closure of a field. Let F be a valued field and G a valued subfield of F. We say that F is a *henselization* of G iff F is an algebraic extension of G, F is a Hensel field, and for any other algebraic extension H of G which is a Hensel field, there is an isomorphic embedding of F into H which is the identity on G.

The following lemma contains the purely algebraic facts needed to prove Theorem 5.4.12.

LEMMA 5.4.13.

(i). *Let F be a valued field. If* $x, y \in F$ *and* $\text{val}(x) < \text{val}(y)$, *then*

$$\text{val}(x+y) = \text{val}(x),$$

$$\text{val}(-x) = \text{val}(x).$$

Moreover, if $x \in \text{val}(F)$ *and* $0 < n < \omega$, *then there is at most one* $y \in \text{val}(F)$ *such that* $y^n = x$.

(ii). *Let F be a Hensel field whose residue class field* F^* *has characteristic zero. Then there is a valued subfield* $F_0 \subset F$ *such that* $F_0 \subset R(F)$ *and* $*$ *maps* F_0 *isomorphically onto* F^*.

(iii). *Let* F_1 *and* G_1 *be valued fields with Hensel subfields* F_0 *and* G_0, *respectively, and suppose* $f : F_0 \cong G_0$ *and* F_1, G_1 *are algebraic extensions of* F_0, G_0, *respectively. If f can be extended to a field isomorphism* $g : F_1 \cong G_1$, *then f can be extended to a valued field isomorphism* $g : F_1 \cong G_1$. *Moreover,* *if σ is a field automorphism of F_1 which is the identity on F_0, and $x \in F_1$, then* $\text{val}(x) = \text{val}(\sigma x)$.

(iv). *Every valued field has a henselization. If F_0 and G_0 are valued fields, F_1 and G_1 are henselizations of them, and $f : F_0 \cong G_0$, then f can be extended to an isomorphism $g : F_1 \cong G_1$ (i.e., the henselization is unique).*

(v). *If F_1 is a henselization of a valued subfield F_0, then $F_0^* = F_1^*$ and* $\text{val}(F_0) = \text{val}(F_1)$.

(vi). *If F is a valued field, and F_0 is a valued subfield, then* $\text{val}(\bar{F}_0)$ *is the closure of* $\text{val}(F_0)$ *under roots in* $\text{val}(F)$. *If F is a Hensel field, F_0 a valued subfield, $F^* = F_0^*$, F^* has characteristic zero, and* $\text{val}(F_0)$ *is closed under roots in* $\text{val}(F)$, *then $\bar{F}_0$ is a henselization of F_0.*

(vii). *Suppose F_1 and G_1 are Hensel fields, F and G are Hensel subfields of F_1 and G_1, respectively, $x \in F_1$ and $y \in G_1$ are transcendental over F and G, f is an isomorphism $f : F \cong G$,*

$$\text{val}(F(x)) = \text{val}(F), \qquad F(x)^* = F^*,$$

$$\textit{and for all} \quad a \in F, \quad f(\text{val}(x-a)) = \text{val}(y-f(a)).$$

Then

$$\text{val}(G(y)) = \text{val}(G), \qquad G(y)^* = G^*,$$

and f can be extended to an isomorphism

$$g : F(x) \cong G(y).$$

(viii). *Suppose F_1 is a Hensel field, F is a Hensel subfield, $x \in F_1$ is transcendental over F, and $F(x)^* = F^*$. If* $\text{val}(F)$ *has more than one element, then*

$$|\text{val}(F(x))| = |\text{val}(F)|.$$

PROOF. (i). This is elementary.

(ii). We have

$$\text{val}(1) = \text{val}(1 \cdot 1) = \text{val}(1) \cdot \text{val}(1),$$

and $\text{val}(1) \neq 0$; hence $\text{val}(1) = 1$. If n is any positive integer, then

$$\text{val}(n) = \text{val}(1 + \dots + 1) \geqslant \text{val}(1) = 1,$$

so $n \in R(F)$. Since F^* has characteristic 0, $n^* = n \neq 0$; hence $\text{val}(n) = 1$. For any positive rational number m/n, $\text{val}(m/n) = \text{val}(m)/\text{val}(n) = 1$. By (i), the value of any negative rational number is 1. Thus the field Q of rational numbers is a subfield of the ring $R(F)$. The union of a chain of fields is a field, so by Zorn's lemma there is a maximal subfield $F_0 \subset R(F)$.

For any nonzero element $x \in F_0$, val (x) = val $(1/x) = 1$. The mapping * embeds F_0 isomorphically onto a subfield G_0 of F^*. We shall show that $G_0 = F^*$.

First suppose $a \in F^* \setminus G_0$ is algebraic over G_0. Then there is an irreducible polynomial $p_1(t) \in G[t]$ which has leading coefficient 1 and $p_1(a) = 0$. Taking inverse images of the coefficients under $* \restriction F_0$, we obtain a polynomial $p(t) \in F_0[t]$ with leading coefficient 1 such that $p^*(t) = p_1(t)$. Since G_0 has characteristic 0, $p^*(t)$ has no multiple roots, and therefore in F^* we may write

$$p^*(t) = q_1(t)(t-a),$$

and $q_1(t)$ and $t-a$ are relatively prime. Now by Hensel's lemma, there exist polynomials $q(t), r(t) \in R(F)[t]$ such that

$$p(t) = q(t)r(t), \qquad q^*(t) = q_1(t), \qquad r^*(t) = t-a.$$

Since the leading coefficients of $q(t)$, $r(t)$ have product 1 and values $\geqslant 1$, they have value 1. Therefore $q(t)$ has the same degree as $q^*(t)$, and $r(t)$ has degree 1. Say $r(t) = b_0 + b_1 t$. Then the element $y = -b_0/b_1$ is a root of $r(t)$ and therefore a root of $p(t)$. Since $p^*(t)$ is irreducible over G_0, $p(t)$ is irreducible over F_0, whence $y \notin F_0$. But val $(b_1) = 1$ and val $(b_0) \geqslant 1$, so val $(y) \geqslant 1$. We have $y^* = -b_0^*/b_1^* = a$. For any polynomial $s(t) \in F_0[t]$ with $\deg(s(t)) < \deg(p(t))$, we have $s(y)^* = s^*(a) \neq 0$, whence val $(s(y)) = 1$. But $F_0(y)$ is the set of all such $s(y)$. Therefore $F_0(y) \subset R(F)$, contradicting the maximality of F_0.

Now suppose $a \in F^* \setminus G_0$ is transcendental over G_0. Choose any $y \in R(F)$ such that $y^* = a$. Then for any nonzero polynomial $p(t) \in F_0[t]$,

$$p(y)^* = p^*(y) \neq 0,$$

so val $(p(y)) = 1$. Hence if $p(t), q(t) \in F_0[t]$ are two nonzero polynomials, then val $(p(y)/q(y)) = 1$. It follows that $F_0(y) \subset R(F)$, again contradicting the maximality of F_0. Therefore $G_0 = F^*$ and (ii) is proved.

(iii). This is proved in [R] Chapter F, Theorem 4. The equivalence of the definition of henselization given here with the one in [R] is proved in Chapter F, Corollary 2 of Theorem 2, and Theorem 4.

(iv). This is proved in [R] Chapter F, Theorem 2.

(v). This is proved in [R] Chapter F, Corollary 1 of Theorem 3.

(vi). Let $\bar{V}$ be the closure of val (F_0) under roots. If $x \in \bar{F}_0$, then by [R] Chapter F, Theorem 1, val $(x) \in \bar{V}$. Thus val $(\bar{F}_0) \subset \bar{V} \subset \bar{F}_0$. But val $(\bar{F}_0)$ is closed under roots, so val $(\bar{F}_0) = \bar{V}$. Now suppose F is a Hensel field, val $(F_0) = \bar{V}$, and $F^* = F_0^*$ has characteristic zero. Let $p(t) \in R(\bar{F}_0)[t]$ have leading coefficient 1. Suppose that in $F^*[t]$ we have $p^*(t) = q'(t)r'(t)$, where $q'(t)$ and $r'(t)$ are relatively prime. Since F is a Hensel field, there exist $q(t), r(t) \in R(F)[t]$ such that $p(t) = q(t)r(t)$, $q^*(t) = q'(t)$, and $r^*(t) = r'(t)$. We may assume that $q(t)$ and $r(t)$ have leading coefficients 1. Because $\bar{F}_0$ is relatively algebraically closed in F, we already have $q(t), r(t) \in \bar{F}_0[t]$. This shows that $\bar{F}_0$ is a Hensel field. By the definition of henselization and by (iv), there is a henselization F_1 of F_0 such that $F_0 \subset F_1 \subset \bar{F}_0$. Then $\bar{F}_0$ is an algebraic extension of F_1 such that

$$\text{val}(\bar{F}_0) = \text{val}(F_1), \qquad \bar{F}_0^* = F_1^*.$$

It is shown in [R] Chapter G, Theorem 2, that if F_1 is a Hensel field and F_1^* has characteristic zero, then F_1 has no proper algebraic valued extension with the same value group and residue class field. It follows that $\bar{F}_0 = F_1$.

(vii). Let F_1' be an algebraic closure of F_1, and choose G_1' similarly. It is proved in [R] Chapter B, Theorem 5, that F_1' can be made into a valued field with the valued subfield F_1, and similarly for G_1'. Let F', G' be the algebraic closures of F, G in F_1', G_1'. By [R] Chapter F, Theorem 1, val (F') is the closure of val (F) under roots in val (F_1'), and similarly for G'. Therefore F', G' are valued subfields of F_1', G_1'. Then by (iii), f can be extended to a valued field isomorphism

$$f' : F' \cong G'.$$

We shall show that for every $b \in F(x)$ and $\alpha \in F' \setminus F$,

(1) there exists $a \in F$ such that val $(b-a) \neq$ val $(\alpha - a)$.

Suppose (1) fails for some b and α. Let $p(t) \in F[t]$ be an irreducible polynomial with leading coefficient 1 such that $p(\alpha) = 0$. Let $\alpha = \alpha_1, \ldots, \alpha_n$ be all the roots of $p(t)$. Thus

$$p(t) = (t-\alpha_1) \ldots (t-\alpha_n) = c_0 + c_1 t + \ldots + c_{n-1} t^{n-1} + t^n.$$

Let

$$e = \text{val}(b - c_{n-1} n^{-1}).$$

Since (1) fails, and by (iii), we have

$$e = \text{val}(\alpha_i - c_{n-1} n^{-1}), \qquad i = 1, \ldots, n.$$

Now let

$$\alpha_i' = e^{-1}(\alpha_i - c_{n-1} n^{-1}), \qquad i = 1, \ldots, n.$$

Thus val $(\alpha_i') = 1$ for each i. Consider the polynomial

$$p_1(t) = (t-\alpha_1') \dots (t-\alpha_n') = d_0 + d_1 t + \dots + d_{n-1} t^{n-1} + t^n.$$

Each coefficient d_i is a symmetric function of $\alpha_1, \dots, \alpha_n$ and hence belongs to F, whence $p(t) \in F[t]$. Moreover, $p_1(t)$ is irreducible over F, for $F(\alpha) = F(\alpha')$, whence α and α' have the same degree n over F. We have

$$\text{(2)} \qquad \text{val}(d_0) = \text{val}(\alpha_1' \dots \alpha_n') = \text{val}(\alpha_1') = \dots = \text{val}(\alpha_n') = 1,$$

and

$$\text{(3)} \qquad d_{n-1} = \alpha_1' + \dots + \alpha_n' = e^{-1}(\alpha_1 + \dots + \alpha_n - c_{n-1}) = 0,$$

and for each $i \leqslant n$, val $(d_i) \geqslant 1$. Let

$$b' = e^{-1}(b - c_{n-1} n^{-1}).$$

Then val $(b') = 1$. Hence $b' \in R(F(x))$, and because $F(x)^* = F^*$, there exists $a \in F$ such that val $(b'-a) > 1$. Then since (1) fails, val $(\alpha'-a) > 1$. By (iii),

$$\text{val}(\alpha_i' - a) = \text{val}(\alpha' - a), \qquad i = 1, \dots, n,$$

whence

$$\text{val}(p(a)) = \text{val}(\alpha'-a)^n > 1.$$

Therefore

$$p^*(a^*) = 0.$$

and hence $p^*(t)$ has the linear factor $(t-a^*)$. By Hensel's lemma, in F, $p^*(t)$ cannot be the product of two relatively prime factors, because $p(t)$ is irreducible over F. Therefore we must have

$$\text{(4)} \qquad p^*(t) = (t-a^*)^n.$$

It follows from (2), (3), and (4) that

$$(a^*)^n = d_0^* \neq 0.$$

However,

$$n(a^*)^{n-1} = d_{n-1}^* = 0,$$

but this is impossible because F^* has characteristic zero. We conclude that (1) does hold for all $b \in F(x)$ and $\alpha \in F' \setminus F$.

Let us now define the function g on $F(x)$ into $G(y)$ by

$$g\left(\frac{a_0 + a_1 x + \dots + a_m x^m}{b_0 + b_1 x + \dots + b_n x^n}\right) = \frac{fa_0 + fa_1 y + \dots + fa_m y^m}{fb_0 + fb_1 y + \dots + fb_n y^n}.$$

Then g is an extension of f and g is a field isomorphism of $F(x)$ onto $G(y)$.

It suffices to show that for all $b \in F(x)$,

$$\operatorname{val}(g(b)) = f(\operatorname{val}(b)).$$

Since f is an isomorphism and the value function preserves products, we need only show that for every irreducible polynomial $p(t) \in F[t]$ with leading coefficient 1,

$$\operatorname{val}(g(p(x))) = f(\operatorname{val}(p(x))). \tag{5}$$

By hypothesis, (5) holds if $p(t)$ has degree $\leqslant 1$. Suppose p has degree > 1 and let α be a root of $p(t)$ in F'. Then $\alpha \in F' \setminus F$, so (1) holds for x and α. Let $a \in F$ be as in (1). We have

$$f(\operatorname{val}(x-a)) = \operatorname{val}(y-f(a)),$$

$$f'(\operatorname{val}(\alpha-a)) = \operatorname{val}(f'(\alpha)-f(a)).$$

In the case that

$$\operatorname{val}(x-a) < \operatorname{val}(\alpha-a),$$

we see that

$$\operatorname{val}(x-\alpha) = \operatorname{val}(x-a),$$

$$\operatorname{val}(y-f'(\alpha)) = \operatorname{val}(y-f(a)),$$

whence

$$\begin{aligned} f(\operatorname{val}(p(x))) &= f(\operatorname{val}(x-\alpha)^n) = f(\operatorname{val}(x-a)^n) \\ &= \operatorname{val}(y-f(a))^n = \operatorname{val}(y-f'(\alpha))^n = \operatorname{val}(g(p(x))), \end{aligned}$$

so (5) holds. In the other case

$$\operatorname{val}(\alpha-a) < \operatorname{val}(x-a),$$

we have

$$\operatorname{val}(x-\alpha) = \operatorname{val}(a-\alpha),$$

$$\operatorname{val}(y-f'(\alpha)) = \operatorname{val}(f(a)-f'(\alpha)),$$

and a similar computation shows that (5) again holds.

(viii). As in (vii), we may let F_1' be an algebraic closure of F_1 and make F_1' into a valued field extension of F_1. Let F' be the algebraic closure of F in F_1'. Again, val (F') is the closure under roots of val (F) in val (F_1'). Since val (F) has more than one element, it follows from the axioms that val (F) is infinite. Therefore, in view of (i), we have

$$|\operatorname{val}(F')| = |\operatorname{val}(F)|.$$

We shall show that

(6) $$|\text{val}(F')| = |\text{val}(F'(x))|,$$

from which it will follow that

$$|\text{val}(F)| \leqslant |\text{val}(F(x))| \leqslant |\text{val}(F'(x))| = |\text{val}(F)|.$$

Each element of $F'(x)$ is a quotient of products of elements of the form $x-\alpha$, $\alpha \in F'$. We observe that if $\alpha, \beta \in F'$ and $\text{val}(x-\alpha) < \text{val}(x-\beta)$, then by (i),

$$\text{val}(x-\alpha) = \text{val}((x-\alpha)-(x-\beta)) = \text{val}(\beta-\alpha),$$

whence $\text{val}(x-\alpha) \in \text{val}(F')$. Therefore the set

$$\{\text{val}(x-\alpha) : \alpha \in F'\} \setminus \text{val}(F')$$

contains at most one element. It follows that

$$|\text{val}(F'(x))| \leqslant \omega + |\text{val}(F')|;$$

thus (6) holds. ⊣

Proof of Theorem 5.4.12. The proof parallels the completeness proof for the theory of real closed fields, using Proposition 5.4.1(i). Let $\hat{F}$ and $\hat{G}$ be saturated models of power ω_1 such that $\hat{F} \equiv F$ and $\hat{G} \equiv G$. By Lemma 5.4.11, $\hat{F}$ and $\hat{G}$ are Hensel fields. It is easy to check that $\hat{F}^* \equiv F^* \equiv G^* \equiv \hat{G}^*$ and $\text{val}(\hat{F}) \equiv \text{val}(F) \equiv \text{val}(G) \equiv \text{val}(\hat{G})$ (see Exercise 5.4.16). Thus, given the continuum hypothesis, we may assume that F and G are saturated models of power ω_1, and we shall prove that $F \cong G$. We note first that the residue class fields F^* and G^* and the value groups $\text{val}(F)$ and $\text{val}(G)$ are ω_1-saturated models (Exercise 5.4.16). Except in the trivial case where $\text{val}(F) = \{1\}$, and F^*, G^*, $\text{val}(F)$ and $\text{val}(G)$ have power ω_1.

By Lemma 5.4.13 (ii) we may identify F^* and G^* with subfields of F and G in such a way that the * mapping is the identity on F^* and G^*. Since $F^* \equiv G^*$, there is an isomorphism $f_0 : F^* \cong G^*$. In the trivial case where $\text{val}(F) = \{1\}$, we have $\text{val}(G) = \{1\}$, and $F = F^*$, $G = G^*$, so $f_0 : F \cong G$.

Suppose now that $\text{val}(F)$ is not trivial. If F_1, G_1 are valued subfields of F, G, we shall write $f_1 : F_1 \leftrightarrow G_1$ iff $f_1 : F_1 \cong G_1$ and

$$(\text{val}(F), x)_{x \in \text{val}(F_1)} \equiv (\text{val}(G), f_1 x)_{x \in \text{val}(F_1)}.$$

We shall prove the following:

(i). F^* and G^* are relatively algebraically closed valued subfields of F and G, respectively.

(ii). Let F_1 and G_1 be relatively algebraically closed valued subfields of F and G, respectively, such that $F^* \subset F_1$ and $G^* \subset G_1$, and val (F_1) is countable. Suppose that $f_0 \subset f_1$ and $f_1 : F_1 \leftrightarrow G_1$. Then for every $x \in F$, there exist relatively algebraically closed valued subfields F_2 and G_2 and a mapping $f_2 : F_2 \leftrightarrow G_2$ such that val (F_2) is countable and

$$x \in F_2, \qquad F_1 \subset F_2, \qquad G_1 \subset G_2, \qquad f_1 \subset f_2.$$

(iii). Property (ii) above holds with the roles of F and G interchanged.

By the usual back and forth argument, (i)–(iii) imply that $F \cong G$. We note that $f_0 : F^* \leftrightarrow G^*$ because val $(F^*) = \{1\} =$ val (G^*), so (i) allows us to start the induction.

Proof of (i). Let $p(t) \in F^*[t]$. Then since $F^* \subset R(F)$, $p(t) \in R(F)[t]$. If $x \in F$ and val $(x) < 1$, then all the nonzero terms $a_m x^m$ of $p(x)$ have different values, val $(x)^m < 1$. Hence val $(p(x)) < 1$, and x is not a root of $p(x)$. Suppose $x \in F$ is a root of $p(t)$. Then val $(x) \geqslant 1$, whence x^* is defined and $p(x^*) = 0$. Thus $p(t)$ already has the root x^* in F^*, whence $F^* = \overline{F^*}$. The same proof shows that $G^* = \overline{G^*}$.

Proof of (ii). We shall first prove a special case:

(iia). If val $(F_1(x)) =$ val (F_1), then (ii) holds.

We wish to use Lemma 5.4.13 (vii). We have

$$F_1(x)^* = F^* = F_1^*.$$

If $x \in F_1$, then (ii) is trivial, so let $x \notin F_1$. Then $F_1(x)$ is a simple transcendental extension of F_1. By Lemma 5.4.13 (vi), $\overline{F_1} = F_1$ is a Hensel field, and similarly G_1 is a Hensel field. We must find $y \in G$ such that for all $a \in F_1$,

$$(1) \qquad f_1(\text{val}\,(x-a)) = \text{val}\,(y-f_1(a)).$$

We claim that:

(2) For every finite $A \subset F_1$ there exists $y \in G$ such that for all $a \in A$, (1) holds.

Let us prove (2). Let $A \subset F_1$ be finite, and let $b \in A$ be such that $c =$ val $(x-b)$ is a maximum. Since val $(F_1) =$ val $(F_1(x))$, $c \in$ val (F_1). For each positive $n < \omega$, val $(n) = 1$ because F^* has characteristic zero, whence val $(nc) = c$. It follows that for all n and all $a \in A$,

$$\text{val}\,(b-nc-a) \geqslant \min\,(\text{val}\,(b-x),\ \text{val}\,(nc), \text{val}\,(x-a)) = \text{val}\,(x-a).$$

Moreover, for each $a \in A$, there is at most one n such that

$$\text{val}\,(b-nc-a) > \text{val}\,(x-a).$$

For if $m < n$, $a \in A$, and

$$\min\,(\text{val}\,(b-mc-a), \text{val}\,(b-nc-a)) > \text{val}\,(x-a),$$

then

$$\text{val}\,(x-a) < \text{val}\,((n-m)c) = c,$$

whence, by Lemma 5.4.13 (i),

$$\text{val}\,(b-nc-a) = \text{val}\,(x-a).$$

But A is finite, so there is a positive $n < \omega$, where

$$\text{val}\,(b-nc-a) = \text{val}\,(x-a) \quad \text{for all} \quad a \in A.$$

Let $y = f(b-nc)$. Then for all $a \in A$,

$$f(\text{val}\,(x-a)) = f(\text{val}\,(b-nc-a)) = \text{val}\,(y-f(a)),$$

and our claim (2) is proved.

Because val $(F_1(x))$ is countable, there is a countable set $A_1 \subset F_1$ such that:

(3) For all $b \in F_1$, there exists $a \in A_1$ with val $(x-a) =$ val $(x-b)$.

Using (2) and the fact that G is ω_1-saturated, we can choose an element $y \in G$ such that (1) holds for all $a \in A_1$. Let $b \in F_1$. Using the equation $F_1(x)^* = F_1^*$, we see that there exists $b' \in F_1$ such that, with $c =$ val $(x-b)$,

$$\text{val}\,((x-b)c^{-1}-b') > 1,$$

whence

$$\text{val}\,(x-(b+cb')) > \text{val}\,(x-b).$$

Hence there exists $a \in A_1$ such that

$$\text{val}\,(x-b) < \text{val}\,(x-a).$$

Therefore

$$\text{val}\,(x-b) = \text{val}\,(a-b) < \text{val}\,(x-a),$$

and applying f,

$$\text{val}\,(fa-fb) = f(\text{val}\,(a-b)) < f(\text{val}\,(x-a)) = \text{val}\,(y-fa).$$

It follows that

$$\begin{aligned}\text{val}\,(y-fb) &= \text{val}\,(y-fa+fa-fb) = \text{val}\,(fa-fb)\\ &= f(\text{val}\,(a-b)) = f(\text{val}\,(x-b)).\end{aligned}$$

So (1) holds for b.

We have verified all the hypothesis of Lemma 5.4.13 (vii). Hence

$$\text{val}\,(G_1(y)) = \text{val}\,(G_1),$$

and f can be extended to an isomorphism

$$g_1 : F_1(x) \cong G_1(y).$$

Since $F_1 = \bar{F}_1$ and $G_1 = \bar{G}_1$, the value groups

(4) $$\text{val}\,(F_1(x)) = \text{val}\,(F_1), \qquad \text{val}\,(G_1(y)) = \text{val}\,(G_1)$$

are closed under roots in val (F) and val (G). Hence, by Lemma 5.4.13 (vi), $\overline{F_1(x)}$ and $\overline{G_1(y)}$ are henselizations of $F_1(x)$ and $G_1(y)$, respectively. Whence by Lemma 5.4.13 (iv), g_1 can be extended to an isomorphism

$$g_2 : \overline{F_1(x)} \cong \overline{G_1(y)}.$$

By Lemma 5.4.13 (v) and (4)

$$\text{val}\,(\overline{F_1(x)}) = \text{val}\,(F_1), \qquad \text{val}\,(\overline{G_1(y)}) = \text{val}\,(G_1).$$

It follows that condition (ii) is satisfied by

$$F_2 = \overline{F_1(x)}, \qquad G_2 = \overline{G_1(y)}, \qquad f_2 = g_2.$$

Thus we have proved the special case (iia).

We now shall prove (ii) in another special case:

(iib). If $x \in$ val (F), then (ii) holds.

Again, we may assume $x \notin F_1$. Since $f_1 : F_1 \leftrightarrow G_1$, and val (F_1) is countable, we may choose $y \in$ val (G) with

(5) $$(\text{val}\,(F), x, a)_{a \in \text{val}(F_1)} \equiv (\text{val}\,(G), y, f_1 a)_{a \in \text{val}(F_1)}.$$

Let V be the subgroup of val (F) generated by val $(F_1) \cup \{x\}$ and W the subgroup of val (G) generated by val $(G_1) \cup \{y\}$. Then V and W are countable. Consider any polynomial

$$p(t) = e_0 + e_1 t + \ldots + e_n t^n \in F_1[t].$$

If $r < s \leqslant n$ and $e_r, e_s \neq 0$, then we must have

$$\text{val}\,(e_r x^r) \neq \text{val}\,(e_s x^s).$$

For otherwise

$$\text{val}\,(e_r x^r) = \text{val}\,(e_r) x^r = \text{val}\,(e_s) x^s,$$

and thus

$$x^{s-r} = \text{val}(e_r)/\text{val}(e_s) \in \text{val}(F_1),$$

and, since val (F_1) is closed under roots, $x \in \text{val}(F_1)$, contradicting $x \notin F_1$. This means that there is a term $e_r x^r$ of $p(x)$ with smallest value, and, by Lemma 5.4.13 (i),

(6) $$\text{val}(p(x)) = \text{val}(e_r)x^r \in V.$$

Similarly, if we let $q(t) \in G_1[t]$ be the image polynomial

$$q(t) = f_1(e_0) + f_1(e_1)t + \ldots + f_1(e_n)t^n,$$

then it follows from (5) that for the same $r \leqslant n$ as before,

(7) $$\text{val}(q(y)) = \text{val}(f_1(e_r))y^r \in W.$$

It follows that val $(F_1(x)) = V$, val $(G_1(y)) = W$. Thus $F_1(x)$ and $G_1(y)$ are valued subfields of F and G, respectively.

Let us define the mapping g_1 from $F_1(x)$ onto $G_1(y)$ by

$$g_1\left(\frac{d_0 + d_1 x + \ldots + d_m x^m}{e_0 + e_1 x + \ldots + e_n x^n}\right) = \frac{f_1(d_0) + \ldots + f_1(d_m)y^m}{f_1(e_0) + \ldots + f_1(e_n)y^n}.$$

Then g_1 maps the field $F_1(x)$ isomorphically onto $G_1(y)$ and $f_1 \subset g_1$. By (6) and (7), g_1 is an isomorphism between the valued fields $F_1(x)$ and $G_1(y)$. From (5),

(8) $$(\text{val}(F), c)_{c \in V} \equiv (\text{val}(G), g_1 c)_{c \in W}.$$

By Lemma 5.4.13 (iv), $F_1(x)$ and $G_1(y)$ have henselizations F^2 and G^2, respectively. It follows from Lemma 5.4.13 (vi) that $\overline{F_1(x)}$ and $\overline{G_1(y)}$ are Hensel fields. Therefore, by the definition of henselization, we may choose F^2 and G^2 so that

$$F_1(x) \subset F^2 \subset \overline{F_1(x)}, \qquad G_1(y) \subset G^2 \subset \overline{G_1(y)}.$$

Using the uniqueness part of Lemma 5.4.13 (iv), we may extend g_1 to an isomorphism

$$g_2 : F^2 \cong G^2.$$

By Lemma 5.4.13 (v),

$$\text{val}(F^2) = V, \qquad \text{val}(G^2) = W.$$

Let $\bar{V}$, $\bar{W}$ be the closures of V, W under roots in val (F), val (G). In view of Lemma 5.4.13 (i), $\bar{V}$ and $\bar{W}$ are countable. Using (8), $g_2 \restriction V$ can be extended in a unique way to an isomorphism $h : \bar{V} \cong \bar{W}$. Indeed,

$$(9) \qquad (\mathrm{val}(F), c)_{c \in V} \equiv (\mathrm{val}(G), hc)_{c \in \bar{V}},$$

and for each $c \in \bar{V}$, c is a root of some irreducible polynomial over F^2 and hc is the root of the image polynomial over G^2. Let $F^3 = F^2(\bar{V})$ and $G^3 = G^2(\bar{W})$ be the subfields generated by $F^2 \cup \bar{V}$ and $G^2 \cup \bar{W}$, respectively. Then F^3 and G^3 are algebraic extensions of F^2 and G^2, respectively. It follows from Lemma 5.4.13 (vi) that val $(F^3) = \bar{V}$, val $(G^3) = \bar{W}$, so F^3 and G^3 are valued subfields. In view of (9), we may extend g_2 to a field isomorphism on F^3 onto G^3. Then it follows from Lemma 5.4.13 (iii) that g_2 can be extended to a valued field isomorphism, $g_3 : F^3 \cong G^3$. Since h is unique, $h = g_3 \restriction \bar{V}$.

We recall that

$$F^* \subset F_1 \subset F^3, \qquad G^* \subset G_1 \subset G^3,$$

whence

$$F^* = F^{3*}, \qquad G^* = G^{3*}.$$

Therefore, by Lemma 5.4.13 (vi), $\overline{F^3}$ and $\overline{G^3}$ are henselizations of F^3 and G^3, respectively. Hence, by Lemma 5.4.13 (iv), we may extend g_3 to an isomorphism $g_4 : \overline{F^3} \cong \overline{G^3}$. By Lemma 5.4.13 (vi) or (v),

$$\mathrm{val}(\overline{F^3}) = \mathrm{val}(F^3) = \bar{V},$$

$$\mathrm{val}(\overline{G^3}) = \mathrm{val}(G^3) = \bar{W}.$$

Thus by (9), $g_4 : \overline{F^3} \leftrightarrow \overline{G^3}$. Also, we have seen that $\bar{V}$ is countable. Hence (ii) holds with $F_2 = \overline{F^3}$, $G_2 = \overline{G^3}, f_2 = g_4$. This proves (iib).

We now prove (ii) in the general case. Since $F_1 = \overline{F_1}$, x is transcendental over F_1. By Lemma 5.4.13 (viii), the value group val $(F_1(x))$ is countable. Hence, applying (iib) countably many times and taking the union, we can find valued subfields F^2, G^2 and a mapping g^2 such that val (F^2) is countable, and

$$F_1 \subset F^2 = \overline{F^2}, \qquad G_1 \subset G^2 = \overline{G^2}, \qquad f_1 \subset g^2 : F^2 \leftrightarrow G^2,$$

$$\mathrm{val}(F_1(x)) \subset \mathrm{val}(F^2).$$

We may then iterate this process and obtain for $2 \leqslant n < \omega$, valued subfields F^{n+1}, G^{n+1} and a mapping g^{n+1} such that val (F^{n+1}) is countable and

$$F^n \subset F^{n+1} = \overline{F^{n+1}}, \qquad G^n \subset G^{n+1} = \overline{G^{n+1}}, \qquad g^n \subset g^{n+1} : F^{n+1} \leftrightarrow G^{n+1},$$

$$\mathrm{val}(F^n(x)) \subset \mathrm{val}(F^{n+1}).$$

Let

$$F^{\omega} = \bigcup_{2 \leqslant n < \omega} F^{n}, \qquad G^{\omega} = \bigcup_{2 \leqslant n < \omega} G^{n}, \qquad g^{\omega} = \bigcup_{2 \leqslant n < \omega} g^{n}.$$

Then all the hypotheses of (ii) hold with F^{ω}, G^{ω}, g^{ω} instead of F_1, G_1, f_1. Moreover,

$$\mathrm{val}\,(F^{\omega}(x)) = \bigcup_{2 \leqslant n < \omega} \mathrm{val}\,(F^{n}(x)) = \mathrm{val}\,(F^{\omega}).$$

Thus we may apply the special case (iia), and we conclude that (ii) holds in general.

Proof of (iii). This is exactly the same as the proof of (ii), with F and G interchanged, and the desired conclusion follows. ⊣

COROLLARY 5.4.14. *If F and G are fields of characteristic zero and $F \equiv G$ then $F((t)) \equiv G((t))$.*

PROOF. By Lemmas 5.4.8 and 5.4.10, Theorem 5.4.12, and the remark that

$$\mathrm{val}\,(F((t))) \cong \langle \mathbb{Z}, +, \cdot, 0, \leqslant \rangle \cong \mathrm{val}\,(G((t))). \dashv$$

Another classical example of a valued field is the field of p-adic numbers, where p is a prime. We first describe the *p-adic valuation* on the field $\mathbb{Q}$ of rational numbers. Every rational number $r \neq 0$ can be written uniquely in the form $r = p^n s$, where $n \in \mathbb{Z}$ is an integer and s is a quotient of two integers not divisible by p. We then define $\mathrm{val}\,(r) = p^n$. In particular, $\mathrm{val}\,(p^n) = p^n$. Let

$$V = \{p^n : n \in \mathbb{Z}\},$$

and put $p^m \leqslant p^n$ iff $m \leqslant n$. Setting $\mathrm{val}\,(0) = 0$ and $r \leqslant 0$ for all $r \in V$, we make $\mathbb{Q}$ into a valued field

$$\langle \mathbb{Q}, +, \cdot, 0, 1, V, \leqslant, \mathrm{val} \rangle,$$

called the field of rational numbers with the p-adic valuation. Its valuation group is isomorphic to $\langle \mathbb{Z}, +, 0, \leqslant \rangle$. Its residue class field is the prime field $\mathbb{Z}_p$ of characteristic p, which has the elements $\mathbb{Z}_p = \{0, 1, ..., p-1\}$. Indeed, if $m = qp+k \in \mathbb{Z}$ and $0 \leqslant k < p$, then $m^* = k$, and if $n^* \neq 0$, then $(m/n)^* = m^*/n^*$ in the sense of $\mathbb{Z}_p$.

The field of *p-adic numbers*, denoted by $\mathbb{Q}_p$, is defined as the *completion* of the field $\mathbb{Q}$ of rational numbers with the p-adic valuation. Its construction is the precise analogue of the construction of $H((t))$ from $H(t)$ described above.

Lemma 5.4.15. *$\mathbb{Q}_p$ is a Hensel field and* $\mathbb{Q}_p^* = \mathbb{Z}_p$, val $(\mathbb{Q}_p) \cong \langle \mathbb{Z}, +, 0, \leqslant \rangle$.

The following lemma is known from the literature.

Lemma 5.4.16.

(i). (Chevalley.) *Suppose $f(t_1 \dots t_n)$ is a polynomial with coefficients in $\mathbb{Z}_p$, zero constant term, and degree $d < n$. Then f has a nontrivial root in $\mathbb{Z}_p$ (the trivial root is $(0, 0, \dots, 0)$).*

(ii). (Lang.) *Let $f(t_1 \dots t_n)$ be a polynomial over $\mathbb{Z}_p((t))$ with zero constant term and degree d such that $d^2 < n$. Then f has a nontrivial zero in $\mathbb{Z}_p((t))$.*

Corollary 5.4.17. *Let $f(t_1 \dots t_n)$ be a polynomial with coefficients in $\mathbb{Z}$, zero constant term, and degree $d < n$. Then there is a finite set Y of primes such that whenever $p \notin Y$, f has a nontrivial zero in the field $\mathbb{Q}_p$ of p-adic numbers.*

Proof. Suppose there is an infinite sequence of primes $p_0 < p_1 < p_2 < \dots$ such that for each $n < \omega$, f has only the trivial root in $\mathbb{Q}_{p_n}$. Let D be a nonprincipal ultrafilter over ω, and form the ultraproduct

$$F = \prod_D \mathbb{Q}_{p_n}.$$

Then f has only the trivial root in F. Since each $\mathbb{Q}_{p_n}$ is a Hensel field, F is a Hensel field. Also,

$$F^* \cong \prod_D \mathbb{Z}_{p_n},$$

whence F^* has characteristic zero. Thus by Lemma 5.4.13 (ii), F^* is isomorphic to a subfield of F. By Lemma 5.4.16 (i), f has a nontrivial root in each field $\mathbb{Z}_{p_n}$, and hence f has a nontrivial root in F^* and thus in F. This contradiction completes the proof. ⊣

The following corollary is an application of Theorem 5.4.12. It is a precise statement of the intuitive principle that the Hensel fields $\mathbb{Q}_p$ and $\mathbb{Z}_p((t))$ are very much alike, even though $\mathbb{Q}_p$ has characteristic zero and $\mathbb{Z}_p((t))$ has characteristic p.

Corollary 5.4.18. *Let φ be any sentence in the language of valued fields. Then for all but finitely many primes p, we have*

$$\mathbb{Q}_p \models \varphi \text{ if and only if } \mathbb{Z}_p((t)) \models \varphi.$$

Proof. By Lemma 5.4.10, each $\mathbb{Z}_p((t))$ is a Hensel field. By definition, each $\mathbb{Q}_p$ is a Hensel field. Furthermore,

$$\mathbb{Z}_p((t))^* = \mathbb{Z}_p = \mathbb{Q}_p^*,$$

$$\mathrm{val}\,(\mathbb{Z}_p((t))) \cong \langle \mathbb{Z}, +, 0, \leqslant \rangle \cong \mathrm{val}\,(\mathbb{Q}_p).$$

Let D be any nonprincipal ultrafilter over the set of all primes, and form ultraproducts

$$F = \prod_D \mathbb{Z}_p((t)), \qquad G = \prod_D \mathbb{Q}_p.$$

Then F and G are Hensel fields. We have

$$F^* \cong \prod_D \mathbb{Z}_p \cong G^*,$$

$$\mathrm{val}\,(F) \cong \prod_D \langle \mathbb{Z}, +, 0, \leqslant \rangle \cong \mathrm{val}\,(G),$$

and F^* has characteristic 0. Thus by Theorem 5.4.12, F and G are elementarily equivalent. Whence, for each sentence φ,

$$S_\varphi = \{p : Q_p \models \varphi \text{ iff } \mathbb{Z}_p((t)) \models \varphi\} \in D.$$

Since this holds for any nonprincipal D, all but finitely many primes must belong to the set S_φ. ⊣

The next corollary looks very similar to Corollary 5.4.17 above, but it depends on Theorem 5.4.12 and hence is much deeper.

Corollary 5.4.19 (Artin's Conjecture). *For each positive integer d there exists a finite set Y of primes such that for every prime $p \notin Y$, every polynomial $f(t_1 \dots t_n)$ of degree d over $\mathbb{Q}_p$ with zero constant term and $n > d^2$ has a nontrivial zero in $\mathbb{Q}_p$.*

Proof. For each pair of positive integers d, n, let $\varphi_{d,n}$ be the sentence in the language of field theory which says:

Every polynomial of degree d with n variables and zero constant term has a nontrivial zero.

We note that if $n > d^2$, then the sentence

$$\varphi_{d,d^2+1} \rightarrow \varphi_{d,n}$$

holds in all fields, because the extra $n-(d^2+1)$ variables can be set equal to zero. By Lemma 5.4.16 (ii), the sentence φ_{d,d^2+1} holds in every field

$\mathbb{Z}_p((t))$. Then by Corollary 5.4.18, φ_{d,d^2+1} holds in $\mathbb{Q}_p$ for all but finitely many primes p. ⊣

We shall see in the exercises that Theorem 5.4.12 also holds when the hypothesis that F^* has characteristic zero is replaced by the hypothesis $F = \mathbb{Q}_p$. This result gives a set of axioms for the complete theory of $\mathbb{Q}_p$, just as Theorem 5.4.4 provided a set of axioms for the complete theory of the field of real numbers.

Instead of valued fields with cross section, one sometimes considers the slightly more general notion of a valued field (F, V, val), where F is a field, V an ordered Abelian group (disjoint from F), and val a function on F onto $V \cup \{0\}$, such that the obvious analogues of the axioms 5.4.7(a)–(f) above hold. With some extra complications, it is possible to generalize Theorem 5.4.12 to Hensel fields without cross section.

EXERCISES

5.4.1. Every η_1 set has power at least 2^ω.

5.4.2. A real closed field F is said to be an η_α field iff $\langle F, \leqslant \rangle$ is an η_α set. Prove that any two η_α fields of power ω_α are isomorphic.

5.4.3. Assume the continuum hypothesis. Prove that a theory T for a countable language $\mathscr{L}$ is model complete iff for every countable model $\mathfrak{A}$ of T and any two saturated models $\mathfrak{B}$, $\mathfrak{C}$ of T of power either ω_1 or finite, if $\mathfrak{A} \subset \mathfrak{B}$ and $\mathfrak{A} \subset \mathfrak{C}$, then $(\mathfrak{B}, a)_{a \in A} \cong (\mathfrak{C}, a)_{a \in A}$.

5.4.4. (i). Prove that the theory of real closed fields is model complete.

(ii). Prove that the theory of real closed ordered fields admits elimination of quantifiers.

5.4.5. Prove that a real closed field F is an η_α field iff it is ω_α-saturated.

5.4.6*. Prove that the theory of infinite atomic Boolean algebras is complete, but is not model complete.

5.4.7. Prove that the theory of infinite atomic Boolean algebras with an extra predicate $\text{At}(x)$ for 'x is an atom' is model complete.

5.4.8. Show that for each complete extension T of the theory of fields, there is a complete theory T' in a language with an extra 1-placed relation symbol U such that (F, U) is a model of T' iff F is an algebraically closed field, U is a subfield of F, which is a model of T, and F is not an algebraic extension of U of degree $\leqslant 2$.

5.4.9. Suppose that F, G are algebraically closed fields, U, V are subfields, $X \subset F$ and $Y \subset G$ are nonempty transcendental sets over U and V, and f is a one–one function from X into Y. Prove that if $U \equiv V$, then

$$(F, U, x)_{x \in X} \equiv (G, V, fx)_{x \in X}.$$

5.4.10*. Prove that the theory of divisible ordered Abelian groups is complete and model complete. It has the following axioms, in the language $\{+, 0, \leqslant\}$:

(i). the axioms for divisible Abelian groups;

(ii). $(\forall xyz)(x \leqslant y \to x+z \leqslant y+z)$;

(iii). the axioms for simple order.

5.4.11. The theory of *Z-groups* has the following axioms in the language $\{+, 0, 1, \leqslant\}$:

(i). the axioms for Abelian groups with identity element 0;

(ii). the axioms for simple order;

(iii). $(\forall xyz)(x \leqslant y \to x+z \leqslant y+z)$;

(iv). 1 is the least element greater than 0;

(v). $(\forall x \exists y)(\bar{n} \cdot y \equiv x \vee \bar{n} \cdot y \equiv x+1 \vee \ldots \vee \bar{n} \cdot y \equiv x+(\bar{n}-1))$, for each positive integer n, where $\bar{n}$ is the term $1+ \ldots +1$, n times. Show that $\langle \mathbb{Z}, +, 0, 1, \leqslant \rangle$ is a Z-group.

5.4.12* (Pressburger). Show that the theory of Z-groups is complete (using the method of this section).

5.4.13. Use Exercise 5.4.12 to obtain a set of axioms for the complete theory of the model $\langle \omega, +, 0, 1, \leqslant \rangle$.

5.4.14. Let T be the theory of Z-groups with extra relations $P_n(x)$, $n = 1, 2, \ldots$, and axioms

$$(\forall x)(P_n(x) \leftrightarrow (\exists y)(n \cdot y \equiv x)).$$

Prove that T is model complete.

5.4.15. Supply the proofs of Lemmas 5.4.8 and 5.4.11.

5.4.16. Let F and G be valued fields such that $F \equiv G$. Show that $F^* \equiv G^*$ and val $(F) \equiv$ val (G). Also, if F is α-saturated, then F^* and val (F) are α-saturated.

5.4.17. Give direct proofs of Theorems 5.4.4, 5.4.6 and 5.4.12 which do not use the continuum hypothesis.

5.4.18. Let T_1 be a theory in the language $\{+, \cdot, 0, 1\}$ and T_2 a theory in the language $\{\cdot, 1, \leqslant\}$. Show that the class of all valued fields F such that $F^* \models T_1$, $\mathrm{val}(F) \models T_2$ is an elementary class, and show how to construct a set of axioms for this class from sets of axioms for T_1 and T_2.

5.4.19*. A *Z-valued field* is a valued field F such that for some $c \in F$, $(\mathrm{val}(F), c)$ is a Z-group. Thus for each field H, $H((c))$ is a Z-valued Hensel field, and for each prime p, the field $\mathbb{Q}_p$ of p-adic numbers is a Z-valued Hensel field. Show that if T_1 is a complete extension of the theory of fields of characteristic zero, then the theory of Z-valued Hensel fields whose residue class fields are models of T_1 is complete.

5.4.20*. Let F be a Hensel field such that F^* has characteristic zero. Let G be a Hensel subfield of F such that $\mathrm{val}(G) \prec \mathrm{val}(F)$ and $G^* = F^*$. Prove that $G \prec F$.

[*Hint*: Reduce the problem to the case where F is saturated, of power ω_1, and $\mathrm{val}(G)$ is countable. Then use the proof of Theorem 5.4.12 to show that any elementary embedding of G into F which is the identity on F^* can be extended to an automorphism of F.]

5.4.21*. Prove the result in the above exercise with the hypothesis $G^* = F^*$ weakened to $G^* \prec F^*$.

5.4.22. Let C be the field of complex numbers. Using partial fractions, the nonzero elements of $C((t))$ can be identified in a natural way with formal power series of the form

$$(1) \qquad a_m t^m + a_{m+1} t^{m+1} + \ldots,$$

where $m \in \mathbb{Z}$, $a_n \in C$ for all $n \geqslant m$, and $a_m \neq 0$. A series (1) is called a *germ of a meromorphic function* iff there is a nonempty neighborhood U of 0 such that for all $z \in U \setminus \{0\}$, the series (1) converges at $t = z$. The set M of all germs of meromorphic functions, together with 0, forms a relatively algebraically closed valued subfield of $C((t))$. Use this fact and Exercise 5.4.20 to show that $M \prec C((t))$.

5.4.23. A valued field F is said to have *rank* 1 iff for all $x, y \in \mathrm{val}(F)$ such that $y > 1$, there is a positive integer n such that $x < y^n$. The *completion* of a valued field F of rank 1 is defined as the set of all equivalence classes of Cauchy sequences in F, just as in the special cases $H((t))$ and $\mathbb{Q}_p$. It is known that the completion of F is a Hensel field which has F as a valued subfield and has the same value group and residue class field as F. Use this to show that if F and G are valued groups of rank 1, F^* has characteristic zero, and $F \equiv G$, then the completions of F and G are elementarily equivalent.

5.4.24. Suppose F is a Hensel field such that for every Hensel field G, $F^* \equiv G^*$ and val $(F) \equiv$ val (G) implies $F \equiv G$. Prove that F has no proper algebraic valued extension with the same value group and residue class field.

Note: In [R] Chapter G, there are examples of Hensel fields F, with F^* of prime characteristic, which do have proper algebraic valued extensions with the same value group and residue class field.

5.4.25*. Let p be a prime. Let F be a Z-valued Hensel field of characteristic zero such that $F^* = \mathbb{Z}_p$ and p is the least element of val (F) such that $p > 1$. Show that $F \equiv \mathbb{Q}_p$, the field of p-adic numbers. This gives a set of axioms for the complete theory of $\mathbb{Q}_p$.

[*Hint*: Similar to Theorem 5.4.12, but Lemma 5.4.13 must be modified.]

5.4.26*. Let p be a prime and $F \equiv \mathbb{Q}_p$. If G is a Hensel subfield of F and val (G) is closed under roots in val (F), then $G \prec F$. Thus the complete theory of the model

$$(\mathbb{Q}_p, P_2, P_3, P_4, \ldots)$$

is model complete, where

$$P_n = \{x \in \text{val}(\mathbb{Q}_p) : x \text{ has an } n\text{th root}\}.$$

5.4.27. Prove Tarski's Theorem 5.4.4 using countable recursively saturated model pairs instead of saturated models.

5.4.28. Prove Theorem 5.4.6 using countable recursively saturated model pairs instead of saturated models.

5.5. Application to Boolean algebras

All of our applications in Section 5.4 were to complete theories. The method of saturated models in some cases can also be applied to incomplete theories, where the aim is to give a useful characterization of all the complete extensions of the given theory. As an illustration, we now give a classification of the complete extensions of the theory of Boolean algebras. The axioms of this theory are given in Example 1.4.3, and some simple definitions and results are found in Example 1.4.3 and the exercises of Section 1.4. Apart from these facts in Section 1.4, we shall assume that the reader has some previous, although not necessarily extensive, experience with Boolean algebras. The classification given here will be used in Section 6.3 in the study of reduced products.

We begin with a list of simple properties of Boolean algebras. Let $\mathscr{L} = \{+, \cdot, ^-, 0, 1\}$ be the language of Boolean algebras. In our axioms we have assumed that $0 \not\equiv 1$. It will be convenient to discard this axiom in this section, and we say that the Boolean algebra

$$\mathfrak{B} = \langle B, +, \cdot, ^-, 0, 1\rangle$$

is *trivial* iff $0 = 1$; otherwise, we say that $\mathfrak{B}$ is *nontrivial*. Observe that according to our definitions, the trivial Boolean algebra is both atomic and atomless. Let a be an element of a Boolean algebra $\mathfrak{B}$. We let

$$B|a = \{x \in B : x \leqslant a\},$$

and we define

$$(x)^{-a} = a \cdot \bar{x} \qquad \text{for all} \quad x \in B|a.$$

Then

$$\mathfrak{B}|a = \langle B|a, +, \cdot, ^{-a}, 0, a\rangle$$

is a Boolean algebra with the operations $\cdot +$, $\cdot$ inherited from $\mathfrak{B}$, $^{-a}$, and constants 0, a. $\mathfrak{B}|a$ is a subalgebra of $\mathfrak{B}$ iff $a = 1$, and $\mathfrak{B}|a$ is the trivial algebra iff $a = 0$. Every element $x \in B$ can be written uniquely as the sum of two elements y and z, where $y \leqslant a$ and $z \leqslant \bar{a}$; in fact, $y = x \cdot a$ and $z = x \cdot \bar{a}$. This decomposition of x into two parts gives rise to a one-to-one mapping of B onto the direct product $B|a \times B|\bar{a}$, which can be shown to be an isomorphism

$$\mathfrak{B} \cong \mathfrak{B}|a \times \mathfrak{B}|\bar{a}. \tag{5.5.1}$$

An element a is said to be *atomic* iff $\mathfrak{B}|a$ is atomic, and, similarly, a is said to be *atomless* iff $\mathfrak{B}|a$ is atomless. Note that the zero element 0 is both atomic and atomless. By (5.5.1), we see that for arbitrary $a \in B$,

(5.5.2)
$\mathfrak{B}$ is atomic iff $\mathfrak{B}|a$ and $\mathfrak{B}|\bar{a}$ are atomic,
$\mathfrak{B}$ is atomless iff $\mathfrak{B}|a$ and $\mathfrak{B}|\bar{a}$ are atomless.

An *ideal* on $\mathfrak{B}$ is a subset $I \subset \mathfrak{B}$ such that for all $x, y, z \in B$,

$$\text{if} \quad x, y \in I, \quad \text{then} \quad x+y \in I \quad \text{and} \quad x \cdot z \in I.$$

If an ideal $I \neq 0$, then $0 \in I$. The set of all complements of elements in an ideal is a filter on $\mathfrak{B}$. If I is an ideal on $\mathfrak{B}$, then the relation (I) defined by

$$x(I)y \quad \text{if and only if} \quad x \cdot \bar{y} + \bar{x} \cdot y \in I$$

is a congruence relation on $\mathfrak{B}$. For $x \in B$, let x/I denote the congruence class of x under (I), and let

$$B/I = \{x/I : x \in B\}.$$

The *quotient algebra*

$$\mathfrak{B}/I = \langle B/I, +, \cdot, ^-, 0, 1\rangle$$

carries the natural extensions of $+, \cdot, ^-$ to B/I, and has as zero element the set I and unit element the filter corresponding to I. $\mathfrak{B}$ is homomorphic to the quotient algebra $\mathfrak{B}/I$ under the homomorphism $x \to x/I$. The kernel of this homomorphism is exactly the set I. The quotient algebra $\mathfrak{B}/I$ is trivial iff $I = B$. An ideal I is *principal* iff $I = B|a$ for some $a \in B$. If $I = B|a$, then the quotient $\mathfrak{B}/I$ is isomorphic to $\mathfrak{B}|\bar{a}$, and hence by (5.5.1), we have

$$\mathfrak{B} \cong \mathfrak{B}/I \times \mathfrak{B}|a.$$

A very special ideal $I(\mathfrak{B})$ on $\mathfrak{B}$ is defined as follows:

$$I(\mathfrak{B}) = \{x \in B : x \text{ can be written as a sum } x = y+z, \text{ where } y \text{ is atomic and } z \text{ is atomless}\}.$$

If $x_1 = y_1+z_1$, $x_2 = y_2+z_2$, y_1 and y_2 are atomic, and z_1 and z_2 are atomless, then $x_1+x_2 = (y_1+y_2)+(z_1+z_2)$, where y_1+y_2 is atomic and z_1+z_2 is atomless. If $u \in B$, then $x_1 \cdot u = y_1 \cdot u + z_1 \cdot u$, where $y_1 \cdot u$ is atomic and $z_1 \cdot u$ is atomless. So $I(\mathfrak{B})$ is an ideal on $\mathfrak{B}$. Note that if $\mathfrak{B}$ has at most a finite number of atoms, then $I(\mathfrak{B}) = B$. If $\mathfrak{B}$ is atomic, then also $I(\mathfrak{B}) = B$. It is easy to check that for an arbitrary element $a \in B$, we have

$$I(\mathfrak{B}|a) = I(\mathfrak{B}) \cap B|a,$$

$$I(\mathfrak{B}|\bar{a}) = I(\mathfrak{B}) \cap B|\bar{a},$$

$$I(\mathfrak{B}) = \{x \in B : x \text{ can be written uniquely as a sum } x = y+z, \text{ where } y \in I(\mathfrak{B}|a) \text{ and } z \in I(\mathfrak{B}|\bar{a})\}.$$

From this we easily obtain

(5.5.3) $$\mathfrak{B}/I(\mathfrak{B}) \cong (\mathfrak{B}|a)/I(\mathfrak{B}|a) \times (\mathfrak{B}|\bar{a})/I(\mathfrak{B}|\bar{a}).$$

We define by induction a sequence

$$\langle \mathfrak{B}^{(k)} \rangle_{k<\omega}$$

of quotient algebras, a sequence

$$\langle \{\langle x, x^{(k)}\rangle : x \in B,\ x^{(k)} \in \mathfrak{B}^{(k)}\}\rangle_{k<\omega}$$

of homomorphisms of $\mathfrak{B}$ onto $\mathfrak{B}^{(k)}$ and a sequence

$$\langle I^{(k)} \rangle_{k<\omega}$$

of ideals on $\mathfrak{B}$ as follows:

$$\text{(5.5.4a)} \qquad \begin{aligned} x^{(0)} &= x, \\ I^{(0)} &= \{0\}, \\ \mathfrak{B}^{(0)} &= \mathfrak{B}; \end{aligned}$$

$$\text{(5.5.4b)} \qquad \begin{aligned} x^{(k+1)} &= x^{(k)}/I(\mathfrak{B}^{(k)}), \\ I^{(k+1)} &= \{x \in B : x^{(k+1)} = 0\}, \\ \mathfrak{B}^{(k+1)} &= \mathfrak{B}^{(k)}/I(\mathfrak{B}^{(k)}). \end{aligned}$$

It follows trivially from the definition that for each $k < \omega$,

(i) the mapping $x \to x^{(k)}$ is a homomorphism of $\mathfrak{B}$ onto $\mathfrak{B}^{(k)}$;

(ii) the set $I^{(k)}$ is an ideal on $\mathfrak{B}$;

(iii) $\mathfrak{B}^{(k)} \cong \mathfrak{B}/I^{(k)}$.

Note that each $I^{(k+1)}$ can also be described as the set of all $x \in B$ such that $x^{(k)}$ is a sum $x^{(k)} = y^{(k)} + z^{(k)}$, where $y^{(k)}$ is atomic in $\mathfrak{B}^{(k)}$ and $z^{(k)}$ is atomless in $\mathfrak{B}^{(k)}$.

PROPOSITION 5.5.5. *For each $k, l < \omega$, there are formulas φ_k, ψ_k, ρ_k, $\eta_{k,l}$, $\sigma_{k,l}$ of $\mathscr{L}$ in the free variable x such that for every Boolean algebra $\mathfrak{B}$ and every $a \in B$:*

(i). $\mathfrak{B} \models \varphi_k[a]$ *iff* $a \in I^{(k)}(a^{(k)} = 0)$.

(ii). $\mathfrak{B} \models \psi_k[a]$ *iff* $a^{(k)}$ *is atomic in* $\mathfrak{B}^{(k)}$.

(iii). $\mathfrak{B} \models \rho_k[a]$ *iff* $a^{(k)}$ *is atomless in* $\mathfrak{B}^{(k)}$.

(iv). $\mathfrak{B} \models \eta_{k,l}[a]$ *iff* $a^{(k)}$ *contains at most l atoms in* $\mathfrak{B}^{(k)}$.

(v). $\mathfrak{B} \models \sigma_{k,l}[a]$ *iff* $a^{(k)}$ *contains at least l atoms in* $\mathfrak{B}^{(k)}$.

The proof, which is not difficult, is left as an exercise.

Let $\mathfrak{B}$ be a nontrivial Boolean algebra. Consider the sequence

$$\mathfrak{B}^{(0)}, \mathfrak{B}^{(1)}, \ldots, \mathfrak{B}^{(k)}, \ldots$$

of quotient algebras defined in (5.5.4). Either some $\mathfrak{B}^{(k)}$ is trivial, in which case all $\mathfrak{B}^{(l)}$, $l \geqslant k$, are trivial, or else no $\mathfrak{B}^{(k)}$ is trivial. In the first case, we can find the least k such that $\mathfrak{B}^{(k+1)}$ is trivial. This means that $\mathfrak{B}^{(k)}$ is nontrivial and each element of $\mathfrak{B}^{(k)}$ is in the ideal $I(\mathfrak{B}^{(k)})$. We now can ask whether $\mathfrak{B}^{(k)}$ is atomless or atomic and whether $\mathfrak{B}^{(k)}$ has a finite or infinite number of atoms. The main result of this section states that the answers to these questions determine the elementary type (the theory) of $\mathfrak{B}$. We assign a pair of *invariants* $(m(\mathfrak{B}), n(\mathfrak{B}))$ to each nontrivial Boolean algebra

$\mathfrak{B}$ as follows:

$$(5.5.6a)\qquad m(\mathfrak{B}) = \begin{cases} \text{the least } k < \omega \text{ such that } \mathfrak{B}^{(k+1)} \text{ is trivial, if such a } k \text{ exists,} \\ \infty, \text{ otherwise;} \end{cases}$$

$$(5.5.6b)\qquad n_0(\mathfrak{B}) = \begin{cases} \infty \text{ if } m(\mathfrak{B}) = k \text{ and } \mathfrak{B}^{(k)} \text{ has infinitely many atoms,} \\ l \text{ if } m(\mathfrak{B}) = k \text{ and } \mathfrak{B}^{(k)} \text{ has } l < \omega \text{ atoms;} \end{cases}$$

$$(5.5.6c)\qquad n(\mathfrak{B}) = \begin{cases} 0 & \text{if } m(\mathfrak{B}) = \infty, \\ n_0(\mathfrak{B}) & \text{if } m(\mathfrak{B}) = k \text{ and } \mathfrak{B}^{(k)} \text{ is atomic,} \\ -n_0(\mathfrak{B}) & \text{if } m(\mathfrak{B}) = k \text{ and } \mathfrak{B}^{(k)} \text{ is not atomic.} \end{cases}$$

Thus $m(\mathfrak{B})$ indicates when $\mathfrak{B}^{(k+1)}$ becomes trivial, the sign of $n(\mathfrak{B})$ indicates whether or not $\mathfrak{B}^{(m(\mathfrak{B}))}$ is atomic, and $n_0(\mathfrak{B}) = |n(\mathfrak{B})|$ indicates the number of atoms in $\mathfrak{B}^{(m(\mathfrak{B}))}$. $m(\mathfrak{B})$ is either a natural number or ∞, and $n(\mathfrak{B})$ is either an integer or $\pm\infty$. When $m(\mathfrak{B}) = k < \omega$ and $n(\mathfrak{B}) = 0$, then $\mathfrak{B}^{(k)}$ is a nontrivial atomless Boolean algebra. The following proposition follows easily from Proposition 5.5.5.

PROPOSITION 5.5.7.

(i). *For each $k, l < \omega$, the following can be expressed by single sentences of $\mathscr{L}$:*

$$m(\mathfrak{B}) = k,$$
$$m(\mathfrak{B}) = k \quad \textit{and} \quad n(\mathfrak{B}) = l,$$
$$m(\mathfrak{B}) = k \quad \textit{and} \quad n(\mathfrak{B}) = -l.$$

(ii). *For each $k < \omega$, the following are expressible by sets of sentences of $\mathscr{L}$:*

$$m(\mathfrak{B}) = \infty,$$
$$m(\mathfrak{B}) = k \textit{ and } n(\mathfrak{B}) = \infty,$$
$$m(\mathfrak{B}) = k \textit{ and } n(\mathfrak{B}) = -\infty.$$

The proof is left as an exercise.

An iteration of (5.5.3) shows that for $k < \omega$ and $a \in B$,

$$\mathfrak{B}^{(k)} \cong (\mathfrak{B}|a)^{(k)} \times (\mathfrak{B}|\bar{a})^{(k)}.$$

The observation (5.5.2) enables us to prove:

PROPOSITION 5.5.8. *Let $\mathfrak{B}$ be a nontrivial Boolean algebra and let $a \in B$. Then:*

(i). $m(\mathfrak{B}) = \max(m(\mathfrak{B}|a), m(\mathfrak{B}|\bar{a}))$.

(ii). *If* $m(\mathfrak{B}|a) < m(\mathfrak{B}|\bar{a})$, *then* $m(\mathfrak{B}) = m(\mathfrak{B}|\bar{a})$ *and* $n(\mathfrak{B}) = n(\mathfrak{B}|\bar{a})$.

(iii). *If* $m(\mathfrak{B}|a) = m(\mathfrak{B}|\bar{a}) < \infty$, *then*

$n(\mathfrak{B}) = 0$ *iff* $n(\mathfrak{B}|a) = n(\mathfrak{B}|\bar{a}) = 0$,

$n(\mathfrak{B}) > 0$ *iff* $n(\mathfrak{B}|a) > 0$ *and* $n(\mathfrak{B}|\bar{a}) > 0$,

$n(\mathfrak{B}) < 0$ *iff either* $n(\mathfrak{B}|a) \leqslant 0$ *or* $n(\mathfrak{B}|\bar{a}) \leqslant 0$, *and not both* $n(\mathfrak{B}|a) = 0$, $n(\mathfrak{B}|\bar{a}) = 0$.

(iv). *If* $m(\mathfrak{B}|a) = m(\mathfrak{B}|\bar{a}) < \infty$, *then* $n_0(\mathfrak{B}) = n_0(\mathfrak{B}|a) + n_0(\mathfrak{B}|\bar{a})$.

□ The lemma below provides the key step to the proof of Theorem 5.5.10.

LEMMA 5.5.9. *Let* $\mathfrak{A}$, $\mathfrak{B}$ *be nontrivial Boolean algebras such that*

$$(m(\mathfrak{A}), n(\mathfrak{A})) = (m(\mathfrak{B}), n(\mathfrak{B})).$$

Assume that $a \in A$, $0 \neq a \neq 1$, *and* $\mathfrak{B}$ *is* ω*-saturated. Then there exists* $b \in B$ *such that* $0 \neq b \neq 1$ *and*

$$(m(\mathfrak{A}|a), n(\mathfrak{A}|a)) = (m(\mathfrak{B}|b), n(\mathfrak{B}|b)),$$
$$(m(\mathfrak{A}|\bar{a}), n(\mathfrak{A}|\bar{a})) = (m(\mathfrak{B}|\bar{b}), n(\mathfrak{B}|\bar{b})).$$

PROOF. First suppose that

$$m(\mathfrak{A}|a) = m(\mathfrak{A}|\bar{a}) = \infty.$$

We have $m(\mathfrak{A}) = \infty$, so $m(\mathfrak{B}) = \infty$. This means that for each $k < \omega$, there exists an element $b \in B$ such that $b^{(k)} \neq 0$ in $\mathfrak{B}^{(k)}$ and $b^{(k)} \neq 1$ in $\mathfrak{B}^{(k)}$. By Proposition 5.5.5 and the ω-saturatedness of $\mathfrak{B}$, there exists $b \in B$ such that $b^{(k)} \neq 0$ in $\mathfrak{B}^{(k)}$ and $b^{(k)} \neq 1$ in $\mathfrak{B}^{(k)}$, for all k. Clearly, for such $b \in B$ we have

$$m(\mathfrak{B}|b) = m(\mathfrak{B}|\bar{b}) = \infty.$$

Next suppose that

$$m(\mathfrak{A}|a) < m(\mathfrak{A}|\bar{a}) = m(\mathfrak{A}).$$

We claim that we only need to find $b \in B$ such that

(1) $$m(\mathfrak{B}|b) = m(\mathfrak{A}|a) \quad \text{and} \quad n(\mathfrak{B}|b) = n(\mathfrak{A}|a).$$

This is because by Proposition 5.5.8 (i) and (ii),

$$m(\mathfrak{A}|\bar{a}) = m(\mathfrak{A}) = m(\mathfrak{B}) = \max(m(\mathfrak{B}|b), m(\mathfrak{B}|\bar{b})) = m(\mathfrak{B}|\bar{b}),$$

and similarly,

$$n(\mathfrak{A}|\bar{a}) = n(\mathfrak{A}) = n(\mathfrak{B}) = n(\mathfrak{B}|\bar{b}).$$

Let $k = m(\mathfrak{A}|a)$. We verify (1) for all possible values of $n(\mathfrak{A}|a)$. Since $k < m(\mathfrak{B})$, we know that

$$(2)\qquad \begin{array}{l} \mathfrak{B}^{(k)} \text{ has more than a finite number of atoms,} \\ \mathfrak{B}^{(k)} \text{ is not atomic.} \end{array}$$

For otherwise $\mathfrak{B}^{(k+1)}$ is trivial, contradicting the definition of $m(\mathfrak{B})$. By (2), Proposition 5.5.5, and the ω-saturatedness of $\mathfrak{B}$, it is easy to find an element $b \in B$ such that

$$m(\mathfrak{B}|b) = k \quad \text{and} \quad n(\mathfrak{B}|b) = n(\mathfrak{A}|a).$$

The only two slightly nontrivial cases are when $n(\mathfrak{A}|a) = \pm\infty$. This is where we need that $\mathfrak{B}$ is ω-saturated.

The case

$$m(\mathfrak{A}|\bar{a}) < m(\mathfrak{A}|a) = m(\mathfrak{A})$$

is handled in exactly the same way as the previous case.

Suppose that

$$m(\mathfrak{A}|a) = m(\mathfrak{A}|\bar{a}) < \infty.$$

Then let $k = m(\mathfrak{A}) = m(\mathfrak{A}|a) = m(\mathfrak{A}|\bar{a})$. If $n(\mathfrak{A}) = 0$, then, by Proposition 5.5.8 (iii), $n(\mathfrak{B}) = n(\mathfrak{A}|a) = n(\mathfrak{A}|\bar{a}) = 0$. In this case, let $b \in B$ be any element such that $b^{(k)} \neq 0$ and $b^{(k)} \neq 1$ in $\mathfrak{B}^{(k)}$. Then clearly,

$$m(\mathfrak{B}|b) = m(\mathfrak{B}|\bar{b}) = k$$

and

$$n(\mathfrak{B}|b) = n(\mathfrak{B}|\bar{b}) = 0.$$

If $n(\mathfrak{A}) > 0$, then by Proposition 5.5.8 (iii) and (iv), we have $n(\mathfrak{B}) > 0$ and

$$n(\mathfrak{A}|a) > 0, \quad n(\mathfrak{A}|\bar{a}) > 0,$$

$$n_0(\mathfrak{B}) = n_0(\mathfrak{A}) = n_0(\mathfrak{A}|a) + n_0(\mathfrak{A}|\bar{a}).$$

In this case, by Proposition 5.5.5 and the ω-saturatedness of $\mathfrak{B}$, we find $b \in B$ such that

$$b^{(k)} \text{ is atomic and contains exactly } n_0(\mathfrak{A}|a) \text{ atoms in } \mathfrak{B}^{(k)},$$

$$\bar{b}^{(k)} \text{ is atomic and contains exactly } n_0(\mathfrak{A}|\bar{a}) \text{ atoms in } \mathfrak{B}^{(k)}.$$

Then

$$m(\mathfrak{B}|b) = m(\mathfrak{B}|\bar{b}) = k,$$

and

$$n(\mathfrak{B}|b) = n(\mathfrak{A}|a), \qquad n(\mathfrak{B}|\bar{b}) = n(\mathfrak{A}|\bar{a}).$$

Finally, if $n(\mathfrak{A}) \leqslant 0$, then by Proposition 5.5.8 (iii) and (iv), we have $n(\mathfrak{B}) \leqslant 0$ and

$$\text{not both } n(\mathfrak{A}|a) = 0, \qquad n(\mathfrak{A}|\bar{a}) = 0,$$

$$\text{either } \; n(\mathfrak{A}|a) \leqslant 0 \;\text{ or }\; n(\mathfrak{A}|\bar{a}) \leqslant 0,$$

$$n_0(\mathfrak{B}) = n_0(\mathfrak{A}) = n_0(\mathfrak{A}|a)+n_0(\mathfrak{A}|\bar{a}).$$

The method should be clear enough by now, so that we can leave the latter case as an exercise. In all cases, we only need the ω-saturatedness of $\mathfrak{B}$ in the case where one of the invariants is $\pm\infty$. ⊣

Let $a_1, \ldots, a_n$ be elements of a Boolean algebra $\mathfrak{B}$. By a *bit* of $a_1, \ldots, a_n$ we mean any one of the 2^n products

$$s = s_1 \cdot s_2 \cdot \ldots \cdot s_n,$$

where each $s_i = a_i$ or $s_i = \bar{a}_i$. If s and t are two bits of $a_1, \ldots, a_n$ and $s \neq t$, then $s \cdot t = 0$. The set of all sums from

$$\{s : s \text{ is a bit of } a_1, \ldots, a_n\}$$

is precisely the finite subalgebra of $\mathfrak{B}$ generated by $a_1, \ldots, a_n$. If $b_1, \ldots, b_n \in B$ and s is a bit of $a_1, \ldots, a_n$, then by the *corresponding bit* t of $b_1, \ldots, b_n$ we mean the product

$$t = t_1 \cdot t_2 \cdot \ldots \cdot t_n,$$

where $t_i = b_i$ if $s_i = a_i$ and $t_i = \bar{b}_i$ if $s_i = \bar{a}_i$. Let $\mathfrak{A}$ and $\mathfrak{B}$ be Boolean algebras and let $a_1, \ldots, a_n \in A$, $b_1, \ldots, b_n \in B$. We say that $(\mathfrak{A}, a_1 \ldots a_n)$ is *similar to* $(\mathfrak{B}, b_1 \ldots b_n)$ iff for every bit s of $a_1, \ldots, a_n$ and the corresponding bit t of $b_1, \ldots, b_n$, the two Boolean algebras $\mathfrak{A}|s$ and $\mathfrak{B}|t$ either are both trivial or have the same pair of invariants. We use the symbols

$$(\mathfrak{A}, a_1 \ldots a_n) \approx (\mathfrak{B}, b_1 \ldots b_n)$$

to denote that they are similar. $\mathfrak{A} \approx \mathfrak{B}$ means simply that either they are both trivial or they have the same invariants. If $a \in {}^{\omega}A$, $b \in {}^{\omega}B$, we write

$$(\mathfrak{A}, a_n)_{n<\omega} \approx (\mathfrak{B}, b_n)_{n<\omega},$$

if and only if for all $n < \omega$,

$$(\mathfrak{A}, a_m)_{m<n} \approx (\mathfrak{B}, b_m)_{m<n}.$$

THEOREM 5.5.10. *A necessary and sufficient condition for two Boolean algebras $\mathfrak{A}$ and $\mathfrak{B}$ to be elementarily equivalent is that either they are both trivial or* *they have the same invariants,*

$$(m(\mathfrak{A}), n(\mathfrak{A})) = (m(\mathfrak{B}), n(\mathfrak{B})).$$

PROOF. The necessity follows from Proposition 5.5.7.

The sufficiency can be proved in any one of several ways. For example, it could be done by the method of elimination of quantifiers (cf. Section 1.5), or using either part of Proposition 5.4.1. We shall use here a simpler method than Proposition 5.4.1. The proof will be a familiar back and forth construction which has only countably many steps, rather than ω_1 steps. The reason we can do this here is because finitely generated Boolean algebras are finite and easily described. The proof uses ω-saturated models and does not depend on the continuum hypothesis.

If $\mathfrak{A}$ and $\mathfrak{B}$ are both nontrivial and finite, then since $\mathfrak{A} \approx \mathfrak{B}$, we have $\mathfrak{A} \cong \mathfrak{B}$, and so $\mathfrak{A}$ and $\mathfrak{B}$ are elementarily equivalent. We may assume that $\mathfrak{A}$ and $\mathfrak{B}$ are both infinite and ω-saturated; in particular, we may assume that $\mathfrak{A}$ and $\mathfrak{B}$ are of power 2^ω, although the continuum hypothesis will in no way be involved in what follows. Let us first prove the following:

(1) For all $a_1, \ldots, a_{n+1} \in A$ and $b_1, \ldots, b_n \in B$, if

$$(\mathfrak{A}, a_1 \ldots a_n) \approx (\mathfrak{B}, b_1 \ldots b_n),$$

then there exists $b_{n+1} \in B$ such that

$$(\mathfrak{A}, a_1 \ldots a_{n+1}) \approx (\mathfrak{B}, b_1 \ldots b_{n+1}).$$

To prove (1), we first prove:

(2) Assume that $(\mathfrak{A}, a_1 \ldots a_n) \approx (\mathfrak{B}, b_1 \ldots b_n)$ and $a_{n+1} \in A$. Then, for each bit s of $a_1, \ldots, a_n$ and corresponding bit t of $b_1, \ldots, b_n$, there exists $c \leqslant t$ such that

$$(\mathfrak{A}|s, a_{n+1} \cdot s) \approx (\mathfrak{B}|t, c).$$

By the assumption of (2), $\mathfrak{A}|s \approx \mathfrak{B}|t$. If they are both trivial, then let $c = 0$. If they are nontrivial, then they have the same invariants. If $a_{n+1} \cdot s = 0$, then let $c = 0$. If $a_{n+1} \cdot s = s$, then let $c = t$. So assume that $0 \neq a_{n+1} \cdot s \neq s$. Clearly, $\mathfrak{B}|t$ is ω-saturated, because all of the operations on $\mathfrak{B}|t$ are definable in $\mathfrak{B}$ using t as a parameter. Whence, by Lemma 5.5.9, there exists $c \in B|t$ such that the conclusion of (2) holds. So (2) is proved.

Now let b_{n+1} be the sum of all c's obtained in (2), one for each bit of $a_1, \ldots, a_n$. It follows that the conclusion of (1) holds.

We arrange all of the terms from the Skolem expansion $\mathscr{L}^*$ of $\mathscr{L}$ in a simple infinite sequence

$$t_0, t_1, \ldots, t_n, \ldots$$

(possibly with repetitions) in such a way that the following hold:

(3) Each Skolem term t_n has at most the free variables $v_0, \ldots, v_{n-1}$.

(4) If $a \in {}^{\omega}A$, $b \in {}^{\omega}B$ are such that for each $n < \omega$,

$$a_{2n} = t_{n\mathfrak{A}^*}[a_0 \ldots a_{n-1}],$$
$$b_{2n+1} = t_{n\mathfrak{B}^*}[b_0 \ldots b_{n-1}],$$

then the ranges of a and b determine elementary submodels of $\mathfrak{A}$ and $\mathfrak{B}$, respectively.

Here $t_{n\mathfrak{A}^*}$ and $t_{n\mathfrak{B}^*}$ are the interpretations of t_n in the Skolem expansions $\mathfrak{A}^*$ and $\mathfrak{B}^*$, respectively. Such an arrangement of the terms t_n can easily be made. Combining (1) with its dual, i.e., $\mathfrak{A}$ and $\mathfrak{B}$ are interchanged, we can find by the familiar back and forth argument, $a \in {}^{\omega}A$ and $b \in {}^{\omega}B$ such that

(5) $$(\mathfrak{A}, a_n)_{n<\omega} \approx (\mathfrak{B}, b_n)_{n<\omega},$$

and the hypothesis of (4) holds. Using (5), a simple argument in Boolean algebras will show that the mapping

$$h : a_n \to b_n$$

is an isomorphism between the two Boolean algebras

$$\mathfrak{A}_0 = \langle \{a_n : n \in \omega\}, +, \cdot, ^-, 0, 1 \rangle,$$

and

$$\mathfrak{B}_0 = \langle \{b_n : n \in \omega\}, +, \cdot, ^-, 0, 1 \rangle.$$

Here we use the fact that a bit of $a_0, \ldots, a_n$ is 0 if and only if the corresponding bit of $b_0, \ldots, b_n$ is 0. The conclusion of (4) says that

$$\mathfrak{A}_0 \prec \mathfrak{A} \quad \text{and} \quad \mathfrak{B}_0 \prec \mathfrak{B}.$$

So $\mathfrak{A} \equiv \mathfrak{B}$. $\dashv$

Exercises

5.5.1. Supply the proofs of (5.5.1)–(5.5.3).

5.5.2. Prove Proposition 5.5.5.

5.5.3. Determine the invariants of the following particular Boolean algebras:
 (i). a finite Boolean algebra with n atoms;
 (ii). a nontrivial atomless Boolean algebra;
 (iii). an infinite atomic Boolean algebra.

5.5.4*. Prove that for each possible pair of invariants (m_1, n_1) among $(\infty, 0)$ and (m, n), $m < \omega$, $-\infty \leqslant n \leqslant +\infty$, there exists a Boolean algebra $\mathfrak{B}$ with invariants (m_1, n_1).

5.5.5. Prove Proposition 5.5.7.

5.5.6. Prove Proposition 5.5.8. More generally, try to compute the invariants of $\mathfrak{A} \times \mathfrak{B}$ from the invariants of $\mathfrak{A}$ and $\mathfrak{B}$.

5.5.7. Give a detailed proof of Lemma 5.5.9.

5.5.8. Using the technique of Theorem 5.5.10, prove that any two saturated Boolean algebras of power ω_1 with the same invariants are isomorphic.

5.5.9. The theory of Boolean algebras has only countably many complete extensions, and each complete extension has only countably many finite types. Thus every Boolean algebra $\mathfrak{A}$ is elementarily equivalent to a countably saturated Boolean algebra $\mathfrak{B}$.

5.5.10*. (a). By the theory of *inductive order* we mean the theory in the language $\mathscr{L} = \{\leqslant, 0\}$ whose axioms are the axioms for the theory of simple order, an axiom stating that 0 is the least element, and the induction scheme

$$(\exists x)\varphi(x) \rightarrow (\exists x \forall y)(\varphi(x) \wedge (\varphi(y) \rightarrow x \leqslant y)),$$

where $\varphi(x)$ is a formula which may possibly have additional free variables $z_1 \dots z_n$. For example, for each ordinal $\alpha > 0$, the model $\langle \alpha, \leqslant, 0 \rangle$ is a model of the theory of inductive order. Let $\mathfrak{A}$ be a model for the theory of inductive order. An element $x \in A$ is said to be a 1-*limit point* iff there does not exist a greatest element $y < x$. x is said to be an $(n+1)$-*limit point* iff there is no greatest n-limit point $y < x$. Every point is a 0-limit point. Thus 0 is an n-limit point for all n. Show that the notion of an n-limit point can be expressed by a single formula of $\mathscr{L}$.

(b). For each model $\mathfrak{A}$ of the theory of inductive order, assign invariants $m(\mathfrak{A})$, $n(\mathfrak{A})$ and $p_k(\mathfrak{A})$, $k = 0, 1, 2, \dots$, as follows:

$$m(\mathfrak{A}) = \begin{cases} k & \text{if } k \text{ is the least natural number such that } \mathfrak{A} \text{ has a greatest} \\ & k\text{-limit point,} \\ \infty & \text{if for all } k < \omega,\ \mathfrak{A} \text{ has arbitrarily large } k\text{-limit points.} \end{cases}$$

$$n(\mathfrak{A}) = \begin{cases} k & \text{if } k \text{ is the greatest natural number such that } \mathfrak{A} \text{ has a} \\ & k\text{-limit point greater than 0,} \\ \infty & \text{if for all } k < \omega,\ \mathfrak{A} \text{ has a } k\text{-limit point greater than 0.} \end{cases}$$

$$p_k(\mathfrak{A}) = \begin{cases} \infty \text{ if } k < m(\mathfrak{A}), \\ \infty \text{ if } k \geqslant m(\mathfrak{A}) \text{ and there are infinitely many } k\text{-limit points} \\ \quad \text{greater than the greatest } (k+1)\text{-limit point}, \\ l \;\; \text{if } k \geqslant m(\mathfrak{A}) \text{ and there are exactly } l < \omega \; k\text{-limit points} \\ \quad \text{greater than the greatest } (k+1)\text{-limit point.} \end{cases}$$

Prove that if $\mathfrak{A}$, $\mathfrak{B}$ are models of the theory of inductive order, then $\mathfrak{A} \equiv \mathfrak{B}$ if and only if $\mathfrak{A}$ and $\mathfrak{B}$ have the same invariants.

[*Hint*: Use the method of Theorem 5.5.10.]

5.5.11. Determine exactly which invariants are the invariants of some model of inductive order. Also, determine which invariants correspond to well ordered models. Use your answer to show that the theory of inductive order has 2^ω complete extensions.

[*Hint*: Consider the normal form of ordinals in descending powers of ω.]

5.5.12*. Which complete extensions of the theory of inductive order have uncountably many types of elements?

5.5.13*. Prove that every sentence which holds in every model $\langle \alpha, \leqslant, 0 \rangle$, where $0 < \alpha < \omega^\omega$ (ordinal exponentiation), is a consequence of the theory of inductive order. Thus the theory of inductive order is the theory of the class of all well ordered models, and is also the theory of the set of models $\{\langle \alpha, \leqslant, 0 \rangle : 0 < \alpha < \omega^\omega\}$.

5.5.14*. Prove that every well-ordered model $\langle \alpha, \leqslant, 0 \rangle$ is elementarily equivalent to a unique model $\langle \beta, \leqslant, 0 \rangle$, where $0 < \beta < \omega^\omega + \omega^\omega$.

5.5.15*. Let Or denote the class of all ordinals. For the purposes of this exercise, we may consider $\langle \mathrm{Or}, \leqslant \rangle$, $\langle \mathrm{Or}, \leqslant, + \rangle$, $\langle \mathrm{Or}, \leqslant, +, \cdot \rangle$ as models where $\leqslant, +, \cdot$ are the usual ordering, addition and multiplication of ordinals. Prove the following:

(i). $\langle \omega^\omega, \leqslant \rangle \prec \langle \mathrm{Or}, \leqslant \rangle$;

(ii). $\langle \omega^{\omega^\omega}, \leqslant, + \rangle \prec \langle \mathrm{Or}, \leqslant, + \rangle$;

(iii). $\langle \omega^{\omega^{\omega^\omega}}, \leqslant, +, \cdot \rangle \prec \langle \mathrm{Or}, \leqslant, +, \cdot \rangle$.

Prove also that in each case every element in the left-hand model is definable in that model by a single formula.

[*Hint*: The proofs of (ii) and (iii) depend on the technique introduced in Exercise 1.3.15 and on the normal form theorem for ordinals.]

CHAPTER 6

MORE ABOUT ULTRAPRODUCTS AND GENERALIZATIONS

In this chapter we shall continue the study of ultraproducts begun in Chapter 4, but we may now make use of saturated models. Section 6.1 ties together the notions of an ultraproduct and of a saturated model, and contains a very neat characterization of elementary classes. The remaining sections deal with various generalizations of the ultraproduct construction.

6.1. Ultraproducts which are saturated

We shall show that certain ultraproducts of models are saturated models. The main theorem in this section is the 'isomorphism theorem' for ultraproducts: Two models are elementarily equivalent if and only if they have isomorphic ultrapowers (Theorem 6.1.15). Some of the theorems in this section and Section 6.3 depend on the continuum hypothesis, or the generalized continuum hypothesis. We begin with a theorem about countably incomplete ultraproducts. Recall that an ultrafilter D is said to be countably incomplete iff D is not closed under countable intersections.

THEOREM 6.1.1. *Let $\mathscr{L}$ be countable, and let D be a countably incomplete ultrafilter over a set I. Then for every family $\mathfrak{A}_i$, $i \in I$, of models for $\mathscr{L}$, the ultraproduct $\prod_D \mathfrak{A}_i$ is ω_1-saturated.*

PROOF. We must show that for every countable sequence a_m, $m < \omega$, of elements of $\prod_D A_i$, and every set $\Sigma(x)$ of formulas of $\mathscr{L} \cup \{c_0, c_1, \ldots\}$, if each finite subset of $\Sigma(x)$ is satisfiable in $(\prod_D \mathfrak{A}_i, a_m)_{m<\omega}$, then $\Sigma(x)$ is satisfiable in $(\prod_D \mathfrak{A}_i, a_m)_{m<\omega}$. Note that if $a_m = \langle a_m(i) : i \in I \rangle_D$, then

$$(\prod_D \mathfrak{A}_i, a_m)_{m<\omega} = \prod_D ((\mathfrak{A}_i, a_m(i))_{m<\omega}).$$

Since $\mathscr{L}$ is an arbitrary countable language and $\mathscr{L} \cup \{c_0, c_1, \ldots\}$ is also countable, it suffices to prove the following:

(1) For every set $\Sigma(x)$ of formulas of $\mathscr{L}$, if each finite subset of $\Sigma(x)$ is satisfiable in $\prod_D \mathfrak{A}_i$, then $\Sigma(x)$ is satisfiable in $\prod_D \mathfrak{A}_i$.

Suppose every finite subset of $\Sigma(x)$ is satisfiable in $\prod_D \mathfrak{A}_i$. Since $\mathscr{L}$ is countable, $\Sigma(x)$ is countable, and we may write

$$\Sigma(x) = \{\sigma_1(x), \sigma_2(x), \ldots\}.$$

As D is countably incomplete, we find a descending chain

$$I = I_0 \supset I_1 \supset I_2 \supset \ldots$$

such that each $I_n \in D$ and $\bigcap_{n<\omega} I_n = 0$. Let $X_0 = I$ and for each positive $n < \omega$, let

$$X_n = I_n \cap \{i \in I : \mathfrak{A}_i \models (\exists x)(\sigma_1(x) \wedge \ldots \wedge \sigma_n(x))\}.$$

Then by the fundamental Theorem 4.1.19, each $X_n \in D$. Moreover, $\bigcap_{n<\omega} X_n = 0$, and $X_n \supset X_{n+1}$. It follows that for each $i \in I$ there is a greatest $n(i) < \omega$ such that $i \in X_{n(i)}$.

We choose a function $f \in \prod_{i \in I} A_i$ in the following way: If $n(i) = 0$, choose $f(i)$ to be an arbitrary element of A_i. If $n(i) > 0$, choose $f(i) \in A_i$ so that

$$\mathfrak{A}_i \models \sigma_1 \wedge \ldots \wedge \sigma_{n(i)}[f(i)].$$

Then whenever $0 < n$ and $i \in X_n$, we have $n \leqslant n(i)$, whence $\mathfrak{A}_i \models \sigma_n[f(i)]$. It follows from the fundamental theorem that $\prod_D \mathfrak{A}_i \models \sigma_n[f_D]$ for all $n > 0$, and therefore the element f_D satisfies $\Sigma(x)$ in $\prod_D \mathfrak{A}_i$. This proves (1). ⊣

COROLLARY 6.1.2. *Assume the continuum hypothesis. Suppose $\mathfrak{A}$, $\mathfrak{B}$ are two models for a countable language $\mathscr{L}$ and $|A|$, $|B| \leqslant \omega_1$. Then the following are equivalent:*

(i). $\mathfrak{A} \equiv \mathfrak{B}$.

(ii). *For all nonprincipal ultrafilters D, E over ω, $\prod_D \mathfrak{A} \cong \prod_E \mathfrak{B}$ (and $\prod_D \mathfrak{A} \cong \prod_D \mathfrak{B}$).*

(iii). *There exist ultrafilters D and E such that $\prod_D \mathfrak{A} \cong \prod_E \mathfrak{B}$.*

PROOF. (i) ⇒ (ii). Assume (i). By Theorem 6.1.1, $\prod_D \mathfrak{A}$ and $\prod_E \mathfrak{B}$ are ω_1-saturated models. Since $|A|$, $|B| \leqslant \omega_1$ and $2^\omega = \omega_1$, the ultrapowers $\prod_D \mathfrak{A}$ and $\prod_E \mathfrak{B}$ have power at most ω_1. By (i) and the fundamental theorem, they are elementarily equivalent. Hence by the uniqueness theorem for saturated models, they are isomorphic.

It is obvious that (ii) implies (iii), and (iii) implies (i) because by the fundamental theorem, every model is elementarily equivalent to its ultrapower. ⊣

Roitman (1982) showed that the continuum hypothesis is needed in Corollary 6.1.2. In an extension of a model of ZFC with α Cohen reals, she proved that for each uncountable regular $\beta \leqslant \alpha$ there is an ultrapower D over ω such that $\prod_D \langle \omega, \leqslant \rangle$ has cofinality β. See also Canjar (1988).

The rest of this section is devoted to the generalization of Theorem 6.1.1 to arbitrary cardinals. We can see from cardinality considerations that the condition that D is countably incomplete is not sufficient for $\prod_D \mathfrak{A}$ to be an ω_2-saturated model. In fact, for $\alpha > \omega$ the assumption that D is α-regular, or even α^+-regular, is not sufficient for $\prod_D \mathfrak{A}$ to be α^+-saturated (see Exercise 6.1.6). We shall need a totally new kind of ultrafilter, the 'good ultrafilters', in order to get a generalization of Theorem 6.1.1.

We first need some notation about functions. Let I be a nonempty set, and β a cardinal. We consider functions f, g on the set $S_\omega(\beta)$ of all finite subsets of β into the set $S(I)$ of all subsets of I. We say that $g \leqslant f$ iff for all $u \in S_\omega(\beta)$, $g(u) \subset f(u)$. Thus $g \leqslant f$ means that each value of g is included in the corresponding value of f. We shall say that f is *monotonic* iff

$$u, w \in S_\omega(\beta) \text{ and } u \subset w \text{ imply } f(u) \supset f(w).$$

Notice that the direction of the inclusion reverses. Thus the larger u is, the smaller $f(u)$ is. Strictly speaking, 'antimonotonic' would be a better name, but it is also longer. The function g is said to be *additive* iff

$$u, w \in s_\omega(\beta) \quad \text{implies} \quad g(u \cup w) = g(u) \cap g(w).$$

Again, notice that there is a union on the left and an intersection on the right. A better but longer name would be 'antiadditive'. It is obvious that:

LEMMA 6.1.3. *Every additive function on $S_\omega(\beta)$ into $S(I)$ is monotonic.*

Since an ultrafilter D over I is a subset of $S(I)$, any function f on $S_\omega(\beta)$ into D is a function on $S_\omega(\beta)$ into $S(I)$. We are interested in that case.

Now for the main definition. Let α be an infinite cardinal. An ultrafilter D over I is said to be *α-good* iff it has the following property:

For every cardinal $\beta < \alpha$ and every monotonic function f on $S_\omega(\beta)$ into D, there exists an additive function g on $S_\omega(\beta)$ into D such that $g \leqslant f$.

Note that if D is α-good, then D is β-good for all infinite cardinals $\beta < \alpha$.

Let us give one example of a monotonic function f on $S_\omega(\beta)$ into D. Suppose D is a countably incomplete ultrafilter over I, and let $I = I_0 \supset I_1 \supset I_2 \supset \ldots$ be a decreasing chain of sets $I_n \in D$ such that $\bigcap_n I_n = 0$. Then the function $f: S_\omega(\beta) \to D$ such that for each $u \in S_\omega(\beta)$, $f(u) = I_{|u|}$ is monotonic. However, f is not additive, because

$$f(u \cap w) = I_{|u \cup w|}, \qquad f(u) \cap f(w) = I_{\max(|u|, |w|)}.$$

The power of the α-good ultrafilters lies in the fact that monotonic functions f like the one above, which are very nonadditive, can be 'refined' to additive functions $g \leqslant f$ on $S_\omega(\beta)$ into D.

In Exercise 6.1.2, we shall see that every countably incomplete ultrafilter is ω_1-good. In order to generalize Theorem 6.1.1 to larger cardinals, we must first prove the existence of countably incomplete ultrafilters which are α-good for a given α. We shall first state the existence theorem, and then prove a series of lemmas which will be used to prove the existence of α-good ultrafilters.

THEOREM 6.1.4. *Let I be a set of power α. Then there exists an α^+-good countably incomplete ultrafilter D over I.*

LEMMA 6.1.5. *For an ultrafilter D to be α^+-good it is necessary and sufficient that for every monotonic function f on $S_\omega(\alpha)$ into D there is an additive function g on $S_\omega(\alpha)$ into D with $g \leqslant f$.*

PROOF. The necessity is obvious.

To prove the sufficiency, let $\beta \leqslant \alpha$, and let $f: S_\omega(\beta) \to D$ be monotonic. Define a function $f': S_\omega(\alpha) \to D$ by

$$f'(u) = f(u \cap \beta), \quad \text{for all} \quad u \in S_\omega(\alpha).$$

Then f' is monotonic. By hypothesis there exists an additive function $g' \leqslant f'$ on $S_\omega(\alpha)$ into D. Now let g be the restriction of g' to $S_\omega(\beta)$. Then g maps $S_\omega(\beta)$ into D, and it is easy to check that g is additive and $g \leqslant f$. This proves that the condition in the lemma is sufficient for D to be α^+-good. ⊣

LEMMA 6.1.6. *Let α be an infinite cardinal. Suppose that X is a set of power α and let Y_x, $x \in X$, be a family of sets each of which has power α. Then there exists a family of sets Z_x, $x \in X$, such that for all $x, y \in X$:*

(i). $Z_x \subset Y_x$;

(ii). Z_x *has power* α;

(iii). *if* $x \neq y$, *then* $Z_x \cap Z_y = 0$.

That is, any family of α sets of power α can be refined to a family of α disjoint sets of power α.

Proof. We may assume without loss of generality that $X = \alpha$. For each ordinal $\beta \leqslant \alpha$, let X_β be the set

$$X_\beta = \{\langle \gamma, \delta \rangle : \gamma \leqslant \delta \quad \text{and} \quad \delta < \beta\}.$$

Thus X_β is a subset of $\beta \times \beta$. (If $\alpha \times \alpha$ is drawn on a graph with two axes, X_β will be a right triangle with sides of length β.) Since α is a limit ordinal, $X_\alpha = \bigcup_{\beta<\alpha} X_\beta$. We shall find a function f with domain X_α such that f is one–one and

$$\text{(1)} \qquad \text{whenever} \quad \gamma \leqslant \delta < \alpha, \quad f(\gamma, \delta) \in Y_\gamma.$$

Once the function f is found, we may define

$$Z_\gamma = \{f(\gamma, \delta) : \gamma \leqslant \delta < \alpha\},$$

and the family Z_γ, $\gamma < \alpha$, clearly has the desired properties (i)–(iii).

The function f is defined by transfinite induction. Let $\beta < \alpha$ and suppose that we already have a one–one function f_β with domain X_β such that (1) holds for β. Since $|X_\beta| < \alpha$, and $|Y_\gamma| = \alpha$ for all $\gamma < \alpha$, we may extend f_β to a one–one function $f_{\beta+1}$ with domain $X_{\beta+1}$ such that (1) holds for $\beta+1$. Simply choose, for each $\gamma \leqslant \beta$, a value $f_{\beta+1}(\gamma, \beta) \in Y_\gamma$ which is different from all the previously chosen values of $f_{\beta+1}$. By taking unions at the limit ordinals, we obtain a chain of functions f_β with domain X_β such that each f_β is one–one and has the property (1). Then the union $f = \bigcup_{\beta<\alpha} f_\beta$ is a one–one function with domain X_α satisfying (1). ⊣

The next definition is the key to the proofs of Lemma 6.1.7 and Theorem 6.1.4.

Let Π be a nonempty collection of partitions of α such that each partition has exactly α equivalence classes, and let F be a nontrivial filter over α. We say that the pair (Π, F) is *consistent* iff given any $X \in F$ and any $X_1, \ldots, X_n$, $n < \omega$, each X_i belonging to a distinct partition $P_i \in \Pi$, $X \cap \bigcap_{1 \leqslant i \leqslant n} X_i \neq 0$. The next lemma contains all the important information about this notion of consistency. If F is a filter and $F \cup E$ has the finite intersection property, then we let (F, E) denote the filter generated by $F \cup E$.

Lemma 6.1.7. *Let α be an infinite cardinal.*

(i). *Let F be a uniform filter over α generated by a subset $E \subset F$ of power at most α. There exists a collection Π of partitions of α such that $|\Pi| = 2^\alpha$ and (Π, F) is consistent.*

(ii). *Suppose that (Π, F) is consistent. Let $J \subset \alpha$. Then either $(\Pi, (F, \{J\}))$ is consistent, or else $(\Pi', (F, \{\alpha \setminus J\}))$ is consistent for some cofinite $\Pi' \subset \Pi$.*

(iii). *Suppose that (Π, F) is consistent. Let p be any monotonic mapping of $S_\omega(\alpha)$ into F and let $P \in \Pi$. Then there exist an extension F' of F and an additive function $q : S_\omega(\alpha) \to F'$ such that $q \leqslant p$ and $(\Pi \setminus \{P\}, F')$ is consistent.*

PROOF. (i). Let J_β, $\beta < \alpha$, be a list of all finite intersections of members of E. Each $|J_\beta| = \alpha$. By Lemma 6.1.6, there are I_β, $\beta < \alpha$, such that $|I_\beta| = \alpha$, $I_\beta \subset J_\beta$, and $I_\beta \cap I_{\beta'} = 0$ if $\beta \neq \beta'$. Consider the set

$$B = \{\langle s, r\rangle : s \in S_\omega(\alpha) \text{ and } r : S(s) \to \alpha\}.$$

Clearly, $|B| = \alpha$. Let $\langle s_\xi, r_\xi\rangle$, $\xi < \alpha$, be an enumeration of B (with possible repetitions) in such a way that

$$B = \{\langle s_\xi, r_\xi\rangle : \xi \in I_\beta\} \text{ for each } \beta < \alpha.$$

For each $J \subset \alpha$ define the function $f_J : \alpha \to \alpha$ as follows:

$$f_J(\xi) = r_\xi(J \cap s_\xi) \quad \text{if} \quad \xi \in \bigcup_{\beta<\alpha} I_\beta,$$

$$f_J(\xi) = 0 \quad \text{otherwise.}$$

We first establish that there are 2^α such functions f_J. Suppose that $J_1 \neq J_2$. We may suppose, by symmetry that there is an $x \in J_1$ and $x \notin J_2$. Let $s = \{x\}$ and $r = \{\langle\{x\}, 0\rangle, \langle 0, 1\rangle\}$. Then $\langle s, r\rangle \in B$, so $\langle s, r\rangle = \langle s_\xi, r_\xi\rangle$ for some ξ. Now $f_{J_1}(\xi) = r(J_1 \cap s) = 0$ and $f_{J_2}(\xi) = r(J_2 \cap s) = 1$. So $f_{J_1} \neq f_{J_2}$. Next, let $\beta, \gamma_1, \ldots, \gamma_n$ be ordinals in α, and let $J_1, \ldots, J_n$ be distinct subsets of α. Then we claim that there is a $\xi \in I_\beta$ such that

$$f_{J_i}(\xi) = \gamma_i \quad \text{for} \quad 1 \leqslant i \leqslant n.$$

To see this, let s be any finite subset of α such that

$$s \cap J_i \neq s \cap J_j \text{ for } 1 \leqslant i < j \leqslant n.$$

Now let $r : S(s) \to \alpha$ be defined in such a way that

$$r(J_i \cap s) = \gamma_i, \qquad 1 \leqslant i \leqslant n.$$

There is a $\xi \in I_\beta$ such that $\langle s_\xi, r_\xi\rangle = \langle s, r\rangle$. Whence

$$f_{J_i}(\xi) = r_\xi(J_i \cap s_\xi) = r(J_i \cap s) = \gamma_i.$$

We have, incidentally, shown that the range of each f_J is α.

Finally, let

$$\Pi = \{\{f_J^{-1}(\gamma) : \gamma < \alpha\} : J \subset \alpha\}.$$

Clearly Π is a collection of 2^α partitions of α. It is obvious that (Π, F) is consistent.

(ii). Suppose that $(\Pi, (F, \{J\}))$ is not consistent. Then there are $X \in F$, $X_i \in P_i \in \Pi$, $1 \leqslant i \leqslant n$, the P_i's distinct, such that

$$J \cap X \cap \bigcap_{1 \leqslant i \leqslant n} X_i = 0. \tag{1}$$

Let $\Pi' = \Pi \setminus \{P_1 \ldots P_n\}$. Let Q_j, $1 \leqslant j \leqslant m$, be distinct elements of Π' and $Y_j \in Q_j$. Then by hypothesis,

$$X \cap \bigcap_{1 \leqslant i \leqslant n} X_i \cap \bigcap_{1 \leqslant j \leqslant m} Y_j \neq 0. \tag{2}$$

It is immediate from (1) and (2) that

$$(\alpha \setminus J) \cap X \cap \bigcap_{1 \leqslant j \leqslant m} Y_j \neq 0.$$

So $(\Pi', (F, \{\alpha \setminus J\}))$ is consistent.

(iii). Let $X_\delta, \delta < \alpha$, be an enumeration of P without repetition. Let $S_\omega(\alpha) = \{t_\delta : \delta < \alpha\}$. For each $\delta < \alpha$, we define the function $q_\delta : S_\omega(\alpha) \to S(\alpha)$ as follows:

$$q_\delta(s) = p(t_\delta) \bigcap X_\delta \quad \text{if} \quad s \subset t_\delta,$$

$$q_\delta(s) = 0 \quad \text{if} \quad s \not\subset t_\delta.$$

Note that $q_\delta(s) \subset p(t_\delta)$, $q_\delta(s) \neq 0$ if $s \subset t_\delta$, and $q_\delta(s_1 \cup s_2) = q_\delta(s_1) \cap q_\delta(s_2)$. This last is because

$$s_1 \cup s_2 \subset t_\delta \text{ iff both } s_1 \subset t_\delta \text{ and } s_2 \subset t_\delta.$$

Define the function $q : S_\omega(\alpha) \to S(\alpha)$ as follows:

$$q(s) = \bigcup_{\delta < \alpha} q_\delta(s), \qquad s \in S_\omega(\alpha).$$

As p is monotone, we see easily that $q(s) \subset p(s)$, so $q \leqslant p$. Since $q_\delta(s) \cap q_{\delta'}(s) = 0$ if $\delta \neq \delta'$, we have that $q(s)$ is a disjoint union of subsets of X_δ. Using the fact that each q_δ is additive and that

$$X_\delta \cap X_{\delta'} \neq 0 \qquad \text{iff} \quad \delta = \delta',$$

we check easily that q is additive. Now let $F' = (F, \text{Rng } q)$. We claim that $(\Pi \setminus \{P\}, F')$ is consistent. Let $X \in F$, $s \in S_\omega(\alpha)$, $X_i \in P_i \in \Pi$, $1 \leqslant i \leqslant n$, the P_i distinct and different from P. Since $s = t_\delta$ for some $\delta < \alpha$, we have $q(s) \supset q_\delta(s) = p(t_\delta) \cap X_\delta$, and

$$X \cap p(t_\delta) \cap X_\delta \cap \bigcap_{1 \leqslant i \leqslant n} X_i \neq 0.$$

Whence

$$X \cap q(s) \cap \bigcap_{1 \leqslant i \leqslant n} X_i \neq 0. \dashv$$

PROOF OF THEOREM 6.1.4. We may assume that $I = \alpha$. Let I_n, $n < \omega$, be a sequence of subsets of α of power α such that $I_{n+1} \subset I_n$ and $\bigcap_{n<\omega} I_n = 0$. Let F_0 be the uniform filter generated by the set $\{I_n : n \in \omega\}$. By Lemma 6.1.7 (i), let Π_0 be any collection of partitions of α such that $|\Pi_0| = 2^\alpha$ and (Π_0, F_0) is consistent. We shall define by transfinite induction two sequences Π_ξ, $\xi < 2^\alpha$, F_ξ, $\xi < 2^\alpha$ such that

$$\Pi_\xi \subset \Pi_\eta, F_\xi \supset F_\eta \quad \text{if} \quad \eta \leqslant \xi < 2^\alpha,$$

$$|\Pi_\xi| = 2^\alpha, \qquad |\Pi_\xi \setminus \Pi_{\xi+1}| < \omega, \qquad \Pi_\lambda = \bigcap_{\eta<\lambda} \Pi_\eta, \qquad \lambda \text{ limit},$$

$$(\Pi_\xi, F_\xi) \text{ is consistent for } \xi < 2^\alpha.$$

The construction is as follows: Let p_ξ, $\xi < 2^\alpha$, be an enumeration of all monotone functions mapping $S_\omega(\alpha)$ into $S(\alpha)$, and let J_ξ, $\xi < 2^\alpha$, be an enumeration of $S(\alpha)$. Suppose that Π_η, F_η for $\eta < \xi < 2^\alpha$ have been defined satisfying all the inductive hypotheses. If ξ is a limit ordinal, then simply let

$$\Pi_\xi = \bigcap_{\eta<\xi} \Pi_\eta \quad \text{and} \quad F_\xi = \bigcup_{\eta<\xi} F_\eta.$$

It is clear that (Π_ξ, F_ξ) is consistent and $|\Pi_\xi| = 2^\alpha$. If $\xi = \lambda+2n+1$, λ a limit ordinal and $n < \omega$, then let J be the first element of $S(\alpha)$ not already in $F_{\xi-1}$. By Lemma 6.1.7 (ii), we can find Π_ξ, F_ξ such that

$$|\Pi_{\xi-1} \setminus \Pi_\xi| < \omega, \qquad |\Pi_\xi| = 2^\alpha,$$

$$J \in F_\xi \quad \text{or} \quad (\alpha \setminus J) \in F_\xi,$$

$$(\Pi_\xi, F_\xi) \text{ is consistent.}$$

If $\xi = \lambda+2n+2$, λ a limit ordinal and $n < \omega$, then let $p : S_\omega(\alpha) \to F_{\xi-1}$ be the first function in the list p_η, $\eta < 2^\alpha$, which we have not already dealt with. By Lemma 6.1.7 (iii), we can find Π_ξ, F_ξ, $q : S_\omega(\alpha) \to F_\xi$ such that

$$|\Pi_{\xi-1} \setminus \Pi_\xi| = 1, \qquad |\Pi_\xi| = 2^\alpha,$$

$$q \leqslant p, q \text{ is additive},$$

$$F_\xi = (F_{\xi-1}, \operatorname{Rng} q),$$

$$(\Pi_\xi, F_\xi) \text{ is consistent.}$$

Let $F = \bigcup_{\xi<2^\alpha} F_\xi$. Because of our construction and cf $(2^\alpha) > \alpha$, we see that F is a countably incomplete α^+-good ultrafilter over α. ⊣

Notice that the full strength of Lemmas 6.1.6 and 6.1.7 allows us to prove that any uniform filter F over α generated from a set of at most α elements can be extended to an α^+-good ultrafilter over α (Exercise 6.1.4).

We now prove the generalization of Theorem 6.1.1.

THEOREM 6.1.8. *Let α be an infinite cardinal and let D be a countably incomplete α-good ultrafilter over a set I. Suppose $\|\mathscr{L}\| < \alpha$. Then for any family $\mathfrak{A}_i$, $i \in I$, of models for $\mathscr{L}$, the ultraproduct $\prod_D \mathfrak{A}_i$ is α-saturated.*

PROOF. In exactly the same way as in the proof of Theorem 6.1.1, we see that it is sufficient to prove:

(1) For every set $\Sigma(x)$ of formulas of $\mathscr{L}$, if every finite subset of $\Sigma(x)$ is satisfiable in $\prod_D \mathfrak{A}_i$, then $\Sigma(x)$ is satisfiable in $\prod_D \mathfrak{A}_i$.

Suppose every finite subset of $\Sigma(x)$ is satisfiable in $\prod_D \mathfrak{A}_i$. Since D is countably incomplete, we may choose a descending chain

$$I = I_0 \supset I_1 \supset I_2 \supset \ldots$$

such that each $I_n \in D$ and $\bigcap_{n<\omega} I_n = 0$. We have $|\Sigma| < \alpha$ because $\|\mathscr{L}\| < \alpha$. Let us define a function

$$f : S_\omega(\Sigma) \to D$$

as follows. For each finite subset σ of Σ, let

$$f(\sigma) = I_{|\sigma|} \cap \{i \in I : \mathfrak{A}_i \models (\exists x) \wedge \sigma\}, \tag{2}$$

with the understanding that $f(\emptyset) = I$. Each $\sigma \in S_\omega(\Sigma)$ is finite and is satisfiable in $\prod_D \mathfrak{A}_i$, whence $\prod_D \mathfrak{A}_i \models (\exists x) \wedge \sigma$. By the fundamental theorem, $f(\sigma) \in D$. Whenever $\sigma \subset \tau \in S_\omega(\Sigma)$, we have

$$I_{|\tau|} \subset I_{|\sigma|}, \qquad \vdash (\exists x) \wedge \tau \to (\exists x) \wedge \sigma,$$

so $f(\tau) \subset f(\sigma)$ and f is monotonic. Now we use the fact that D is α-good. Since D is α-good, there is an additive function $g \leqslant f$ on $S_\omega(\Sigma)$ into D. For each $i \in I$, let

$$\sigma(i) = \bigcup \{\theta \in \Sigma : i \in g(\{\theta\})\}. \tag{3}$$

If $|\sigma(i)| \geqslant n$, then $i \in I_n$. Because if $\sigma(i)$ has at least n distinct elements $\theta_1, \ldots, \theta_n$, then for $s \leqslant n$ we have

$$i \in g(\theta_s),$$

whence using the additivity of g,

$$i \in g(\{\theta_1\}) \cap \ldots \cap g(\{\theta_n\}) = g(\{\theta_1, \ldots, \theta_n\}) \subset f(\{\theta_1, \ldots, \theta_n\}) \subset I_n.$$

We recall that $\bigcap_{n<\omega} I_n = 0$, and therefore

$$\text{for each } i \in I, \quad \sigma(i) \text{ is finite.}$$

We now choose an element h_D which satisfies $\Sigma(x)$ in $\prod_D \mathfrak{A}_i$. For each $i \in I$, we have by (2), (3), and additivity,

$$i \in \bigcap\{g(\{\theta\}) : \theta \in \sigma(i)\} = g(\sigma(i)) \subset f(\sigma(i)),$$

and therefore $i \in f(\sigma(i))$. Then by (2) we may choose an element $h(i) \in A_i$ such that

(4) $$\mathfrak{A}_i \models \bigwedge \sigma(i)[h(i)].$$

Now, whenever $\theta \in \Sigma$ and $i \in g(\{\theta\})$, we have $\theta \in \sigma(i)$ and by (4), $\mathfrak{A}_i \models \theta[h(i)]$. But $g(\{\theta\}) \in D$, so by the fundamental theorem $\prod_D \mathfrak{A}_i \models \theta[h_D]$ for all $\theta \in \Sigma$. This shows that h_D satisfies Σ in $\prod_D \mathfrak{A}_i$. ⊣

Using Theorems 6.1.4 and 6.1.8, and an instance of the GCH, let us quickly draw a conclusion which is stronger than the isomorphism theorem 6.1.15, proved later in this section.

Theorem 6.1.9. *Let* $\|\mathscr{L}\| \leqslant \alpha$ *and* $\mathfrak{A}, \mathfrak{B}$ *be models for* $\mathscr{L}$ *with* $|A|, |B| \leqslant \alpha^+$. *Assume that* $2^\alpha = \alpha^+$. *Let* D *be an* α^+*-good countably incomplete ultrafilter over a set* I *of power* α. *Then the following are equivalent*:

(i). $\mathfrak{A} \equiv \mathfrak{B}$.

(ii). $\prod_D \mathfrak{A} \cong \prod_D \mathfrak{B}$.

Proof. (i) ⇒ (ii). Assume (i). By Theorem 6.1.8, the ultrapowers $\prod_D \mathfrak{A}$ and $\prod_D \mathfrak{B}$ are both α^+-saturated. Moreover, they both have cardinality at most $(\alpha^+)^\alpha = 2^\alpha = \alpha^+$. Using the equivalences $\prod_D \mathfrak{A} \equiv \mathfrak{A} \equiv \mathfrak{B} \equiv \prod_D \mathfrak{B}$ and the uniqueness theorem for saturated models, we have the isomorphism $\prod_D \mathfrak{A} \cong \prod_D \mathfrak{B}$.

(ii) ⇒ (i). This is obvious. ⊣

It follows from Theorem 6.1.9 that if we assume the GCH, then for any two models $\mathfrak{A}, \mathfrak{B}$ for an arbitrary $\mathscr{L}$, $\mathfrak{A}$ and $\mathfrak{B}$ are elementarily equivalent if and only if they have isomorphic ultrapowers. This isomorphism theorem gives us a purely algebraic characterization of the notion of elementary equivalence. It turns out that while it is not known whether Theorem 6.1.9 is true without the assumption $2^\alpha = \alpha^+$, the above consequence of Theorem 6.1.9 is always true without the GCH. The next sequence of lemmas is designed to prove the isomorphism theorem without the GCH.

We first give a crucial definition. Let λ, κ be infinite cardinals and let μ be the least cardinal α such that $\lambda^\alpha > \lambda$. It is easily seen that $\mu \leqslant \lambda$ and μ is a regular cardinal. Let F be a set of functions $f : \lambda \to \mu$ and let G be a set of functions $g : \lambda \to \beta(g)$, where $\beta(g)$ is a cardinal less than μ. Let D be a filter over λ. We say that the triple (F, G, D) is *κ-consistent* iff the following hold:

(i). D is generated by a subset $E \subset D$ of power at most κ. This means that $E \subset D$, $|E| \leqslant \kappa$, E is closed under finite intersections, and every element of D is a superset of some element in E.

(ii). Whenever we are given a cardinal $\beta < \mu$, a β-termed sequence f_ρ, $\rho < \beta$, of distinct functions in F, a β-termed sequence of ordinals σ_ρ, $\rho < \beta$, less than μ, and two functions $f \in F$ and $g \in G$, then the set

$$\{\xi < \lambda : f_\rho(\xi) = \sigma_\rho \text{ for all } \rho < \beta \text{ and } f(\xi) = g(\xi)\}$$

together with D generate a nontrivial filter over λ.

A consequence of this definition is that D is a nontrivial filter over λ.

The next lemma is a straightforward generalization of Lemma 6.1.7(i). Indeed, the definition of the triple (F, G, D) being κ-consistent is a generalization of the notion of the pair (Π, F) being consistent.

LEMMA 6.1.10. *There is a family F of 2^λ functions from λ to μ such that the triple $(F, 0, \{\lambda\})$ is μ-consistent.*

PROOF. Let

$$H = \{(A, S, h) : A \subset \lambda,\ |A| < \mu,\ S \subset S(A),\ |S| < \mu, \text{ and } h : S \to \mu\}.$$

By the definition of μ, $|H| = \lambda$. Whence we may enumerate $H = \{(A_\xi, S_\xi, h_\xi) : \xi < \lambda\}$. For $B \subset \lambda$, and $\xi < \lambda$, define

$$f_B(\xi) = \begin{cases} h_\xi(B \cap A_\xi) & \text{if } B \cap A_\xi \in S_\xi, \\ 0 & \text{if } B \cap A_\xi \notin S_\xi. \end{cases}$$

Finally, let $F = \{f_B : B \subset \lambda\}$. It is easy to see that if $B \neq C$, then $f_B \neq f_C$. Let $\beta < \mu$ and let B_ρ, $\rho < \beta$, be distinct subsets of λ and let σ_ρ, $\rho < \beta$, be a sequence of ordinals less than μ. We show that the set

$$\{\xi < \lambda : f_{B_\rho}(\xi) = \sigma_\rho \quad \text{for all} \quad \rho < \beta\}$$

is not empty. Let A be a set of power less than μ such that $B_\rho \cap A \neq B_{\rho'} \cap A$

for all $\rho, \rho' < \beta, \rho \neq \rho'$. Let $S = \{B_\rho \cap A : \rho < \beta\}$ and let $h(A \cap B_\rho) = \sigma_\rho$. Then $(A, S, h) \in H$, and whence $(A, S, h) = (A_\xi, S_\xi, h_\xi)$ for some $\xi < \lambda$. Hence

$$f_{B_\rho}(\xi) = h_\xi(B_\rho \cap A_\xi) = h(B_\rho \cap A) = \sigma_\rho \quad \text{for all} \quad \rho < \beta. \dashv$$

LEMMA 6.1.11.

(i). *If (F, G, D) is κ-consistent and $\kappa < \gamma$, then (F, G, D) is γ-consistent.*

(ii). *Suppose that (F_ξ, G_ξ, D_ξ) is κ_ξ-consistent for every $\xi < \delta$. Suppose further that $F_\xi \supset F_\eta$, $G_\xi \subset G_\eta$, $D_\xi \subset D_\eta$ whenever $\xi < \eta < \delta$, and $\kappa_\xi \leqslant \kappa$ for each $\xi < \delta$, and* cf $(\delta) \leqslant \kappa$. *Then $(\bigcap_{\xi<\delta} F_\xi, \bigcup_{\xi<\delta} G_\xi, \bigcup_{\xi<\delta} D_\xi)$ is κ-consistent.*

(iii). *If (F, G, D) is κ-consistent and $F' \subset F$, $G' \subset G$, then (F', G', D) is κ-consistent.*

PROOF. The proofs of all parts of this lemma consist of simply checking the definitions. We leave it as an exercise for the reader. $\dashv$

The next lemma contains the heart of the matter.

LEMMA 6.1.12. *Let G be a set of functions from λ to cardinals less than μ such that $\mu + |G| \leqslant \kappa$. Suppose that $(F, 0, D)$ is κ-consistent. Then there is an $F' \subset F$ such that $|F \setminus F'| \leqslant \kappa$ and the triple (F', G, D) is κ-consistent.*

PROOF. Since $|G| \leqslant \kappa$, it is clearly sufficient to prove that:

(1) For every element $g \in G$, there is a subset $F_g \subset F$ such that $|F_g| \leqslant \kappa$ and $(F - F_g, \{g\}, D)$ is κ-consistent.

This is because from (1) it follows that the set $F' = F - \bigcup_{g \in G}\{F_g\}$ is the required set. To prove (1), let $g \in G$. Suppose that (1) fails to hold for g. Then

(2) For every subset $\bar{F} \subset F$ of power at most κ, $(F \setminus \bar{F}, \{g\}, D)$ is not κ-consistent.

We now define by induction sets F_ξ, $\bar{F}_\xi$, $\xi < \kappa^+$, such that (making use of (2))

$F_0 = F$;

$\bar{F}_\xi$ is a subset of F_ξ of power at most κ such that $(F_\xi \setminus \bar{F}_\xi, \{g\}, D)$ is not κ-consistent;

$F_{\xi+1} = F_\xi \setminus \bar{F}_\xi$;

$F_\eta = \bigcap_{\xi<\eta} F_\xi$ if η is a limit ordinal, $\eta < \kappa^+$.

Furthermore, since $\mu \leqslant \kappa$, if we examine what it really means for $(F_\xi, \{g\}, D)$ to be not κ-consistent, we can make sure that for each $\xi < \kappa^+$, there are a cardinal $\beta_\xi < \mu$, a sequence of ordinals σ_ρ^ξ, $\rho < \beta_\xi$, less than μ, a sequence of distinct functions $f_\rho^\xi \in \bar{F}_\xi$, $\rho < \beta_\xi$, and a function $f^\xi \in \bar{F}_\xi$ such that the set

$$A_\xi = \{\nu < \lambda : f_\rho^\xi(\nu) = \sigma_\rho^\xi \quad \text{for all} \quad \rho < \beta_\xi \quad \text{and} \quad f^\xi(\nu) = g(\nu)\}$$

is inconsistent with D. Remember now that D is generated by a set E of power at most κ. Since each A_ξ is inconsistent with D, we have that,

$$\text{for each } \xi < \kappa^+, \text{ there is an } X_\xi \in E \text{ such that } A_\xi \cap X_\xi = 0.$$

Since $|E| \leqslant \kappa$, this means that there is a set $X \in E$ and there are κ^+ ordinals ξ such that $A_\xi \cap X = 0$. Since $\mu \leqslant \kappa$, we may assume without loss of generality that for all $\xi \in \kappa^+$,

$$\beta_\xi = \beta < \mu \quad \text{and} \quad A_\xi \cap X = 0.$$

Let $\gamma < \mu$ be a cardinal such that $g : \lambda \to \gamma$.

It is clear that $\gamma < \kappa^+$. Consider now the functions f_ρ^ξ, $\xi < \gamma$, $\rho < \beta$, and the functions f^ξ, $\xi < \gamma$. Consider also the ordinals σ_ρ^ξ, $\xi < \gamma$, $\rho < \beta$, and the ordinals $\xi < \gamma$. Clearly, we may well-order all the functions and the ordinals by the cardinal $\beta + \gamma < \mu$. Since the triple $(F, 0, D)$ is κ-consistent, the set

$$A = \{\nu < \lambda : f_\rho^\xi(\nu) = \sigma_\rho^\xi, \quad \text{for all} \quad \xi < \gamma \quad \text{and} \quad \rho < \beta, \quad \text{and} \quad f^\xi(\nu) = \xi, \quad \text{for all} \quad \xi < \gamma\}$$

is consistent with D. So, in particular, $A \cap X \neq 0$. Now let $\nu \in A \cap X$. We have that $\nu \notin A_\xi$ for all $\xi < \kappa^+$, because $A_\xi \cap X = 0$. But now, for some $\xi < \gamma$, $g(\nu) = \xi$. Whence $\nu \in A_\xi$. This is a contradiction. So (1) is proved. ⊣

LEMMA 6.1.13

(i). *Suppose that* $(F, 0, D)$ *is* κ*-consistent and* $A \subset \lambda$. *Then there is an* $F' \subset F$, $|F \setminus F'| < \mu$, *such that either* $(F', 0, D')$ *is* κ*-consistent, where* D' *is the filter generated by* D *and* $\{A\}$, *or else* $(F', 0, D'')$ *is* κ*-consistent, where* D'' *is the filter generated by* D *and* $\{\lambda \setminus A\}$.

(ii). *Suppose that* $(F, 0, D)$ *is* κ*-consistent,* $\mu \leqslant \kappa$, *and* $A_\xi \subset \lambda$, *for* $\xi < \kappa$. *Then there are* $F' \subset F$ *and a filter* $D' \supset D$ *such that* $|F \setminus F'| \leqslant \kappa$, $(F', 0, D')$ *is* κ*-consistent, and for all* $\xi < \kappa$, *either* $A_\xi \in D'$ *or* $(\lambda \setminus A_\xi) \in D'$.

PROOF. (i). The proof of (i) is an analogue of the proof of Lemma 6.1.7 (ii). First of all, it is clear that both D' and D'' are generated by a subset of power at most κ. Suppose that $(F, 0, D')$ is not κ-consistent. Then there are a cardinal $\beta < \mu$, distinct functions f_ρ, $\rho < \beta$, and ordinals σ_ρ, $\rho < \beta$, such that the set

$$B = \{\xi < \lambda : f_\rho(\xi) = \sigma_\rho \quad \text{for all} \quad \rho < \beta\}$$

is inconsistent with D'. Whence there is a set $X \in E$ such that $B \cap X \cap A = 0$. Let $F' = F \setminus \{f_\rho : \rho < \beta\}$. We show that $(F', 0, D'')$ is κ-consistent. Let $\beta' < \mu$ and let the sequences f'_ρ, $\rho < \beta'$ and σ'_ρ, $\rho < \beta'$, be given. Consider the set

$$B' = \{\xi < \lambda : f'_\rho(\xi) = \sigma'_\rho \quad \text{for all} \quad \rho < \beta'\}.$$

Since $(F, 0, D)$ is κ-consistent, the set $B \cap B'$ is consistent with D. Let Y be any set in E. We have that $B \cap B' \cap Y \cap X \neq 0$. Since $B \cap X \cap A = \emptyset$, it follows that $B' \cap Y \cap (\lambda \setminus A) \neq 0$. So B' is consistent with D''.

(ii). The proof of (ii) is a simple iteration of (i). ⊣

The following lemma is a key step in constructing the isomorphism.

LEMMA 6.1.14. *Suppose that $\mathfrak{A}$ is a model with $|A| < \mu$, and $(F, 0, D)$ is κ-consistent. Let φ_ξ, $\xi < \kappa$, be formulas of $\mathscr{L}$. We assume that the set $\{\varphi_\xi : \xi < \kappa\}$ is closed under conjunction. We write each $\varphi_\xi = \varphi_\xi(x, y)$. For each $\xi < \kappa$, let $a^\xi : \lambda \to A$ be a function mapping λ into A. Suppose that for each $\xi < \kappa$,*

$$\{\nu < \lambda : \mathfrak{A} \models (\exists x)\varphi_\xi(x, a^\xi(\nu))\} \in D.$$

Then there are $a : \lambda \to A$, $F' \subset F$, $D' \supset D$, such that $|F \setminus F'| \leqslant \kappa$, $(F', 0, D')$ is κ-consistent, and, for every $\xi < \kappa$,

$$\{\nu < \lambda : \mathfrak{A} \models \varphi_\xi(a(\nu), a^\xi(\nu))\} \in D'.$$

PROOF. Let $|A| = \alpha$ and let $\{a_\xi : \xi < \alpha\}$ be an enumeration of A. We define functions $g_\xi : \lambda \to \alpha$ for $\xi < \kappa$ as follows: for each $\nu < \lambda$,

$$g_\xi(\nu) = \begin{cases} \text{the first ordinal } \eta \text{ such that } \mathfrak{A} \models \varphi_\xi(a_\eta a^\xi(\nu)) \text{ if such an } a_\eta \text{ exists,} \\ 0 \text{ otherwise.} \end{cases}$$

Let $G = \{g_\xi : \xi < \kappa\}$. Note that $\mu + |G| \leqslant \kappa$. So, by Lemma 6.1.12, there is $\bar{F} \subset F$ such that $|F \setminus \bar{F}| \leqslant \kappa$ and $(\bar{F}, G, D)$ is κ-consistent. Now let f be any function in $\bar{F}$. Define $a : \lambda \to A$ as follows:

$$a(\nu) = \begin{cases} a_{f(\nu)} & \text{if } f(\nu) < \alpha, \\ a_0 & \text{otherwise.} \end{cases}$$

For each $\xi < \kappa$, define

$$B_\xi = \{\nu < \lambda : \mathfrak{A} \models \varphi_\xi(a(\nu), a^\xi(\nu))\}.$$

Let D' be the filter generated by D and $\{B_\xi : \xi < \kappa\}$. Note that D' is generated by a subset of power at most κ. Let $F' = \bar{F} \setminus \{f\}$. We now show that the conclusion of the theorem holds for F', D' and a. We simply have to show that $(F', 0, D')$ is κ-consistent.

Let $\beta < \mu$, and suppose we are given the sequences f_ρ, $\rho < \beta$, and σ_ρ, $\rho < \beta$. Let

$$B = \{\nu < \lambda : f_\rho(\nu) = \sigma_\rho \quad \text{for every } \rho < \beta\}.$$

If B were inconsistent with D', then there are $X \in D$ and some B_ξ, $\xi < \kappa$, such that

$$B \cap X \cap B_\xi = 0.$$

Now consider the function $f \in \bar{F}$ together with the sequence f_ρ, $\rho < \beta$, and the function $g_\xi \in G$. Since $(\bar{F}, G, D)$ is κ-consistent, the set

$$\bar{B} = \{\nu < \lambda : f_\rho(\nu) = \sigma_\rho \text{ for all } \rho < \beta \text{ and } f(\nu) = g_\xi(\nu)\}$$

is consistent with D. If we look at the definition of g_ξ and of B_ξ, we see that

$$\bar{B} \subset B \cap B_\xi.$$

Since $\bar{B} \cap X \neq 0$, it follows that $B \cap B_\xi \cap X \neq 0$. This contradicts the assumption that B is inconsistent with D'. ⊣

We observe that only notational difficulties prevented us from proving the above lemma with the set of formulas $\varphi_\xi(x, y_1, \ldots, y_{n_\xi})$, $\xi < \kappa$, $n_\xi < \omega$. Then we would have to deal with the functions $a_1^\xi, \ldots, a_{n_\xi}^\xi$, and the sets

$$\{\nu < \lambda : \mathfrak{A} \models (\exists x)\varphi_\xi(x, a_1^\xi(\nu), \ldots, a_{n_\xi}^\xi(\nu))\}.$$

However, nothing will be gained in the proof. So we shall assume that Lemma 6.1.14 holds in this more general form.

THEOREM 6.1.15 (Isomorphism Theorem). *Let $\mathfrak{A}$ and $\mathfrak{B}$ be models for $\mathscr{L}$. Then $\mathfrak{A}$ and $\mathfrak{B}$ are elementarily equivalent if and only if they have isomorphic ultrapowers.*

Proof. One direction of the theorem is trivial, so let us concentrate on the hard direction.

Let $\mathfrak{A}$ and $\mathfrak{B}$ be equivalent models for $\mathscr{L}$. We may assume that μ and λ are cardinals such that μ is the least cardinal such that $\lambda^\mu > \lambda$ and both $|A| < \mu$ and $|B| < \mu$. Since $2^{|A|} \leqslant \lambda^{|A|} = \lambda$, we may further assume without loss of generality that $||\mathscr{L}|| \leqslant \lambda$. We shall now construct by transfinite induction on ordinals $\rho < 2^\lambda$ an ultrafilter D over λ and an isomorphism between the ultrapowers $\prod_D \mathfrak{A}$ and $\prod_D \mathfrak{B}$. We start out with

$$|F_0| = 2^\lambda \text{ and } D_0 = \{\lambda\}, \text{ and } (F_0, 0, D_0) \text{ is } \lambda\text{-consistent.}$$

This is clearly possible by Lemmas 6.1.10 and 6.1.11(i) (as $\mu \leqslant \lambda$). We now look for a decreasing sequence of the F_ρ, $\rho < 2^\lambda$, and an increasing sequence of the D_ρ, $\rho < 2^\lambda$, satisfying, among other conditions:

(1) $$|F_0 - F_\rho| \leqslant \lambda + |\rho|, \quad \text{so} \quad |F_\rho| = 2^\lambda;$$

$$(F_\rho, 0, D_\rho) \text{ is } \lambda + |\rho|\text{-consistent;}$$

$$F_\eta = \bigcap_{\rho<\eta} F_\rho, \quad D_\eta = \bigcup_{\rho<\eta} D_\rho, \quad \text{for } \eta \text{ a limit ordinal;}$$

eventually, every subset of λ either belongs to some D_ρ or its complement belongs to some D_ρ, $\rho < 2^\lambda$, so $D = \bigcup_{\rho<2^\lambda} D_\rho$ is the required ultrafilter. We also want to construct two sequences of elements of $a_\rho : \lambda \to A$ and $b_\rho : \lambda \to B$ so that the sequences a_ρ, $\rho < 2^\lambda$ and b_ρ, $\rho < 2^\lambda$ exhaust all the elements of A^λ and B^λ, respectively, and for each $\xi < 2^\lambda$ the following conditions hold:

(2) for any formula $\varphi(x_1, \ldots, x_n)$ of $\mathscr{L}$ and any elements $a_{\rho_1}, \ldots, a_{\rho_n}$, each $\rho_i < \xi$, either

$$\{\nu < \lambda : \mathfrak{A} \models \varphi[a_{\rho_1}(\nu), \ldots, a_{\rho_n}(\nu)]\} \in D_\xi$$

or

$$\{\nu < \lambda : \mathfrak{A} \models \neg\varphi[a_{\rho_1}(\nu), \ldots, a_{\rho_n}(\nu)]\} \in D_\xi;$$

(3) for any formula $\varphi(x_1, \ldots, x_n)$ of $\mathscr{L}$ and any elements $a_{\rho_1}, \ldots, a_{\rho_n}$ and $b_{\rho_1}, \ldots, b_{\rho_n}$, each $\rho_i < \xi$, we have

$$\{\nu < \lambda : \mathfrak{A} \models \varphi[a_{\rho_1}(\nu) \ldots a_{\rho_n}(\nu)]\} \in D_\xi \quad \text{if and only if}$$

$$\{\nu < \lambda : \mathfrak{B} \models \varphi[b_{\rho_1}(\nu) \ldots b_{\rho_n}(\nu)]\} \in D_\xi.$$

Using Lemmas 6.1.13(i) and 6.1.11, it is easy to satisfy all the requirements of (1). Also, if conditions (2) and (3) hold for all $\xi < \eta < 2^\lambda$ for some limit ordinal η, then they hold automatically for η. Whence, using the by now familiar back and forth technique, we only need to satisfy (1)–(3) when $\xi = \sigma+1$, in which case we either pick a_σ arbitrarily (in order to exhaust A^λ) and find b_σ, or vice versa. Since the situation is symmetrical, let us pick a_σ to be the first element of A^λ which has not yet been put in the list $a_\rho, \rho < \sigma$. We now seek to find $F_{\sigma+1}$, $D_{\sigma+1}$ and b_σ so that (1)–(3) hold.

For each formula $\varphi(xy_1 \ldots y_n)$ of $\mathscr{L}$ and ordinals $\rho_1, \ldots, \rho_n < \sigma$, consider the set X, depending on $\varphi, \rho_1, \ldots, \rho_n$,

$$X(\varphi, \rho_1, \ldots, \rho_n) = \{\nu < \lambda : \mathfrak{A} \models \varphi[a_\sigma(\nu)a_{\rho_1}(\nu) \ldots a_{\rho_n}(\nu)]\}.$$

There are altogether $\lambda+|\sigma|$ such sets X. Since $(F_\sigma, 0, D_\sigma)$ is $\lambda+|\sigma|$-consistent, by Lemma 6.1.13(ii) we can find $F' \subset F_\sigma$, $D' \supset D_\sigma$ such that

$|F_\sigma \setminus F'| \leqslant \lambda+|\sigma|$,

$(F', 0, D')$ is $\lambda+|\sigma|$-consistent,

each $X(\varphi, \rho_1, \ldots, \rho_n)$ is in D' or its complement is in D'.

Let Γ be the set of all formulas $\varphi(xa_{\rho_1} \ldots a_{\rho_n})$ such that the corresponding set

$$X(\varphi, \rho_1, \ldots, \rho_n) \in D'.$$

Note that if $\varphi(xa_{\rho_1} \ldots a_{\rho_n}) \notin \Gamma$, then $\neg\varphi(xa_{\rho_1} \ldots a_{\rho_n}) \in \Gamma$. Furthermore, for each $\varphi(xa_{\rho_1} \ldots a_{\rho_n}) \in \Gamma$, the set Y defined below (Y depending again on $\varphi, \rho_1, \ldots, \rho_n$) belongs to D':

$$Y(\varphi, \rho_1, \ldots, \rho_n) = \{\nu < \lambda : \mathfrak{A} \models (\exists x)\varphi(xa_{\rho_1}(\nu) \ldots a_{\rho_n}(\nu))\} \in D'.$$

Note that there are at most $\lambda+|\sigma|$ such sets Y. We now claim that the set $Z(\varphi, \rho_1, \ldots, \rho_n)$ depending on $\varphi(xa_{\rho_1} \ldots a_{\rho_n}) \in \Gamma$ defined below is in D':

$$Z(\varphi, \rho_1, \ldots, \rho_n) = \{\nu < \lambda : \mathfrak{B} \models (\exists x)\varphi(xb_{\rho_1}(\nu) \ldots b_{\rho_n}(\nu))\} \in D'.$$

The reason for this is that if $Z(\varphi, \rho_1, \ldots, \rho_n) \notin D'$, then $Z(\varphi, \rho_1, \ldots, \rho_n) \notin D_\sigma$. Using (3), we see that this implies $Y(\varphi, \rho_1, \ldots, \rho_n) \notin D_\sigma$. Now, by (2), $[\lambda \setminus Y(\varphi, \rho_1, \ldots, \rho_n)] \in D_\sigma \subset D'$. This contradicts that D' is a nontrivial filter. We can now apply Lemma 6.1.14 to get a $b_\sigma : \lambda \to B$, $F_{\sigma+1} \subset F'$, $D_{\sigma+1} \supset D'$ such that $|F' \setminus F_{\sigma+1}| \leqslant \lambda+|\sigma|$, $(F_{\sigma+1}, 0, D_{\sigma+1})$ is $\lambda+|\sigma|$-consistent, and for each $\varphi(xa_{\rho_1}, \ldots a_{\rho_n}) \in \Gamma$,

$$\{\nu < \lambda : \mathfrak{B} \models \varphi[b_\sigma(\nu)b_{\rho_1}(\nu) \ldots b_{\rho_n}(\nu)]\} \in D_{\sigma+1}.$$

We now have our $F_{\sigma+1}$, $D_{\sigma+1}$ and b_σ. First of all, by our choice of D', it is clear that $D_{\sigma+1}$ satisfies (2). To show that (3) holds, we only note that if $\varphi(xa_{\rho_1}, \ldots a_{\rho_n}) \in \Gamma$, then (3) will hold for φ and $a_\sigma, a_{\rho_1}, \ldots, a_{\rho_n}$. If $\varphi(xa_{\rho_1} \ldots a_{\rho_n}) \notin \Gamma$, then $\neg\varphi(xa_{\rho_1} \ldots a_{\rho_n}) \in \Gamma$. Then (3) holds for $\neg \varphi$, $a_\sigma, a_{\rho_1}, \ldots, a_{\rho_n}$, and hence also for $\varphi, a_\sigma, a_{\rho_1}, \ldots, a_{\rho_n}$. So the induction is complete.

Let $D = \bigcup_{\rho < 2^\lambda} D_\rho$. Then D is an ultrafilter over λ. Furthermore, the mapping $a_{\rho D} \to b_{\rho D}$ is the required isomorphism of $\prod_D \mathfrak{A}$ onto $\prod_D \mathfrak{B}$. ⊣

We shall see in some of the exercises that Theorem 6.1.15 can be improved in various ways. However, no new ideas are involved in these improvements.

As a corollary we give an algebraic characterization of the notion of an elementary class.

Corollary 6.1.16. *Let K be any class of models for $\mathscr{L}$.*

(i). *K is an elementary class if and only if K is closed under ultraproducts and isomorphisms, and the complement of K is closed under ultrapowers.*

(ii). *K is a basic elementary class if and only if both K and its complement are closed under ultraproducts and isomorphisms.*

Proof. By Theorems 6.1.15 and 4.1.12. ⊣

The next corollary is a strong form of the Craig interpolation theorem.

Corollary 6.1.17 (Separation Theorem). *Let K, L be two classes of models for $\mathscr{L}$ such that $K \cap L = 0$ and both K and L are closed under isomorphisms and ultraproducts. Then there exists a basic elementary class M such that $K \subset M$ and $L \cap M = 0$.*

Proof. Let K' be the class of all models $\mathfrak{A}$ for $\mathscr{L}$ such that some $\mathfrak{B} \in K$ is elementarily equivalent to $\mathfrak{A}$. Define L' similarly. Then $K \subset K'$, $L \subset L'$ and K', L' are closed under elementary equivalence. It follows from the fundamental theorem 4.1.9 that the ultraproduct construction preserves elementary equivalence, that is, if D is an ultrafilter over I and $\mathfrak{A}_i \equiv \mathfrak{B}_i$ for all $i \in I$, then $\prod_D \mathfrak{A}_i \equiv \prod_D \mathfrak{B}_i$. Therefore the classes K', L' are also closed under ultraproducts. By Theorem 4.1.12, K' and L' are elementary classes. We claim that $K' \cap L' = 0$. For if $\mathfrak{A} \in K' \cap L'$, then there exist $\mathfrak{B} \in K$, $\mathfrak{C} \in L$ such that $\mathfrak{B} \equiv \mathfrak{A} \equiv \mathfrak{C}$, and by the isomorphism theorem there is an ultrafilter D such that $\prod_D \mathfrak{B} \cong \prod_D \mathfrak{C}$; since K, L are closed under isomorphisms and ultrapowers, this implies that $\prod_D \mathfrak{B} \in K \cap L$, contradicting $K \cap L = 0$.

Now, since $K' \cap L' = 0$ and K', L' are elementary classes, it follows from the compactness theorem that there is a basic elementary class M with $K' \subset M$, $L' \cap M = 0$. Since $K \subset K'$, $L \subset L'$, we have $K \subset M$, $L \cap M = 0$. ⊣

EXERCISES

6.1.1*. Give examples for the following:

(i). Theorem 6.1.1 is false when $\mathscr{L}$ is uncountable.

(ii). A countable model $\mathfrak{A}$ for a language $\mathscr{L}$ of power 2^ω and nonprincipal ultrafilters D, E over ω such that $\prod_D \mathfrak{A} \not\cong \prod_E \mathfrak{A}$.

(iii). Like (ii), but with $\prod_D \mathfrak{A} \not\cong \prod_D(\prod_D \mathfrak{A})$.

(iv). For every nonprincipal ultrafilter D over ω, there are countable models $\mathfrak{A}, \mathfrak{B}$ with 2^ω symbols such that $\mathfrak{A} \equiv \mathfrak{B}$ but $\prod_D \mathfrak{A} \not\cong \prod_D \mathfrak{B}$.

6.1.2. Prove that every ultrafilter is ω_1-good. (Thus Theorem 6.1.8 implies Theorem 6.1.1.)

6.1.3*. Every countably incomplete α^+-good ultrafilter is α-regular. Hence the index set has power at least α.

6.1.4. Let $|I| = \alpha$. Every set $E \subset S(I)$ such that $|E| \leqslant \alpha$, every element of E has power α, and E is closed under finite intersections can be extended to an α^+-good ultrafilter over I.

6.1.5*. There are 2^{2^α} different α^+-good ultrafilters over α.

6.1.6*. There exists an α-regular ultrafilter over α which is not ω_2-good. If $\omega_1 \leqslant \beta < \alpha$, then there exists an α-regular ultrafilter over α which is β^+-good but not β^{++}-good.

6.1.7. If D is an α-complete ultrafilter, and if $\mathfrak{A}$ is a model for a language of cardinal $||\mathscr{L}|| < \alpha$, then $\prod_D \mathfrak{A}$ is α-saturated if and only if $\mathfrak{A}$ is α-saturated. (Note that $\alpha > \omega$ because $||\mathscr{L}|| \geqslant \omega$.)

6.1.8. Let D be an α-good countably incomplete ultrafilter over a set I. Then for any family of nonempty finite sets A_i, $i \in I$, the ultraproduct $\prod_D A_i$ is either finite or has cardinality $\geqslant \alpha$.

6.1.9. Every model $\mathfrak{A}$ of power $\leqslant 2^\alpha$ has an elementary extension $\mathfrak{B}$ of power $\leqslant 2^\alpha$ such that every reduct of $\mathfrak{B}$ to a language with at most α symbols is α^+-saturated.

6.1.10*. Assume $\alpha^+ = 2^\alpha$ and let $\mathfrak{A}, \mathfrak{B}$ be models of power $\leqslant \alpha^+$ such that

$\mathfrak{A} \equiv \mathfrak{B}$. Let $||\mathscr{L}|| = \beta$. Prove that if D, E are α^+-good countably incomplete ultrafilters over α and F is any β-regular ultrafilter, then

$$\prod_F (\prod_D \mathfrak{A}) \cong \prod_F (\prod_E \mathfrak{B}).$$

[*Hint*: All reducts of $\prod_D \mathfrak{A}$, $\prod_E \mathfrak{B}$ to finite sublanguages of $\mathscr{L}$ are isomorphic.]

6.1.11*. Let μ be the least cardinal α such that $\lambda^\alpha > \lambda$. Then there exists an ultrafilter D over λ such that for any two models $\mathfrak{A}$, $\mathfrak{B}$ of power less than μ, $\mathfrak{A} \equiv \mathfrak{B}$ if and only if $\prod_D \mathfrak{A} \cong \prod_D \mathfrak{B}$.

6.1.12. Let μ and λ be as in Exercise 6.1.11. Then there exists an ultrafilter D over λ (in fact the same one as in Exercise 6.1.11) such that for every model $\mathfrak{A}$ for $\mathscr{L}$ with $|A| < \mu$ and $||\mathscr{L}|| \leqslant \lambda$, the ultrapower $\prod_D \mathfrak{A}$ is λ^+-saturated.

6.1.13*. Again let μ, λ be as in Exercise 6.1.11. Then one can find an ultrafilter D over λ with the additional property that whenever we are given two sequences of models $\mathfrak{A}_\xi$, $\xi < \lambda$, and $\mathfrak{B}_\xi$, $\xi < \lambda$ for the same $\mathscr{L}$ such that $|A_\xi| \leqslant \beta < \mu$, $|B_\xi| \leqslant \beta < \mu$ for all $\xi < \lambda$, and $\prod_D \mathfrak{A}_\xi \equiv \prod_D \mathfrak{B}_\xi$, then $\prod_D \mathfrak{A}_\xi \cong \prod_D \mathfrak{B}_\xi$.

6.1.14 (Consistency Theorem). Let K, L be two classes of models and let T_1, T_2 be the theories of K, L. Prove that $T_1 \cup T_2$ is consistent if and only if some ultraproduct of members of K is isomorphic to some ultraproduct of members of L (cf. Exercise 4.1.23).

6.1.15*. Let $\mathfrak{A}$ be an infinite simply ordered set, and D a nonprincipal ultrafilter over ω. Prove that $\prod_D \mathfrak{A}$ is not $(2^\omega)^+$-saturated. Also prove that $\prod_D \langle \omega_1, < \rangle$ is not ω_2-saturated.

6.1.16**. Let D be an α-regular ultrafilter. Prove that if $\mathfrak{A} = \langle \omega, +, \cdot, 0, 1 \rangle$ is the standard model for number theory, and $\prod_D \mathfrak{A}$ is α^+-saturated, then D is α^+-good. A similar result holds for other models, e.g., $\mathfrak{A} = \langle S_\omega(\omega), \subset \rangle$.

6.1.17*. Prove that for each ultrafilter D and each cardinal $\alpha > \omega$, (i)–(iv) below are equivalent:

(i). D is countably incomplete and α-good.

(ii). For every model $\mathfrak{A}$ for a countable language, $\prod_D \mathfrak{A}$ is α-saturated.

(iii). For each family $\mathfrak{A}_i$, $i \in I$, of models for a language with fewer than α symbols, $\prod_D \mathfrak{A}_i$ is α-universal.

(iv). Same as (iii) with α-saturated.

6.1.18. Let D be an α-regular ultrafilter and let $\mathfrak{A}$, $\mathfrak{B}$ be models for a language of power $\leqslant \alpha$ with $\mathfrak{A} \equiv \mathfrak{B}$. Then $\prod_D \mathfrak{A}$ is α^+-saturated if, and only if, $\prod_D \mathfrak{B}$ is α^+-saturated.

[*Hint*: See Exercise 4.3.36.]

6.1.19. Let D be an ω_2-good countably incomplete ultrafilter over ω_1, and let E be a countably incomplete ultrafilter over ω. Let $\mathfrak{A}$ be the standard model for number theory, $\mathfrak{A} = \langle \omega, +, \cdot, 0, 1 \rangle$. Prove that

$$\prod_D (\prod_E \mathfrak{A}) \not\cong \prod_E (\prod_D \mathfrak{A}).$$

6.1.20*. Let $\omega \leqslant \alpha < \beta$, let D be an α^+-good ultrafilter over α, let E be a β^+-good ultrafilter over β, and let D, E be countably incomplete. Let $\mathfrak{A}$ be an infinite simply ordered set, and form the double ultrapower $\mathfrak{B} = \prod_D(\prod_E \mathfrak{A})$. Prove that:

(i). $\mathfrak{B}$ is α^+-saturated;

(ii). $\mathfrak{B}$ is not α^{++}-homogeneous (hence not α^{++}-saturated);

(iii). $\mathfrak{B}$ is β^+-universal.

6.1.21*. Let $S_n(\alpha)$ denote the set of all subsets of α of power $< n$. An ultrafilter D is $S_n(\alpha)$-good iff for every monotonic function $f: S_n(\alpha) \to D$ there is an additive function $g : S_n(\alpha) \to D$ such that $g \leqslant f$. Every ultrafilter is trivially $S_0(\alpha)$, $S_1(\alpha)$, and $S_2(\alpha)$-good. Show that if $3 \leqslant n < \omega$, then D is $S_n(\alpha)$-good if and only if D is $S_3(\alpha)$-good.

6.1.22**. Let D be a countably incomplete ultrafilter. Then D is $S_3(\alpha)$-good if and only if D is α^+-good.

6.1.23. Let $\mathfrak{A}$ be a model and $\varphi(uv_1 \dots v_n)$ a formula. A set $S \subset A$ is said to be *φ-definable* iff there are $a_1, \dots, a_n$ in A such that

$$S = \{b \in A : \mathfrak{A} \models \varphi[ba_1 \dots a_n]\}.$$

By a (φ, m)-cover of $\mathfrak{A}$ we mean a collection of m φ-definable sets $S_1, \dots, S_m \subset A$ such that

$$S_1 \cup \dots \cup S_m = A,$$

but no proper subfamily of $S_1, \dots, S_m$ covers A. A model $\mathfrak{A}$ has the *finite cover property* iff there is a formula $\varphi(uv_1 \dots v_n)$ such that for arbitrarily large $m < \omega$, $\mathfrak{A}$ has a (φ, m)-cover. Show that if $\mathfrak{A} \equiv \mathfrak{B}$ and $\mathfrak{A}$ has the finite cover property, then so does $\mathfrak{B}$.

6.1.24**. Show that the following are equivalent:

(i). $\mathfrak{A}$ does not have the finite cover property.

(ii). There is a cardinal $\alpha \geqslant 2^\omega$ such that for every α-regular ultrafilter D, $\prod_D \mathfrak{A}$ is α^+-saturated.

(iii). For all α and α-regular D, $\prod_D \mathfrak{A}$ is α^+-saturated.

6.1.25*. If $||\mathscr{L}|| = \omega$, $\mathfrak{A}$ is α-saturated, D is an ultrafilter over a set of power β, and $\prod_D \mathfrak{A}$ is $(2^\beta)^+$-saturated, then $\prod_D \mathfrak{A}$ is α-saturated.

6.1.26. Prove that for every countably incomplete ultrafilter D, every bounded ultrapower of a superstructure $V(X)$ modulo D is ω_1-saturated.

6.1.27. Prove that a countably incomplete ultrafilter D is α-good if and only if for every superstructure $V(X)$, the bounded ultrapower of $V(X)$ modulo D is α-saturated.

6.2. Direct products, reduced products, and Horn sentences

This and the next section contain a thorough study of the notions listed in the above title. One of the results is a preservation theorem of the type encountered in Section 5.2, namely: A sentence φ is preserved under reduced products if and only if it is equivalent to a Horn sentence. There are a number of equally interesting results concerning direct products, for example: (i) a sentence φ is preserved under direct products of two models if and only if it is preserved under arbitrary direct products; (ii) every sentence is equivalent to a Boolean combination of sentences preserved under reduced products.

Recall from Chapter 4 that if D is a proper filter over a nonempty index set I, then the reduced product of the models $\mathfrak{A}_i$, $i \in I$, modulo D is denoted by $\prod_D \mathfrak{A}_i$. The reduced power of $\mathfrak{A}$ modulo D is denoted by $\prod_D \mathfrak{A}$. In the case where the filter D consists only of the set $\{I\}$, then we call $\prod_{\{I\}} \mathfrak{A}_i$ the *direct product* of the models $\mathfrak{A}_i$, $i \in I$. (See Exercise 4.1.12 for a more explicit definition of the direct product.) We henceforth agree to drop the subscript $\{I\}$ on $\prod$ and write $\prod_{i \in I} \mathfrak{A}_i$ for the direct product. Our understanding in this section is that I shall always be nonempty, and D shall always be a proper filter over I. If the index set I has exactly two members, $I = \{1, 2\}$, then the reduced product $\prod_D \mathfrak{A}_i$ degenerates into a model which is either isomorphic to $\mathfrak{A}_1$ or to $\mathfrak{A}_2$, or is the direct product of the two models $\mathfrak{A}_1$ and $\mathfrak{A}_2$, which we shall denote by $\mathfrak{A}_1 \times \mathfrak{A}_2$. Considered as a binary operation on models, the operation $\times$ is (up to isomorphism) commutative and associative. There is a unique (up to isomorphism) one-element model $\mathfrak{A}$ for $\mathscr{L}$ which acts like an identity for $\times$, namely the one-element model in which all relations are nonempty. For this model $\mathfrak{A}$, we have obviously $\mathfrak{B} \cong \mathfrak{B} \times \mathfrak{A} \cong \mathfrak{A} \times \mathfrak{B}$ for all $\mathfrak{B}$. We refer to $\mathfrak{A}_1 \times \mathfrak{A}_2 \times \ldots \times \mathfrak{A}_n$ as a *finite direct product* of $\mathfrak{A}_1, \ldots, \mathfrak{A}_n$, and to $\prod_{i \in I} \mathfrak{A}_i$ as an *arbitrary direct product* of models $\mathfrak{A}_i$. It will be clear from the context whether we have the finite or arbitrary direct product in mind. We keep a similar convention as regards finite or arbitrary direct powers.

Since D is a filter over I, it is a filter in the complete Boolean algebra $S(I)$, the set of all subsets of I, under the usual Boolean operations. Thus we may define the quotient Boolean algebra $S(I)/D$ in the usual way. If we let $\mathbf{1}$ denote the two-element Boolean algebra (1 here denotes the fact that this Boolean algebra has exactly one atom), then there is a natural isomorphism of $S(I)/D$ onto the reduced product $\prod_D \mathbf{1}$ (see Exercise 6.2.1). Much of the study of reduced products $\prod_D \mathfrak{A}_i$ depends on a knowledge of the theory of the Boolean algebra $\prod_D \mathbf{1}$.

At this point we need some general combinatorial lemmas on direct and reduced products. The student already knows that the operation $\times$ on models is commutative and associative (up to isomorphism). Furthermore, it is easy to verify that a direct product $\prod_{i \in I} \mathfrak{A}_i$ of direct products $\mathfrak{A}_i = \prod_{j \in J_i} \mathfrak{B}_j$ is again (isomorphic to) a direct product of models with a suitably chosen index set. Also a direct product $\prod_{i \in I} \mathfrak{A}_i$ can always be regrouped, depending on a partition $I = \bigcup_{j \in J} I_j$ of I, so that it is a direct product $\prod_{j \in J} \mathfrak{B}_j$ of direct products $\mathfrak{B}_j = \prod_{i \in I_j} \mathfrak{A}_i$. These same facts for reduced products (and also for ultraproducts) are slightly less trivial and we shall state them carefully below, leaving most of the proofs to the exercises. (This topic is treated in a more thorough manner again at the beginning of Section 6.4.) Let D and E be filters over the sets I and J, respectively. We define $D \times E$ to be the collection of all subsets Y of $I \times J$ such that

$$\{j \in J : \{i \in I : \langle ij \rangle \in Y\} \in D\} \in E.$$

The reader is asked to verify that $D \times E$ is a filter over $I \times J$. In particular, if D and E are proper filters, then so is $D \times E$. Let D be a filter over I, and let $I = I_1 \cup \ldots \cup I_n$ be a partition of I into disjoint sets I_j. Suppose that,

$$\text{for all } j,\ 1 \leqslant j \leqslant n, \text{ for all } X \in D,\ X \cap I_j \neq 0.$$

Then the sets

$$D_j = \{X \cap I_j : X \in D\} \text{ for } 1 \leqslant j \leqslant n$$

are proper filters over I_j. The following proposition contains essentially all that we need to know about combinations of reduced products.

PROPOSITION 6.2.1.

(i). *Suppose that D and E are filters over I and J. Then for all models* $\mathfrak{A}_{ij}$, $\langle ij\rangle \in I\times J$,

$$\prod_{D\times E} \mathfrak{A}_{ij} \cong \prod_{E} \big(\prod_{D} \mathfrak{A}_{ij}\big).$$

(ii). *Suppose* $I = I_1 \cup \ldots \cup I_n$, *D is a filter over I, and* D_j *are filters over* I_j *defined as above. Then for all models* $\mathfrak{A}_i$, $i \in I$,

$$\prod_{D} \mathfrak{A}_i \cong \prod_{D_1} \mathfrak{A}_i \times \prod_{D_2} \mathfrak{A}_i \times \ldots \times \prod_{D_n} \mathfrak{A}_i .$$

PROOF. We give a precise definition of the isomorphisms and leave the rest as an exercise. Let $a \in \prod_{\langle ij\rangle \in I\times J} A_{ij}$. For $j \in J$, let a^j be the restriction of a to the set $I\times\{j\}$; a^j is essentially a member of $\prod_{i\in I} A_{ij}$, whence a^j_D is in $\prod_D A_{ij}$. Let f be the function defined by

$$f(a_{D\times E}) = (\lambda j \in J(a^j_D))_E .$$

Then f is the desired isomorphism. The required isomorphism for (ii) is constructed in very much the same manner. For $1 \leqslant j \leqslant n$, let a_j be the restriction of $a \in \prod_{i\in I} A_i$ to the set I_j. Then the mapping

$$g(a_D) = \langle a^1_{D_1} \ldots a^n_{D_n}\rangle$$

will work. ⊣

A formula φ of $\mathscr{L}$ is said to be a *basic Horn formula* iff φ is a disjunction of formulas θ_i,

$$\varphi = \theta_1 \vee \ldots \vee \theta_m ,$$

where at most one of the formulas θ_i is an atomic formula, the rest being negations of atomic formulas. If $m = 1$, then φ is either an atomic or the negation of an atomic formula. If $m > 1$ and θ_m is an atomic formula, say, then φ is equivalent to a formula

$$(\psi_1 \wedge \psi_2 \wedge \ldots \wedge \psi_{m-1} \rightarrow \theta_m),$$

where each ψ_i, $1 \leqslant i < m$, is also an atomic formula. A *Horn formula* is built up from basic Horn formulas with the connectives $\wedge$, $\exists$ and $\forall$. A *Horn sentence* is a Horn formula with no free variables.

Proposition 6.2.2.

(i). *Let* $\varphi(x_1 \ldots x_n)$ *be a Horn formula and let* $\mathfrak{A}_i$, $i \in I$, *be models for* $\mathscr{L}$. *Let D be a (proper) filter over I, and let* $a^1, \ldots, a^n \in \prod_{i \in I} A_i$. *If*

$$\{i \in I : \mathfrak{A}_i \models \varphi[a^1(i) \ldots a^n(i)]\} \in D,$$

then

$$\prod_D \mathfrak{A}_i \models \varphi[a^1_D \ldots a^n_D].$$

(ii). *Every Horn sentence is preserved under reduced products.*

(iii). *Every Horn sentence is preserved under direct (finite or arbitrary) products.*

Proof. We shall sketch a proof of (i). Then (ii) and (iii) follow easily. Suppose $\varphi = \theta_1 \vee \ldots \vee \theta_m$ is a basic Horn formula. Assume first that all θ_j are negations of atomic formulas ψ_j, and

$$X = \{i \in I : \mathfrak{A}_i \models (\neg \psi_1 \vee \ldots \vee \neg \psi_m)[a^1(i) \ldots a^n(i)]\} \in D.$$

This means that for some j, $1 \leqslant j \leqslant m$, the set

$$X_j = \{i \in I : \mathfrak{A}_i \models \psi_j[a^1(i) \ldots a^n(i)]\} \notin D.$$

For otherwise $X \cap X_1 \cap \ldots \cap X_m = 0 \in D$, which is a contradiction. So, by the definition of reduced products, for some j,

$$\prod_D \mathfrak{A}_i \models \neg \psi_j[a^1_D \ldots a^n_D],$$

and hence

$$\prod_D \mathfrak{A}_i \models \varphi[a^1_D \ldots a^n_D].$$

Next suppose that $\theta_i = \neg \psi_j$, $1 \leqslant j < m$, and ψ_j and θ_m are atomic formulas. Let us disregard the trivial case $m = 1$, and assume that for $1 \leqslant j < m$

$$\prod_D \mathfrak{A}_i \models \psi_j[a^1_D \ldots a^n_D].$$

Then each

$$X_j = \{i \in I : \mathfrak{A}_i \models \psi_j[a^1(i) \ldots a^n(i)]\} \in D.$$

So $X = X_1 \cap \ldots \cap X_{m-1} \in D$. Now if

$$i \in X \cap \{i \in I : \mathfrak{A}_i \models \varphi[a^1(i) \ldots a^n(i)]\},$$

then $\mathfrak{A}_i \models \theta_m[a^1(i) \ldots a^n(i)]$. So

$$\{i \in I : \mathfrak{A}_i \models \theta_m[a^1(i) \ldots a^n(i)]\} \in D,$$

and

$$\Pi_D \mathfrak{A}_i \models \theta_m[a_D^1 \dots a_D^n].$$

Now (i) is proved for basic Horn formulas. We leave the induction steps for $\varphi_1 \wedge \varphi_2$, $(\exists x)\varphi$, and $(\forall x)\varphi$ to the reader. ⊣

We turn next to the converse of the above proposition. There are two natural preservation problems: Which sentences are preserved under reduced products, and which sentences are preserved under direct products? The following example shows that the two questions have different answers.

EXAMPLE 6.2.3. There is a Σ_2^0 sentence φ which is preserved under arbitrary direct products but not under reduced products. Hence the sentence φ is not logically equivalent to a Horn sentence.

Let φ be the conjunction of the axioms for Boolean algebras plus the sentence 'there is at least one atom', i.e.

$$(\exists x \forall y)[x \not\equiv 0 \wedge (x \cdot y \equiv y \rightarrow y \equiv x \vee y \equiv 0)].$$

Since the axioms for Boolean algebras are universal, φ is Σ_2^0. To see that φ is preserved under direct products, we note that if an element $f \in \prod_{i \in I} \mathfrak{A}_i$ is an atom at one coordinate i and zero everywhere else, then f is an atom in $\prod_{i \in I} \mathfrak{A}_i$. However, φ is not preserved under reduced products because if D is the Frechét filter of all cofinite subsets of ω, then any reduced product $\prod_D \mathfrak{A}_n$ of Boolean algebras is atomless.

Let us take up the preservation problem for reduced products.

Let I be a set of infinite power α. The phrase 'for almost all $i \in I$', shall mean 'for all but fewer than α elements $i \in I$'.

LEMMA 6.2.4. *Suppose that* $\|\mathscr{L}\| \leqslant |I| = \alpha$, $2^\alpha = \alpha^+$, *and that* $\mathfrak{A}_i$, $i \in I$, *are models for* $\mathscr{L}$ *with* $|A_i| \leqslant \alpha^+$. *Let* $\mathfrak{B}$ *be either a finite model or a saturated model of power* α^+. *If every Horn sentence holding on almost all* $\mathfrak{A}_i$ *holds on* $\mathfrak{B}$, *then* $\mathfrak{B}$ *is isomorphic to a reduced product* $\prod_D \mathfrak{A}_i$ *of the models* $\mathfrak{A}_i$ *modulo a filter* D *over* I.

PROOF. Let $A = \prod_{i \in I} A_i$. Since $2^\alpha = \alpha^+$, $|A| \leqslant \alpha^+$. We first find a mapping h of A onto B such that

(1) for all Horn formulas $\varphi(x_1 \dots x_n)$ and all $y^1, \dots, y^n \in A$, if for almost all $i \in I$, $\mathfrak{A}_i \models \varphi[y^1(i) \dots y^n(i)]$, then $\mathfrak{B} \models \varphi[hy^1 \dots hy^n]$.

Let $a \in {}^{\alpha^+}A$ and $b \in {}^{\alpha^+}B$ enumerate A and B, respectively. We shall construct by transfinite induction two sequences $\bar{a} \in {}^{\alpha^+}A$ and $\bar{b} \in {}^{\alpha^+}B$ such that for each $\nu < \alpha^+$,

(2) every Horn sentence (in $\mathscr{L} \cup \{c_\xi : \xi < \nu\}$) holding in almost all models $(\mathfrak{A}_i, \bar{a}_\xi(i))_{\xi<\nu}$ holds in $(\mathfrak{B}, \bar{b}_\xi)_{\xi<\nu}$,

and

(3) range $(a) \subset$ range $(\bar{a})$, range $(b) \subset$ range $(\bar{b})$.

Inductions of this form are by now so familiar that we shall skip the formalities and settle down on the hard parts. Suppose that (2) holds for ν. We shall take the two possibilities of $\nu = \eta+2k$ and $\nu = \eta+2k+1$ (η limit ordinal, $k < \omega$) separately.

Case 1. $\nu = \eta+2k$. Define $\bar{a}_\nu = a_{\eta+k}$. Let

$\Sigma = \{\varphi(x) : \varphi(x)$ is a Horn formula of $\mathscr{L} \cup \{c_\xi : \xi < \nu\}$ and for almost all $i \in I$, $(\mathfrak{A}_i, \bar{a}_\xi(i))_{\xi<\nu} \models \varphi[\bar{a}_\nu(i)]\}$.

It is immediate that Σ is closed under conjunction. By inductive hypothesis, because $(\exists x)\varphi$ is a Horn sentence of $\mathscr{L} \cup \{c_\xi : \xi < \nu\}$,

$$\text{for every } \varphi \in \Sigma,\ (\mathfrak{B}, \bar{b}_\xi)_{\xi<\nu} \models (\exists x)\varphi(x).$$

This shows that Σ can be extended to a type of $(\mathfrak{B}, \bar{b}_\xi)_{\xi<\nu}$. Since the latter model is α^+-saturated, some element, say $\bar{b}_\nu \in B$ is such that

$$(\mathfrak{B}, \bar{b}_\xi)_{\xi<\nu} \models \Sigma[\bar{b}_\nu].$$

Clearly (2) now holds for $\nu+1$.

Case 2. $\nu = \eta+2k+1$. Let $\bar{b}_\nu = b_{\eta+k}$. Let

$\Sigma = \{\varphi(x) : \varphi$ is a Horn formula of $\mathscr{L} \cup \{c_\xi : \xi < \nu\}$ and $(\mathfrak{B}, \bar{b}_\xi)_{\xi<\nu} \models \neg\varphi[\bar{b}_\nu]\}$.

Let $\varphi \in \Sigma$. Then $(\mathfrak{B}, \bar{b}_\xi)_{\xi<\nu} \models (\exists x)\neg\varphi$, so by predicate logic $(\mathfrak{B}, \bar{b}_\xi)_{\xi<\nu}$ does not satisfy the Horn sentence $(\forall x)\varphi$. By our inductive hypothesis (2), we see that the set I_φ defined by

$$I_\varphi = \{i \in I : (\mathfrak{A}_i, \bar{a}_\xi(i))_{\xi<\nu} \text{ does not satisfy } (\forall x)\varphi\}$$

is a subset of I of power α. Since $\|\mathscr{L}\| \leqslant \alpha$, the set Σ also has power at most α. Applying Lemma 6.1.6, we have that

(4) for each $\varphi \in \Sigma$, there is a subset $J_\varphi \subset I_\varphi$ of power α such that $J_\varphi \cap J_\psi = 0$ if $\varphi, \psi \in \Sigma$ and $\varphi \neq \psi$.

Now define $a_\nu \in \prod_{i\in I} A_i$ as follows:

if $i \in J_\varphi$, pick $a_\nu(i)$ so that $(\mathfrak{A}_i, \bar{a}_\xi(i))_{\xi<\nu} \models \neg\varphi[a_\nu(i)]$,

if $i \notin \bigcup_{\varphi\in\Sigma} J_\varphi$, pick $a_\nu(i)$ arbitrarily.

The definition of a_ν is such that, given any Horn sentence φ such that φ holds for almost all $(\mathfrak{A}_i, \bar{a}_\xi(i))_{\xi\leqslant\nu}$, then φ must hold on $(\mathfrak{B}, b_\xi)_{\xi\leqslant\nu}$, for otherwise φ would have already failed to hold on at least α models $(\mathfrak{A}_i, \bar{a}_\xi(i))_{\xi\leqslant\nu}$. So again (2) holds for $\nu+1$. After the induction is complete, clearly (3) will hold. The mapping h is defined by $h(\bar{a}_\nu) = b_\nu$, $\nu < \alpha^+$. It is well-defined because if $\bar{a}_\xi = \bar{a}_\eta$, then for all $i \in I$, $\bar{a}_\xi(i) = \bar{a}_\eta(i)$ whence, using the fact that $c_\xi \equiv c_\eta$ is a Horn sentence, we have $b_\xi = b_\eta$. So (1) is proved.

For each atomic formula $\varphi(x_1 \ldots x_n)$ of $\mathscr{L}$ and each sequence $y = (y^1 \ldots y^n)$ of elements of A, define

$$K_{\varphi,y} = \{i \in I : \mathfrak{A}_i \models \varphi[y^1(i) \ldots y^n(i)]\}.$$

If $\mathfrak{B} \models \varphi[hy^1 \ldots hy^n]$, then $|K_{\varphi,y}| = \alpha$; if otherwise, then, since $\neg\varphi$ is a Horn formula, $\mathfrak{B} \models \neg\varphi[hy^1 \ldots hy^n]$. Let E be the collection of all $K_{\varphi,y}$ such that $\mathfrak{B} \models \varphi[hy^1 \ldots hy^n]$. Since the disjunction of negations of atomic formulas is a Horn formula, the same reasoning as above will show that

(5) every finite intersection of elements of E has power α.

From (5) we see easily that E can be extended to the (proper) filter D over I such that for $X \subset I$,

$X \in D$ iff X contains a finite intersection of members of E.

The final step of the proof consists in showing that $\mathfrak{B}$ is isomorphic to $\prod_D \mathfrak{A}_i$. Let us first show that

$$\text{for all } y^1, y^2 \in A, y^1_D = y^2_D \text{ iff } hy^1 = hy^2.$$

If $y^1_D = y^2_D$, then $K = \{i \in I : y^1(i) = y^2(i)\} \in D$. So

(6) $$K \supset K_{\varphi_1,z_1} \cap \ldots \cap K_{\varphi_n,z_n}, \quad \text{for} \quad K_{\varphi_i,z_i} \in E.$$

Consider now the Horn formula (roughly speaking)

$$\psi = (\varphi_1 \wedge \ldots \wedge \varphi_n \rightarrow y^1 \equiv y^2).$$

By (6), ψ holds in all models $\mathfrak{A}_i$ under the obvious interpretations of $z_1, \ldots, z_n, y^1, y^2$. So ψ holds in $\mathfrak{B}$ under the interpretation $hz_1, \ldots, hz_n, hy^1, hy^2$. By the definition of E, we have $\mathfrak{B} \models \varphi_1[hz_1], \ldots, \mathfrak{B} \models \varphi_n[hz_n]$. So $\mathfrak{B} \models hy^1 \equiv hy^2$ and $hy^1 = hy^2$. On the other hand, if $hy^1 = hy^2$, then for the formula (roughly speaking) $\varphi(y^1y^2) = (y^1 \equiv y^2)$ there will be a $K \in E$, thus $K \in D$, such that

$$K = \{i \in I : \mathfrak{A}_i \models y^1(i) \equiv y^2(i)\}.$$

Whence $y^1_D = y^2_D$. We leave the remaining analogous arguments showing that h preserves all atomic and negations of atomic formulas to the reader. So h induces an isomorphism of $\prod_D \mathfrak{A}_i$ onto $\mathfrak{B}$ and the proposition is proved. ⊣

The preceding lemma is similar to a result from Section 6.1 on ultraproducts: Under the hypotheses of the lemma, if every sentence holding in almost all $\mathfrak{A}_i$ holds in $\mathfrak{B}$, then $\mathfrak{B}$ is isomorphic to an ultraproduct $\prod_D \mathfrak{A}_i$ of the models $\mathfrak{A}_i$ modulo an ultrafilter D over I. Recall that in Section 6.1 we proved that an ultrafilter D has the property that every ultrapower $\prod_D \mathfrak{A}$ is α-saturated if and only if D is α-good. We mention in passing that there is an analogous theorem for reduced products. Benda and Shelah (1972) proved the following.

Let $\alpha > \omega$. A filter D has the property that every reduced power $\prod_D \mathfrak{A}$ is α-saturated if and only if D is countably incomplete, α-good, and the Boolean algebra $\prod_D \mathbf{1}$ is α-saturated.

We say that φ is a *reduced product sentence* iff it is preserved under reduced products. We similarly define *reduced power sentences*, *finite direct power sentences*, and so forth. We now prove a preservation theorem for reduced products. Our first proof uses the continuum hypothesis, $2^\omega = \omega_1$. We shall then show how to eliminate the continuum hypothesis using one result which will be proved in the next section.

THEOREM 6.2.5. *Assume the continuum hypothesis, $2^\omega = \omega^+$. Then φ is a reduced product sentence if and only if φ is equivalent to a Horn sentence.*

PROOF. The fact that Horn sentences are reduced product sentences was shown in Proposition 6.2.2, and its proof does not require the continuum hypothesis. Assume that φ is a reduced product sentence. If φ is inconsistent then it is equivalent to the Horn sentence $(\forall x)(x \not\equiv x)$. So assume φ is consistent. Since we are dealing with a single sentence φ, we may assume that $\mathscr{L}$ is countable. Let

$$\Sigma = \{\psi : \psi \text{ is a Horn sentence and } \vdash \varphi \rightarrow \psi\}.$$

Σ is not empty and is closed under conjunctions. We prove that

(1) $\qquad$ for some $\psi \in \Sigma$, $\vdash \psi \rightarrow \varphi$.

By the compactness theorem, it is sufficient to prove $\Sigma \models \varphi$. Let $\mathfrak{B}$ be a model of Σ, and we may assume that either $\mathfrak{B}$ is finite or $\mathfrak{B}$ is saturated of power

ω_1. Let

$$\Psi = \{\psi : \psi \text{ is a Horn sentence and } \varphi \wedge \neg \psi \text{ is consistent}\}.$$

If $\Psi = 0$, then every Horn sentence is in Σ, and again φ is inconsistent. So $\Psi \neq 0$. For each $\psi \in \Psi$, pick a model $\mathfrak{A}_\psi$ of $\varphi \wedge \neg \psi$. Each $\mathfrak{A}_\psi$ is countable. Let $I = \omega \times \Psi$. For each $i = \langle n, \psi \rangle \in I$, let $\mathfrak{A}_i = \mathfrak{A}_\psi$. Now let ψ be an arbitrary Horn sentence and assume that ψ holds in almost all $\mathfrak{A}_i$. Then $\psi \notin \Psi$, so $\psi \in \Sigma$ and $\mathfrak{B} \models \psi$. By Lemma 6.2.4, $\mathfrak{B}$ is isomorphic to a reduced product $\prod_D \mathfrak{A}_i$. Since $\mathfrak{A}_i \models \varphi$ and φ is a reduced product sentence, we have $\mathfrak{B} \models \varphi$. So (1) is true. ⊣

We ask the reader to prove in an exercise that it is sufficient in the hypothesis of Theorem 6.2.5 to assume that $2^\alpha = \alpha^+$ holds for some infinite cardinal α.

Proposition 6.2.6. *Assume* $2^\omega = \omega_1$. *Then we have*:

(i). φ *is a reduced power sentence if and only if it is equivalent to a disjunction of Horn sentences.*

(ii). φ *is a finite direct product and a reduced power sentence if and only if it is equivalent to a Horn sentence.*

Proof. (i). Let φ be a reduced power sentence. Let

$$\Sigma = \{\psi_1 \vee \ldots \vee \psi_m : \psi_1, \ldots, \psi_m \text{ are Horn sentences and } \vdash \psi_1 \vee \ldots \vee \psi_m \rightarrow \varphi\}.$$

Σ is closed under disjunctions. Suppose

(1) for no $\psi \in \Sigma$ is $\vdash \varphi \rightarrow \psi$.

Then by compactness, $\{\varphi\} \cup \{\neg \psi : \psi \in \Sigma\}$ is consistent. Let $\mathfrak{A}$ be any countable model of $\{\varphi\} \cup \{\neg \psi : \psi \in \Sigma\}$. Note that any Horn sentence ψ which holds on $\mathfrak{A}$ must also have a model which does not satisfy φ. Since Horn sentences are closed under conjunctions, by compactness once more, we can find a model $\mathfrak{B}$ such that $\mathfrak{B} \models \neg \varphi$ and

every Horn sentence holding in $\mathfrak{A}$ holds in $\mathfrak{B}$.

We may assume that $\mathfrak{B}$ is either finite or saturated of power ω_1. Let $\mathfrak{A} = \mathfrak{A}_i$, $i \in \omega$. Then clearly every Horn sentence holding on almost all $\mathfrak{A}_i$ will hold in $\mathfrak{B}$. By Lemma 6.2.4, $\mathfrak{B}$ is isomorphic to some reduced power $\prod_D \mathfrak{A}$ of $\mathfrak{A}$. This contradicts the fact that φ is a reduced power sentence and that $\mathfrak{B} \models \neg \varphi$. So (1) is false.

(ii). Here we argue in a similar manner. Suppose that φ is a finite direct product sentence and a reduced power sentence. Then by (i), we already know that φ is a disjunction of Horn sentences. For simplicity, let $\varphi = \psi_1 \vee \psi_2$. Let

$$\Sigma = \{\psi : \psi \text{ a Horn sentence and } \vdash \varphi \rightarrow \psi\}.$$

We prove that

(2) $$\text{for some } \psi \in \Sigma, \quad \vdash \psi \rightarrow \varphi.$$

By a similar argument as in Theorem 6.2.5, for every $\mathfrak{B} \models \Sigma$ there are models $\mathfrak{A}_i, i \in I$, of φ such that $\mathfrak{B} \equiv \prod_D \mathfrak{A}_i$. We now break up the index set I into two pieces I_1 and I_2 such that

$$\text{if } i \in I_1, \quad \text{then} \quad \mathfrak{A}_i \models \psi_1,$$
$$\text{if } i \in I_2, \quad \text{then} \quad \mathfrak{A}_i \models \psi_2.$$

We also assume without loss of generality that

$$X \cap I_1 \neq 0 \text{ and } X \cap I_2 \neq 0 \text{ for all } X \in D.$$

Then, by Proposition 6.2.1, $\prod_D \mathfrak{A}_i \cong \prod_{D_1} \mathfrak{A}_i \times \prod_{D_2} \mathfrak{A}_i$. Clearly $\prod_{D_1} \mathfrak{A}_i \models \psi_1$, $\prod_{D_2} \mathfrak{A}_i \models \psi_2$, and because φ is a direct product sentence, $\prod_D \mathfrak{A}_i \models \varphi$ and hence $\mathfrak{B} \models \varphi$. So (2) holds. Notice that the proof does not claim that either $\varphi = \psi_1$ or $\varphi = \psi_2$. $\dashv$

The latter two results, Theorem 6.2.5 and Proposition 6.2.6, were first proved by Keisler (1965d) using the continuum hypothesis. Galvin (1965) found an indirect way to eliminate the continuum hypothesis from these results. Then Shelah (1972a) obtained a direct proof without the continuum hypothesis along the lines of his proof of the isomorphism theorem for ultrapowers. We shall present Galvin's indirect method of eliminating the continuum hypothesis. It takes a circuitous journey through recursive functions, Boolean algebras and the constructible universe, but offers an example of a method which has proved useful in other circumstances.

Let us assume that the language $\mathscr{L}$ is countable. We may assign suitable Gödel numbers to the symbols and expressions of $\mathscr{L}$ so that all of the usual syntactical properties and transformations will become, under the correspondence given by the Gödel numbers, recursive predicates and functions. Thus, for example, the predicates

φ is a sentence of $\mathscr{L}$;
φ is a Horn sentence of $\mathscr{L}$;
φ is an axiom of $\mathscr{L}$;
n is a proof of φ from the axioms of $\mathscr{L}$;

are all recursive predicates. As is well known, the predicate

$$\varphi \text{ is a theorem of } \mathscr{L}$$

is not recursive if $\mathscr{L}$ has at least one binary relation symbol; however, it is also well known that this predicate is a recursively enumerable predicate, i.e., a predicate obtained from a recursive predicate by prefixing a single existential (number) quantifier. A predicate $P(x)$ is said to be *arithmetical* iff it is obtained from some recursive predicate $R(xy_1 \ldots y_n)$ by prefixing a sequence of quantifiers $(Q_1 y_1 \ldots Q_n y_n)$(Q_i existential or universal) such that for all x,

$$P(x) \text{ iff } (Q_1 y_1 \ldots Q_n y_n) R(xy_1 \ldots y_n).$$

An *arithmetical statement* is a zero-placed arithmetical predicate. Let ZF denote Zermelo–Fraenkel set theory (see Appendix), and let ZFL denote ZF plus the axiom of constructibility (for an esoteric form of this axiom see Chapter 4; for the more usual form see Section 7.4). The following are implicit in Gödel (1940):

(1) Let Γ be an arithmetical statement. Then Γ is provable in ZF if and only if Γ is provable in ZFL; or, in symbols, $\vdash_{ZF} \Gamma$ iff $\vdash_{ZFL} \Gamma$.

(2) In ZFL we can prove the axiom of choice and the generalized continuum hypothesis (GCH).

From (1) and (2), it is immediate that:

(3) Let Γ be an arithmetical statement; if Γ is provable in ZF with the axiom of choice and the GCH, then Γ is provable in ZF.

Let us now consider the statement:

(4) For all sentences φ, φ is a reduced product sentence iff φ is equivalent to a Horn sentence.

There is no question that a part of (4), namely, the predicate

$$\varphi \text{ is equivalent to a Horn sentence,}$$

is an arithmetical predicate.

We shall prove in the next section that the other part of (4) is also an arithmetical predicate. That is, we have the following:

LEMMA 6.2.7. *The predicate*

$$\varphi \textit{ is a reduced product sentence}$$

is an arithmetical predicate.

Using this lemma we see that (4) is an arithmetical statement. Theorem 6.2.5 states that (4) can be proved in ZFC with the GCH. We conclude by (3) that the theorem can be proved in ZFC alone. So the one fact needed to eliminate the continuum hypothesis is Lemma 6.2.7 above.

We now consider the preservation problem for direct products. A syntactical characterization of direct product sentences was given by Weinstein (1965). Since his characterization is very complicated, we shall not give it here. Instead we concentrate on the simple cases of universal and existential direct product sentences.

PROPOSITION 6.2.8. *Let φ be a universal sentence. Then φ is a (finite) direct product sentence if and only if φ is equivalent to a universal Horn sentence.*

PROOF. We shall illustrate the method of proof by an example. Suppose that φ is a direct product sentence of the following form:

$$\varphi = (\forall x)[(\neg P_1(x) \vee P_2(x) \vee P_3(x)) \wedge (\neg Q_1(x) \vee Q_2(x) \vee Q_3(x))].$$

We have assumed for simplicity that x is the only free variable occurring in the atomic formulas $P_1, P_2, P_3, Q_1, Q_2, Q_3$. We assert that φ is equivalent to one of the four universal Horn sentences below:

$$\varphi_1 = (\forall x)((P_1 \to P_2) \wedge (Q_1 \to Q_2)),$$
$$\varphi_2 = (\forall x)((P_1 \to P_3) \wedge (Q_1 \to Q_2)),$$
$$\varphi_3 = (\forall x)((P_1 \to P_2) \wedge (Q_1 \to Q_3)),$$
$$\varphi_4 = (\forall x)((P_1 \to P_3) \wedge (Q_1 \to Q_3)).$$

Clearly $\models \varphi_i \to \varphi$ for $1 \leqslant i \leqslant 4$. Suppose that not $\models \varphi \to \varphi_i$ for $1 \leqslant i \leqslant 4$. Then there are four models $\mathfrak{A}_i$, $1 \leqslant i \leqslant 4$, of φ, and four elements a_i, $1 \leqslant i \leqslant 4$, $a_i \in A_i$, such that

$$\mathfrak{A}_1 \models (P_1(a_1) \wedge \neg P_2(a_1)) \vee (Q_1(a_1) \wedge \neg Q_2(a_1)), \text{ etc.}$$

A simple combinatorial argument will show that some direct product $\mathfrak{A}_i \times \mathfrak{A}_j$ of these four models will contain an element $\langle a_i a_j \rangle$ which will not satisfy the matrix of φ. This contradicts $\mathfrak{A}_i \times \mathfrak{A}_j \models \varphi$. ⊣

PROPOSITION 6.2.9. *Let φ be an existential sentence. Then φ is a (finite) direct product sentence if φ is equivalent to an existential Horn sentence.*

PROOF. We here give a proof assuming $2^\omega = \omega_1$. This assumption can be eliminated using 6.3.7. Assume φ is an existential sentence preserved under $\mathfrak{A} \times \mathfrak{B}$. Note that the natural embedding maps $\mathfrak{A}$ isomorphically into $\prod_D \mathfrak{A}$. Since φ is preserved under extensions, this remark shows that φ is a reduced power sentence. Hence, by 6.2.6, φ is a reduced product sentence. Let

$$\Sigma = \{\psi : \psi \text{ is an existential Horn sentence and } \vdash \varphi \rightarrow \psi\}.$$

Σ is closed under conjunctions. If

(1) $$\text{for no } \psi \in \Sigma, \text{ is } \vdash \psi \rightarrow \varphi,$$

then by compactness find a model $\mathfrak{B}$ of $\{\neg\varphi\} \cup \Sigma$. For each existential Horn sentence ψ such that $\varphi \wedge \neg\psi$ is consistent, find a countable model $\mathfrak{A}_\psi$ of $\varphi \wedge \neg\psi$. By the technique of Theorem 6.2.5, we have a system of models $\mathfrak{A}_i$, $i \in I$, with I countable and such that

every existential Horn sentence holding on almost all $\mathfrak{A}_i$ holds on $\mathfrak{B}$.

Assume now $\mathfrak{B}$ is at least 2^ω-saturated. This is always possible, regardless of whether $\mathfrak{B}$ is finite or infinite, and it does not use any form of the continuum hypothesis. Inspecting the argument of Lemma 6.2.4, we see that a reduced product $\prod_D \mathfrak{A}_i$ of the models $\mathfrak{A}_i$ is isomorphically embeddable into $\mathfrak{B}$. Since each $\mathfrak{A}_i \models \varphi$, we have $\prod_D \mathfrak{A}_i \models \varphi$. Since φ is existential, $\mathfrak{B} \models \varphi$. This contradicts $\mathfrak{B} \models \neg\varphi$. So (1) fails. ⊣

Weinstein (1965) has shown that a universal–existential sentence is a (finite) direct product sentence if and only if it is logically equivalent to a universal–existential Horn sentence. Example 6.2.3 shows that we cannot go further, since it gives an existential–universal direct product sentence which is not equivalent to a Horn sentence.

EXERCISES

6.2.1. Let D be a filter over I. Show that the relation

$$X \equiv_D Y \text{ iff } X \cap Z = Y \cap Z \text{ and } \bar{X} \cap Z = \bar{Y} \cap Z \text{ for some } Z \in D$$

is a congruence relation on the field of sets $S(I)$. Thus we can define in the natural way the quotient Boolean algebra $S(I)/D$, with the elements of the Boolean algebra being equivalence classes X/D for $X \subset I$. If we identify a subset $X \subset I$ with its characteristic function f_X ($f_X(i) = 1$ if $i \in X$, $f_X(i) = 0$ if $i \notin X$), then there is a natural isomorphism of $S(I)/D$ onto the reduced power $\prod_D \mathbf{1}$, where $\mathbf{1}$ is the 2-element Boolean algebra.

6.2.2. Show that any reduced product $\prod_D \mathfrak{A}_i$ over a finite index set I is isomorphic to a direct product $\prod_{i\in J} \mathfrak{A}_i$ over a subset J of I. This is true more generally whenever D is a principal filter over I.

6.2.3. Let D_j be filters over the sets I_j, and let D be a filter over J. Then there is a filter E over the set

$$K = \bigcup_{j\in J} (I_j \times \{j\}),$$

so that

$$\prod_E \mathfrak{A}_{ij} \cong \prod_D \prod_{D_j} \mathfrak{A}_{ij}.$$

6.2.4*. Using Section 6.1, prove Theorem 6.2.5 directly without the CH.

6.2.5. Prove an analogue of Theorem 6.2.5 for elementary classes: that is, an elementary class K is closed under reduced products if and only if K is the class of all models of some set of Horn sentences.

6.2.6*. Use Lemma 6.2.4 to prove the following separation theorem. Let K_1 and K_2 be arbitrary classes of models, each closed under reduced products and isomorphisms. Suppose that $K_1 \cap K_2 = 0$. Then there are two elementary classes L_1 and L_2 each determined by a set of Horn sentences such that

$$K_1 \subset L_1, \qquad K_2 \subset L_2, \qquad L_1 \cap L_2 = 0.$$

6.2.7. Let T be a theory in $\mathscr{L}$. Then the following are equivalent, using $2^\omega = \omega_1$:

(i). Every formula is equivalent under T to a Horn formula.

(ii). Every reduced product $\prod_D \mathfrak{A}_i$ of models $\mathfrak{A}_i$, $i \in I$, of T which is itself a model of T is an ultraproduct of models of T; in fact, D will be an ultrafilter over I.

Examples of theories T which satisfy (ii) are: theory of simple order, theory of fields, theory of division rings. (See Exercise 4.1.30.)

6.2.8. We say that a basic Horn formula $\varphi = \theta_1 \vee \ldots \vee \theta_m$ is a *strict basic Horn formuls* iff some θ_i is an atomic formula. A *strict Horn formula* is built up from strict basic Horn formulas by using $\wedge$, $\exists$ and $\forall$. Prove the following using $2^\omega = \omega_1$. A sentence φ is preserved under reduced products $\prod_D \mathfrak{A}_i$, whether or not D is an improper filter over I, iff φ is equivalent to a strict Horn sentence.

6.2.9. A disjunction of atomic and negations of atomic formulas $\theta_1 \vee \ldots \vee \theta_m$ is said to be *weakly Horn* iff at most two of the terms θ_i, θ_j are atomic formulas, the rest being negations of atomic formulas. The closure of these formulas under $\wedge$, $\exists$, $\forall$ is called the set of *weak Horn* sentences. Show by the technique of Lemma 6.2.4 ($2^{\omega} = \omega_1$), that if every weak Horn sentence holding on all models of a theory T holds on $\mathfrak{B}$, then $\mathfrak{B}$ is a model of T. Whence every sentence is equivalent to a weak Horn sentence, a result which also has a proof without the continuum hypothesis.

[*Hint*: Show that the filter D constructed in Lemma 6.2.4 so that $\mathfrak{B} \cong \prod_D \mathfrak{A}$ is actually an ultrafilter.]

6.2.10*. φ is said to be a *special Horn sentence* iff φ is a conjunction of sentences of the form

$$(\forall x_1 \ldots x_n)(\psi \to \theta),$$

where θ is an atomic formula, and ψ is a positive formula. Prove that φ is preserved under subdirect products if and only if φ is equivalent to a special Horn sentence. [A submodel $\mathfrak{B}$ of a direct product $\prod_{i \in I} \mathfrak{A}_i$ of models of φ is a *subdirect product of models* $\mathfrak{A}_i$ iff for all $i \in I$, $\{b(i) : b \in B\} = A_i$.] Also, an example of a special Horn sentence is the sentence that expresses the condition: a group has a trivial center, i.e.,

$$(\forall x)[(\forall y)x + y \equiv y + x \to x \equiv 0].$$

Find an example of a Horn sentence which is not equivalent to any special Horn sentence.

6.2.11. A sentence φ is *preserved by reduced factors* iff whenever $\prod_D \mathfrak{A}_i \models \varphi$, then $\{i : \mathfrak{A}_i \models \varphi\} \in D$. Prove that every positive sentence is preserved by reduced factors. Also prove directly without using Lyndon's theorem that every sentence preserved under homomorphic images is preserved by reduced factors. Find a reduced factor sentence which is not equivalent to any positive sentence.

6.2.12. Prove that reduced products lead to an easy proof of the following compactness theorem for Horn sentences. Let Σ be a set of Horn sentences. If every finite subset of Σ has a model, then Σ has a model.

6.2.13*. An early result in the subject of universal algebras is the following theorem of Birkhoff. Let K be an arbitrary class of algebras. Then for K to be an equational class, i.e., a class of models of some set of equations, it is necessary and sufficient that K be closed under subalgebras, homomorphisms and arbitrary direct products. Using the results of Sections 6.1 and 6.2, give a proof of this theorem.

[*Hint*: First show that K is closed under ultraproducts and the complement of K is closed under ultrapowers. Thus K is an elementary class. Since K is closed under homomorphisms, K is determined by a set of positive sentences. Since K is closed under subalgebras, K is determined by a set of universal positive sentences. Since K is closed under direct products, K is determined by a set of universal positive Horn sentences. Such sentences are equations.]

6.2.14. By examining the proofs of Lemma 6.2.4 and Theorem 6.2.5, deduce that a sentence φ is preserved under countably indexed reduced products iff it is preserved under arbitrary reduced products. Do the same for reduced powers.

6.2.15*. The definition of α-good makes sense for filters as well as ultrafilters. Suppose D is a filter over α and D contains an α^+-good countable incomplete filter. Assume the GCH. Show that, if $||\mathscr{L}|| \leqslant \alpha$, $\mathfrak{A}$ and $\mathfrak{B}$ are models of power at most α^+, and $\mathfrak{A} \equiv \mathfrak{B}$, then $\prod_D \mathfrak{A} \cong \prod_D \mathfrak{B}$.

6.3. Direct products, reduced products, and Horn sentences (continued)

This section is a continuation of Section 6.2. Consider an arbitrary reduced product $\prod_D \mathfrak{A}_i$. We ask the following question: given the models $\mathfrak{A}_i$, $i \in I$, and the filter D, how can we determine which sentences are true in the reduced product $\prod_D \mathfrak{A}_i$? We shall see that the truth value of a formula φ in $\prod_D \mathfrak{A}_i$ is in a sense determined by the truth values of certain other formulas in the models $\mathfrak{A}_i$ and in the Boolean algebra $S(I)/D$. This fact will be developed to give a great deal of information about direct and reduced product sentences. We begin with a key definition.

6.3.1. A formula $\varphi(x_1 \dots x_n)$ of $\mathscr{L}$ is *determined* by the sequence of formulas $(\sigma; \psi_1, \dots, \psi_m)$ iff:

(i). Each ψ_j is a formula of $\mathscr{L}$ with at most the variables $x_1, \dots, x_n$ free.

(ii). $\sigma(y_1 \dots y_m)$ is a formula in the language of Boolean algebras which is monotonic, i.e.

$$T_{\mathrm{BA}} \vdash (\forall z_1 \dots z_m y_1 \dots y_m)(\sigma(y_1 \dots y_m) \wedge \bigwedge_{1 \leqslant j \leqslant m} y_j \leqslant z_j \rightarrow \sigma(z_1 \dots z_m)).$$

(Here, and hereafter, T_{BA} shall denote the theory of Boolean algebras.)

(iii). Let I be an arbitrary index set, D a filter over I, and $\mathfrak{A}_i$, $i \in I$, be models of $\mathscr{L}$. Let $a^1, \dots, a^n \in \prod_{i \in I} A_i$, and for each j, $1 \leqslant j \leqslant m$, let

$$X^j = \{i \in I : \mathfrak{A}_i \models \psi_j[a^1(i) \dots a^n(i)]\}.$$

Then

$$\prod_D \mathfrak{A}_i \models \varphi[a_D^1 \dots a_D^n] \quad \text{iff} \quad S(I)/D \models \sigma[X^1/D \dots X^m/D].$$

We say that φ is *determined* iff it is determined by some sequence.

PROPOSITION 6.3.2. *Every formula φ is determined.*

PROOF. This is proved by induction on the formula φ. Suppose that $\varphi(x_1 \dots x_n)$ is an atomic formula. We show that φ is determined by $(\sigma : \varphi)$, where $\sigma(y)$ is the formula $y \equiv 1$ (in the language of T_{BA}). Clearly 6.3.1 (i) and (ii) hold. Assume the hypothesis of 6.3.1(iii). Then by the definition of reduced products, the following are equivalent:

$$\prod_D \mathfrak{A}_i \models \varphi[a_D^1 \dots a_D^n],$$

$$X = \{i \in I : \mathfrak{A}_i \models \varphi[a^1(i) \dots a^n(i)]\} \in D,$$

$$X/D = 1 \text{ in } S(I)/D,$$

$$S(I)/D \models \sigma[X/D].$$

Suppose that φ is determined by the sequence $(\sigma; \psi_1 \dots \psi_m)$. Say $\sigma = \sigma(y_1 \dots y_m)$, with at most $y_1, \dots, y_m$ free. Let $\tau = \neg\sigma(-y_1 \dots -y_m)$. ($-$ is the Boolean complement.) Then $\neg\varphi$ is determined by $(\tau; \neg\psi_1 \dots \neg\psi_m)$. Clearly, 6.3.1 (i) and (ii) hold. For $1 \leqslant j \leqslant m$, let

$$X_j = \{i \in I : \mathfrak{A}_i \models \neg\psi_j[a^1(i) \dots a^n(i)]\}.$$

Then the following are equivalent:

$$\prod_D \mathfrak{A}_i \models \neg\varphi[a_D^1 \dots a_D^n],$$

$$\text{not } \prod_D \mathfrak{A}_i \models \varphi[a_D^1 \dots a_D^n],$$

$$\text{not } S(I)/D \models \sigma[(I \setminus X_1)/D \dots (I \setminus X_m)/D],$$

$$S(I)/D \models \neg\sigma[-(X_1/D) \dots -(X_m/D)],$$

$$S(I)/D \models \tau[X_1/D \dots X_m/D].$$

Suppose that φ is determined by the sequence $(\sigma; \psi_1 \dots \psi_m)$, and χ is determined by the sequence $(\tau; \theta_1 \dots \theta_l)$. Let $y_1, \dots, y_m$ and $z_1, \dots, z_l$ be distinct vaiiables and let

$$\rho(y_1 \dots y_m z_1 \dots z_l) = \sigma(y_1 \dots y_m) \wedge \tau(z_1 \dots z_l).$$

We may assume that $\{x_1, \ldots, x_n\}$ contains all the free variables in φ or χ. Then $\varphi \wedge \chi$ is determined by the sequence $(\rho; \psi_1 \ldots \psi_m \theta_1 \ldots \theta_l)$. Again 6.3.1 (i) and (ii) hold trivially. Assume the hypothesis of 6.3.1(iii), with the sets X^j, $1 \leqslant j \leqslant m$, and Y^k, $1 \leqslant k \leqslant l$. Then the following are equivalent:

$$\prod_D \mathfrak{A}_i \models \varphi \wedge \chi[a_D^1 \ldots a_D^n],$$

$$\prod_D \mathfrak{A}_i \models \varphi[a_D^1 \ldots a_D^n] \quad \text{and} \quad \prod_D \mathfrak{A}_i \models \chi[a_D^1 \ldots a_D^n],$$

$$S(I)/D \models \sigma[X^1/D \ldots X^m/D] \quad \text{and} \quad S(I)/D \models \tau[Y^1/D \ldots Y^l/D],$$

$$S(I)/D \models \rho[X^1/D \ldots X^m/D Y^1/D \ldots Y^l/D].$$

Finally, suppose that $\varphi(xx_1 \ldots x_n)$ is determined by the sequence $(\sigma; \psi_1 \ldots \psi_m)$. Let $l = 2^m$ and let $s_1, \ldots, s_l$ be a listing of all subsets of $\{1, \ldots, m\}$ with $s_k = \{k\}$ for $1 \leqslant k \leqslant m$. For $1 \leqslant k \leqslant l$, let $\theta_k = \exists x \bigwedge_{j \in s_k} \psi_j$. (The empty conjunction shall be some true formula.) We see that $\theta_k = \exists x \psi_k$ for $1 \leqslant k \leqslant m$. Let

$$\tau(y_1 \ldots y_l) = (\exists z_1 \ldots z_l)[\bigwedge_{1 \leqslant k \leqslant l} (z_k \leqslant y_k) \wedge \bigwedge_{s_i \cup s_j = s_k} (z_i \cdot z_j \equiv z_k) \wedge \sigma(z_1 \ldots z_m)].$$

We show that $(\exists x)\varphi$ is determined by $(\tau; \theta_1 \ldots \theta_l)$. 6.3.1(i) holds trivially, and 6.3.1(ii) follows simply because of the syntactical form of τ. Let us first suppose that $\prod_D \mathfrak{A}_i \models (\exists x)\varphi[a_D^1 \ldots a_D^n]$. Then there exists an $a \in \prod_{i \in I} A_i$ such that $\prod_D \mathfrak{A}_i \models \varphi[a_D a_D^1 \ldots a_D^n]$. For $1 \leqslant k \leqslant l$, let

$$Y^k = \{i \in I : \mathfrak{A}_i \models \theta_k[a^1(i) \ldots a^n(i)]\}$$

and let

$$Z^k = \{i \in I : \mathfrak{A}_i \models \bigwedge_{j \in s_k} \psi_j[a(i)a^1(i) \ldots a^n(i)]\}.$$

It is clear that $Z^k \subset Y^k$

$$\bigwedge_{s_i \cup s_j = s_k} (Z^i \cap Z^j = Z^k),$$

and

$$S(I)/D \models \sigma[Z^1/D \ldots Z^m/D].$$

Whence $S(I)/D \models \tau[Y^1/D \ldots Y^l/D]$. On the other hand, let Y^k, $1 \leqslant k \leqslant l$, be defined as above, and suppose that there are sets Z^k, $1 \leqslant k \leqslant l$, such that for all k, $1 \leqslant k \leqslant l$,

$$S(I)/D \models Z^k/D \leqslant Y^k/D,$$

$$(1) \qquad S(I)/D \models \bigwedge_{s_i \cup s_j = s_k} (Z^i/D \cdot Z^j/D \equiv Z^k/D),$$

$$S(I)/D \models \sigma[Z^1/D \ldots Z^m/D].$$

Since the number of conditions in (1) is finite, we can find a set $X \in D$ such that

$$(2) \qquad \begin{aligned} &Z^k \cap X \subset Y^k && \text{for} \quad 1 \leqslant k \leqslant l, \\ &Z^i \cap Z^j \cap X = Z^k \cap X && \text{for} \quad s_i \cup s_j = s_k. \end{aligned}$$

For each $i \in X$, let $s = \{j : i \in Z^j, 1 \leqslant j \leqslant m\}$. Then because of (2) and the definition of Y^k, we see that

$$\mathfrak{A}_i \models (\exists x) \bigwedge_{j \in s} \psi_j[a^1(i) \dots a^n(i)].$$

So pick an $a(i) \in A_i$ such that

$$\mathfrak{A}_i \models \bigwedge_{j \in s} \psi_j[a(i)a^1(i) \dots a^n(i)].$$

Define $a(i) \in A_i$ arbitrarily if $i \notin X$. For $1 \leqslant j \leqslant m$, let

$$W^j = \{i \in I : \mathfrak{A}_i \models \psi_j[a(i)a^1(i) \dots a^n(i)]\}.$$

Note that by our choice of a,

$$Z^j \cap X \subset W^j, \qquad 1 \leqslant j \leqslant m.$$

Whence $Z^j/D \leqslant W^j/D$. Since σ satisfies 6.3.1 (ii), we have from (1),

$$S(I)/D \models \sigma[W^1/D \dots W^m/D].$$

But, since φ is determined by $(\sigma; \psi_1 \dots \psi_m)$, this implies

$$\prod_D \mathfrak{A}_i \models \varphi[a_D a_D^1 \dots a_D^n],$$

so

$$\prod_D \mathfrak{A}_i \models (\exists x)\varphi[a_D^1 \dots a_D^n]. \dashv$$

Inspecting the proof of Proposition 6.3.2, we see that the following additional remarks can be made. If a sentence or formula φ is given explicitly, then by the procedure described in Proposition 6.3.2, we can find explicit formulas $\sigma, \psi_1, \dots, \psi_m$ such that φ is determined by $(\sigma; \psi_1 \dots \psi_m)$. In other words, if we knew the precise shape of φ as a concatenation of symbols of $\mathscr{L}$, then we can exhibit specific formulas $\sigma, \psi_1, \dots, \psi_m$ which determine φ. We call such a process *effective*. Another way of putting it is that a process is effective if and only if a machine can be programmed so that when we feed in the expression for φ, it will automatically and mechanically give us the sequence $(\sigma; \psi_1 \dots \psi_m)$ of formulas with each symbol printed on a piece of paper. Such processes are also called *recursive*; however, since this is a book on model theory, we shall not go into recursion theory here. So whatever we shall need later about recursion theory, we shall assume it outright and ask the interested students to look up the matter in some references. There should, however, be no question about what we mean when we say that something is effective. The following proposition summarizes these remarks.

PROPOSITION 6.3.3.

(i). *Given a formula φ, there is an effective way of finding a sequence* $(\sigma; \psi_1 \ldots \psi_m)$ *which determines* φ.

(ii). *If* φ *is a* $\mathbf{\Sigma}_n^0$ *(resp.* $\mathbf{\Pi}_n^0$*) formula then there is an effective way of finding a sequence* $(\sigma; \psi_1 \ldots \psi_m)$ *which determines* φ *and such that each* ψ_j *is a* $\mathbf{\Sigma}_n^0$ *(resp.* $\mathbf{\Pi}_n^0$*) formula.*

(iii). *If* φ *is a* $\mathbf{\Sigma}_n^0$ *(resp.* $\mathbf{\Pi}_n^0$*) sentence, then each of the* ψ_j *in the determining sequence is a* $\mathbf{\Sigma}_n^0$ *(resp.* $\mathbf{\Pi}_n^0$*) sentence.*

The proof of this proposition is straightforward. We notice that the notion of determining sequences is closed under logical equivalence. That is, if φ is determined by $(\sigma; \psi_1 \ldots \psi_m)$ and if the formulas $\varphi', \sigma', \psi'_1, \ldots, \psi'_m$ are term-by-term equivalent to the formulas $\varphi, \sigma, \psi_1, \ldots, \psi_m$, then φ' is determined by the sequence $(\sigma'; \psi'_1, \ldots, \psi'_m)$. This remark is predicated on the assumption that the formulas $\varphi', \sigma', \psi'_1, \ldots, \psi'_m$ have the correct number of free variables as required in 6.3.1.

An immediate application is the following:

THEOREM 6.3.4. *Reduced products, reduced powers, direct products and direct powers all preserve elementary equivalence.*

PROOF. Suppose that $\mathfrak{A}_i \equiv \mathfrak{B}_i$ for $i \in I$. Consider the reduced products $\mathfrak{A} = \prod_D \mathfrak{A}_i$ and $\mathfrak{B} = \prod_D \mathfrak{B}_i$. Let φ be any sentence of $\mathscr{L}$ and let $(\sigma; \psi_1 \ldots \psi_m)$ be a determining sequence of φ. Each ψ_j is a sentence of $\mathscr{L}$. By our assumption, for $1 \leqslant j \leqslant m$,

$$\{i \in I : \mathfrak{A}_i \models \psi_j\} = X^j = \{i \in I : \mathfrak{B}_i \models \psi_j\}.$$

From this it follows that

$$\mathfrak{A} \models \varphi \quad \text{iff} \quad S(I)/D \models \sigma[X^1/D \ldots X^m/D] \quad \text{iff} \quad \mathfrak{B} \models \varphi.$$

Since reduced powers, direct products and direct powers are all special cases of reduced products, the proposition is proved. ⊣

Let S be a finite set of formulas, and let

$\neg S$ be the set of negations of formulas in S;
$\bigwedge S$ be the set of all finite conjunctions of formulas in S;
$(\exists x)S$ be the set of all formulas $(\exists x)\varphi$ with $\varphi \in S$;
$B(S)$ be the set of all Boolean combinations of formulas in S.

Since in what follows we are only interested in formulas up to logical equivalence, we may assume that the sets $\bigwedge S$ and $B(S)$ are also finite. In fact, in our discussions hereafter we shall frequently identify a formula φ with the equivalence class of all formulas ψ logically equivalent to φ. We shall try to be careful in checking that a representative φ can always be chosen so that it has the required (small) number of free variables, or that it is a formula of a particular form Σ_n^0 or Π_n^0. Some words about the notion of effectiveness when applied to sets of the form $\bigwedge S$ or $B(S)$. In the first case, we can take for $\bigwedge S$ all finite conjuncts of members of S up to $|S|$ of such members. Then we can effectively find, given φ, ψ in $\bigwedge S$, an element θ in $\bigwedge S$ such that θ is equivalent to $\varphi \wedge \psi$. In the second case, we let $B(S)$ be the set of all disjunctions (of length $\leqslant 2^{|S|}$) of conjunctions of the form

$$\bigwedge_{1 \leqslant i \leqslant |S|} \varepsilon_i \varphi_i,$$

where $S = \{\varphi_i : 1 \leqslant i \leqslant |S|\}$, and each ε_i is either the empty sequence or the symbol $\neg$. Clearly, $S \subset B(S)$, because $\bigvee \bigwedge \varphi$ is the same as φ. Furthermore, given φ, ψ in $B(S)$, we can effectively find $\theta, \rho, \chi \in B(S)$ such that $\vdash \theta \leftrightarrow \neg \varphi$, $\vdash \rho \leftrightarrow \varphi \vee \psi$, $\vdash \chi \leftrightarrow \varphi \wedge \psi$. Also, we can effectively find an inconsistent formula and a valid formula in $B(S)$. In passing from S to $B(S)$, we see that we do not increase the number of free variables in the formulas, and also if each $\varphi \in S$ is equivalent to a Boolean combination of Σ_n^0 formulas, then each $\psi \in B(S)$ is equivalent to a Σ_{n+1}^0 and also to a Π_{n+1}^0 formula. When we write the expression $S = B(S)$, we shall mean that S is closed under all Boolean operations, and that, given φ, ψ in S, we can effectively find formulas in S equivalent to an inconsistent formula, a valid formula, $\neg \varphi$, $\varphi \vee \psi$ and $\varphi \wedge \psi$.

A set S of formulas is said to be *self-determining* iff S is finite and for each $\varphi \in S$, there is a sequence $(\sigma; \psi_1 \ldots \psi_m)$ which determines φ with all $\psi_j \in S$. Since the property of being a determining sequence for φ cannot be decided effectively, we say that we can *effectively find* a self-determining set S iff S is explicitly given, and for each $\varphi \in S$, the determining sequence $(\sigma; \psi_1 \ldots \psi_m)$ is also explicitly given.

PROPOSITION 6.3.5. *Let S and T be self-determining sets. Then each of the following sets is self-determining*:

$$S \cup T, \quad \neg S, \quad \bigwedge S, \quad (\exists x) \bigwedge S, \quad B(S).$$

PROOF. The fact that $S \cup T$ is self-determining follows from the definition. The sets $\neg S$, $\bigwedge S$, $\exists x \bigwedge S$ are self-determining by examining the proof of Proposition 6.3.2 and the remarks preceding this proposition. By combining the argument for $\neg S$ and for $\bigwedge S$, and by examining the specific formulas in $B(S)$, we see that $B(S)$ is self-determining. ⊣

REMARK. In Proposition 6.3.5, if S and T are explicitly given, as well as the correspondence between φ and its determining sequence $(\sigma; \psi_1 \ldots \psi_m)$, then we can effectively find the sets in the conclusion.

A set S of formulas is *autonomous* iff S is self-determining and $S = B(S)$, i.e., the Boolean operations on S are closed and effective. The following is the fundamental theorem of autonomous sets. We state it in several parts for emphasis.

THEOREM 6.3.6.

(i). *Given a formula φ of $\mathscr{L}$, we can effectively find an autonomous set S_φ to which φ belongs.*

(ii). *Given a finite set Φ of formulas of $\mathscr{L}$, we can effectively find an autonomous set S_Φ which extends Φ.*

(iii). *If φ is equivalent to a Boolean combination of Σ_n^0 formulas (resp. Σ_n^0 sentences), then each member of S_φ is equivalent to a Boolean combination of Σ_n^0 formulas (resp. Σ_n^0 sentences) which has at most the free variables that are in φ.*

(iv). *This part extends* (ii) *in the same way that* (iii) *extends* (i).

PROOF. The proof depends heavily on the preceding results. We sketch a proof of part (i) and the rest will follow by a careful inspection. It is sufficient to prove that every formula φ belongs to a self-determining set S, for then $B(S)$ is autonomous and $\varphi \in B(S)$. Clearly, if φ is an atomic formula, the set $\{\varphi\}$ is self-determining. If φ belongs to a self-determining set S, then $\neg \varphi$ belongs to the self-determining set $\neg S$. If φ and ψ belong, respectively, to self-determining sets S and T, then $\varphi \wedge \psi$ belongs to the self-determining set $\bigwedge (S \cup T)$. If φ belongs to a self-determining set S, then $(\exists x)\varphi$ belongs to the self-determining set $(\exists x) \bigwedge S$. ⊣

We point out here a fact which we shall need later: If φ belongs to an autonomous set S, then $(\exists x)\varphi$ belongs to the autonomous set $B((\exists x)S)$.

Theorem 6.3.4 motivates the following series of definitions. Let S be an autonomous set of sentences. S clearly gives rise to a finite Boolean algebra under the logical operations of $\neg, \wedge, \vee, 0, 1$, where 0 and 1 are identified, respectively, with an inconsistent sentence in S and a valid sentence in S. We shall use the symbol S to denote also the Boolean algebra. Since S is finite, S is a Boolean algebra with exactly n atoms $\varphi_1, \ldots, \varphi_n$, $n \geqslant 1$. We refer to the set $\{\varphi_1 \ldots \varphi_n\}$ as the set of atoms of S, and we write

$$\mathrm{at}(S) = \{\varphi_1 \ldots \varphi_n\}.$$

Note that for $1 \leqslant i, j \leqslant n$, we have:

6.3.7.

(i). $\varphi_1 \vee \ldots \vee \varphi_n = 1$.

(ii). $\varphi_i \neq 0$, $\varphi_i \wedge \varphi_j = 0$.

(iii). If $\varphi \in S$ *and* $\varphi \leqslant \varphi_i$, *then* $\varphi = 0$ *or* $\varphi = \varphi_i$.

(iv). If $\varphi \in S$, *then* φ *is a disjunction (possibly empty) of elements of* $\mathrm{at}(S)$.

(v). *Every model satisfies some* φ_i.

(vi). φ_i *is consistent and so has a model.*

(vii). *If* $\mathfrak{A}$ *is a model of* φ_i *and* $\mathfrak{B}$ *is a model of* φ_j, *then*

$$i = j \textit{ iff for all } \varphi \in S,\ \mathfrak{A} \models \varphi \textit{ iff } \mathfrak{B} \models \varphi.$$

Let I be an index set and let D be a filter over I. Suppose that $\varphi_i \in \mathrm{at}(S)$ for $i \in I$. We shall show that there is a $\sigma \in \mathrm{at}(S)$ such that

$$(*) \qquad \text{whenever} \quad \mathfrak{A}_i \models \varphi_i, \quad \text{then} \quad \prod_D \mathfrak{A}_i \models \sigma.$$

It is sufficient for this purpose to show that if we have also $\mathfrak{B}_i \models \varphi_i$, then

$$\prod_D \mathfrak{A}_i \models \varphi \quad \text{iff} \quad \prod_D \mathfrak{B}_i \models \varphi, \quad \text{for all} \quad \varphi \in S.$$

So let $\varphi \in S$ and let it be determined by a sequence $(\sigma; \psi_1, \ldots, \psi_m)$ with the $\psi_j \in S$. Since $\mathfrak{A}_i$ and $\mathfrak{B}_i$ are equivalent under all sentences of S, we see that for $1 \leqslant j \leqslant m$,

$$\{i \in I : \mathfrak{A}_i \models \psi_j\} = X^j = \{i \in I : \mathfrak{B}_i \models \psi_j\}.$$

Whence

$$\prod_D \mathfrak{A}_i \models \varphi \quad \text{iff} \quad S(I)/D \models \sigma[X^1/D \ldots X^m/D] \quad \text{iff} \quad \prod_D \mathfrak{B}_i \models \varphi.$$

So we can now define, without ambiguity,

$$\prod_D \varphi_i = \sigma$$

in precisely the sense of (*) above. We agree to call $\prod_D \varphi_i$ the *reduced product of the atoms* φ_i, $i \in I$, *modulo the filter* D. Our notation extends immediately to

reduced powers: $\prod_D \varphi$,

direct products: $\prod_{i \in I} \varphi_i$,

finite direct products: $\varphi_i \times \varphi_j$.

Ordinarily, D is taken to be a proper filter over I. In the case where D is improper, then $\prod_D \mathfrak{A}_i$ degenerates into the trivial 1-element model with relations and functions defined everywhere. This trivial 1-element model is, of course, a model of some $\varphi_i \in \mathrm{at}(S)$. Let us denote this particular element of $\mathrm{at}(S)$ by ζ, or, more precisely, ζ_S. From our definitions, we see that all of the combinatorial properties of reduced and direct products of models carry over to the analogous operations on $\mathrm{at}(S)$. In particular, to mention just a few examples, $\times$ is commutative and associative, and

$$\varphi_i \times \zeta = \varphi_i = \zeta \times \varphi_i \text{ for } 1 \leqslant i \leqslant n.$$

Thus we shall not bother to put parentheses in expressions like $\varphi_{i_1} \times \ldots \times \varphi_{i_m}$.

Let ψ_i, $i \in I$, now be arbitrary elements of S; we define the *reduced product of* ψ_i *modulo* D as

$$\prod_D \psi_i = \bigvee \{\sigma \in \mathrm{at}(S) : \text{there are } \quad \varphi_i \in \mathrm{at}(S), \quad \varphi_i \leqslant \psi_i \quad \text{and} \quad \sigma = \prod_D \varphi_i\}.$$

Note that if some $\psi_i = 0$, then $\prod_D \psi_i = 0$. Also, the same property expressed in (*) still holds, namely

$$\text{whenever} \quad \mathfrak{A}_i \models \psi_i, \quad \text{then} \quad \prod_D \mathfrak{A}_i \models \prod_D \psi_i.$$

Analogously, we have

the *reduced power*: $\prod_D \psi$,

the *direct product*: $\prod_{i \in I} \psi_i$,

the *finite direct product*: $\psi_i \times \psi_j$,

of elements of S. Needless to say, all combinatorial properties of reduced products for models carry over to S.

Some interesting consequences can now be deduced from the fact that $\prod_D$ can be defined on at(S) and on S. Many of the consequences depend on the fact that S is finite, and others depend on the fact that S can be effectively formed. First, the following remarks are practically obvious and are left as an exercise:

6.3.8. *Let $\varphi \in S$. Then*:

(i). *φ is a reduced product sentence iff $\prod_D \varphi \leqslant \varphi$ for all I, D.*

(ii). *φ is a reduced power sentence iff for all atoms $\psi \leqslant \varphi$ and all I, D, $\prod_D \psi \leqslant \varphi$.*

(iii). *φ is a direct product sentence iff $\prod_{i \in I} \varphi \leqslant \varphi$ for all I, D.*

(iv). *φ is a direct power sentence iff for all atoms $\psi \leqslant \varphi$ and all I, $\prod_{i \in I} \psi \leqslant \varphi$.*

(v). *φ is a finite direct product sentence iff $\varphi \times \varphi \leqslant \varphi$.*

(vi). *φ is a finite direct power sentence iff for all atoms $\psi \leqslant \varphi$ and all n, $\psi \times \ldots \times \psi$ (n times) $\leqslant \varphi$.*

Secondly, we should make some remarks concerning the effectiveness of the operations $\times$ and $\prod_D$ on S. We have already shown that given a formula φ, we can effectively find an autonomous set S such that $\varphi \in S$. By the definition of autonomous sets we know that given $\varphi, \psi \in S$, we can effectively find ρ, χ and θ in S such that

$$\vdash \rho \leftrightarrow \neg \varphi, \qquad \vdash \chi \leftrightarrow \varphi \vee \psi, \qquad \vdash \theta \leftrightarrow \varphi \wedge \psi.$$

Furthermore, we can effectively find sentences in S that are inconsistent and valid. All this will not guarantee that we can effectively tell which sentences in S give rise to elements of at(S). This is because to say that $\varphi \neq 0$ requires essentially that φ be consistent, and we have no effective way of determining that. Thus it is impossible to make the operations $\times$ and $\prod_D$ effective on S. On the other hand, even though we cannot distinguish effectively the atoms of S from the nonatoms, we can, by looking at the definition of $B(S)$, always find effectively a subset S' of S such that at$(S) \subset S'$ and every sentence in S' is either an atom of S or is inconsistent. A similar consideration tells us that given $\varphi \in S$, we can always find effectively a subset S' of S such that every atom of S included in φ belongs to S' and every sentence of S' is either an atom included in φ or is inconsistent.

THEOREM 6.3.9.

(i). *φ is a finite direct power sentence iff φ is equivalent to a disjunction of finite direct product sentences.*

(ii). *φ is a direct power sentence iff φ is equivalent to a disjunction of direct product sentences.*

(iii). *φ is a reduced power sentence iff φ is equivalent to a disjunction of reduced product sentences.*

(iv). *φ is a reduced product sentence iff φ is a finite direct product sentence and a reduced power sentence.*

PROOF. (i). Let φ be finite direct power sentence. Find autonomous S so that $\varphi \in S$. Consider all $\psi \in S$ such that

$$\psi \leqslant \varphi \text{ and } \psi \text{ is a finite direct product sentence.}$$

We claim that φ is (equivalent to) a disjunction of such sentences ψ. Let $\tau \in \text{at}(S)$ and $\tau \leqslant \varphi$. Then the set of all finite direct powers of τ is a finite set of atoms each of which is included in φ. Let ψ be the disjunction of all such atoms. Clearly, $\tau \leqslant \psi \leqslant \varphi$ and ψ is a finite direct product sentence. So φ is a disjunction of such sentences.

(ii). This proof is similar to that of (i).

(iii). Let φ be a reduced power sentence, and let S be an autonomous set containing φ. Again consider all $\psi \in S$ such that

$$\psi \leqslant \varphi \text{ and } \psi \text{ is a reduced product sentence.}$$

Let $\tau \in \text{at}(S)$ and $\tau \leqslant \varphi$. Then the set of all reduced products of τ is a finite set of atoms included in φ. Let ψ be the disjunction of such atoms; then, because a reduced product of reduced products of τ is again a reduced product of τ (see Exercise 6.2.3), ψ is a reduced product sentence and $\tau \leqslant \psi$. So φ is a disjunction of such sentences.

(iv). Suppose that φ is a finite direct product sentence and a reduced power sentence. Let $\varphi \in S$ and S be autonomous. Let ψ_i, $i \in I$, be atoms of S such that $\psi_i \leqslant \varphi$, and we wish to show that the reduced product $\prod_D \psi_i$ is an atom included in φ. We may assume without loss of generality that

$$\{\psi_i : i \in I\} = \{\varphi_j : 1 \leqslant j \leqslant l\}, \text{ each } \varphi_j \in \text{at}(S)$$

and that for each j and each $X \in D$,

$$I_j = \{i \in I : \psi_i = \varphi_j\}$$

and

$$X \cap I_j \neq 0.$$

Defining the filters D_j on I_j as in Proposition 6.2.1, we see that

$$\prod_D \psi_i = \prod_{D_1} \varphi_1 \times \prod_{D_2} \varphi_2 \times \dots \times \prod_{D_l} \varphi_l.$$

Since φ is a reduced power sentence, each $\prod_{D_j} \varphi_j$ is an atom included in φ, and since φ is a finite direct product sentence $\prod_D \psi_i$ is included in φ. ⊣

The technique used above also leads to an interpolation lemma (see Exercise 6.3.2). A further refinement of the technique will lead to the next main result, namely, that a finite direct product sentence is always a direct product sentence.

Let S be an autonomous set. The (arbitrary) direct power operation $\varphi^I = \prod_{i \in I} \varphi$ on S is said to be *essentially finite, with index m*, iff for all $\varphi \in S$ and all (nonempty) index sets I with at least m elements, there is a subset J of I with exactly m elements such that

$$\varphi^J = \varphi^K = \varphi^I \quad \text{for all} \quad K \quad \text{such that} \quad J \subset K \subset I.$$

If m is understood, then we say that (arbitrary) direct power is *essentially finite* on S.

A refinement of the argument of Theorem 6.3.6 leads to the following:

PROPOSITION 6.3.10. *Given a sentence φ of $\mathscr{L}$, then we can effectively find an autonomous set of sentences S such that $\varphi \in S$. Furthermore, the direct power on S is essentially finite with an index m which can also be effectively determined from φ.*

We should mention here that while the operations $\times$ and $\prod_D$ on S are not effective, we can, nevertheless, determine the bound m effectively. The proof of the proposition depends on the following lemma. First a convention: Suppose S is an autonomous set of formulas in the free variables $x_1, \dots, x_n$. We can find a corresponding autonomous set of sentences S' simply by replacing all free occurrences of $x_1, \dots, x_n$ in $\varphi \in S$ by new constants $c_1, \dots, c_n$. In this way, our definitions of reduced products, etc., on S' carry over in a natural way to S. Also, we may, of course, speak of S having an essentially finite direct power, etc.

LEMMA 6.3.11. *Every finite set Φ of Boolean combinations of Σ_n^0 formulas can be effectively extended to an autonomous set S of Boolean combinations of Σ_n^0 formulas (every free variable occurring in a formula of S occurs in some* *formula in Φ) having an essentially finite direct power and such that the index m is effectively calculable.*

Proof (in outline). We prove this by induction on n. For $n = 0$, let $S = B(S)$ be the set of all Boolean combinations of atomic formulas occurring in Φ. The number m is simply the total number of different atomic formulas (not predicate symbols) occurring in Φ. This can be seen as follows: Let $(\mathfrak{A}_i, a_1(i), \ldots, a_n(i))$, $i \in I$, be models for $\mathscr{L} \cup \{c_1 \ldots c_n\}$, and suppose that $|I| \geqslant m$. For each atomic formula $\varphi(x_1 \ldots x_n)$ in Φ, pick an index $i \in I$ such that

$$\mathfrak{A}_i \models \neg\, \varphi[a_1(i) \ldots a_n(i)]$$

if there is such an i. Collecting all these indices together in a set J with at most m elements, we see that for all K, $J \subset K \subset I$, and all $\varphi \in S$,

$$\prod_{i \in J} (\mathfrak{A}_i, a_1(i), \ldots, a_n(i)) \models \varphi \quad \text{iff} \quad \prod_{i \in K} (\mathfrak{A}_i, a_1(i), \ldots, a_n(i)) \models \varphi$$

$$\text{iff} \quad \prod_{i \in I} (\mathfrak{A}_i, a_1(i), \ldots, a_n(i)) \models \varphi.$$

This proves the assertion for $n = 0$. Assume the lemma is true for n. Let Φ be a finite subset of Boolean combinations of Σ^0_{n+1} formulas. We may without loss of generality assume that there is a finite set Ψ of Boolean combinations of Σ^0_n formulas and a sequence $u = (y_1 \ldots y_m)$ of variables such that each $\varphi \in \Phi$ is a Boolean combination of formulas in $(\exists u)\Psi$. By induction hypothesis, Ψ can be extended to an autonomous set T of Boolean combinations of Σ^0_n formulas such that the direct power on T is essentially finite with index m. Let S be the autonomous set $S = B((\exists u)T)$. (Here we use the remark following Theorem 6.3.6, and the observation that in the proof of Proposition 6.3.2 it is inessential whether we quantify on a single variable or a finite sequence of variables simultaneously.) Thus every formula in S is a Boolean combination of Σ^0_{n+1} formulas and S extends Φ. We now show that the direct power on S is essentially finite. Let $\theta_1, \ldots, \theta_l$ be an effective listing of formulas of T which contains all atoms of T and such that each θ_i is an atom of T or is inconsistent. We now claim that the set S' of all conjunctions

$$(1) \qquad \varphi = \bigwedge_{1 \leqslant i \leqslant l} \varepsilon_i (\exists u)\theta_i, \qquad \varepsilon_i \text{ is empty or is } \neg,$$

contains all atoms of S and that each such conjunction is either an atom of S or is inconsistent. This is evident because, for each $\theta \in T$, $(\exists u)\theta$ is equivalent to a disjunction of formulas of the form $(\exists u)\theta_i$, and $\neg\,(\exists u)\theta$ is equivalent to a conjunction of formulas of the form $\neg (\exists u)\theta_i$. By the disjunctive normal form theorem, every atom of S has a representative in S', and clearly every conjunct in S' is either an atom or is inconsistent. We shall show that $m \cdot |S'|$ is the required index. We now argue informally. For each atom φ of S', let $g(\varphi)$ be the disjunction of all atoms θ_i of T which appears in the representation of φ in (1) as $(\exists u)\theta_i$, without the $\neg$ symbol. Then we can establish the following:

(2) g is a one-to-one mapping of $\mathrm{at}(S)$ into T,

and

(3) g preserves direct products, i.e. $g(\prod_{i\in I}\varphi_i) = \prod_{i\in I} g(\varphi_i)$.

We leave the verification of (2) and (3) as an exercise. Since direct power on T is essentially finite with index m, (2) and (3) show that direct power on $\mathrm{at}(S)$ is essentially finite with index m. Remember that the total number of atoms of S is at most $|S'|$; we see that the product of any $k \geqslant m \cdot |S'|$ number of atoms of S can always be regrouped into a subproduct of exactly $m \cdot |S'|$ terms. (Here we use strongly the fact that m is an index for direct powers on $\mathrm{at}(S)$.) Furthermore, in any direct product of $m \cdot |S'|$ atoms of S, some atom must repeat at least m times. So we can add this atom any number of times to the product without changing the result of the product. Since direct product on S is a natural extension of the product of atoms of S, the above remarks show that direct power on S has index $m \cdot |S'|$. ⊣

Let S be an autonomous set of sentences. We define the relation $\leqslant_\times$ on $\mathrm{at}(S)$ as follows: for $\varphi, \psi \in \mathrm{at}(S)$,

$$\varphi \leqslant_\times \psi \text{ iff for some } \theta \in \mathrm{at}(S),\ \varphi \times \theta = \psi.$$

Since there is an identity ζ_S in $\mathrm{at}(S)$, $\leqslant_\times$ is a reflexive relation. If $\varphi = \theta \times \psi$ and $\psi = \chi \times \tau$, then $\varphi = (\theta \times \chi) \times \tau$. So $\leqslant_\times$ is transitive.

PROPOSITION 6.3.12. *Let S be an autonomous set. If the direct power on S is essentially finite, then $\leqslant_\times$ on* $\mathrm{at}(S)$ *is a partial ordering of* $\mathrm{at}(S)$.

PROOF. Suppose that $\varphi, \psi, \theta, \chi \in \mathrm{at}(S)$, and $\varphi = \theta \times \psi$ and $\psi = \chi \times \varphi$. Then

$$\varphi = \theta \times \psi = (\theta \times \chi) \times \varphi = (\theta \times \chi)^n \times \varphi \quad \text{for all} \quad n < \omega.$$

Let m be the index of direct power on S, then by commutative and associative properties of $\times$, we have

$$\varphi = \theta^m \times \chi^m \times \varphi = \theta^{m+1} \times \chi^m \times \varphi = \theta \times (\theta^m \times \chi^m \times \varphi) = \theta \times \varphi = \psi.$$

So $\leqslant_\times$ is antisymmetric. ⊣

Let $\varphi \in \mathrm{at}(S)$, and consider the set

$$(*) \qquad \Phi = \{\psi \in \mathrm{at}(S) : \varphi \leqslant_\times \psi\}.$$

It is immediate that if $\psi_1, \psi_2 \in \Phi$, then $\psi_1 \times \psi_2 \in \Phi$. Thus the sentence $\bigvee \Phi$ is a finite direct product sentence. Similar considerations show that the sentence

$$(**) \qquad \bigvee \{\psi \in \mathrm{at}(S) : \psi \leqslant_\times \varphi\}$$

is preserved under direct factors. It is known that if $\leqslant_\times$ is a (finite) partial ordering relation, then every $\varphi \in \mathrm{at}(S)$ is a Boolean combination of sentences of the form (*), as well as sentences of the form (**). Thus we have:

Proposition 6.3.13.

(i). *Every sentence is equivalent to a Boolean combination of finite direct product sentences.*

(ii). *Every sentence is equivalent to a Boolean combination of sentences preserved under direct factors.*

Proof. Given φ, find by Theorem 6.3.9 and Proposition 6.3.10 an autonomous set S such that $\varphi \in S$ and such that $\leqslant_\times$ is a partial ordering of $\mathrm{at}(S)$. The proof now follows from the remarks above. ⊣

We shall see later that part (i) can be improved.

Theorem 6.3.14.

(i). *Every finite direct product sentence is an (arbitrary) direct product sentence.*

(ii). *Every finite direct power sentence is an (arbitrary) direct power sentence.*

(iii). *More generally, given a sentence φ we can effectively find a number n such that for all index sets I and all models $\mathfrak{A}_i$, $i \in I$, there is a subset J of I with at most n elements such that for all K, $J \subset K \subset I$,*

$$\prod_{i \in K} \mathfrak{A}_i \models \varphi \quad \textit{iff} \quad \prod_{i \in I} \mathfrak{A}_i \models \varphi.$$

PROOF. Both (i) and (ii) are consequences of (iii) (in fact, by Theorem 6.3.9, part (ii) already follows from part (i)), so we shall only prove (iii).

(iii). Given φ, find effectively an autonomous set S on which direct power is essentially finite with index m. Let $n = m \cdot |S|$. (This is really a waste, since we can find a much better upper bound to the number of atoms of S than $|S|$.) Suppose that $\mathfrak{A}_i$, $i \in I$, are given. Suppose that $\mathfrak{A}_i \models \psi_i$, where ψ_i are atoms of S. If an atom ψ of S occurs in the list ψ_i, $i \in I$, less than m times, then we put into J all indices $i \in I$ such that $\psi = \psi_i$. If an atom ψ of S occurs in the list ψ_i, $i \in I$, m or more times, we simply select any m indices i such that $\psi = \psi_i$ and put them in J. At the end, J will have at most n elements.

To prove the conclusion of (iii), it is sufficient to do it for all atoms $\psi \leqslant \varphi$. Let K be such that $J \subset K \subset I$, and let $\psi \in \mathrm{at}(S)$, $\psi \leqslant \varphi$. Then using the fact that direct power on S has index m, we see that, by a suitable regrouping of the atoms ψ_i, $i \in I$,

$$\prod_{i \in J} \psi_i = \prod_{i \in K} \psi_i = \prod_{i \in I} \psi_i.$$

So $\prod_{i \in K} \psi_i = \psi$ iff $\prod_{i \in I} \psi_i = \psi$, and

$$\prod_{i \in K} \mathfrak{A}_i \models \psi \quad \text{iff} \quad \prod_{i \in I} \mathfrak{A}_i \models \psi. \dashv$$

Pursuing our line of investigation even further, we shall next use this same sort of method to show that every sentence of $\mathscr{L}$ is equivalent to a Boolean combination of reduced product sentences. Notice that by Theorem 6.3.9 (iii), it is clearly sufficient to show that every sentence is equivalent to a Boolean combination of reduced power sentences.

Let us now return to the definition of the reduced power $\prod_D \varphi$ defined on the atoms of an autonomous set. We recall that $\prod_D \varphi = \psi$ iff whenever $\mathfrak{A}_i$, $i \in I$, are models for $\mathscr{L}$ and each $\mathfrak{A}_i \models \varphi$, then $\prod_D \mathfrak{A}_i \models \psi$. Since S is autonomous, there is a sequence $(\sigma; \psi_1 \ldots \psi_m)$ with $\psi_j \in S$ which determines ψ. This means that, letting

$$X^j = \{i \in I : \mathfrak{A}_i \models \psi_j\}, \qquad 1 \leqslant j \leqslant m,$$

we have

$$\prod_D \mathfrak{A}_i \models \psi \quad \text{iff} \quad S(I)/D \models \sigma[X^1/D \ldots X^m/D].$$

Notice that since $\mathfrak{A}_i \models \varphi$, the sets X^j do not depend on $\mathfrak{A}_i$, $i \in I$, but only depend on ψ_j, and hence on ψ. Furthermore, depending on whether the atom φ is included in ψ_j or excluded from ψ_j, the sets X^j are, respectively, I or 0. Whence, whether or not $\prod_D \varphi = \psi$ *depends only on* whether or not $S(I)/D$

$\models \sigma'$, where σ' is a sentence in the language of Boolean algebras depending only on ψ. From this we draw the following simple but important conclusion:

If $S(I)/D$ and $S(J)/E$ are elementarily equivalent Boolean algebras, then $\prod_D \varphi = \psi$ iff $\prod_E \varphi = \psi$.

This implies that the reduced power $\prod_D \varphi$ on at(S) depends only on the (Boolean) elementary type of $S(I)/D$. Let the letters $a, b, c, \dots$ range over elementary types of Booleans algebras of the form $S(I)/D$, i.e., those Boolean algebras which are (proper) homomorphic images of a complete atomic Boolean algebra. We define, for $\varphi \in$ at(S) and a the type of $S(I)/D$,

$$a \cdot \varphi = \text{the unique } \psi \in \text{at}(S) \text{ such that } \prod_D \varphi = \psi.$$

Let Q denote the set of elementary types of quotient Boolean algebras of the form $S(I)/D$. We see from Theorem 6.3.4 that the direct product of two such algebras $S(I)/D$ and $S(J)/E$ is determined, up to equivalence, by the types of $S(I)/D$ and $S(J)/E$. Also, it follows from the above discussion that a reduced power $\prod_E S(I)/D$ of such an algebra modulo a filter E over J is determined, up to equivalence, by the types of $S(I)/D$ and $S(J)/E$. From these remarks we can introduce meaningful definitions of $a \times b$ and $a \cdot b$ for $a, b \in Q$. We now introduce certain specific types of Boolean algebras, namely,

$\bar{n}$ = the type of the Boolean algebra with exactly n atoms, $n \geq 1$;
$\overline{\infty}$ = the type of the (complete) theory of atomless Boolean algebras.

The next two lemmas tell us all we need to know about the three operations $a \times b$, $a \cdot b$ and $a \cdot \varphi$.

Lemma 6.3.15. *Let a, b range in Q. Then*:

(i). $a \times b \in Q$, $a \cdot b \in Q$, $\bar{n} \in Q$, $\overline{\infty} \in Q$.
(ii). $a \cdot \overline{\infty} = \overline{\infty} \cdot a = \overline{\infty}$.
(iii). *Either* $a = \overline{\infty}$ *or* $a = c \times \bar{1}$ *for some* $c \in Q$.

Proof. We briefly sketch proofs of parts of the lemma, leaving the detailed verification to the reader. Let a be the type of $S(I)/D$ and b be the type of $S(J)/E$. We may assume that $I \cap J = 0$. Then $a \times b$ is the type of

$$S(I \cup J)/\{X \cup Y : X \in D, Y \in E\},$$

and $a \cdot b$ is the type of

$$S(I \times J)/D \times E. \tag{1}$$

Clearly $\bar{n} \in Q$ for each $n < \omega$. To obtain $\overline{\infty} \in Q$, consider the type of

$$S(\omega)/\{X \subset \omega : (\omega \setminus X) \text{ is finite}\}.$$

We point out that in general, $a \cdot b \neq b \cdot a$, however, it is easy to verify that $a \cdot \overline{\infty} = \overline{\infty} \cdot a = \overline{\infty}$, either in a straightforward manner, or using the definition of $a \cdot b$ given in (1). For the last statement, if a is not atomless nor trivial, then a has an atom. It is then easy to 'factor' this atom out. ⊣

LEMMA 6.3.16. *Let S be an autonomous set and let φ, ψ range in* at(S) *and a, b range in Q. Then*:

(i). $(a \cdot b) \cdot \varphi = a \cdot (b \cdot \varphi)$.
(ii). $(a \times b) \cdot \varphi = a \cdot \varphi \times b \cdot \varphi$.
(iii). $a \cdot (\varphi \times \psi) = a \cdot \varphi \times a \cdot \psi$.
(iv). $\zeta_S = a \cdot \zeta_S$.
(v). $\bar{1} \cdot \varphi = \varphi$.

We leave the easy proof of Lemma 6.3.16 as an exercise. For $\varphi, \psi \in \text{at}(S)$, we introduce another ordering relation $\leqslant_Q$ defined by

$$\varphi \leqslant_Q \psi \text{ iff } a \cdot \varphi = \psi \text{ for some } a \in Q.$$

The next two results are analogues of Propositions 6.3.12 and 6.3.13.

PROPOSITION 6.3.17. *Let S be an autonomous set of sentences. If the relation $\leqslant_\times$ on* at(S) *is a partial ordering of* at(S), *then so is the relation $\leqslant_Q$.*

PROOF. By Lemma 6.3.16 (i) and (v), $\leqslant_Q$ is reflexive and transitive. So we must show that $\leqslant_Q$ is antisymmetric. So let $\varphi = a \cdot \psi$ and $\psi = b \cdot \varphi$ with $a, b \in Q$, $\varphi, \psi \in \text{at}(S)$. If one of a or b is $\overline{\infty}$, say $a = \overline{\infty}$, then by Lemma 6.3.15 (ii),

$$\varphi = \overline{\infty} \cdot \psi = (b \cdot \overline{\infty}) \cdot \psi = b \cdot \varphi = \psi.$$

If $a \neq \overline{\infty}$, then by Lemma 6.3.15 (iii), $a = c \times \bar{1}$ for some $c \in Q$. Then, by Lemma 6.3.16 (ii) and (v),

$$\varphi = (c \times \bar{1}) \cdot \psi = c \cdot \psi \times \bar{1} \cdot \psi = c \cdot \psi \times \psi.$$

So $\psi \leqslant_\times \varphi$. Similarly, if $b \neq \overline{\infty}$, then $\varphi \leqslant_\times \psi$. Whence, by hypothesis, $\varphi = \psi$. ⊣

THEOREM 6.3.18. *Every sentence φ is equivalent to a Boolean combination of reduced product sentences.*

PROOF. We find an autonomous set S containing φ such that direct power is essentially finite on S. By Proposition 6.3.12, $\leqslant_\times$ is a partial ordering of at(S). By Proposition 6.3.17 $\leqslant_Q$ is also a partial ordering of at(S). This means that every sentence in S is a Boolean combination of sentences of the form

$$\varphi_\theta = \bigvee\{\psi \in \mathrm{at}(S) : \theta \leqslant_Q \psi\} \text{ with } \theta \text{ ranging in } \mathrm{at}(S).$$

Clearly, if $\theta \leqslant_Q \psi$, then $\theta \leqslant_Q \prod_D \psi$ for all reduced powers $\prod_D \psi$. So each φ_θ is a reduced power sentence. The result now follows from Theorem 6.3.9 (iii). ⊣

In order to be quite specific in what follows, let us assume that S is an autonomous set. Let

$$T = \{\bigwedge_{\varphi \in S} (\varepsilon_\varphi)\varphi : \varepsilon_\varphi \text{ is either the empty sequence or is } \neg\},$$

and, as before, let $B(S)$ be the set of disjunctions of sentences in T up to $l = |T|$ terms. As we noted earlier, $B(S)$ is autonomous. But the more important thing is that T consists of all atoms of $B(S)$ plus, possibly, a few other inconsistent sentences. Every $\varphi \in S$, of course, is equivalent to the disjunction of the subset T_φ of T defined by

$$T_\varphi = \{\theta \in T : \varphi \text{ occurs positively in } \theta\}.$$

From this, since each $\varphi \in S$ is determined by a sequence $(\sigma; \psi_1 \ldots \psi_m)$ with $\psi_j \in S$, we see that, with a sufficient amount of writing, we can show that, letting $T = \{\theta_1 \ldots \theta_l\}$,

> for each k, $1 \leqslant k \leqslant l$, we can find effectively a formula σ_k in the variables $y_1, \ldots, y_l$, so that the sequence $(\sigma_k; \theta_1 \ldots \theta_l)$ determines θ_k.

Let $J \subset l$ and let $1 \leqslant k \leqslant l$. Define the sentence $\rho_{J,k}$, in the language of Boolean algebras, as follows:

$$\rho_{J,k} = (\exists y_1 \ldots y_l)(\sigma_k(y_1 \ldots y_l) \wedge (\sum_{j \in J} y_j) = 1$$
$$\wedge \bigwedge_{\substack{i,j \in J \\ i \neq j}} (y_i \cdot y_j = 0) \wedge \bigwedge_{j \notin J} (y_j = 0)).$$

For $J \subset l$, let $\Phi(J)$ be the following statement:

> for every k such that $k \notin J$ and $1 \leqslant k \leqslant l$, the Boolean sentence $\rho_{J,k}$ holds in no Boolean algebra $S(I)/D$.

For $\varphi \in S$, let $J_\varphi = \{k : \theta_k \in T_\varphi\}$.

Proposition 6.3.19. *φ is a reduced product sentence if and only if for some $J \subset J_\varphi$, $\Phi(J)$ and φ is equivalent to $\bigvee_{k\in J}\theta_k$.*

Proof. (i). Assume first that φ is a reduced product sentence. Let

$$J = \{k \in J_\varphi : \theta_k \text{ is an atom in } B(S)\}.$$

Clearly, $J \subset J_\varphi$ and φ is equivalent to $\bigvee_{k\in J}\theta_k$. We now prove $\Phi(J)$. If φ is inconsistent, then $J = 0$ and ρ_{Jk} can never hold in any Boolean algebra. So $\Phi(J)$. Now assume that φ is consistent. Suppose, on the contrary, that

$\rho_{J,k}$ holds in some Boolean algebra $S(I)/D$ for some $k \notin J$.

Thus there are sets $X^1/D, \dots, X^l/D$ such that

$$S(I)/D \models \sigma_k[X^1/D \dots X^l/D] \wedge \sum_{j\in J} X^j/D = 1 \wedge \bigwedge_{1\leqslant i<j\leqslant l} (X^i/D \cdot X^j/D = 0).$$

Let $X \in D$ be such that

$$\bigcup_{j\in J} X^j \supset X,$$

and

$$X^i \cap X^j \cap X = 0 \qquad \text{for } 1 \leqslant i < j \leqslant l.$$

Notice that each X^j/D for $j \notin J$ is forced to be equal to 0. For $j \in J$ and $i \in X^j \cap X$, let $\mathfrak{A}_i$ be a model of θ_j. For $i \in I \setminus X$, pick $\mathfrak{A}_i$ to be any model of φ (we have assumed that φ is consistent). It should be clear that if $j \notin J$, then either θ_j is inconsistent, or θ_j is an atom not included in φ. Define for $1 \leqslant j \leqslant l$,

(1) $$Y^j = \{i \in I : \mathfrak{A}_i \models \theta_j\}.$$

It is clear that

$$Y^j = 0 \quad \text{if} \quad j \notin J,$$

and

(2) $$Y^j/D = X^j/D, \qquad 1 \leqslant j \leqslant l,$$
$$\sum_{j\in J} Y^j/D = 1,$$
$$Y^i/D \cdot Y^j/D = 0, \qquad 1 \leqslant i < j \leqslant l.$$

Since $S(I)/D \models \sigma_k[X^1/D \dots X^l/D]$, we have $S(I)/D \models \sigma_k[Y^1/D \dots Y^l/D]$. Since $(\sigma_k; \theta_1 \dots \theta_l)$ determines θ_k, we see that

(3) $$\prod_D \mathfrak{A}_i \models \theta_k.$$

This is a contradiction to the fact that φ is preserved under reduced products, as θ_k is an atom not included in φ.

(ii). In the other direction, suppose that for some $J \subset J_\varphi$, $\Phi(J)$ and φ is equivalent to $\bigvee_{k \in J} \theta_k$. Let I, D, and $\mathfrak{A}_i$, $i \in I$, be given such that each $\mathfrak{A}_i \models \varphi$. Define the sets Y^j, $1 \leqslant j \leqslant l$, as in (1). Then it is clear that (2) holds. There is some atom θ_k such that (3) holds. Whence

(4) $$S(I)/D \models \sigma_k[Y^1/D \ldots Y^l/D].$$

Putting (2) and (4) together, we see that

$$\rho_{J,k} \text{ holds in the Boolean algebra } S(I)/D.$$

This implies by $\Phi(J)$ that $k \in J$. So $\prod_D \mathfrak{A}_i \models \varphi$. ⊣

Let us now examine the statement $\Sigma(\varphi)$ about the sentence φ: for some $J \subset J_\varphi$, $\Phi(J)$ and φ is equivalent to $\bigvee_{k \in J} \theta_k$. Given an arbitrary sentence φ, we see that the sentences $\theta_1, \ldots, \theta_l$ can be found effectively; similarly, the formulas $\sigma_1, \ldots, \sigma_l$ can also be found effectively. The same remark applies to the sets T_φ and J_φ. Since J_φ is finite, we can examine in turn all J such that $J \subset J_\varphi$. Given a $J \subset J_\varphi$, the sentence $\bigvee_{k \in J} \theta_k$ can be constructed explicitly. Whence we can ask if $\vdash \varphi \leftrightarrow \bigvee_{k \in J} \theta_k$; this involves finding a proof of the sentence $\varphi \leftrightarrow \bigvee_{k \in J} \theta_k$. This then turns out to be a recursively enumerable condition. Turning now to the statement $\Phi(J)$, we see first of all that the construction of $\rho_{J,k}$ for each $J \subset J_\varphi$ and $k \notin J$ is quite explicit. However, the statement

$$\rho_{J,k} \text{ holds in no Boolean algebra } S(I)/D$$

can in no way be interpreted as an effective statement. We shall see later that $\Sigma(\varphi)$ is an arithmetical statement.

Finally, the last result in this section uses the analysis of Boolean algebras in Section 5.5.

THEOREM 6.3.20 (Ershov's Theorem). *Let σ be a sentence in the language of Boolean algebras, and let T_{BA} be the axioms of the theory of Boolean algebras. Then the following are equivalent:*

(i). *$T_{\mathrm{BA}} \cup \{\sigma\}$ is consistent.*

(ii). *σ holds in some quotient Boolean algebra $S(\omega)/D$.*

PROOF. We first observe that (i) follows immediately from (ii).

(i) → (ii). In the other direction, we shall need several intermediate steps.

First observe that (i) is equivalent to the assertion that

$$\sigma \text{ holds in some Boolean algebra } \mathfrak{B}.$$

Using the notation from Section 5.5, let $m(\mathfrak{B})$ and $n(\mathfrak{B})$ be the pair of invariants assigned to $\mathfrak{B}$. If we could find a quotient algebra $S(\omega)/D$ such that

(1) $$m(S(\omega)/D) = m(\mathfrak{B}) \quad \text{and} \quad n(S(\omega)/D) = n(\mathfrak{B}),$$

then we could conclude from 5.5.10 that σ holds in $S(\omega)/D$, whence (ii) would be proved. To prove (1), we shall show that

(2) for every pair of possible invariants (m, n) for Boolean algebras, there exists a quotient algebra $S(\omega)/D$ such that $m(S(\omega)/D) = m$ and $n(S(\omega)/D) = n$.

We divide the proof of (2) into two cases: m is finite, or $m = \infty$.

Let us first deal with the case when m is finite. Recall from Section 5.5 that n is either an integer or is $= \pm\infty$, depending on the number of atoms $\mathfrak{A}^{(m)}$ has and on whether $\mathfrak{A}^{(m)}$ is or is not atomic. We shall first prove the following lemma by induction on the finite values of m.

LEMMA 6.3.21. *Let n be either an integer or $\pm\infty$. Then there is a quotient algebra $S(\omega)/D$ such that:*

(i). *D contains all cofinite subsets of ω;*

(ii). *$m(S(\omega)/D) = m$ and $n(S(\omega)/D) = n$.*

PROOF. First consider the case $m = 0$ and $n = 1$. Let D be any nonprincipal ultrafilter over ω. Then $S(\omega)/D$ has exactly one atom, and D satisfies (i). For any positive value of n, with m remaining 0, we use the construction in Lemma 6.3.15 to find a D over ω satisfying (i) and (ii). If n is negative, we again appeal to the construction in Lemma 6.3.15, piecing together an atomic part with an atomless part. Thus the case $m = 0$ is proved.

Consider now any pair of invariants $(m+1, n)$, n either an integer or $\pm\infty$. By induction hypothesis, we already have a quotient algebra $S(\omega)/D$ satisfying (i) and (ii). We shall now construct another quotient algebra $\mathfrak{B} = S(\omega \times \omega)/F$ such that

(3) $$\text{(i) holds for } F, \quad \text{and} \quad \mathfrak{B}^{(1)} \cong S(\omega)/D.$$

($\mathfrak{B}^{(1)}$ is defined in Section 5.5.) From this it follows readily that $m(\mathfrak{B}) = m+1$ and $n(\mathfrak{B}) = n$. To construct F, first let E be any nonprincipal ultrafilter on ω. Let F be the following collection of subsets of $\omega \times \omega$: for

$X \subset \omega \times \omega$, let $X_i = X \cap \{i\} \times \omega$. Then (in what follows we shall frequently identify X_i with the set $\{n : \langle in \rangle \in X_i\}$)

(4) $X \in F$ iff $X_i \in E$ for all $i \in \omega$, and $\{i \in \omega : X_i \text{ is cofinite}\} \in D$.

It is quite clear that F is a filter over $\omega \times \omega$; furthermore, F contains every cofinite subset of $\omega \times \omega$. Let $\mathfrak{B} = S(\omega \times \omega)/F$ and we assert that $\mathfrak{B}$ is the desired quotient algebra satisfying (3).

Let I be any subset of ω. In view of the definition (4) of F, we see that the following pair of equivalences are true:

(5) $I/D = 0$ iff there is an $X \in F$ such that $I \cap \{i \in \omega : X_i \text{ is cofinite}\} = 0$;
$I/D \neq 0$ iff for all $X \in F$, $I \cap \{i \in \omega : X_i \text{ is cofinite}\} \neq 0$.

Since, by assumption, D contains all cofinite subsets of ω, we see that $\{i\}/D = 0$ for al $i \in \omega$, and whence, by (5),

(6) for all $i \in \omega$, there is an $X \in F$ such that $X_i \in E$ but is not cofinite.

One more definition: for each $X \subset \omega \times \omega$, let

$$(X) = \{i \in \omega : X_i \in E\}.$$

The atoms of $\mathfrak{B}$ are characterized by the following: Let $X \subset \omega \times \omega$. Then

(7) X/F is an atom of $\mathfrak{B}$ if and only if (X) is a singleton $\{i\}$ and $X/F = X_i/F$.

To prove (7), suppose first that (X) is a singleton $\{i\}$, and that $X/F = X_i/F$. It is sufficient to prove that X_i/F is an atom of $\mathfrak{B}$. Notice that $\omega \setminus X_i \notin E$; therefore $\omega \times \omega \setminus X_i \notin F$, so $X_i/F \neq 0$. Suppose that $Y \subset X_i$. Then either $Y \in E$ or $X_i \setminus Y \in E$; so either

$$Y/F = X/F, \text{ or } Y/F = 0.$$

On the other hand, suppose that X/F is an atom of $\mathfrak{B}$. We first show that

(8) $$(X) \neq 0.$$

Suppose that (8) fails. Let

$$J = \{i \in \omega : X_i \text{ is infinite}\}.$$

Since (8) fails to hold, we see that each $X_i \notin E$ for $i \in J$. From this it follows that $(\omega \setminus J) \notin D$, for otherwise $X/F = 0$. Now for each $i \in J$, let Y_i and Z_i be disjoint infinite subsets of X_i, and let

$$W = \bigcup_{i \in J} Y_i \quad \text{and} \quad U = \bigcup_{i \in J} Z_i.$$

It follows from (5) that

$$0 \neq W/F \leqslant X/F, \qquad 0 \neq U/F \leqslant X/F, \qquad W/F \neq U/F.$$

This contradicts X/F is an atom. So (8) holds. If $i, j \in (X)$ and $i \neq j$, then it is evident that

$$0 \neq X_i/F \leqslant X/F, \qquad 0 \neq X_j/F \leqslant X/F, \qquad X_i/F \neq X_j/F.$$

So it follows that $(X) = \{i\}$ for some $i \in \omega$. By the direction of (7) that we have already proved, we have that X_i/F is an atom of $\mathfrak{B}$ and $X_i/F \leqslant X/F$. It follows that $X_i/F = X/F$. So (7) is proved.

It follows easily from (7) and the arguments leading to it that atomless elements of $\mathfrak{B}$ are characterized as follows: Let $X \subset \omega \times \omega$. Then

(9) X/F is atomless in $\mathfrak{B}$ if and only if $X_i \notin E$ for each $i \in (X)$.

We leave the easy proof of (9) to the reader. The next two assertions characterize those elements of $\mathfrak{B}$ which belong to $I(\mathfrak{B})$, i.e., those elements of $\mathfrak{B}$ which can be decomposed into an atomic part and an atomless part (see Section 5.5). Let $X < \omega \times \omega$.

(10) If $(X)/D = 0$, then in $\mathfrak{B}$ the sum of all atoms covered by X/F exists and is equal to $(\bigcup_{i \in (X)} X_i)/F$. So in this case $X/F \in I(\mathfrak{B})$.

(11) If $(X)/D \neq 0$, then in $\mathfrak{B}$ the sum of all atoms covered by X/F does not exist. So in this case $X/F \notin I(\mathfrak{B})$.

Let us first prove (10). We have already seen by (7) that each X_i/F is an atom covered by X/F. Let $Y = \bigcup_{i \in (X)} X_i$. It is immediate that $X_i/F \leqslant Y/F \leqslant X/F$ for $i \in (X)$. Furthermore, it is easy to show that

every atom covered by X/F is one of X_i/F with $i \in (X)$.

Thus Y/F covers all atoms covered by X/F. Suppose that Z/F is an atomless element such that $Z/F \leqslant Y/F$. We may suppose that $Z \subset Y$, and, by (9), $Z_i \notin E$ for all $i \in (X)$. Since $(X)/D = 0$, it follows that $Z/F = 0$. So (10) is proved.

Assume now $(X)/D \neq 0$. By previous considerations we know that X_i/F with $i \in (X)$ is the set of all atoms covered by X/F. Also, these atoms are covered by Y/F, where $Y = \bigcup_{i \in (X)} X_i$. Suppose the sum of these atoms exists in $\mathfrak{B}$; let it be denoted by Z/F. We may suppose that $Z \subset Y$. Since $X_i/F \leqslant Z/F$, we see that for each $i \in (X)$, $Z_i \in E$, for otherwise $X_i/F \cdot Z/F = 0$. Since E is nonprincipal, Z_i is infinite. Whence we pick a subset $W_i \subset Z_i$ such that W_i is infinite and $W_i \notin E$. Let $W = \bigcup_{i \in (X)} W_i$. By (9), W is an atomless element of $\mathfrak{B}$ such that $0 \neq W/F \leqslant Z/F$. This contradicts the definition of Z/F. So (11) is proved.

By (10) and (11), we see that

$$(X)/D = 0 \text{ if, and only if, } (X/F)/I(\mathfrak{B}) = 0.$$

Furthermore, the mapping $X \to (X)$ is a homomorphism of $S(\omega \times \omega)$ onto $S(\omega)$. Whence we have

$$S(\omega)/D \cong (S(\omega \times \omega)/F)/I(\mathfrak{B}),$$

which proves (3) and completes the induction. (We may, of course, identify $\omega \times \omega$ with ω and redefine F, so that $\mathfrak{B}$ is isomorphic to a quotient algebra $S(\omega)/F$.) The lemma is proved. ⊣

Finally, to construct a quotient algebra $S(\omega)/D$ such that $m(S(\omega)/D) = \infty$, we proceed as follows: Let I_n, $n < \omega$, be a partition of ω into disjoint infinite sets. On each I_n, let D_n be a filter over I_n such that $m(S(I_n)/D_n) = n$. This is possible by the lemma. Now let

$$D = \{X \subset \omega : X \cap I_n \in D_n \text{ for all } n < \omega\}.$$

We leave it to the reader to verify that $m(S(\omega)/D) = \infty$. This concludes the proof of Ershov's theorem. ⊣

Using Proposition 6.3.19 and Ershov's theorem we can at last fill the gap needed to eliminate the continuum hypothesis in the preceding section. We must prove:

LEMMA 6.2.7. *The predicate*

φ is a reduced product sentence

is an arithmetical predicate.

PROOF. Using Ershov's theorem, we see that the following three statements are equivalent:

$\rho_{J,k}$ holds in no Boolean algebra $S(I)/D$;
$\rho_{J,k}$ holds in no Boolean algebra $S(\omega)/D$;
$\rho_{J,k}$ is inconsistent with the theory of Boolean algebras.

Notice that this last statement is quite obviously an arithmetical statement about $\rho_{J,k}$. Whence we have, by Proposition 6.3.19 and the discussion following it: the predicate

$$\varphi \text{ is a reduced product sentence}$$

is an arithmetical predicate. ⊣

Now, we can restate Theorem 6.2.5 as follows:

THEOREM 6.2.5′. *φ is a reduced product sentence if and only if φ is equivalent to a Horn sentence.*

We can, of course, apply the same technique to eliminate the continuum hypothesis from Proposition 6.2.6.

EXERCISES

6.3.1. Let $\mathscr{L}$ be a countable language, and let there be a suitable Gödel numbering of the formulas of $\mathscr{L}$. Show that the set of direct product sentences of $\mathscr{L}$ is a recursively enumerable set.

[*Hint*: Show that in $\mathscr{L} \cup \{U, V\}$, where U and V are 1-placed relation symbols, one can express that φ is a direct product sentence by considering the validity of a suitable sentence which expresses $(\varphi^{(U)} \wedge \varphi^{(V)} \to \varphi^{(U \times V)})$. Will the same method work for direct power, reduced power or reduced product sentences?]

6.3.2*. Prove by the technique of Theorem 6.3.9 the following interpolation lemma. Suppose φ and ψ are sentences, with ψ true in every finite direct product (with one or more factors) of models of φ. Then there exists a finite direct product sentence θ such that $\varphi \vdash \theta$ and $\theta \vdash \psi$.

6.3.3*. Prove by the methods of Theorem 6.3.6 the following result: Every Π^0_{n+1} reduced (direct) product sentence φ is equivalent to a sentence of the form

$$(\forall x_1 \dots x_k)\psi,$$

where ψ is a Σ^0_n reduced (direct) product formula. (φ is a reduced product formula if $\varphi(c_1 \dots c_k)$, obtained from φ by replacing x_i by constants c_i, is a reduced product sentence.)

[*Hint*: Let S be any self-determining set of Σ^0_n formulas such that S is closed, up to equivalence, under both $\wedge$ and $\vee$. Suppose, for simplicity, that x is the only free variable occurring in S. For each $\psi \in S$, let ψ' be a formula of S equivalent to the disjunction of all formulas $\varphi_{\mathfrak{A},a} \in S$, where

$$\varphi_{\mathfrak{A},a} = \bigwedge \{\sigma \in S : \mathfrak{A} \models \sigma[a],\ a \in A, \text{ and } \mathfrak{A} \models (\forall x)\psi\}.$$

The mapping $\psi \to \psi'$ of S into S satisfies the following:

(i). $\vdash (\forall x)\psi \leftrightarrow (\forall x)\psi'$;

(ii). for every set I and filter D over I, and all formulas $\psi, \psi_i, i \in I$, of S, if

every reduced product $\prod_D \mathfrak{A}_i$ of models $\mathfrak{A}_i \models (\forall x)\psi_i$ is a model of $(\forall x)\psi$,

then

every reduced product $\prod_D \mathfrak{A}_i$ of models $\mathfrak{A}_i \models \psi_i'(c)$ is a model of $\psi'(c)$.]

6.3.4. Combining Exercise 6.3.3 and Proposition 6.2.9, show that every $\mathbf{\Pi}_2^0$ sentence that is preserved under direct products is equivalent to a $\mathbf{\Pi}_2^0$ Horn sentence.

6.3.5. Prove that every sentence φ containing only the identity symbol and which is closed under direct products is equivalent to a Horn sentence containing only the identity symbol.

[*Hint*: Use the results of Section 1.5 to show that every sentence φ containing only the identity symbol is equivalent to a $\mathbf{\Pi}_2^0$ sentence. Then apply Exercise 6.3.4.]

In an entirely analogous manner, every direct product sentence containing only $\equiv$ and 1-placed relation symbols is equivalent to a Horn sentence.

6.3.6. If $\mathscr{L}$ is countable, then for every I and models $\mathfrak{A}_i$, $i \in I$, there is a countable subset $K \subset I$ such that for all J, $K \subset J \subset I$, $\prod_{i \in J} \mathfrak{A}_i \equiv \prod_{i \in I} \mathfrak{A}_i$.

6.3.7*. Use Theorem 6.3.14 to show the following Cantor–Bernstein type result: Let $\mathfrak{A}, \mathfrak{B}, \mathfrak{C}$ be three models for $\mathscr{L}$. If $\mathfrak{A} \equiv \mathfrak{A} \times \mathfrak{B} \times \mathfrak{C}$, then $\mathfrak{A} \equiv \mathfrak{A} \times \mathfrak{B}$. (In general, we cannot replace $\equiv$ by $\cong$.)

6.3.8. Use Theorem 6.3.18 to show that: If two model $\mathfrak{A}$ and $\mathfrak{B}$ are equivalent under all Horn sentences of $\mathscr{L}$, then they are equivalent.

6.3.9. φ is said to be closed under *reduced roots* iff $\mathfrak{A} \models \varphi$ whenever $\prod_D \mathfrak{A} \models \varphi$. Prove the following consequence of Theorem 6.3.18: Every sentence is equivalent to a Boolean combination of reduced root sentences.

6.3.10*. A class K of models is said to be *compact* iff for all sets of sentences Σ, Σ has a model in K iff every finite subset if Σ has a model in K. Evidently, K is compact iff the closure of K under elementary equivalence is an elementary class. Prove that if K is compact, then the class of all arbitrary direct products of members of K is again compact.

6.3.11. Use the results of Section 5.5 to prove that the condition

$$\{\sigma\} \cup T_{\mathbf{BA}} \text{ is consistent (or is inconsistent)}$$

is a recursive predicate.

6.3.12. Show the following sharper form of Proposition 6.3.19: Let φ be an arbitrary sentence of $\mathscr{L}$. Then we can effectively find reduced product sentences $\psi_1, \ldots, \psi_m$ such that

(i). $\vdash \psi_j \to \varphi$ for $1 \leqslant j \leqslant m$;

(ii). for every reduced product sentence ψ, if $\vdash \psi \to \varphi$, then $\vdash \psi \to \psi_j$ for some j, $1 \leqslant j \leqslant m$.

6.3.13. Let φ be the sentence which says that there are not exactly three elements in the universe. Then φ is equivalent both to a Σ_2^0 sentence and to a Π_2^0 Horn sentence. However, φ is not equivalent to any Σ_2^0 Horn sentence.

6.3.14*. Prove that every Horn sentence φ which holds in the two-element Boolean algebra holds in every Boolean algebra.

[*Hint*: First eliminate the existential quantifiers in φ by using the fact that every Boolean function on $\{0, 1\}$ can be built up from $+$, $\cdot$, and $-$. Then use the fact that universal Horn sentences are preserved under subdirect products.]

6.3.15*. Assuming the continuum hypothesis, give a shorter proof of Ershov's theorem.

[*Hint*: Use Lemma 6.2.4 and the preceding exercise.]

6.4. Limit ultrapowers and complete extensions

The limit ultrapower construction is a generalization of the ultrapower construction. It shares many of the desirable properties of ultrapower; in particular, the fundamental theorem, the expansion theorem, and the existence of a natural embedding. In a number of problems in model theory, the ultrapower construction is not broad enough, and the limit ultrapower construction gives us the model we want. While the definition of an ultrapower is simpler, we shall see that the class of limit ultrapowers of a model $\mathfrak{A}$ is in some ways a much more natural class than the class of ultrapowers of $\mathfrak{A}$ (cf. Corollary 6.4.11). Before we come to the definition of limit ultrapower we shall study in some detail complete extensions of models. They are a very strong kind of elementary extension, and are intimately connected with limit ultrapowers.

Consider a model $\mathfrak{A}$ for $\mathscr{L}$ of infinite power α. There are exactly 2^α different (finitary) relations and functions on the set A. Let us expand the language $\mathscr{L}$ by adding a new symbol for each relation and function over A, and a constant symbol for each element of A. The new language $\mathscr{L}^{\#}$ will have 2^α new relation and function symbols, and α new constant symbols. Let $\mathfrak{A}^{\#}$ be the expansion of $\mathfrak{A}$ to $\mathscr{L}^{\#}$, where each new symbol is given its natural interpretation. We shall call $\mathfrak{A}^{\#}$ the *completion of* $\mathfrak{A}$.

Let $\mathfrak{A}$, $\mathfrak{B}$ be two models for $\mathscr{L}$ and let $\mathfrak{A}^{\#}$ be the completion of $\mathfrak{A}$. We shall say that a mapping f is a *complete embedding* of $\mathfrak{A}$ into $\mathfrak{B}$ iff there exists an expansion $\mathfrak{C}$ of $\mathfrak{B}$ to $\mathscr{L}^{\#}$ such that $f: \mathfrak{A}^{\#} \prec \mathfrak{C}$. In other words, f is an elementary embedding, and we can simultaneously extend all relations and functions on A to relations and functions on B such that f is still an elementary embedding from the expansion of $\mathfrak{A}$ into the expansion of $\mathfrak{B}$. Thus a complete embedding is an elementary embedding of a very special kind.

The natural embedding for ultrapowers gives us one example of a complete embedding.

PROPOSITION 6.4.1. *Let D be an ultrafilter and $\mathfrak{A}$ a model. Then the natural embedding d is a complete embedding of $\mathfrak{A}$ into the ultrapower $\prod_D \mathfrak{A}$.*

PROOF. Let $\mathfrak{A}^{\#}$ be the completion of $\mathfrak{A}$. By the fundamental theorem, $d: \mathfrak{A}^{\#} \prec \prod_D \mathfrak{A}^{\#}$. But by the expansion theorem, $\prod_D \mathfrak{A}^{\#}$ is an expansion of $\prod_D \mathfrak{A}$ to $\mathscr{L}^{\#}$. ⊣

We shall see more examples of complete embeddings later on. However, it may be instructive to mention one such example now.

EXAMPLE 6.4.2. Let $\mathfrak{A}$ be a model, let D be an ultrafilter, and let β be an infinite cardinal. Let

$$B = \{g_D : g \in {}^I A \quad \text{and} \quad |\text{range}(g)| < \beta\},$$

and let $\mathfrak{B}$ be the submodel of $\prod_D \mathfrak{A}$ with universe B. Then, by repeating the proof of the fundamental theorem, we can show that d is a complete embedding of $\mathfrak{A}$ into $\mathfrak{B}$. We leave the details as an exercise.

It is often easier to talk about extensions instead of embeddings. We shall say that $\mathfrak{B}$ is a *complete extension of* $\mathfrak{A}$ iff $\mathfrak{A} \subset \mathfrak{B}$ and the identity function is a complete embedding of $\mathfrak{A}$ into $\mathfrak{B}$. Let us call $\mathfrak{B}$ an *ultrapower extension of*

$\mathfrak{A}$ iff $\mathfrak{A} \subset \mathfrak{B}$ and there is an ultrapower $\prod_D \mathfrak{A}$ of $\mathfrak{A}$ such that

$$(\mathfrak{B}, a)_{a \in A} \cong (\prod_D \mathfrak{A}, d(a))_{a \in A}.$$

Thus, by Proposition 6.4.1, we have the following:

COROLLARY 6.4.3. *Every ultrapower extension of a model $\mathfrak{A}$ is a complete extension of $\mathfrak{A}$.*

The converse of the above corollary is false (cf. Exercise 6.4.6). However, the following is a weak converse.

THEOREM 6.4.4. *Let $\mathfrak{B}$ be a complete extension of $\mathfrak{A}$. Then for each $b \in \mathfrak{B}$, there exists an ultrapower extension $\mathfrak{C}$ of $\mathfrak{A}$ such that $\mathfrak{C} \prec \mathfrak{B}$ and $b \in C$.*

PROOF. Let $b \in B$. Let $\mathfrak{A}^{\#}$ be the completion of $\mathfrak{A}$, let $\mathscr{L}^{\#}$ be the language of $\mathfrak{A}^{\#}$, and let $\mathfrak{B}^{\#}$ be an expansion of $\mathfrak{B}$ to $\mathscr{L}^{\#}$ such that $\mathfrak{A}^{\#} \prec \mathfrak{B}^{\#}$. For each relation R on A, let R' be the corresponding relation in $\mathfrak{B}^{\#}$, and use a similar notation for functions. Consider the set

$$D = \{R \in S(A) : b \in R'\}.$$

Since $\mathfrak{A}^{\#} \prec \mathfrak{B}^{\#}$, it is easily seen that D is an ultrafilter over the set A. Consider the ultrapower $\prod_D \mathfrak{A}^{\#}$. To prove the theorem, it suffices to find an elementary embedding $g : \prod_D \mathfrak{A}^{\#} \prec \mathfrak{B}^{\#}$ such that $b \in$ range (g). It suffices because each element $a \in A$ is a constant of $\mathfrak{A}^{\#}$, whence $g(d(a)) = a$, and thus the range $\mathfrak{C}$ of g is an ultrapower extension of $\mathfrak{A}$ containing b, and $\mathfrak{C} \prec \mathfrak{B}$.

Let i be the identity function on A. Consider any formula $\varphi(x)$ of $\mathscr{L}^{\#}$, and let

$$R_\varphi = \{a \in A : \mathfrak{A}^{\#} \models \varphi[a]\}.$$

Then R_φ is a relation of $\mathfrak{A}^{\#}$, and

$$\mathfrak{A}^{\#} \models (\forall x)(\varphi(x) \leftrightarrow R_\varphi(x)). \tag{1}$$

Because of the definition of D, the following are equivalent:

$$\prod_D \mathfrak{A}^{\#} \models \varphi[i_D];$$

$$R_\varphi \in D;$$

$$b \in R'_\varphi;$$

$$\mathfrak{B}^{\#} \models \varphi[b].$$

The last two lines are equivalent because $\mathfrak{A}^{\#} \prec \mathfrak{B}^{\#}$, whence (1) holds for $\mathfrak{B}^{\#}$. Then

(2) $$(\prod_D \mathfrak{A}^{\#}, i_D) \equiv (\mathfrak{B}^{\#}, b).$$

Now consider an arbitrary element f_D of $\prod_D \mathfrak{A}^{\#}$. f is a function of $\mathfrak{A}^{\#}$; let f'' be the corresponding function of $\prod_D \mathfrak{A}^{\#}$. Then

$$f''(i_D) = \langle f(i(a)) : a \in A \rangle_D = f_D.$$

Let us add a new constant symbol c_f to $\mathscr{L}^{\#}$ for each $f \in {}^A A$. Expand the models $\prod_D \mathfrak{A}^{\#}$, $\mathfrak{B}^{\#}$ by interpreting c_f by f_D in the former and by $f'(b)$ in the latter. Then for each sentence $\varphi(c_{f_1} \dots c_{f_n})$ of the new language, the following are equivalent:

$$(\prod_D \mathfrak{A}^{\#}, f_D)_{f \in {}^A A} \models \varphi(c_{f_1} \dots c_{f_n});$$
$$\prod_D \mathfrak{A}^{\#} \models \varphi[f_1''(i_D) \dots f_n''(i_D)];$$
$$\prod_D \mathfrak{A}^{\#} \models \varphi(f_1(x) \dots f_n(x))[i_D];$$
$$\mathfrak{B}^{\#} \models \varphi(f_1(x) \dots f_n'(x))[b];$$
$$\mathfrak{B}^{\#} \models \varphi[f_1'(b) \dots f_n'(b)];$$
$$(\mathfrak{B}^{\#}, f'(b))_{f \in {}^A A} \models \varphi(c_{f_1} \dots c_{f_n}).$$

Thus

$$(\prod_D \mathfrak{A}^{\#}, f_D)_{f \in {}^A A} \equiv (\mathfrak{B}^{\#}, f'(b))_{f \in {}^A A}.$$

Therefore the map $f_D \to f'(b)$ is an elementary embedding on $\prod_D \mathfrak{A}^{\#}$ into $\mathfrak{B}^{\#}$ which maps i_D to b. ⊣

THEOREM 6.4.5 (Keisler-Rabin Theorem). *Let α be an infinite cardinal, and suppose that ω is the only measurable cardinal $\leqslant \alpha$. (That is, either α is less than the first measurable cardinal $> \omega$, or else ω is the only measurable cardinal.) Then the following are equivalent:*

(i). $\alpha = \alpha^{\omega}$.

(ii). *Every model $\mathfrak{A}$ of power α (with any number of relations) has a proper elementary extension of power α.*

(iii). *Every model $\mathfrak{A}$ of power α has a proper complete extension of power α.*

PROOF. (i) ⇒ (ii). Let D be a nonprincipal ultrafilter over ω. Then, for any model $\mathfrak{A}$ of power α, d is an elementary embedding of $\mathfrak{A}$ into $\prod_D \mathfrak{A}$.

Since A is infinite and D is not ω_1-complete, d is a proper embedding (Proposition 4.2.4). Moreover,

$$\alpha \leqslant |\prod_D A| \leqslant \alpha^\omega = \alpha.$$

Thus $\prod_D \mathfrak{A}$ is isomorphic to a proper elementary extension of $\mathfrak{A}$, of power α, and (ii) holds.

(ii) $\Rightarrow$ (iii). Let $\mathfrak{A}$ be a model of power α. Form the completion $\mathfrak{A}^\#$ of $\mathfrak{A}$. Applying (ii) to $\mathfrak{A}^\#$, there is a proper elementary extension $\mathfrak{B}^\#$ of power α. Then the reduct $\mathfrak{B}$ of $\mathfrak{B}^\#$ to $\mathscr{L}$ is a proper complete extension of $\mathfrak{A}$, so (iii) holds.

(iii) $\Rightarrow$ (i). Let $\mathfrak{A}$ be a model of power α and let $\mathfrak{B}$ be a proper complete extension of power α. Then there exists $b \in B \setminus A$. By Theorem 6.4.4, there is an ultrapower extension $\mathfrak{C}$ of $\mathfrak{A}$ with $b \in C$, $\mathfrak{C} \prec \mathfrak{B}$. Thus $\mathfrak{C}$ is isomorphic to some ultrapower $\prod_D \mathfrak{A}$, say $f: \mathfrak{C} \cong \prod_D \mathfrak{A}$, and for all $a \in A, f(a) = d(a)$. Since $b \in C \setminus A, f(b) \notin d(A)$, whence d maps A properly into $\prod_D A$. Then, by Proposition 4.2.4, D is not an α^+-complete ultrafilter. But since there is no measurable cardinal β such that $\omega_1 \leqslant \beta \leqslant \alpha$, it follows from Proposition 4.2.7 that D is not ω_1-complete, i.e. D is countably incomplete. Therefore D is ω-regular; so by Proposition 4.3.9, $|\prod_D A| = |\prod_D A|^\omega$. But $\alpha \leqslant |\prod_D A|$, so

$$\alpha^\omega \leqslant |\prod_D A|^\omega = |\prod_D A| = |C| \leqslant |B| = \alpha,$$

whence (i) holds. ⊣

Other applications of Theorem 6.4.4 are given as exercises.

We now come to the definition of the limit ultrapower. Roughly, a limit ultrapower of $\mathfrak{A}$ is a submodel of an ultrapower $\prod_D \mathfrak{A}$ made up of elements f_D, where f is 'almost constant', that is, the set of pairs $\langle i,j \rangle$ such that $f(i) = f(j)$ belongs to a given filter over $I \times I$.

Here is the formal definition: For each $g \in {}^I A$, the *equivalence relation determined* by g, denoted by eq (g) is defined as

$$\text{eq}(g) = \{\langle i,j \rangle \in I \times I : g(i) = g(j)\}.$$

Now let D be an ultrafilter over I and let V be a filter over $I \times I$. For a set A, the *limit ultrapower* $\prod_{D|V} A$ is the set

$$\prod_{D|V} A = \{g_D : g \in {}^I A \quad \text{and} \quad \text{eq}(g) \in V\}.$$

Before defining the limit ultrapower of a model we need the proposition below.

PROPOSITION 6.4.6. *Let $\mathfrak{A}$ be a model, D an ultrafilter over I, and V a filter over $I \times I$. Then $\prod_{D|V} A$ is a nonempty subset of $\prod_D A$ which is closed under the functions and constants of the model $\prod_D \mathfrak{A}$.*

PROOF. If f is a constant function in ${}^I A$, then $\mathrm{eq}(f) = I \times I$, so $\mathrm{eq}(f) \in V$ and $f_D \in \prod_{D|V} A$. This shows that, for all $a \in A$, $d(a) \in \prod_{D|V} A$. Therefore $\prod_{D|V} A$ is nonempty and contains all the constants of $\prod_D \mathfrak{A}$. Let G be an n-placed function of $\mathfrak{A}$, and H the corresponding function of $\prod_D \mathfrak{A}$. We show that if

(1) $$f_D^1, \ldots, f_D^n \in \prod_{D|V} A,$$

then

(2) $$H(f_D^1 \ldots f_D^n) \in \prod_{D|V} A.$$

Assume (1). Then we may suppose that

$$\mathrm{eq}(f^1) \in V, \ldots, \mathrm{eq}(f^n) \in V.$$

Hence

$$\mathrm{eq}(f^1) \cap \ldots \cap \mathrm{eq}(f^n) \in V.$$

But for all $\langle i, j \rangle \in I \times I$,

$$f^1(i) = f^1(j), \ldots, f^n(i) = f^n(j) \text{ implies } G(f^1(i) \ldots f^n(i)) = G(f^1(j) \ldots f^n(j)).$$

Thus

$$\mathrm{eq}(f^1) \cap \ldots \cap \mathrm{eq}(f^n) \subset \mathrm{eq}(\langle G(f^1(i) \ldots f^n(i)) : i \in I \rangle) \in V.$$

Since

$$H(f_D^1 \ldots f_D^n) = \langle G(f^1(i) \ldots f^n(i)) : i \in I \rangle_D,$$

the formula (2) follows. ⊣

We may now define the *limit ultrapower* $\prod_{D|V} \mathfrak{A}$ of the model $\mathfrak{A}$ to be the submodel of $\prod_D \mathfrak{A}$ whose universe is the set $\prod_{D|V} A$.

The model $\mathfrak{B}$ of Example 6.4.2 is just the limit ultrapower $\mathfrak{B} = \prod_{D|V} \mathfrak{A}$, where V is the filter over $I \times I$ generated by the set of all equivalence relations on I having $< \beta$ equivalence classes.

There is also such a thing as a limit ultraproduct which has been studied in the literature. We shall not take it up here. We shall next concern ourselves with a study of some model-theoretic properties of the limit ultrapower.

PROPOSITION 6.4.7 (Expansion Theorem). *If* $\mathfrak{A}'$ *is an expansion of* $\mathfrak{A}$ *to a language* $\mathscr{L}'$, *then a limit ultrapower* $\prod_{D|V}(\mathfrak{A}')$ *is an expansion of* $\prod_{D|V}\mathfrak{A}$ *to* $\mathscr{L}'$.

PROPOSITION 6.4.8 (Fundamental Theorem). *Let* $\prod_{D|V}\mathfrak{A}$ *be a limit ultrapower of* $\mathfrak{A}$. *Then* $\prod_{D|V}\mathfrak{A}$ *is an elementary submodel of the ultrapower* $\prod_D\mathfrak{A}$.

We leave the proofs of the two propositions above as exercises. Proposition 6.4.8, combined with the fundamental theorem for ultraproducts, shows that we have available the same criterion for truth in $\prod_{D|V}\mathfrak{A}$ as we have in $\prod_D\mathfrak{A}$.

COROLLARY 6.4.9. *Let* $\prod_{D|V}\mathfrak{A}$ *be a limit ultrapower of* $\mathfrak{A}$. *Then the natural embedding d of* $\mathfrak{A}$ *into* $\prod_D\mathfrak{A}$ *is also a complete embedding of* $\mathfrak{A}$ *into* $\prod_{D|V}\mathfrak{A}$. *Moreover, there is a complete extension* $\mathfrak{B}$ *of* $\mathfrak{A}$ *such that*

$$(\mathfrak{B}, a)_{a\in A} \cong (\prod_{D|V}\mathfrak{A}, d(a))_{a\in A}.$$

This time it turns out that the converse is also true.

THEOREM 6.4.10. *For all models* $\mathfrak{A}$, $\mathfrak{B}$, *the following are equivalent*:

(i). $\mathfrak{B}$ *is a complete extension of* $\mathfrak{A}$.

(ii). *There exists a limit ultrapower* $\prod_{D|V}\mathfrak{A}$ *of* $\mathfrak{A}$ *such that*

$$(\mathfrak{B}, a)_{a\in A} \cong (\prod_{D|V}\mathfrak{A}, d(a))_{a\in A}.$$

PROOF. (ii) $\Rightarrow$ (i). This is immediate from Corollary 6.4.9.

(i) $\Rightarrow$ (ii). Let $\mathfrak{A}^{\#}$ be the completion of $\mathfrak{A}$ in a language $\mathscr{L}^{\#}$, and let $\mathfrak{B}^{\#}$ be an expansion of $\mathfrak{B}$ to $\mathscr{L}^{\#}$ such that $\mathfrak{A}^{\#} \prec \mathfrak{B}^{\#}$. Then $\mathfrak{A}^{\#} \equiv \mathfrak{B}^{\#}$, whence by Corollary 4.3.13, $\mathfrak{B}^{\#}$ is elementarily embeddable in some ultrapower $\prod_D\mathfrak{A}^{\#}$. Let $\pi : \mathfrak{B}^{\#} \prec \prod_D\mathfrak{A}^{\#}$ be such an elementary embedding. Since each $a \in A$ is a constant of $\mathfrak{A}^{\#}$ and its interpretation in $\prod_D\mathfrak{A}^{\#}$ is $d(a)$, we have

(1) $$\pi(a) = d(a) \text{ for all } a \in A.$$

Let C be the range of π. We shall find a filter V over $I \times I$, where $I = \bigcup D$, such that $C = \prod_{D|V}A$.

Define V to be the filter over $I \times I$ generated by the sets

$$\{\mathrm{eq}(f) : f \in {}^I A \quad \text{and} \quad f_D \in C\}.$$

Then obviously

$$(2) \qquad C \subset \prod_{D|V} A.$$

Let $g_D \in \prod_{D|V} A$. Then there exists $f =_D g$ such that $\mathrm{eq}(f) \in V$. This means that there exist $h_D^1, \ldots, h_D^n \in C$ such that

$$\mathrm{eq}(h^1) \cap \ldots \cap \mathrm{eq}(h^n) \subset \mathrm{eq}(f).$$

It follows that there is an n-placed function G over A such that for all $i \in I$,

$$G(h^1(i) \ldots h^n(i)) = f(i).$$

Since $\mathfrak{A}^{\#}$ is the completion of $\mathfrak{A}$, G is a function of $\mathfrak{A}^{\#}$. Let G'' be the corresponding function of $\prod_D \mathfrak{A}^{\#}$. Then

$$G''(h_D^1 \ldots h_D^n) = f_D.$$

Since C is the image of π, C must be closed under all the functions of $\prod_D \mathfrak{A}^{\#}$. Thus $f_D \in C$. We have shown that

$$(3) \qquad \prod_{D|V} A \subset C.$$

From (2) and (3) we see that

$$(4) \qquad \pi : \mathfrak{B}^{\#} \cong \prod_{D|V} \mathfrak{A}^{\#}.$$

The desired conclusion (ii) now follows from (1) and (4), taking reducts to $\mathscr{L}$. ⊣

Corollary 6.4.11. *$\mathfrak{B}$ is isomorphic to a complete extension of $\mathfrak{A}$ iff $\mathfrak{B}$ is isomorphic to a limit ultrapower of $\mathfrak{A}$.*

The following corollary is an application of limit ultrapowers to give a result involving complete embeddings only.

Corollary 6.4.12. *Suppose f is a complete embedding of $\mathfrak{A}$ into $\mathfrak{B}$. Then for any extension $\mathfrak{A}' \supset \mathfrak{A}$, there exists an extension $\mathfrak{B}' \supset \mathfrak{B}$ such that f can be extended to a complete embedding f' of $(\mathfrak{A}', A)$ into $(\mathfrak{B}', B)$.*

Proof. By Theorem 6.4.10, there is a limit ultrapower $\prod_{D|V} \mathfrak{A}$ and an isomorphism

$$\pi : (\mathfrak{B}, f(a))_{a \in A} \cong \Big(\prod_{D|V} \mathfrak{A}, d(a)\Big)_{a \in A}.$$

Form the limit ultrapower $\prod_{D|V}(\mathfrak{A}', A)$. Let us identify, for $f \in {}^I A$, the equivalence class f_D in the sense of $\prod_{D|V} A$ with the equivalence class f_D in the sense of $\prod_{D|V} A'$. With this identification, $\prod_{D|V} \mathfrak{A}$ is a submodel of $\prod_{D|V} \mathfrak{A}'$, and $\prod_{D|V}(\mathfrak{A}', A) = (\prod_{D|V} \mathfrak{A}', \prod_{D|V} A)$. It follows that there is an extension $\mathfrak{B}' \supset \mathfrak{B}$ and an isomorphism

$$\pi' : (\mathfrak{B}', B) \cong \prod_{D|V} (\mathfrak{A}', A)$$

such that $\pi \subset \pi'$. Define the map $f' : A' \to B'$ by

$$f'(a) = \pi'^{-1} d(a).$$

Since d is a complete embedding of $(\mathfrak{A}', A)$ into its limit ultrapower, f' is a complete embedding of $(\mathfrak{A}', A)$ into $(\mathfrak{B}', B)$. Finally, f' is an extension of f, for, whenever $a \in A$,

$$f'(a) = \pi'^{-1} d(a) = \pi^{-1} d(a) = f(a). \dashv$$

Notice that in the above corollary there is no limit on the size of the model $\mathfrak{A}'$. For example, we have the following:

COROLLARY 6.4.13. *Suppose α is an ordinal and $f : \langle R(\alpha), \in \rangle \prec \mathfrak{B}$. If there exist extensions $\mathfrak{B}' \supset \mathfrak{B}$ and $f' \supset f$ such that $f' : \langle R(\alpha+1), \in \rangle \prec \mathfrak{B}'$, then for all ordinals $\beta > \alpha$ there exist extensions $\mathfrak{B}'' \supset \mathfrak{B}$ and $f'' \supset f$ such that $f'' : \langle R(\beta), \in \rangle \prec \mathfrak{B}''$.*

PROOF. Every relation, function and constant of $R(\alpha)$ is an element of $R(\alpha+1)$. Therefore f is a complete embedding of $\langle R(\alpha), \in \rangle$ into $\mathfrak{B}$, and the result follows. $\dashv$

COROLLARY 6.4.14. *Suppose f is a complete embedding of $\langle \alpha, < \rangle$ into a model $\mathfrak{B}$. Then there exist an extension $\mathfrak{B}' \supset \mathfrak{B}$ and $f' \supset f$ such that:*

(i). *f' is a complete embedding of $\langle R(\alpha), \in \rangle$ into $\mathfrak{B}'$;*

(ii). *B is the set of all ordinals of the model $\mathfrak{B}'$.*

PROOF. By Corollary 6.4.12, there is a complete embedding $f' \supset f$ of $(\langle R(\alpha), \in \rangle, \alpha)$ into $(\mathfrak{B}', \mathfrak{B})$ for some $\mathfrak{B}' \supset \mathfrak{B}$. Then f' is a complete embedding of $\langle R(\alpha), \in \rangle$ into $\mathfrak{B}'$. Since α is the set of ordinals of $(\langle R(\alpha), \in \rangle, \alpha)$, B is the set of ordinals of $(\mathfrak{B}', B)$, whence (ii) holds. $\dashv$

We conclude this section with some applications of the limit ultrapower construction to the nonstandard universes studied in Section 4.4. We begin with an improvement of Proposition 4.4.10.

PROPOSITION 6.4.15. *Let* $\langle V(X), V(Y), * \rangle$ *be a nonstandard universe and let* $\mathfrak{A}$ *be a model whose universe* A *is a set relative to* $V(X)$. *Then the restriction of* $*$ *to* A *is a complete embedding of* $\mathfrak{A}$ *into* $^*\mathfrak{A}$.

PROOF. The completion $\mathfrak{B} = \mathfrak{A}^{\#}$ of $\mathfrak{A}$ also has the universe A which is a set relative to $V(X)$. By Proposition 4.4.10, the restriction of $*$ to A is an elementary embedding of $\mathfrak{B}$ into $^*\mathfrak{B}$, and $^*\mathfrak{B}$ is an expansion of $^*\mathfrak{A}$. ⊣

If we define a complete bounded elementary embedding in a manner analogous to our definition of a complete elementary embedding, then Theorem 4.4.16 shows that for any nonstandard universe $\langle V(X), V(Y), * \rangle$, $*$ is a complete bounded elementary embedding of $\langle V(X), \in \rangle$ into $\langle {}^*V(X), \in \rangle$.

It follows from 6.4.11 and 6.4.15 and $^*\mathfrak{A}$ is isomorphic to a limit ultrapower of $\mathfrak{A}$. We now wish to obtain an analogous result for the whole superstructure $\langle V(X), \in \rangle$. Our plan is to introduce a notion of a bounded limit ultrapower and show that every nonstandard universe can be "represented" as a bounded limit ultrapower.

Let D be an ultrafilter over an index set I and let U be a filter over $I \times I$. Assume that D and U have the property that for every infinite set A, the natural embedding $d : A \to \prod_{D|U} A$ is a proper embedding. We imitate the definition of a bounded ultrapower in Section 4.4 but restrict ourselves to functions $f : I \to V(X)$ such that $\mathrm{eq}(f) \in U$ when appropriate. Let Y be the ordinary limit ultrapower $\prod_{D|U} X$ of X. For each natural number n, let

$$W_n = \{f \in {}^I V(X) : \{i : f(i) \in V_n(X)\} \in D \text{ and } \mathrm{eq}(f) \in U\}.$$

Let $W = \bigcup_{n<\omega} W_n$. For each $x \in V(X)$, let $c(x)$ be the constant function at x. As in Section 4.4, there is a unique function $/D : W \to V(Y)$ such that

$$f/D = \{g \in {}^I X : \mathrm{eq}(g) \in U \text{ and } f =_D g\}$$

if $f \in W_0$, and

$$f/D = \{g/D : g \in W \text{ and } \{i : g(i) \in f(i)\} \in D\}$$

if $f \in W \setminus W_0$.

Now define the embedding $* : V(X) \to V(Y)$ by $^*A = c(A)/D$. We call the triple $\langle V(X), V(Y), * \rangle$ the *bounded limit ultrapower* of $V(X)$ modulo $D|U$.

THEOREM 6.4.16. *Each bounded limit ultrapower of $V(X)$ is a nonstandard universe.*

The proof of Theorem 6.4.16 follows the same pattern as the proof in Section 4.4 that a bounded ultrapower is a nonstandard universe, and requires an analogue of Łoś' theorem for bounded limit ultrapowers. We now state a converse result.

THEOREM 6.4.17 (Representation Theorem for Nonstandard Universes). *Every nonstandard universe $\langle V(X), V(Y), * \rangle$ is a bounded limit ultrapower of $V(X)$.*

PROOF. By Exercise 4.4.29, there is an ultrafilter D with bounded ultrapower $\langle V(X), V(Y), \rho \rangle$ and a bounded elementary embedding $h : \langle {}^*V(X), \in \rangle \to \langle {}^{\rho}V(X), \in \rangle$ such that $h({}^*A) = {}^{\rho}A$ for all $A \in V(X)$. By the argument used to prove Theorem 6.4.10, one can find a filter U over $I \times I$ such that $\langle V(X), V(Y), * \rangle$ is the bounded limit ultrapower of $V(X)$ modulo $D|U$. ⊣

COROLLARY 6.4.18. *For every nonstandard universe $\langle V(X), V(Y), * \rangle$ there is a model $\mathfrak{A}$ for $\mathscr{L} = \{\in\}$ such that $\langle {}^*V(X), \in \rangle$ is the truncation of $\mathfrak{A}$ and $*$ is an ordinary elementary embedding of $\langle V(X), \in \rangle$ into $\mathfrak{A}$.*

PROOF. By Theorem 6.4.17, there exist D and U such that $\langle V(X), V(Y), * \rangle$ is a bounded limit ultrapower of $V(X)$ modulo $D|U$. Let h be the Mostowski collapse of the ordinary limit ultrapower $\prod_{D|U} \langle V(X), \in \rangle$. Proposition 4.4.18 on bounded ultrapowers can readily be extended to the case of bounded limit ultrapowers, so that ${}^*V(X)$ is the range of h and for all $f \in W$, $h(f_D) = f/D$. For each $x \in V(X)$, ${}^*x = h(d(x))$. By Corollary 6.4.8, d is an elementary embedding of $\langle V(X), \in \rangle$ into $\prod_{D|U} \langle V(X), \in \rangle$. By renaming elements and replacing $d(x)$ by ${}^*x$ for each $x \in V(X)$, we may replace $\prod_{D|U} \langle V(X), \in \rangle$ by an isomorphic model $\mathfrak{A}$ such that $\langle {}^*V(X), \in \rangle$ is the truncation of $\mathfrak{A}$ and $*$ is an elementary embedding of $\langle V(X), \in \rangle$ into $\mathfrak{A}$. ⊣

EXERCISES

6.4.1. If $\mathfrak{A}^{\#}$ is the completion of $\mathfrak{A}$, then the theory of $\mathfrak{A}^{\#}$ has built-in Skolem functions.

6.4.2. $\mathfrak{B}$ belongs to every pseudoelementary class which contains $\mathfrak{A}$ iff $\mathfrak{B}$ is isomorphic to a complete extension of $\mathfrak{A}$. (See Exercise 4.1.17.)

6.4.3. Let f be a complete embedding of $\mathfrak{A}$ into $\mathfrak{B}$. Then:

(i). If $\mathscr{L}' \subset \mathscr{L}$, then f is a complete embedding from the reduct of $\mathfrak{A}$ to $\mathscr{L}'$ into the reduct of $\mathfrak{B}$ to $\mathscr{L}'$.

(ii). If $\mathscr{L} \subset \mathscr{L}'$ and $\mathfrak{A}'$ is an expansion of $\mathfrak{A}$ to $\mathscr{L}'$, then there exists an expansion $\mathfrak{B}'$ of $\mathfrak{B}$ to $\mathscr{L}'$ such that f is a complete embedding of $\mathfrak{A}'$ into $\mathfrak{B}'$.

6.4.4. (i). If $\mathfrak{C}$ is a complete extension of $\mathfrak{B}$ and $\mathfrak{B}$ is a complete extension of $\mathfrak{A}$, then $\mathfrak{C}$ is a complete extension of $\mathfrak{A}$.

(ii). If $\mathfrak{A}_{n+1}$ is a complete extension of $\mathfrak{A}_n$ for all $n < \omega$, then $\bigcup_{m<\omega} \mathfrak{A}_m$ is a complete extension of each $\mathfrak{A}_n$. Generalize to well ordered chains of arbitrary length.

6.4.5*. Let $\mathfrak{A} = \langle A, L, \ldots \rangle$ be a model such that L is a simple ordering of cofinality $> \omega$. Then $\mathfrak{A}$ has complete extensions $\mathfrak{B} = \langle B, M, \ldots \rangle$ of arbitrarily large cardinality such that A is cofinal in $\langle B, M \rangle$.

6.4.6. Let $\omega \leqslant \alpha$, $2^\alpha \leqslant \beta$. Then every model $\mathfrak{A}$ of power α has a complete extension $\mathfrak{B}$ of power β. If ω is the only measurable cardinal $\leqslant \alpha$ and if $\mathrm{cf}(\beta) = \omega$, then the complete extension $\mathfrak{B}$ is not an ultrapower extension of $\mathfrak{A}$.

6.4.7* (i). No proper ultrapower extension of $\langle \omega, < \rangle$ has cofinality ω.

(ii). $\langle \omega, < \rangle$ has a proper complete extension of cofinality ω.

6.4.8. Prove the assertion in Example 6.4.2.

6.4.9. Let D be an ultrafilter over a set I and let T be a topology on I. Let $\mathfrak{A}$ be a model and give A the discrete topology. Let

$$B = \{f_D : f : I \to A \text{ and } f \text{ is continuous}\}.$$

Let $\mathfrak{B}$ be the submodel of $\prod_D \mathfrak{A}$ with universe B. Prove that $\mathfrak{B} \prec \prod_D \mathfrak{A}$ and d is a complete embedding of $\mathfrak{A}$ into $\mathfrak{B}$.

6.4.10. Suppose ω is the only measurable cardinal $\leqslant \alpha$ and let $\mathfrak{A}$ be an (α, ω)-model. Show that there is a pseudoelementary class K such that $\mathfrak{A} \in K$, and every (α, ω)-model $\mathfrak{B} \in K$ is isomorphic to $\mathfrak{A}$.

6.4.11. Suppose ω is the only measurable cardinal $\leqslant \alpha$, and let $\mathfrak{A}$ be a model of power α. Let $\Sigma(x)$ be a countable set of formulas such that $\mathfrak{A}$ omits Σ and Σ is consistent with the theory of $\mathfrak{A}$. Find a pseudoelementary class K such that $\mathfrak{A} \in K$ and every model $\mathfrak{B} \in K$ which omits Σ is isomorphic to $\mathfrak{A}$.

6.4.12*. Suppose there exists an uncountable measurable cardinal $\beta \leqslant \alpha$. Assume the GCH. Prove that every model of power α has a proper complete extension of power α.

[*Hint*: Show that if $\beta < \alpha < \alpha^\omega$ and D is a countably complete ultrafilter over β, then $|\prod_D \alpha| = \alpha$.]

6.4.13. Let α be an ordinal, and suppose ω is the only measurable cardinal $\leqslant \alpha$. Assume that $\langle R(\alpha), \epsilon\rangle \prec \langle B, E\rangle$ and ω is 'fixed' in the sense that $\{b \in B : bE\omega\} = \omega$. Then:

(i). If α is a successor ordinal $\alpha = \gamma+1$, then $\langle B, E\rangle = \langle R(\alpha), \epsilon\rangle$.

(ii). In general, $\langle B, E\rangle$ is an 'end extension' of $\langle R(\alpha), \epsilon\rangle$ in the sense that

$$\{b \in B : \text{for some } a \in R(\alpha),\ bEa\} = R(\alpha).$$

6.4.14. Let α be a cardinal and suppose $\langle B, L\rangle$ is a complete extension of $\langle \alpha, <\rangle$ such that α is not cofinal in $\langle B, L\rangle$. Then there exists an ultrafilter D which is not descendingly α-complete (see Exercise 4.3.10), such that the model $(\prod_D\langle \alpha, <\rangle, d(\beta))_{\beta\in\alpha}$ is elementarily embeddable in $(\langle B, L\rangle, \beta)_{\beta\in\alpha}$.

[*Hint*: Use Exercise 4.3.34.]

6.4.15. Let $\langle B', B, L\rangle$ be a complete extension of the model $\langle \alpha^+, \alpha, <\rangle$. If α is cofinal in $\langle B, L\rangle$, then α^+ is cofinal in $\langle B', L\rangle$.

[*Hint*: Use Exercise 4.3.12.]

6.4.16. If V is the improper filter $V = S(I\times I)$, then $\prod_{D|V}\mathfrak{A} = \prod_D \mathfrak{A}$. Thus every ultrapower is a limit ultrapower.

6.4.17. If V is the trivial filter $V = \{I\times I\}$, then $\prod_{D|V}\mathfrak{A} = d(\mathfrak{A})$, where d is the natural embedding.

6.4.18. The limit ultrapower $\prod_{D|V}\mathfrak{A}$ depends only on the equivalence relations in V. That is, if $\mathscr{E}$ is the set of all equivalence relations on I, and $V \cap \mathscr{E} = W \cap \mathscr{E}$, then $\prod_{D|V}\mathfrak{A} = \prod_{D|W}\mathfrak{A}$.

6.4.19. Prove the results Proposition 6.4.7, Proposition 6.4.8, and Corollary 6.4.9.

6.4.20. Suppose V, W are filters over $I\times I$ and $V \subset W$. Then $\prod_{D|V}\mathfrak{A} \prec \prod_{D|W}\mathfrak{A}$.

6.4.21. In Exercise 6.4.9, $\mathfrak{B}$ is a limit ultrapower of $\mathfrak{A}$.

6.4.22. When D is a proper filter over I and V is a filter over $I \times I$, define the *limit reduced power* $\prod_{D|V} \mathfrak{A}$, and verify Proposition 6.4.6 for it.

6.4.23*. For any two models $\mathfrak{A}$, $\mathfrak{B}$ such that $\mathfrak{A} \equiv \mathfrak{B}$, there exists a model $\mathfrak{C}$ such that both $\mathfrak{A}$ and $\mathfrak{B}$ are completely embeddable in $\mathfrak{C}$.

[*Hint*: Use Exercise 4.3.29.]

Furthermore, $\mathfrak{C}$ may be chosen of power $2^{|A|} \cup 2^{|B|}$.

6.4.24. $\mathfrak{A} \equiv \mathfrak{B}$ iff there exist limit ultrapowers $\prod_{D|V} \mathfrak{A} \cong \prod_{E|W} \mathfrak{B}$. Further, we may take the limit ultrapowers to be of power $2^{|A|} \cup 2^{|B|}$.

6.4.25. If V is a principal filter over $I \times I$, then $\prod_{D|V} \mathfrak{A}$ is isomorphic to an ultrapower of $\mathfrak{A}$.

6.4.26. Suppose $\mathfrak{B}$ is a complete extension of $\mathfrak{A}$. Then there exists a set M of models such that:

(i). For each $\mathfrak{C} \in M$, $\mathfrak{C}$ is an ultrapower extension of $\mathfrak{A}$ and $\mathfrak{C} \prec \mathfrak{B}$, and $|C| \leqslant 2^{|A|}$.

(ii). For all $\mathfrak{C}_1, \mathfrak{C}_2 \in M$, there exists $\mathfrak{C} \in M$ such that $\mathfrak{C}_1 \prec \mathfrak{C}$, $\mathfrak{C}_2 \prec \mathfrak{C}$ (i.e., M is directed).

(iii). $\mathfrak{B} = \bigcup M$.

This justifies the term 'limit ultrapower'.

[*Hint*: Use Theorem 6.4.10 and Exercises 6.4.20 and 6.4.25.]

6.4.27. K is an elementary class iff K is closed under ultraproducts and both K and its complement are closed under limit ultrapowers and isomorphism.

6.4.28*. Let α be a regular cardinal. Let $\mathfrak{A}$ be a model $\mathfrak{A} = \langle \alpha, <, \ldots \rangle$ and suppose $\mathfrak{B} = \langle B, L, \ldots \rangle$ is a complete extension of $\mathfrak{A}$. Let

$$B_0 = \{b \in B : bLa \text{ for some } a \in \alpha\},$$

and let $\mathfrak{B}_0$ be the submodel of $\mathfrak{B}$ with universe B_0. Prove that $\mathfrak{B}_0$ is also a complete extension of $\mathfrak{A}$.

[*Hint*: Show that if $\mathfrak{B} \cong \prod_{D|V} \mathfrak{A}$, then $\mathfrak{B}_0 \cong \prod_{D|W} \mathfrak{A}$, where W is the filter such that $\text{eq}(f) \in W$ iff $\text{eq}(f) \in V$ and $|\text{range} f| < \alpha$.]

6.4.29*. Suppose α is an inaccessible cardinal, $\langle R(\alpha), \in \rangle \prec \langle B, E \rangle$, and for all $b \in B$ there exists $a \in R(\alpha)$ such that bEa. Then $\langle B, E \rangle$ is a complete extension of $\langle R(\alpha), \in \rangle$.

[*Hint*: Every relation and function on $R(\alpha)$ of power $< \alpha$ belongs to $R(\alpha)$. Use this fact and the method proof of Theorem 6.4.10 to show that for some limit ultrapower $\prod_{D|V}\langle R(\alpha), \in\rangle$, we have

$$\langle B, E, a\rangle_{a\in R(\alpha)} \cong (\prod_{D|V} \langle R(\alpha), \in\rangle, d(a))_{a\in R(\alpha)}.]$$

6.4.30*. Let $\mathfrak{A}$ be a model with built-in Skolem functions. A function $f \in {}^A A$ is said to be *definable* iff there exist a formula $\varphi(xyz_1 \dots z_n)$ and elements $a_1, \dots, a_n \in A$ such that for all $a, b \in A$,

$$\mathfrak{A} \models \varphi[aba_1 \dots a_n] \text{ iff } f(a) = b.$$

Let D be any ultrafilter over A, and let

$$B = \{f_D \in \prod_D A : f \text{ is definable}\}.$$

Let $\mathfrak{B}$ be the submodel $\mathfrak{B} \subset \prod_D \mathfrak{A}$ with universe B. Show that:

(i). $d(\mathfrak{A}) \prec \mathfrak{B} \prec \prod_D \mathfrak{B}$.

(ii). If D is nonprincipal, then $d(\mathfrak{A}) \neq \mathfrak{B}$.

(iii). If $\mathfrak{A}$, $\mathscr{L}$ are countable, so is $\mathfrak{B}$. In general, $|B| \leqslant |A| \cup ||\mathscr{L}||$.

(iv). If $\mathfrak{A}$ is the completion of model $\mathfrak{C}$, then $\mathfrak{B} = \prod_D \mathfrak{A}$.

6.4.31. Generalize the construction of the preceding exercise by considering a definable subset $U \subset A$, taking an ultrafilter D over U and

$$B = \{f_D : f \in {}^U A \quad \text{and} \quad f \text{ is definable in } \mathfrak{A}\}.$$

6.4.32. Let $\mathfrak{A} = \langle A, U, \dots\rangle$ be a model for a countable language which has built-in Skolem functions. Let D be an ultrafilter over A such that for every definable function $f \in {}^A A$, either

$$\{a \in A : f(a) \notin U\} \in D,$$

or there exists $u \in U$ such that

$$\{a \in A : f(a) = u\} \in D.$$

If $\mathfrak{B} = \langle B, V, \dots\rangle$ is the model constructed in Exercise 4.3.30, show that $V = d(U)$.

6.4.33. Use the preceding exercises to give a new proof of Theorem 3.2.14.

6.4.34*. Let $\mathfrak{A}$ be a (not necessarily countable) model of ZF, let x be a regular cardinal of $\mathfrak{A}$, and suppose the set

$$\{y \in A : \mathfrak{A} \models y \in 2^{2^x}\}$$

is countable. Then $\mathfrak{A}$ has an elementary extension $\mathfrak{B}$ such that x is fixed and

2^x is enlarged, i.e.

$$\{y \in B : \mathfrak{B} \models y \in x\} \subset A, \qquad \{y \in B : \mathfrak{B} \models y \in 2^x\} \setminus A \neq 0,$$

and, furthermore,

$$\{y \in B : \mathfrak{B} \models y \in 2^{2^x}\}$$

is still countable. Iterating this, show that there is an elementary extension $\mathfrak{C}$ of $\mathfrak{A}$ such that

$$\{y \in C : \mathfrak{C} \models y \in x\} \subset A, \quad |\{y \in C : \mathfrak{C} \models y \in 2^x\}| = \omega_1.$$

6.4.35 (GCH). Let $\mathfrak{A}$ be a saturated model of power α^+ for a language $\mathscr{L}$, where $||\mathscr{L}|| \leqslant \alpha$. Then every model $\mathfrak{B} \equiv \mathfrak{A}$ of power $\leqslant \alpha^+$ has a complete embedding into $\mathfrak{A}$. If $\mathfrak{B} \prec \mathfrak{A}$ and $|B| \leqslant \alpha$, then $\mathfrak{A}$ is a complete extension of $\mathfrak{B}$.

6.4.36*. Give an example of a set Σ of sentences in an uncountable language $\mathscr{L}$ such that:

(i). Σ has a countable model.

(ii). Every countable subset Σ_0 of Σ has a countable model $\mathfrak{A}$, for the language $\mathscr{L}_0$ having just the symbols occurring in Σ_0, such that $\mathfrak{A}$ cannot be expanded to a model of Σ.

[*Hint*: Use Theorem 6.4.5 with $\alpha = \omega$, and pick Σ so that the successor function symbol occurs in every member of Σ.]

6.4.37*. In the above exercise, find a Σ which also has the property:

(iii). For each pair S, T of symbols of $\mathscr{L}$, there are only countably many $\sigma \in \Sigma$ such that both S and T occur in σ.

6.4.38*. Let Σ be a set of sentences in $\mathscr{L}$ such that each symbol of $\mathscr{L}$ occurs in only countably many members of Σ. Then there is a countable subset Σ_0 of Σ such that every model $\mathfrak{A}$ of Σ_0, for the language $\mathscr{L}_0$ having just the symbols occurring in Σ_0, can be expanded to a model of Σ.

6.4.39*. Let $V(X)$ be a superstructure with infinite base set X. Prove that for each uncountable cardinal $\alpha \leqslant |X|$ there are nonstandard universes $\langle V(X), V(Y), * \rangle$ and $\langle V(X), V(Z), \rho \rangle$ such that ${}^*A \subset {}^{\rho}A$ for all $A \subset X$, and ${}^*A = {}^{\rho}A$ if and only if $|A| < \alpha$. [*Hint*: Use an appropriate bounded ultrapower and bounded limit ultrapower.]

6.4.40*. Let $V(X)$ be a superstructure with infinite base set X. Prove that for each $n > 1$ there are nonstandard universes $\langle V(X), V(Y), * \rangle$ and $\langle V(X), V(Y), \rho \rangle$ such that ${}^*A = {}^{\rho}A$ for all $A \in V_n(X)$, and ${}^*V_n(X)$ is a proper subset of ${}^{\rho}V_n(X)$.

6.5. Iterated ultrapowers

We now take up the following question: Can we construct additional models by iterating the ultrapower construction? The answer is yes, and the models so constructed are useful. The iterated ultrapowers will always be limit ultrapowers, but the particular way in which they are constructed can give them desirable properties which are not shared by arbitrary limit ultrapowers. Our starting point is a result showing that finite iterations of ultrapowers give nothing new.

Consider a pair of ultrafilters D, E over sets I, J. We define $D \times E$ to be the set of all $Y \in S(I \times J)$ such that

$$\{j \in J : \{i \in I : \langle i, j\rangle \in Y\} \in D\} \in E.$$

Suppose now that $D_1, \ldots, D_{n+1}$ are ultrafilters over sets $I_1, \ldots, I_{n+1}$; then we define inductively

$$D_1 \times \ldots \times D_{n+1} = (D_1 \times \ldots \times D_n) \times D_{n+1}.$$

PROPOSITION 6.5.1. *In the above notation, $D \times E$ is an ultrafilter over $I \times J$, and $D_1 \times \ldots \times D_n$ is an ultrafilter over $I_1 \times \ldots \times I_n$.*

PROPOSITION 6.5.2 (Finite Iteration Theorem). *Let D and E be ultrafilters. Then for any model $\mathfrak{A}$,*

$$\prod_{D\times E} \mathfrak{A} \cong \prod_{E} \Big(\prod_{D} \mathfrak{A}\Big).$$

If $D_1, \ldots, D_n$ are ultrafilters, $n > 0$, then

$$\prod_{D_1 \times \ldots \times D_n} \mathfrak{A} \cong \prod_{D_n} \Big(\ldots \prod_{D_2} \Big(\prod_{D_1} \mathfrak{A}\Big) \ldots \Big).$$

PROOF. The second statement follows from the first by induction, so we need only prove the first statement. Let $f \in {}^{I\times J}A$. For each $j \in J$, let $f_j \in {}^{I}A$ be the function

$$f_j = \langle f(ij) : i \in I\rangle.$$

We shall show that the relation associating $f_{D\times E}$ with $f^* = \langle f_{jD} : j \in J\rangle_E$ is an isomorphism of $\prod_{D\times E} \mathfrak{A}$ onto $\prod_E (\prod_D \mathfrak{A})$. Consider any formula $\varphi(x_1 \ldots x_n)$ of $\mathscr{L}$ and any $f^1, \ldots, f^n \in {}^{I\times J}A$. It suffices to prove

(1) $$\prod_{D\times E} \mathfrak{A} \models \varphi[f^1_{D\times E} \ldots f^n_{D\times E}]$$

iff

(2) $$\prod_{E} \Big(\prod_{D} \mathfrak{A}\Big) \models \varphi[f^{1*} \ldots f^{n*}].$$

By the fundamental theorem, (1) is equivalent to

(3) $$\{\langle i,j\rangle : \mathfrak{A} \models \varphi[f^1(ij) \ldots f^n(ij)]\} \in D \times E.$$

But (3) in turn is equivalent to (4) and (5):

(4) $$\{j : \{i : \mathfrak{A} \models \varphi[f^1(ij) \ldots f^n(ij)]\} \in D\} \in E,$$

(5) $$\{j : \prod_D \mathfrak{A} \models \varphi[f^1_{jD} \ldots f^n_{jD}]\} \in E.$$

If the fundamental theorem is involved a third time, (5) is equivalent to (2). ⊣.

Examples show that in general, $\prod_{D \times E}\mathfrak{A}$ and $\prod_{E \times D}\mathfrak{A}$ are not isomorphic. Thus the order in which we iterate the ultrapower is important. For the general case, suppose we are given a simply ordered (nonempty) set $\langle Y, <\rangle$ and for each $y \in Y$ an ultrafilter D_y over a set I_y. When the ordering $<$ is understood, we shall also let Y denote the ordered set $\langle Y, <\rangle$. We shall define an iterated ultrapower of $\mathfrak{A}$ modulo D_y. Roughly speaking, it is the smallest model $\mathfrak{B}$ in which all the finite iterations $\prod_{D_{y_1} \times \ldots \times D_{y_n}} \mathfrak{A}$, $y_1 < \ldots < y_n$ in Y, can be embedded in a natural way. When Y is a finite set, say $Y = \{y_1, \ldots, y_n\}$ in increasing order, then the iterated ultrapower of $\mathfrak{A}$ modulo D_y will coincide with the ultrapower $\prod_{D_{y_1} \times \ldots \times D_{y_n}} \mathfrak{A}$.

Let $K = \prod_{y \in Y} I_y$ be the Cartesian product of the index sets I_y. Let $Z \subset Y$. We say that a function f with domain K *lives on* Z iff $f(i)$ depends only on $i \restriction Z$, that is, for all $i, j \in K$,

$$i \restriction Z = j \restriction Z \text{ implies } f(i) = f(j).$$

A subset $s \subset K$ is said to *live on* Z iff the characteristic function of s lives on Z. Thus s lives on Z iff for all $i, j \in K$,

$$i \restriction Z = j \restriction Z \text{ implies } i \in s \text{ iff } j \in s.$$

By the *trace* of s on $\{y_1, \ldots, y_n\}$, denoted by $s_{y_1 \ldots y_n}$, we mean the set

$$s_{y_1 \ldots y_n} = \{\langle i(y_1) \ldots i(y_n)\rangle : i \in s\}.$$

Thus $s_{y_1 \ldots y_n}$ is a subset of $I_{y_1} \times \ldots \times I_{y_n}$. We now define a *product* $\times_Y D_y$ of the ultrafilters D_y as follows:

$\times_Y D_y$ is the set of all $s \subset K$ such that, for some $y_1 < \ldots < y_n$ in Y:

(1) s lives on $\{y_1, \ldots, y_n\}$;

(2) $s_{y_1 \ldots y_n} \in D_{y_1} \times \ldots \times D_{y_n}$.

Thus $\bigtimes_Y D_y$ is a set of subsets of K, and each of its elements lives on a finite subset of K. Before we can go any further, we need a lemma which is harder to prove than we would expect. For this reason we are giving the proof in some detail, even though it makes terrible reading.

LEMMA 6.5.3. *Suppose $s \subset K$, and s lives on both the sets $Y_0 = \{y_1, \ldots, y_m\}$ and $Z_0 = \{z_1, \ldots, z_n\}$, where $y_1 < \ldots < y_m$, $z_1 < \ldots < z_n$ in Y. Then*

$$s_{y_1 \ldots y_m} \in D_{y_1} \times \ldots \times D_{y_m}$$

iff

$$s_{z_1 \ldots z_n} \in D_{z_1} \times \ldots \times D_{z_n}.$$

PROOF. If $t \in S(K)$ and $j \in I_{u_1} \times \ldots \times I_{u_p}$, let $t|j = \{i \in t : \langle i(u_1) \ldots i(u_p)\rangle = j\}$. Note that s also lives on $Y_0 \cup Z_0$. Therefore it suffices to prove the lemma for the case $Y_0 \subset Z_0$. In this case, the statement $i \in s$ depends only on $i(y_1), \ldots, i(y_m)$ and is independent of $i(z)$ for $z \in Z_0 \setminus Y_0$. We may assume further that $n = m+1$, for we can get from the set Y_0 to the set Z_0 by adding one element at a time. Thus there is a unique element $z \in Z_0 \setminus Y_0$. It follows by induction on m that for each r, $1 \leqslant r < m$, and each set $t \subset I_{y_1} \times \ldots \times I_{y_m}$, we have

$$t \in D_{y_1} \times \ldots \times D_{y_m}$$

iff

$$\{j \in I_{y_{r+1}} \times \ldots \times I_{y_m} : (t|j)_{y_1 \ldots y_r} \in D_{y_1} \times \ldots \times D_{y_r}\} \in D_{y_{r+1}} \times \ldots \times D_{y_m}.$$

Therefore

$$\text{(1)}\quad \begin{array}{l} s_{y_1 \ldots y_m} \in D_{y_1} \times \ldots \times D_{y_m} \quad \text{iff} \\ \{j \in I_{y_{r+1}} \times \ldots \times I_{y_m} : (s|j)_{y_1 \ldots y_r} \in D_{y_1} \times \ldots \times D_{y_r}\} \in D_{y_{r+1}} \times \ldots \times D_{y_m}. \end{array}$$

A corresponding statement also holds for Z_0. We have three cases.

Case 1. z is the last element of Z_0, i.e., $y_m < z$. Then the following are equivalent, since s lives on Y_0:

$$s_{y_1 \ldots y_m} \in D_{y_1} \times \ldots \times D_{y_m};$$

$$\{i(z) : s_{y_1 \ldots y_m} \in D_{y_1} \times \ldots \times D_{y_m}\} = I_z;$$

$$\{i(z) : s_{y_1 \ldots y_m} \in D_{y_1} \times \ldots \times D_{y_m}\} \in D_z;$$

$$s_{y_1 \ldots y_m z} \in D_{y_1} \times \ldots \times D_{y_m} \times D_z.$$

Case 2. z is the first element of Z_0, i.e., $z < y_1$. Then the following are equivalent, since s lives on Y_0:

$$s_{y_1 \ldots y_m} \in D_{y_1} \times \ldots \times D_{y_m};$$

$$\{j \in I_{y_1} \times \ldots \times I_{y_m} : (s|j)_z = I_z\} \in D_{y_1} \times \ldots \times D_{y_m};$$

$$\{j \in I_{y_1} \times \ldots \times I_{y_m} : (s|j)_z \in D_z\} \in D_{y_1} \times \ldots \times D_{y_m};$$

$$s_{zy_1 \ldots y_m} \in D_z \times D_{y_1} \times \ldots \times D_{y_m}.$$

The last step used the statement (1) for Z_0.

Case 3. Otherwise. There is a greatest element $y_z \in Y_0$ such that $y_r < z$, and $r < m$. Then again using the assumption that s lives on Y_0, and using (1) for both Y_0 and Z_0, we see that the following are equivalent, where

$$D_0 = D_{y_1} \times \ldots \times D_{y_r}, \qquad D_1 = D_{y_{r+1}} \times \ldots \times D_{y_m},$$

and I_0, I_1 are defined analogously:

$$s_{y_1 \ldots y_m} \in D_0 \times D_1;$$

$$\{j \in I_1 : (s|j)_{y_1 \ldots y_r} \in D_0\} \in D_1;$$

$$\{j \in I_1 : \{h \in I_z : (s|h^\frown j)_{y_1 \ldots y_r} \in D_0\} = I_z\} \in D_1;$$

$$\{j \in I_1 : \{h \in I_z : (s|h^\frown j)_{y_1 \ldots y_r} \in D_0\} \in D_z\} \in D_1;$$

$$\{j \in I_1 : (s|j)_{y_1 \ldots y_r z} \in D_0 \times D_z\} \in D_1;$$

$$s_{z_1 \ldots z_{m+1}} \in D_0 \times D_z \times D_1.$$

We have proved the desired result in each of the three cases. ⊣

PROPOSITION 6.5.4. *Let S be the set of all subsets s of K which live on a finite subset of Y. Then S is closed under finite unions, finite intersections, and complementation relative to K. Moreover, there exists an ultrafilter D over K such that*

$$D \cap S = \times_Y D_y.$$

PROOF. The first statement is easy. To prove the second statement, it suffices to prove:

(1) For all $s \in S$, either $s \in \times_Y D_y$ or $(K \setminus s) \in \times_Y D_y$.

(2) The set $\times_Y D_y$ has the finite intersection property.

First we prove (1). Let $s \in S$. Then s lives on some finite set $Y_0 = \{y_1, \ldots, y_m\}$, where $y_1 < \ldots < y_m$. Since $D_{y_1} \times \ldots \times D_{y_m}$ is an ultrafilter over $I_{y_1} \times \ldots \times I_{y_m}$, we have either

(3) $$s_{y_1 \ldots y_m} \in D_{y_1} \times \ldots \times D_{y_m},$$

or

(4) $$I_{y_1} \times \ldots \times I_{y_m} \setminus s_{y_1 \ldots y_m} \in D_{y_1} \times \ldots \times D_{y_m}.$$

But since s lives on Y_0,

$$(K \setminus s)_{y_1 \dots y_m} = I_{y_1} \times \dots \times I_{y_m} \setminus s_{y_1 \dots y_m}.$$

Hence, in the case (3), $s \in \bigtimes_Y D_y$, while in the case (4), $K \setminus s \in \bigtimes_Y D_y$.

We need Lemma 6.5.3 to prove (2). It suffices to prove that $\bigtimes_Y D_y$ is closed under finite intersection and does not contain the empty set. The latter is true because for any $y_1 < \dots < y_m$ in Y,

$$0_{y_1 \dots y_m} = 0 \notin D_{y_1} \times \dots \bigtimes D_{y_m}.$$

Suppose that $s,t \in \bigtimes_Y D_y$. Say s lives on Y_0 and t lives on Z_0, where Y_0, Z_0 are finite. Then $s \cap t$ lives on the finite set $Y_0 \cup Z_0$. Let

$$Y_0 \cup Z_0 = \{y_1, \dots, y_m\}, \qquad y_1 < \dots < y_m.$$

Then, by Lemma 6.5.3, both of the sets

$$A = s_{y_1 \dots y_m}, \qquad B = t_{y_1 \dots y_m}$$

belong to the ultrafilter $D_{y_1} \times \dots \times D_{y_m}$. Hence the intersection $A \cap B$ belongs to the ultrafilter. But since s and t live on $\{y_1, \dots, y_m\}$, we have

$$i \in s \quad \text{iff} \quad \langle i(y_1) \dots i(y_m) \rangle \in A,$$

$$i \in t \quad \text{iff} \quad \langle i(y_1) \dots i(y_m) \rangle \in B,$$

whence

$$i \in s \cap t \quad \text{iff} \quad \langle i(y_1) \dots i(y_m) \rangle \in A \cap B.$$

It follows that

$$(s \cap t)_{y_1 \dots y_m} = A \cap B \in D_{y_1} \times \dots \times D_{y_m}$$

and therefore $s \cap t \in \bigtimes_y D_y$. ⊣

We are now over the hump. We continue to assume that $\langle Y, < \rangle$ is a simply-ordered set, D_y is an ultrafilter over I_y for each $y \in Y$, K is the Cartesian product $K = \prod_{y \in Y} I_y$, and S is the set of all $s \subset K$ which live on a finite subset of Y. Now let $\mathfrak{A}$ be a model for $\mathscr{L}$. Let C be the set of all functions $f : K \to A$ such that f lives on a finite subset of Y. Let us also write $E = \times_Y D_y$, for brevity. We shall say that two functions $f,g \in C$ are *equivalent modulo* E, in symbols $f =_E g$, iff

$$\{i \in K : f(i) = g(i)\} \in E.$$

As usual, the equivalence class of f is written

$$f_E = \{g \in C : f =_E g\}.$$

The *iterated ultrapower* $\prod_E A$ of the set A is defined as the set of all equivalence classes:

$$\prod_E A = \{f_E : f \in C\}.$$

Finally, the *iterated ultrapower* $\prod_E \mathfrak{A}$ of the model $\mathfrak{A}$ is described in the following way. The universe of $\prod_E \mathfrak{A}$ is the set $\prod_E A$. For each n-placed relation R of $\mathfrak{A}$, the corresponding relation R' of $\prod_E \mathfrak{A}$ is determined by

$$R'(f_E^1 \ldots f_E^n) \quad \text{iff} \quad \{i \in K : R(f^1(i) \ldots f^n(i))\} \in E.$$

For each n-placed function G of $\mathfrak{A}$, the corresponding function G' of $\prod_E \mathfrak{A}$ is given by

$$G'(f_E^1 \ldots f_E^n) = \langle G(f^1(i) \ldots f^n(i)) : i \in K \rangle_E .$$

For each constant a of $\mathfrak{A}$, the corresponding constant a' of $\prod_E \mathfrak{A}$ is the element

$$a' = \langle a : i \in K \rangle_E .$$

The similarity between this definition and the definition of the ultrapower is obvious. The difference is that we are now dealing with subsets $S \subset S(K)$, $C \subset {}^K A$, $E \subset D$, rather than the whole sets. As in defining the ultraproduct, there are certain details which must be verified before we can be sure that the definition is meaningful. We list these details in a proposition.

PROPOSITION 6.5.5.

(i). *If* $f^1 \ldots f^n \in C$, *then*

$$\{i \in K : f^1(i) = f^2(i)\} \in S;$$

$$\{i \in K : R(f^1(i) \ldots f^n(i))\} \in S;$$

$$\langle G(f^1(i) \ldots f^n(i)) : i \in K \rangle \in C.$$

(ii). *The relation* $=_E$ *is an equivalence relation over* C.

(iii). *If* $f^1 =_E g^1, \ldots, f^n =_E g^n$, *then*

$$\{i \in K : R(f^1(i) \ldots f^n(i))\} \in E \quad \textit{iff} \quad \{i \in K : R(g^1(i) \ldots g^n(i))\} \in E,$$

and

$$\langle G(f^1(i) \ldots f^n(i)) : i \in K \rangle =_E \langle G(g^1(i) \ldots g^n(i)) : i \in K \rangle.$$

The basic results for iterated ultrapowers are similar to the corresponding results for ultraproducts and limit ultrapowers, and have essentially the same proofs.

PROPOSITION 6.5.6 (Expansion Theorem). *Let* $E = \times_Y D_y$. *If* $\mathfrak{A}'$ *is an ex-*

pansion of $\mathfrak{A}$ to a language $\mathscr{L}'$, then an iterated ultrapower $\prod_E(\mathfrak{A}')$ is an expansion of the iterated ultrapower $\prod_E \mathfrak{A}$ to $\mathscr{L}'$.

PROPOSITION 6.5.7 (Fundamental Theorem). *Let $E = \times_Y D_y$, and form the iterated ultrapower $\prod_E \mathfrak{A}$. Then for each formula $\varphi(x_1 \dots x_n)$ of $\mathscr{L}$ and all $f^1, \dots, f^n \in {}^K A$ which live on finite subsets of Y, we have*

$$\prod_E \mathfrak{A} \models \varphi[f_E^1 \dots f_E^n]$$

iff

$$\{i \in K : \mathfrak{A} \models \varphi[f^1(i) \dots f^n(i)]\} \in E.$$

The corresponding statement for terms also holds, and as before, the fundamental theorem must first be proved for terms. The natural embedding $d : A \to \prod_E A$ is defined as before by putting $d(a)$ equal to the equivalence class of the constant function with value a. Then, we have the following:

COROLLARY 6.5.8. *Let $E = \times_Y D_y$ and form the iterated ultrapower $\prod_E \mathfrak{A}$. Then the natural embedding d is a complete embedding of $\mathfrak{A}$ into $\prod_E \mathfrak{A}$.*

The iterated ultrapower abounds with natural embeddings, in addition to the original natural embedding of $\mathfrak{A}$ into its limit ultrapower. Let us consider a finite subset $Z \subset Y$, and let $Z = \{z_1, \dots, z_n\}$ be arranged in increasing order. It is then natural to associate the functions

$$g : I_{z_1} \times \dots \times I_{z_n} \to A$$

with the functions $f : K \to A$ which live on Z. With $E = \times_Y D_y$, $E_Z = D_{z_1} \times \dots \times D_{z_n}$, we may define the *natural embedding*

$$d_Z : \prod_{E_Z} \mathfrak{A} \to \prod_E \mathfrak{A}$$

by

$$d_Z(g_{E_Z}) = \langle g(i(z_1) \dots i(z_n)) : i \in K \rangle_E .$$

For $Z = 0$, we adopt the convention $d_0 = d$, the natural embedding of $\mathfrak{A}$ into $\prod_E \mathfrak{A}$. The embedding d_Z depends on Y, D_y, and the set A, as well as on Z.

PROPOSITION 6.5.9. *Let Z be a finite subset of Y. Then:*

(i). *d_Z is a one–one function on $\prod_{E_Z} A$ into $\prod_E A$.*

(ii). $\text{range}(d_Z) = \{f_E \in \prod_E A : f \text{ lives on } Z\}$.

(iii). $\prod_E A = \bigcup \{\text{range}(d_W) : W \in S_\omega(Y)\}$.

(iv). *If $y \neq z$ in Y, then*

$$\text{range}\,(d_{\{y\}}) \cap \text{range}\,(d_{\{z\}}) = \text{range}\,(d_0).$$

PROOF. (i). The following string of equivalent statements, for all $g,h : I_{z_1} \times \ldots \times I_{z_n} \to A$, show both that d_Z is a function and that it is one–one:

$$g =_{E_Z} h;$$
$$\{j \in I_{z_1} \times \ldots \times I_{z_n} : g(j) = h(j)\} \in E_Z;$$
$$\{i \in K : g(i(z_1) \ldots i(z_n)) = h(i(z_1) \ldots i(z_n))\} \in E;$$
$$\langle g(i(z_1) \ldots i(z_n)) : i \in K\rangle =_E \langle h(i(z_1) \ldots i(z_n)) : i \in K\rangle.$$

(ii). This follows at once from the fact that a function $f: K \to A$ lives on Z iff there is a $g : I_{z_1} \times \ldots \times I_{z_n} \to A$ such that

$$f = \langle g(i(z_1) \ldots i(z_n)) : i \in K\rangle.$$

(iii). This follows from (ii).

(iv). Obviously, range$(d_0) \subset$ range$(d_{\{y\}})$ and similarly for z. Let

$$b \in \text{range}\,(d_{\{y\}}) \cap \text{range}\,(d_{\{z\}}).$$

Then by (iii) there exist f, $g \in {}^K A$ such that $b = f_E = g_E$ and f lives on $\{y\}$, g lives on $\{z\}$. Suppose, for example, $y < z$. Let $f'(i(y)) = f(i)$, $g'(i(z)) = g(i)$. Then

$$\{i \in K : f'(i(y)) = g'(i(z))\} = \{i \in K : f(i) = g(i)\} \in E,$$

whence

$$\{j \in I_z : \{i \in I_y : f'(i) = g'(j)\} \in D_y\} \in D_z.$$

Thus there exists $j_0 \in I_z$ such that

$$\{i \in I_y : f'(i) = g'(j_0)\} \in D_y.$$

Let $a = g'(j_0)$. Then $a \in A$, and

$$\{i \in I_y : f'(i) = a\} \in D_y,$$

whence

$$\{i \in K : f(i) = a\} \in E,$$

so $f_E = d_0(a) \in$ range (d_0). ⊣

PROPOSITION 6.5.10. *Whenever* $W \subset Z \in S_\omega(Y)$, *we have*

$$d_Z : \prod_{E_Z} \mathfrak{A} \prec \prod_E \mathfrak{A}$$

and

$$d_W(\prod_{E_W} \mathfrak{A}) \prec d_Z(\prod_{E_Z} \mathfrak{A}) \prec \prod_E \mathfrak{A}.$$

PROOF. Let $\varphi(x_1 \dots x_m)$ be a formula of $\mathscr{L}$, and let $f^1, \dots, f^m$ be functions on $I_{z_1} \times \dots \times I_{z_n}$ into A which live on a finite subset of Z. Using the fundamental theorem we see that the following are equivalent:

$$\prod_{E_Z} \mathfrak{A} \models \varphi[f^1_{E_Z} \dots f^m_{E_Z}];$$

$$s = \{i \in I_{z_1} \times \dots \times I_{z_n} : \mathfrak{A} \models \varphi[f^1(i) \dots f^m(i)]\} \in E_Z;$$

$$\{i \in K : \mathfrak{A} \models \varphi[f^1(i(z_1) \dots i(z_n)) \dots f^m(i(z_1) \dots i(z_n))]\} \in E;$$

$$\{i \in K : \mathfrak{A} \models \varphi[\bar{f}^1(i) \dots \bar{f}^m(i)]\} \in E.$$

Here

$$\bar{f}^r = \langle f^r(i(z_1) \dots i(z_n)) : i \in K \rangle;$$

$$\prod_E \mathfrak{A} \models \varphi[\bar{f}^1_E \dots \bar{f}^m_E];$$

$$\prod_E \mathfrak{A} \models \varphi[d_Z(f^1_{E_Z}) \dots d_Z(f^m_{E_Z})].$$

This shows that $d_Z : \prod_{E_Z} \mathfrak{A} \prec \prod_E \mathfrak{A}$. The second assertion follows from the first and the fact that range$(d_W) \subset$ range (d_Z). ⊣

The most useful case of the iterated ultrapower is the case where all the ultrafilters D_y, $y \in Y$, are the same, say $D_y = D$. In this case we shall write $\times_Y D = \times_Y D_y$. The next two results are applications of iterated ultrapowers to the Löwenheim–Skolem problem (cf. Section 4.3).

THEOREM 6.5.11 (Narrowing a Gap). *Let $\mathscr{L}$ have a 1-placed relation symbol U.*

(i). *If $\alpha^\omega \geq \beta' \geq \beta^\omega$ and $\alpha \geq \beta \geq \omega$, then every (α, β)-model has a complete extension which is an (α^ω, β')-model.*

(ii). *Suppose $\alpha \geq \alpha' \geq \beta' \geq \beta^\omega$, $\beta \geq \omega$ and $\alpha' \geq ||\mathscr{L}||$. Then every theory which admits (α, β) admits (α', β').*

PROOF. (ii) follows easily from (i) and the downward Löwenheim–Skolem theorem.

(i). Choose a countably incomplete ultrafilter D over ω. Let $\langle Y, < \rangle$ be any linearly ordered set of power β'. Let $\mathfrak{A} = \langle A, V, \dots \rangle$ be an (α, β)-model, where $|A| = \alpha$, $|V| = \beta$. Form the iterated ultrapower

$$\mathfrak{A}' = \langle A', V', \dots \rangle = \prod_E \mathfrak{A}, \qquad E = \times_Y D.$$

Since d is a complete embedding of $\mathfrak{A}$ into $\mathfrak{A}'$, $\mathfrak{A}'$ is isomorphic to a complete extension of $\mathfrak{A}$. For each finite $Z \subset Y$, let $\prod_{E_Z} \mathfrak{A} = \langle A_Z, V_Z, \dots \rangle$. Then, by Proposition 6.5.10, d_Z is an elementary embedding of $\prod_{E_Z}\mathfrak{A}$ into $\mathfrak{A}'$. By Proposition 6.5.9,

$$A' = \bigcup \{\text{range}(d_Z) : Z \in S_\omega(Y)\}.$$

When Z is finite, E_Z is just an ultrafilter over a finite number of copies of ω. If $Z \neq 0$, then E_Z is also countably incomplete, since D is. It follows that when Z is a nonempty finite set, $|\prod_{E_Z} A| = \alpha^\omega$(by Proposition 4.3.7). Then

$$\alpha^\omega \leqslant |A'| \leqslant \alpha^\omega \cdot |Y| = \alpha^\omega \cdot \beta' = \alpha^\omega,$$

so $|A'| = \alpha^\omega$. We also have

(1) $$V' = \bigcup \{d_Z(V_Z) : Z \in S_\omega(Y)\},$$

whence $\beta = |V| \leqslant |\prod_{E_Z} V| \leqslant \beta^\omega$, and $|V_Z| = |\prod_{E_Z} V|$. Thus

(2) $$|V_Z| \leqslant \beta^\omega.$$

Since V is infinite and D is countably incomplete, d maps V properly into $\prod_D V$. It follows that for each $y \in Y$,

(3) $$d_0(V) \text{ is a proper subset of } d_{\{y\}}(V_{\{y\}}).$$

But by Proposition 6.5.9(iv), for $y \neq z$ in Y we have

(4) $$d_{\{y\}}(V_{\{y\}}) \cap d_{\{z\}}(V_{\{z\}}) = d_0(V).$$

It follows from (1), (3) and (4) that

$$\beta' = |Y| \leqslant |V'|.$$

Also, from (2),

$$|V'| \leqslant |Y| \cdot \beta^\omega = \beta' \cdot \beta^\omega = \beta'.$$

Therefore $|V'| = \beta'$. This shows that $\mathfrak{A}'$ is an (α^ω, β')-model, and (i) is proved. ⊣

The above theorem eliminates the continuum hypothesis from the earlier result, Corollary 4.3.11.

Ths next result complements Theorem 6.4.5. It is of interest when $\alpha < \alpha^\omega < 2^\alpha$.

COROLLARY 6.5.12. *Let α, β be cardinals, $\alpha > \beta \geqslant \omega$, and suppose ω is the only measurable cardinal $\leqslant \beta$. Then the following are equivalent*:

(i). $\alpha \geqslant \beta^\omega$.

(ii). *Every model $\mathfrak{A}$ of power β has an elementary extension of power α.*

(iii). *Every model $\mathfrak{A}$ of power β has a complete extension of power α.*

Proof. Assume (i). Let D be a countably incomplete ultrafilter over ω, let Y be a simply ordered set of power α, and form the iterated ultrapower $\mathfrak{A}' = \prod_{\times YD} \mathfrak{A}$. The proof of Theorem 6.5.11 shows that $\mathfrak{A}'$ has power α and is isomorphic to a complete extension of $\mathfrak{A}$. Thus (ii) and (iii) follow from (i).

By applying (ii) to the completion of $\mathfrak{A}$, we see that (ii) implies (iii).

So far we have not used the assumption that ω is the only measurable cardinal $\leqslant \alpha$. It is used only to go from (iii) to (i).

Assume (iii), and let $\mathfrak{A}$ have power β and $\mathfrak{B}$ be a complete extension of $\mathfrak{A}$ of power α. Since $\alpha > \beta$, there exists $b \in B \setminus A$. By Theorem 6.4.4, there is an ultrapower extension $\mathfrak{C}$ of $\mathfrak{A}$ such that $b \in C$ and $\mathfrak{C} \prec \mathfrak{B}$. Then $(\mathfrak{C}, a)_{a \in A}$ is isomorphic to an ultrapower $(\prod_D \mathfrak{A}, d(a))_{a \in A}$. Since $b \notin A$, there must be an element of $\prod_D A$ not in $d(A)$. Thus D is not β^+-complete, whence by our measurability assumption D is countably incomplete. Then, by Proposition 4.3.7, $|\prod_D A| = \beta^\omega$, so $\beta^\omega = |C| \leqslant |B| = \alpha$, and (i) holds. ⊣

What happens if we form an iterated ultrapower where all the ultrafilters D_y are α-complete? We would like such iterated ultrapowers to preserve sentences of the infinitary language $\mathscr{L}_\alpha$, but unfortunately this is not the case (see Exercise 6.5.20). However, in one special situation, they do preserve the property of well foundedness. We shall exploit that fact to get more information about measurable cardinals. The rest of this section depends heavily on Section 4.2.

Lemma 6.5.13. *Suppose $\langle Y, <' \rangle$ is an inversely well ordered set, and, for each $y \in Y$, D_y is an α-complete ultrafilter. Then the intersection of any subset of $\times_Y D_y$ of power less than α is nonempty.*

Proof. Let $E \subset \times_Y D_y$, $|E| < \alpha$. We may assume that $\langle Y, <' \rangle = \langle \beta, > \rangle$, where β is an ordinal. We assume that β is a limit ordinal and leave the other case for the reader. For each $s \in E$, choose a finite set $L(s) \subset Y$ on which s lives. Suppose $s \in E$, $0 < \gamma \langle \beta$, and $L(s) = \{\delta_1, \ldots, \delta_n\}$, where $\delta_1 > \ldots > \delta_n = 0$. We shall define a set $F_\gamma(s)$. There are two cases, depending on the position of γ.

Case 1: $\delta_p \geqslant \gamma > \delta_{p+1}$. Let $F_\gamma(s)$ be the set of all $j \in I_{\delta_{p+1}} \times \ldots \times I_{\delta_n}$ such that $(s|j)_{\delta_1 \ldots \delta_p} \in D_{\delta_1} \times \ldots \times D_{\delta_p}$.

Case 2 : $\gamma > \delta_1$. Let $F_\gamma(s) = s_{\delta_1 \ldots \delta_n}$.

Putting $p = 0$ in Case 2, we always have $F_\gamma(s) \in D_{\delta_{p+1}} \times \ldots \times D_{\delta_n}$. Let us say that a function j *passes through* $F_\gamma(s)$ iff

$$\langle j(\delta_{p+1}) \dots j(\delta_n)\rangle \in F_\gamma(s).$$

Now for each $\gamma \in \beta$, $0 < \gamma$, we may define

$$F_\gamma = \{j \in \prod_{\delta \in \gamma} I_\delta : \text{for all } s \in E,\ j \text{ passes through } F_\gamma(s)\}.$$

Let $F_0 = \{0\}$. We claim that:

(1) If $j_\delta \in F_\delta$ for all $\delta < \gamma$ and if $\delta < \eta$ implies $j_\delta \subset j_\eta$, then $\bigcup_{\delta<\gamma} j_\delta \in F_\gamma$.

(2) If $j \in F_\gamma$, then j can be extended to a function $k \in F_{\gamma+1}$.

(1) follows because each set $L(s)$ is finite. To see (2), we first note that for each $s \in E$, the set

$$X_s = \{i \in I_\gamma : j^\frown i \text{ passes through } F_{\gamma+1}(s)\} \in D_\gamma.$$

Since $|E| < \alpha$ and D_γ is α-complete, the set $\bigcap_{s\in E} X_s$ belongs to D_γ, hence is nonempty. Choosing $i \in \bigcap_{s\in E} X_s$, we find that $j^\frown i = k$ belongs to $F_{\gamma+1}$, and (2) is proved. Using Zorn's lemma, (1) and (2) imply that there exists a function $i \in K$ such that for all $s \in E$,

$$\langle i(\delta) : \delta \in L(s)\rangle \in s_{\delta_1 \dots \delta_n},$$

that is, $i \in s$. Therefore $i \in \bigcap E$, and $\bigcap E \neq 0$. ⊣

THEOREM 6.5.14. *Let α be an uncountable measurable cardinal. Then:*

(i). *For all $\beta \geqslant \alpha$, $\langle \beta, <\rangle$ has well ordered complete extensions of arbitrarily large power.*

(ii). *For each cardinal $\gamma > 2^\alpha$, $\langle \gamma, <\rangle$ is a complete extension of $\langle \alpha, <\rangle$.*

PROOF. (i). Let D be a nonprincipal α-complete ultrafilter over α. Given a cardinal γ, we wish to find a well ordered complete extension of $\langle \beta, <\rangle$ of power at least γ. Let $\langle Y, <'\rangle$ be an inversely well ordered set of order type $\langle \gamma, >\rangle$, and form the iterated ultrapower $\langle B, L\rangle = \prod_E \langle \beta, <\rangle$, where $E = \mathsf{X}_Y D$. Since D is nonprincipal, it is not α^+-complete, and therefore $d(\beta)$ is a proper subset of $\prod_D \beta$. It follows that for each $y \in \gamma$, range (d_0) is a proper subset of range$(d_{\{y\}})$. Then by Proposition 6.5.9(iv), it follows that B has power at least γ. Suppose $\langle B, L\rangle$ is not well ordered. Then there is a countable sequence $f_E^0, f_E^1, \dots$ of elements of B such that

$$f_E^{n+1} L f_E^n, \qquad n = 0, 1, 2, \dots.$$

Then for each n,

$$X_n = \{i \in K : f^{n+1}(i) < f^n(i)\} \in E.$$

Since $\alpha > \omega$, we may apply Lemma 6.5.13, and $\bigcap_{n<\omega} X_n$ is nonempty. Let $i \in \bigcap_{n<\omega} X_n$. Then in $\langle \beta, < \rangle$ we have the decreasing sequence

$$f^0(i) > f^1(i) > \ldots .$$

But this contradicts the fact that $\langle \beta, < \rangle$ is well-ordered, whence $\langle B, L \rangle$ must be well-ordered. Finally, by Corollary 6.5.8, $\langle B, L \rangle$ is isomorphic to a complete extension of $\langle \beta, < \rangle$.

(ii). Let D and E be as in the proof of (i), where γ is the given cardinal, $\gamma > \alpha$. Form the iterated ultrapower $\langle A, L \rangle = \prod_E \langle \alpha, < \rangle$. Then $\langle A, L \rangle$ is well-ordered and $|A| \geq \gamma$. We claim that

(1) $d(\alpha)$ is an initial segment of $\langle A, L \rangle$.

For let $f_E \in A$ and $f_E L d(\delta)$ for some $\delta < \alpha$. Let f live on $\{y_1, \ldots, y_n\}$, where $y_1 > \ldots > y_n$. Then (with $D_y = D$ for all y),

$$\{i \in {}^Y\alpha : f(i) < \delta\}_{y_1 \ldots y_n} \in D_{y_1} \times \ldots \times D_{y_n}.$$

Since D is α-complete, $D_{y_1} \times \ldots \times D_{y_n}$ is α-complete, whence there exists $\eta < \delta$ such that

$$\{i \in {}^Y\alpha : f(i) = \eta\}_{y_1 \ldots y_n} \in D_{y_1} \times \ldots \times D_{y_n}.$$

So $f_E = d(\eta)$. This proves (1). Let γ' be the order type of $\langle A, L \rangle$. Then $\langle \gamma', < \rangle$ is a complete extension of $\langle \alpha, < \rangle$. By (i), $|A| \geq \gamma$, whence we have $\gamma' \geq \gamma$.

It remains to prove $\gamma' \leq \gamma$. Let $f_E \in A$, and suppose that f lives on $y_1, \ldots, y_n$, where $\gamma > y_1 > \ldots > y_n$. Let $Z = \{y \in \gamma : y_1 \geq y\}$, and let

$$C = \{g_E \in A : g_E L f_E \text{ and } g \text{ lives on a finite subset of } Z\}.$$

Since there are at most $|Z| \cdot \alpha^\alpha$ functions which live on a finite subset of Z, and α^α, $|Z| < \gamma$, we have $|C| < \gamma$. Consider an arbitrary $h_E \in A$ such that $h_E L f_E$. There is a finite number of elements $z_1 > \ldots > z_m$ such that h lives on $\{z_1, \ldots, z_m\}$, and $\{y_1, \ldots, y_n\} \subset \{z_1, \ldots, z_m\}$. Suppose $z_1 > y_1$. Then f lives on $\{z_2, \ldots, z_m\}$. Since $h_E L f_E$, the set

$$s = \{i \in {}^I Y : h(i) < f(i)\} \in E.$$

For $j \in I_{z_1} \times \ldots \times I_{z_m}$, define $h'(j) = h(i)$ and $f'(j) = f(i)$, where $\langle i(z_1) \ldots i(z_m) \rangle = j$. Then

$$\{j \in I_{z_1} \times \ldots \times I_{z_m} : h'(j) < f'(j)\} \in D_{z_1} \times \ldots \times D_{z_m}.$$

Hence the set

$$U = \{j_1 \in I_{z_2} \times \ldots \times I_{z_m} : \{j_0 \in I_{z_1} : h'(j_0{}^\frown j_1) < f'(j_0{}^\frown j_1)\} \in D_{z_1}\}$$

belongs to $D_{z_2} \times \ldots \times D_{z_m}$. But $f'(j_0{}^\frown j_1)$ depends only on j_1 and not on j_0, since f lives on $\{z_2, \ldots, z_m\}$. Thus, since D is α-complete and $f'(j_0{}^\frown j_1) < \alpha$, there is for each $j_1 \in U$ an element $l'(j_1) < f'(j_0{}^\frown j_1)$ such that

$$\{j_0 \in I_{z_1} : h'(j_0{}^\frown j_1) = l'(j_1)\} \in D_z.$$

It follows that if we define $l : {}^I Y \to \alpha$ by $l(i) = l'\langle i(z_2) \ldots i(z_m)\rangle$, then l lives on $\{z_2, \ldots, z_m\}$ and $h_E = l_E$. For some $r \leqslant m$, $y_1 = z_r$. Continuing the above process, we obtain a function k living on $\{z_r, \ldots, z_m\}$ such that $k_E = h_E$. But then k lives on a finite subset of Z, so $k_E \in C$, whence $h_E \in C$. This means that f_E has $|C| < \gamma$ predecessors in $\langle A, L\rangle$. We conclude that $\gamma' \leqslant \gamma$. ⊣

The theorem can also be stated in terms of models of the form $\langle R(\alpha), \in\rangle$ instead of $\langle \alpha, <\rangle$.

COROLLARY 6.5.15. *Let α be an uncountable measurable cardinal. Then:*

(i). *For all $\beta \geqslant \alpha$, the model $\langle R(\beta), \in\rangle$ has well founded complete extensions of arbitrarily large power.*

(ii). *For each cardinal $\gamma > 2^\alpha$, the model $\langle R(\alpha), \in\rangle$ has a complete extension $\langle B, \in\rangle$ such that $\gamma \subset B \subset R(\gamma)$ and B is a transitive set.*

PROOF. (i). Let γ be a cardinal and let $\langle B_0, <\rangle$ be a complete well ordered extension of $\langle \beta, <\rangle$ of power $\geqslant \gamma$. By Corollary 6.4.14, $\langle B_0, <\rangle$ can be extended to a complete extension $\langle B, E\rangle$ of $\langle R(\beta), \in\rangle$ which has the same ordinals as $\langle B_0, <\rangle$. It follows that $\langle B, E\rangle$ is well founded.

We leave the proof of (ii) as an exercise. ⊣

EXERCISES

6.5.1. Prove that if D, E are ultrafilters, then so is $D \times E$. Moreover, if D, E are proper filters, so is $D \times E$. Prove a similar result for $D_1 \times \ldots \times D_n$.

6.5.2. Let D, E be proper filters over sets I, J. For each $\langle i, j\rangle \in I \times J$ let $\mathfrak{A}_{ij}$ be a model. Prove that

$$\prod_{D \times E} \mathfrak{A}_{ij} \cong \prod_E \big(\prod_D \mathfrak{A}_{ij}\big).$$

Thus an ultraproduct of ultraproducts is isomorphic to an ultraproduct.

6.5.3. Let E be a proper filter over J and for each $j \in J$ let D_j be a proper filter over I_j. Let $K = \bigcup_{j \in J} I_j$. Prove that the set

$$F = \{X \in S(K) : \{j \in J : X \cap I_j \in D_j\} \in E\}$$

is a proper filter over K. Moreover, if E and every D_j is an ultrafilter, then F is an ultrafilter.

6.5.4. In the above exercise assume that the sets I_j are pairwise disjoint. Prove that

$$\prod_F \mathfrak{A}_k \cong \prod_E (\prod_{D_j} \mathfrak{A}_i).$$

6.5.5. Let D, E be ultrafilters over I, J and let V, W be filters over $I \times I, J \times J$. Give an appropriate definition of $V \times W$ and show that for any model $\mathfrak{A}$,

$$\prod_{D \times E | V \times W} \mathfrak{A} \cong \prod_{E|W} (\prod_{D|V} \mathfrak{A}).$$

6.5.6. Suppose D is an α-regular filter and E is a proper filter. Show that $D \times E$ and $E \times D$ are α-regular filters.

6.5.7. Suppose D is a uniform filter over I, E is a proper filter over J, and $|J| \leqslant |I|$. Then $D \times E$ and $E \times D$ are uniform filters.

6.5.8. If D and E are α-complete filters, then $D \times E$ is α-complete. The same holds for descendingly α-complete.

6.5.9. Suppose D is a uniform ω_1-complete ultrafilter, and E is a countably incomplete ultrafilter over ω. Then $D \times E$ and $E \times D$ are uniform countably incomplete ultrafilters which are not ω_1-regular.

6.5.10. Supply the proofs of the results Propositions 6.5.5–6.5.7 and Corollary 6.5.8.

6.5.11. Let $\langle Y, < \rangle$ be a finite ordered set, where $Y = \{y_1, \ldots, y_n\}$ in increasing order. Given ultrafilters $D_{y_1}, \ldots, D_{y_n}$, let $E = \times_Y D_y$. Show that for any model $\mathfrak{A}$,

$$\prod_E \mathfrak{A} \cong \prod_{D_{y_1} \times \ldots \times D_{y_n}} \mathfrak{A}.$$

6.5.12. Given $\langle Y, < \rangle$ and ultrafilters D_y, $y \in Y$, let $E = \times_Y D_y$ and $K = \prod_{y \in Y} I_y$. Find a filter V over $K \times K$ such that for all filters $D \supset E$ over K,

$$\prod_E \mathfrak{A} \cong \prod_{D|V} \mathfrak{A}.$$

Thus iterated ultrapowers are limit ultrapowers.

6.5.13. Show that if $E = \times_Y D_y$, then for any nonempty set A

$$|\prod_E A| \leqslant |Y| \cup \sup \{|A|^{|I_y|} : y \in Y\}.$$

If each D_y is $|I_y|$-regular and the set A is infinite, then equality holds.

6.5.14. Let $\mathfrak{A}_0 \prec \mathfrak{A}_1 \prec \ldots$ be an elementary chain of length ω such that each $\mathfrak{A}_{n+1}$ is an ultrapower extension of $\mathfrak{A}_n$; specifically,

$$(\mathfrak{A}_{n+1}, a)_{a \in A_n} \cong (\prod_{D_n} \mathfrak{A}_n, d(a))_{a \in A_n}.$$

Let $\langle Y, <\rangle = \langle \omega, <\rangle$ and $E = \times_Y D_n$. Prove that

$$\bigcup_{n<\omega} \mathfrak{A}_n \cong \prod_E \mathfrak{A}_n.$$

6.5.15. Show that the result of the above exercise does not hold for elementary chains of length $\omega+1$.

[*Hint*: Consider the cardinalities.]

6.5.16. $\mathfrak{A} \equiv \mathfrak{B}$ iff they have isomorphic iterated ultrapowers of type $\langle Y, <\rangle = \langle \omega, <\rangle$.

6.5.17. Show that reduced powers and limit ultrapowers can also be iterated. That is, given $\langle Y, <\rangle$ and filters D_y, define

$$\times_Y D_y \quad \text{and} \prod_{\times_Y D_y} \mathfrak{A};$$

and, given ultrafilters D_y and filters V_y over $I_y \times I_y$, define

$$\times_Y (D_y | V_y) \quad \text{and} \prod_{\times_Y (D_y|V_y)} \mathfrak{A}.$$

6.5.18. Let $\langle Z, <\rangle$ be a finite initial segment of the simply ordered set $\langle Y, <\rangle$, and let $W = Y \setminus Z$. Let D_y, $y \in Y$, be ultrafilters. Show that

$$\prod_{\times_Y D_y} \mathfrak{A} \cong \prod_{\times_W D_w} (\prod_{\times_Z D_z} \mathfrak{A}).$$

Give an example showing that the result is not always true for infinite initial segments Z of Y.

6.5.19*. Given an iterated ultrapower $\prod_E A$, where $E = \times_Y D_y$, let $W, Z \in S_\omega(Y)$. Show that

$$\text{range}(d_Z) \cap \text{range}(d_W) = \text{range}(d_{Z \cap W}).$$

6.5.20. Show that Lemma 6.5.13 is not true for the ordered set $\langle Y, <\rangle =$

$\langle\omega, <\rangle$. Also show that if α is measurable and D is a nonprincipal α-complete ultrafilter over α, then $\prod_{X_\omega D}\langle\alpha, <\rangle$ is not well ordered.

6.5.21. Let α be an uncountable measurable cardinal and let $\mathfrak{A}$ be a model of power $\geqslant \alpha$. Prove that $\mathfrak{A}$ has complete extensions $\mathfrak{B}$ of arbitrarily large power such that every sentence of the infinitary language $\mathscr{L}_\alpha$ of the form $(\forall U)\bigvee\Phi$ which holds in $\mathfrak{A}$ holds in $\mathfrak{B}$ (where Φ is a set of finite formulas). This is a more general form of Theorem 6.5.14.

6.5.22. Let α be an uncountable measurable cardinal. Then for each cardinal $\gamma > 2^\alpha$, the model $\langle R(\alpha), \in\rangle$ has a complete extension $\langle B, \in\rangle$ such that $\gamma \subset B \subset R(\gamma)$ and B is a transitive set. (Use Exercise 1.4.18 on the existence of the transitive realization of a well founded model.)

The next few exercises show how iterated ultrapowers can be used to give another proof of the results on indiscernible elements in Section 3.3.

6.5.23. Let $\langle Y, <\rangle$ and $\langle Y', <'\rangle$ be simply ordered sets and let σ be an ordermorphism of Y into Y'. Let $E = \times_Y D$, $E' = \times_{Y'} D$, where D is an ultrafilter over I. Define the map $\sigma^* : {}^{Y'}I \to {}^{Y}I$ by $\sigma^*(j) = j \circ \sigma$. Then define a *natural embedding* $d_\sigma : \prod_D A \to \prod_E A$ by $d_\sigma(g_E) = (g \circ \sigma^*)_{E'}$. Prove that for any model $\mathfrak{A}$, d_σ is an elementary embedding of $\prod_E \mathfrak{A}$ into $\prod_{E'} \mathfrak{A}$. Moreover, if σ maps Y onto Y', then d_σ maps $\prod_E \mathfrak{A}$ isomorphically onto $\prod_{E'} \mathfrak{A}$.

6.5.24. Let Y, Z be simply ordered sets and let D_y, E_z be ultrafilters for each $y \in Y$, $z \in Z$. Let $\sigma : Y \to Z$ be an ordermorphic embedding such that for all $y \in Y$, $D_y = E_{\sigma(y)}$. Find a natural elementary embedding

$$d_\sigma : \prod_{\times_Y D_y} \mathfrak{A} \prec \prod_{\times_Z E_z} \mathfrak{A}.$$

6.5.25. Let D be an ultrafilter, let Y, Z be simply ordered sets, and let σ be an ordermorphic embedding of Y into Z. Let $\mathfrak{A}$ be a model. Show that

$$\text{range}(d_\sigma) = \{f_{\times_Z D} : f \text{ lives on range}(\sigma)\}.$$

6.5.26. Let D be an ultrafilter, Y an ordered set, and $E = \times_Y D$. For each $b \in \prod_D A$, define $e_b : Y \to \prod_E A$ by $e_b(y) = d_{\{y\}}(b)$ for each $y \in Y$. Show that:

(i). If $b \in \prod_D A \setminus d(A)$, then e_b is a one–one function.

(ii). For all $b \in \prod_D A$ and all ordermorphisms $\sigma : Y \to Y$, $d_\sigma \circ e_b = e_b \circ \sigma$.

6.5.27. Let $\mathfrak{A}$ be a model, let D be an ultrafilter such that $\prod_D A \setminus d(A) \neq 0$, and let $\langle Y, <\rangle$ be a simply ordered set. Let $\mathfrak{B}$ be the iterated ultrapower

$\mathfrak{B} = \prod_{\times_Y D} \mathfrak{A}$. Prove that there exists a set $C \subset B$ and an ordering $<'$ of C such that:

(i). $\langle C, <' \rangle \cong \langle Y, < \rangle$;

(ii). every automorphism of $\langle C, <' \rangle$ can be extended to an automorphism of $\mathfrak{B}$;

(iii). C is a set of indiscernibles in $\mathfrak{B}$.

6.5.28. Let Y, Y', Y'' be simply ordered sets and let $\sigma : Y \to Y'$, $\tau : Y' \to Y''$ be ordermorphic embeddings. Then for any model $\mathfrak{A}$ and ultrafilter D, the natural embeddings satisfy

$$d_{\tau \circ \sigma} = d_\tau \circ d_\sigma .$$

6.5.29. Let Y be a simply ordered set, D an ultrafilter, $E = \times_Y D$, $\mathfrak{A}$ a model. Then the mapping $\sigma \to d_\sigma$ is an isomorphic embedding of the group of automorphisms of Y into the group of automorphisms of $\prod_E \mathfrak{A}$. It also maps the semigroup of ordermorphic embeddings of Y into Y into the semigroup of elementary embeddings of $\prod_E \mathfrak{A}$ into itself.

6.5.30*. Let $E = \times_Y D_y$ and form the iterated ultrapower

$$\langle B, M \rangle = \prod_E \langle A, L \rangle,$$

where L is a well ordering of A. Suppose that $f_E M g_E$ in $\langle B, M \rangle$, and f lives on W and g lives on Z, where $W, Z \in S_\omega(Y)$. Suppose further that, in Y, $W \setminus Z$ is an initial segment of $W \cup Z$. Then there exists h_E in B such that $f_E M h_E$, $h_E M g_E$, and h lives on $W \cap Z$.

6.5.31*. Let α be a regular cardinal, $\alpha > 2^\omega$. Let $\langle Y, < \rangle$ be ordermorphic to $\langle \alpha, > \rangle$. For each $y \in Y$, let D_y be an ultrafilter over ω. Let $\mathfrak{A} = \langle A, L \rangle$ be a model such that L well orders A. Then the iterated ultrapower $\prod_{\times_Y D_y} \mathfrak{A}$ has no strictly descending sequences of length α.

6.5.32. Let D, E be ultrafilters, and let E be countably incomplete. Then $D \times E$ is α-good if and only if E is α-good.

6.5.33. Use the above exercise to do Exercise 6.1.6.

6.5.34 (GCH). Let D be an α^+-good countably incomplete ultrafilter over α, and E a β^+-good countably incomplete ultrafilter over β. Form the ultrafilter $D \times E$ over the set $I = \alpha \times \beta$. Find two families of finite nonempty sets $A_i, B_i, i \in I$, such that

$$|\prod_{D \times E} A_i| = \alpha^+, \qquad |\prod_{D \times E} B_i| = \beta^+.$$

6.5.35. Reformulate Exercise 6.1.19 to show that if $2^{\omega_1} = \omega_2$, then there exist ultrafilters D, E and a countable model $\mathfrak{A}$ such that

$$\prod_{D \times E} \mathfrak{A} \not\cong \prod_{E \times D} \mathfrak{A}.$$

6.5.36. A filter D over a set I or power α is said to be *selective* iff for every partition $I = \bigcup_{\beta<\alpha} X_\beta$ of I into α nonempty sets X_β such that $\bigcup_{\beta<\gamma} X_\beta \notin D$ for all $\gamma < \alpha$, there exists $Y \in D$ such that for all $\beta < \alpha$, $|Y \cap X_\beta| = 1$. (That is, Y is a choice function for the partition.) Let D be a selective ultrafilter over α. Show that there cannot exist nonprincipal ultrafilters E, F over sets J, K such that F is uniform, $|K| = \alpha$, and there is a one–one mapping of $J \times K$ onto α which maps $E \times F$ onto D.

6.5.37. Let D be a countably incomplete, selective, α^+-good ultrafilter over α. Show that there cannot exist countably incomplete ultrafilters E, F over sets J, K such that some one–one function f of $J \times K$ onto α maps $E \times F$ onto D.

6.5.38* (GCH). Show that there exists an ultrafilter D over α which is countably incomplete, selective, and α^+-good.

6.5.39. Let D be a countably incomplete ultrafilter and E an α-complete ultrafilter. Show that $D \times E$ is α-good if and only if D is α-good. (Cf. Exercise 6.5.32.)

CHAPTER 7

SELECTED TOPICS

7.1. Categoricity in power

Let us recall that a theory T in $\mathscr{L}$ is said to be *categorical in power* α, or *α-categorical*, iff T has a model of power α and every two models of T of power α are isomorphic. (In other words, T has exactly one model of power α, up to isomorphism.) In Section 2.3 we obtained a necessary and sufficient condition for a theory to be categorical in power ω. In this section we shall study theories which are categorical in uncountable powers.

We shall assume throughout this section that the language $\mathscr{L}$ is countable, $||\mathscr{L}|| = \omega$. We shall be interested only in complete theories, since every α-categorical theory having no finite models is complete.

In Section 1.4 we have seen examples of theories of each of the following kinds:

categorical in every infinite power;
categorical in power ω but not in any uncountable power;
categorical in every uncountable power but not in power ω;
not categorical in any infinite power.

It was conjectured by Łoś that these are the only possibilities. Łoś's conjecture may be stated as follows: If a theory T in $\mathscr{L}$ is categorical in some uncountable power, then T is categorical in every uncountable power. This conjecture was proved by Morley (1965a), and our chief aim in this section is to prove Morley's theorem (Theorem 7.1.14). First, we give a list of examples of theories which are categorical in every uncountable power. Only a few natural examples of such theories are known:

(i). Pure identity theory.
(ii). Infinite Abelian groups in which all elements have order p (p prime).
(iii). Divisible torsion-free Abelian groups.

(iv). Algebraically closed fields of characteristic p (zero or prime).

(v). The theory of all models $\langle A, G\rangle$, where A is an infinite set and G is a permutation of A with no finite cycles.

(vi). The theory of all models $\langle A, U, G\rangle$, where G is one–one function with n places mapping U^n onto $A \setminus U$.

(vii). The theory in the language $\mathscr{L} = \{c_n : n < \omega\}$ with the axioms $\neg\, c_n \equiv c_m$, where $n < m < \omega$.

(viii). The complete theory of the model $\langle \omega, S\rangle$, where S is the successor function.

Our proof of Morley's theorem will put together several of the methods developed earlier in this book, including indiscernible sets, saturated models, and one of the Löwenheim–Skolem–Tarski theorems for two cardinals, namely Theorem 3.2.11.

LEMMA 7.1.1. *Let T be a complete theory in $\mathscr{L}$ such that every model of T of power ω_1 is saturated. Then every uncountable model of T is saturated. Hence if T has infinite models, T is categorical in every uncountable power.*

PROOF. We shall use Theorem 3.2.11. Suppose that $\alpha > \omega_1$ and T has a model $\mathfrak{A}$ of power α which is not saturated. Then there exists a set $X \subset A$ of power $|X| < \alpha$ and a set $\Sigma(v)$ of formulas in the expanded language $\mathscr{L}_X = \mathscr{L} \cup \{c_x : x \in X\}$ such that every finite subset of Σ is satisfiable in the model $(\mathfrak{A}, x)_{x\in X}$, but Σ is not satisfiable in $(\mathfrak{A}, x)_{x\in X}$. Since $||\mathscr{L}|| = \omega$, we have $||\mathscr{L}_X|| = |X| \cup \omega < \alpha$; hence $|\Sigma| < \alpha$. We may therefore choose a subset $U \subset A$ of power $|U| = |\Sigma|$ and a one–one function φ of U onto Σ. We shall now expand the model $\mathfrak{A}$ so that we can say that an element satisfies the formula corresponding to some $u \in U$. Besides the new one-placed relation U, we shall introduce two two-placed relations R and S.

The mapping φ associates with each element $a \in U$ a formula φ_a in Σ, whence φ_a is a formula in the language $\mathscr{L}_X$ in the variable v. Let R be the set of all pairs $\langle a, x\rangle$ in $A \times A$ such that $a \in U$, $x \in X$, and the constant c_x occurs in the formula φ_a. Thus for each $a \in A$ there are only finitely many, and possibly zero, $x \in A$ such that $R(a, x)$. Let S be the set of all pairs $\langle a, b\rangle \in A \times A$ such that $a \in U$ and $(\mathfrak{A}, x)_{x\in X} \models \varphi_a[b]$. We form the expanded model

$$\mathfrak{A}^* = (\mathfrak{A}, U, R, S).$$

For each $a \in U$, the following formula $\psi(u)$ is satisfied by a in $(\mathfrak{A}^*, x)_{x\in X}$:

$$(1) \qquad (\forall v)(S(uv) \leftrightarrow \varphi_a(v)).$$

Furthermore, for each $n < \omega$, the following sentence, which says 'every n

elements of Σ are simultaneously satisfied', holds in $\mathfrak{A}^*$:

(2) $$(\forall u_1 \ldots u_n)[U(u_1) \wedge \ldots \wedge U(u_n) \rightarrow (\exists v)(S(u_1 v) \wedge \ldots \wedge S(u_n v))].$$

Finally, the sentence below, which says 'Σ is not satisfied', holds in $\mathfrak{A}^*$:

(3) $$\neg (\exists v \forall u)(U(u) \rightarrow S(uv)).$$

By Theorem 3.2.14, since $|U| < |A|$, there exist two models $\mathfrak{B}^*$ and $\mathfrak{C}^*$ such that $\mathfrak{B}^* \prec \mathfrak{A}^*$, $\mathfrak{B}^* \prec \mathfrak{C}^*$, $|B| = \omega$, $|C| = \omega_1$, and the interpretation of U is the same in $\mathfrak{B}^*$ as in $\mathfrak{C}^*$. Thus we may write

$$\mathfrak{B}^* = (\mathfrak{B}, U', R', S'), \qquad \mathfrak{C}^* = (\mathfrak{C}, U', R'', S''),$$

and U', being a subset of B, is countable. It follows that, with $X' = X \cap B$,

(4) $$(\mathfrak{B}^*, x)_{x \in X'} \prec (\mathfrak{A}^*, x)_{x \in X'},$$

and

(5) $$(\mathfrak{B}^*, x)_{x \in X'} \prec (\mathfrak{C}^*, x)_{x \in X'}.$$

For each $a \in U'$, the set $\{x \in A : R(ax)\}$ is finite, and therefore

$$\{x \in A : R(ax)\} = \{x \in B : R'(ax)\} \subset X'.$$

It follows that for each $a \in U'$, φ_a is a formula in the language $\mathscr{L}_{X'}$ of the model $(\mathfrak{B}^*, x)_{x \in X'}$. Therefore, by (4) and (5), for each $a \in U'$ the formula (1) is satisfied by a in both of the models $(\mathfrak{B}^*, x)_{x \in X'}$ and $(\mathfrak{C}^*, x)_{x \in X'}$. Also, the sentence (2) for each $n < \omega$ and the sentence (3) hold in $\mathfrak{B}^*$ and in $\mathfrak{C}^*$. We conclude that in the model $(\mathfrak{C}, x)_{x \in X'}$ the set of formulas

$$\Sigma'(v) = \{\varphi_a(v) : a \in U'\}$$

has the property that every finite subset of $\Sigma'(v)$ is satisfied but $\Sigma'(v)$ is not satisfied. Since $X' \subset B$, X' is countable. Thus the model $\mathfrak{C}$ is not ω_1-saturated. But $\mathfrak{C}$ does have power ω_1, and by (4) and (5), $\mathfrak{C}$ is a model of T. This shows that T has a model $\mathfrak{C}$ of power ω_1 which is not saturated, contrary to our hypothesis. So every uncountable model of T is saturated.

Finally, if T is complete and has infinite models, then any two models of T of the same infinite power are elementarily equivalent and saturated, whence by the uniqueness theorem for saturated models they are isomorphic and T is categorical in every uncountable power. ⊣

In the above lemma, the countability of the language $\mathscr{L}$ was used very strongly, because it was needed in the proof of Theorem 3.2.11.

We need yet another application of Theorem 3.2.11. This application concerns the notion of a theory being stable in a power, which is of considerable interest in its own right as well as an important tool in the proof of Morley's theorem. Let us recall that a type $\Sigma(v)$ is a maximal consistent set of formulas in the single variable v. Thus each element of a model realizes a unique type.

7.1.2. A theory T is *stable in power* α iff for every model $\mathfrak{A}$ of T and every set $X \subset A$ of power α, the simple expansion $(\mathfrak{A}, x)_{x \in X}$ realizes exactly α types in a single variable v.

Here is our second application of Theorem 3.2.11.

LEMMA 7.1.3. *Suppose T is a theory in $\mathscr{L}$ which is stable in power ω. Then T is stable in every infinite power.*

PROOF. Suppose $\alpha \geqslant \omega$ and T is not stable in power α. Then T has a model $\mathfrak{A}$ with a set $X \subset A$ of power α such that $(\mathfrak{A}, x)_{x \in X}$ realizes at least α^+ types in v. By the Löwenheim–Skolem–Tarski theorem, we may assume without loss of generality that $\mathfrak{A}$ has power exactly α^+ and $(\mathfrak{A}, x)_{x \in X}$ realizes exactly α^+ types in v. As in Lemma 7.1.1, we form an appropriate expansion of the model $\mathfrak{A}$.

Since $\mathscr{L}$ is countable, the language $\mathscr{L}_X = \mathscr{L} \cup \{c_x : x \in X\}$ has power α. Therefore the set Σ of *all* formulas $\varphi(v)$ of $\mathscr{L}_X$ in the variable v has power α. Choose a subset $U \subset A$ of power α with a one–one function φ on U onto Σ. As in Lemma 7.1.1, let R and S be binary relations on A such that

$$R(ax) \quad \text{iff} \quad a \in U \quad \text{and} \quad c_x \text{ occurs in } \varphi_a;$$

$$S(ab) \quad \text{iff} \quad a \in U \quad \text{and} \quad (\mathfrak{A}, x)_{x \in X} \models \varphi_a[b].$$

Let $V \subset A$ be a set of power $|V| = \alpha^+$ such that any two distinct elements of V realize different types in $(\mathfrak{A}, x)_{x \in X}$, and let G be a one–one function of A into V. Now form the expanded model

$$\mathfrak{A}^* = (\mathfrak{A}, U, V, R, S, G).$$

For each $a \in U$, in the model $(\mathfrak{A}^*, x)_{x \in X}$ a satisfies the formula

$$(1) \qquad (\forall v)(S(uv) \leftrightarrow \varphi_a(v)).$$

Furthermore, in the model $\mathfrak{A}^*$ the following sentences hold:

$$(2) \qquad (\forall vw)[v \not\equiv w \wedge V(v) \wedge V(w) \rightarrow (\exists u)(U(u) \wedge \neg (S(uv) \leftrightarrow S(uw)))],$$

$$(3) \qquad (\forall vw)(v \not\equiv w \rightarrow V(G(v)) \wedge G(v) \not\equiv G(w)).$$

The first one says 'two distinct elements of V realize different types', and the second one says 'G maps A one–one into V'. Now, using Theorem 3.2.11 and the fact that $|U|<|A|$, we obtain models

$$\mathfrak{B}^* = (\mathfrak{B}, U', V', R', S', G'),$$
$$\mathfrak{C}^* = (\mathfrak{C}, U', V'', R'', S'', G''),$$

such that $\mathfrak{B}^* \prec \mathfrak{A}^*$, $|B| = \omega$, $\mathfrak{B}^* \prec \mathfrak{C}^*$, and $|C| = \omega_1$. Since (3) holds in $\mathfrak{C}^*$, it follows that

(4) $$|V''| = \omega_1.$$

As before, for all $a \in U'$, every $x \in X$ such that $R(ax)$ belongs to $X' = X \cap B$, whence $\varphi_a(v)$ is a formula in $\mathscr{L}_{X'}$. Thus for each $a \in U'$, the formula (1) is satisfied in $(\mathfrak{B}^*, x)_{x\in X'}$ and in $(\mathfrak{C}^*, x)_{x\in X'}$ by the element a. Since (2) holds in $\mathfrak{C}^*$, and U is interpreted by U' in $\mathfrak{C}^*$, any two distinct elements of V'' realize different types in the model $(\mathfrak{C}, x)_{x\in X'}$. Also $X' \subset B$, so X' is countable. Thus, by (4), the model $(\mathfrak{C}, x)_{x\in X'}$ realizes uncountably many types, whence T is not stable in power ω. ⊣

The next lemma ties the notion of stability in with the notion of categoricity.

Lemma 7.1.4. *If a theory T is categorical in some uncountable power α, then T is stable in power ω.*

Proof. Suppose T is not stable in power ω. Then T has a model $\mathfrak{A}$ of power ω_1 with a countable set $X \subset A$ such that the model $(\mathfrak{A}, x)_{x\in X}$ realizes ω_1 types. Now we shall use a result in Section 3.3 on indiscernible sets. By Corollary 3.3.14, T has a model $\mathfrak{B}$ of power α with the property that for all $Y \subset B$, the expanded model $(\mathfrak{B}, y)_{y\in Y}$ realizes at most $|Y| \cup \omega$ types. But since T is α-categorical, every model $\mathfrak{B}$ of T of power α has the above property. By the Löwenheim–Skolem–Tarski theorem, the model $\mathfrak{A}$ has an elementary extension $\mathfrak{B}$ of power α. But then the model $(\mathfrak{B}, x)_{x\in X}$ realizes only $|X| \cup \omega = \omega$ types, while the model $(\mathfrak{A}, x)_{x\in X}$ realizes ω_1 types. This is a contradiction, because every element of A realizes the same type in $(\mathfrak{A}, x)_{x\in X}$ as it realizes in $(\mathfrak{B}, x)_{x\in X}$. ⊣

In view of Lemma 7.1.4, we wish to make a further study of theories which are stable in power ω. There are complete theories which are stable in power ω but are not categorical in power ω_1. The simplest example is the theory of all models $\langle A, U\rangle$, where U and $A\setminus U$ are infinite. Another example is the theory of all equivalence relations $\langle A, E\rangle$, where E has infinitely many equivalence classes and each class is infinite. Both of these examples are ω-categorical.

Here is an example of a complete theory which is ω-stable but not categorical in any power. T is the theory of all models $\langle A, U_0, U_1, U_2, \ldots\rangle$, where each U_n is a unary relation such that the U_n are infinite disjoint subsets of A. This theory has countably many countable models. If we add to T constant symbols c_{mn} and axioms stating that the constants are distinct and $U_n(c_{mn})$, we obtain a complete ω-stable theory with 2^ω countable models.

The corollary below gives a whole class of non-ω-stable theories.

Corollary 7.1.5. *Let* $\mathfrak{A} = \langle A, \leqslant, R_0, \ldots\rangle$ *be an infinite model such that* $\leqslant$ *is a simple ordering. Then* $\mathrm{Th}(\mathfrak{A})$ *is not* ω*-stable and is not* ω_1*-categorical.*

Proof. By the compactness theorem, there is a model $\mathfrak{B} \equiv \mathfrak{A}$ which has a subset $X \subset B$ such that $\langle X, \leqslant\rangle$ is isomorphic to the rationals. For each initial segment $Y \subset X$, the set of formulas

$$\{c_y \leqslant v : y \in Y\} \cup \{v \leqslant c_x : x \in X\setminus Y\}$$

is consistent with $\mathrm{Th}(\mathfrak{B}_X)$. But X has 2^ω initial segments, so $\mathrm{Th}(\mathfrak{B}_X)$ has 2^ω types. ⊣

Lemma 7.1.6. *Suppose T is a theory which has infinite models and is stable in power* ω. *Then, for every regular cardinal* $\alpha > \omega$, T *has an* α*-saturated model of every power* $\beta \geqslant \alpha$.

Proof. Consider an arbitrary model $\mathfrak{A}$ of T of power β. Form the expanded model $(\mathfrak{A}, a)_{a\in A}$, and let $T_\mathfrak{A}$ be the complete theory of this model. By the compactness theorem, there is a model $(\mathfrak{B}, a)_{a\in A}$ in which every type consistent with $T_\mathfrak{A}$ is realized. By Lemma 7.1.3, T is stable in power β. Hence $(\mathfrak{B}, a)_{a\in A}$ realizes only $|A| = \beta$ different types. Therefore by the Löwenheim–Skolem–Tarski theorem there is an elementary submodel $(\mathfrak{C}, a)_{a\in A} \prec (\mathfrak{B}, a)_{a\in A}$ of power β in which every type consistent with $T_\mathfrak{A}$ is realized. From Propositions 3.1.1 and 3.1.3 on elementary extensions, $\mathfrak{C}$ is an elementary extension of $\mathfrak{A}$. We have shown:

(1) Every model $\mathfrak{A}$ of T of power β has an elementary extension $\mathfrak{A}'$ of power β such that every type which is consistent with the theory of $(\mathfrak{A}, a)_{a\in A}$ is realized in $(\mathfrak{A}', a)_{a\in A}$.

Now we use (1) α times, and form an elementary chain $\mathfrak{A}_\gamma$, $\gamma < \alpha$, such that each $\mathfrak{A}_\gamma$ is a model of T of power β,

$$\mathfrak{A}_\gamma = \bigcup_{\delta<\gamma}\mathfrak{A}_\delta \quad \text{when} \quad \gamma \text{ is a limit ordinal,}$$

and for all $\gamma < \alpha$, the condition (1) holds for $\mathfrak{A} = \mathfrak{A}_\gamma$, $\mathfrak{A}' = \mathfrak{A}_{\gamma+1}$. Let $\mathfrak{B}$ be the union $\mathfrak{B} = \bigcup_{\gamma<\alpha}\mathfrak{A}_\gamma$. Then $\mathfrak{B}$ is a model of T, and since $\alpha \leqslant \beta$, $\mathfrak{B}$ has power β.

We show that $\mathfrak{B}$ is α-saturated. Consider any set $X \subset B$ of power $|X| < \alpha$. Since α is regular, $|X| < \mathrm{cf}(\alpha)$, and it follows that there exists $\gamma < \alpha$ such that $X \subset \mathfrak{A}_\gamma$. Then every type $\Sigma(v)$ consistent with the complete theory of the model $(\mathfrak{A}_\gamma, x)_{x\in X}$ is realized in $(\mathfrak{A}_{\gamma+1}, x)_{x\in X}$, and hence in $(\mathfrak{B}, x)_{x\in X}$. But since $\mathfrak{A}_\gamma \prec \mathfrak{B}$, the models $(\mathfrak{A}_\gamma, x)_{x\in X}$ and $(\mathfrak{B}, x)_{x\in X}$ have the same complete theories. This shows that $\mathfrak{B}$ is α-saturated. ⊣

At this point we can already prove the 'upward' half of Morley's categoricity theorem.

LEMMA 7.1.7 (Upward Morley Theorem). *Every complete theory T which is categorical in power ω_1 is categorical in every uncountable power.*

PROOF. Let T be categorical in power ω_1. Then T is stable in power ω. Since ω_1 is regular, T has a saturated model of power ω_1. Therefore every model of T of power ω_1 is saturated. By Lemma 7.1.1, all uncountable models of T are saturated. Since T is complete, any two saturated models of T of the same power are isomorphic. Hence T is categorical in every uncountable power. ⊣

Before continuing towards Morley's theorem, we can at this point derive two corollaries about categoricity in power which are of interest.

COROLLARY 7.1.8. *Let α be an uncountable cardinal and let T be a complete theory which has infinite models. Then a necessary and sufficient condition for T to be categorical in power α is that every model of T of power α is saturated.*

PROOF. The sufficiency follows at once from the uniqueness theorem for saturated models. To prove the necessity, we assume that T is α-categorical. Let $\mathfrak{A}$ be a model of T of power α. By Lemma 7.1.4, T is stable in power ω. Thus, by Lemma 7.1.6, for every regular cardinal $\omega < \alpha' \leqslant \alpha$, T has an

α'-saturated model of power α. But since $\mathfrak{A}$ is the *only* model of T of power α (up to isomorphism), $\mathfrak{A}$ is α'-saturated for all regular α', $\omega < \alpha' \leqslant \alpha$. If α is a successor cardinal, then α is itself regular, so $\mathfrak{A}$ is saturated. On the other hand, if α is singular, then $\mathfrak{A}$ is γ^+-saturated for all infinite $\gamma < \alpha$, and so $\mathfrak{A}$ is saturated by Proposition 5.1.1. ⊣

COROLLARY 7.1.9. *Suppose T is categorical in an uncountable cardinal α. Then for every model $\mathfrak{A}$ of T and every countable set $X \subset A$, the complete theory of the expanded model $(\mathfrak{A}, x)_{x\in X}$ is α-categorical.*

PROOF. Let $(\mathfrak{B}, y_x)_{x\in X} \equiv (\mathfrak{A}, x)_{x\in X}$ and let $\mathfrak{B}$ have power α. Then the models $\mathfrak{B}$ and $(\mathfrak{B}, y_x)_{x\in X}$ are both saturated models of power α.⊣

The next two lemmas concern atomic models of ω-stable theories. Let us recall that a *complete formula*, or *atom*, of T is a formula $\varphi(x_1 \dots x_n)$ which is consistent with T and is such that for every formula $\psi(x_1 \dots x_n)$, either $T \models \varphi \to \psi$ or $T \models \varphi \to \neg\psi$. If $\varphi(x_1 \dots x_n)$ is complete and $T \models \varphi(x_1 \dots x_n) \to \psi(x_1 \dots x_n)$, we call φ an atom of ψ. If $\psi(x_1 \dots x_n)$ has an atom in T, it is said to be *completable* in T. The theory T is said to be *atomic* iff every formula consistent with T is completable in T.

In Section 2.3, we only considered atomic theories in countable languages. Now we shall need to study atomic theories in uncountable languages formed by adding constants to $\mathscr{L}$.

LEMMA 7.1.10. *Suppose T is a theory in a countable language $\mathscr{L}$ and T is stable in power ω. Then for every model $\mathfrak{A}$ of T and every subset $X \subset A$, the complete theory of the expanded model $(\mathfrak{A}, x)_{x\in X}$ is atomic.*

PROOF. Let $\mathfrak{A}_X = (\mathfrak{A}, x)_{x\in X}$, let T_X be the complete theory of $\mathfrak{A}_X$, and let $\mathscr{L}_X = \mathscr{L} \cup \{c_x : x \in X\}$ be the language of $\mathfrak{A}_X$. Assume that for some $X \subset A$, T_X is not atomic. We shall obtain a contradiction. There is a smallest positive integer n such that there exists an $X \subset A$ and a formula $\psi(v_1 \dots v_n)$ which is consistent with T_X and is incompletable. We shall first reduce the lemma to the case $n = 1$.

Suppose for the moment that $n > 1$. Then the formula $(\exists v_n)\psi(v_1 \dots v_n)$ has an atom $\varphi(v_1 \dots v_{n-1})$. Add new constants $c_1, \dots, c_{n-1}$ to $\mathscr{L}_X$. Then the set of sentences

$$T_X \cup \{\varphi(c_1 \dots c_{n-1})\} = T_X'$$

determines a complete theory in $\mathscr{L}_X \cup \{c_1 \dots c_{n-1}\}$. We claim that the formula $\psi(c_1 \dots c_{n-1} v_n)$ has no atoms in T'_X. For assume that $\theta(c_1 \dots c_{n-1} v_n)$ is an atom of $\psi(c_1 \dots c_{n-1} v_n)$. Then

$$T'_X \vdash \theta(c_1 \dots c_{n-1} v_n) \rightarrow \psi(c_1 \dots c_{n-1} v_n),$$

whence

$$T_X \vdash \varphi(v_1 \dots v_{n-1}) \wedge \theta(v_1 \dots v_n) \rightarrow \psi(v_1 \dots v_n).$$

However, it is easily seen that the formula $\varphi(v_1 \dots v_{n-1}) \wedge \theta(v_1 \dots v_n)$ is an atom of T_X, contradicting the assumption that $\psi(v_1 \dots v_n)$ has no atoms. This proves the claim that $\psi(c_1 \dots c_{n-1} v_n)$ has no atoms in T'_X. Now let $a_1, \dots, a_{n-1}$ be elements of A such that $\mathfrak{A}_X \models \varphi[a_1 \dots a_{n-1}]$. Then $(\mathfrak{A}_X, a_1 \dots a_{n-1})$ is a model of T'_X, so the formula $\psi(c_1 \dots c_{n-1} v_n)$ has no atoms in the complete theory of $(\mathfrak{A}_X, a_1 \dots a_{n-1})$.

We may assume hereafter that $n = 1$, so $\psi(v)$ is consistent with T_X and has no atoms.

We shall use a binary tree argument. For every formula $\theta(v)$ which is consistent with and incompletable in T_X, we may choose two consistent formulas $\theta_0(v)$ and $\theta_1(v)$ such that

$$T_X \models \theta_0 \rightarrow \theta, \qquad T_X \models \theta_1 \rightarrow \theta, \qquad T_X \models \neg(\theta_0 \wedge \theta_1).$$

The formulas $\theta_0(v)$ and $\theta_1(v)$ are then still incompletable in T_X. Repeating the process with $\psi(v)$ as a starting point, we obtain a tree of consistent incompletable formulas,

$$\psi \begin{cases} \psi_0 \begin{cases} \psi_{00} \cdots \\ \psi_{01} \cdots \end{cases} \\ \psi_1 \begin{cases} \psi_{10} \cdots \\ \psi_{11} \cdots \end{cases} \end{cases}$$

Let Y be the set of all constants $x \in X$ which occur in some formula of this tree. Then Y is a countable subset of X. Let T_Y be the complete theory of $(\mathfrak{A}, x)_{x \in Y}$. The tree has 2^ω branches, and each branch is a set of formulas consistent with T_Y. It follows that T_Y has 2^ω types in v. By the compactness theorem, T_Y has a model which realizes 2^ω types. Therefore T is not stable in power ω, contrary to the hypothesis. ⊣

Let us recall that a model $\mathfrak{A}$ is said to be *atomic* iff every n-tuple of elements of A satisfies a complete formula in $\mathrm{Th}(\mathfrak{A})$. Also, $\mathfrak{A}$ is said to be *prime* iff $\mathfrak{A}$ is elementarily embeddable in every model of $\mathrm{Th}(\mathfrak{A})$. In Section 2.3, we proved that every atomic theory in a countable language has a model which is atomic and prime. We now generalize this result to uncountable languages. Later on we shall discuss the question of uniqueness of prime models.

LEMMA 7.1.11. *Let T be a theory such that for every model $\mathfrak{A}$ of T and every subset $X \subset A$, the complete theory of the expanded model $\mathfrak{A}_X = (\mathfrak{A}, x)_{x \in X}$ is atomic. Then for every model $\mathfrak{A}$ of T and every subset $X \subset A$, the complete theory of $\mathfrak{A}_X$ has an atomic prime model.*

PROOF. We shall construct an elementary submodel of $\mathfrak{A}_X$ which is atomic and prime. Let $\mathscr{L}_X$ be the language of $\mathfrak{A}_X$ and let $||\mathscr{L}_X|| = \alpha$. We shall find a sequence a_β, $\beta < \alpha \cdot \omega$, of elements of A such that:

(1) For all $\beta < \alpha \cdot \omega$, a_β realizes an atom in the complete theory of $(\mathfrak{A}_X, a_\gamma)_{\gamma < \beta}$.

(2) For all $n < \omega$ and every atom $\varphi(v)$ in the complete theory of $(\mathfrak{A}_X, a_\gamma)_{\gamma < \alpha \cdot n}$, there exists $\delta < \alpha \cdot (n+1)$ such that a_δ realizes $\varphi(v)$.

This is done inductively. Suppose $m < \omega$ and we have a sequence a_β, $\beta < \alpha \cdot m$ such that (1) holds for all $\beta < \alpha \cdot m$ and (2) holds for all $n < m$. Let $\varphi_\delta(v)$, $\delta < \alpha$, be a list of all the atoms in the variable v for the complete theory of the model $(\mathfrak{A}_X, a_\gamma)_{\gamma < \alpha \cdot m}$. Choose for $a_{\alpha \cdot m}$ an element of A which satisfies $\varphi_0(v)$ in $(\mathfrak{A}_X, a_\gamma)_{\gamma < \alpha \cdot m}$. For the next step, we use the fact that the complete theory of $(\mathfrak{A}_X, a_\gamma)_{\gamma < \alpha \cdot m+1}$ is atomic. Therefore the formula $\varphi_1(v)$ has an atom $\varphi_1'(v)$ in the complete theory of $(\mathfrak{A}_X, a_\gamma)_{\gamma < \alpha \cdot m+1}$. (Note that $\varphi_1(v)$ itself is an atom for the complete theory of $(\mathfrak{A}_X, a_\gamma)_{\gamma < \alpha \cdot m}$, but perhaps not for $(\mathfrak{A}_X, a_\gamma)_{\gamma < \alpha \cdot m+1}$.) Now choose $a_{\alpha \cdot m+1}$ satisfying $\varphi_1'(v)$. We may continue in this manner to obtain $a_{\alpha \cdot m+\delta}$, $\delta < \alpha$, such that (1) holds for $\beta < \alpha \cdot (m+1)$ and (2) holds for $n = m$. This completes the induction.

Now let $\mathfrak{B}_X$ be the submodel of $\mathfrak{A}_X$ such that $B = \{a_\beta : \beta < \alpha \cdot \omega\}$. For any formula $\varphi(v)$ which is consistent with the complete theory of $(\mathfrak{A}_X, a_\beta)_{\beta < \alpha \cdot \omega}$, there is an $n < \omega$ such that all the constants c_β occurring in $\varphi(v)$ appear before $\alpha \cdot n$, $\beta < \alpha \cdot n$. Since the theory of $(\mathfrak{A}_X, a_\beta)_{\beta < \alpha \cdot n}$ is atomic, $\varphi(v)$ contains an atom $\varphi'(v)$ in that theory. Then by (2) there is a $\delta < \alpha$ such that $a_{\alpha \cdot n+\delta}$ satisfies $\varphi'(v)$, and hence satisfies $\varphi(v)$. But $a_{\alpha \cdot n+\delta}$ is an element of B. This shows that

(3) $$\mathfrak{B}_X \prec \mathfrak{A}_X.$$

We first show that $\mathfrak{B}_X$ is prime. Let $\mathfrak{C}_X \equiv \mathfrak{B}_X$. Since each a_β realizes an atom in the complete theory of $(\mathfrak{B}_X, a_\gamma)_{\gamma<\beta}$, we see by induction that there are elements $d_\gamma \in C$, $\gamma < \alpha \cdot \omega$, such that

$$(\mathfrak{B}_X, a_\gamma)_{\gamma<\alpha\cdot\omega} \equiv (\mathfrak{C}_X, d_\gamma)_{\gamma<\alpha\cdot\omega}.$$

It follows that the mapping $a_\gamma \to d_\gamma$ is an elementary embedding of $\mathfrak{B}_X$ into $\mathfrak{C}_X$, whence $\mathfrak{B}_X$ is prime.

It remains to be proved that $\mathfrak{B}_X$ is an atomic model. To do this, we prove by induction that for all $\beta < \alpha \cdot \omega$,

(4) for all $\delta_1, ..., \delta_n < \beta$, the n-tuple $a_{\delta_1}, ..., a_{\delta_n}$ satisfies an atom in $\mathfrak{A}_X$.

Suppose (4) holds for all $\beta < \gamma$. If γ is a limit ordinal, (4) obviously holds for $\beta = \gamma$. Suppose γ is a successor ordinal, $\gamma = \eta + 1$. Consider $\delta_1, ..., \delta_n < \eta$, and look at the $(n+1)$-tuple $a_{\delta_1}, ..., a_{\delta_n}, a_\eta$. By (1), a_η satisfies an atom $\varphi(v)$ in the complete theory of $(\mathfrak{A}_X, a_\beta)_{\beta<\eta}$. Let $c_{\lambda_1}, ..., c_{\lambda_m}$ be all the constants appearing in $\varphi(v)$. Then by our inductive hypothesis, the $(n+m)$-tuple $a_{\delta_1}, ..., a_{\delta_n}, a_{\lambda_1}, ..., a_{\lambda_m}$ satisfies an atom $\theta(v_1 ... v_n u_1 ... u_m)$ in the theory of $\mathfrak{A}_X$. Moreover, the formula $\varphi(v)$ may be written in the form $\varphi(vc_{\lambda_1} ... c_{\lambda_m})$, where $\varphi(vu_1 ... u_m)$ is a formula in the theory of $\mathfrak{A}_X$.

For any formula $\psi(v)$ of $(\mathfrak{A}_X, a_\beta)_{\beta<\eta}$, one of the formulas

$$(\forall v)(\varphi(v) \to \psi(v)), \quad (\forall v)(\varphi(v) \to \neg\psi(v))$$

holds in that model. Hence for any formula $\psi(vv_1 ... v_n u_1 ... u_m)$ of $\mathscr{L}_X$, either

$$\mathfrak{A}_X \models \theta(v_1 ... v_n u_1 ... u_m) \to (\forall v)(\varphi(vu_1 ... u_m) \to \psi),$$

or else

$$\mathfrak{A}_X \models \theta(v_1 ... v_n u_1 ... u_m) \to (\forall v)(\varphi(vu_1 ... u_m) \to \neg\psi).$$

It follows that the formula

$$\theta(v_1 ... v_n u_1 ... u_m) \wedge \varphi(vu_1 ... u_m)$$

is an atom in the theory of $\mathfrak{A}_X$, for the variables $v, v_1, ..., v_n, u_1, ..., u_m$. Hence the formula

$$(\exists u_1 ... u_m)[\theta(v_1 ... v_n u_1 ... u_m) \wedge \varphi(vu_1 ... u_m)]$$

is an atom in the theory of $\mathfrak{A}_X$, and this atom is satisfied by the elements $a_{\delta_1}, ..., a_{\delta_n}, a_\eta$. This completes the proof of (4) by induction. ⊣

Combining the last two lemmas, we have the following:

Corollary 7.1.12. *Let T be a theory in a countable language such that T is stable in power ω. Then for every model $\mathfrak{A}$ of T and every subset $X \subset A$, the complete theory of $\mathfrak{A}_X$ has an atomic prime model.*

Lemma 7.1.13. *Suppose T is a complete ω-stable theory and $\mathfrak{A}$ is an uncountable model of T. Then there is a proper elementary extension $\mathfrak{B} \succ \mathfrak{A}$ such that every countable set $\Gamma(v)$ of formulas of $\mathscr{L}_A$ which is realized in $\mathfrak{B}_A$ is realized in $\mathfrak{A}_A$.*

Proof. We first find a formula $\psi(v)$ of $\mathscr{L}_A$ such that:

(1) $\{b \in A : \mathfrak{A}_A \models \psi[b]\}$ is uncountable.

(2) For any formula $\varphi(v)$ of $\mathscr{L}_A$, at least one of the sets

$$\{b \in A : \mathfrak{A}_A \models \psi \wedge \varphi[b]\}, \qquad \{b \in A : \mathfrak{A}_A \models \psi \wedge \neg\varphi[b]\}$$

is countable.

Suppose there is no such formula $\psi(v)$. That is, for any formula $\varphi(v)$ satisfied by uncountably many elements in $\mathfrak{A}_A$, there are two formulas $\varphi_0(v)$ and $\varphi_1(v)$ such that

$$\mathfrak{A}_A \models \varphi_0 \to \varphi, \qquad \mathfrak{A}_A \models \varphi_1 \to \varphi, \qquad \mathfrak{A}_A \models \neg(\varphi_0 \wedge \varphi_1),$$

and $\varphi_0(v)$, $\varphi_1(v)$ are each satisfied by uncountably many elements in $\mathfrak{A}_A$. Using a binary tree argument, we could then find 2^ω types consistent with $\text{Th}(\mathfrak{A}_X)$ for some countable $X \subset A$. But this contradicts the hypothesis that T is ω-stable. Therefore there is a formula $\psi(v)$ of $\mathscr{L}_A$ satisfying (1) and (2).

Now let c be a new constant symbol and let Δ be the set of all sentences $\varphi(c)$ in $\mathscr{L}_A \cup \{c\}$ such that all but countably many elements which satisfy $\psi(v)$ satisfy $\varphi(v)$ in $\mathfrak{A}_A$. Then by (1) and (2), Δ is a complete theory in $\mathscr{L}_A \cup \{c\}$. Also, $\text{Th}(\mathfrak{A}_A) \subset \Delta$. By the preceding corollary, Δ has an atomic model

$$(\mathfrak{B}, a, c)_{a \in A},$$

and $\mathfrak{A} \prec \mathfrak{B}$. $\mathfrak{B}$ is a proper extension of $\mathfrak{A}$ because for each $a \in A$,

$$\Delta \models \neg c \equiv a.$$

We show that every countable set $\Gamma(v)$ of formulas of $\mathscr{L}_A$ which is realized in $\mathfrak{B}_A$ is realized in $\mathfrak{A}_A$. Suppose $b \in B$ realizes $\Gamma(v)$. Then b satisfies a complete formula $\varphi(cv)$ in the atomic model $\mathfrak{B}_{A \cup \{c\}}$. It follows by the completeness of Δ that

$$\Delta \models (\exists v)\varphi(cv), \qquad \Delta \models \varphi(cv) \rightarrow \gamma(v),$$

for all $\gamma(v) \in \Gamma(v)$. Since $\Gamma(v)$ is countable, there is a countable subset $\Delta_0 \subset \Delta$ such that

$$\Delta_0 \models (\exists v)\varphi(cv), \qquad \Delta_0 \models \varphi(cv) \rightarrow \gamma(v)$$

for all $\gamma(v) \in \Gamma(v)$. Since each $\delta(v)$, where $\delta(c) \in \Delta_0$, is satisfied by all but countably many elements which satisfy $\psi(v)$ in $\mathfrak{A}_A$, there is an element $c_0 \in A$ such that $(\mathfrak{A}_A, c_0)$ is a model of Δ_0. But then there is an element $d_0 \in A$ such that $\mathfrak{A}_A \models \varphi[c_0 d_0]$, and hence d_0 realizes $\Gamma(v)$ in $\mathfrak{A}_A$. This completes the proof. ⊣

THEOREM 7.1.14 (Morley Categoricity Theorem). *Suppose T is a complete theory in a countable language. If T is categorical in some uncountable power, then T is categorical in every uncountable power.*

PROOF. Let T be categorical in the uncountable power α. By Lemma 7.1.1, it suffices to prove that every model of T of power ω_1 is saturated. Let $\mathfrak{A}$ be a model of T of power ω_1. Since T is α-categorical, T is ω-stable. Using the preceding lemma α times and taking unions at limit ordinals, we obtain an elementary extension $\mathfrak{B} \succ \mathfrak{A}$ of power α such that every countable set $\Gamma(v)$ of formulas of $\mathscr{L}_A$ which is realized in $\mathfrak{B}_A$ is realized in $\mathfrak{A}_A$. Since the ω-stable theory T has an ω_1-saturated model of power α and is α-categorical, $\mathfrak{B}$ must be ω_1-saturated. Consider any countable set $Y \subset A$ and any set $\Gamma(v)$ of formulas of $\mathscr{L}_Y$ consistent with $\mathrm{Th}(\mathfrak{A}_Y)$. Since $\mathfrak{B}$ is ω_1-saturated, $\Gamma(v)$ is realized in $\mathfrak{B}_Y$ and therefore is realized in $\mathfrak{A}_Y$. It follows that $\mathfrak{A}$ is ω_1-saturated. ⊣

The above proof of Morley's theorem actually shows a bit more, namely:

COROLLARY 7.1.15. *Let T be a complete theory in a countable language and let $\alpha > \omega$. If every model of T of power α is ω_1-saturated, then every uncountable model of T is saturated.*

The proof we have given for Morley's theorem is considerably simpler than Morley's original proof and is due to Baldwin and Lachlan. The two key lemmas which were not available to Morley were the Löwenheim–Skolem–Tarski Theorem for two cardinals and Lemma 7.1.13. On the other hand, Morley's original proof gave a great deal of additional information on ω-stable theories which is of independent interest. One of Morley's observations is not hard to prove and concerns indiscernibles in stable theories.

A subset X of a model $\mathfrak{A}$ is said to be *totally indiscernible* iff for any two n-tuples $x_1, \ldots, x_n$ and $y_1, \ldots, y_n$ in X such that $x_i \neq x_j$ and $y_i \neq y_j$, whenever $i \neq j$, we have

$$(\mathfrak{A}, x_1 \ldots x_n) \equiv (\mathfrak{A}, y_1 \ldots y_n).$$

Note that if X is totally indiscernible in $\mathfrak{A}$, then any simple ordering $\langle X, < \rangle$ is indiscernible in $\mathfrak{A}$.

For example, in an algebraically closed field any transcendental basis is a set of total indiscernibles.

THEOREM 7.1.16. *Suppose* $\mathfrak{A}$ *is a model of an* ω*-stable theory. Then any infinite set* $\langle X, < \rangle$ *of indiscernibles in* $\mathfrak{A}$ *is totally indiscernible.*

PROOF. Suppose $\langle X, < \rangle$ is indiscernible but not totally indiscernible. Then for some formula $\varphi(v_1 \ldots v_n)$ and some permutation π of $\{1, \ldots, n\}$, we have

$$\mathfrak{A} \models \varphi[x_1 \ldots x_n], \qquad \mathfrak{A} \models \neg \varphi[x_{\pi 1} \ldots x_{\pi n}],$$

for all increasing n-tuples $x_1 < \ldots < x_n$ in X. Since any permutation is a product of transpositions of the form $(i, i+1)$, we may assume that π is such a transposition. By the stretching theorem we may assume that $\langle X, < \rangle$ is a countable dense ordering. Fix elements

$$x_1 < \ldots < x_{i-1} < x_{i+2} < \ldots < x_n$$

in X, and let Y be the interval of $\langle X, < \rangle$ between x_{i-1} and x_{i+2}. Let

$$\varphi(yz) = \varphi(x_1 \ldots x_{i-1} y z x_{i+2} \ldots x_n).$$

For each initial segment S of Y, the set of formulas

$$\Gamma_S(v) = \{\varphi(yv) : y \in S\} \cup \{\neg \varphi(yv) : y \in Y \setminus S\}$$

is consistent with $\mathrm{Th}(\mathfrak{A}_X)$. Therefore $\mathfrak{A}_X$ has 2^ω types in v, contradicting ω-stability. ⊣

In his original proof of the categoricity theorem, Morley developed a way to classify formulas $\varphi(v)$ in $\mathfrak{A}_A$ by giving each formula an ordinal called its *transcendence rank*. The higher the transcendence rank, the greater the 'complexity' of the formula. The notion of transcendence rank has proved useful for other purposes. We present here the basic results of Morley and then discuss some later applications.

We begin with the transcendence rank zero. Consider a model $\mathfrak{A}$. A formula $\varphi(x)$ in $\mathscr{L}_A$ is said to be *algebraic*, or of *transcendence rank zero*, in $\mathfrak{A}$ iff $\varphi(x)$ is satisfied by at least one but only finitely many elements in $\mathfrak{A}$. The *degree n* of $\varphi(v)$ is the number of elements of A which satisfy $\varphi(x)$.

Notice that each algebraic formula $\varphi(x)$ of degree n is equivalent in $\mathrm{Th}(\mathfrak{A}_A)$ to a finite disjunction

$$x \equiv a_1 \vee \ldots \vee x \equiv a_n.$$

In particular, the algebraic formulas of degree 1 are equivalent to equations $x \equiv a$.

Now let us consider a subset $X \subset A$. An algebraic formula $\varphi(x)$ in $\mathscr{L}_X$ is said to be *irreducible over X* iff the set of consequences of $\varphi(x)$ in $\mathrm{Th}(\mathfrak{A}_X)$ is a type in $\mathscr{L}_X$, i.e., is maximal consistent in $\mathscr{L}_X$. A type $\Sigma(x)$ in $\mathrm{Th}(\mathfrak{A}_X)$ is said to be *algebraic over X* iff it is the set of all consequences of some irreducible algebraic formula over X. Similarly, an element $a \in A$ is *algebraic over X* iff a satisfies some irreducible algebraic formula over X.

LEMMA 7.1.17. *Let $X \subset A$ and let $\varphi(x)$ be an algebraic formula in $\mathscr{L}_X$. Then $\varphi(x)$ is equivalent in* $\mathrm{Th}(\mathfrak{A}_X)$ *to a finite disjunction of algebraic irreducible formulas over X.*

PROOF. Let $a_1, \ldots, a_n$ be the elements of A which satisfy $\varphi(x)$. Choose a maximal subset $\{b_1, \ldots, b_m\} \subset \{a_1, \ldots, a_n\}$ such that the b_i all realize different types in $\mathfrak{A}_X$. For each i, choose a formula $\psi_i(x)$ in $\mathscr{L}_X$ which is satisfied by b_i but not by any $b_j, j \neq i$. Then for each i, $\varphi_i(x) = \varphi(x) \wedge \psi_i(x)$ is algebraic and irreducible over X, and

$$\mathfrak{A}_X \models \varphi \leftrightarrow \varphi_1 \vee \ldots \vee \varphi_m. \dashv$$

For example, let $\mathfrak{A}$ be an algebraically closed field and let X be a subfield of $\mathfrak{A}$. If $p(x)$ is a polynomial over X with degree n and no multiple roots, then the equation $p(x) \equiv 0$ is an algebraic formula of degree n. The equation $p(x) \equiv 0$ is irreducible over X if and only if the polynomial $p(x)$ is irreducible over X. An element $a \in A$ is algebraic over X if and only if a belongs to the algebraic closure of X. Because any two elements not in the algebraic closure of X realize the same type in $\mathfrak{A}_X$, and if there is one such element, then there are infinitely many.

As a second example, let $\mathfrak{A}$ be a dense ordering without endpoints and let $X \subset A$. Then, using the analysis of dense ordering by elimination of quantifiers in Section 1.5, we see that an element a is algebraic over X if and only if $a \in X$.

We now introduce the higher transcendence ranks. The idea is to 'throw out' the algebraic formulas and look at what is left. It is convenient to work with ω_1-saturated models.

Let $\mathfrak{A}$ be an ω_1-saturated model. The *transcendence rank*, or Morley rank, of a formula $\varphi(x)$ of $\mathscr{L}_A$ is defined recursively as follows: $\varphi(x)$ has *rank* α in $\mathfrak{A}$, where α is an ordinal, if and only if the set of formulas

$$(1) \qquad \{\varphi(x)\} \cup \{\sigma(x) : \neg\sigma(x) \text{ has rank } < \alpha\}$$

of $\mathscr{L}_A$ is consistent and has finitely many maximal consistent extensions in $\text{Th}(\mathfrak{A}_A)$. The number of maximal consistent extensions of the set of formulas (1) is called the *degree* of a formula $\varphi(x)$ of rank α.

The degree is thus always a positive integer. Algebraic is the same as of rank zero, because a consistent formula $\varphi(x)$ is satisfied by finitely many elements $a_1, \ldots, a_n$ in $\mathfrak{A}$ if and only if it has finitely many maximal consistent extensions determined by the formulas $x \equiv a_1, \ldots, x \equiv a_n$.

If $\varphi(x)$ is inconsistent with $\text{Th}(\mathfrak{A}_A)$ we give it rank -1. If $\varphi(x)$ is consistent with $\text{Th}(\mathfrak{A}_A)$ but has no ordinal as a rank we give it rank ∞. Thus each formula $\varphi(x)$ has a unique rank.

If we let

$$S_\alpha(x) = \{\sigma(x) : \neg\sigma(x) \text{ has rank } < \alpha\}$$

for each ordinal α, then the transcendence rank may be described as follows: $S_0(x)$ is the set of all consequences of $\text{Th}(\mathfrak{A}_A)$. A formula $\varphi(x)$ is algebraic if it is consistent with $S_0(x)$ and has only finitely many maximal consistent extensions. $S_1(x)$ is equivalent to the set $\{x \not\equiv a : a \in A\}$. $\varphi(x)$ has rank 1 if $\varphi(x) \cup S_1(x)$ is consistent and has only finitely many maximal consistent extensions. In general, we see that:

(i). If $\varphi(x) \cup S_\alpha(x)$ is inconsistent, $\varphi(x)$ has rank $< \alpha$.

(ii). If $\varphi(x) \cup S_\alpha(x)$ is consistent and has finitely many maximal consistent extensions, $\varphi(x)$ has rank α.

(iii). If $\varphi(x) \cup S_\alpha(x)$ has infinitely many maximal consistent extensions, $\varphi(x)$ has rank $> \alpha$.

(iv). If $\varphi(x) \cup S_\alpha(x)$ is consistent for all α, then $\varphi(x)$ has rank ∞.

(v). $S_\alpha(x)$ is closed under consequences, and increases with α.

From the definition and intervening discussion, we see that:

LEMMA 7.1.18. *Let $\varphi(x)$ and $\psi(x)$ be formulas of $\mathscr{L}_A$.*

(i). *The rank of* $\varphi(x) \vee \psi(x)$ *is the maximum of the rank of* $\varphi(x)$ *and the rank of* $\psi(x)$.

(ii). *If* $\varphi(x)$ *and* $\psi(x)$ *have rank* α *and* $\mathfrak{A}_A \models \neg(\varphi(x) \wedge \psi(x))$, *then the degree of* $\varphi(x) \vee \psi(x)$ *is the sum of the degree of* $\varphi(x)$ *and the degree of* $\psi(x)$.

(iii). *Every formula of rank* α *is equivalent in* $\mathrm{Th}(\mathfrak{A}_A)$ *to a finite disjunction of formulas of rank* α *and degree* 1.

(iv). *If there is a formula of rank* α, α *an ordinal, then there are formulas of every rank* $\beta < \alpha$.

It follows that the valid formula $x \equiv x$ has the largest rank. We call the rank of $x \equiv x$ in $\mathfrak{A}$ the *rank of the model* $\mathfrak{A}$.

To complete our series of definitions, we let X be a subset of A. A formula $\varphi(x)$ in $\mathscr{L}_X$ of rank α is said to be *irreducible* over X iff the set of consequences of

$$\{\varphi(x)\} \cup (S_\alpha(x) \cap \mathscr{L}_X)$$

is a type in $\mathscr{L}_X$. (Thus irreducible over A is the same thing as degree 1.) A type $\Sigma(x)$ in $\mathrm{Th}(\mathfrak{A}_X)$ is of *rank* α over X iff it is the set of consequences of

$$\{\varphi(x)\} \cup (S_\alpha(x) \cap \mathscr{L}_X),$$

where $\varphi(x)$ is irreducible over X and of rank α. An element $a \in A$ has *rank* α *over* X iff its type in $\mathfrak{A}_X$ does.

LEMMA 7.1.19. *Let* $X \subset A$ *and let* $\varphi(x)$ *be a formula in* $\mathscr{L}_X$ *of rank* α. *Then* $\varphi(x)$ *is equivalent in* $\mathrm{Th}(\mathfrak{A}_X)$ *to a finite disjunction of irreducible formulas of rank* α *over* X. *Moreover,* $\varphi(x)$ *is irreducible over* X *if and only if* $\varphi(x)$ *is irreducible over every finite* $X_0 \subset X$, *where* $\varphi(x)$ *is in* $\mathscr{L}_{X_0}$.

PROOF. Like Lemma 7.1.17. ⊣

Let us return to our examples. First let $\mathfrak{A}$ be an algebraically closed field and X a subfield of $\mathfrak{A}$. The model $\mathfrak{A}$ has transcendence rank 1, because the set of formulas

$$\{x \not\equiv a : a \in A\}$$

determines a type in $\mathrm{Th}(\mathfrak{A}_A)$. The formula $x \equiv x$ is irreducible of rank 1 over X. An element $a \in A$ has rank 1 over X if and only if a is not in the algebraic closure of X.

The example where $\mathfrak{A}$ is a dense linear order without endpoints is quite different. In this model there are no formulas of rank 1, because if $\varphi(x)$

is satisfied by infinitely many elements, then

$$\{\varphi(x)\} \cup \{x \not\equiv a : a \in A\}$$

has infinitely many maximal consistent extensions in $\mathrm{Th}(\mathfrak{A}_A)$. Therefore φ has rank ∞ and every element of $A \setminus X$ has rank ∞ over X.

Here is an example of a model with rank 2. Let $\mathfrak{A} = \langle A, E\rangle$, where E is an equivalence relation with ω_1 equivalence classes, each of power ω_1. The formula $E(av)$ has rank 1 for each $a \in A$, because the set

$$\{E(av)\} \cup \{b \not\equiv v : b \in A\}$$

determines a type $\Sigma(v)$ in $\mathrm{Th}(\mathfrak{A}_A)$. The formula $v \equiv v$ has rank 2 because

$$\{v \equiv v\} \cup \{\neg E(av) : a \in A\}$$

determines a type in $\mathrm{Th}(\mathfrak{A}_A)$, but

$$\{v \equiv v\} \cup \{b \not\equiv v : b \in A\}$$

does not. Thus $\mathfrak{A}$ has transcendence rank 2.

For each $\alpha < \omega_1$, there are examples of theories whose models have transcendence rank α.

We now prove that in two elementarily equivalent ω_1-saturated models, the same formula has the same transcendence rank. This is the reason that ω_1-saturated models are more convenient than arbitrary models.

Lemma 7.1.20. *Let $\mathfrak{A}, \mathfrak{B}$ be ω_1-saturated models, let X be a subset of $A \cap B$, and suppose $\mathfrak{A}_X \equiv \mathfrak{B}_X$. Then:*

(i). *For any formula $\varphi(x)$ in $\mathscr{L}_X$, $\varphi(x)$ has the same rank in $\mathfrak{A}$ as in $\mathfrak{B}$. Thus $\mathfrak{A}$ and $\mathfrak{B}$ have the same rank.*

(ii). *If the rank of $\varphi(x)$ is an ordinal, then $\varphi(x)$ has the same degree in $\mathfrak{A}$ as in $\mathfrak{B}$, and $\varphi(x)$ is irreducible over X in $\mathfrak{A}$ if and only if it is irreducible over X in $\mathfrak{B}$.*

Proof. It suffices to prove the result in the case that X is finite.

(i). We prove by induction on α that for all ω_1-saturated models $\mathfrak{A}$ and $\mathfrak{B}$ and finite sets X, if $\mathfrak{A}_X \equiv \mathfrak{B}_X$ and $\varphi(x)$ in $\mathscr{L}_X$ has rank α in $\mathfrak{A}$, then it has rank α in $\mathfrak{B}$. We temporarily write $S_\alpha(\mathfrak{A}, x)$ for $S_\alpha(x)$ in $\mathfrak{A}$.

Assume the result for all $\beta < \alpha$. Suppose $\varphi(x)$ has rank α in $\mathfrak{A}$. Then $\varphi(x)$ cannot have rank $< \alpha$ in $\mathfrak{B}$. We assume $\varphi(x)$ has rank $> \alpha$ in $\mathfrak{B}$ and obtain a contradiction. The set of formulas

$$\{\varphi(x)\} \cup S_\alpha(\mathfrak{B}, x)$$

has infinitely many maximal consistent extensions. Therefore for some countable set C with $X \subset C \subset B$, the set

$$\{\varphi(x)\} \cup (S_\alpha(\mathfrak{B}, x) \cap \mathscr{L}_C)$$

has infinitely many maximal consistent extensions in $\mathscr{L}_C$. Since $\mathfrak{A}$ is ω_1-saturated, there is a mapping $f : C \to A$ such that $f(b) = b$ for $b \in X$ and

$$(\mathfrak{A}, fb)_{b \in C} \equiv (\mathfrak{B}, b)_{b \in C}.$$

To simplify the notation, we may assume that $C \subset A \cap B$ and $\mathfrak{A}_C \equiv \mathfrak{B}_C$. By inductive hypothesis, the set of formulas $S_\alpha(x) \cap \mathscr{L}_C$ is the same with respect to $\mathfrak{A}$ as with respect to $\mathfrak{B}$. Since $S_\alpha(x)$ is closed under consequence, it follows that

$$\{\varphi(x)\} \cup S_\alpha(\mathfrak{A}, x)$$

has infinitely many maximal consistent extensions. This contradicts the assumption that $\varphi(x)$ has rank α in $\mathfrak{A}$. Therefore $\varphi(x)$ also has rank α in $\mathfrak{B}$.

(ii). The proof is similar to (i). ⊣

COROLLARY 7.1.21. *If the rank of $\varphi(x)$ in $\mathfrak{A}$ is an ordinal α, then $\alpha < (2^\omega)^+$.*

PROOF. There are only 2^ω theories of saturated models $\mathfrak{A}_X$, X countable. Therefore there are only 2^ω different ranks of formulas. ⊣

Lachlan (1971) has improved the above result to $\alpha < \omega_1$.

We now characterize ω-stable theories in terms of transcendence rank. By the above lemma, any two ω_1-saturated models of a complete theory T have the same rank. We call this rank the *rank of* T. T is called *totally transcendental* iff its rank is an ordinal, rather than ∞.

THEOREM 7.1.22. *Let T be a complete theory.*

(i). *T is ω-stable if and only if T is totally transcendental.*

(ii). *If T is totally transcendental, then its rank is a countable ordinal.*

PROOF. (i). First assume T has rank ∞. Let $\mathfrak{A}$ be an ω_1-saturated model of T and let $S(x) = \bigcup_\alpha S_\alpha(x)$. Then $S(x)$ is consistent and $S(x) = S_\beta(x)$ for some β. For every formula $\varphi(x)$ consistent with $S(x)$, the set

$$\{\varphi(x)\} \cup S(x)$$

has infinitely many maximal consistent extensions. Using a tree argument, we can then find a countable set $C \subset A$ such that $\mathrm{Th}(\mathfrak{A}_C)$ has 2^ω types. Therefore T is not ω-stable.

Now assume T has rank $< \infty$, and let $\mathfrak{A}$ be a model of T and X a countable subset of A. We may assume $\mathfrak{A}$ is ω_1-saturated.

Let $\Sigma(x)$ be a type in $\mathrm{Th}(\mathfrak{A}_X)$. Each formula $\sigma(x)$ in $\Sigma(x)$ has an ordinal rank. Let $\sigma(x)$ be a formula in $\Sigma(x)$ of minimal rank α. Since $\sigma(x)$ is a finite disjunction of irreducible formulas over X, we may assume that $\sigma(x)$ is irreducible over X. Then $\Sigma(x)$ is the set of all consequences of

$$\sigma(x) \cup (S_\alpha(x) \cap \mathscr{L}_X)$$

Since there are only countably many formulas in $\mathscr{L}_X$ and each type is determined by a formula as above, there are only countably many types in $\mathrm{Th}(\mathfrak{A}_X)$. Hence T is ω-stable.

(ii). Let T be ω-stable. Then for each n, T has only countably many complete extensions in the language $\mathscr{L} \cup \{c_1, \dots, c_n\}$. The rank of a formula $\varphi(xa_1 \dots a_n)$ in an ω_1-saturated model $\mathfrak{A}$ of T is completely determined by the theory of $(\mathfrak{A}, a_1 \dots a_n)$. Therefore there are only countably many ranks of formulas in $\mathfrak{A}$. Thus T has rank less than ω_1. ⊣

As an application of transcendence rank, we show that every uncountable model of an ω-stable theory has indiscernibles.

THEOREM 7.1.23. *Let T be a complete ω-stable theory. Then every model $\mathfrak{A}$ of T of regular power $\alpha > \omega$ has a set of total indiscernibles of power α. In fact, if $Y, Z \subset A$ and $|Y| < |Z| = \alpha$, then $\mathfrak{A}_Y$ has a totally indiscernible set $X \subset Z$ of power α.*

PROOF. We may assume that $\mathfrak{A}$ is ω_1-saturated, since $\mathfrak{A}$ has an ω_1-saturated elementary extension of power α. Let β be the least ordinal such that there is a formula $\varphi(x)$ in $\mathscr{L}_A$ of rank β which is satisfied by α elements of Z. Thus $0 < \beta < \omega_1$. There is a formula $\varphi(x)$ of rank β and degree 1 which is satisfied by α elements of Z. For any set $Y' \subset A$ of power $|Y'| < \alpha$, there are fewer than α formulas in $\mathscr{L}_{Y'}$ of rank $< \beta$, and each one is satisfied by fewer than α elements of Z. Therefore the set

$$S_{Y'}(x) = \{\varphi(x)\} \cup (S_\beta(x) \cap \mathscr{L}_{Y'})$$

is satisfied by α elements of Z. We may assume that all the constants of $\varphi(x)$ belong to Y. Then for each $Y' \supset Y$, the set $S_{Y'}(x)$ determines a type of Th($\mathfrak{A}_{Y'}$). We may choose elements

$$a_0, a_1, \ldots, a_\gamma, \ldots, \qquad \gamma < \alpha,$$

such that each a_γ belongs to Z and realizes the set

$$S_{Y_\gamma}(x), \qquad Y_\gamma = Y \cup \{a_\delta : \delta < \gamma\}.$$

Let X be the set $\{a_\gamma : \gamma < \alpha\}$. We shall prove that X is a set of indiscernibles in $\mathfrak{A}_Y$. It suffices to show that $\langle X, <\rangle$ is indiscernible, where $a_\delta < a_\gamma$ iff $\delta < \gamma$. We show by induction on n that for all $\gamma_1 < \ldots < \gamma_n < \alpha$ and $\delta_1 < \ldots < \delta_n < \alpha$, we have

$$(\mathfrak{A}_Y a_{\gamma_1} \ldots a_{\gamma_n}) \equiv (\mathfrak{A}_Y a_{\delta_1} \ldots a_{\delta_n}).$$

Assume this holds for n and let

$$Y' = Y \cup \{a_{\gamma_1}, \ldots, a_{\gamma_n}\},$$

$$Y'' = Y \cup \{a_{\delta_1}, \ldots, a_{\delta_n}\}.$$

Then

$$\theta(a_{\gamma_1} \ldots a_{\gamma_n} x) \in S_{Y'}(x)$$

if and only if

$$\theta(a_{\delta_1} \ldots a_{\delta_n} x) \in S_{Y''}(x).$$

Thus, for each $\gamma_{n+1} > \gamma_n$ and $\delta_{n+1} > \delta_n$, the following are equivalent:

$$\mathfrak{A}_Y \models \theta(a_{\gamma_1} \ldots a_{\gamma_n} a_{\gamma_{n+1}});$$

$$S_{Y'}(x) \models \theta(a_{\gamma_1} \ldots a_{\gamma_n} x);$$

$$S_{Y''}(x) \models \theta(a_{\delta_1} \ldots a_{\delta_n} x);$$

$$\mathfrak{A}_Y \models \theta(a_{\delta_1} \ldots a_{\delta_n} a_{\delta_{n+1}}).$$

This shows that for all $n < \omega$,

$$(\mathfrak{A}_Y a_{\gamma_1} \ldots a_{\gamma_n}) \equiv (\mathfrak{A}_Y a_{\delta_1} \ldots a_{\delta_n}). \dashv$$

In the years since the writing of the second edition of this book, the methods used by Morley to prove his categoricity theorem have developed into a major new branch of model theory, which goes by the names of Classification Theory, Stability Theory, or Forking. Several books on the subject are now available, including Baldwin (1988), Lascar (1986), Pillay (1983), Poizat (1985), and Shelah (1978), (19??). An excellent survey of the subject is given by Baldwin (1987a).

The aim of the theory is to isolate classes of models which can be classified up to isomorphism using invariants like the Morley rank. The theory has been most successful in the case of stable theories, that is, theories which are α-stable for some α. Instead of the original Morley rank, the theory stresses an analogue of the Morley rank with respect to a finite set of formulas, and a powerful generalization of the notion of algebraic dependence called forking. There has been progress both in the general study of stable theories and in the application of the method to particular classes of algebras.

We shall not go further into the subject in detail here. However, to give a feeling for the progress which has been made, we shall state a selection of results about categoricity and stability.

Shelah used transcendence degrees to study prime and atomic models of ω-stable theories with extra constants. We proved in Corollary 7.1.12 that if T is ω-stable, then for every model $\mathfrak{A}$ of T and every set $X \subset A$, $\text{Th}(\mathfrak{A}_X)$ has a prime atomic model. Shelah proved the following, which answered a question of Sacks.

7.1.24. *Suppose T is ω-stable, $\mathfrak{A}$ is a model of T, and $X \subset A$.*

(i). $\text{Th}(\mathfrak{A}_X)$ *has a unique prime model up to isomorphism.*

(ii). *A model $\mathfrak{B}_X$ of* $\text{Th}(\mathfrak{A}_X)$ *is prime if and only if it is atomic and has no uncountable set of indiscernibles.*

The rough idea of the proof is to build an isomorphism between two prime models by a back and forth construction using an induction on the ranks of elements over X.

Shelah (1978) proved the following extension of 7.1.24 to stable theories.

7.1.24′. *If T is a stable theory in a countable language and* $\text{Th}(\mathfrak{A}_X)$ *has a prime model for every model $\mathfrak{A}$ of T and countable subset $X \subset A$, then the prime model of* $\text{Th}(\mathfrak{A}_X)$ *is unique for every such $\mathfrak{A}$ and X.*

Morley's theorem was extended to uncountable languages by Shelah (1978), extending various partial results of Rowbottom, Ressayre, and Keisler. Shelah proved that:

7.1.25. *If a complete theory T is categorical in one cardinal greater than $||\mathscr{L}||$, then T is categorical in every cardinal greater than $||\mathscr{L}||$. If $||\mathscr{L}|| > \omega$ and T is categorical in cardinal $||\mathscr{L}||$, then T has fewer than $||\mathscr{L}||$ nonequivalent formulas and T is categorical in every cardinal greater than $||\mathscr{L}||$.*

Returning to countable languages, Baldwin improved Morley's result that every ω_1-categorical theory has countable rank.

7.1.26. *Every ω_1-categorical theory T has finite Morley rank.*

Baldwin and Lachlan obtained the following result concerning countable models of ω_1-categorical theories.

7.1.27. *Suppose T is an ω_1-categorical theory in a countable language. Then T is either ω-categorical or T has exactly countably many countable models up to isomorphism. Moreover, all countable models of T are ω-homogeneous.*

As a by-product of the above theorem, Baldwin and Lachlan gave a characterization of ω_1-categorical theories in terms of countable models. To state the result, we introduce some notation. Given a model $\mathfrak{A}$ and a formula $\varphi(x)$ in the language $\mathscr{L}_A$, let $\varphi(\mathfrak{A})$ be the set

$$\varphi(\mathfrak{A}) = \{a \in A\colon a \text{ satisfies } \varphi(x) \text{ in } \mathfrak{A}_A\}.$$

7.1.28. *Let T be a complete theory in a countable language. Then T is ω_1-categorical if and only if the following two conditions hold:*

(i). *T is stable in power ω.*

(ii). *For every pair of countable models $\mathfrak{A}$, $\mathfrak{B}$ of T with $\mathfrak{A} \precneqq \mathfrak{B}$, if $\varphi(x)$ is a formula of $\mathscr{L}_A$ and $\varphi(\mathfrak{A})$ is infinite, then $\varphi(\mathfrak{A}) \subsetneqq \varphi(\mathfrak{B})$.*

The necessity of (i) and (ii) for ω_1-categoricity was already known, and the sufficiency was proved by Baldwin and Lachlan. The necessity of (i) is given in Lemma 7.1.4 (due to Morley), while the necessity of (ii) follows from Vaught's two-cardinal theorem in Section 3.2.

The notion of a stable theory in powers other than ω has also proved to be fruitful. Shelah and Harnik (see Shelah, 1971d) showed that:

7.1.29. *If a complete theory T is stable in power α, then T has a saturated model of power α.*

Shelah has classified theories by the cardinals in which they are stable.

7.1.30. *For every complete theory T in a countable language, exactly one of the following happens:*

(i). *T is stable in every cardinal $\alpha \geqslant 2^\omega$.*

(ii). *T is stable in power α if and only if $\alpha = \alpha^\omega$.*

(iii). *T is not stable in any infinite cardinal.*

T is called *superstable* if (i) holds, *stable* if (i) or (ii) holds (i.e., T is stable in some α), and *unstable* if (iii) holds. Note that ω-stable implies superstable.

EXAMPLE 7.1.31. The theory of countably many independent unary relations is stable in power α if and only if $\alpha \geqslant 2^\omega$. So this theory is superstable but not ω-stable.

EXAMPLE 7.1.32 (Shelah). Here is a complete theory which is stable but not superstable, i.e. has property (ii): The language has countably many binary relations $E_0, E_1, E_2, \ldots$, and the axioms state that:

(i). Each E_n is an equivalence relation with infinitely many classes, $E_{n+1}(x, y) \to E_n(x, y)$.

(ii). Each class of E_n is the union of infinitely many classes of E_{n+1}.

The following theorem of Shelah gives a characterization of unstable theories which does not involve cardinals.

7.1.33. *A complete theory T is unstable if and only if there is a formula $\varphi(x_1 \ldots x_n y_1 \ldots y_n) = \varphi(\vec{x}\vec{y})$, a model $\mathfrak{A}$ of T, and an infinite set $X \subset A^n$ such that the relation*

$$\{\langle \vec{x}, \vec{y} \rangle : \vec{x}, \vec{y} \in X \text{ and } \mathfrak{A} \models \varphi[\vec{x}, \vec{y}]\}$$

simply orders X.

For example, we see that the theory of dense order without end-points is unstable, and so is the theory of atomless Boolean algebras.

One of the test problems which has been a driving force for classification theory is known as the spectrum problem. For a theory T, let $I(T, \alpha)$ be the number of models of T of power α up to isomorphism. The spectrum problem is:

Which functions $f(\alpha) = I(T, \alpha)$, α an infinite cardinal, are possible where T is a complete theory for a countable language?

To give the reader an idea of the progress in classification theory without giving any more definitions, we shall state a sampling of newer results on categoricity and the spectrum problem.

It is clear that $I(T, \alpha)$ is at most 2^α. Morley's theorem shows that if $I(T, \alpha) = 1$ for some uncountable α, then $I(T, \alpha) = 1$ for all uncountable α. Theorem 2.3.15 of Vaught shows that $I(T, \omega) \neq 2$. Theorem 7.1.26 of Baldwin and Lachlan shows that if $I(T, \omega_1) = 1$ then either $I(T, \omega) = 1$ or

$I(T, \omega) = \omega$. At the opposite extreme from Morley's theorem, Shelah (1978) has proved

7.1.34. *Every complete theory in a countable language which is not superstable has* 2^α *models in each cardinal* $\alpha > \omega$.

There are a number of open questions involving stability and the spectrum for small cardinals. A problem which has attracted particular interest is Vaught's conjecture that if $I(T, \omega) > \omega$ then $I(T, \omega) = 2^\omega$. Several special cases of Vaught's conjecture have been proved. Steel (1978) proved Vaught's conjecture for theories of trees (extending earlier work of M. Rubin, L. Marcus, and A. Miller). Shelah, Harrington and Makkai (1984) proved Vaught's conjecture for ω-stable theories. Lo (1985) proved Vaught's conjecture for homogeneous models.

Extending earlier work of Lachlan and of Buechler (1984), Hrushovsky (19??) proved Kueker's conjecture for stable theories, that every complete stable theory such that all uncountable models are ω-saturated is either ω_0-categorical or ω_1-categorical.

Another problem concerns finite axiomatizability of categorical theories. Peretyatkin (1980) gave an example of a finitely axiomatizable complete ω_1-categorical theory. Zilber (1980, 1984) and Cherlin, Harrington and Lachlan (1985) proved that there is no finitely axiomatizable complete theory which is categorical in all infinite powers, and the latter paper improved this to show that there is no complete finitely axiomatizable ω-categorical ω-stable theory.

Lachlan (1973) improved Theorem 7.1.27 by proving that a superstable theory has either one or infinitely many countable models. The corresponding problem for stable theories is open.

Classification theory has yielded a nearly complete solution of the spectrum problem for sufficiently large cardinals α. Shelah (1985), (19??) has proved the following results. The first result was a conjecture of Morley, and the second result is the solution of the spectrum problem.

7.1.35. *Let* T *be a complete theory in a countable language and suppose that* T *is not* ω_1*-categorical. Then for all infinite cardinals* $\alpha \leq \beta$ *we have* $I(T, \alpha) \leq I(T, \beta)$.

7.1.36. *Let* T *be a complete theory in a countable language. Then either*

(a) $I(T, \alpha) = 2^\alpha$ *for all uncountable cardinals* α, *or for all ordinals* ξ,

(b) $$I(T, \aleph_\xi) < \beth_{\omega_1}(|\xi|).$$

Moreover, case (b) *splits into three subcases:*
(b1) *There is an ordinal* $\zeta < \omega_1$ *such that for all ordinals* $\xi > 0$,

$$I(T, \aleph_\xi) = \beth_\zeta(|\omega + \xi|);$$

(b2) *There exist* $n < \omega$ *and a cardinal* $1 \leq \beta \leq 2^\omega$ *such that for all ordinals* $\xi \geq 2^\omega$,

$$I(T, \aleph_\xi) = \beth_n(|\xi|^\beta);$$

(b3) *There exist* $k, m < \omega$ *and a cardinal* $\beta \leq (2^\omega)^+$ *such that for all ordinals* $\xi \geq 2^\omega$,

$$I(T, \aleph_\xi) = \beth_k(\Sigma_{\gamma<\beta}\beth_m(|\xi|^\gamma)).$$

The dichotomy between (a) and (b) is called the *main gap*. It shows that $I(T, \alpha)$ for $\alpha > \omega$ is either the largest possible function 2^α or is much smaller than 2^α.

EXERCISES

Unless we specify otherwise, assume the language $\mathscr{L}$ is countable.

7.1.1. Prove the following directly from Lemmas 7.1.1, Corollary 7.1.5, and Lemma 7.1.6: Any complete theory T which is ω_1-categorical is categorical in every uncountable power.

7.1.2. A class K of models for $\mathscr{L}$ is said to be *α-categorical* iff K has a model $\mathfrak{A} \in K$ of power α, and any two models of power α in K are isomorphic. Prove that if T' is categorical in power α in an expansion $\mathscr{L}'$ of the language $\mathscr{L}$, then the class K of all reducts to $\mathscr{L}$ of models of T' is α-categorical.

7.1.3*. Give an example of a PC_Δ class of models of $\mathscr{L}$ (see Chapter 4) which is categorical in power $\beth_\omega$ but not in power ω_1. ($\mathscr{L}$ is countable.)

7.1.4*. Prove that every PC_Δ class (in a countable language) which is ω_1-categorical is categorical in every uncountable power.

[*Hint*: First try assuming $2^\omega = \omega_1$ and use the proof of Lemma 7.1.1. In general, also use the proofs of Corollary 7.1.5 and Lemma 7.1.6.]

7.1.5. Prove the following stronger form of Lemma 7.1.1. Let T be a complete theory of $\mathscr{L}$ and let $\Sigma_0(v), \Sigma_1(v), \ldots$ be sets of formulas in v. If every model of T of power ω_1 which omits $\Sigma_0(v), \Sigma_1(v), \ldots$ is homogeneous, then every uncountable model of T which omits $\Sigma_0(v), \Sigma_1(v), \ldots$ is homogeneous. (Compare with Exercise 5.1.15.)

7.1.6. Let T be a complete theory and let K be the class of all models of T which omit $\Sigma_0(v), \Sigma_1(v), \ldots$. Prove that if K is ω_1-categorical and every $\mathfrak{A} \in K$ of power ω_1 is homogeneous, then K is categorical in every uncountable power.

7.1.7*. Give an example of a complete theory in a language with $\aleph_{\omega+1}$ symbols which is categorical in power $\aleph_\omega$ but not in any larger power.

7.1.8. Give an example of a complete theory in a language with α symbols which is categorical in every power $\beta > \alpha$ but is not categorical in power α (for each infinite cardinal α).

7.1.9. Suppose that a theory T in $\mathscr{L}$ is stable in power ω and $\mathfrak{A}$ is a model of T. Prove that for every countable set $X \subset A$, the complete theory of $(\mathfrak{A}, x)_{x \in X}$ is stable in power ω.

7.1.10. Prove that Example 7.1.31 gives a complete theory which is stable in every power $\alpha \geqslant 2^\omega$ but is not stable in power ω. Thus the converse of Lemma 7.1.4 fails.

7.1.11. Show that a theory T is stable for types in one variable iff it is stable for types in finitely many variables. That is, if T is stable in power α, then for every model $\mathfrak{A}$ of T, every $X \subset A$ of power α, and every $n < \omega$, $\mathfrak{A}_X$ realizes exactly α types in $v_1, \ldots, v_n$.

7.1.12. For each of the examples of ω_1-categorical theories given in the text, find an infinite set of indiscernibles in a model of power ω_1.

7.1.13*. Let $\varphi(v_1 \ldots v_n)$ be a formula, $\mathfrak{A}$ a model, and X an infinite subset of $\mathfrak{A}$. Suppose that for every sequence $x_1, \ldots, x_n$ of distinct elements of X there is a permutation π of $\{1, \ldots, n\}$ such that exactly one of

$$\mathfrak{A} \models \varphi[x_1 \ldots x_n], \qquad \mathfrak{A} \models \varphi[x_{\pi 1} \ldots x_{\pi n}]$$

holds. Then the complete theory T of $\mathfrak{A}$ is not stable in power ω, whence T is not ω_1-categorical.

[*Hint*: Use Ramsey's theorem.]

7.1.14*. Let T be a complete theory and let K be the class of all models of T which omit $\Sigma_0(v), \Sigma_1(v), \ldots$. Assume that $\alpha > \omega$ and:

(i). K is α-categorical;

(ii). every countable model in K has arbitrarily large elementary extensions;

(iii). every model of K of power α is ω_1-homogeneous.

Prove that K is categorical in every uncountable power, and every uncountable model of K is homogeneous.

[*Hint*: Use the methods of this section and Section 7.2.]

7.1.15*. Let T be a complete theory and let $\varphi(u, v)$ be a formula. Let $\mathfrak{A}$ be a model of T. Suppose that for all $n < \omega$ there exist elements $u_1, \ldots, u_n \in A$ such that

$$\mathfrak{A} \models (\forall v)(\varphi(u_1 v) \vee \ldots \vee \varphi(u_n v)),$$

but for all $m \leqslant n$,

$$\mathfrak{A} \models \neg (\forall v) \bigvee_{\substack{1 \leqslant p \leqslant n \\ p \neq m}} \varphi(u_p v).$$

Prove that T is not ω_1-categorical.

[*Hint*: Use Ramsey's theorem.]

7.1.16*. Let T be a complete theory in $\mathscr{L}$ and α an uncountable cardinal. Suppose that for every model $\mathfrak{A}$ of T of power α and every finite set $X \subset \mathfrak{A}$, the model $(\mathfrak{A}, x)_{x \in X}$ is α^+-universal. Prove that T is categorical in every uncountable power.

7.1.17*. Suppose T is categorical in power ω_1. Then every countable model $\mathfrak{A}$ of T has a proper elementary extension $\mathfrak{B}$ such that:

(i). $\mathfrak{B}$ is prime over $\mathfrak{A}$, i.e., for every proper elementary extension $\mathfrak{C}$ of $\mathfrak{A}$, $(\mathfrak{B}, a)_{a \in A}$ is elementarily embeddable in $(\mathfrak{C}, a)_{a \in A}$;

(ii). $\mathfrak{B}$ is minimal over $\mathfrak{A}$, i.e., if $\mathfrak{A} \prec \mathfrak{C} \prec \mathfrak{B}$, then $\mathfrak{A} = \mathfrak{C}$ or $\mathfrak{B} = \mathfrak{C}$.

7.1.18*. If T is ω_1-categorical, then T has at most ω nonisomorphic countable models.

7.1.19**. If T is a complete theory which has uncountably many types, then T has 2^{ω_1} nonisomorphic models of power ω_1.

7.2. An extension of Ramsey's theorem and applications; some two-cardinal theorems

We introduced the notion of indiscernibles in Section 3.3, and have made use of this notion in Section 7.1. The main results on indiscernibles were stated and proved in Section 3.3, and they depend, at least for the existence of models with indiscernibles, on Ramsey's theorem (Theorem 3.3.7).

The omitting types theorem (Theorem 2.2.3) gives a sufficient condition for a (countable) theory to have a (countable) model omitting a type Σ. This theorem is definitely false if the language $\mathscr{L}$ is not countable or if we require that T has a noncountable model omitting Σ. It was first discovered by Morley (1965b) that the method of indiscernibles allows us to construct arbitrarily large models of T which have indiscernibles and omit a type Σ, provided that T has sufficiently large models omitting Σ to start with. He observed that the simple Ramsey theorem for countable sets no longer suffices, and what is needed is an extension of Ramsey's theorem first proved by Erdös and Rado (1956). This new method also enabled him to deduce rather easily a two-cardinal theorem of Vaught. This new line of approach of constructing models using the result of Erdös and Rado will form the substance of this section. We have decided to include in this section a two-cardinal theorem of Chang, which does not use indiscernibles, but does require the notion of saturated models introduced in Section 5.1.

Our first task is to state and prove the partition theorem of Erdös and Rado. Recall from Section 3.3 that $[X]^n$ is the set of all n-element subsets of X. If α is an infinite cardinal, then, as has been introduced elsewhere in the book,

$$\beth_0(\alpha) = \alpha,$$

$$\beth_{\xi+1}(\alpha) = 2^{\beth_\xi(\alpha)},$$

$$\beth_\eta(\alpha) = \bigcup_{\xi<\eta}\beth_\xi(\alpha),$$

if η is a limit ordinal different from 0.
When $\alpha = \omega$, we simply write $\beth_\xi$ for $\beth_\xi(\omega)$. The generalized continuum hypothesis is, of course, equivalent to the assertion $\beth_\xi = \aleph_\xi$ for all ordinals ξ.

THEOREM 7.2.1 (Erdös and Rado). *Let α be an infinite cardinal and let $n \in \omega$. Suppose that*

$$|X| > \beth_n(\alpha), \qquad [X]^{n+1} \subset \bigcup_{i\in I} C_i, \qquad |I| \leqslant \alpha.$$

Then there are a subset $Y \subset X$ and an $i \in I$ such that

$$|Y| > \alpha \quad \text{and} \quad [Y]^{n+1} \subset C_i.$$

PROOF. The theorem is proved by induction on n. The case that $n = 0$ follows from the pigeonhole principle: If more than α elements are divided into α classes, then some class must contain more than α elements. Let $n > 0$ and assume the theorem is true for $n-1$. We may suppose that $I \subset X$ and that

$C_i \cap C_j = 0$ if $i \neq j$. Let R be the $(n+2)$-placed relation on X defined by

$$R(x_0 \ldots x_n i) \text{ iff } i \in I \text{ and } \{x_0 \ldots x_n\} \in C_i.$$

Form the model

$$\mathfrak{A} = \langle X, R, i \rangle_{i \in I}.$$

Let us say that an elementary submodel $\mathfrak{B} \prec \mathfrak{A}$ is *β-saturated relative to* $\mathfrak{A}$ iff for every $Z \subset B$ of power $|Z| < \beta$, every type $\Sigma(v)$ which is realized in $(\mathfrak{A}, b)_{b \in Z}$ is realized in $(\mathfrak{B}, b)_{b \in Z}$.

We claim that $\mathfrak{A}$ has an elementary submodel $\mathfrak{B}$ of power $\beth_n(\alpha)$ such that $I \subset B$ and $\mathfrak{B}$ is $\beth_{n-1}(\alpha)^+$-saturated relative to $\mathfrak{A}$. The proof is like the proof of the existence of β^+-saturated models of power 2^β. We need only form an elementary chain

$$\mathfrak{B}_\gamma \prec \mathfrak{A}, \qquad \gamma < \beth_{n-1}(\alpha)^+$$

such that $I \subset B_0$, each $\mathfrak{B}_\gamma$ has power $\beth_n(\alpha)$, and for all $Z \subset B_\gamma$ of power $\beth_{n-1}(\alpha)$, every type $\Sigma(v)$ realized in $(\mathfrak{A}, b)_{b \in Z}$ is realized in $(\mathfrak{B}_{\gamma+1}, b)_{b \in Z}$. This is possible because there are at most $\beth_n(\alpha)^{\beth_{n-1}(\alpha)} \cdot 2^{\beth_{n-1}(\alpha)} = \beth_n(\alpha)$ such sets Z and types $\Sigma(v)$ at stage γ.

Since $\mathfrak{A}$ has power $> \beth_n(\alpha)$, $\mathfrak{B}$ is a proper submodel of $\mathfrak{A}$. Choose an element $c \in A \setminus B$. We then form a sequence

$$b_\gamma, \qquad \gamma < \beth_{n-1}(\alpha)^+$$

of elements of B such that for all γ, b_γ realizes the same type as c in $(\mathfrak{A}, b_\delta)_{\delta < \gamma}$. All the b_γ are distinct because $c \notin B$. Let U be the set of all b_γ. Then U has power $\beth_{n-1}(\alpha)^+$. Partition $[U]^n$ into disjoint sets D_i, $i \in I$, as follows: for $\gamma_0 < \ldots < \gamma_{n-1} < \beth_{n-1}(\alpha)^+$,

$$\{b_{\gamma_0}, \ldots, b_{\gamma_{n-1}}\} \in D_i \quad \text{iff} \quad \{b_{\gamma_0}, \ldots, b_{\gamma_{n-1}}, c\} \in C_i.$$

By the induction hypothesis for $n-1$, there is a subset $Y \subset U$ of power $|Y| > \alpha$ such that for some $i \in I$,

$$[Y]^n \subset D_i.$$

Then for any $\gamma_0 < \ldots < \gamma_{n-1} < \gamma_n$,

$$\{b_{\gamma_0}, \ldots, b_{\gamma_{n-1}}, c\} \in C_i.$$

Since b_{γ_n} realizes the same type as c in $(\mathfrak{A}, b_\delta)_{\delta < \gamma}$, we have

$$\{b_{\gamma_0}, \ldots, b_{\gamma_n}\} \in C_i.$$

This shows that

$$[Y]^{n+1} \subset C_i. \dashv$$

We next state and prove Morley's theorem on omitting types of elements in its simplest form, leaving various ramifications of the result to the exercises. Let $\mathscr{L}$ be a countable language, T be a theory in $\mathscr{L}$, and $\Sigma(v)$ be a set of formulas of T in the single free variable v. Recall that a model $\mathfrak{A}$ of T omits Σ iff for all $a \in A$ it is not the case that $\mathfrak{A} \models \Sigma[a]$. Or, equivalently, for all $a \in A$, there is a $\sigma(v) \in \Sigma$ such that $\mathfrak{A} \models \neg\, \sigma[a]$.

THEOREM 7.2.2 (Morley's Theorem on Omitting Types). *Let $\mathscr{L}$, T, Σ be as in the above. If for each cardinal $\beth_\xi$, $\xi < \omega_1$, T has a model of power greater than $\beth_\xi$ omitting Σ, then T has a model omitting Σ in each infinite power.*

PROOF. For $\xi < \omega_1$, let $\mathfrak{A}_\xi = \langle A_\xi, \ldots \rangle$ be a model of T of power greater than $\beth_\xi$ omitting Σ. Let $\mathscr{L}^*$ be an expansion of $\mathscr{L}$ with built-in Skolem functions and let T^* be the extension of T to $\mathscr{L}^*$ by adding all axioms for the Skolem theory. Clearly, by results of Chapter 3, each $\mathfrak{A}_\xi$ has a corresponding expansion to $\mathfrak{A}_\xi^*$. Then for all $\xi < \omega_1$,

(1) for all $a \in A_\xi$, there is a $\sigma(v) \in \Sigma$, such that $\mathfrak{A}_\xi^* \models \neg\, \sigma[a]$.

Since $\mathscr{L}$ is countable, $\mathscr{L}^*$ is still countable. Thus, let

$$t_0, t_1, t_2, \ldots, t_n, \ldots, \quad n \in \omega,$$

be an enumeration of all the terms of $\mathscr{L}^*$. We may assume that t_n has p_n places, and that each term t_n is in the variables $v_1, \ldots, v_{p_n}$. Let $c_1, c_2, \ldots$ be new constant symbols, and let $\overline{\mathscr{L}} = \mathscr{L}^* \cup \{c_1, c_2, \ldots\}$. We wish to prove the following: There is a sequence

(2) $$\sigma_0(v), \ldots, \sigma_n(v), \ldots, \quad n \in \omega$$

of formulas of Σ such that the following set $\overline{T}$ of sentences of $\overline{\mathscr{L}}$ is consistent:

(3) the set T^*,
and the sentences
$\neg\, \sigma_n(t_n(c_{i_1} \ldots c_{i_{p_n}}))$, whenever $i_1 < i_2 < \ldots < i_{p_n} < \omega$,
$\neg\, (c_i \equiv c_j)$, whenever $i < j < \omega$.

The proof of the existence of the sequence (2) and the consistency of (3) is by induction on n, and follows from an auxiliary induction given below. We assume that each set A_ξ is simply ordered in some manner by $<$. (The context will make it clear whether we are referring to $<_\xi$ or to $<_\eta$.) We prove by induction on n:

(4) There are formulas $\sigma_0(v), \ldots, \sigma_{n-1}(v) \in \Sigma$, and a cofinal subset F_n of ω_1, and a sequence of subsets $X_\xi \subset A_\xi$, $\xi \in F_n$, such that, for each $\xi \in F_n$, we have

(i). $|X_\xi| > \beth_\lambda$, where ξ is the λth-element of F_n, and
(ii). for each $m < n$, $\mathfrak{A}_\xi \models \neg\, \sigma_m[t_m(x_1 \dots x_{p_m})]$ whenever $x_1 < \dots < x_{p_m}$ are in X_ξ.

The statement (4) is obviously true if we take $n = 0$, $F_0 = \omega_1$, and $X_\xi = A_\xi$. Then (4) (i) holds, and (4) (ii) holds vacuously. Assume therefore that (4) holds for n, namely, we have found $\sigma_0, \dots, \sigma_{n-1}$, F_n, and X_ξ, $\xi \in F_n$, that satisfy the conclusions (4) (i) and (ii). Consider the term $t_n(v_1 \dots v_{p_n})$. Assume first that $p_n \geqslant 1$. We first form a subset G_n of F_n by omitting all elements of F_n which cannot be written as $\eta + (p_n - 1)$(in the sense of the natural ordering on F_n). Thus

$$G_n = \{\xi \in F_n : \text{there is an } \eta \in F_n \text{ such that } (\xi \setminus \eta) \cap F_n \text{ has exactly } p_n - 1 \text{ members}\}.$$

G_n is still cofinal in ω_1, and, furthermore, for $\xi \in G_n$,

(5) $$|X_\xi| > \beth_{\lambda + p_n - 1} \text{ if } \xi \text{ is the } \lambda\text{th element of } G_n.$$

Since $\mathfrak{A}_\xi^*$ omits Σ, we see that whenever $x_1 < \dots < x_{p_n}$ are from X_ξ,

$$\text{it is not the case that } \mathfrak{A}_\xi^* \models \Sigma(t_n(x_1 \dots x_{p_n})).$$

So there is a $\sigma(v) \in \Sigma$ such that

(6) $$\mathfrak{A}_\xi^* \models \neg\, \sigma[t_n(x_1 \dots x_{p_n})].$$

Remember that Σ is countable, so (6) represents a partition of $[X_\xi]^{p_n}$ into a countable number of pieces. Since (5) holds, we may use Theorem 7.2.1 to find a subset $Y_\xi \subset X_\xi$ such that

(7) $$|Y_\xi| > \beth_\lambda \text{ if } \xi \text{ is the } \lambda\text{th element of } G_n,$$

and

(8) there is a $\sigma \in \Sigma$, call it σ_ξ, such that
$\mathfrak{A}_\xi^* \models \neg\, \sigma_\xi[t_n(x_1 \dots x_{p_n})]$ whenever $x_1 < \dots < x_{p_n}$ are from Y_ξ.

(7) and (8), of course, hold for each $\xi \in G_n$. Since $|G_n| = \omega_1$ and $|\{\sigma_\xi : \xi \in G_n\}| = \omega$, we see that by (8),

(9) there is a formula $\sigma \in \Sigma$, call it σ_n, and a cofinal subset F_{n+1} of G_n such that for each $\xi \in F_{n+1}$,

$$\mathfrak{A}_\xi^* \models \neg\, \sigma_n[t_n(x_1 \dots x_{p_n})], \text{ where } x_1 < \dots < x_{p_n} \text{ are from } Y_\xi.$$

Since $F_{n+1} \subset G_n$, it follows easily from (7) that

(10) $$|Y_\xi| > \beth_\lambda \text{ if } \xi \text{ is the } \lambda\text{th element of } F_{n+1}.$$

Since each $Y_\xi \subset X_\xi$, we see that (9) and (10) show that the formulas $\sigma_0, \ldots, \sigma_n$, the set F_{n+1}, and the sequence Y_ξ, $\xi \in F_{n+1}$, satisfy the conclusions of (4). If $p_n = 0$, then t_n is the interpretation of a constant element in each $\mathfrak{A}^*_\xi$, $\xi \in F_n$. From this we can easily find a cofinal subset $F_{n+1} \subset F_n$ such that (9) holds. So (4) is proved.

We now take the sequence of formulas in (2) simply as the sequence of formulas given to us by (4). The consistency of the set (3) now follows easily by compactness. For suppose Δ is any finite subset of the set (3). It will contain some term t_n for a highest n. Now take any $\xi \in F_{n+1}$. Since X_ξ is an infinite subset of A_ξ, the model $\mathfrak{A}^*_\xi$ is more than adequate to show the consistency of Δ.

Finally, let α be any infinite cardinal, and let $\mathscr{L}'$ be the language $\mathscr{L}^* \cup \{c_\xi : \xi < \alpha\}$ with new constant symbols c_ξ, $\xi < \alpha$. Let T' be the set of sentences of $\mathscr{L}'$ given by:

the set T^*;
all sentences $\neg\, \sigma_n(t_n(c_{\xi_1} \ldots c_{\xi_{p_n}}))$ with $\xi_1 < \ldots < \xi_{p_n} < \alpha$;
all sentences $\neg\, (c_\xi \equiv c_\eta)$ with $\xi < \eta < \alpha$.

Then a simple argument using the consistency of (3) shows that T' is consistent. Let $\mathfrak{B}' = \langle B', \ldots \rangle$ be any model of T', and let C be the set of all interpretations of c_ξ, $\xi < \alpha$, in $\mathfrak{B}'$. Let $\mathfrak{B}$ be the model for $\mathscr{L}$ (the original language) obtained from the Skolem hull $\mathfrak{H}(C)$ of the set C in $\mathfrak{B}'$. Several facts are evident. Since $|C| = \alpha$ and $\mathscr{L}^*$ is countable, the model $\mathfrak{B}$ has power exactly α. $\mathfrak{B}$ is an elementary submodel of $\mathfrak{B}'$ (in the sense of $\mathscr{L}$), whence $\mathfrak{B}$ is a model of T. We now prove that $\mathfrak{B}$ omits Σ. Each element $b \in B$ is obtained from the set C by applying a (Skolem) term t_n to a sequence $x_1, \ldots, x_{p_n}$ from C. We may suppose that $b = t_n(x_1 \ldots x_{p_n})$ and $x_1, \ldots, x_{p_n}$ are interpretations of $c_{\xi_1}, \ldots, c_{\xi_{p_n}}$, with $\xi_1 < \ldots < \xi_{p_n} < \alpha$. From the definition of T', we see that

$$\mathfrak{B}' \models \neg\, \sigma_n[t_n(x_1 \ldots x_{p_n})],$$

whence

$$\mathfrak{B} \models \neg\, \sigma_n[b]. \dashv$$

The idea behind the proof of Theorem 7.2.2, once understood, is quite simple. We shall now amplify the basic idea in a number of remarks. The reader is asked to check these remarks in the exercises.

The first remark is that Theorem 7.2.2 remains true if the set of formulas $\Sigma(v)$ is replaced by a set of formulas $\Sigma(v_1 \ldots v_m)$ in the variables $v_1, \ldots, v_m$.

The required changes in the proof are minimal. We simply replace the sequence

$$t_0, t_1, \ldots, t_n, \ldots, \qquad n \in \omega$$

of all terms of $\mathscr{L}^*$ by a sequence enumerating all m-tuples of terms of $\mathscr{L}^*$. The variables occurring in an m-tuple of terms may be ordered in an arbitrary fashion. From this point on, the proof may be repeated almost word for word.

Secondly, we could have easily required that the constants c_n occurring in (3) be indiscernible in the sense of Section 3.3. To do this would only require one or two additional steps and will not complicate the proof very much. Thus the conclusion of Theorem 7.2.2 can be strengthened by saying that in each infinite power α, T has a model of power α which omits Σ and has a set of α indiscernibles.

The third remark is that in Theorem 7.2.2, we may replace the single set of formulas Σ by any countable set S of sets of formulas of $\mathscr{L}$. If we specify that $\mathfrak{A}$ omits S is to mean that $\mathfrak{A}$ omits Σ for each $\Sigma \in S$, then Theorem 7.2.2 can be reworded, with S replacing Σ. The sets of formulas in S may be in different numbers of free variables. The truth of this remark depends on the fact that S is countable and that the proof of condition (4) in Theorem 7.2.2 involves only a finite induction.

What happens to Theorem 7.2.2 when $\mathscr{L}$ is no longer countable, or when $\mathscr{L}$ is replaced by languages broader than the ordinary first-order language, will be touched upon in the exercises. There is, however, one case of the nonclassical languages which deserves to be pointed out here. Suppose that $\mathscr{L}$ is a countable language which has among its symbols a unary relation $\underline{\omega}$ and constants $\underline{0}, \underline{1}, \ldots, \underline{n}, \ldots$, $n \in \omega$. An *ω-model* is a model $\mathfrak{A}$ for $\mathscr{L}$, in which $\underline{\omega}$ is interpreted by ω and each $\underline{n}$ by n.

COROLLARY 7.2.3. *Let T be a theory of $\mathscr{L}$. Suppose that for each cardinal $\beth_\xi$, $\xi < \omega_1$, T has an ω-model of power greater than $\beth_\xi$. Then T has an ω-model in each infinite power.*

The proof of Corollary 7.2.3 is immediate when we consider the set of formulas

$$\Sigma = \{\underline{\omega}(v), v \not\equiv \underline{0}, \ldots, v \not\equiv \underline{n}, \ldots\}.$$

If we introduce the ordering relation $\leqslant$ on ω into an ω-model and require that $\leqslant$ be standard, then we obtain the notion of an ω-standard model. For ω-standard models, Corollary 7.2.3 can be improved if T is a single sentence. (See Exercise 7.2.7.)

Before we go on to exploit the idea of Theorem 7.2.2 in an application on two-cardinal theorems, we give two counterexamples in the next two propositions. The first example shows that Theorem 7.2.2 cannot be improved, and the second example shows that the exact analogue of Theorem 7.2.2 for languages with α symbols, where $\alpha \geqslant \omega$ and $\mathrm{cf}(\alpha) > \omega$, will fail. Thus Theorem 7.2.2 is false if we replace ω, ω_1 by ω_1, ω_2 everywhere.

PROPOSITION 7.2.4. *Let ξ be an infinite ordinal. Then there are a language $\mathscr{L}$ with $|\xi|$ symbols, and Σ and T of $\mathscr{L}$ such that T has a model of power $\beth_\xi$ omitting Σ but has no model of power exceeding $\beth_\xi$ omitting Σ.*

PROOF. This proof is a generalization of the idea behind Proposition 3.2.11 (iii). Let $\mathscr{L}$ contain the following symbols:

constant symbols c_λ for $\lambda \leqslant \xi$;

two binary relation symbols $<$ and P, and a unary symbol U;

a 3-placed relation symbol Q.

Let $\Sigma(v)$ be the following set of formulas:

$$\Sigma = U(v) \cup \{v \not\equiv c_\lambda : \lambda \leqslant \xi\}.$$

Let T be a set of sentences of $\mathscr{L}$ which expresses the following:

(i). $c_\lambda < c_\mu$ if $\lambda < \mu \leqslant \xi$; $U(c_\lambda)$ if $\lambda \leqslant \xi$;

(ii). $<$ is a simple ordering relation on U

(for convenience, let P_x denote the set $\{y : P(xy)\}$);

(iii). $P_x \subset P_y$ if $x < y$;

(iv). P_{c_0} is the set $\{x : x < c_\omega\}$;

(v). P_{c_ξ} is the universe of the model;

(vi). if $x \in S$ and x has no immediate predecessor, then $P_x = \bigcup_{y<x} P_y$;

(vii). if $x \in S$ and x has an immediate predecessor y, then

$$(\forall uv)[P(xu) \wedge P(xv) \wedge u \not\equiv v \rightarrow (\exists w)(P(yw) \wedge \neg (Q(ywu) \leftrightarrow Q(ywv)))].$$

Note that except for the infinite number of sentences of condition (i), the rest of the sentences of T, namely (ii)–(vii), can be put in a single sentence. Let $\mathfrak{A}$ now be any model of T omitting Σ. This means that the interpretation of $<$ on $\mathfrak{A}$ is isomorphic to the natural $<$ relation on the ordinal $\xi+1$. Using conditions (iv)–(vii), we can prove by induction on $\lambda \leqslant \mu$ that

$$|\text{the interpretation of } P_{c_\lambda} \text{ in } \mathfrak{A}| \leqslant \beth_\lambda.$$

From this and (v) we see that $|A| \leqslant \beth_\xi$. By interpreting c_ξ by ξ and $P_{\xi+1}$ by $S(P_\xi)$, we can obtain a model $\mathfrak{A}$ of T such that $\mathfrak{A}$ has cardinal exactly $\beth_\xi$ and $\mathfrak{A}$ omits $\Sigma(v)$. ⊣

The ideas of Proposition 7.2.4 can be sharpened and refined to obtain the following:

PROPOSITION 7.2.5. *Let α be an infinite cardinal such that $\operatorname{cf}(\alpha) > \omega$. Then there is a language $\mathscr{L}$ with α symbols, and Σ, T of $\mathscr{L}$ such that T has a model of power $\beth_{\alpha^+}$ omitting Σ, but has no model of power greater than $\beth_{\alpha^+}$ omitting Σ.*

PROOF (in outline). Let $\mathscr{L}$ contain at least the symbols c_λ, $\lambda < \alpha$, U, and a binary $<$. As before, let

$$\Sigma = U(v) \cup \{v \not\equiv c_\lambda : \lambda < \alpha\}.$$

Before describing T in detail, we let T have at least the sentences

(i). $c_\lambda < c_\mu$ if $\lambda < \mu < \alpha$; $U(c_\lambda)$ if $\lambda < \alpha$;

(ii). $<$ is a simple ordering relation on U.

Thus any model $\mathfrak{A}$ of T which omits Σ will contain the standard ordering $<$ on α. To complete the description of T, we first assert that the following ideas are expressible in $\mathscr{L}$ (with the addition of a number of finitary relation symbols):

$\varphi_1(r)$: r is a simple ordering relation on α;
$\varphi_2(r)$: r is a well ordering relation on α.

Let R denote the collection of those r such that $\varphi_2(r)$.

R is well ordered by the initial segment relation;
to each $r \in R$ there corresponds a subset P_r;
the sets P_r, $r \in R$, satisfy conditions (vi) and (vii) of Proposition 7.2.4.

All of these, except for $\varphi_2(r)$, are routine, and we ask the reader to provide a proof of Proposition 7.2.5 in Exercise 7.2.9. To show that $\varphi_2(r)$ is expressible in $\mathscr{L}$, we need to know the following special fact depending very much on $\operatorname{cf}(\alpha) > \omega$ and α being a cardinal:

r is a well ordering of α iff r restricted to any $\lambda < \alpha$ is order isomorphic to some ordinal $\mu < \alpha$.

This should be enough of a hint. ⊣

We now turn our attention to some more results for the two-cardinal problem, first introduced in Section 3.2. Recall that if $\alpha \geqslant \beta \geqslant \omega$, then a theory T is said to admit the pair (α, β) iff T has a model $\mathfrak{A} = \langle A, V, \ldots \rangle$, where $|A| = \alpha$ and $|V| = \beta$. We remind the reader that, apart from the collection of results in Section 3.2 (in particular, Proposition 3.2.11, Theorem 3.2.12, Theorem 3.2.14, and some exercises), we have also dealt with the two-cardinal problem in Section 4.3 (in particular, Theorem 4.3.10 and Corollary 4.3.11) and in Section 6.5 (Theorem 6.5.11). The idea behind Theorem 7.2.2 is used to prove the following theorem of Vaught; the proof is essentially due to Morley.

THEOREM 7.2.6. *Suppose $\mathscr{L}$ is a countable language, T is a theory of $\mathscr{L}$, and for each $n \in \omega$, T admits some pair $(\beth_n(\gamma_n), \gamma_n)$ of infinite cardinals. Then T admits every pair of infinite cardinals (α, β).*

PROOF. For each n, let $\mathfrak{A}_n = \langle A_n, V_n, <_n, \ldots \rangle$ be a model of T such that $|A_n| > \beth_n(|V_n|)$; we may assume that $<_n$ is an ordering of A_n. As in the proof of Theorem 7.2.2, we first consider the expansion $\mathscr{L}^*$ of $\mathscr{L}$ with built-in Skolem functions, and the corresponding expansions $\mathfrak{A}_n^*$ of $\mathfrak{A}_n$. $\mathscr{L}^*$ is still countable, so it has only a countable number of terms. For later purposes, we shall consider all terms $t(x_1 \ldots x_n; y_1 \ldots y_m)$ of $\mathscr{L}^*$ with the (distinct) variables $x_1, \ldots, x_n$ and $y_1, \ldots, y_m$. Each term $t(v_1 \ldots v_l)$ of $\mathscr{L}^*$ can, of course, be considered as many terms when we separate $v_1, \ldots, v_l$ into two sets of variables. Nevertheless, $\mathscr{L}^*$ will only have a countable number of such terms, and we shall arrange them in the sequence

$$t_0, t_1, \ldots, t_n, \ldots, \qquad n \in \omega.$$

We assume that t_n is in terms of the variables $x_1, \ldots, x_{q_n}$ and $y_1, \ldots, y_{p_n}$. Consider the language $\mathscr{L}^* \cup \{c_1, c_2, \ldots\}$, and the set $\bar{T}$ of sentences consisting of:

$c_i \not\equiv c_j$ if $i < j$;
the sentences of the Skolem expansion T^* of T;
the sentences which express that U is infinite;
all sentences of the following form:

$$\begin{aligned}\sigma_n(c_{i_1} \ldots c_{i_{p_n}}; c_{j_1} \ldots c_{j_{p_n}}) = (\forall x_1 \ldots x_{q_n})[&U(x_1) \wedge \ldots \wedge U(x_{q_n}) \\ &\wedge x_1 < \ldots < x_{q_n} \wedge U(t_n(x_1 \ldots x_{q_n}; c_{i_1} \ldots c_{i_{p_n}})) \\ &\rightarrow t_n(x_1 \ldots x_{q_n}; c_{i_1} \ldots c_{i_{p_n}}) \equiv t_n(x_1 \ldots x_{q_n}; c_{j_1} \ldots c_{j_{p_n}})]\end{aligned}$$

where $n \in \omega$, and $i_1 < \ldots < i_{p_n} < \omega, j_1 < \ldots < j_{p_n} < \omega$.

We wish to show first that

(1) $\bar{T}$ is consistent.

This is shown by an induction on the following (we have assumed that each A_n is ordered by $<_n$):

(2) There are an infinite subset $F_n \subset \omega$, and a sequence $X_m \subset A_m$, $m \in F_n$, such that for each $m \in F_n$ we have:
 (i). $|X_m| > \beth_l(|V_m|)$ if m is the lth element of F_n;
 (ii). for all $k < n$ and all $u_1 < \dots < u_{pk}$ and $w_1 < \dots < w_{pk}$ from X_m,

$$\mathfrak{A}_m^* \models \sigma_k[u_1 \dots u_{pk};\ \ w_1 \dots w_{pk}].$$

For $n = 0$, (2) holds trivially, with $F_0 = \omega$, and $X_m = A_m$ for all $m \in F_0$. Assume (2) holds for n. Consider the term $t_n(x_1 \dots x_{q_n}; y_1, \dots y_{p_n})$; for simplicity, and without loss of generality, let us only consider the case where q_n and p_n are both nonzero. Let

$$F_{n+1} = F_n \setminus \text{the first } p_n \text{ elements of } F_n.$$

Clearly, F_{n+1} is an infinite subset of ω, and

(3) $|X_m| > \beth_{l+p_n}(|V_m|)$ if m is the lth element of F_{n+1}.

We now partition $[X_m]^{p_n}$, for $m \in F_{n+1}$, into $2^{|V_m|}$ classes as follows: We say that $u_1 < \dots < u_{p_n}$ and $w_1 < \dots < w_{p_n}$ from X_m are equivalent iff

$$\mathfrak{A}_m^* \models \sigma_n[u_1 \dots u_{p_n};\ \ w_1 \dots w_{p_n}].$$

It is clear that this is indeed an equivalence relation. Furthermore, the equivalence class of any sequence $u_1 < \dots < u_{p_n}$ is determined by a mapping f of the q_n-tuples of V_m into $V_m \cup \{a_m\}$, where a_m is some arbitrary but fixed element of $A_m \setminus V_m$; i.e., the mapping is given by

$$f(u_1 \dots u_{p_n}) = \begin{cases} t_n[x_1 \dots x_{q_n}; u_1 \dots u_{p_n}] & \text{if this element is in } V_m; \\ a_m & \text{otherwise.} \end{cases}$$

From this, we see that the total number of equivalence classes on X_m is at most $2^{|V_m|}$. Whence, by (3) and Theorem 7.2.1, if m is the lth element of F_{n+1},

(4) there is a subset $Y_m \subset X_m$ such that $|Y_m| > \beth_l(|V_m|)$ and all members of $[Y_m]^{p_n}$ are equivalent.

The set F_{n+1} and the sequence Y_m, $m \in F_{n+1}$, satisfy the conclusions of (2) for $n+1$. So (2) holds for all n. Since each V_n is infinite, we see that, by (2), a simple compactness argument proves (1).

Let now $\alpha > \beta$ be infinite cardinals. Consider the language $\mathscr{L}^* \cup \{c_\xi : \xi < \alpha\}$ and a set T' of this language by imitating the definition of $\bar{T}$, replacing, of course, $i_1 < \ldots < i_{p_n} < \omega$ by $\xi_1 < \ldots < \xi_{p_n} < \alpha$, etc. T' is clearly consistent. Hence it has an (α, α)-model. Call this model $\mathfrak{B} = \langle B, V, \ldots \rangle$. Let X be any subset of V of power β, and let Y be X union the set of interpretations of all c_ξ, $\xi < \alpha$, in $\mathfrak{B}$. Y has power α. We claim that

(5) the Skolem hull of Y in $\mathfrak{B}$ with respect to the language $\mathscr{L}^*$ is an (α, β)-model of T.

Let A be the Skolem hull of Y in $\mathfrak{B}$. It is clear that A determines a model for $\mathscr{L}$ which is a model of T of power α. It only remains to show that $|A \cap V| = \beta$. Suppose that $a \in A \cap V$. Then either (i) $a \in X$, or else (ii) for some $n \in \omega$, $\xi_1 < \ldots < \xi_n < \alpha$, and $x_1 < \ldots < x_{q_n}$ from X, we have

$$a = t_n[x_1 \ldots x_{q_n};\ d_{\xi_1} \ldots d_{\xi_n}].$$

By our choice of T' and $\mathfrak{B}$ we can never get more than a countable number of elements $a \in A \cap V$ by (ii). Whence $\beta = |X| \leqslant |A \cap V| \leqslant \beta$. Thus $\mathfrak{A} = \langle A, A \cap V, \ldots \rangle$ is an (α, β)-model of T and the theorem is proved. ⊣

For our next and last theorem of this section we shall not use the method originating with Theorem 7.2.1, but shall return to the method of saturated models introduced in Section 5.1.

THEOREM 7.2.7 (Chang's Two-Cardinal Theorem). *Let $\mathscr{L}$ be a countable language and let T be a theory of $\mathscr{L}$. Assume the generalized continuum hypothesis. Suppose that T admits a pair (α, β), with $\alpha > \beta \geqslant \omega$. Then T admits every pair (δ^+, δ), where δ is a regular infinite cardinal.*

PROOF. First note that the theorem for the case $\delta = \omega$ is already known, namely Theorem 3.2.12, and, in fact, its proof does not require the GCH. Henceforth, let us assume that δ is a regular cardinal greater than ω. Let $\mathfrak{A} = \langle A, V, \ldots \rangle$ be a model of T such that $|A| = \alpha$ and $|V| = \beta$. Let R be a new binary relation over V and such that R indexes all the finite subsets of V; this is possible because V is an infinite set. Thus we have:

(1) for any $u_1, \ldots, u_n \in V$, there exists $u \in V$ such that

$$(\mathfrak{A}, R) \models (\forall t)(U(t) \to (P(ut) \leftrightarrow t \equiv u_1 \vee \ldots \vee t \equiv u_n)).$$

(P is a new binary relation symbol whose interpretation is R.) By the downward Löwenheim–Skolem theorem applied to the language $\mathscr{L} \cup \{P\}$, and the model $(\mathfrak{A}, R)$, there exists an elementary submodel $\langle B, V, R, ...\rangle$ of $(\mathfrak{A}, R)$ such that $V \subset B$ and $|B| = \beta$. Let Q be a new 1-placed relation symbol, and consider the model

$$(\mathfrak{A}, B, R) = \langle A, B, V, R, ...\rangle$$

for $\mathscr{L} \cup \{P, Q\}$. Let T' be the theory of the model $(\mathfrak{A}, B, R)$. Note that T' contains all closures of formulas of the form (by (1))

$$(2)\quad U(z_1) \wedge ... \wedge U(z_n) \wedge \varphi(z_1 x_1 \ldots x_m) \wedge ... \wedge \varphi(z_n x_1 \ldots x_m) \rightarrow$$
$$\rightarrow (\exists y)(P(yz_1) \wedge ... \wedge P(yz_n) \wedge (\forall t)(P(yt) \rightarrow \varphi(tx_1 \ldots x_m))),$$

where φ is an arbitrary formula of $\mathscr{L} \cup \{P, Q\}$. By Proposition 5.1.5 (i), T' has a saturated model

$$\langle A_0, B_0, V_0, R_0, ...\rangle$$

of power δ. Since both B and V are infinite, it follows easily that $|A_0| = |B_0| = |V_0| = \delta$. Furthermore, there are enough sentences in the theory T' to ensure that the model generated by the set B_0 in the model $\langle A_0, B_0, V_0, R_0, ...\rangle$ has universe B_0; in fact, let it be denoted by

$$\mathfrak{B}_0 = \langle B_0, V_0, R_0, ...\rangle.$$

$\mathfrak{B}_0$ is a proper elementary submodel (in the language $\mathscr{L} \cup \{P\}$) of the model

$$\mathfrak{A}_0 = \langle A_0, V_0, R_0, ...\rangle.$$

It is obvious that both $\mathfrak{B}_0$ and $\mathfrak{A}_0$ are saturated models of power δ. Since they are also equivalent, by the uniqueness of saturated models, Theorem 5.1.13, they are isomorphic. Because B_0 is a proper subset of A_0, and because the interpretations of U and P in $\mathfrak{B}_0$ and $\mathfrak{A}_0$ are identical, we see that the following is true:

(3) If $\mathfrak{A}' = \langle A', V', R', ...\rangle$ is any model for $\mathscr{L} \cup \{P\}$, which is elementarily embeddable into $\mathfrak{A}_0$ by an isomorphism which maps the set V' onto the set V_0, then $\mathfrak{A}'$ has a proper elementary extension $\mathfrak{A}'' = \langle A'', V'', R'', ...\rangle$ such that

$$V'' = V', \qquad R'' = R', \qquad \mathfrak{A}'' \cong \mathfrak{A}_0.$$

The point of (3) is that in going from $\mathfrak{A}'$ to $\mathfrak{A}''$, we do not increase the set V' nor the relation R', but we do add at least one element that is in $A'' \setminus A'$. We shall now define by transfinite induction an elementary chain of models

$$\mathfrak{A}_0 = \langle A_0, V_0, R_0, ...\rangle \prec ... \prec \mathfrak{A}_\nu = \langle A_\nu, V_0, R_0, ...\rangle \prec ..., \qquad \nu < \delta^+,$$

satisfying the following: for every $\nu < \delta^+$,

(4) if $\xi < \eta \leqslant \nu$, then $\mathfrak{A}_0 \cong \mathfrak{A}_\eta$ and $\mathfrak{A}_\xi$ is a proper elementary submodel of $\mathfrak{A}_\eta$; furthermore, the interpretations of P and U on $\mathfrak{A}_\xi$ and $\mathfrak{A}_\eta$ are the same.

Let μ be an ordinal such that $0 < \mu < \delta^+$. Assume that the models $\mathfrak{A}_\nu$, $\nu < \mu$, have been found such that (4) holds for all $\nu < \mu$. If $\mu = \nu+1$, then by (3), we can easily extend the elementary chain by finding a model $\mathfrak{A}_\mu = \langle A_\mu, V_0, R_0, \ldots\rangle$ such that

$$\mathfrak{A}_\mu \cong \mathfrak{A}_0 \text{ and } \mathfrak{A}_\mu \text{ is a proper elementary extension of } \mathfrak{A}_\nu.$$

Hence (4) holds with ν replaced by μ. So far, as the reader can see, the proof follows the idea of Theorem 3.2.12. It is in the limit case that the difficulty occurs.

Suppose that μ is a limit ordinal. Let $\mathfrak{A}' = \bigcup_{\xi<\mu}\mathfrak{A}_\xi$. Thus $\mathfrak{A}' = \langle A', V_0, R_0, \ldots\rangle$, where $A' = \bigcup_{\xi<\mu} A_\xi$. Since $\mu < \delta^+$, we see that $|A'| = \delta$, and that A' is a proper elementary extension of each $\mathfrak{A}_\xi$, $\xi < \mu$. Thus $\mathfrak{A}' \equiv \mathfrak{A}_0$. We shall show in the next few paragraphs that

(5) $\mathfrak{A}'$ is elementarily embeddable into $\mathfrak{A}_0$ in such a way that the set V_0 is mapped onto itself.

The proof of (5) is by a back and forth argument, which should be quite familiar to the student by now. Since the model $\mathfrak{A}_0$ is saturated, there will never be any difficulty in finding the value of the correspondence in the set A_0. Observe that the model $\mathfrak{A}'$ is not necessarily saturated. Thus in order to prove (5), we need to show that:

(6) the model $\mathfrak{A}'$ is V_0-saturated, i.e., for all $\lambda < \delta$ and all $a \in {}^\lambda A'$, the model $(\mathfrak{A}', a)$ realizes every type $\Sigma(v)$ over it which contains the formula $U(v)$.

To prove (6), let λ, a, Σ be as in the above. Since Σ is a type of $(\mathfrak{A}', a)$ containing the formula $U(v)$,

(7) Σ is finitely satisfiable in $(\mathfrak{A}', a)$ by elements of V_0.

We simply have to prove that Σ is simultaneously satisfiable in $(\mathfrak{A}', a)$. Let s be a finite subset of Σ, and let

$$\sigma_s(v) = (\bigwedge_{\sigma\in s} \sigma(v)) \wedge U(v).$$

By (7), we know that for every $s \in S_\omega(\Sigma)$,

$$(\mathfrak{A}', a) \models (\exists v)\sigma_s(v).$$

The formula σ_s will contain a finite number of constants $c_{\xi_1}, \ldots, c_{\xi_n}$, for the elements $a_{\xi_1}, \ldots, a_{\xi_n}$ of a, with $\xi_i < \lambda$. Hence we may write (the $a_{\xi_1}, \ldots, a_{\xi_n}$, of course, depend on s)

$$\mathfrak{A}' \models (\exists v)\sigma_s(va_{\xi_1} \ldots a_{\xi_n}).$$

Now the formula $\sigma_s(vv_1v_2 \ldots v_n)$ is a formula of $\mathscr{L} \cup^{\cdot} \{P\}$. The elements $a_{\xi_1}, \ldots, a_{\xi_n}$ all occur in the set A', whence, for some ordinal, say $\nu_s < \mu$, we have

$$a_{\xi_1}, \ldots, a_{\xi_n} \in A_{\nu_s}.$$

Since $\mathfrak{A}_{\nu_s} \prec \mathfrak{A}'$, we have

$$\mathfrak{A}_{\nu_s} \models (\exists v)\sigma_s(va_{\xi_1} \ldots a_{\xi_n}).$$

Let $u_s \in V_0$ be an element such that

$$\mathfrak{A}_{\nu_s} \models \sigma_s(u_s a_{\xi_1} \ldots a_{\xi_n}).$$

Suppose s' is another finite subset of Σ such that $s \subset s'$. Then we see easily that $u_{s'} \in V_0$ and $u_{s'} \in A_{\nu_s}$, and

$$\mathfrak{A}_{\nu_s} \models \sigma_s(u_{s'}\, a_{\xi_1} \ldots a_{\xi_n}).$$

Now for a fixed $s \in S_\omega(\Sigma)$, consider the set of conditions

$$\Gamma_s(v) = \{P(vu_{s'}) : s \subset s' \in S_\omega(\Sigma)\}$$
$$\cup \{(\forall t)(P(vt) \to \sigma_s(va_{\xi_1} \ldots a_{\xi_n}))\}.$$

Since $|\Sigma| < \delta$, we see easily that this set of conditions involves less than δ elements from $\mathfrak{A}_{\nu_s}$. Furthermore, by the observation of (2), this set of conditions is finitely satisfiable in $\mathfrak{A}_{\nu_s}$. Since $\mathfrak{A}_{\nu_s}$ is saturated, the set of conditions $\Gamma_s(v)$ is simultaneously satisfiable in $\mathfrak{A}_{\nu_s}$ by an element, say, $w_s \in V_0$. So, in particular, we may select for each $s \in S_\omega(\Sigma)$ an element $w_s \in V_0$ such that

(8) $$\mathfrak{A}_{\nu_s} \models (\forall t)(P(w_s t) \to \sigma_s(ta_{\xi_1} \ldots a_{\xi_n})),$$

and

(9) $$\mathfrak{A}_{\nu_s} \models P(w_s u_{s'}), \quad \text{for all } s' \text{ such that } s \subset s' \in S_\omega(\Sigma).$$

Consider now the set of conditions

$$\Delta(v) = \{P(w_s v) : s \in S_\omega(\Sigma)\}.$$

Notice that $|\Delta| < \delta$ and all elements $w_s \in V_0$; hence they all belong to A_0. Given $w_{s_1}, \ldots, w_{s_n}$, and letting $s = s_1 \cup \ldots \cup s_n$, we see that, from (9),

$$\mathfrak{A}_{v_s} \models P(w_{s_1} u_s) \wedge \ldots \wedge P(w_{s_n} u_s),$$

and thus

$$\mathfrak{A}_0 \models P(w_{s_1} u_s) \wedge \ldots \wedge P(w_{s_n} u_s).$$

So the set of conditions $\Delta(v)$ is finitely satisfiable in $\mathfrak{A}_0$. Since $\mathfrak{A}_0$ is saturated, $\Delta(v)$ is satisfiable in $\mathfrak{A}_0$ by an element, say, u. By (8), we see that

$$\mathfrak{A}_{v_s} \models \sigma_s(ua_{\xi_1} \ldots a_{\xi_n}) \quad \text{for all} \quad s \in S_\omega(\Sigma).$$

This implies immediately that

$$(\mathfrak{A}', a) \models \Sigma(u).$$

So (6) is proved. We leave the straightforward proof of (5) from (6) to the reader as an exercise. Once we have (5), then it is trivial to find a model $\mathfrak{A}_\mu = \langle A_\mu, V_0, R_0, \ldots \rangle$ such that (4) holds with v replaced by μ. The induction is now complete. The model $\bigcup_{v<\delta^+} \mathfrak{A}_v$ is now a (δ^+, δ) model of T (ignoring the superfluous relation R_0). ⊣

As the reader can see, this proof leans heavily on the existence of saturated models, and hence on the GCH. If δ happens to be inaccessible, then, of course, we may apply Proposition 5.1.5 (ii) and avoid the GCH.

At this point we would like to discuss the current status of two-cardinal problems and other associated problems. The reader will see that two-cardinal problems in model theory are intimately tied to various problems in set theory.

The so-called *gap n conjecture* can be stated as follows (assuming the language $\mathscr{L}$ is countable): every theory T of $\mathscr{L}$ which admits some pair $(\aleph_n(\alpha), \alpha)$, $\alpha \geqslant \omega$, admits every pair $(\aleph_n(\beta), \beta)$, $\beta \geqslant \omega$. The gap n conjecture for $n \geqslant 2$ easily implies a weak form of the GCH which is known to be unprovable in ZFC (see Exercise 7.2.32).

Jensen has proved the following (unpublished) theorem.

7.2.8. *The axiom of constructibility implies the gap n conjecture for all* $n < \omega$.

This result shows that the gap n conjecture is consistent and improves some earlier consistency results of Silver (1971).

Before proving the above Theorem 7.2.8, Jensen obtained the weaker result that the gap 1 conjecture follows from the axiom of constructibility.

Theorem 7.2.7 above shows that the gap 1 conjecture, where β is a regular cardinal, follows from the GCH. It is still an open question whether the gap 1 conjecture, where β is singular, can be proved from the GCH.

The gap 1 conjecture is related to the problem of the existence of special Aronsajn trees in set theory. The notion of a tree over a set X was defined in the proof of Theorem 4.2.23.

DEFINITION 7.2.9. Let α be an infinite cardinal. An *α-Aronszajn tree* is a tree over the set α which has order α, fewer than α elements of each order, and no branch of order α. α has the *tree property* if there is no α-Aronszajn tree.

A *special α^+-Aronszajn tree* is an α^+-Aronszajn tree T which has a function $f:\alpha^+ \to \alpha$ such that whenever $x\ T\ y$ and $x \neq y$, $f(x) \neq f(y)$.

Exercise 4.2.21 showed that an inaccessible cardinal has the tree property if and only if it is weakly compact. The following result on successors of regular cardinals is due to Aronszajn and Specker.

7.2.10. (i). *There is a special ω_1-Aronszajn tree.*

(ii). (GCH) *For every regular α, there is a special α^+-Aronszajn tree.*

It follows that ω_1 does not have the tree property, and under the GCH, no successor of a regular cardinal has the tree property. The connection between special Aronszajn trees and the gap 1 conjecture is given by the following fact which was observed independently by Rowbottom and Silver.

7.2.11. *There is a sentence σ in a finite language $\mathscr{L}$ such that for all infinite cardinals α, σ admits (α^+, α) if and only if there exists a special α^+-Aronszajn tree.*

PROOF. Let $\mathscr{L} = \{U, T, <, r, f, g, h\}$ where U is a unary relation, T and $\leqslant$ are binary relations, r and f are unary functions, and g and h are binary functions. Let σ be a sentence of $\mathscr{L}$ which says the following:

T is a strict tree, that is, a strict partial order which is a strict linear order below each element.

$<$ is a strict linear order.

U is an initial segment for $<$: $(\exists x)(\forall y)[U(y) \leftrightarrow y < x]$.

r acts like the tree order function:

$$x \; T \; y \rightarrow r(x) < r(y);$$

$$(\forall z)(\exists x) r(x) \equiv z;$$

$$z < r(y) \rightarrow (\exists x)[x \; T \; y \wedge r(x) \equiv z].$$

f acts like the function for a special Aronszajn tree:

$$(\forall x) U(f(x));$$

$$x \; T \; y \rightarrow f(x) \neq f(y).$$

For each x, $\{y : y < x\}$ has power at most $|U|$ (using g).
For each x, $\{y : r(y) = x\}$ has power at most $|U|$ (using h).

Each special α^+-Aronszajn tree T with the function f can be made into a model of σ of type (α^+, α) where U is α, $<$ is the usual order on α^+, and $r(x)$ is the order of x. On the other hand, given a model of σ of type (α^+, α), we can build a special α^+-Aronszajn tree by choosing a subset S of the universe which has order type α^+ with respect to $<$, and restricting T to $r^{-1}(S)$. The sentence σ guarantees that the tree has order α^+ and at most α elements of each order. There is no branch of order α^+ because f maps each branch one to one into the set U of power α. ⊣

By 7.2.10, the sentence σ of 7.2.11 admits (ω_1, ω). By 7.2.11, if every sentence which admits (ω_1, ω) admits (α^+, α), then there exists a special α^+-Aronszajn tree. Thus the full gap 1 conjecture implies that there exists a special α^+-Aronszajn tree for every α, and hence no successor cardinal has the tree property. By 7.2.8, the axiom of constructibility implies that a cardinal has the tree property if and only if it is weakly compact.

On the other hand, Mitchell, Silver, and Shelah respectively have proved the following consistency results relative to large cardinals. Mahlo numbers are defined in Exercise 4.2.12, where it is stated that every uncountable weakly compact cardinal is Mahlo, and every Mahlo number is inaccessible. The notion of a supercompact cardinal is stronger than a strongly compact cardinal, and can be found in Kanamori, Reinhardt, and Solovay (1978).

7.2.12. (i). ZFC + [There is a Mahlo number] *is consistent if and only if* ZFC + [There are no special ω_2-Aronszajn trees] *is consistent.*

(ii). ZFC + [There is a weakly compact cardinal $> \omega$] *is consistent if and only if* ZFC + [ω_2 has the tree property] *is consistent.*

(iii). *If* ZFC + [There is a supercompact cardinal] *is consistent, then so is* ZFC + GCH + [There are no special $\omega_{\omega+1}$-Aronszajn trees].

By (i), if ZFC + [There is a Mahlo number] is consistent, then there is a model of ZFC in which the sentence σ admits (ω_1, ω) but does not admit (ω_2, ω_1). This shows that the assumption GCH is necessary in Theorem 7.2.7.

By (iii), if ZFC + [There is a supercompact cardinal] is consistent, then there is a model of ZFC + GCH in which the sentence σ admits (α^+, α) for all regular α but does not admit $(\omega_{\omega+1}, \omega_\omega)$. Thus the full gap 1 conjecture cannot even be proved from the GCH.

Turning now to the gap 2 conjecture, the related set-theoretical problem turns out to involve the existence of Kurepa families or trees. We say that $\alpha \geqslant \omega$ has the property K_1, $K_1(\alpha)$, iff there is a (Kurepa) family F of subsets of α^+ such that $|F| = \alpha^{++}$ and for every $\xi < \alpha^+$,

$$|\{X \cap \xi : X \in F\}| \leqslant \alpha.$$

We say that $\alpha \geqslant \omega$ has the property K_2, $K_2(\alpha)$, iff there is a (Kurepa) tree T on α^+ of height α^+ having at most α elements at each level and at least α^{++} branches of length α^+. (See also Exercise 4.3.17.) It is again easy to see that:

7.2.13. *There is a sentence σ in a suitable language such that for all infinite cardinals α, σ admits (α^{++}, α) if and only if $K_1(\alpha)$ and also if and only if $K_2(\alpha)$.*

Thus the gap 2 conjecture would imply a transfer principle for Kurepa families or trees. The property $K_1(\omega)$ (or $K_2(\omega)$) is known as the Kurepa hypothesis. It was known that (a result of Lévy and Rowbottom, later improved by Stewart):

If ZFC is consistent, then so is ZFC + GCH + $K_2(\omega)$.

This result has been further improved by Solovay as follows, where $V = L$ denotes the axiom of constructibility.

7.2.14. ZFC + $V = L \vdash [K_2(\alpha)$ for every infinite cardinal $\alpha]$.

Concerning the negation of the Kurepa hypothesis, Silver and Solovay proved that

$$\text{ZFC} + \neg K_2(\omega) \text{ is consistent}$$

if and only if

$$\text{ZFC} + [\text{there exists an uncountable inaccessible cardinal}]$$

is consistent.

By a result of Silver, we can never hope to prove the gap 2 conjecture in ZFC + GCH. He has shown the following:

7.2.15. *If*

$$\text{ZFC} + [\text{there are two uncountable inaccessible cardinals}]$$

is consistent, then so is each of the following:

(a). $\text{ZFC} + \text{GCH} + K_2(\omega_1) + \neg K_2(\omega)$;
(b). $\text{ZFC} + \text{GCH} + K_2(\omega) + \neg K_2(\omega_1)$;
(c). $\text{ZFC} + \text{GCH} + K_2(\omega_1) + \neg K_2(\omega_5)$;
(d). $\text{ZFC} + \text{GCH} + K_2(\omega_5) + \neg K_2(\omega_1)$.

These results together with 7.2.13 show that we can never hope to prove certain instances of the gap 2 conjecture. For example, from 7.2.15(a) and 7.2.13 we can never prove that whenever a theory T has an (ω_3, ω_1) model, then it has an (ω_2, ω) model.

A conjecture related to the gap 1 conjecture, known as Chang's conjecture, will be discussed in the next Section 7.3, where again we see that set theory plays a dominant role.

EXERCISES

7.2.1. Prove that Theorem 7.2.2 remains true if the type $\Sigma(v)$ is replaced by a type $\Sigma(v_1 \ldots v_m)$ in the variables $v_1, \ldots, v_m$.

7.2.2. Strengthen the conclusion of Theorem 7.2.2 by requiring that the models of T of arbitrary infinite powers omitting Σ be generated from indiscernibles.

7.2.3. Prove that Theorem 7.2.2 remains true if the type Σ is replaced by a countable set S of types.

7.2.4*. Let $\mathscr{L}$ have $\alpha \geqslant \omega$ symbols and let Σ be a type of $\mathscr{L}$. Prove, by using the ideas of Theorem 7.2.2, that if for each $\xi < (2^\alpha)^+$, T has a model of power exceeding $\beth_\xi$ omitting Σ, then T has models in arbitrarily large infinite powers omitting Σ. Also, T has a model which omits Σ and has an infinite set of indiscernibles.

7.2.5*. Show that Exercise 7.2.4 remains true if Σ is replaced by an arbitrary set S of types of $\mathscr{L}$. Thus S may have as many as 2^α types of $\mathscr{L}$. Show also by means of counterexamples, like Proposition 7.2.4, that $\beth_{(2^\alpha)^+}$ is the least cardinal enjoying this property.

[*Hint*: Let $\mathscr{L}$ contain the 2-placed relation symbols $\leqslant$ and E, and constants c_ξ, $\xi < \alpha$. Let β be an ordinal less than $(2^\alpha)^+$, and let W be a well ordering of $S(\alpha)$ of type β. For $X, Y \subset \alpha$, let

$$\Sigma_{XY}(v_0v_1) = \{c_\xi E v_0 : \xi \in X\} \cup \{\neg\, c_\xi E v_0 : \xi \notin X\}$$
$$\cup \{c_\xi E v_1 : \xi \in Y\} \cup \{\neg\, c_\xi E v_1 : \xi \notin Y\} \cup \{\neg\,(v_0 \leqslant v_1)\}.$$

Let $S = \{\Sigma_{XY} : W(XY)\} \cup \{(v_0 \leqslant v_1 \wedge v_0 \not\equiv v_1 \wedge v_1 \leqslant v_0)\}$.]

7.2.6*. Let $\mathscr{L}$ have $\alpha \geqslant \omega$ symbols and suppose that $\operatorname{cf}(\alpha) = \omega$. Let Σ be a type of $\mathscr{L}$. Assume the GCH. Prove that if for each $\xi < \alpha^+$, T has a model of power exceeding $\beth_\xi$ omitting Σ, then T has models in arbitrarily large infinite powers omitting Σ. (Again, Σ may be replaced by a set S of types, with $|S| \leqslant \alpha$.)

7.2.7**. Let $\mathscr{L}$ have $\alpha \geqslant \omega$ symbols. Suppose that $\mathscr{L}$ has at least the symbols c_λ, $\lambda < \alpha$, and the binary predicate symbol $<$. A model $\mathfrak{A}$ for $\mathscr{L}$ is said to be *α-standard* iff the interpretation of $<$ in $\mathfrak{A}$ is precisely the natural ordering of $\xi < \alpha$. Clearly ω-standard models are also ω-models. Let m_α be the least ordinal λ such that for every sentence φ of $\mathscr{L}$, if φ has an α-standard model of power $\beth_\lambda$, then φ has arbitrarily large α-standard models. Prove the following:

(i). $\alpha < m_\alpha < \alpha^+$ if $\operatorname{cf}(\alpha) = \omega$ (GCH if $\alpha > \omega$).

(ii). $\alpha^+ < m_\alpha < \alpha^{++}$ if $\operatorname{cf}(\alpha) > \omega$.

(iii). m_ω is the first nonrecursive ordinal, i.e., the Church–Kleene ω_1.

There are results in the literature relating m_α and the notion of α^+-recursiveness, for $\alpha > \omega$.

7.2.8**. Let T be a complete theory of $\mathscr{L}$ and let Σ be a type of $\mathscr{L}$. We say that T and Σ characterize an infinite cardinal α iff:

(i). T and Σ have at most α symbols;

(ii). there are models of T omitting Σ in just those infinite powers $\beta < \alpha$.

A formula $\varphi(v)$ of $\mathscr{L}$ is said to be *algebraic in T* iff in every model $\mathfrak{A}$ of T

there are only a finite number of elements of $\mathfrak{A}$ satisfying φ. An element a of a model $\mathfrak{A}$ of T is said to be *algebraic* iff a satisfies some algebraic formula $\varphi(v)$ in T. Disprove the following conjecture of Morley: If T is complete and T and Σ characterize $\alpha > \omega$, then every model of T contains an infinite number of algebraic elements.

7.2.9. Give a complete proof of Proposition 7.2.5. Make sure that other than the sentences of (i), all other requirements on the theory T can be expressed by a single sentence of $\mathscr{L}$.

7.2.10. Let Σ be a set of formulas of a countable language $\mathscr{L}$ in the variable v. We say that (T, Σ) admits the pair (α, β) iff T has a model $\mathfrak{A}$ such that $|A| = \alpha$ and

$$|\{a \in A : \mathfrak{A} \models \Sigma[a]\}| = \beta.$$

Use the technique of Theorems 7.2.2 and 7.2.6 to establish the following: If for each $\xi < \omega_1$, (T, Σ) admits some pair $(\beth_\xi(\gamma_\xi), \gamma_\xi)$ of infinite cardinals, then (T, Σ) admits all pairs (α, β) of infinite cardinals.

7.2.11. Find a version of Theorem 7.2.6 for noncountable languages. Do the same for Exercise 7.2.10. (See Exercise 7.2.4.)

7.2.12. Prove that the following property of a set Σ of sentences is compact: Σ admits all pairs (α, β) of infinite cardinals $\alpha \geqslant \beta$. That is, prove that Σ admits all pairs (α, β) if and only if every finite subset of Σ admits all pairs (α, β).

7.2.13*. Recall from Section 4.2 (in particular, 4.2.9 and 4.2.10) that we have already defined the infinitary language $\mathscr{L}_\alpha$, where α is an infinite cardinal. We shall now describe a much more restricted sublanguage of $\mathscr{L}_\alpha$, but one which is nevertheless richer than the classical language $\mathscr{L}_\omega$. The language $\mathscr{L}_{\alpha\omega}$ is obtained by adding to the rules of formation of $\mathscr{L}$ the following two new rules: the first new rule is already given in 4.2.9, namely:

> If Φ is a set of formulas of $\mathscr{L}_{\alpha\omega}$ of power $|\Phi| < \alpha$, then $\bigwedge\Phi$ is a formula of $\mathscr{L}_{\alpha\omega}$.

The second new rule is:

> If φ is a formula of $\mathscr{L}_{\alpha\omega}$ and V is a finite set of variables, then $(\forall V)\varphi$ is a formula of $\mathscr{L}_{\alpha\omega}$.

Notice that $\mathscr{L}_{\alpha\omega}$ still only has a countable number of individual variables, and that any sentence of $\mathscr{L}_{\alpha\omega}$ will only contain subformulas each having only a finite number of free variables. (Why?)

Let α be an infinite cardinal and let $\mathscr{L}$ have at most α symbols. Consider the language $\mathscr{L}_{\alpha^+\omega}$ built up from the symbols of $\mathscr{L}$. We define the Hanf number of $\mathscr{L}_{\alpha^+\omega}$ to be the least cardinal β such that every sentence φ of $\mathscr{L}_{\alpha^+\omega}$ having a model of power β also has models of arbitrarily high powers. We define the Morley number of $\mathscr{L}$ to be the least cardinal γ such that for any type Σ ($|\Sigma| \leqslant \alpha$) and theory T of $\mathscr{L}$, if T has a model of power γ omitting Σ, then T has models of arbitrarily high powers omitting Σ. Prove the following:

(i). The Hanf number of $\mathscr{L}_{\alpha^+\omega}$ exists.

(ii). For some language $\mathscr{L}'$, with $||\mathscr{L}'|| = \alpha$, the Hanf number of $\mathscr{L}_{\alpha^+\omega}$ is exactly the Morley number of $\mathscr{L}'$. Hence, by Exercise 7.2.4 and Proposition 7.2.4, the Hanf number of $\mathscr{L}_{\alpha^+\omega}$ is trapped between $\beth_{\alpha^+}$ and $\beth_{(2^\alpha)^+}$.

(iii). If $\mathrm{cf}(\alpha) = \omega$ and we assume GCH, then the Hanf number of $\mathscr{L}_{\alpha^+\omega}$ is exactly $\beth_{\alpha^+}$.

(iv). The Hanf number of $\mathscr{L}_{\omega_1\omega}$ is $\beth_{\omega_1}$.

(v). If $\mathrm{cf}(\alpha) > \omega$, then the Hanf number of $\mathscr{L}_{\alpha^+\omega}$ is greater than $\beth_{\alpha^+}$.

This exercise shows that for these suitably restricted sublanguages of $\mathscr{L}_\alpha$, we can find Hanf numbers of these languages of reasonable size. As soon as we consider the full language $\mathscr{L}_\alpha$, say, even for $\mathscr{L}_{\omega_1}$, the Hanf number of $\mathscr{L}_{\omega_1}$ exists, but is incredibly large.

7.2.14. Complete the proof of Theorem 7.2.7 by proving statement (5).

7.2.15. Suppose that $\mathscr{L}$ has $\gamma \geqslant \omega$ symbols. Show that Theorem 7.2.7 remains true, without any essential changes in the proof, if we require that δ be a regular cardinal greater than γ.

7.2.16*. Suppose that $\mathscr{L}$ has $\gamma > \omega$ symbols, and suppose that γ is regular. Prove that Exercise 7.2.15 can be strengthened by letting $\delta \geqslant \gamma$.

[*Hint*: Replace all n-placed relations (or functions) in a model for $\mathscr{L}$ by an $(n+1)$-placed relation (or function) which indexes, by means of elements of the model, all the n-placed relations (or functions) of the model. Thus given any model for $\mathscr{L}$ of power at least γ, we may pass to an 'equivalent' model of the same power which has only a countable number of relations and functions. Next observe that the tower constructed in Theorem 7.2.7 can be made so that $\mathfrak{B} \prec \mathfrak{A}_0$, where $\mathfrak{B}$ is any model of T' of power δ.]

7.2.17. Suppose that $\mathscr{L}$ has $\gamma > \omega$ symbols. Assume the GCH. Prove that if a theory T admits the pair (α^+, α), then it admits the pair $(\alpha^{+++}, \alpha^{++})$. Notice that no assumption is made on the relative sizes of α and γ. It is an open problem whether T admits the pair (α^{++}, α^+).

7.2.18*. Combine Theorems 7.2.6 and 7.2.7 to obtain the following three-cardinal theorem: We say that T *admits* (α, β, γ), with $\alpha \geqslant \beta \geqslant \gamma$, iff T has a model

$$\mathfrak{A} = \langle A, V_1, V_2, \ldots \rangle$$

such that $|A| = \alpha$, $|V_1| = \beta$, $|V_2| = \gamma$. Let $\mathscr{L}$ be a countable language and let T be a theory of $\mathscr{L}$. Assume the GCH. Suppose that for each $n \in \omega$, T admits a triple $(\aleph_{n+1}(\gamma), \gamma^+, \gamma)$, $\gamma \geqslant \omega$. Then T admits all triples (α, β^+, β), with $\alpha \geqslant \beta^+$, $\beta \geqslant \omega$, and β regular.

7.2.19. Prove that the symmetric version of Exercise 7.2.18 does not hold. Namely, there is a theory T which admits $(\beth_\omega^+, \beth_\omega, \beth_0)$, but does not admit $(\beth_n^+, \beth_n, \beth_0)$ for any $0 < n < \omega$.

7.2.20. Prove that the gap 2 conjecture is equivalent to the following three-cardinal problem: If T admits $(\alpha^{++}, \alpha^+, \alpha)$, then T admits all $(\beta^{++}, \beta^+, \beta)$, $\alpha, \beta \geqslant \omega$.

7.2.21. Suppose that $\omega \leqslant \gamma$, $\beth_\omega(\gamma) \leqslant \beta$, $\beth_\omega(\beta) \leqslant \alpha$, and T is a theory in a countable language which admits (α, β, γ). Prove that T admits every triple of infinite cardinals $(\alpha', \beta', \gamma')$, where $\alpha' \geqslant \beta' \geqslant \gamma'$. (This is a generalization of Theorem 7.2.6.)

7.2.22*. Let T be a theory in a countable language. If $\omega \leqslant \gamma < \beta$ and $\beth_\omega(\beta) \leqslant \alpha$ and T admits (α, β, γ), then for any $\alpha' \geqslant \beta$, T admits $(\alpha', \beta^\omega, \gamma^\omega)$.

7.2.23. Let T be a theory in a countable language and $\Sigma(v)$ a set of formulas. We say that T *admits a pair* (α, β) *omitting* Σ iff T has a model $\langle A, V, \ldots \rangle$ which omits Σ and where $|A| = \alpha$, $|V| = \beta$. Prove that if $\beth_{\omega_1} \leqslant \beta$ and $\beth_{\omega_1}(\beta) \leqslant \alpha$ and if T admits (α, β) omitting Σ, then T admits every pair of infinite cardinals (α', β') omitting Σ. (Compare with Theorem 7.2.6 and Exercise 7.2.21.)

7.2.24*. Let T, Σ be as above. If $\omega \leqslant \gamma < \beta$ and $\beth_{\omega_1}(\beta) \leqslant \alpha$ and T admits (α, β, γ) omitting Σ, then for any cardinal $\alpha' \geqslant \omega_1$, T admits the triple $(\alpha', \omega_1, \omega)$ omitting Σ. (Compare with Exercise 7.2.18.)

7.2.25*. State and prove the analogues of Exercises 7.2.21 and 7.2.23 for countable sequences of cardinals.

7.2.26. Show that in each of the exercises 7.2.21–7.2.24 the hypothesis can be weakened in a way analogous to Theorem 7.2.6.

7.2.27*. Prove the assertion 7.2.10.

7.2.28**. Prove that ω_1 does not have the tree property by finding a tree as described in 7.2.8 and 7.2.9. Next use Theorem 7.2.7 to establish 7.2.9.

7.2.29*. Prove the assertion 7.2.13.

7.2.30*. Recall that in Section 4.2 we introduced the notion of a weakly compact cardinal and the infinitary logic $\mathscr{L}_\alpha$. Prove the following theorem: Let α be a weakly compact cardinal and let $\mathfrak{A}$ be a model of the form $\mathfrak{A} = \langle R(\alpha), \in, R_0, R_1, ...\rangle$. Then there exist elementary extensions $\mathfrak{B} = \langle B, E, S_0, S_1, ...\rangle$ of $\mathfrak{A}$ of arbitrarily large power such that for all $b \in B$ and $a \in R(\alpha)$, bEa implies $b \in R(\alpha)$. (This improves Exercise 4.2.21 (iii).)

[*Hint*: We may assume that $\mathfrak{A}$ has built-in Skolem functions. Let T be a theory in the language $\mathscr{L}_\alpha$, with constants c_a for each $a \in R(\alpha)$ and new constants $d_0, d_1, ...,$ with the axioms:

(1) the elementary diagram of $\mathfrak{A}$;

(2) $\varphi(d_{n_0} \dots d_{n_r}) \leftrightarrow \varphi(d_{m_0} \dots d_{m_r})$, where $n_0 < \dots < n_r$, $m_0 < \dots < m_r$, and $\varphi(v_0 \dots v_r)$ is a formula in $\mathscr{L} \cup \{c_a : a \in R(\alpha)\}$;

(3) $(\forall x)(x \in c_a \leftrightarrow \bigvee_{b\in a} x \equiv c_b)$ for each $a \in R(\alpha)$;

(4) d_m, d_n are ordinals and $d_m < d_n$, where $m < n$.

Use Exercise 7.2.4 to show that every subset of T of power $< \alpha$ has a model, whence by weak compactness, T has a model. Then form the desired model $\mathfrak{B}$ by replacing $d_0, d_1, ...$ by a generating set of indiscernibles of any power.]

The following related problem is open: Does every model of ZFC which has an end elementary extension have arbitrarily large end elementary extensions? For a partial solution, see the next exercise.

7.2.31**. Prove that every model of Zermelo–Fraenkel set theory whose ordinals have cofinality ω has arbitrarily large elementary extensions in which all new ordinals occur after all the old ordinals.

[*Hint*: Try it first for models of the axiom of choice, using the Erdös–Rado theorem within the model. For the general case, prove a version of the Erdös–Rado theorem without the axiom of choice.]

7.2.32. Show that the gap n conjecture implies the following weak form of the GCH: If $m < n$ and there exists an infinite cardinal α such that $2^\alpha = \aleph_m(\alpha)$, then $2^\beta = \aleph_m(\beta)$ for all infinite cardinals β. In particular, the gap 2 conjecture implies that if $2^\alpha = \alpha^+$ for some infinite cardinal α, then the GCH holds.

7.2.33*. Let $\mathscr{L}$ be a countable language and let P be a 1-placed relation symbol of $\mathscr{L}$. A model $\mathfrak{A} = \langle A, R, ...\rangle$ for $\mathscr{L}$ is a *two-cardinal model* iff $|A| > |R| \geqslant \omega$. A set Σ of sentences of $\mathscr{L}$ is a *set of axioms for two-cardinal* *models* iff $\mathfrak{A} \models \Sigma$ if and only if $\mathfrak{A}$ is elementarily equivalent to a two-cardinal model. Prove that the following is a set of axioms for two-cardinal models: for each n and each finite sequence of formulas $\varphi_0, \ldots, \varphi_m$ of $\mathscr{L}$ in the variables $x_0, \ldots, x_n$,

$$(\exists v_0 \forall x_0 \exists y_0 z_0 \ldots \forall x_n \exists y_n z_n)[\bigwedge_{0 \leq i \leq n} v_0 \neq y_i$$

$$\wedge \bigwedge_{0 \leq i, j \leq n} (P(x_j) \wedge x_i = z_j \rightarrow y_i = x_j) \wedge \bigwedge_{0 \leq j \leq m} (\varphi_j(x_0 \ldots x_n) \rightarrow \varphi_j(y_0 \ldots y_n))].$$

7.3. Models of large cardinality

In this section we shall investigate three closely related problems involving models and cardinal numbers. To state the first problem, recall that a model of type (α, β) is a model $\mathfrak{A} = \langle A, U, \ldots \rangle$ such that $|A| = \alpha$, $|U| = \beta$.

7.3.1. *Which pairs of cardinals (α, β) and (γ, δ) have the property that every model $\mathfrak{A}$ of type (α, β) for a countable language has an elementary submodel $\mathfrak{B}$ of type (γ, δ)?*

We say that *Chang's conjecture* holds for the pairs (α, β), (γ, δ) iff they have the above property.

Chang's conjecture is a stronger form of the Löwenheim–Skolem problem for pairs of cardinals; if Chang's conjecture holds for (α, β), (γ, δ), then every theory which admits (α, β) admits (γ, δ). Chang's conjecture is interesting only when $\alpha > \beta > \delta$ and $\alpha \geqslant \gamma > \delta$; the other cases are trivial. When $\alpha > \gamma > \beta = \delta$, Chang's conjecture is an easy corollary of the Löwenheim–Skolem–Tarski theorem. The 'dual' case, $\alpha = \gamma > \beta > \delta$, is a difficult problem which leads to consistency results in set theory.

Here is the second problem.

7.3.2. *Which cardinals α have the property that every model of power α for a countable language has a proper elementary submodel of power α?*

A model $\mathfrak{A}$ for a countable language which does not have any proper elementary submodel of power $|A|$ is called a *Jónsson model*. Thus the problem may be restated: For which cardinals α does there exist no Jónsson model of power α?

It is not hard to see that if $\alpha > \beta > \delta$ and Chang's conjecture holds for the pairs (α, β), (α, δ), then there is no Jónsson model of power α (Proposition 7.3.4). Thus an affirmative answer to problem 7.3.1, with $\alpha = \gamma > \beta > \delta$, implies an affirmative answer to problem 7.3.2.

For the third problem, we need the notion of a well ordered model. A model $\mathfrak{A} = \langle A, <, \ldots \rangle$ is said to be *well ordered of type* α iff $<$ is a well ordering of the set A of order type α.

7.3.3. *Which cardinals* α, β, *with* $\beta < \alpha$, *have the property that every well ordered model of type* α (*for a countable language*) *has an elementary submodel which is well ordered of type* β?

This is a Löwenheim–Skolem problem for well ordered models. There is a natural similar question for $\beta > \alpha$, which will be dealt with in the exercises (cf. Exercise 7.3.34).

We shall begin with a few elementary negative results concerning these three problems. Then we shall proceed to some deeper positive results which concern large cardinals, the so-called Ramsey cardinals. In Section 4.2, we studied large cardinals, the measurable cardinals, with the aid of the ultraproduct construction. We now resume the study begun there, but this time we shall apply other model-theoretic constructions, especially Skolem functions and indiscernible elements. We shall first show that the class of Ramsey cardinals lies between the classes of weakly compact cardinals and measurable cardinals. Then it will be shown that the three questions above have affirmative answers for Ramsey cardinals. In the next section, 7.4, the results concerning Ramsey cardinals and the three questions 7.3.1–7.3.3 will be applied to the notion of a constructible set, and Scott's Theorem 4.2.18 (that the axiom of constructibility and the axiom of measurability contradict each other) will be vastly improved. The standard facts about constructible sets which we shall need will be stated there without proofs.

The following result gives relationships between problems 7.3.1–7.3.3.

PROPOSITION 7.3.4.

(i). *If* $\alpha > \beta > \gamma$ *and Chang's conjecture holds for the pairs* (α, β), (α, γ), *then there is no Jónsson model of power* α.

(ii). *Chang's conjecture holds for the pairs* (α^+, α), (β^+, β) *if and only if every well ordered model of type* α^+ *has an elementary submodel of type* β^+ (*for a countable language*).

PROOF. (i). Let $\mathfrak{A}$ be a model of power α (with $\mathscr{L}$ countable). Let $U \subset A$ have power β, whence $(\mathfrak{A}, U)$ is a model of type (α, β). By Chang's conjecture, there is a model $(\mathfrak{B}, V) \prec (\mathfrak{A}, U)$ of type (α, γ). But then $V = U \cap B$ and V is a proper subset of U, so B is a proper subset of A.

Thus $\mathfrak{B}$ is a proper elementary submodel of $\mathfrak{A}$ of power α.

(ii). Assume Chang's conjecture for $(\alpha^+, \alpha), (\beta^+, \beta)$. Let $\mathfrak{A} = \langle A, <, \ldots\rangle$ be well ordered of type α^+. Let U be an initial segment of $\langle A, <\rangle$ of type α. Each $a \in A$ has at most α predecessors, so there is a binary function $F: A \times A \to A$ such that for each $a \in A$,

$$\{F(a, b) : b \in U\} = \{c \in A : c < a\}.$$

The model $(\mathfrak{A}, U, F)$ of type (α^+, α) has an elementary submodel $(\mathfrak{B}, V, G)$ of type (β^+, β). In each of these models the sentence

$$(1) \qquad (\forall x \forall y)[y < x \to (\exists z)(U(z) \wedge F(x, z) \equiv y)]$$

holds. It follows that every element of B has at most $|V| = \beta$ predecessors. Moreover, $|B| = \beta^+$ and $\mathfrak{B} \prec \mathfrak{A}$, so $\langle B, <\rangle$ is well ordered of type $\geqslant \beta^+$. Therefore $\langle B, <\rangle$ has exactly type β^+.

Now assume every well ordered model of type α^+ has an elementary submodel of type β^+. Let $\mathfrak{A} = \langle A, U, \ldots\rangle$ be of type (α^+, α). Choose a well ordering $<$ of A of type α^+ such that U is an initial segment of type α. As before, choose $F: A \times A \to A$ such that (1) holds in $(\mathfrak{A}, <, F)$. Let $(\mathfrak{B}, <, G) \prec (\mathfrak{A}, <, F)$ be of type β^+. Since U is a proper initial segment of $\langle A, <\rangle$, V is a proper initial segment of $\langle B, <\rangle$. But $\langle B, <\rangle$ has order type exactly β^+, so $|V| \leqslant \beta$. However, by (1) each element of B has at most $|V|$ predecessors, whence $|V| = \beta$. Thus $\mathfrak{B}$ is of type (β^+, β). ⊣

From Proposition 3.2.11 (iv) and (v), we see that Chang's conjecture fails for all pairs $(\beth_n(\alpha), \alpha), (\beta, \delta)$ when $\beta > \beth_n(\delta)$, and $(\aleph_n(\alpha), \alpha), (\beta, \delta)$ when $\beta > \aleph_n(\delta)$, because the corresponding two-cardinal problems fail. That is, finite gaps cannot be stretched. Chang's conjecture leads once more to the problem of the existence of Kurepa families, which we encountered in the preceding section. We introduced in Section 7.2 the notation $K_1(\alpha)$ for 'there exists a Kurepa family of subsets of α^+'. It is not hard to see that:

7.3.5. *If* $K_1(\alpha)$, *then Chang's conjecture fails for* $(\alpha^{++}, \alpha^+), (\alpha^+, \alpha)$.

Thus it follows from 7.2.14 that the axiom of constructibility implies that Chang's conjecture fails for all $(\alpha^{++}, \alpha^+), (\alpha^+, \alpha)$. On the other hand, Silver has shown the following:

7.3.6. *If* ZFC + 'there is a Ramsey cardinal' *is consistent, then so is*

ZFC + 'Chang's conjecture holds for $(\omega^{++}, \omega^+), (\omega^+, \omega)$'.

Ramsey cardinals are defined just before Proposition 7.3.9. Actually, Silver proved the conclusion of 7.3.6 assuming the weaker hypothesis that the existence of the cardinal $\kappa(\omega_1)$ defined in Exercise 7.3.23 is consistent.

Chang's conjecture for (ω_3, ω_2), (ω_2, ω_1) is related to stronger large cardinal notions, such as the huge cardinals in Kanamori, Reinhart, and Solovay (1978). Laver proved that:

7.3.6′. *For each* $n < \omega$, *if* ZFC + 'there is a huge cardinal' *is consistent, then so is* ZFC + 'Chang's conjecture holds for $(\omega_{n+2}, \omega_{n+1})$, (ω_{n+1}, ω_n)'.

Relative consistency results involving three-cardinal forms of Chang's conjecture were obtained by Foreman (1982).

Results in the other direction can be stated using the notion of an *inner model* of set theory, which is a model of the form $\langle A, \in \rangle$ where A is a transitive proper class. Examples of inner models of ZFC are the model $\langle L, \in \rangle$ of constructible sets (defined in Section 7.3) and the core model of Jensen and Dodd (1981). Koepke (1988) used the core model to prove:

7.3.6″. *Chang's conjecture for* (ω_3, ω_2), (ω_2, ω_1) *implies that for each ordinal* α *there is an inner model of ZFC whose set of uncountable measurable cardinals has order type* α.

Let us now look for cases in which the answer to questions 7.3.2 and 7.3.3 are negative. To give a negative answer to 7.3.2, we must give an example of a Jónsson model of power α. There are obvious examples of Jónsson models of power ω: e.g., the standard model of arithmetic, $\langle \omega, +, \cdot, 0 \rangle$, is a Jónsson model. The next result gives further examples.

PROPOSITION 7.3.7. *Assume the generalized continuum hypothesis. Then for every successor cardinal* α^+, *there exists a Jónsson model of power* α^+.

PROOF. Let the universe set be $A = \alpha^+$. By the GCH, $(\alpha^+)^\alpha = 2^\alpha = \alpha^+$, so the set A has exactly α^+ subsets of power α. Let X_β, $\beta < \alpha^+$, be an enumeration of all subsets $X \subset A$ of power $|X| = \alpha$. Consider any $\gamma < \alpha^+$, with $\alpha \leqslant \gamma$. Then $|\gamma| = \alpha$. From Lemma 6.1.6, any collection of α sets of power α can be refined to a collection of α disjoint sets of power α. That is, there are subsets $Y_\beta \subset X_\beta$, $\beta < \gamma$, such that each $|Y_\beta| = \alpha$ and the sets Y_β

are pairwise disjoint. For each $\beta < \gamma$, we may choose a one–one function $f_{\beta\gamma}$ on Y_β onto γ. The sets Y_β, $\beta < \gamma$, are pairwise disjoint, so the $f_{\beta\gamma}$ may be 'extended' to a function f on $A \times A$ into A such that

$$\text{if } \beta < \gamma < \alpha^+ \quad \text{and} \quad y \in Y_\beta, \quad \text{then} \quad f(y\gamma) = f_{\beta\gamma}(y).$$

Let $\mathfrak{B}$ be any submodel of $\mathfrak{A}$ of power α^+, and let $\xi \in A$. We wish to prove that $\xi \in B$. Take any subset of B of power α, say X_β. Then there exists $\gamma \in B$ such that $\xi < \gamma$ and $X_\beta \subset \gamma$. Since $f_{\beta\gamma}$ maps Y_β onto γ, and $Y_\beta \subset X_\beta$, there exists $\eta \in X_\beta \subset B$ such that $f(\eta\gamma) = f_{\beta\gamma}(\eta) = \xi$. But $\eta \in B$ and $\gamma \in B$, whence $\xi \in B$. ⊣

In Exercise 7.3.15 we see that without the GCH it can be proved that there exist Jónsson models of power ω_n, that is, ω_n is not a Jónsson cardinal, for each finite n. Tryba (1984) proved without the GCH that for each regular cardinal α there exist Jónsson models of power α^+, that is, no successor of a regular cardinal is Jónsson. Recently, Shelah proved without the GCH that there exists a Jónsson model of power $\omega_{\omega+1}$. His proof works for certain other small successors of singular cardinals, but it is open whether one can prove in ZFC that no successor cardinal is Jónsson.

Using the core model, Koepke (1988) proved that if there exists a Jónsson cardinal ω_ξ such that $\xi < \omega_\xi$, then for each α there is an inner model of ZFC whose set of uncountable measurable cardinals has order type α. He also showed that if there exists a singular Jónsson cardinal of uncountable cofinality α, then there is an inner model of ZFC with α measurable cardinals.

In the other direction, Prikry (1970) proved that a singular cardinal which is the union of a strictly increasing sequence of measurable cardinals is a Jónsson cardinal. Devlin (1973) showed that if

$$\text{ZFC} + [\text{there exists a measurable cardinal} > \omega]$$

is consistent, then so is

$$\text{ZFC} + [\text{there is no Jónsson model of power } 2^\omega].$$

Combining the results of Koepke and Prikry, it follows that for each $n > 0$,

$$\text{ZFC} + \text{'There is a Jónsson cardinal of cofinality } \omega_n\text{'}$$

is consistent if and only if

$$\text{ZFC} + \text{'There are } \omega_n \text{ measurable cardinals'}$$

is consistent.

The above proposition shows that, assuming the GCH, the Chang conjecture fails for (α^+, β), (α^+, γ) whenever $\alpha^+ > \beta > \gamma$. Here are some examples for problem 7.3.3.

PROPOSITION 7.3.8. *Suppose that every theory T for a countable language which has a well ordered model of type α has a well ordered model of type β. Then*:

(i). *If $\alpha > \omega$, then $\beta > \omega$.*

(ii). *If α is singular, then β is singular.*

(iii). *If α is a successor cardinal, then so is β.*

(iv). *If there exists γ such that $\gamma < \alpha \leqslant 2^\gamma$, then there exists δ such that $\delta < \beta \leqslant 2^\delta$.*

PROOF. (i). If $\alpha > \omega$, then $\langle \alpha, < \rangle$ is not elementarily equivalent to $\langle \omega, < \rangle$.

(ii). The theory of all models $\langle A, <, a, F \rangle$, where F is a function, with

$$(\forall x \exists y)(y < a \wedge x < F(y)),$$

has well ordered models of singular types only.

(iii). The theory of all models $\langle A, <, a, F \rangle$, where

$$(\forall x \forall y)[y < x \rightarrow (\exists z)(z < a \wedge F(x, z) \equiv y)],$$

has well ordered models of successor types only.

(iv). Consider the theory of all models $\langle A, <, a, E \rangle$ satisfying

$$(\forall x \forall y)[x \not\equiv y \rightarrow (\exists z)(z < a \wedge \neg (zEx \leftrightarrow zEy))]. \dashv$$

Let us now define Ramsey cardinals. Recall that the set of all finite subsets of X is denoted by $S_\omega(X)$, and the set of all subsets of X of power n is denoted by $[X]^n$. Thus $S_\omega(X) = \bigcup_{n<\omega}[X]^n$.

An infinite cardinal α is said to be a *Ramsey cardinal* iff for every set X of power α, if

$$S_\omega(X) \subset \bigcup_{i \in I} C_i \quad \text{and} \quad |I| < \alpha,$$

then there are a subset $Y \subset X$ of power α and elements $i_1, i_2, \ldots \in I$ such that

$$[Y]^n \subset C_{i_n}, \qquad n = 1, 2, \ldots .$$

Ramsey cardinals are so named because the partition property defining them is suggested by Ramsey's theorem.

PROPOSITION 7.3.9. *If α is a Ramsey cardinal, then $\alpha > \omega$ and α is inaccessible.*

PROOF. First let $\alpha = \omega$ and take $X = \alpha$. Let C_0 be the set of all $x \in S_\omega(X)$ such that $|x| \in x$. Let $C_1 = S_\omega(X) \setminus C_0$. Consider any infinite subset $Y \subset X$, and write $Y = \{y_0, y_1, ...\}$ in increasing order. Let $n = y_0$. Then the sets $\{y_0, ..., y_{n-1}\}$ and $\{y_1, ..., y_n\}$ do not both belong to the same C_i. Thus ω is not a Ramsey cardinal.

Now let α be singular and let $\beta = \mathrm{cf}(\alpha) < \alpha$. Take $X = \alpha$, and let I be a cofinal subset of α of power β. Then put $x \in C_i$ iff i is the least element of I which is an upper bound of x. Clearly, for any $Y \subset X$ of power α, we cannot even have an $i \in I$ such that $[Y]^1 \subset C_i$.

Finally, let α be a cardinal such that for some $\gamma < \alpha$, $\alpha \leqslant 2^\gamma$. Take $I = \gamma$ and let X be a subset of $S(\gamma)$ of power α. For each $\{a, b\} \in [X]^2$, put $\{a, b\} \in C_i$, where i is the least element of I which belongs to exactly one of a, b. Define C_i arbitrarily outside $[X]^2$. Then for any $Y \subset X$ of power $|Y| \geqslant 3$ and any $i \in I$, we cannot have $[Y]^2 \subset C_i$. Therefore α is not Ramsey. ⊣

Ramsey cardinals can also be defined model-theoretically:

PROPOSITION 7.3.10. *α is a Ramsey cardinal if and only if every model $\mathfrak{A}$ of power α for a countable language has a set of indiscernibles of power α.*

PROOF. Suppose α is a Ramsey cardinal. Let I be the set of all sets of formulas of $\mathscr{L}$. Since $||\mathscr{L}|| = \omega$, $|I| = 2^\omega$, and because Ramsey cardinals are inaccessible and $> \omega$, $|I| < \alpha$. Choose any simple ordering $<$ of A. Partition $S_\omega(A)$ by putting $\{a_1, ..., a_n\} \in C_i$ iff $a_1 < ... < a_n$ and

$$i = \{\varphi(v_1 \ldots v_n) : \mathfrak{A} \models \varphi[a_1 \ldots a_n]\}.$$

Since α is Ramsey, there is a set $X \subset A$ of power α such that for each n, all increasing n-tuples in $\langle X, < \rangle$ belong to the same C_i. This means that $\langle X, < \rangle$ is a set of indiscernibles in $\mathfrak{A}$.

For the converse, let $S_\omega(\alpha) \subset \bigcup_{i \in I} C_i$ and $|I| < \alpha$. We may assume without loss of generality that I is a proper initial segment of $\langle \alpha, < \rangle$. Form the model

$$\mathfrak{A} = \langle \alpha, <, I, F_1, F_2, ... \rangle,$$

where for each $n < \omega$, F_n is a function on A^n into I such that if $x_1 < ... < x_n$,

$$F_n(x_1 \ldots x_n) = i \quad \text{implies} \quad \{x_1, ..., x_n\} \in C_i.$$

Then $\mathfrak{A}$ has a set $\langle X, <' \rangle$ of indiscernibles of power α. Since $<$ is already a relation of $\mathfrak{A}$, $<'$ must be either $<$ or $>$, and we may assume that $<'$ is $<$ (restricted to X). Since α is a cardinal, $\langle X, < \rangle$ has order type α.

Consider two n-tuples $x_1 < \dots < x_n$ and $y_1 < \dots < y_n$ in X. Suppose

(1) $$F_n(x_1 \dots x_n) \neq F_n(y_1 \dots y_n).$$

Then choose $z_1 < \dots < z_n$ in X such that $x_n < z_1$ and $y_n < z_1$. We cannot have

$$F_n(x_1 \dots x_n) = F_n(z_1 \dots z_n) = F_n(y_1 \dots y_n),$$

so we may assume

(2) $$F_n(x_1 \dots x_n) \neq F_n(z_1 \dots z_n).$$

There exists a sequence of n-tuples

$$u_1^\xi < \dots < u_n^\xi, \qquad \xi < \alpha,$$

of elements of X such that

$$\xi < \zeta < \alpha \quad \text{implies} \quad u_n^\xi < u_1^\zeta.$$

But by (2) and indiscernibility,

$$F_n(u_1^\xi \dots u_n^\xi), \qquad \xi < \alpha,$$

are all different elements of I, contradicting $|I| < \alpha$. This shows that (1) is never the case, and it follows that any two elements of $[X]^n$ belong to the same C_i (with $i = F_n(x_1 \dots x_n)$). Thus α is a Ramsey cardinal. ⊣

We shall now show that Ramsey cardinals are weakly compact. Recall that in Section 4.2 we defined a cardinal α to be weakly compact iff α is inaccessible, and in the infinitary language $\mathscr{L}_\alpha$ the following weak compactness theorem holds: If Σ is a set of sentences of $\mathscr{L}_\alpha$ of power $|\Sigma| = \alpha$, and if every subset of Σ of power $< \alpha$ has a model, then Σ has a model. Furthermore, we proved in 4.2 that an inaccessible cardinal α is weakly compact iff α has the tree property: For every tree T on α of order α such that for all $\beta < \alpha$, fewer than α elements have order β, T has a branch of order α. We show that Ramsey cardinals have the tree property and hence are weakly compact.

LEMMA 7.3.11. *Every Ramsey cardinal α has the following property: Every simply ordered set $\langle A, L \rangle$ of power α has a subset of power α, which is either well ordered or inversely well ordered by L.*

PROOF. We choose any well ordering $<$ of A of type α. We then partition $[A]^2$ into two parts as follows:

$$\{a, b\} \in C_1 \quad \text{iff} \quad (a < b \text{ iff } aLb),$$

$$\{a, b\} \in C_2 \quad \text{otherwise.}$$

That is, C_1 contains the pairs on which the two orderings L, $<$ coincide. We have assumed that α is a Ramsey cardinal, so there is a set $X \subset A$ such that $|X| = \alpha$ and either $[X]^2 \subset C_1$ or $[X]^2 \subset C_2$. In the former case, X is well ordered by L, and in the latter case it is inversely well ordered by L. ⊣

PROPOSITION 7.3.12. *Every Ramsey cardinal is weakly compact.*

PROOF. Let α be a Ramsey cardinal. We show that α has the tree property. Let T be a tree on α of order α such that for all $\beta < \alpha$, fewer than α elements have order β. We define a linear ordering xLy on α which extends the ordering T as follows: Let x, y be distinct elements of α.

Case 1. If xTy, then xLy.

Case 2. If yTx, then yLx.

Case 3. Otherwise: both the sets

$$\{x\} \cup \{z : zTx \quad \text{and not} \quad zTy\} = X,$$

$$\{y\} \cup \{z : zTy \quad \text{and not} \quad zTx\} = Y$$

are nonempty sets well ordered by T. Let $x' \in X$, $y' \in Y$ be the least elements of these two sets. Then xLy iff $x' < y'$ (as elements of α).

It is easy to check that L linearly orders α. By Lemma 7.3.11, there is a set $Z \subset \alpha$ of power α which is either well ordered or inversely well ordered by L. Suppose Z is well ordered by L. We may assume the order type is α (otherwise take a subset of Z of order type α). Let

$$B = \{x \in \alpha : \text{for some } y \in Z \text{ and all } z \in Z,\ yLz \text{ implies } xTz\},$$

that is, the elements of Z are eventually above x in the tree T. If $x \in B$ and uTx, then obviously $u \in B$. For each $\beta < \alpha$, there is at most one $x \in B$ of order β, because no element is above two elements of order β in T. On the other hand, for each $\beta < \alpha$ there exists $x \in B$ of order β. To see this, note that α is inaccessible by Proposition 7.3.9 and there are $< \alpha$ elements of order $\leqslant \beta$, whence Z contains α elements of order $> \beta$, so there is an x_0 of order β such that Z contains α elements above x_0 in T. Since $\langle Z, L\rangle$ has order type α, the elements of Z above x_0 in T are cofinal in $\langle Z, L\rangle$.

However, if $x_0 Ty$, $x_0 Tz$, and yLu, uLz, then from the definition of L it follows that $x_0 Tu$. Therefore the elements of Z are eventually above x_0 in T, i.e. $x_0 \in B$. It follows that B is a branch of T of order α. ⊣

Let us recall from Section 4.2 that a cardinal α is said to be measurable iff there exists a nonprincipal α-complete ultrafilter over α. Moreover, an ultrafilter D over α is said to be normal iff $\alpha > \omega$, D is α-complete and nonprincipal, and in the ultrapower $\prod_D \langle \alpha, < \rangle$, the αth element is the equivalence class f_D of the identity function f. It was shown in Proposition 4.2.20 that every measurable cardinal $\alpha > \omega$ has a normal ultrafilter over α.

Theorem 7.3.13. *Let α be an uncountable measurable cardinal and let D be a normal ultrafilter over α. If I is a set of power $|I| < \alpha$ and if $S_\omega(\alpha) \subset \bigcup_{i \in I} C_i$, then there exists a set $Y \in D$ and elements $i_1, i_2, i_3, \ldots \in I$ such that*

$$[Y]^n \subset C_{i_n}, \quad n = 1, 2, 3, \ldots.$$

Proof. Since D is countably complete, it suffices to prove that for each positive integer n there is an $i \in I$ and $Y \in D$ such that $[Y]^n \subset C_i$. For then the intersection of these Y's will have the required property. Fix n. Well order I and let f be the function on $[\alpha]^n$ into I such that $f(x)$ is the least $i \in I$ with $x \in C_i$. We extend f to a function on $\bigcup_{m<n}[\alpha]^m$ into I as follows. Assuming that $m < n$ and f is already defined on $[\alpha]^{m+1}$, we define f on $[\alpha]^m$ by the condition

(1) $$f(x) = i \quad \text{iff} \quad \{\beta \in \alpha : \beta \notin x \text{ and } f(x \cup \{\beta\}) = i\} \in D.$$

There is a unique $i \in I$ satisfying (1) because D is an α-complete ultrafilter. Ultimately, we obtain by induction a value $f(0) = i_0$ when $m = 0$.

By transfinite induction define elements $\beta_\xi \in \alpha$ and sets $X_\xi, Y_\xi \subset \alpha$ as follows, for $\xi < \alpha$:

(2) $$X_\xi = \{\beta_\zeta : \zeta < \xi\},$$

(3) $$Y_\xi = \{\gamma \in \alpha : \gamma \geqslant \xi \text{ and for all } m \leqslant n \text{ and } x \in [X_\xi \cup \{\gamma\}]^m,\ f(x) = i_0\},$$

(4) $$\beta_\xi \text{ is the least } \gamma \in Y_\xi.$$

From the definition of f and the α-completeness of D, we see that for all $\xi \in \alpha$,

(5) $$\text{for all } m \leqslant n \text{ and } x \in [X_\xi]^m,\ f(x) = i_0,$$

(6) $$Y_\xi \in D, \text{ whence } \beta_\xi \text{ exists.}$$

Now for each $\gamma \in \alpha$, let $g(\gamma)$ be the least $\xi \in \alpha$ such that $\gamma \notin Y_{\xi+1}$. From (3) we see that $\gamma \notin Y_{\gamma+1}$, whence $g(\gamma) \leqslant \gamma$. Since each $Y_{\xi+1} \in D$, we have for all ξ,

$$\{\gamma \in \alpha : g(\gamma) = \xi\} \notin D.$$

Now we use the fact that D is normal. By Proposition 4.2.19,

$$Y = \{\gamma \in \alpha : g(\gamma) = \gamma\} \in D.$$

We claim that

$$\text{(7)} \qquad \text{for all } \gamma \in Y, \gamma \in Y_\gamma.$$

If $\gamma = \eta+1$, then $g(\gamma) > \eta$, whence $\gamma \in Y_{\eta+1} = Y_\gamma$. If γ is a limit ordinal, then $\gamma \in \bigcap_{\xi<\gamma} Y_\xi$, and it follows from (3) that $\gamma \in Y_\gamma$. We conclude from (7), (3) and (4) that for all $\gamma \in Y$, $\gamma = \beta_\gamma$. Therefore, by (2) and (5), we have $[Y]^n \subset C_{i_0}$. ⊣

Observe the similarity between the above proof and the proof of Ramsey's Theorem 3.3.7.

COROLLARY 7.3.14. *Every uncountable measurable cardinal is a Ramsey cardinal.*

PROOF. Let α be an uncountable measurable cardinal and let D be a normal ultrafilter over α. Let $|I| < \alpha$ and let $S_\omega(\alpha) = \bigcup_{i\in I} C_i$. By Theorem 7.3.13, there is a set $Y \in D$ and elements $i_1, i_2, i_3, \ldots \in I$ such that

$$[Y]^n \subset C_{i_n}, \qquad n = 1, 2, 3, \ldots.$$

Since D is normal, every element of D has power α; in particular, $|Y| = \alpha$. ⊣

Actually, the property of being a measurable cardinal is much stronger than that of being a Ramsey cardinal.

PROPOSITION 7.3.15. *Let α be an uncountable measurable cardinal, and let D be a normal ultrafilter over α. Then the set of all Ramsey cardinals $\beta < \alpha$ belongs to D. Hence α is the αth Ramsey cardinal.*

PROOF. The proof is like Theorem 4.2.23. From the definition of Ramsey cardinals, we see that there is a sentence φ such that for all ordinals β, $\langle R(\beta+1), \in\rangle \models \varphi$ iff β is a Ramsey cardinal. By Corollary 7.3.14, φ holds in $\langle R(\alpha+1), \in\rangle$, and then, by Corollary 4.2.22, $\{\beta \in \alpha : \langle R(\beta+1), \in\rangle \models \varphi\} \in D$. ⊣

The following theorem applies Ramsey cardinals to the problems of Chang and Jónsson described in 7.3.1 and 7.3.2.

THEOREM 7.3.16 (Rowbottom's Theorem). *Let α be a Ramsey cardinal. Then*:
(i). *If $\omega \leqslant \gamma < \beta < \alpha$, then Chang's conjecture holds for (α, β), (α, γ).*
(ii). *There are no Jónsson models of power α.*

PROOF. From Proposition 7.3.4, (ii) follows from (i). We prove only (i).
Let $\mathfrak{A} = \langle A, U, \ldots\rangle$ be a model with $|A| = \alpha$, $|U| = \beta$. We may assume without loss of generality that $\mathfrak{A}$ has built-in Skolem functions. Choose a linear ordering $<$ of A, an element $a \in A \setminus U$, and a subset $W \subset U$ of power γ. Form a simple expansion $\mathfrak{A}^*$ of $\mathfrak{A}$ in the language $\mathscr{L}^*$ by adding a constant c_w for each $w \in W$. Let T be the set of all Skolem terms of the language $\mathscr{L}^*$. We shall let I be the set $I = {}^T(U \cup \{a\})$ of all functions on T into $U \cup \{a\}$. Since T has power γ and U has power β, I has power β^γ. But α is inaccessible by Proposition 7.3.9, so $|I| < \alpha$.
We make a partition $S_\omega(A) = \bigcup_{i \in I} C_i$ as follows: Call a term $t \in T$ an *n-term* iff all the variables occurring in t are among $v_1, \ldots, v_n$ (whence we can write t in the form $t(v_1 \ldots v_n)$). Let $x = \{a_1, \ldots, a_n\} \in S_\omega(A)$, where $a_1 < \ldots < a_n$. Then we put $x \in C_i$, where $i \in I$ is given by:

$i(t) = a$ if t is not an n-term;
$i(t) = a$ if t is an n-term and $t(a_1 \ldots a_n) \in A \setminus U$;
$i(t) = u$ if t is an n-term and $t(a_1 \ldots a_n) = u \in U$.

Since α is a Ramsey cardinal, there exists a set $Y \subset A$ of power α and elements $i_1, i_2, i_3, \ldots \in I$ such that

$$[Y]^n \subset C_{i_n}, \qquad n = 1, 2, 3, \ldots .$$

Let $\mathfrak{B}^*$ be the Skolem hull $\mathfrak{H}(Y)$ in the model $\mathfrak{A}^*$, and let $\mathfrak{B} = \langle B, V, \ldots\rangle$ be the reduct of $\mathfrak{B}^*$ to the language $\mathscr{L}$. Then $\mathfrak{B}$ is an elementary submodel of $\mathfrak{A}$ of power α. Since each element of W is a constant of $\mathfrak{B}^*$ and belongs to U, we have $W \subset V$, whence

$$|W| = \gamma \leqslant |V|.$$

On the other hand, each element $b \in V$ is of the form $b = t(y_1 \ldots y_n)$ for some Skolem term $t \in T$ and some $y_1 < \ldots < y_n$ in Y. Since $\{y_1, \ldots, y_n\} \in [Y]^n \subset C_{i_n}$, we have $i_n(t) = b$. Therefore

$$b \in \bigcup_{n<\omega} \operatorname{range}(i_n) \cap U,$$

hence

$$V \subset \bigcup_{n<\omega} \text{range}(i_n) \cap U,$$

and it follows that

$$|V| \leqslant \omega \cdot |T| = \omega \cdot \gamma = \gamma.$$

Thus $|V| = \gamma$. ⊣

A cardinal α is said to be a *Rowbottom cardinal* iff it has the property:

For all $\beta < \alpha$, Chang's conjecture holds for (α, β), (α, ω).

Thus the above theorem shows that all Ramsey cardinals are Rowbottom cardinals. Exercise 7.3.7 shows that all Rowbottom cardinals are either weakly inaccessible or have cofinality ω. Prikry has obtained the following consistency results:

7.3.17a. *If*

ZFC+'there exists a measurable cardinal $> \omega$'

is consistent, then so are

ZFC+GCH+'there exists a Rowbottom cardinal $\alpha > \omega$ of cofinality ω',

and

ZFC+'there exists a Rowbottom cardinal α of cofinality ω such that $\omega < \alpha < 2^{\omega}$'.

Analogous consistency results for weakly inaccessible cardinals were obtained by Solovay:

7.3.17b. *If*

ZFC+'there is a measurable cardinal $> \omega$'

is consistent, then so are

ZFC+'2^{ω} is a Rowbottom cardinal',

and

ZFC+'there exists a weakly inaccessible Rowbottom cardinal α with $\omega < \alpha < 2^{\omega}$'.

We shall see in Section 7.4 that the axiom of constructibility implies that there are no Rowbottom cardinals $> \omega$. Kunen has shown that:

7.3.17c. *If*

ZFC+'there is a measurable cardinal $> \omega$'

is consistent, then so is

ZFC+'there is a measurable cardinal $> \omega$'
+'for all $\alpha > \omega$, α is Rowbottom iff α is Ramsey, and also iff there is no Jónsson model of power α'.

Kleinburg (1972) has announced the following result:

ZFC+'there exists a Rowbottom cardinal'

is consistent if and only if

ZFC+'there is a cardinal α in which there is no Jónsson model'

is consistent.

It is not known whether or not the following are consistent:

ZFC+'$\aleph_\omega$ is a Rowbottom cardinal',

ZFC+'the first inaccessible cardinal exists and is Rowbottom',

ZFC+'every cardinal which is of cofinality ω or weakly inaccessible is Rowbottom'.

Apter (1983) has obtained a partial solution of the first of these problems, showing that if ZFC + 'there are infinitely many measurable cardinals' is consistent, so is ZF + [ω_n-dependent choice] + [ω_ω is a Rowbottom cardinal].

We next apply Ramsey cardinals to problem 7.3.3. We shall here use the notion of a set of indiscernibles which was introduced in Section 3.3 and used heavily in the preceding sections.

THEOREM 7.3.18 (Silver's Theorem). *Let α be a Ramsey cardinal, and let $\mathscr{L}$ be countable.*

(i). *If β is a cardinal with $\omega < \beta \leqslant \alpha$, then every well ordered model of type α has an elementary submodel which is well ordered of type β.*

(ii). *Let $\mathfrak{A}$ be a well ordered model of type α. Then there is an elementary chain*

$$\mathfrak{A}_\xi = \langle A_\xi, <, \ldots\rangle, \ \xi \text{ an ordinal} > 0,$$

with the following properties:

(a) $\mathfrak{A}_\alpha \prec \mathfrak{A}$ *(note that $\alpha = \omega_\alpha$);*
(b) *each $\mathfrak{A}_\xi$ is a well ordered model of type ω_ξ;*
(c) *whenever $\xi < \eta$, A_ξ is an initial segment of $\langle A_\eta, <\rangle$;*
(d) *if u_ξ is defined as the ω_ξth element with respect to $<$, then for each η the set $\{u_\xi : \xi < \eta\}$ is indiscernible in $\mathfrak{A}_\eta$.*

Note that (ii) obviously implies (i).

Before proving the theorem we shall prove two lemmas. At the outset, observe that if $\mathfrak{A}^*$ is a Skolem expansion of $\mathfrak{A}$, then $\mathfrak{A}$ is well ordered of type β iff $\mathfrak{A}^*$ is. Therefore, when we prove Theorem 7.3.18, we may assume without loss of generality that $\mathfrak{A}$ has built-in Skolem functions.

LEMMA 7.3.19. *Let α be an infinite cardinal, let $\mathfrak{A} = \langle A, <, ...\rangle$ be a well ordered model of type α with built-in Skolem functions, and suppose that $\langle X, <\rangle$ is a set of indiscernibles in $\mathfrak{A}$ of power α. Then every increasing sequence $x_1 < x_2 < ...$ of elements of X satisfies the following formulas in $\mathfrak{A}$:*

(i). $v_i < v_j$, *where* $i < j < \omega$;

(ii). $t(v_1 ... v_m) < v_n$, *where t is a term and* $m < n < \omega$;

(iii). $t(v_1 ... v_m v_{j_1} ... v_{j_n}) \leqslant v_m \rightarrow t(v_1 ... v_m v_{j_1} ... v_{j_n}) \equiv t(v_1 ... v_m v_{k_1} ... v_{k_n})$, *where t is a term and*

$$m < j_1 < ... < j_n < \omega, \qquad m < k_1 < ... < k_n < \omega.$$

PROOF. Increasing sequences from X obviously satisfy (i).

For (ii), note that $\langle A, <\rangle$ has type α and $|X| = \alpha$, so X is cofinal in $\langle A, <\rangle$. Therefore, given $x_1 < ... < x_m$ in X, the element $t(x_1 ... x_m)$ must be less than some $y \in X$, with $x_m < y$. But then, by the indiscernibility of X, we always have $t(x_1 ... x_m) < x_n$ when $x_m < x_n \in X$. Thus (ii) holds.

The proof of (iii) depends on the fact that each element of $\langle A, <\rangle$ has fewer than α predecessors. Let $x_1 < ... < x_m < y_1 < ... y_n$ and $x_m < z_1 < ... < z_n$ in $\langle X, <\rangle$, and suppose $t(x_1 ... x_m y_1 ... y_n) \leqslant x_m$, but

$$(1) \qquad t(x_1 ... x_m y_1 ... y_n) \neq t(x_1 ... x_m z_1 ... z_n).$$

Choose $w_1 < ... < w_n$ in X such that $y_n < w_1$, $z_n < w_1$. Since we cannot have

$$t(x_1 ... x_m y_1 ... y_n) = t(x_1 ... x_m w_1 ... w_n),$$

and

$$t(x_1 ... x_m z_1 ... z_n) = t(x_1 ... x_m w_1 ... w_n),$$

in view of (1), we must have

$$(2) \qquad \begin{aligned} t(x_1 ... x_m y_1 ... y_n) &\neq t(x_1 ... x_m w_1 ... w_n), \\ t(x_1 ... x_m z_1 ... z_n) &\neq t(x_1 ... x_m w_1 ... w_n). \end{aligned}$$

Moreover, by indiscernibility,

$$(3) \qquad t(x_1 ... x_m w_1 ... w_n) \leqslant x_m.$$

Now choose α different n-tuples

$$w_1^\beta < \dots < w_n^\beta, \qquad \beta < \alpha,$$

in X, such that $\gamma < \beta < \alpha$ implies

$$x_1 < \dots < x_m < w_1^\gamma < \dots < w_n^\gamma < w_1^\beta < \dots < w_n^\beta.$$

Then by (2), (3), and indiscernibility, the elements $t(x_1 \dots x_m w_1^\beta \dots w_n^\beta)$, $\beta < \alpha$, are all different and all less than x_m. This is impossible since x_m has fewer than α predecessors. We conclude that (iii) holds for increasing sequences from X. ⊣

LEMMA 7.3.20. *Suppose* $\mathfrak{B} = \langle B, <, \dots\rangle$ *is a well-ordered model for a countable language,* $\mathfrak{B}$ *has built-in Skolem functions,* $\langle X, <\rangle$ *is an uncountable set of indiscernibles in* $\mathfrak{B}$, *and every increasing sequence of elements of* X *satisfies the formulas* (i)–(iii) *of Lemma* 7.3.19 *in* $\mathfrak{B}$. *Then there is an elementary chain*

$$\mathfrak{A}_\xi = \langle A_\xi, <, \dots\rangle, \ \xi \text{ an ordinal} > 0,$$

of models $\mathfrak{A}_\xi \equiv \mathfrak{B}$ *such that conditions* (b)–(d) *of Theorem* 7.3.18 *hold and, moreover,*

(e) *each* $\mathfrak{A}_\xi$ *is generated by a set* X_ξ *of indiscernibles such that increasing sequences of* X_ξ *and* X *satisfy the same formulas.*

PROOF. For each $\xi > 0$, let $\langle X_\xi, <\rangle = \langle \omega_\xi, <\rangle$. By the stretching theorem 3.3.11(b), there is a model $\mathfrak{A}_\xi$ which is generated by the set X_ξ of indiscernibles, and the sets of formulas satisfied by increasing sequences from X in $\mathfrak{B}$ and from X_ξ in $\mathfrak{A}_\xi$ are the same. By the elementary embedding theorem 3.3.11(d), whenever $\xi < \eta$, there is a unique elementary embedding of $\mathfrak{A}_\xi$ into $\mathfrak{A}_\eta$, whose restriction to X_ξ is the identity mapping. We may therefore identify the elements of $\mathfrak{A}_\xi$ with their images in $\mathfrak{A}_\eta$, and we obtain an elementary chain $\mathfrak{A}_\xi$, $\xi > 0$. By our construction, the condition (e) holds.

Since each φ true in $\mathfrak{B}$ is satisfied by increasing sequences of X in $\mathfrak{B}$, and hence, by increasing sequences of X_ξ in $\mathfrak{A}_\xi$, φ is true in $\mathfrak{A}_\xi$. Thus $\mathfrak{A}_\xi \equiv \mathfrak{B}$. In particular, the relation $<$ simply orders each $\mathfrak{A}_\xi$. Moreover, each countable subset of $\mathfrak{A}_\xi$ is included in the Skolem hull $\mathfrak{H}(Z)$ of a countable subset $Z \subset X_\xi$. But X is uncountable, and therefore there is a one–one ordermorphic mapping f of $\langle Z, <\rangle$ into $\langle X, <\rangle$ (recall that both $\langle X, <\rangle$ and $\langle Z, <\rangle$ are well ordered). By the elementary embedding theorem 3.3.11(d), f can be extended to an elementary embedding of $\mathfrak{H}(Z)$ into $\mathfrak{B}$.

Since $\mathfrak{B}$ is well ordered, $\mathfrak{H}(Z)$ is well ordered. Thus every countable subset of $\mathfrak{A}_\xi$ is well ordered; hence $\mathfrak{A}_\xi$ contains no infinite decreasing sequence, and $\mathfrak{A}_\xi$ is well ordered.

The order type of each $\langle A_\xi, <\rangle$ is at least ω_ξ because $X_\xi \subset A_\xi$ and $|X_\xi| = \omega_\xi$. All formulas of the form (ii) in Lemma 7.3.19 are satisfied by increasing sequences in X_ξ, and X_ξ generates $\mathfrak{A}_\xi$. This guarantees that X_ξ is a cofinal subset of $\langle A_\xi, <\rangle$. The formulas (iii) of Lemma 7.3.19 guarantee that, for each $x \in X_\xi$, every element $y \in A_\xi$ which precedes x in the ordering $<$ is determined uniquely by a term $t(v_{i_1} \dots v_{i_m} v_{j_1} \dots v_{j_n})$ and an m-tuple $x_1 < \dots < x_m$ of elements of X_ξ, with $x_m = x$. (The variable v_{i_m} does not necessarily occur in t.) We have assumed that $\mathscr{L}$ is a countable language, so there are only ω terms. Since $\langle X_\xi, <\rangle = \langle \omega_\xi, <\rangle$, the element x has fewer than ω_ξ predecessors in the ordering $\langle X_\xi, <\rangle$. Therefore there are fewer than ω_ξ finite sequences $x_1 < \dots < x_m = x$ in X_ξ. Also, $\xi > 0$, so $\omega < \omega_\xi$. It follows that the element x has fewer than ω_ξ predecessors in the ordering $\langle A_\xi, <\rangle$. We have shown that X_ξ is cofinal in $\langle A_\xi, <\rangle$, and it follows that every element of A_ξ has fewer than ω_ξ predecessors. Therefore $\langle A_\xi, <\rangle$ has order type exactly ω_ξ, whence $\mathfrak{A}_\xi$ is a well ordered model of type ω_ξ. This verifies condition (b).

Now let $0 < \xi < \eta$. We prove (c). Let $a \in A_\eta$, $b \in A_\xi$, $a < b$. We must show that $a \in A_\xi$. The element a may be written

$$a = t(x_1 \dots x_{m-1} y_1 \dots y_n),$$

where $x_1 < \dots < x_{m-1} < y_1 < \dots < y_n$, the x's are in X_ξ, and the y's are in $X_\eta \setminus X_\xi$. For some $x_m \in X_\xi$, we have $x_{m-1} < x_m$ and $b < x_m$. Adding an inessential variable v_m, we may write

$$a = t(x_1 \dots x_m y_1 \dots y_n),$$

and we also have

$$t(x_1 \dots x_m y_1 \dots y_n) < x_m.$$

We may choose $w_1, \dots, w_n \in X_\xi$ such that

$$x_1 < \dots < x_m < w_1 < \dots < w_n.$$

We also have $x_m < y_1$ because $x_m \in X_\xi$ and $y_1 \in X_\eta - X_\xi$. Using Lemma 7.3.19 (iii), we have

$$a = t(x_1 \dots x_m y_1 \dots y_n) = t(x_1 \dots x_m w_1 \dots w_n).$$

This shows that $a \in A_\xi$, and (c) holds.

It remains to prove (d). We let u_ξ be the ω_ξth element in the ordering $<$. Let $0 < \xi < \eta$. We have already established that A_ξ is a proper initial segment of A_η of order type ω_ξ, and it follows that u_ξ is the least element of $A_\eta \setminus A_\xi$ in the ordering $<$. Moreover, if $\xi < \eta < \zeta$, then the ω_ξth elements of $\mathfrak{A}_\eta$ and $\mathfrak{A}_\zeta$ are the same, as u_ξ is defined independently of η. Our plan is to show that all the u_ξ 'come from the same term'.

Consider first the case $\xi = 1 < \eta$. Recall from our construction that $\langle X, < \rangle = \langle \omega_1, < \rangle$, $\langle X_\eta, < \rangle = \langle \omega_\eta, < \rangle$. Therefore ω_1 is the first element of X_η which does not belong to $\mathfrak{A}_1$. For all $x \in X_1$, we have

$$x < u_1 \quad \text{and} \quad u_1 \leqslant \omega_1.$$

The element u_1 is expressible as a term in $\mathfrak{A}_\eta$:

$$u_1 = t(x_1 \ldots x_m \omega_1 y_1 \ldots y_n)$$

where the x's are in X_1, the y's are in $X_\eta \setminus X_1$, and

$$x_1 < \ldots < x_m < \omega_1 < y_1 < \ldots < y_n.$$

It follows from Lemma 7.3.19 (iii) that

$$u_1 = t(x_1 \ldots x_m \omega_1 \omega_1 + 1 \ldots \omega_1 + n).$$

Thus for all $x \in X_1$,

$$x < t(x_1 \ldots x_m \omega_1 \omega_1 + 1 \ldots \omega_1 + n) \leqslant \omega_1. \tag{1}$$

Now consider an arbitrary ξ, $1 \leqslant \xi < \eta$. As before, ω_ξ is the first element of X_η which is not in A_ξ, and for all $x \in X_\xi$, we have

$$x < u_\xi \quad \text{and} \quad u_\xi \leqslant \omega_\xi.$$

Express u_ξ as a term in $\mathfrak{A}_\eta$,

$$u_\xi = s(w_1 \ldots w_p \omega_\xi z_1 \ldots z_q)$$

where the w's are in X_ξ, the z's are in $X_\eta - X_\xi$, and

$$w_1 < \ldots < w_p < \omega_\xi < z_1 < \ldots < z_q.$$

Again, using Lemma 7.3.19 (iii),

$$u_\xi = s(w_1 \ldots w_p \omega_\xi \omega_\xi + 1 \ldots \omega_\xi + q).$$

Then whenever $w_p < w \in X_\xi$,

$$w < s(w_1 \ldots w_p \omega_\xi \omega_\xi + 1 \ldots \omega_\xi + q) \leqslant \omega_\xi. \tag{2}$$

Choose $w_1' < \dots < w_p'$ in X_1 such that for all $i \leqslant m$ and $j \leqslant p$,

(3) $$x_i < w_j \text{ iff } x_i < w_j', \qquad x_i > w_j \text{ iff } x_i > w_j'.$$

By indiscernibility and (2), whenever $w_p' < x \in X_1$, we have

$$x < s(w_1' \dots w_p' \omega_1 \omega_1 + 1 \dots \omega_1 + q) \leqslant \omega_1 .$$

Since X_1 is cofinal in $\langle A_1, < \rangle$, it follows that

$$s(w_1' \dots w_p' \omega_1 \omega_1 + 1 \dots \omega_1 + q) \in A_\eta \setminus A_1 ,$$

whence

(4) $$u_1 = t(x_1 \dots x_m \omega_1 \omega_1 + 1 \dots \omega_1 + n) \leqslant s(w_1' \dots w_p' \omega_1 \omega_1 + 1 \dots \omega_1 + q).$$

Using indiscernibility and (3), (4),

(5) $$t(x_1 \dots x_m \omega_\xi \omega_\xi + 1 \dots \omega_\xi + n) \leqslant s(w_1 \dots w_p \omega_\xi \omega_\xi + 1 \dots \omega_\xi + q) = u_\xi .$$

Moreover, using indiscernibility and (1), we see that for all $w \in X_\xi$,

(6) $$w < t(x_1 \dots x_m \omega_\xi \omega_\xi + 1 \dots \omega_\xi + n).$$

Since X_ξ is cofinal in $\langle A_\xi, < \rangle$, (5) and (6) imply that

(7) $$u_\xi = t(x_1 \dots x_m \omega_\xi \omega_\xi + 1 \dots \omega_\xi + n).$$

Note that whenever $1 \leqslant \xi < \zeta < \eta$,

(8) $$x_1 < \dots < x_m < \omega_\xi < \dots < \omega_{\xi+n} < \omega_\zeta < \dots < \omega_{\zeta+n} .$$

From (7), (8), and the indiscernibility of $\langle X_\eta, < \rangle$ in $\mathfrak{A}_\eta$, it follows that the set $\langle \{u_\xi : \xi < \eta\}, < \rangle$ is indiscernible in $\mathfrak{A}_\eta$, and this is the required condition (d). ⊣

Proof of Theorem 7.3.18. As we observed already, we need only prove (ii) and we may assume that $\mathfrak{A}$ has built-in Skolem functions. By Proposition 7.3.10, $\mathfrak{A}$ has a set $\langle X, < \rangle$ of indiscernibles of power α. Thus the hypotheses of Lemma 7.3.19 hold. X is uncountable, so by Lemma 7.3.19, the hypotheses of the next lemma, 7.3.20, are satisfied by $\mathfrak{A} = \mathfrak{B}$. Then by Lemma 7.3.20, there is an elementary chain

$$\mathfrak{A}_\xi, \ \xi \text{ is an ordinal} > 0,$$

which satisfies conditions (b)–(d) of the Theorem and condition (e) of Lemma 7.3.20. Finally, by (e) the model $\mathfrak{A}_\alpha$ is generated by a set $\langle X_\alpha, < \rangle$ of indiscernibles whose increasing sequences satisfy the same formulas as the increasing sequences from X. Furthermore, α is inaccessible, so $\omega_\alpha = \alpha$ and $\mathfrak{A}_\alpha$ has order type $\omega_\alpha = \alpha$, whence $\langle X_\alpha, < \rangle$ has order type $\leqslant \alpha$.

Therefore $\langle X_\alpha, <\rangle$ can be ordermorphically embedded one–one into $\langle X, <\rangle$. Then by the elementary embedding theorem, $\mathfrak{A}_\alpha$ is elementarily embeddable in $\mathfrak{A}$, so (a) can also be satisfied. ⊣

Exercises

7.3.1. Investigate what happens to Chang's conjecture in the 'trivial' cases where $\alpha > \beta > \delta$ and $\alpha \geqslant \gamma > \delta$ are not both true.

7.3.2. If Chang's conjecture holds for (α, β), (γ, δ) and $\alpha' \geqslant \alpha \geqslant \beta$, then Chang's conjecture holds for (α', β), (γ, δ).

7.3.3. If Chang's conjecture holds for (α, β), (γ, δ) and also for (γ, δ), (μ, ν), then Chang's conjecture holds for (α, β), (μ, ν).

7.3.4. Let us say that Chang's conjecture holds for (α, β), $(\gamma, < \delta)$ iff every model of type (α, β) for a countable language has an elementary submodel of some type (γ, μ), with $\mu < \delta$. Prove that if every well ordered model of type α for a countable language has an elementary submodel which is well ordered of type γ, then for all $\beta < \alpha$, Chang's conjecture holds for (α, β), $(\gamma, < \gamma)$.

7.3.5. If Chang's conjecture holds for (α, β), $(\gamma, < \delta)$ and if $\alpha \geqslant \beta \geqslant \beta'$, then Chang's conjecture holds for (α, β'), $(\gamma, < \delta)$. If Chang's conjecture holds for (α, β), (γ, ω) and $\alpha \geqslant \beta \geqslant \beta'$, then it holds for (α, β'), (γ, ω).

7.3.6. If $\delta < \mathrm{cf}(\alpha) \leqslant \beta \leqslant \alpha$, then Chang's conjecture fails for (α, β), (α, δ).

7.3.7. If α is a Rowbottom cardinal, then either α is weakly inaccessible or $\mathrm{cf}(\alpha) = \omega$.

7.3.8. Prove the assertion 7.3.5.

7.3.9*. Assume $\alpha > \beta \geqslant \delta$ and $\alpha \geqslant \gamma \geqslant \delta$. Then Chang's conjecture for (α, β), $(\gamma, < \delta)$ is equivalent to the following partition property:

For every set X of power $|X| = \alpha$ and I of power $|I| = \beta$, and every partition $S_\omega(X) \subset \bigcup_{i\in I} C_i$, there exist subsets $Y \subset X$ of power $|Y| = \gamma$ and $J \subset I$ of power $|J| < \delta$ such that $S_\omega(Y) \subset \bigcup_{j\in J} C_j$.

7.3.10*. Let $\alpha > \beta \geqslant \delta$, $\alpha > \gamma \geqslant \delta$. Assume that for all $\mu < \delta$, $\beta^\mu = \beta$. If Chang's conjecture holds for (α, β), $(\gamma, < \delta)$, then it still holds for languages with fewer than δ symbols. If $\mu' \leqslant \mu < \gamma$ and $\beta = \beta^\mu$ and Chang's conjecture holds for (α, β), (γ, μ'), then it holds for (α, β), (γ, μ).

[*Hint*: Use Exercise 7.3.9.]

7.3.11. $\mathfrak{A}$ is said to be a ***Jónsson algebra*** of power α iff $|A| = \alpha$ and $\mathfrak{A}$ has no proper submodel of power α (and $\mathfrak{A}$ is a model for a countable language having only function symbols). Every Jónsson algebra is obviously a Jónsson model. Prove that if there exists a Jónsson model of power α, then there exists a Jónsson algebra of power α.

7.3.12*. Prove that if there exists a Jónsson algebra of power α, then there exists one which has a single binary commutative function.

7.3.13*. Prove that for every cardinal α there is a Jónsson algebra $\langle A, F\rangle$ of power α, where F is an infinitary function with ω places.

7.3.14. Let $\beta < \alpha$. Prove that there exists a Jónsson model of power α with β relations iff there exists a model $\mathfrak{A}$ of power α with countably many relations and a set $X \subset A$ of power β such that $\mathfrak{A}$ has no proper elementary submodel of power α which includes X.

7.3.15*. Prove without the GCH that for each $n < \omega$, there is a Jónsson model of power $\aleph_n$. More generally, for all cardinals α, if there is a Jónsson model of power α, then there is a Jónsson model of power α^+.

7.3.16. Prove that α is a Ramsey cardinal iff $\alpha > \omega$ and every model $\mathfrak{A}$, for a language $\mathscr{L}$ of power $||\mathscr{L}|| < \alpha$, has a set of indiscernibles of power α. (Cf. Proposition 7.3.10.)

7.3.17. The 'Erdös notation'. Let α, β, γ be cardinals and n be a natural number. The notation

$$\alpha \to (\beta)^n_\gamma$$

is defined to mean that:

For every partition $[\alpha]^n = \bigcup_{i\in\gamma} C_i$ of $[\alpha]^n$ into γ parts, there exists $X \subset \alpha$ of power $|X| = \beta$ and $i \in \gamma$ such that $[X]^n \subset C_i$.

Similarly,

$$\alpha \to (\beta)^{<\omega}_\gamma$$

means that:

For every partition $S_\omega(\alpha) = \bigcup_{i\in\gamma} C_i$ of $S_\omega(\alpha)$ into γ parts, there exist $X \subset \alpha$ of power α and elements $i_1, i_2, \ldots \in \gamma$ such that $[X]^n \subset C_{i_n}$, $n = 1, 2, \ldots$.

Check the following:

(i). Ramsey's theorem (Theorem 3.3.7) says that for all $n < \omega$, $\alpha \geqslant \omega$,

$$\alpha \to (\omega)^n_2.$$

(ii). Erdös and Rado's theorem 7.2.1 says that for all $n < \omega$ and $\alpha \geqslant \omega$,

$$\beth_n(\alpha)^+ \rightarrow (\alpha^+)^{n+1}_{\alpha}.$$

(iii). α is a Ramsey cardinal iff for all $\beta < \alpha$,

$$\alpha \rightarrow (\alpha)^{<\omega}_{\beta}.$$

(iv). The relations $\alpha \rightarrow (\beta)^n_\gamma$ and $\alpha \rightarrow (\beta)^{<\omega}_\gamma$ remain true when the cardinal α is made larger and when the cardinals on the right of the arrow are made smaller.

7.3.18. Show that if $\alpha \rightarrow (\gamma)^{<\omega}_{\beta}$, then Chang's conjecture holds for (α, β), (γ, ω).

7.3.19*. Show that α is a Ramsey cardinal iff

$$\alpha \rightarrow (\alpha)^{<\omega}_2.$$

[*Hint*: Use Proposition 7.3.10.]

7.3.20. $\alpha \rightarrow (\beta)^{<\omega}_2$ if and only if every model of power α for a countable language has a set of indiscernibles of power β.

7.3.21*. If $\alpha \rightarrow (\alpha)^2_2$, then α is weakly compact.

[*Hint*: Show that $\alpha \rightarrow (\alpha)^2_2$ implies the condition in Lemma 7.3.11 and then use the proof of Proposition 7.3.12. Also it must be shown that $\alpha \rightarrow (\alpha)^2_2$ implies that α is inaccessible.]

7.3.22*. If α is weakly compact, then for all $\beta < \alpha$ and $n < \omega$, $\alpha \rightarrow (\alpha)^n_\beta$.

[*Hint*: Use an argument like the proof of Ramsey's theorem and Theorem 7.3.13, but where D is an α-complete ultrafilter in an α-complete field of sets of power α.]

7.3.23*. For each infinite cardinal β, let $\kappa(\beta)$ be the least cardinal α such that

$$\alpha \rightarrow (\beta)^{<\omega}_2$$

(if such an α exists). Thus α is Ramsey iff $\alpha = \kappa(\alpha)$. Prove that if $\kappa(\beta)$ exists, then

$$\text{for all} \quad \gamma < \kappa(\beta), \kappa(\beta) \rightarrow (\beta)^{<\omega}_\gamma.$$

7.3.24. If $\kappa(\beta)$ exists, then it is inaccessible.

[*Hint*: Use above exercise.]

($\kappa(\beta)$ is called the *Erdös cardinal* for β).

7.3.25*. If $\beta < \alpha$ and $\kappa(\alpha)$ exists, then $\kappa(\beta) < \kappa(\alpha)$ (α, β are infinite cardinals).

7.3.26**. If $\kappa(\omega)$ exists, there is a cardinal $\omega < \beta < \kappa(\omega)$ such that β is $\mathbf{\Pi}_n^m$-indescribable for all $m, n < \omega$.

[*Hint*: Let α be a limit ordinal $> \kappa(\omega)$, and consider a model $\langle R(\alpha), \in, R, \ldots\rangle$ with built-in Skolem functions. This model has a set $\langle X, <\rangle$ of indiscernibles of order type ω, with $X \subset \kappa(\omega)$. The mapping $x_n \to x_{n+1}$ determines an elementary embedding $f : \mathfrak{H}(X) \prec \mathfrak{H}(X)$. Take for β the first ordinal moved by f.]

This exercise shows that there are weakly compact cardinals $\omega < \beta < \kappa(\omega)$; hence the first uncountable weakly compact cardinal is not Ramsey.

7.3.27*. If $\alpha < \kappa(\alpha)$, then $\kappa(\alpha)$ is not weakly compact.

[*Hint*: Show that $\kappa(\alpha)$ is $\mathbf{\Pi}_1^1$-describable, using the equation $\kappa(\alpha)^\alpha = \kappa(\alpha)$ from Exercise 7.3.24.]

7.3.28*. Let D be an α-complete nonprincipal ultrafilter over α. Then D has the partition property in Theorem 7.3.13 iff there is a permutation σ of α such that $\sigma(D) = \{\sigma(X) : X \in D\}$ is a normal ultrafilter over α.

7.3.29. State and prove a version of Rowbottom's theorem for models of languages $\mathscr{L}$ of power $||\mathscr{L}|| < \alpha$.

7.3.30*. Suppose that α is a limit of measurable cardinals and $\mathrm{cf}(\alpha) \leqslant \gamma \leqslant \beta < \alpha$. Then Chang's conjecture holds for (α, β), (α, γ). Thus every limit α of measurable cardinals such that $\mathrm{cf}(\alpha) = \omega$ is a Rowbottom cardinal.

7.3.31. State and prove a generalization of Silver's theorem for languages of power less than α.

7.3.32. A model $\mathfrak{A} = \langle A, U, <, \ldots\rangle$ is said to be well ordered on U of type α iff $<$ is a well ordering of the set U of order type α. Generalize Silver's theorem for well ordered models on U.

7.3.33**. Show that Silver's theorem is still true when the hypothesis 'α is a Ramsey cardinal' is weakened to 'for some $\gamma \geqslant \omega_1$, $\alpha = \kappa(\gamma)$'.

[*Hint*: Verify the hypothesis of Lemma 7.3.20.]

7.3.34*. Suppose α is a Ramsey cardinal and $\mathfrak{A}$ is a well ordered model of type α for a language of power $\leqslant \alpha$. Then for every cardinal $\beta > \alpha$, $\mathfrak{A}$ has an elementary extension which is well ordered of type β. In fact, there is an elementary chain $\mathfrak{A}_\beta$, β a cardinal $\geqslant \alpha$, such that $\mathfrak{A}_\alpha = \mathfrak{A}$ and conditions (b)–(d) of Silver's theorem hold.

[*Hint*: Use the fact that α is weakly compact.]

7.3.35*. A model $\mathfrak{A} = \langle A, <, \ldots\rangle$ is said to be ω_1-*well ordered* iff $<$ is a simple ordering of A and there is no decreasing sequence of length ω_1. If T is a theory in a countable language and T has an ω_1-well ordered model of power $\beth_\omega$, then T has ω_1-well ordered models of all infinite powers.
[*Hint*: Use the Erdös–Rado theorem.]

7.3.36. Let h_{wo} be the Hanf number for well ordered models, i.e., the least cardinal $\alpha \geqslant \omega$ such that for every theory T in a countable language, if T has a well ordered model of power α, then T has well ordered models of all infinite powers. Prove that h_{wo} exists.

7.3.37*. Prove that if $\kappa(\omega_1)$ exists, then $h_{wo} < \kappa(\omega_1)$.

7.3.38*. Show that $\omega_1 \leqslant \mathrm{cf}\,(h_{wo}) \leqslant 2^\omega$.
[*Hint*: Prove that if a theory T has well ordered models of every power $\beta < \alpha$, but does not have well ordered models of every infinite power, then there exists another theory T' which has a well ordered model of power α but does not have well ordered models of all infinite powers.]

7.3.39*. Show that $h_{wo} = \beth_{h_{wo}}$. Thus $\omega_1 < h_{wo}$, $\beth_{\omega_1} < h_{wo}$, $\beth_{\aleph_{\omega_1}} < h_{wo}$, etc.

7.3.40*. If the first inaccessible cardinal $\alpha > \omega$ exists, then $\alpha < h_{wo}$.

7.3.41*. If $\kappa(\omega)$ exists, then $\kappa(\omega) < h_{wo}$.

7.3.42. $\mathfrak{A} = \langle A, <, \ldots\rangle$ is said to be an α-*like* model iff $<$ simply orders A, A has power α, and every element of A has fewer than α predecessors. (Thus a well ordered model is α-like iff it is of type α.) Give examples of the following:

(i). A theory which has an α-like model for all $\alpha > \omega$, but not for $\alpha = \omega$.
(ii). A theory which has α-like models iff α is a successor cardinal.
(iii). A theory which has α-like models iff α is singular.

For the following exercises assume that $\mathscr{L}$ is countable.

7.3.43 (GCH). If T has an α^+-like model, then for all regular β, T has a β^+-like model. The GCH is not needed when $\beta^+ = \omega_1$.
[*Hint*: Use the two-cardinal theorems.]

7.3.44*. If T has an ω-like model, then for all α, T has an α-like model.
[*Hint*: Use indiscernible elements.]

7.3.45* (GCH). If α is inaccessible and T has α-like models, then for all regular β, T has β^+-like models.
[*Hint*: Without using the GCH, show that T has an ω_1-like model, and then use Exercise 7.3.43.]

7.3.46**. Let α be a strong limit cardinal. If T has an α-like model, then for all singular cardinals β, T has a β-like model.

[*Hint*: Use 'doubly-indexed' indiscernibles.]

7.3.47. In Exercise 4.2.12, we defined α to be a *Mahlo number* iff every closed unbounded subset of α contains an inaccessible cardinal. Mahlo numbers are also called ρ_0 *cardinals*. We now define α to be a ρ_{n+1} *cardinal* iff every closed unbounded subset of α contains a ρ_n cardinal. Show that if α is a ρ_{n+1} cardinal, then α is the αth ρ_n cardinal. Also, if $\alpha > \omega$ and α is weakly compact, then α is a ρ_n cardinal.

7.3.48**. Suppose that for each $n < \omega$, there is a ρ_n cardinal α_n such that T has an α_n-like model. Then for every cardinal $\beta > \omega$, T has a β-like model.

7.3.49*. For each $n < \omega$, there is a sentence σ_n such that for all α, σ_n has an α-like model iff α is not a ρ_n cardinal.

7.3.50**. A *semigroup* $\langle G, \cdot\rangle$ is a set G together with an associative binary operation $\cdot$. A semigroup is a group if $\cdot$ is both left and right solvable, i.e.,

$$\text{for all } x, y \in G, \qquad (\exists z)(x \cdot z = y) \quad \text{and} \quad (\exists z)(z \cdot x = y).$$

Let α be an infinite cardinal. Prove that if a semigroup $\langle G, \cdot\rangle$ of power α is a Jónsson semigroup, i.e., it has no proper sub-semigroups of the same power, then either $\langle G, \cdot\rangle$ is a group or else

$$\mathrm{cf}(\alpha) = \omega \quad \text{and} \quad (\exists\beta)(\beta < \alpha \wedge \alpha < 2^\beta).$$

Hence, if $\mathrm{cf}(\alpha) > \omega$ or if $(\beta < \alpha \rightarrow 2^\beta \leqslant \alpha)$, then every Jónsson semigroup of power α is a Jónsson group.

7.4. Large cardinals and the constructible universe

In Section 4.2, Scott's theorem states that the existence of a measurable cardinal $> \omega$ implies that the axiom of constructibility is false. We were able to give a proof there with a minimum knowledge of constructible sets. Using the results of the preceding section, Scott's theorem can be vastly improved. From the weaker hypothesis of the existence of a Ramsey cardinal, it will be shown that the axiom of constructibility becomes false at the earliest possible stage, i.e., there are nonconstructible sets of natural numbers (Theorem 7.4.7). The even weaker hypothesis that at least one 'nontrivial' case of one of the three questions 7.3.1–7.3.3 has an affirmative answer, still implies that the axiom of constructibility fails (Theorem 7.4.10). It also implies that there are nonconstructible subsets of ω (cf. Theorem 7.4.11), but we shall not give the (long) proof here.

This section will depend heavily on the basic properties of constructible sets. They are available in several books, for example, Cohen (1966), Gödel (1940), or Shoenfield (1967). We shall not attempt here to give a complete introduction to constructible sets. Instead we give only a brief outline which will serve to fix notation and point out what is needed for the theorems that follow. We advise the reader to learn the basic properties of constructible sets elsewhere before attempting to read this material.

In Section 4.2 we gave the following 'model-theoretic' statement of the axiom of constructibility:

For every regular cardinal $\alpha > \omega$, there is no proper subset $M \subset H(\alpha)$ such that $\alpha \subset M$ and $\langle M, \in \rangle$ is a model of ZF$-$P.

This definition is not convenient for our present purposes, and we shall have to turn to the usual statement of the axiom of constructibility which is familiar from the literature.

Given a set M of sets, a subset $X \subset M$ is said to be *definable in M* (with parameters) iff there exists a formula $\varphi(uv_1 \dots v_n)$ of the language $\mathscr{L} = \{\in\}$ and elements $y_1, \dots, y_n \in M$ such that

$$X = \{x \in M : \langle M, \in \rangle \models \varphi[xy_1 \dots y_n]\}.$$

Note that each formula $\varphi(uv_1 \dots v_n)$ and n-tuple $y_1, \dots, y_n \in M$ defines a unique subset $X \subset M$. We now define by induction the chain of sets $L(\alpha)$, α an ordinal.

7.4.1. If $\alpha = 0$, then $L(\alpha) = 0$. If $\alpha > 0$ is a limit ordinal, then

$$L(\alpha) = \bigcup_{\beta < \alpha} L(\beta).$$

If $\alpha = \beta + 1$, then

$$L(\alpha) = \{X \subset L(\beta) : X \text{ is definable in } L(\beta)\}.$$

It follows that in all cases $L(\alpha) = \bigcup_{\beta < \alpha} L(\beta+1)$.

A set x is said to be *constructible* iff there exists an α such that $x \in L(\alpha)$. The class of all constructible sets is called the *constructible universe* and is denoted by L. Thus

$$L = \bigcup_{\alpha} L(\alpha).$$

PROPOSITION 7.4.2. *The axiom of constructibility holds iff every set is constructible.*

The usual statement of the axiom of constructibility is that every set is constructible, and the above proposition states that this usual statement is equivalent to the one given in Section 4.2.

PROPOSITION 7.4.3.

(i). *If* $\alpha < \beta$, *then* $L(\alpha) \subset L(\beta)$ *and* $L(\alpha) \in L(\beta)$.

(ii). $\alpha \subset L(\alpha) \subset R(\alpha)$ *and* $\alpha \in L(\alpha+1) \setminus L(\alpha)$.

(iii). $L(\alpha)$ *is a transitive set, i.e.* $x \in y \in L(\alpha)$ *implies* $x \in L(\alpha)$. *Consequently, L is a transitive class.*

(iv). *If* α *is an infinite ordinal, then* $|L(\alpha)| = |\alpha|$.

(v). *If* α *is a cardinal,* $\langle B, E\rangle \equiv \langle L(\alpha), \in\rangle$, *and* $\langle B, E\rangle$ *is well-founded, then there are a unique* η *and function* f *such that*

$$f : \langle B, E\rangle \cong \langle L(\eta), \in\rangle.$$

(vi). *If* $x \in L$ *and* $x \subset L(\xi)$, *then* $x \in L(|\xi|^+)$. *More generally, if* $x \in L$ *and* x *is a finitary relation over* $L(\xi)$, *then* $x \in L(|\xi|^+)$.

(vii). *If* $\alpha = \beth_\alpha$, *then* $L(\alpha) = R(\alpha) \cap L$.

One important result is that the constructible universe is a 'model' of ZF+axiom of constructibility.

Before we can state this and further results clearly, we must be more specific about the nature of the intuitive underlying set theory (UST) which we are operating in. Up to this point we have been able to get by without making any definite commitment to whether the UST is like ZFC, Bernays–Gödel set theory, or Bernays–Morse set theory. But results such as the one above about the constructible universe must be stated one way when the UST is ZFC and another way when it is Bernays–Morse. So from now on we shall work under the definite understanding that our UST is Bernays–Morse set theory. This makes it possible for us to study models whose universes are classes rather than sets.

The usual ordered pair construction

$$\langle x, y\rangle = \{\{x\}, \{x, y\}\}$$

does not work for proper classes. However, it is easy to find a substitute, for instance

$$\langle X, Y\rangle = (X \times \{0\}) \cup (Y \times \{1\}).$$

Consider a class A and a binary relation E over A. We may then form the model $\langle A, E\rangle$ for the language $\mathscr{L}_\in = \{\in\}$. The notion of satisfaction of a formula φ of $\mathscr{L}_\in$ in the model $\langle A, E\rangle$ may then be defined recursively in the usual way. Since A might be a proper class, the Bernays–Morse axioms are needed to make such a definition. We shall extend the usual model-theoretic notation to such models, e.g.,

$$\langle A, E\rangle \models \varphi[a_1 \ldots a_n],$$

$$\langle A, E\rangle \prec \langle B, F\rangle,$$

and so on.

In particular, we may consider the model $\langle V, \in\rangle$, where V is the class of all sets. If $\varphi(x_1 \ldots x_n)$ is a formula of $\mathscr{L}_\in$ and $a_1, \ldots, a_n$ are sets, then

$$\varphi(a_1 \ldots a_n) \text{ is true}$$

if, and only if,

$$\langle V, \in\rangle \models \varphi[a_1 \ldots a_n].$$

If φ is a formula of $\mathscr{L}_\in$ and y is a new variable, then $\varphi^{(y)}$ denotes the relativization of φ to y; it is formed by replacing

$$(\exists x)\sigma \text{ by } (\exists x)(x \in y \wedge \sigma) \text{ and } (\forall x)\sigma \text{ by } (\forall x)(x \in y \to \sigma)$$

everywhere. With this notation we see that for every class A and all $a_1, \ldots, a_n \in A$,

$$\varphi^{(A)}(a_1 \ldots a_n) \text{ is true}$$

if and only if

$$\langle A, \in\rangle \models \varphi[a_1 \ldots a_n].$$

It is convenient to use ZFL to denote the theory

$$\text{ZF} + \text{axiom of constructibility.}$$

With this understanding we can now state the following theorem:

Theorem 7.4.4 (Gödel).
(i). $\langle L, \in\rangle \models$ ZFL.
(ii). (axiom of choice)$^{(L)}$.
(iii). (GCH)$^{(L)}$.

Combining Proposition 7.4.3(vii) with Theorem 7.4.4(i), we obtain the following useful fact:

$$(\alpha = \beth_\alpha \text{ implies } R(\alpha) = L(\alpha))^{(L)}.$$

The formal proof of Theorem 7.4.4 can be carried out in ZF and yields the following:

Theorem 7.4.4′.
(i). ZF $\vdash$ (ZFL)$^{(L)}$.
(ii). ZFL $\vdash$ axiom of choice.
(iii). ZFL $\vdash$ GCH.

Hereafter, by 'formula' we shall mean a formula of the language $\mathscr{L}_\in = \{\in\}$.

Let A be a transitive class. We say that the formula $\varphi(x_1 \dots x_n)$ is *absolute for A* iff for all $a_1, \dots, a_n \in A$,

$$\varphi(a_1 \dots a_n) \leftrightarrow \varphi^{(A)}(a_1 \dots a_n).$$

(*Note*: We can replace the latter line by

$$\langle V, \in\rangle \models \varphi[a_1 \dots a_n] \leftrightarrow \langle A, \in\rangle \models \varphi[a_1 \dots a_n].)$$

Whence a formula $\varphi(x_1 \dots x_n)$ is *absolute for L* iff for all $a_1, \dots, a_n \in L$,

$$\varphi(a_1 \dots a_n) \leftrightarrow \varphi^{(L)}(a_1 \dots a_n).$$

It is obvious that every atomic formula is absolute for L as well as for every transitive class B. We shall be particularly interested in absoluteness for the sets $L(\alpha)$. A convenient sufficient condition for absoluteness is given by the notion of a limited formula.

By a *limited* (or *bounded*) *quantifier* we mean a quantifier of the form $(\forall u \in v)$ or $(\exists u \in v)$, where u and v are set variables. $(\forall u \in v)\psi$ means $\forall u(u \in v \to \psi)$, and $(\exists u \in v)\psi$ means $(\exists u)(u \in v \wedge \psi)$. A formula $\varphi(v_1 \dots v_n)$ is said to be a *limited formula* iff all its quantifiers are limited.

LEMMA 7.4.5. *Let $\varphi(v_1 \dots v_n)$ be a limited formula. Then φ is absolute for L and also for every transitive class B (hence for each set $L(\alpha)$).*

The proof is an easy induction on the length of φ.

Our flexible use of abbreviations (introduced into set-theoretical formulas in the standard way) can lead to trouble when we are dealing with the notion of absoluteness. The difficulty is that an abbreviation might stand for either of two or more formulas which are logically equivalent with respect to ZFC but not logically equivalent; thus if $\langle B, \in\rangle$ is not a model of ZFC, it may well happen that one of these formulas is absolute for B, while the other is not. For example, consider the abbreviation '$x = \langle y, z\rangle$' and the two formulas (1) and (2) below:

(1) $$(\forall u)[u \in x \leftrightarrow (\forall v)(v \in u \leftrightarrow v = y) \vee (\forall v)(v \in u \leftrightarrow v = y \vee v = z)],$$

or

(2) $$(\exists uv)[u \in x \wedge v \in x \wedge (\forall w)(w \in x \to w = u \vee w = v) \wedge y \in u \wedge (\forall w)(w \in u \to w = y) \wedge y \in v \wedge z \in v \wedge (\forall w)(w \in v \to w = y \vee w = z)].$$

The formulas are not logically equivalent, but they are logically equivalent with respect to ZFC. Thus '$x = \langle y, z\rangle$' can stand for either (1) or (2). The formula (2) is a limited formula, so by Lemma 7.4.5, it is absolute for every transitive class B. However, formula (1) is not absolute for the transitive set $B = \{0\}$.

To get around this difficulty, let us say that an abbreviation φ for a formula can be *expressed by* a formula ψ iff ψ is one of the formulas which φ abbreviates. Then we shall tacitly agree that whenever an abbreviation for a formula can be expressed by a limited formula, it really does stand for a limited formula. The following are examples of abbreviations which can be expressed by limited formulas:

'$x \subset y$';

'$x = \langle y, z\rangle$';

'x is a function';

'x is an ordinal'.

We shall also need the following very special absoluteness results:

Lemma 7.4.6. *Suppose $\eta \geqslant |\xi|^+$. Then for all $x \in L(\eta)$ (and for an appropriate formula defining $L(\xi)$):*

(i). *$x = L(\xi)$ iff $(x = L(\xi))^{(L(\eta))}$, and*

(ii). *$\langle x, \in\rangle \models \sigma[a_1 \dots a_n]$ iff $(\langle x, \in\rangle \models \sigma[a_1 \dots a_n])^{(L(\eta))}$ for all formulas σ and $a_1, \dots, a_n \in x$.*

The reader should be warned that some very simple and familiar formulas are not absolute for L or for $L(\alpha)$, for example,

x is a cardinal;

x is countable;

$x = S(y)$;

$x = R(\beta)$.

We now come to the first of the main results of this section.

Theorem 7.4.7. *Suppose there exists a Ramsey cardinal α. Then:*

(i). *There are only countably many constructible subsets of ω.*

(ii). *For any two cardinals β, γ with $\omega < \beta < \gamma$, we have $\langle L(\beta), \in\rangle \prec \langle L(\gamma), \in\rangle$.*

(iii). *If γ is a cardinal, then*

$$\langle\{\beta : \beta \text{ is a cardinal and } \omega < \beta < \gamma\}, <\rangle$$

is a set of indiscernibles in $\langle L(\gamma), \in\rangle$.

PROOF. It is not hard to prove that (ii) implies (i). However, we shall prove (i) directly from Rowbottom's theorem, and then prove (ii) and (iii) from Silver's theorem.

(i). Let U be the set $L \cap S(\omega)$ of all constructible subsets of ω; then $U \subset L(\alpha)$. Consider the model $\langle L(\alpha), \in, U\rangle$. Thus $|L(\alpha)| = \alpha$ and $|U| \leqslant 2^\omega < \alpha$. By Rowbottom's theorem, there is an elementary submodel $\langle B, \in, V\rangle \prec \langle L(\alpha), \in, U\rangle$ such that $|B| = \alpha$, $|V| = \omega$. There is a unique η and f such that $f : \langle B, \in\rangle \cong \langle L(\eta), \in\rangle$. Since $|B| = \alpha$, $|\eta| = |L(\eta)| \geqslant \alpha$, so $\eta \geqslant \alpha$. But every ordinal of $\langle B, \in\rangle$ is an ordinal of $\langle L(\alpha), \in\rangle$, so the ordinals of $\langle B, \in\rangle$ and hence of $\langle L(\eta), \in\rangle$ have order type $\leqslant \alpha$. Then $\eta \leqslant \alpha$, whence

$$f : \langle B, \in\rangle \cong \langle L(\alpha), \in\rangle.$$

The statement 'y is a finite ordinal' can be expressed by a limited formula, and therefore the model $\langle L(\alpha), \in, U\rangle$ satisfies the sentence

$$(\forall x)(U(x) \leftrightarrow (\forall y)(y \in x \to y \text{ is a finite ordinal})).$$

Hence the above sentence also holds in $\langle B, \in, V\rangle$, and it follows that the isomorphism f maps V onto U, whence $|U| = |V| = \omega$.

(ii). We may choose a well ordering $<^*$ of the set $L(\alpha)$ such that for all $\zeta < \alpha$, the set $L(\zeta)$ is an initial segment of $\langle L(\alpha), <^*\rangle$. This is done by first well-ordering $L(1)$, then $L(2) \setminus L(1)$, etc. Since $L(\alpha)$ has power α, and for all $\zeta < \alpha$, $L(\zeta)$ has power $|\zeta| < \alpha$, the well ordering $<^*$ has order type exactly α. This gives us a well ordered model

$$\mathfrak{A} = \langle L(\alpha), <^*, \in\rangle \tag{1}$$

of type α. From our choice of $<^*$ we see that $\mathfrak{A}$ satisfies

$$(\forall x, y)(x \in y \to x <^* y).$$

This is because $x \in y \in L(\zeta)$ implies that $x \in L(\eta)$ for some $\eta < \zeta$, and each $L(\eta)$ is an initial segment of $\langle L(\alpha), <^*\rangle$.

Using Silver's theorem, there is an elementary chain

$$\mathfrak{A}_\xi = \langle A_\xi, <_\xi, \in_\xi\rangle, \qquad \xi > 0,$$

with the properties (a)–(d). By (a) we have

$$\langle A_\xi, \in_\xi\rangle \equiv \langle L(\alpha), \in\rangle,$$

and by (b), $\mathfrak{A}_\xi$ is a well ordered model of type ω_ξ. Using (a) again, (1) holds in $\mathfrak{A}_\xi$, and therefore $\in_\xi$ is a well founded relation. So by 7.4.3, there is a unique ordinal η_ξ and isomorphism

$$f_\xi : \langle A_\xi, \in_\xi \rangle \cong \langle L(\eta_\xi), \in \rangle.$$

Since $|L(\zeta)| = |\zeta|$ for all infinite ζ,

$$\eta_\xi \geqslant |\eta_\xi| = |L(\eta_\xi)| = |A_\xi| = \omega_\xi,$$

so $\eta_\xi \geqslant \omega_\xi$. On the other hand, by (1), the ordinals of $\mathfrak{A}_\xi$ have order type $\leqslant \omega_\xi$, whence the ordinals of $\langle L(\eta_\xi), \in \rangle$ have order type $\leqslant \omega_\xi$. But the formula 'y is an ordinal' is absolute for $L(\eta_\xi)$. It follows that $\eta_\xi \leqslant \omega_\xi$, and thus $\eta_\xi = \omega_\xi$. Hence

(2) $$f_\xi : \langle A_\xi, \in_\xi \rangle \cong \langle L(\omega_\xi), \in \rangle, \qquad \xi > 0.$$

We claim that

(3) $$\text{if } 0 < \zeta < \xi, \text{ then } f_\xi \restriction A_\zeta = f_\zeta.$$

To prove (3), we note first that

$$b \in A_\xi, \qquad a \in A_\zeta, \qquad b \in_\xi a \text{ implies } b \in_\zeta a,$$

because by (1) we have $b <_\xi a$ and by (c) the set A_ζ is an initial segment in $\langle A_\xi, <_\xi \rangle$. Since $L(\omega_\xi)$ is transitive, we have

$$f_\xi(a) = \{f_\xi(b) : b \in_\xi a\}, \qquad a \in A_\xi.$$

Moreover,

$$f_\zeta(a) = \{f_\zeta(b) : b \in_\zeta a\} = \{f_\zeta(b) : b \in_\xi a\}, \qquad a \in A_\zeta.$$

Then by induction on $<_\zeta$, we see that $f_\zeta(a) = f_\xi(a)$ for all $a \in A_\zeta$, and (3) follows.

Combining (2), (3), and the fact that the models $\mathfrak{A}_\xi$ form an elementary chain, we conclude that

$$\langle L(\omega_\zeta), \in \rangle \prec \langle L(\omega_\xi), \in \rangle.$$

This proves (ii).

(iii). Given $\xi > 0$ and $0 < \zeta < \xi$, let u_ζ be the ω_ζth element under $<_\xi$. In our proof of (ii), the well ordering $<^*$ could have been chosen so that for each ordinal $\beta < \alpha$, β is the first element of $L(\alpha) \setminus L(\beta)$. Since the formulas '$y$ is an ordinal' and '$x \in L(y)$' are absolute for $L(\alpha)$, the model $\mathfrak{A}$ satisfies

(4) $$(\forall y)[y \text{ is an ordinal} \rightarrow (\forall x)(x <^* y \leftrightarrow x \in L(y))].$$

Then (4) also holds in $\mathfrak{A}_\xi$. Moreover, for each $0 < \zeta < \xi$, $a \in A_\xi$, we have

$$a \in A_\zeta \quad \text{iff} \quad f_\xi(a) \in L(\omega_\zeta) \quad \text{iff} \quad \mathfrak{A}_\xi \models a \in L(f_\xi^{-1}(\omega_\zeta)).$$

Thus by (4),

$$a \in A_\zeta \quad \text{iff} \quad a <_\xi f_\xi^{-1}(\omega_\zeta).$$

This means that $f_\xi^{-1}(\omega_\zeta)$ is the ω_ζth element in the ordering $<_\xi$, that is,

$$f_\xi(u_\zeta) = \omega_\zeta.$$

Condition (d) of Silver's theorem says that the set

$$\langle \{u_\zeta : 0 < \zeta < \xi\}, <_\xi \rangle$$

is indiscernible in $\mathfrak{A}_\xi$. By (1), we have $u_\zeta <_\xi u_\eta$ iff $\zeta < \eta$. It follows that

$$\langle \{\omega_\zeta : 0 < \zeta < \xi\}, \in \rangle$$

is indiscernible in the model $\langle L(\omega_\xi), \in \rangle$. This proves (iii). ⊣

Theorem 7.4.7 is remarkable, for it shows that the assumption that a large cardinal exists (a Ramsey cardinal) implies something new about sets of natural numbers. Of course, even the existence of an inaccessible cardinal $\alpha > \omega$ implies number-theoretic facts like 'ZF is consistent'. However, Theorem 7.4.7 is more exciting because it is a case where an axiom of infinity implies something about sets of real numbers, which is mathematically interesting and unexpected.

Motivated by Theorem 7.4.7 and the next corollary, we introduce the set $O^\#$ and the set-theoretic sentence "$O^\#$ exists". Intuitively, $O^\#$ is the set of all formulas of $\mathscr{L}_\in$ which are satisfied in $\langle L, \in \rangle$ by any increasing tuple of uncountable cardinal numbers. In ordinary ZFC, this definition of $O^\#$ cannot be formalized because it involves satisfaction in the proper class $\langle L, \in \rangle$ and assumes that the uncountable cardinals are a class of indiscernibles. The sentence "$O^\#$ exists" is defined as the conjunction of parts (ii) and (iii) of Theorem 7.4.7. We shall see in the next corollary that when $O^\#$ exists, the intuitive definition of $O^\#$ can be formalized in the language $\mathscr{L}_\in$ of set theory. We first given an official definition of $O^\#$ within ZFC.

Definition. We say that $O^\#$ *exists* iff

(i). For any two cardinals β, γ with $\omega < \beta < \gamma$, we have $\langle L(\beta), \in \rangle \prec \langle L(\gamma), \in \rangle$,

(ii). If γ is a cardinal, then the set of uncountable cardinals $\beta < \gamma$ with the usual ordering is a set of indiscernibles in $\langle L(\gamma), \in \rangle$.

If $O^{\#}$ exists, then we define $O^{\#}$ to be the set of all formulas $\varphi(x_1 \dots x_n)$ of $\mathscr{L}_\in$ such that

$$\langle L(\omega_\omega), \in \rangle \models \varphi[\omega_1 \dots \omega_n].$$

Here ω_m is the mth infinite cardinal in the ordinary sense, not in the sense of L.

Theorem 7.4.7 may be restated as follows:

THEOREM 7.4.7′. *If there is a Ramsey cardinal, then $O^{\#}$ exists.*

We now show that if $O^{\#}$ exists then we can formulate satisfaction in the 'model' $\langle L, \in \rangle$.

PROPOSITION 7.4.8. *Assume that $O^{\#}$ exists, Then:*

(i). *For all cardinals $\beta > \omega$, $\langle L(\beta), \in \rangle \prec \langle L, \in \rangle$.*

(ii). *There is a formula $\psi(u, v)$ of $\mathscr{L}_\in$ such that for all u and v,*

$\psi(uv)$ iff 'u is a formula of $\mathscr{L}_\in$, $v \in L$, and $\langle L, \in \rangle \models u[v]$'.

(iii). *If $\varphi(v)$ is a formula of $\mathscr{L}_\in$ and $\langle L, \in \rangle \models (\exists v)\varphi$, then there is a countable $x \in L(\omega_1)$ such that $\langle L, \in \rangle \models \varphi[x]$.*

(iv). *Let C be the class of all uncountable cardinals. Then $\langle C, < \rangle$ is a class of indiscernibles in $\langle L, \in \rangle$.*
Moreover, $O^{\#}$ is the set of all formulas $\varphi(x_1 \dots x_n)$ which are satisfied by increasing tuples from C in $\langle L, \in \rangle$.

(v). *$V \neq L$, and in fact $S(\omega)^L$ is countable and belongs to $L(\omega_1)$.*

PROOF. (i). By Theorem 7.4.7(ii), the models $\langle L(\beta), \in \rangle$, β a cardinal $> \omega$, form an elementary chain. The union of this chain is $\langle L, \in \rangle$. The proof of the theorem on unions of elementary chains still goes through in this case, so

$$\langle L(\beta), \in \rangle \prec \langle L, \in \rangle.$$

(ii). Let $\psi(uv)$ be the formula which says:

'u is a formula of $\mathscr{L}_\in$, and there exists a cardinal $\beta > \omega$ such that $v \in L(\beta)$ and v satisfies u in $\langle L(\beta), \in \rangle$'.

Then it follows from (i) that $\psi(uv)$ has the required property.

(iii). If $\langle L, \in \rangle \models (\exists v)\varphi$, then by (i), $\langle L(\omega_1), \in \rangle \models (\exists v)\varphi$, whence there exists $x \in L(\omega_1)$, such that $\langle L(\omega_1), \in \rangle \models \varphi[x]$. Again using (i), $\langle L, \in \rangle \models$

$\varphi[x]$. Since $x \in L(\omega_1)$, $x \subset L(\xi)$ for some $\omega \leqslant \xi < \omega_1$. Thus $|x| \leqslant |L(\xi)| = |\xi| = \omega$, whence x is countable.

(iv). Let $\alpha_1 < \ldots < \alpha_n$ and $\beta_1 < \ldots < \beta_n$ be increasing n-tuples of uncountable cardinals, and let γ be a cardinal with $\alpha_n < \gamma$, $\beta_n < \gamma$. By Theorem 7.4.7(iii), we have

$$\langle L(\gamma), \in, \alpha_1 \ldots \alpha_n \rangle \equiv \langle L(\gamma), \in, \beta_1 \ldots \beta_n \rangle.$$

Then by (i),

$$\langle L, \in, \alpha_1 \ldots \alpha_n \rangle \equiv \langle L, \in, \beta_1 \ldots \beta_n \rangle. \dashv$$

The next Theorem 7.4.10 completely solves the problems 7.3.1–7.3.3 under the assumption of the axiom of constructibility. Lemma 7.4.9 below allows us to handle all three problems at once.

LEMMA 7.4.9. *Let α be a cardinal and assume:*

(i). *For all cardinals $\beta \leqslant \alpha$, if $\langle B, \in \rangle \prec \langle L(\alpha), \in \rangle$ and $|B| = \beta$, then $L(\beta) \subset B$.*

Then each of the following hold:

(ii). *$\langle L(\alpha), \in \rangle$ is a Jónsson model of power α.*

(iii). *For all cardinals $\delta < \beta < \alpha$ and $\delta < \gamma \leqslant \alpha$, Chang's conjecture fails for the pairs (α, β), (γ, δ).*

(iv). *For all cardinals $\beta < \alpha$, there exists a well ordered model of type α which has no elementary submodel of type β.*

PROOF. (ii) follows by taking $\beta = \alpha$.

For (iii), consider the model $\mathfrak{A} = \langle L(\alpha), L(\beta), \in \rangle$ of type (α, β) and suppose there is a $\mathfrak{B} \prec \mathfrak{A}$, $\mathfrak{B} = \langle B, V, \in \rangle$, of type (γ, δ). Then $V = B \cap L(\beta)$. By (i), $L(\gamma) \subset B$. If $\gamma \leqslant \beta$, then also $L(\gamma) \subset L(\beta)$, whence $L(\gamma) \subset V$, and this contradicts $|V| = \delta < \gamma$. Similarly, if $\beta \leqslant \gamma$, then $L(\beta) \subset B$, whence $L(\beta) \subset V$ and this contradicts $|V| = \delta < \beta$. Thus there is no $\mathfrak{B} \prec \mathfrak{A}$ of type (γ, δ).

To prove (iv), choose a well ordering $<^*$ of $L(\alpha)$ of order type α, such that $L(\beta)$ is an initial segment. Then $\langle L(\alpha), <^*, \in, \beta \rangle$ is a well ordered model of type α. Here β is a constant of the model. Suppose $\langle B, <^*, \in, \beta \rangle \prec \langle L(\alpha), <^*, \in, \beta \rangle$ is an elementary submodel of type β. Then $|B| = \beta$, so (i) implies that $L(\beta) \subset B$. But then the element $\beta \notin L(\beta)$ has β predecessors, so $\langle B, <^*, \in, \beta \rangle$ cannot have type β. Thus $\langle L(\alpha), <^*, \in, \beta \rangle$ has no elementary submodels of type β. $\dashv$

THEOREM 7.4.10. *Assume the axiom of constructibility. Then conditions (i)–(iv) of Lemma 7.4.9 hold for all cardinals α.*

PROOF. By Lemma 7.4.9, it suffices to prove (i). To do this, we assume that (i) is false and get a contradiction. Thus we assume that there are cardinals $\alpha > \beta$ and a model $\langle B, \in\rangle \prec \langle L(\alpha), \in\rangle$ of power $|B| = \beta$ such that not $L(\beta) \subset B$. Since $\langle B, \in\rangle$ is well-founded, there is a unique ordinal η and isomorphism

$$f : \langle L(\eta), \in\rangle \cong \langle B, \in\rangle.$$

Then f is also an elementary embedding

$$f : \langle L(\eta), \in\rangle \prec \langle L(\alpha), \in\rangle.$$

We have

$$|\eta| = |L(\eta)| = \beta, \text{ so } \eta \geqslant \beta.$$

For any limited formula $\varphi(v_1 \dots v_n)$ and any $x_1, \dots, x_n \in L(\eta)$, we have

(1) $$\varphi(x_1 \dots x_n) \text{ iff } \varphi(f(x_1) \dots f(x_n)),$$

because limited formulas are absolute for $L(\eta)$ and $L(\alpha)$. Also, by Lemma 7.4.6, for all $\mu < \beta$ we have

(2) $$f(L(\mu)) = L(f(\mu)).$$

This is because $x = L(\mu)$ iff $(x = L(\mu))^{(L(\eta))}$ iff $(f(x) = L(f(\mu)))^{(L(\alpha))}$ iff $f(x) = L(f(\mu))$.

Since $L(\beta)$ is not a subset of B, there exists $x \in L(\beta)$ such that $f(x) \neq x$. Let ξ be the least ordinal such that

(3) $$\text{for some } x \in L(\xi+1), \quad f(x) \neq x.$$

β is a limit ordinal, so $\xi < \beta$. In fact, β is a cardinal, so

(4) $$|\xi|^+ \leqslant \beta \leqslant \eta.$$

It follows from the equation $L(\xi) = \bigcup_{\nu<\xi} L(\nu+1)$ that

(5) $$\text{for all } x \in L(\xi), f(x) = x.$$

Thus ξ is the greatest ordinal with the property (5). We claim that

(6) $$f(\xi) > \xi.$$

To prove (6), assume that $f(\xi) \leqslant \xi$. By (5) and the fact that f is one–one, we have $f(\xi) \geqslant \xi$, so $f(\xi) = \xi$. From (4) we see that $L(\xi) \in L(\eta)$. Let $x \in L(\xi+1)$. Then there are a formula $\varphi(uv_1 \dots v_n)$ and $y_1, \dots, y_n \in L(\xi)$ such that

(7) $$x = \{z \in L(\xi) : \varphi^{(L(\xi))}(zy_1 \dots y_n)\}.$$

The statement

$$x = \{z \in u : \varphi^{(u)}(zy_1 \dots y_n)\}$$

is limited, so by (1),

(8) $$f(x) = \{z \in f(L(\xi)) : \varphi^{(f(L(\xi)))}(zf(y_1) \dots f(y_n))\}.$$

But by (2), $f(L(\xi)) = L(f(\xi)) = L(\xi)$, and by (5), $f(y_1) = y_1, \dots, f(y_n) = y_n$. It follows from (7) and (8) that $f(x) = x$. But x was an arbitrary element of $L(\xi+1)$, so we have contradicted (3). We conclude that $f(\xi) > \xi$, i.e., (6) holds.

We have $\xi > \omega$, because for each $\mu \leqslant \omega$ there is a limited formula $\varphi(v)$ such that μ is the unique set satisfying $\varphi(v)$, whence $f(\mu) = \mu$. Also, it follows from (4) that

(9) $$S(\xi) \cap L \subset L(\eta).$$

Up to this point in the proof, we have not used the axiom of constructibility. Using it now, (9) implies that $S(\xi) \subset L(\eta)$. Define

$$D = \{x \in S(\xi) : \xi \in f(x)\}.$$

We claim that D is a nonprincipal ξ-complete ultrafilter over ξ. This is essentially the result of Exercise 4.2.4, but we shall give a direct proof here. Note that the formulas $x = 0$, $z = x \cap y$, $z = \xi \setminus x$ are limited. Thus $f(0) = 0$, so $\xi \in f(0)$ and $0 \notin D$. Also, for all x, y: $x \in D$ and $y \in D$ implies $\xi \in f(x)$ and $\xi \in f(y)$. So $\xi \in f(x) \cap f(y) = f(x \cap y)$, and $x \cap y \in D$. And, for all $x \subset \xi$, $x \notin D$ implies $\xi \notin f(x)$. So $\xi \in f(\xi) \setminus f(x) = f(\xi \setminus x)$, and $\xi \setminus x \in D$. This makes D an ultrafilter. D is nonprincipal, because for all $\zeta \in \xi$,

$$\xi \notin \{\zeta\} = \{f(\zeta)\} = f(\{\zeta\}),$$

so $\{\zeta\} \notin D$. To prove D that is ξ-complete, let $\zeta \in \xi$ and $\{x_\pi : \pi \in \zeta\} \subset D$. We have a slight difficulty because it is not clear whether $\{x_\pi : \pi \in \zeta\} \in L(\eta)$. However, we get around this by letting

$$r = \{\langle \pi, \mu \rangle \in \zeta \times \xi : \mu \in x_\pi\}.$$

Then $r \in L(\eta)$. Each x_π belongs to $L(\eta)$, and

$$\{\mu \in \xi : \langle \pi, \mu \rangle \in r\} = x_\pi \in D.$$

The formula $v = \{\mu \in \xi : \langle \pi, \mu \rangle \in r\}$ is limited. Therefore for all $\pi < \zeta$,

$$\begin{aligned} \xi \in f(x_\pi) &= \{\mu \in f(\xi) : \langle f(\pi), \mu \rangle \in f(r)\} \\ &= \{\mu \in f(\xi) : \langle \pi, \mu \rangle \in f(r)\}, \end{aligned}$$

whence $\langle \pi, \xi \rangle \in f(r)$. Hence

$$\xi \in \{\mu \in f(\xi) : (\forall \pi \in \zeta)\langle \pi, \mu \rangle \in f(r)\} = z.$$

Since the formula $v = \{\mu \in \xi : (\forall \pi \in \zeta)\langle \pi, \mu \rangle \in r\}$ is also limited,

$$f(\bigcap_{\pi < \zeta} x_\pi) = f(\{\mu \in \xi : (\forall \pi \in \zeta)\langle \pi, \mu \rangle \in r\}) = z.$$

Hence $\bigcap_{\pi < r} x_\pi \in D$.

We have shown that ξ is a measurable cardinal $> \omega$. But by Scott's theorem 4.2.18, this contradicts the axiom of constructibility. ⊣

By 7.4.8, if $V = L$ then $O^\#$ does not exist. By Theorem 7.4.10, if $V = L$ then 7.4.9 (i) holds. Kunen has obtained the following improvement of Theorem 7.4.10.

7.4.11. *If $O^\#$ does not exist then* 7.4.9 (i) *holds.*

The proof combines the methods of the preceding and present sections with a generalization of the iterated ultrapower construction of Section 6.5. Although many of the tools required for the proof have already been introduced in this book, we shall not give the proof here because it is quite long. We shall prove one last result connecting the problems 7.3.1–7.3.3 with the constructible universe.

THEOREM 7.4.12. *Assume that there is no ordinal $\xi > \omega$ such that*

$$(\xi \text{ is a weakly compact cardinal})^{(L)}.$$

Then conditions (i)–(iv) *of Lemma* 7.4.9 *hold.*

PROOF. Assume (i) is false. Let α, η,

$$f : \langle L(\eta), \in \rangle \prec \langle L(\alpha), \in \rangle,$$

and ξ be as in the previous proof. We show first that $(\xi \text{ is inaccessible})^{(L)}$. It is easy to check that ξ is accessible if and only if

(1) there exist $\zeta < \xi$ and $y \subset \xi$ and $r \subset \zeta \times y$ such that y is cofinal in $\langle \xi, < \rangle$ and for all $\mu, \pi \in y$, $\mu \neq \pi$ implies $\{v : r(v, \mu)\} \neq \{v : r(v, \pi)\}$.

Let $\varphi(\zeta, \xi, y, r)$ be an abbreviation for the part of (1) inside the quantifiers; formally,

$$(\forall z \in \xi)(\exists u \in y)[z \in u] \wedge (\forall \mu, \pi \in y)[\mu \not\equiv \pi \rightarrow \\ \neg (\forall v \in \zeta)[(\exists s \in r)(s \equiv \langle v, \mu \rangle) \leftrightarrow (\exists s \in r)(s \equiv \langle v, \pi \rangle)]].$$

Note that all the quantifiers in φ are limited. It follows that φ is absolute for L. Suppose (ξ is accessible)$^{(L)}$. Then there exist $\zeta \in \xi$ and $y \in S(\xi) \cap L$ and $r \in S(\zeta \times y) \cap L$ such that

$$\varphi^{(L)}(\zeta, \xi, y, r).$$

Using absoluteness we have

$$\varphi(\zeta, \xi, y, r).$$

Since $|\xi|^+ \leqslant \eta$, we have $y \in L(\eta)$ and $r \in L(\eta)$. Then as we observed in the previous proof, it follows that

$$\varphi(f(\zeta), f(\xi), f(y), f(r)).$$

But $f(\zeta) = \zeta$, so

(2) $$\varphi(\zeta, f(\xi), f(y), f(r)).$$

The statement (2) implies that $f(y)$ is cofinal in $f(\xi)$, whence there exists $\mu < f(\xi)$ with $\xi \leqslant \mu$ and $\mu \in f(y)$. Let

(3) $$z = \{v \in \zeta : \langle v, \mu \rangle \in f(r)\}.$$

We note that for all $\pi \in \xi$, we have $f(\pi) = \pi$, and $\pi \in y$ iff $f(\pi) \in f(y)$; therefore

(4) $$y = f(y) \cap \xi.$$

For all $\pi \in f(y) \cap \xi$, $\pi \neq \mu$, whence by (2) and (3),

(5) $$z \neq \{v \in \zeta : \langle v, \pi \rangle \in f(r)\}.$$

We have

$$z \subset f(z), \qquad f(z) \subset f(\zeta) = \zeta,$$

$$(\zeta - z) \subset f(\zeta - z) = f(\zeta) - f(z) = \zeta - f(z),$$

and therefore $z = f(z)$. Thus from (3) and (5), respectively, we obtain, using (4),

$$(\exists \mu \in y)(z = \{v \in \zeta : \langle v, \mu \rangle \in r\}),$$

$$(\forall \pi \in y)(z \neq \{v \in \zeta : \langle v, \pi \rangle \in r\}).$$

This is a contradiction. Hence (ξ is an inaccessible cardinal)$^{(L)}$.

To show that (ξ is weakly compact)$^{(L)}$ we shall use the characterization of weakly compact given in Exercise 4.2.2(iii). It says that a cardinal β is weakly compact iff β is inaccessible and every model of the form

$$\langle R(\beta), \in, S_1 \dots S_n \rangle$$

has a proper elementary extension

$$\langle C, E, T_1 \dots T_n \rangle$$

such that

$$c \in C, \qquad b \in R(\beta), \qquad cEb \text{ imply } c \in R(\beta).$$

Note that since (ξ is inaccessible)$^{(L)}$, we have ($\xi = \beth_\xi)^{(L)}$; hence

$$(R(\xi) = L(\xi))^{(L)}.$$

Furthermore, using Lemma 7.4.6, we have

$$f(L(\xi)) = L(f(\xi)).$$

Consider a model $\mathfrak{B} \in L$ of the form

$$\mathfrak{B} = \langle L(\xi), \in, S_1 \dots S_n \rangle.$$

Then $S_1, \dots, S_n \in L$. Since $S_1, \dots, S_n$ are finite-placed relations over $L(\xi)$, they all belong to $L(|\xi|^+)$. Hence $\mathfrak{B} \in L(|\xi|^+)$, and $\mathfrak{B} \in L(\eta)$. Now form the model

$$f(\mathfrak{B}) = \langle f(L(\xi)), \in, f(S_1) \dots f(S_n) \rangle.$$

Recalling that f is an elementary embedding of $\langle L(\eta), \in \rangle$ into $\langle L(\alpha), \in \rangle$, we have for each $\mathscr{L}_\in$ formula $\varphi(v_1 \dots v_n)$ and $x_1, \dots, x_n \in L(\xi)$,

$$\begin{aligned}
&(\mathfrak{B} \models \varphi[x_1 \dots x_n])^{(L)} \\
\text{iff } &\langle L(\eta), \in \rangle \models (\mathfrak{B} \models \varphi[x_1 \dots x_n]) \\
\text{iff } &\langle L(\alpha), \in \rangle \models (f(\mathfrak{B}) \models \varphi[f(x_1) \dots f(x_n)]) \\
\text{iff } &(f(\mathfrak{B}) \models \varphi[f(x_1) \dots f(x_n)])^{(L)}.
\end{aligned}$$

This is because the statement '$\mathfrak{B} \models \varphi[x_1 \dots x_n]$' is absolute by Lemma 7.4.6(ii). But $f(x_1) = x_1, \dots, f(x_n) = x_n$. Therefore $(\mathfrak{B} \prec f(\mathfrak{B}))^{(L)}$. On the other hand, $\xi \subset L(\xi)$, whence $\xi \in f(\xi) \subset f(L(\xi))$, so $\xi \in f(L(\xi))$. But $\xi \notin L(\xi)$. Hence

$$(f(\mathfrak{B}) \text{ is a proper elementary extension of } \mathfrak{B})^{(L)}.$$

If $y \in f(L(\xi))$, $z \in L(\xi)$, and $y \in z$, then $y \in L(\xi)$ because $L(\xi)$ is transitive. Thus $f(\mathfrak{B})$ has the required property. ⊣

COROLLARY 7.4.13. *If there exists a Ramsey cardinal, then there is a countable* *ordinal $\beta > \omega$ such that*

$$(\beta \textit{ is a weakly compact cardinal})^{(L)}.$$

Proof. By Rowbottom's theorem, there is no Jónsson model of power α. Then by Theorem 7.4.11, there is an ordinal $\xi > \omega$ such that

$$(\xi \text{ is a weakly compact cardinal})^{(L)}.$$

The desired conclusion now follows from Corollary 7.4.8(ii). ⊣

The core model of Jensen and Dodd (1981) has many of the structure properties of L but allows measurable cardinals. The methods and results described in this section have been considerably extended using the core model and other inner models instead of L. These methods have led to results on regular ultrafilters, Chang's conjecture and Jónsson models which have been stated without proof in Sections 4.3 and 7.3. See, for example, Devlin (1973), Kanamori and Magidor (1978), Donder, Jensen and Koppelberg (1981), Donder and Koepke (1983), Levinski (1984), Koepke (1988), and Donder (1988).

Exercises

7.4.1. Show that Corollary 7.4.8(i) can be reformulated as the following theorem of ZFC: Let $\mathfrak{A} = \langle A, E\rangle$ be a model of ZFC+ 'there exists a Ramsey cardinal'. Let

$$L^{\mathfrak{A}} = \{a \in A : \mathfrak{A} \models \text{`}a \in L\text{'}\},$$

$$\mathfrak{A} \models \text{`}\beta \text{ is a cardinal} > \omega\text{'},$$

and

$$L(\beta)^{\mathfrak{A}} = \{a \in A : \mathfrak{A} \models \text{`}a \in L(\beta)\text{'}\}.$$

Then

$$\langle L(\beta)^{\mathfrak{A}}, E\rangle \prec \langle L^{\mathfrak{A}}, E\rangle.$$

Give similar formulations for parts (iii) and (iv) of the corollary.

7.4.2*. Assuming that $\kappa(\omega_1)$ exists (cf. Exercise 7.3.23), prove the conclusions of Theorem 7.4.7.

[*Hint*: Use Lemma 7.3.20 or Exercise 7.3.33 instead of Silver's theorem.]

7.4.3. Assume that there exists a cardinal α and a well ordering $<^*$ of $L(\alpha)$ such that the model $\langle L(\alpha), <^*, \in\rangle$ satisfies the sentences (1), (4) of the proof of Theorem 7.4.7 and also has an uncountable set of indiscernibles. Show that the conclusions of Theorem 7.4.7 hold.

7.4.4*. Assume that the class of cardinals $> \omega$ is indiscernible in the model $\langle L, \in \rangle$ (i.e., condition 7.4.8(iv)). Let $O^{\#}$ be the set of all formulas $\varphi(v_1 \dots v_n)$ of $\mathscr{L}_\in$ such that $\langle L, \in \rangle \models \varphi[\omega_1 \dots \omega_n]$. Here $\omega_1, \dots, \omega_n$ are in the ordinary sense, not in the sense of L. Prove that $O^{\#} \notin L$.

[*Hint*: Let α be a cardinal $> \aleph_\omega$. Then $O^{\#}$ is also the set of all $\mathscr{L}_\in$ formulas $\varphi(v_1 \dots v_n)$ such that $\langle L(\alpha), \in \rangle \models \varphi[\omega_1 \dots \omega_n]$. Use the fact that there exists a well ordering $<^*$ of $L(\alpha)$ which is definable in $\langle L(\alpha), \in \rangle$ and such that $\langle L(\alpha), \in \rangle$ satisfies sentences (1), (4) of the proof of Theorem 7.4.7.]

7.4.5. Condition (iv) of Corollary 7.4.8 implies conclusions (i)–(iii) of Corollary 7.4.8.

[*Hint*: Use Exercise 7.4.3.]

7.4.6. Suppose that α is a Rowbottom cardinal. Then for every formula $\varphi(v)$ of $\mathscr{L}_\in$ such that $\langle L(\alpha), \in \rangle \models (\exists! v)\varphi(v)$, there exists a countable $x \in L(\omega_1)$ such that $\langle L(\alpha), \in \rangle \models \varphi[x]$. In particular, if $\alpha > \omega$, then there are only countably many constructible subsets of ω.

7.4.7. Suppose $\alpha > \omega_1$ and Chang's conjecture holds for $(\alpha, \omega_1), (\omega_1, \omega)$. Then there are only countably many constructible subsets of ω. [*Remark*: From exercises in Section 7.3, if Chang's conjecture holds for some pair $(\alpha, \beta), (\gamma, \omega)$ with $\alpha \geqslant \beta > \omega, \alpha \geqslant \gamma > \omega$, then it holds for (α, ω_1), (ω_1, ω).]

7.4.8*. Assume that α is a cardinal and every countable subset of α is constructible. Then conditions (i)–(iv) of Lemma 7.4.9 hold.

[*Hint*: Like the proof of Theorem 7.4.10, but prove that ξ is a Ramsey cardinal instead of proving that ξ is measurable.]

7.4.9. If γ is an inaccessible cardinal, then (γ is an inaccessible cardinal)$^{(L)}$.

7.4.10*. If γ is a weakly compact cardinal, then (γ is a weakly compact cardinal)$^{(L)}$.

[*Hint*: By the preceding exercise, (γ is an inaccessible cardinal)$^{(L)}$. Thus $(R(\gamma) = L(\gamma))^{(L)}$. Let $\gamma > \omega$. Consider a constructible model $\mathfrak{B} = \langle L(\gamma), \in, S_1 \dots S_n \rangle$. Take a cardinal $\alpha > \gamma$ and choose an elementary submodel $\mathfrak{C} \prec \langle L(\alpha), \in \rangle$ of power γ such that $\gamma \subset C, \gamma \in C, \mathfrak{B} \in C$. Then $\mathfrak{C} \cong \langle L(\eta), \in \rangle$ for some η. By weak compactness, there is a $\delta \geqslant \eta$ and an elementary embedding $f: \langle L(\eta), \in \rangle \prec \langle L(\delta), \in \rangle$ such that γ is the least ordinal with $f(\gamma) \neq \gamma$. Now use the method of Theorem 7.4.11 to show that in the constructible universe $f(\mathfrak{B})$ is an elementary extension of $\mathfrak{B}$ of the required kind.]

7.4.11*. Suppose $f: \langle L(\eta), \in\rangle \prec \langle L(\alpha), \in\rangle$, α is a cardinal, ξ is the least ordinal such that $f(\xi) \neq \xi$, and $|\xi|^+ \leqslant \eta$ (cf. the proof of Theorem 7.4.11). Then

$$(\xi \text{ is } \mathbf{\Pi}_1^2\text{-indescribable})^{(L)}.$$

(The notion of a $\mathbf{\Pi}_n^m$-indescribable cardinal is defined in Exercise 4.2.19.) If in addition α is a limit cardinal, then

$$(\xi \text{ is indescribable})^{(L)}.$$

The same conclusion holds if $f: \langle L, \in\rangle \prec \langle L, \in\rangle$ and ξ is the least ordinal such that $f(\xi) \neq \xi$.

7.4.12*. Assume that the class C of all uncountable cardinals is indiscernible in $\langle L, \in\rangle$. Then for each cardinal $\alpha > \omega$, there exists an elementary embedding $f: \langle L, \in\rangle \prec \langle L, \in\rangle$ such that α is the least ordinal such that $f(\alpha) \neq \alpha$. Hence

$$(\alpha \text{ is weakly compact})^{(L)},$$

and

$$(\alpha \text{ is indescribable})^{(L)}.$$

[*Hint*: Use the fact that there is a function F definable in $\langle L, \in\rangle$ such that the sentences (1)–(4) of the proof of Theorem 7.4.7 hold in $\langle L, \in\rangle$. Thus $\langle L, \in\rangle$ has definable Skolem functions, and the Skolem hull of C is isomorphic to $\langle L, \in\rangle$. Next use the elementary embedding theorem for indiscernibles, i.e., any ordermorphism of the class of indiscernibles into itself generates an elementary embedding $f: \langle L, \in\rangle \prec \langle L, \in\rangle$.]

7.4.13**. Assume that $f: \langle L(\eta), \in\rangle \prec \langle L(\alpha), \in\rangle$, α is a cardinal, ξ is the least ordinal such that $f(\xi) \neq \xi$, and $|\xi|^+ \leqslant \eta$. (Same hypotheses as Exercise 7.4.11.) Then there exists an elementary embedding

$$f: \langle L, \in\rangle \prec \langle L, \in\rangle$$

such that ξ is the least ordinal with $f(\xi) \neq \xi$.

[*Hint*: Form an 'ultrapower' $\prod_D \langle L, \in\rangle$. Let

$$D = \{x \in L \cap S(\xi) : \xi \in f(x)\}.$$

Note that $D \notin L$. Let the elements of $\prod_D \langle L, \in\rangle$ be the equivalence classes of *constructible* functions $g \in^{\xi} L$. Show that the ultrapower is well-founded and therefore isomorphic to $\langle L, \in\rangle$. Let f be the composition of the natural embedding and this isomorphism. This exercise is the first step in the proof of the result 7.4.11 of Kunen.]

7.4.14. Using Exercises 7.4.12 and 13, show that if

$$\text{(there is no indescribable cardinal)}^{(L)},$$

then conditions (i)–(iv) of Lemma 7.4.9 hold.

Elaborating on the hint for Exercise 7.4.12, there is a formula $x <^* y$ such that the following are theorems of ZFL:

$$(\forall xy)(x \in y \rightarrow x <^* y),$$
$$(\forall \alpha)(L(\alpha) \text{ is well ordered by } <^*),$$
$$(\forall \alpha)(L(\alpha) \text{ is an initial segment of } <^*).$$

Using $<^*$, we can define Skolem functions in ZFL. For the following problem we shall use formulas containing $<^*$ and Skolem terms as abbreviations for the formulas which define them in ZFL.

7.4.15*. Assume that the class of all uncountable cardinals is indiscernible in $\langle L, \in \rangle$, and let $O^{\#}$ be the set of $\mathscr{L}_{\in}$ formulas defined in Exercise 7.4.4. Show that $O^{\#}$ is the unique set Σ such that:

(1). Σ is a maximal consistent set of formulas of $\mathscr{L}_{\in}$.

(2). All axioms of ZFL belong to Σ.

(3). All the $\mathscr{L}_{\in}$ formulas (i)–(iii) of Lemma 7.3.19 belong to Σ.

(4). All $\mathscr{L}_{\in}$ formulas of the form

$$v_n < t(v_1 \dots v_{n-1} v_{n+1} \dots v_m) \rightarrow v_{n+1} \leqslant t(v_1 \dots v_{n-1} v_{n+1} \dots v_m)$$

belong to Σ.

(5). All $\mathscr{L}_{\in}$ formulas of the form

$$\varphi(v_{i_1} \dots v_{i_n}) \leftrightarrow \varphi(v_{j_1} \dots v_{j_n}), \qquad i_1 < \dots < i_n < \omega, j_1 < \dots < j_n < \omega,$$

belong to Σ.

(6). For any well ordering $<'$ of ω, the model generated by the set $\langle \omega, <' \rangle$ of indiscernibles whose increasing sequences satisfy Σ is well ordered by $<^*$.

Note: It turns out that, with an appropriate Gödel numbering of $\mathscr{L}_{\in}$ formulas, the conditions (1)–(6) can be expressed as a Π^1_2 formula over $\langle R(\omega), \in \rangle$ (that is, second-order universal–existential). It then follows that the statement $\varphi \in O^{\#}$ is Δ^1_3, i.e., can be defined both by a Π^1_3 formula and by a Σ^1_3 formula in $\langle R(\omega), \in \rangle$. This is known to be the simplest possible kind of nonconstructible set, in the Σ and Π hierarchy. Solovay has shown that if ZF is consistent, so is ZF+ 'there is a nonconstructible Δ^1_3 set'.

7.4.16*. Assume the class of cardinals $> \omega$ is indiscernible in $\langle L, \in \rangle$ and define $O^{\#}$ as before. Let M be a transitive set or class, $\langle M, \in \rangle \models$ ZFC,

$C_M = \{\alpha \in M : \langle M, \in\rangle \models \alpha \text{ is a cardinal} > \omega\}$, and

$$L_M = \{x \in M : \langle M, \in\rangle \models x \in L\}.$$

Prove that if $O^{\#} \in M$, then $\langle C_M, <\rangle$ is a set of indiscernibles in $\langle L_M, \in\rangle$. Hence all the conclusions of Corollary 7.4.8 hold in $\langle M, \in\rangle$ (by Exercise 7.4.5). This means that simply by putting the set $O^{\#}$ of natural numbers into the model, we can make every definable element of $\langle L, \in\rangle$ countable!

7.4.17*. Gaifman's construction. Let $\alpha > \omega$ be a measurable cardinal and let D be an α-complete nonprincipal ultrafilter over α. Recall the iterated ultrapower construction of Section 6.5. For each ordinal β, let

$$E_\beta = \times_{\langle \beta, > \rangle} D$$

be the ultrafilter formed by iterating D, β times with the inverse well order. Prove that:

(i). The iterated ultrapower $\prod_{E_\beta} \langle L, \in\rangle$ is well founded and isomorphic to $\langle L, \in\rangle$.

(ii). For each $\gamma \in \beta$, let $d_{\{\gamma\}} : \prod_D \langle L, \in\rangle \prec \prod_{E_\beta} \langle L, \in\rangle$ be the natural embedding. Let $b \in \prod_D \langle L, \in\rangle$ be the equivalence class of the identity function on α. Then $\{d_{\{\gamma\}}(b) : \gamma \in \beta\}$ is indiscernible in $\prod_{E_\beta} \langle L, \in\rangle$ (with the obvious ordering). Hence $\langle L, \in\rangle$ has indiscernibles.

7.4.18*. Generalize Theorem 7.4.7(ii) to the following. Suppose there exists a Ramsey cardinal α. Then for any three infinite cardinals $\alpha < \beta < \gamma$, β and γ regular, we have

$$\langle L(\beta), \in\rangle \prec_\alpha \langle L(\gamma), \in\rangle.$$

($\prec_\alpha$ denotes elementary submodel in the sense of the infinitary language $\mathscr{L}_\alpha$ introduced in Chapter 4.) In particular, we have

$$\langle L(\omega_2), \in\rangle \prec_{\omega_1} \langle L, \in\rangle.$$

This result can be improved in the style of Exercise 7.4.2.

APPENDIX A

SET THEORY

In the first part of this appendix, we shall develop the intuitive set theory which is needed for the theory of models. Our purpose is both to fix notation, and to present some basic results about ordinals, transfinite induction and cardinals. The treatment below will fill the gap between the amount of set theory used in most other branches of mathematics and the slightly larger amount used in model theory.

The last part of this appendix contains the formal lists of axioms for four axiomatic set theories, those of Zermelo, Zermelo–Fraenkel, Bernays, and Bernays–Morse.

The empty set will be denoted by 0. The set of all x such that the condition $\varphi(x)$ holds, if such a set exists, is denoted by $\{x : \varphi(x)\}$. The set of all elements of X which are not elements of Y is denoted by $X \setminus Y$. We write $Y \subset X$ if Y is a subset of X, including the possibility $Y = X$. We use the usual notation for unions and intersections. The *ordered pair* of x and y is defined by

$$\langle x, y\rangle = \{\{x\}, \{x, y\}\}.$$

We define *ordered n-tuples* inductively by

$$\langle x\rangle = x, \qquad \langle x_1, \ldots, x_n, x_{n+1}\rangle = \langle\langle x_1, \ldots, x_n\rangle, x_{n+1}\rangle.$$

The *Cartesian product* $X \times Y$ is the set of all ordered pairs $\langle x, y\rangle$ with $x \in X$, $y \in Y$, and we write

$$X_1 \times \ldots \times X_n \times X_{n+1} = (X_1 \times \ldots \times X_n) \times X_{n+1},$$

$$X^1 = X, \qquad X^{n+1} = X^n \times X.$$

An *n-ary relation* over X is a subset of X^n, and a *function* is a binary many–one relation. If $\langle x, y\rangle \in f$, where f is a function, we may write $y = f(x)$ or $y = f_x$. If R is an n-ary relation we sometimes write $R(x_1, \ldots, x_n)$ for

$\langle x_1, \ldots, x_n \rangle \in R$. If R is binary, we may write xRy for $\langle x, y \rangle \in R$. We shall say that f is a function *on* X *onto* Y if X = domain (f) and Y = range (f), f is a function *on* X *into* Y if X = domain (f) and range (f) $\subset Y$. The *restriction* of a function f to a set $Y \subset$ domain (f) is written $f|Y$. An *n-ary operation over* X is a function on X^n into X. We shall denote the set of all functions on X into Y by ${}^XY$. In particular, note that ${}^0X = \{0\}$. The *Cartesian product* of a collection $\{X_i : i \in I\}$ of sets is denoted by $\prod_{i \in I} X_i$ and is defined to be the set of all functions f with domain I such that $f(i) \in X_i$ for all $i \in I$.

An *equivalence relation* over a set X is a binary relation R over X which is *reflexive* (i.e., xRx for all $x \in X$), *symmetric* (i.e., xRy implies yRx), and *transitive* (i.e., xRy and yRz implies xRz). A *partial ordering* of a set X is a binary relation R over X which is reflexive, transitive and *antisymmetric* (i.e., xRy and yRx implies $x = y$). *A simple ordering* of a set X is a partial ordering R of X which is *connected* (i.e., for all $x, y \in X$, either xRy or yRx). A *well ordering* of a set X is a simple ordering R of X with the property that every nonempty subset Y of X has a least element (i.e., an element $y \in Y$ such that yRz for all $z \in Y$). A *strict well ordering* of X is a relation R over X which is irreflexive ($x \in X$ implies not xRx) and such that the reflexive relation $R \cup \{\langle x, x \rangle : x \in X\}$ is a well ordering of X. It is easily seen that if X is (strictly) well ordered by R and $Y \subset X$, then Y is (strictly) well ordered by $R \cap Y \times Y$.

Any set X of sets is partially ordered by the inclusion relation $\subset$. X is said to be a *chain* iff X is simply ordered by $\subset$, and a *well ordered chain* iff X is well ordered by $\subset$. Two chains X, Y are said to be *isomorphic* iff there is a one–one function f, called an isomorphism on X onto Y, such that for all $x, y \in X$, $x \subset y$ if and only if $f(x) \subset f(y)$.

We now turn to ordinal numbers or ordinals. An *ordinal* is a set α such that $\bigcup \alpha \subset \alpha$ and α is strictly well ordered by the $\in$ relation. As a rule, we shall use lower case Greek letters for ordinals. This definition of ordinals is, of course, quite artificial, although it has by now become fairly standard in the literature. The intuitive idea of an ordinal is not a set at all, but a type of well ordering. Our definition of an ordinal as a certain kind of set is just a trick. However, it has the advantage of making our notation much simpler. We did the same sort of thing in our artificial definitions of ordered pairs and of functions. (If we wanted to do things in the most natural way, we would start with several different kinds of basic objects instead of starting only with sets.) We shall develop the elementary facts about ordinals carefully, so we can see that they really behave the way we want them to.

A.1. *Every element of an ordinal is an ordinal.*

Proof. Let α be an ordinal and $x \in \alpha$. If $y \in x$, then $y \in \bigcup\alpha$, so $y \in \alpha$. Hence $x \subset \alpha$, and it follows that x is strictly well ordered by $\in$.

We now show that $\bigcup x \subset x$, or in other words, $z \in y$, $y \in x$ imply $z \in x$. Let $z \in y$, $y \in x$. Since $\bigcup\alpha \subset \alpha$, we have $y \in \alpha$, and hence $z \in \alpha$. $\in$ is transitive over α, so $z \in x$. ⊣

A.2. *If α, β are ordinals, then $\alpha \subset \beta$ if and only if $\alpha \in \beta$ or $\alpha = \beta$.*

Proof. If $\alpha \in \beta$, then $\alpha \subset \bigcup\beta$ and $\bigcup\beta \subset \beta$, so $\alpha \subset \beta$.

Assume $\alpha \subset \beta$ and $\alpha \neq \beta$. Let γ be the least element of the nonempty set $\beta \setminus \alpha$. To show that $\alpha \in \beta$, we shall prove that $\alpha = \gamma$. If $\delta \in \gamma$, then $\delta \in \beta$, and since γ is the least element of $\beta \setminus \alpha$, $\delta \in \alpha$. Therefore $\gamma \subset \alpha$. Now let $\delta \in \alpha$. Since $\alpha \subset \beta$, $\delta \in \beta$. β is strictly well ordered by $\in$, so we have either $\gamma \in \delta$, $\gamma = \delta$, or $\delta \in \gamma$. Since $\gamma \notin \alpha$ and $\delta \in \alpha$, we conclude that $\gamma \neq \delta$ and $\gamma \notin \delta$. Therefore $\delta \in \gamma$. This shows that $\alpha \subset \gamma$, and completes the proof. ⊣

By combining the two previous results, we see the following:

A.3. *Every ordinal α is a well ordered chain.*

The next result is a considerable improvement over A.3.

A.4.

(i). *Every set of ordinals is strictly well ordered by ε.*
(ii). *Every set of ordinals is a well ordered chain.*

Proof. Since (i) obviously follows from (ii) and A.2, we prove only (ii).

Let X be a set of ordinals. Then $\subset$ partially orders X. Suppose α, $\beta \in X$. If not $\beta \subset \alpha$, then there is a least ordinal $\gamma \in \beta \setminus \alpha$. Since $\delta \in \gamma$ implies $\delta \in \beta \cap \alpha$, we have $\gamma \subset \alpha$. But $\gamma \notin \alpha$, so by A.2, $\gamma = \alpha$. Therefore $\alpha \in \beta$ and hence $\alpha \subset \beta$. This shows that $\subset$ is connected over X, and thus $\subset$ simply orders X.

Now let Y be a nonempty subset of X. Choose $\alpha \in Y$.

Case 1: $\alpha \cap Y = 0$. Let $\beta \in Y$; then $\beta \notin \alpha$. Moreover, $\alpha \subset \beta$ or $\beta \subset \alpha$. If $\beta \subset \alpha$, then $\beta = \alpha$. Hence α is the least element of Y.

Case 2: $\alpha \cap Y \neq 0$. Then $\alpha \cap Y$ has a least element γ. We have $\gamma \cap Y = 0$, because $\delta \in \gamma \cap Y$ implies that $\delta \in \alpha \cap Y$ and $\delta \in \gamma$, contradicting our choice of γ as the least element of $\alpha \cap Y$. Since $\gamma \cap Y = 0$, we see from Case 1 that γ is the least element of Y. Therefore $\subset$ well orders X. ⊣

Now that we have shown that the ordinals are strictly well ordered by $\in$, we shall usually write $\alpha < \beta$ for $\alpha \in \beta$, and $\alpha \leqslant \beta$ for $\alpha \in \beta$ or $\alpha = \beta$. We repeat for emphasis that

$$\alpha < \beta \text{ means the same as } \alpha \in \beta.$$

Notice that A.1 says that each ordinal α is equal to the set of all ordinals $\beta < \alpha$. Moreover, A.2 says that $\alpha \subset \beta$ if and only if $\alpha \leqslant \beta$.

The first three ordinals are: 0, $1 = \{0\}$, $2 = \{0, \{0\}\}$. The *successor of* α is the ordinal $\alpha+1 = \alpha \cup \{\alpha\}$, which is the least ordinal greater than α. α is said to be a *limit ordinal* iff it is the successor of no ordinal. The smallest limit ordinal other than 0 is denoted by ω. The elements of ω are called *finite ordinals*, or *natural numbers*; we use the letters m, n, p, q, r, s for arbitrary natural numbers.

The *infimum* of a nonempty set X of ordinals is the least element of X, and the *supremum* of X is the least ordinal which is greater than or equal to every element of X. The exercise A.5 below provides a very convenient notation for the infimum, $\bigcap X$, and the supremum, $\bigcup X$, of a set X of ordinals.

A.5. *Let X be a nonempty set of ordinals. Then $\bigcap X$ and $\bigcup X$ are ordinals. Indeed, $\bigcap X$ is the infimum of X and $\bigcup X$ is the supremum of X.*

We now come to the very useful principle of transfinite induction.

A.6 (Transfinite Induction). *Let $P(\alpha)$ be a property of ordinals. Assume that for all ordinals β, if $P(\gamma)$ holds for all $\gamma < \beta$, then $P(\beta)$ holds. Then we have $P(\alpha)$ for all ordinals α.*

PROOF. We suppose that $P(\alpha)$ fails for some α, and arrive at a contradiction. Let

$$X = \{\gamma \leqslant \alpha : P(\gamma) \text{ fails}\}.$$

X is not empty, because $\alpha \in X$. Thus X has a least element β. But $P(\gamma)$ holds for all $\gamma < \beta$, so by hypothesis $P(\beta)$ holds. This contradicts $\beta \in X$. ⊣

We shall see many examples of proofs which use transfinite induction in this book. Here is a first example.

A.7. *If α, β are ordinals which are isomorphic well ordered chains, then $\alpha = \beta$. Furthermore, the only isomorphism from the chain α to itself is the identity function.*

PROOF. Let $P(\alpha)$ be the property:

> 'the only ordinal which is isomorphic to α is α, and the only isomorphism from α to α is the identity function'.

Assume that $P(\gamma)$ holds for all $\gamma < \beta$. Let f be an arbitrary isomorphism from β to an ordinal δ. For each $\gamma < \beta$, the restriction of f to γ is an isomorphism from γ to the ordinal $f(\gamma)$. But $P(\gamma)$ holds, so $\gamma = f(\gamma)$ and f is the identity function on β. Therefore $\delta = \beta$, and $P(\beta)$ holds. ⊣

The whole point in studying ordinals is that every well ordering 'looks like' some ordinal. To make this precise we shall say that a well ordering R over a set X has *order type* α iff there is a one–one function f on X onto α such that xRy implies $f(x) \leqslant f(y)$. The function f is called an *isomorphism* from R to α. We see from A.7 that a well ordering R has *at most one* order type, because any two order types of R must be isomorphic and hence equal. Moreover, there is at most one isomorphism f between R and its order type α, because if f, g are two such isomorphisms, then $f^{-1}g$ is an isomorphism from α to α. The next result shows that the order type always exists.

A.8. *Every well ordering has exactly one order type.*

PROOF. The proof of this proposition depends on another version of the principle of transfinite induction which applies to well orderings rather than ordinals. Let R be a well ordering of a set X, and let $x \in X$. By the *initial segment* of R *determined by* x we mean the restriction $R \cap Y \times Y$ of R to the set

$$Y = \{y \in X : yRx \text{ and } y \neq x\}.$$

We use the following principle of transfinite induction – the easy proof is left for the reader.

(1) *Let $P(R)$ be a property of well orderings. Assume that for every well ordering S, if $P(T)$ holds for every initial segment T of S, then $P(S)$. Then $P(R)$ holds for all R.*

We now apply the principle (1), where $P(R)$ is the property: 'R has an order type'. Suppose every initial segment of a well ordering S over X has an order type. It suffices to prove that S has an order type. For each $x \in X$ let f_x be the unique isomorphism from the initial segment of S determined by x to its order type.

Case 1: S has a greatest element x. Then, if β is the order type of the initial segment of S determined by x, the function

$$f = f_x \cup \{\langle x, \beta\rangle\}$$

is an isomorphism from S to $\beta+1$, so $P(S)$ holds.

Case 2: S has no greatest element. We note that if ySx, then the restriction g of f_x to the set

$$\{z : zSy \text{ and } z \neq y\}$$

is an isomorphism from the initial segment of S determined by y to an ordinal. Therefore $g = f_y$ and $f_y \subset f_x$. Now form the union f of the chain $\{f_x : x \in X\}$. f is an isomorphism from S to the supremum of the order types of the initial segments of S. Thus again $P(S)$ holds and our proof is complete. ⊣

A function f whose domain is an ordinal α is called an (α-*termed*) *sequence*. An *enumeration* of a set X is a sequence whose range is X. We sometimes use the notation $\langle f_0, f_1, \ldots, f_\beta, \ldots\rangle$, $\beta < \alpha$, or f_β, $\beta < \alpha$, or even $\langle f_0, f_1, \ldots\rangle$, for an α-termed sequence f.

The *sum* $\alpha+\beta$ of two ordinals is the ordinal $\gamma \geqslant \alpha$ with the property that the chain $\gamma \setminus \alpha$ is isomorphic to the chain β. It takes transfinite induction to show that $\alpha+\beta$ exists and is unique. Intuitively we think of $\alpha+\beta$ as the list of ordinals in α followed by the list of ordinals in β. Let f be an α-termed sequence and g a β-termed sequence. The *concatenation* $f^\frown g$ of f and g is the $(\alpha+\beta)$-termed sequence which is, intuitively, obtained by first listing f_ξ, $\xi < \alpha$, and then listing g_ζ, $\zeta < \beta$. More formally, $f^\frown g$ may be defined by

$$f^\frown g(\xi) = f(\xi) \qquad \text{for} \quad \xi < \alpha,$$
$$f^\frown g(\alpha+\zeta) = g(\zeta) \qquad \text{for} \quad \zeta < \beta.$$

We shall always assume the *axiom of choice*, which states that:

If X_i is a nonempty set for each $i \in I$, then the Cartesian product $\prod_{i\in I} X_i$ is nonempty.

The axiom of choice has a number of equivalent formulations which can be found in almost any advanced textbook in mathematics. For convenience, we shall state a few of them here. The proofs of their equivalence to the axiom of choice can be found, for example, in KELLEY (1955).

A.9 (Well Ordering Principle). *Every set can be well ordered.*

A.10 (Enumeration Principle). *Every set can be enumerated.*

A.11 (Hausdorff Maximal Principle). *Every set X of sets includes a maximal chain Y, i.e., a chain $Y \subset X$ such that if $Y \subset Z \subset X$ and Z is a chain, then $Y = Z$.*

A.12 (Zorn's Lemma). *Let X be a nonempty set of sets which is closed under unions of nonempty chains (i.e., if $0 \neq Y \subset X$ and Y is a chain, then $\bigcup Y \in X$); then X has a maximal element, i.e. an element $x \in X$ such that $x \subset y \in X$ implies $x = y$.*

We shall often wish to introduce a definition by transfinite recursion. This type of definition is made possible by the following important result in set theory.

A.13. *Let G be a function on $\bigcup_{\beta<\alpha} {}^{\beta}X$ into X. Then there exists a unique α-termed sequence f such that $f(\beta) = G(f|\beta)$ for all $\beta < \alpha$.*

PROOF. We argue by transfinite induction, A.6. Assume the proposition is true for all $\alpha < \alpha_0$. We shall prove the proposition for $\alpha = \alpha_0$. Let G be a function on $\bigcup_{\beta<\alpha_0} {}^{\beta}X$ into X. For each $\gamma < \alpha_0$ there is a unique γ-termed sequence f_γ such that $f_\gamma(\beta) = G(f_\gamma|\beta)$ for all $\beta < \gamma$. If α_0 is not a limit ordinal, say $\alpha_0 = \gamma+1$, we let f be the α-termed sequence such that $f|\gamma = f_\gamma$, and $f(\gamma) = G(f_\gamma)$. Then $f(\beta) = G(f|\beta)$ for all $\beta < \alpha_0$.

Suppose now that α_0 is a limit ordinal. We note that if $\delta < \gamma < \alpha_0$, then $f_\gamma|\delta$ is a δ-termed sequence with the property that for all $\beta < \delta$,

$$f_\gamma|\delta(\beta) = f_\gamma(\beta) = G(f_\gamma|\beta) = G((f_\gamma|\delta)|\beta).$$

By uniqueness of f_δ, we have $f_\gamma|\delta = f_\delta$. In other words, $f_\delta \subset f_\gamma$. Now it follows that the union $f = \bigcup_{\gamma<\alpha} f_\gamma$ is an α-termed sequence with the property that $f(\beta) = f_{\beta+1}(\beta) = G(f_{\beta+1}|\beta) = G(f|\beta)$. In both cases we have found the desired function f.

To show uniqueness, let f' be another α_0-termed sequence such that $f'(\beta) = G(f'|\beta)$ for all $\beta < \alpha_0$. Then by the uniqueness of f_β,

$$f'|\beta = f_\beta = f|\beta \quad \text{for all} \quad \beta < \alpha_0.$$

It follows that

$$f'(\beta) = G(f'|\beta) = G(f|\beta) = f(\beta)$$

for all $\beta < \alpha_0$, so $f = f'$. ⊣

By the *power*, or *cardinality*, of a set X, denoted by $|X|$, we mean the least ordinal α such that X is enumerated by an α-termed sequence. (The existence of $|X|$ requires the axiom of choice.) An ordinal α is said to be a *cardinal*, or *initial ordinal*, if $\alpha = |\alpha|$. We shall use lower case Greek letters for cardinals as well as ordinals. The ξth infinite cardinal is denoted by $\aleph_\xi$, or alternatively by ω_ξ.

The *successor* of a cardinal α, denoted by α^+, is the least cardinal greater than α; thus $(\aleph_\xi)^+ = \aleph_{\xi+1}$. A cardinal α is said to be a *limit cardinal* iff it is not the successor of a cardinal. The *cardinal power* of α with *exponent* β is defined by $\alpha^\beta = |{}^\beta\alpha|$. The 'beths' $\beth_\xi$ are defined recursively by

$$\beth_0 = \aleph_0, \qquad \beth_{\xi+1} = 2^{\beth_\xi},$$

and when ξ is a limit ordinal,

$$\beth_\xi = \bigcup_{\zeta<\xi} \beth_\zeta.$$

The 'beths' are closely tied in with the notion of a strong limit cardinal. A cardinal α is called a *strong limit cardinal* iff, for all cardinals $\beta < \alpha$, $2^\beta < \alpha$. We shall let the reader prove the following.

A.14.

(i). *α is an infinite limit cardinal if and only if there is a limit ordinal ξ such that $\alpha = \aleph_\xi$.*

(ii). *α is an infinite strong limit cardinal if and only if there is a limit ordinal ξ such that $\alpha = \beth_\xi$.*

A.15.

(i). $\xi \leqslant \aleph_\xi \leqslant \beth_\xi$.

(ii). *There are arbitrarily large cardinals α such that*

$$\alpha = \aleph_\alpha = \beth_\alpha.$$

The *continuum hypothesis* (CH) states that $\beth_1 = \aleph_1$, and the *generalized continuum hypothesis* (GCH) states that, for all ξ, $2^{\aleph_\xi} = \aleph_{\xi+1}$. Thus the GCH implies that for all ξ, $\beth_\xi = \aleph_\xi$. The GCH also implies that every limit cardinal is a strong limit cardinal.

We do not assume the CH or GCH as part of our intuitive set theory. The GCH is interesting because it dramatically simplifies the arithmetic of cardinals, and we shall sometimes prove theorems which need the GCH as an additional hypothesis.

Some of the basic results in the arithmetic of cardinals are collected in the next three propositions. They are included mainly for reference, so the proofs are not given. Most of them are proved in Kamke (1950).

A.16. *The following three conditions are equivalent*:

(i). $|X| \leqslant |Y|$ (*as ordinals*).

(ii). *There is a one-one function on X into Y.*

(iii). *There is a function on Y onto X.*

A.17. *Let* α, β, γ *be arbitrary cardinals.*

(i). $\alpha < 2^\alpha$.

(ii). *If* $\alpha \leqslant \beta$, *then* $\alpha^\gamma \leqslant \beta^\gamma$ *and* (*if* $0 < \gamma$) $\gamma^\alpha \leqslant \gamma^\beta$.

(iii). $\alpha^0 = 1$; $1^\alpha = 1$; *if* $\alpha > 0$, $0^\alpha = 0$.

(iv). *Let* β *be infinite and* $\gamma > 0$. *Then*

$$(\alpha^\beta)^\gamma = (\alpha^\gamma)^\beta = \alpha^\beta \cup \alpha^\gamma = \alpha^{\beta \cup \gamma}.$$

Special case: $(2^\beta)^\beta = 2^\beta$.

(v). *If* α *is infinite and* $n > 0$, *then* $\alpha^n = \alpha$.

(vi). *If X is a nonempty set of cardinals, then* $\bigcup X$ *and* $\bigcap X$ *are cardinals.*

A.18.

(i). *If* $X \cup Y$ *is infinite, then*

$$|X \cup Y| = |X| \cup |Y|.$$

(ii). *If* $X \cup Y$ *is infinite and X and Y are nonempty, then*

$$|X \times Y| = |X| \cup |Y|.$$

(iii). *If X is an infinite set of sets, then*

$$|\bigcup X| \leqslant |X| \cup \bigcup\{|x| : x \in X\},$$

that is, the cardinal of $\bigcup X$ *is at most the supremum of the cardinal of X and the cardinals of the elements of X.*

The inequality in (iii) above may be replaced by an equality in the following two important cases:

(a) *X is well ordered by inclusion;*
(b) *any two sets in X are disjoint.*

The set of all subsets of X, also called the *power set* of X, is written $S(X)$. If α is a cardinal, we shall write

$$S_\alpha(X) = \{x \subset X : |x| < \alpha\},$$
$$S^\alpha(X) = \{x \subset X : |X \setminus x| < \alpha\}.$$

For example, $S_\omega(X)$ is the set of all finite subsets of X and $S^\omega(X)$ the set of all cofinite subsets of X.

A.19.
(i). $|S(X)| = 2^{|X|}$.
(ii). *If X is infinite, then*

$$|S_\omega(X)| = |X|.$$

Part (i) is obvious, and (ii) follows from A.17 (v).

The *rank function* $R(\xi)$ is obtained by iterating the operation of forming power sets. We define inductively

$$R(0) = 0;$$
$$R(\xi+1) = S(R(\xi));$$
if $\xi > 0$ is a limit ordinal, $R(\xi) = \bigcup_{\eta<\xi} R(\eta)$.

The rank function has the following obvious properties:

A.20.
(i). *If* $\eta < \xi$, *then* $R(\eta) \subset R(\xi)$.
(ii). *For each ordinal* ξ, $|R(\omega+\xi)| = \beth_\xi$.
(iii). $R(\xi) = \bigcup_{\eta<\xi} S(R(\eta))$, *for all* $\xi > 0$.

The *axiom of regularity* is the following statement:

A.21. *For every set* x *there is an ordinal* ξ *such that* $x \in R(\xi)$.

A.22. *The axiom of regularity is equivalent to each of the following:*
(i). *For every nonempty set* x *there exists* $y \in x$ *such that* $x \cap y = 0$.
(ii). *There is no infinite sequence* $x_0, x_1, x_2, \ldots$ *of sets such that* $x_1 \in x_0$, $x_2 \in x_1$, $x_3 \in x_2, \ldots$.

The equivalence of (i) with the axiom of regularity does not require the axiom of choice, but the axiom of choice is needed to prove that (ii) implies the axiom of regularity. The statement (ii) has the consequence that for every set x, $x \notin x$, and more generally there cannot be any finite cycle $x_0 \in x_1 \in \ldots \in x_n \in x_0$.

The axiom of regularity is not used at all in this book. It is, however, usually included among the axioms of set theory, because only the 'regular' sets $x \in R(\xi)$ are considered to be mathematically interesting. For this reason we may as well include the axiom of regularity as part of the intuitive set theory.

One consequence of the axiom of regularity which may be useful in model theory is the following principle of induction (compare A.6):

A.23. *Let $P(x)$ be a property of sets. Assume that for all sets y, if $P(z)$ holds for all $z \in y$, then $P(y)$ holds. Then we have $P(x)$ for all sets x.*

In the next few pages we shall develop in some detail the important notion of cofinality. Let ξ be a limit ordinal. A set X is said to be *cofinal in* ξ iff $X \subset \xi$ and $\xi = \bigcup X$. The *cofinality* of ξ, written $\mathrm{cf}(\xi)$, is the least cardinal α such that a set of power α is cofinal in ξ.

Notice in particular that $\mathrm{cf}(0) = 0$. We define the cofinality of a successor ordinal by simply writing

$$\mathrm{cf}(\xi+1) = 1.$$

How does the cofinality function behave? One thing which we can see at once from the definition is

$$\omega \leqslant \mathrm{cf}(\xi) \leqslant \xi,$$

whenever ξ is an infinite limit ordinal. Putting $\xi = \omega$, we obtain the equation

$$\omega = \mathrm{cf}(\omega).$$

A cardinal α is said to be *regular* iff $\mathrm{cf}(\alpha) = \alpha$, and to be *singular* iff $\mathrm{cf}(\alpha) < \alpha$. So we have the following:

A.24. *ω is regular.*

By a quirk of fate, 0 and 1 are regular, but all finite cardinals $n > 1$ are singular. However, the notions of regular and singular cardinals are important only for infinite cardinals. Are there any other regular cardinals? The next proposition gives us some others.

A.25. *Any infinite successor cardinal α^+ is regular.*

PROOF. Let X be cofinal in α^+. By A.18 (iii),

$$\alpha^+ \leqslant |X| \cup \bigcup\{|\beta| : \beta \in X\}.$$

But $|\beta| \leqslant \alpha$ for all $\beta \in X$, so

$$\bigcup\{|\beta| : \beta \in X\} < \alpha^+.$$

This means that $|X| = \alpha^+$, and $\mathrm{cf}(\alpha^+) = \alpha^+$. ⊣

When we begin looking at limit cardinals other than ω, we usually find that they are singular. For example, it is easy to see that

$$\mathrm{cf}(\aleph_\omega) = \omega, \qquad \mathrm{cf}(\beth_\omega) = \omega.$$

Thus $\aleph_\omega$ and $\beth_\omega$ are singular. More generally, we may prove another result.

A.26. *For any limit ordinal $\xi > 0$, we have*

$$\mathrm{cf}(\aleph_\xi) = \mathrm{cf}(\beth_\xi) = \mathrm{cf}(\xi).$$

PROOF. If X is cofinal in ξ, then the set $Y = \{\aleph_\gamma : \gamma \in X\}$ is cofinal in $\aleph_\xi$, and $|Y| = |X|$. Hence $\mathrm{cf}(\aleph_\xi) \leqslant \mathrm{cf}(\xi)$.

If X' is cofinal in $\aleph_\xi$, then the set

$$Y' = \{\gamma : \text{for some } \beta \in X', |\beta| = \aleph_\gamma\}$$

is cofinal in ξ, and $|Y'| \leqslant |X'|$. So we have $\mathrm{cf}(\xi) \leqslant \mathrm{cf}(\aleph_\xi)$. This proves $\mathrm{cf}(\xi) = \mathrm{cf}(\aleph_\xi)$. The proof of the equation $\mathrm{cf}(\xi) = \mathrm{cf}(\beth_\xi)$ is similar. ⊣

We now, naturally, ask the question: Are any limit cardinals greater than ω regular? From the usual axioms of set theory it is impossible to prove that the answer is yes, and it is probably also impossible to prove that the answer is no. A *strong* limit cardinal which is regular is called an *inaccessible cardinal.* Thus ω is an inaccessible cardinal. Other inaccessible cardinals, if they exist, have many nice properties in common with ω. The *hypothesis of inaccessibility* states that there are inaccessible cardinals greater than ω. We shall not assume the hypothesis of inaccessibility as a part of our intuitive set theory – like the GCH, it is a plausible extra assumption about set theory which can usually be avoided.

Let us turn again to the cofinality function.

A.27. *Let ξ be an infinite limit ordinal. Then there is a cofinal set Y in ξ which is isomorphic as a chain to* $\mathrm{cf}(\xi)$. *If ξ is a limit cardinal, we may choose Y to be a set of cardinals.*

Proof. Let $\alpha = \mathrm{cf}(\xi)$ and let X be a cofinal set in ξ of power α. Let x_β, $\beta < \alpha$, be an enumeration of X. We define a function f on α into ξ by transfinite recursion. Let $f(0) = 0$. Now let $0 < \beta < \alpha$, and suppose we have already chosen $f(\gamma) \in \xi$ for each $\gamma < \beta$. Then the set

$$Z = \{x_\gamma : \gamma < \beta\} \cup \{f(\gamma) : \gamma < \beta\}$$

is a subset of ξ of power less than α, so we cannot have $\xi = \bigcup Z$. The only other possibility is $\bigcup Z < \xi$. We may therefore define $f(\beta)$ to be the least $\zeta < \xi$ such that $\bigcup Z < \zeta$. The function f on α into ξ has the following properties:

(1) if $\gamma < \beta < \alpha$, then $f(\gamma) < f(\beta)$;

(2) $\xi = \bigcup_{\beta<\alpha} f(\beta)$.

It follows that the set $Y = \mathrm{range}\,(f)$ is cofinal in ξ, and Y has the isomorphism f with α. ⊣

A.28. *For every ordinal ξ,* $\mathrm{cf}(\xi)$ *is a regular cardinal.*

Proof. We must show that

$$\mathrm{cf}(\mathrm{cf}(\xi)) = \mathrm{cf}(\xi).$$

This equation is obvious if $\mathrm{cf}(\xi) = 0$ or $\mathrm{cf}(\xi) = 1$, so let us assume that $\mathrm{cf}(\xi)$ is infinite, and let $\alpha = \mathrm{cf}(\xi)$. By A.27, there is a set Y cofinal in ξ which is isomorphic to α, and let f be the isomorphism from α to Y. Let Z be cofinal in α. Then the set

$$W = \{f(\beta) : \beta \in Z\}$$

is cofinal in ξ. Then $\alpha \leqslant |W|$. But $|W| \leqslant |Z|$, so $\alpha \leqslant |Z|$ and $\mathrm{cf}(\alpha) = \alpha$. ⊣

A.29. *If α is an infinite cardinal, then $\alpha < \alpha^{\mathrm{cf}(\alpha)}$.*

Proof. Let Y be a set cofinal in α and f an isomorphism from $\mathrm{cf}\,(\alpha)$ to Y.

Now let g be any function on α into the set ${}^{\mathrm{cf}(\alpha)}\alpha$. We wish to show that g cannot be onto ${}^{\mathrm{cf}(\alpha)}\alpha$. To show this we define a function $h \in {}^{\mathrm{cf}(\alpha)}\alpha$ in the following way. For each $\beta < \mathrm{cf}\,(\alpha)$, $h(\beta)$ is the least $\xi < \alpha$ which does not belong to the set

(1) $$\{g_\gamma(\beta) : \gamma < f(\beta)\}.$$

We know that there is such a $\xi < \alpha$ because the set (1) must have power at most $|f(\beta)|$, and $|f(\beta)| < \alpha$. Now the function h cannot lie anywhere in the list g_γ, $\gamma < \alpha$, because when $f(\beta) > \gamma$ we have $h(\beta) \neq g_\gamma(\beta)$. It follows that g does not map α onto ${}^{\mathrm{cf}(\alpha)}\alpha$, and so

$$\alpha < |{}^{\mathrm{cf}(\alpha)}\alpha| = \alpha^{\mathrm{cf}(\alpha)}. \dashv$$

Incidentally, the result which we have just proved is a form of what is known as Konig's theorem. It is an improvement of Cantor's theorem that $\alpha < 2^\alpha$ (A.17 (i)), because we know that

$$2^{\mathrm{cf}(\alpha)} \leqslant \alpha^\alpha = 2^\alpha.$$

Indeed, the proof of A.29 is a form of Cantor's famous 'diagonal method' which is used to prove that $\alpha < 2^\alpha$.

In this part of the appendix, we shall give a brief description of three of the main systems of axiomatic set theory, namely the set theories of Zermelo–Fraenkel, Bernays, and Bernays–Morse.

A.30. Zermelo–Fraenkel Set Theory, ZF. This theory is formulated in first-order logic with identity and the binary relation symbol $\in$. The axioms are the following, 1–9:

1. *Extensionality*: $(\forall xy)(x \equiv y \leftrightarrow (\forall z)(z \in x \leftrightarrow z \in y))$.
   Intuitively, $x \equiv y$ iff x and y have the same elements.
2. *Null set*: $(\exists x \forall y)(\neg\, y \in x)$.
3. *Pairs*: $(\forall xy \exists z \forall u)(u \in z \leftrightarrow u \equiv x \vee u \equiv y)$.
   Intuitively, if x, y are sets, so is $\{x, y\}$.
4. *Unions*: $(\forall x \exists y \forall z)(z \in y \leftrightarrow (\exists w)(z \in w \wedge w \in x))$.
   Intuitively, if x is a set, so is $\bigcup x$.
5. *Power set*: $(\forall x \exists y \forall z)(z \in y \leftrightarrow (\forall w)(w \in z \rightarrow w \in x))$.
   Intuitively, if x is a set, so is $S(x)$.
6. *Infinity*: $(\exists x)((\exists y)(y \in x) \wedge (\forall y)(y \in x \rightarrow (\exists z)(y \in z \wedge z \in x)))$.
   Intuitively, there exist infinite sets.
7. *Regularity*: $(\forall x)((\exists y)(y \in x) \rightarrow (\exists y)(y \in x \wedge \neg\, (\exists z)(z \in y \wedge z \in x)))$.
   Intuitively, every nonempty set is disjoint from one of its elements.
8. *Subsets*: $(\forall x \exists y \forall z)(z \in y \leftrightarrow z \in x \wedge \varphi(zu_1 \ldots u_n))$, where φ is any formula in which y does not occur.
   Intuitively, $y = \{z \in x : \varphi(zu_1 \ldots u_n)\}$ is a set.

9. *Collection*: $(\forall x)[x \in u \rightarrow (\exists z)\varphi(xzuv_1 \ldots v_n)] \rightarrow (\exists y \forall x)[x \in u \rightarrow (\exists z)(z \in y \wedge \varphi(xzuv_1 \ldots v_n))]$, where φ is a formula in which y does not occur.
Intuitively, if each of the classes $C_x = \{z : \varphi(xz \ldots)\}$, $x \in u$, is non-empty, then there exists a set y which meets each of these classes.

Note that both 8 and 9 are infinite schemes of axioms rather than a single axiom. The subset scheme and the collection scheme are often combined into a single scheme of axioms called the replacement scheme.

Replacement:
$(\forall x \exists! z)\varphi(xzuv_1 \ldots v_n) \rightarrow (\exists y \forall z)[z \in y \leftrightarrow (\exists x)(x \in u \wedge \varphi(xzuv_1 \ldots v_n))]$
where φ is a formula in which y does not occur and $\exists! z$ means 'there exists a unique z'.
Intuitively, if $F(x)$ is the unique z such that $\varphi(xz \ldots)$, then $\{F(x) : x \in u\}$ is a set.

The scheme of replacement can be proved from the axioms of ZF. On the other hand, the subset and collection schemes can be proved from the remaining axioms 1–7 of ZF plus the scheme of replacement. Thus axioms 1–7 plus replacement is an alternative set of axioms for ZF. The advantage of our list of axioms 1–9 is that it is easy to describe the subtheories of ZF below.

Zermelo–Fraenkel set theory with choice, ZFC, has the axioms of ZF plus the following:

10. *Choice*: $(\forall x \exists y)[y$ is a function with domain $x \wedge (\forall z)(z \in x \wedge (\exists u)(u \in z) \rightarrow y(z) \in z)]$.
Intuitively, every set has a choice function.

All of the results in this book, except for a very small number of results which involve proper classes, can be formulated and proved in ZFC.

A.31. Subtheories of ZF. Two important subtheories of ZF are Zermelo set theory and ZF minus the power set axiom.

(i). *Zermelo set theory*, Z, has all the axioms of ZF except the scheme of collection, that is, axioms 1–8. For each limit ordinal $\alpha > \omega$, $\langle R(\alpha), \in \rangle$ is a model of Zermelo set theory. The importance of this theory is mainly historical, for it came before ZF. Zermelo set theory with the axiom of choice is adequate for much of classical mathematics and most of the results of this book. However, it is not adequate for defining the sequences $\aleph_\alpha$ or $R(\alpha)$, or proving that every well ordering has the order type of some ordinal. Thus one has to treat cardinals and well orderings differently in Zermelo set theory.

(ii). *Zermelo–Fraenkel set theory minus the power set axiom*, or ZF − P, has all the axioms of ZF except the power set axiom, namely axioms 1–4 and 6–9. This theory (and some of its subtheories) is important in the recent literature because of its relation to generalized recursion theory and infinitary logic. It is not adequate for classical mathematics because the set of real numbers cannot be shown to exist. On the other hand, much of modern set theory can be done in ZF − P. We use ZF − P in Section 4.2 to give a quick statement of the axiom of constructibility.

(iii). *Set-theoretic arithmetic*. This theory has all the axioms of ZF except that the axiom of infinity is replaced by its negation. Thus the axioms are 1, 2, 3, 4, 5, ¬ 6, 7, 8, 9. The standard model for this theory is $\langle R(\omega), \in\rangle$. Set-theoretic arithmetic is equivalent to Peano arithmetic in the sense that every formula of one language has a simple translation into a formula of the other language such that a sentence is provable in one theory if and only if its translation is provable in the other theory.

In ZF, one can prove the consistency of Zermelo set theory, ZF − P, and set-theoretic arithmetic. The consistency of set-theoretic arithmetic can also be proved in Zermelo set theory and in ZF − P.

A.32. Bernays Set Theory. This theory is formulated in first-order logic with identity and with a binary relation symbol $\in$ and a 1-placed relation symbol V. $V(x)$ is read 'x is a set', and $\neg V(x)$ is read 'x is a proper class'. In order to make the axioms readable, we need to introduce some abbreviations. Given a formula φ, we define the *relativization* of φ to V, in symbols φ^V, inductively as follows:

$$\text{If } \varphi \text{ is an atomic formula, } \varphi^V = \varphi;$$
$$(\varphi_1 \wedge \varphi_2)^V = \varphi_1^V \wedge \varphi_2^V;$$
$$(\neg \varphi_1)^V = \neg(\varphi_1^V);$$
$$((\exists x)\varphi)^V = (\exists x)(V(x) \wedge \varphi^V);$$
$$((\forall x)\varphi)^V = (\forall x)(V(x) \rightarrow \varphi^V).$$

Thus φ^V is interpreted as φ with all the quantified variables restricted to V.

The axioms of Bernays set theory are given by the list of axioms 1*–10* below:

1*. *Extensionality*: Same as axiom 1.

2*–6*. The relativizations to V of the axioms 2–6, i.e. the null set axiom, axiom of pairs, axiom of unions, power set axiom and axiom of infinity.

7*. *Regularity*: Same as axiom 7.

8*. *Replacement*:
$(\forall w)(w \text{ is a function} \rightarrow [(\forall u \exists z \forall y)(y \in z \leftrightarrow (\exists x)(x \in u \wedge y \equiv w(x)))]^V)$, where the notion of a function and the notion $x(z)$ are used in the usual way. Note that this time the axiom of replacement is a single formula rather than an infinite scheme, because the class w takes the place of the formula $\varphi(x, y)$.

9*. *Universe class*: $(\forall x)(V(x) \leftrightarrow (\exists y)(x \in y))$.
Intuitively, x is a set iff x is an element of some class.

10*. *Comprehension*:

$$(\exists x \forall y)(y \in x \leftrightarrow V(y) \wedge \varphi^V(yu_1 \ldots u_n)),$$

where $\varphi(yu_1 \ldots u_n)$ is any formula in which x does not occur.
Intuitively, the class $x = \{y : V(y) \wedge \varphi^V(yu_1 \ldots u_n)\}$ exists. This axiom states that there are lots of classes. Note that $V(y)$ is needed here to avoid Russell's paradox.

There are two natural axioms of choice for Bernays set theory, one of which is stronger than the other. The weaker axiom of choice, the 'axiom of choice for sets', is just the relativization of Axiom 10 to V. The stronger axiom of choice, the 'axiom of choice for classes', is Axiom 10 as it stands. Intuitively, the former says that every set has a choice function and the latter says that every class has a choice function.

A.33. Bernays–Morse Set Theory. This theory has the same axioms as Bernays set theory except that the comprehension scheme 10* is replaced by the stronger scheme

10**. $(\exists x \forall y)(y \in x \leftrightarrow V(y) \wedge \varphi(yu_1 \ldots u_n))$
where φ is any formula in which x does not occur.

The difference is that Bernays set theory has the comprehension scheme only for formulas whose quantifiers range over sets, while in Bernays–Morse set theory the quantifiers in the comprehension scheme are allowed to range over arbitrary classes. Thus Bernays–Morse set theory is an extension of Bernays set theory. The relation between the models of the three set theories described here is discussed in Section 1.4. Bernays–Morse set theory will be especially convenient in the last section of the book, Section 7.4, where the results are most easily expressed in terms of classes. The axiom of choice for sets, or the stronger axiom of choice for classes, may also be added to Bernays–Morse set theory.

APPENDIX B

OPEN PROBLEMS IN CLASSICAL MODEL THEORY

Here is a list of selected open problems. Generally, a problem is selected either because it has historical interest, or because it has intrinsic interest, or because its solution promises interesting or important new methods. The problems are arranged roughly into three groups: old problems (1–7), specific problems (8–21), and general problems (22–24). When a problem can be stated most strikingly as a conjecture, we shall state it that way, even where we believe that the conjecture is not true. Unless otherwise stated, assume that the language $\mathscr{L}$ is countable. Whenever the proposer of a problem is known to us, his name appears in parentheses after the problem.

1. *The finite spectrum problem*: Given a sentence φ, the *finite spectrum* of φ is the set of all $n < \omega$ such that φ has a model of power n.

Conjecture: If $S \subset \omega$ is the finite spectrum of some sentence of $\mathscr{L}$, then $\omega \setminus S$ is also the finite spectrum of some sentence of $\mathscr{L}$ (Scholz).

2. *Conjecture*: There is a Jónsson group of power ω_1 (Kurosh).

3. For each $n \leqslant \omega$, let $\mathfrak{A}_n$ be the free group with n generators x_m, $m < n$.

Conjecture: If $2 \leqslant p < n$, then $\mathfrak{A}_p \equiv \mathfrak{A}_n$, and in fact $\mathfrak{A}_p \prec \mathfrak{A}_n$ (Tarski).

4. *Conjecture*: If D is an uniform countably incomplete ultrafilter over β and A is any infinite set, then $|\prod_D A| = |A|^\beta$. (Known to be true if $\beta = \omega_n$, $n < \omega$, and $V = L$ holds.)

5. *Conjecture*: If T is a complete finitely axiomatizable theory in $\mathscr{L}$, then T is not ω_1-categorical, and T is unstable.

6. *Conjecture*: Without the continuum hypothesis, it can be shown that any theory T with more than ω non-isomorphic models of power ω has 2^ω non-isomorphic models of power ω (Vaught).

7. A model $\mathfrak{A}$ is said to be *rigid* iff the only automorphism of $\mathfrak{A}$ is the identity function. Given a theory T in $\mathscr{L}$, define $R(T)$ to be the class of all infinite cardinals α such that T has a rigid model of power α.

Conjecture (GCH): For every theory T in $\mathscr{L}$, either

$$R(T) = \{\alpha : \alpha \geqslant \omega\},$$

$$R(T) = \{\alpha : \alpha > \omega\},$$

$$R(T) = \{\alpha : \omega \leqslant \alpha \leqslant \omega_\beta\} \text{ for some ordinal } \beta < \omega_1,$$

or

$$R(T) = \{\alpha : \omega < \alpha \leqslant \omega_\beta\} \text{ for some ordinal } \beta < \omega_1$$

(Ehrenfeucht).

8. *Conjecture*: There is a Jónsson model of power $\aleph_\omega$; of power $\beth_\omega$ (Jónsson).

9. For a theory T in $\mathscr{L}$, let $f_T(\alpha)$ be the number of non-isomorphic models of T of power α.

Conjecture: If $\omega < \alpha < \beta$, then $f_T(\alpha) \leqslant f_T(\beta)$ (Morley).

10. *Conjecture* (GCH): If $\alpha, \beta \geqslant \omega$ and β is singular, then every theory which admits (α^+, α) admits (β^+, β). This has been proved by Jensen assuming $V = L$ (Vaught).

11. Find a simple set of axioms for the theory of all models $\mathfrak{A} = \langle A, U, \ldots\rangle$ such that $|A| \geqslant \beth_\omega(|U|)$ (Vaught).

12. *Conjecture*: Every theory T which admits $(\aleph_\omega, \aleph_0)$ admits $(2^{\aleph_0}, \aleph_0)$. This is trivial with the CH (Shelah).

13. 3-*cardinal problem*: For which triples of cardinals (α, β, γ), $(\alpha', \beta', \gamma')$ does every theory which admits (α, β, γ) admit $(\alpha', \beta', \gamma')$? (Vaught).

14. *Conjecture* (GCH or $V = L$): If α is less than the first measurable cardinal $> \omega$, then every uniform ultrafilter over α is regular (Keisler).

15. *Conjecture*: Let $0 < n < \omega$. Every uniform ultrafilter over ω_n is regular (Keisler). This was proved by Prikry and Jensen assuming $V = L$.

16. *Conjecture* (GCH): If there is a uniform ω_1-descendingly complete ultrafilter over α, then α is no smaller than the first uncountable measurable cardinal (Keisler).

17. *Conjecture* (GCH): Let D be an ultrafilter over α such that every ultraproduct $\prod_D A_\beta$ which is infinite has power $\geqslant \alpha^+$. Then D is α^+-good (Keisler).

18. *Conjecture*: Let $|A|, |B|, ||\mathscr{L}|| \leqslant \alpha$ and let D be a regular ultrafilter over α. If $\mathfrak{A} \equiv \mathfrak{B}$, then $\prod_D \mathfrak{A} \cong \prod_D \mathfrak{B}$. (Chang and Keisler)

19. *Conjecture*: If D is a regular ultrafilter over α, then for all infinite $\mathfrak{A}$, $\prod_D \mathfrak{A}$ is α^{++}-universal. (Chang and Keisler)

20. *Conjecture*: Let T be a complete theory having infinite models. Then exactly one of the following will hold:

(i). T is categorical in every power $\aleph_\alpha$, $\alpha \geqslant \beth_1$.

(ii). T has exactly $\beth_2$ non-isomorphic models in every power $\aleph_\alpha$, $\alpha \geqslant \beth_1$.

(iii). T has at least $|\alpha|$ non-isomorphic models in every power $\aleph_\alpha$, $\alpha \geqslant \beth_1$.

(Shelah)

21. Let $\mathscr{L}' \subset \mathscr{L}$ and let T be a complete theory in $\mathscr{L}$. T is said to be *γ-saturated over $\mathscr{L}'$* iff every model of T of power γ with a γ-saturated reduct to $\mathscr{L}'$ is γ-saturated.

Conjecture (GCH): Let $\alpha, \beta > \omega$. If T is α^+-saturated over $\mathscr{L}'$, then T is β^+-saturated over $\mathscr{L}'$. (Chang)

22. Investigate set theory based on the axioms of ZFC plus the gap n conjecture, with or without the GCH.

23. Develop a theory of models which stresses the order type of the model $\mathfrak{A} = \langle A, <, \ldots \rangle$ rather than the cardinality of the set A.

24. Develop the model theory of second- and higher-order logic.

The following discussion surveys the progress which has been made on the above problems between the publication of the first edition of this book in 1973, and the writing of the current edition in 1989.

Conjecture 1 remains open. It is related to the conjecture that exponential time is equal to nondeterministic exponential time in complexity theory.

Conjecture 2 was proved by Shelah (1980). In fact, Shelah constructed a group G of power ω_1 such that for every $X \subset G$ of power ω_1, every $g \in G$ is a product of fewer than 10^6 elements of X. Thus G has no proper subgroups of power ω_1, and even no proper subsemigroups of power ω_1.

Conjecture 3 remains open. The following partial results was proved by Sacerdote (1973). If $2 \leqslant p < n \leqslant \omega$, then for every Σ_3^0 formula $\varphi(x_1 \dots x_m)$ and $a_1, \dots, a_m \in A_p$, if $\mathfrak{A}_p \models \varphi[a_1 \dots a_m]$, then $\mathfrak{A}_n \models \varphi[a_1 \dots a_m]$.

Conjecture 4 should have been stated with the additional hypothesis that β is less than the first uncountable measurable cardinal, for otherwise there is an easy counterexample. Let E be a countably complete uniform ultrafilter over β, let F be a uniform ultrafilter over ω, and let $D = E \times F$. Then $|\prod_d \omega| = 2^\omega$.

Magidor (1976), (1979) obtained counterexamples to Conjecture 4 for $\beta = \omega_3$ and $\beta = \omega_2$ assuming the consistency of ZFC + [There is a huge cardinal]. Foreman, Magidor, and Shelah have announced an analogus result for $\beta = \omega_1$. See the discussion in Section 4.3 following Proposition 4.3.9.

Donder (1988) proved that if ZFC is consistent, then so is ZFC + [Every uniform ultrafilter over an infinite set is regular], and it follows that ZFC + [Conjecture 4] is also consistent.

Conjecture 5: Makowski (1974) disproved half of Conjecture 5, giving an example of a complete finitely axiomatizable superstable theory. Peretyatkin (1980) gave a full counterexample by giving an example of a complete finitely axoimatizable ω_1-categorical theory. The corresponding problem for theories categorical in all infinite powers turned out to have the opposite answer. Zilber (1980, 1984) and Cherlin, Harrington and Lachlan (1985) proved that there is no finitely axiomatizable complete theory which is categorical in all infinite powers, and the latter paper improved this to show that there is no complete finitely axiomatizable ω-categorical ω-stable theory.

Conjecture 6, known as Vaught's Conjecture, remains open. It has been proved in several special cases, which are discussed at the end of Section 7.1.

Conjecture 7 was disproved by Shelah (1976). He showed, assuming the GCH, that a class C of cardinals has the form $R(T)$ for some finitely axiomatizable T if and only if there is a Σ_2^1 sentence φ such that C is the class of all infinite cardinals of models of φ. Even without the GCH, the classes $R(T)$ are much richer than Conjecture 7 suggests.

Conjecture 8 is open.

Conjecture 9 was proved by Shelah (1985), (19??). See the discussion at the end of Section 7.1.

Conjecture 10: It was announced in an abstract by Litman and Shelah

(1976), and proved in Shelah (1978) (as explained by Ben-David 1988), that the negation of Conjecture 10 is consistent relative to a strong axiom of infinity. The result is that if ZFC + [There is a supercompact cardinal] is consistent, then so is ZFC + GCH + [There is a theory which admits (ω_1, ω) but does not admit $(\omega_{\omega+1}, \omega_\omega)$]. See Section 7.2 for further discussion on this problem.

Conjecture 11 was partially solved by Barwise (1975) and Schmerl (1977). They independently gave sets of axioms for the theory in question. However, neither set of axioms is simple enough to be easily read.

Conjecture 12 was proved by Shelah (1977) without the CH.

Conjecture 13 remains open in general. It was solved under the assumption of the axiom of constructibility by Jensen (see Devlin 1984). Foreman (1982) obtained consistency results concerning Chang's conjecture for three cardinals. Partial results in ZFC are indicated in problems 7.2.18–7.2.22.

Conjectures 14 and 15 are closely related to Conjecture 4 on cardinalities of ultrapowers.. Donder (1988) proved that the axiom of constructibility implies Conjecture 14 (and 15), that every uniform ultrafilter is regular. Donder used a weaker hypothesis than $V = L$, and his result improves several earlier theorems, including the result of Ketonen (1976) that if $O^{\#}$ does not exist then every uniform ultrafilter over ω_1 is regular. Magidor (1976), (1979) showed that the negation of Conjecture 15 is consistent relative to a large cardinal axiom. He proved that if ZFC + [There exists a huge cardinal] is consistent, then so are ZFC + [There is a uniform ultrafilter over ω_n which is not regular] where n is either 3 or 2. Foreman, Magidor, and Shelah announced an analogus result for $n = 1$.

Conjecture 16: Jensen, Prikry, and Silver (see Prikry (1973)) prove assuming $V = L$ that for every $\alpha > \omega$ which is regular but not weakly compact, and every $\beta < \alpha$, every uniform D over α is β-descendingly incomplete. D.V. and G. V. Choodnovsky (1974) extend the result to certain weakly compact cardinals. Silver (1974) proves assuming $0^{\#}$ does not exist that for every inaccessible $\alpha > \omega$ and every uniform D over α, there exists $\omega < \beta < \alpha$ such that D is not β-descendingly complete.

Conjecture 17 was disproved by Shelah (1976c). Without the GCH he constructed, over each $\alpha > \omega$, a regular non-good ultrafilter D such that every ultraproduct $\prod_D A_\beta$ is finite or of power $\geqslant 2^\alpha$.

The negation of Conjecture 18 was shown to be consistent by Shelah

(1988a, 19??a). He proved that if ZFC is consistent, then so is ZFC + $2^{\omega} = \omega_2$ + [There are countable graphs $\mathfrak{A}$, $\mathfrak{B}$ such that $\mathfrak{A} \equiv \mathfrak{B}$ but there are no ultrafilters D, E over ω with $\prod_D \mathfrak{A} \cong \prod_E \mathfrak{B}$].

Conjecture 19 is still open.

A modified form of Conjecture 20 was proved by Shelah (1985, 19??). See the discussion at the end of Section 7.1.

Conjecture 21 was solved negatively by Pabion (1982).

Update, October 2011: The current state of Conjecture 1 is surveyed in the paper Durand et.al., "Fifty Years of the Spectrum Problem", submitted to the Bulletin of Symbolic Logic and currently online at
`www.diku.dk/hjemmesider/ansatte/neil/SpectraSubmitted.pdf` .
The question was posed by Asser in 1955. In 1972, Jones and Selman proved that a set is the finite spectrum of some first order sentence if and only if it is computable in nondeterministic exponential time.

Conjecture 3 was solved affirmatively by Kharlampovich and Myasnikov, "Elementary theory of free non-abelian groups". J. Algebra 302 (2006), 451-552, and Sela, "Diophantine geometry over groups. VI. The elementary theory of a free group.", Geom. Funct. Anal. 16 (2006), 707-730.

Kennedy and Shelah, "More on regular reduced products", J. Symb. Logic 69 (2004), 1261-66, answered Conjecture 19 by showing that it is equivalent to a combinatorial condition on the ultrafilter. That paper also improved Shelah's earlier negative answer to Conjecture 18.

The remaining open problems in the list are Conjecture 1, Conjecture 6 of Vaught, Conjecture 8 on Jonsson models, and 13 (the 3-cardinal problem).

HISTORICAL NOTES

Notes for Chapter 1

1.1 For an account of the early development of sentential and predicate logic, see Church (1956). Universal algebra as a separate subject appeared in the works of Whitehead and Russell (1913, 1925, 1927); in more recent years, the books of Birkhoff (1961), Cohn (1965), and Grätzer (1968) are representative of the field. The books of Cohn and Grätzer contain further discussions of the relation between universal algebra and model theory.

1.2 Many results in this section (as well as the exercises) are special cases of more general theorems about the model theory of first-order languages. All of these more general results will appear in later sections of the book.

Exercise 1.2.19: The case $|\mathscr{S}| = \omega_1$ is due to Kreisel and Specker (see Kreisel, 1962); the general case is due to Reznikoff (1965).

1.3 The basic notions of satisfaction and truth are due to Tarski (1935a). Also due to him are some of the other notions such as: elementary equivalence between models, the semantical notion of consequence, submodels, extensions. We again refer the reader to Church (1956) for a complete account of the history of first-order languages.

1.3.11: Lindenbaum (see Tarski, 1930a).

1.3.20: Gödel (1930).

1.3.21–22: Gödel (1930) for $\mathscr{L}$ countable, Malcev (1936) for uncountable languages.

Exercise 1.3.14: Fuhrken (1968).

Exercises 1.3.15–20: These exercises give a general method for proving that two models are equivalent. The method was introduced by Fraïssé (1954) and, independently, by Ehrenfeucht (1961).

1.4 The definition of a theory is due to Tarski. Most of the examples in this section are well-known mathematical facts. Some references on number theory and set theory can be found in the text. For more information on the early development of set theory, see Fraenkel and Bar-Hillel (1958). Sikorski (1964) is a good reference for Boolean algebras.

1.4.2: Cantor (1895).

1.4.4: Stone (1936).

1.4.10: Steinitz (1910).

Exercise 1.4.9: Stone (1936).

Exercise 1.4.10: Lindenbaum (see Tarski, 1935b).

Exercise 1.4.15: Tarski (1938).

Exercise 1.4.18: Mostowski (1949).

1.5 Elimination of quantifiers is an important method in proving positive results in questions involving decidability. This section only contains two simple examples.

1.5.3, 4: Langford (1927).

1.5.7–9: Tarski (1936).

Exercise 1.5.7: (i) Behmann (1922); (ii) Szmielew (1955); (iii) Tarski (1931); (iv) Behmann (1922).

Exercises 1.5.8–9: Pressburger (1930).

Exercise 1.5.11: Szmielew (1955).

Notes for Chapter 2

2.1 The Gödel completeness theorem was first proved for countable languages by Gödel (1930) and in general by Malcev (1936). The proof given here is due to Henkin (1949). Henkin's proof is of particular importance in model theory because it introduces the method of constructing models from individual constants. The Löwenheim–Skolem–Tarski theorem was first proved even before the completeness theorem; the case $\alpha = \omega$ was proved by Löwenheim (1915) and Skolem (1920), and the general case is due to Tarski. The subject of model theory received a great impetus from the early applications of the completeness theorem by Henkin, Malcev, Robinson, and Tarski. The method of diagrams originated in the works of Henkin and Robinson and leads naturally to uncountable languages. Nonstandard analysis was introduced by Robinson (1961a, 1966), and has been developed by several researchers in recent years as a new approach to analysis.

2.2.1–5: Henkin (1949).

2.1.6: Skolem (1920).
2.1.7: Skolem (1934).
2.1.9: Henkin (1953).
2.1.10–11: Robinson (1951).
2.1.13: Marczewski–Szpilrajn (1930).
Exercise 2.1.1: Skolem (1934).
Exercise 2.1.2: Stone (1936), Henkin (1954a).
Exercise 2.1.5: Robinson (1951).
Exercise 2.1.12: Henkin (1953).
Exercises 2.1.13–14: Tarski (1952).
Exercises 2.1.15–19: Rasiowa and Sikorski (1950).

2.2 The ω-completeness theorem is due to Henkin and Orey. Various improvements were later obtained by different authors. One of these improvements is the omitting-types theorem 2.2.3, which is from Grzegorczyk, Mostowski, Ryll-Nardzewski.

Padoa (1901) observed that if a sentence $\varphi(P)$ has two models which are alike except for the interpretation of P, then the relation P is not explicitly definable from $\varphi(P)$. Beth's theorem is the converse of Padoa's method and was proved by Beth (1953). Later, the Craig interpolation theorem and the Robinson consistency theorem were proved independently by different methods, but each one easily implies the other. They form one of the fundamental properties of first-order predicate logic. Craig and Robinson both gave a proof of Beth's theorem. A number of different proofs of these results have appeared in the literature. The proof given here is due to Henkin (1963). Another proof, using special models, will be given in Chapter 5.

The material in this section is covered from a somewhat different viewpoint in Kreisel and Krivine (1967).

2.2.9, 10, 15: Grzegorczyk, Mostowski, Ryll-Nardzewski (1961).
2.2.19: Chang (1964a), Kreisel and Krivine (1967).
2.2.13: Henkin (1954b), Orey (1956).
2.2.18: Keisler and Morley (1968).
2.2.20: Craig (1957).
2.2.22: Beth (1953).
2.2.23: Robinson (1956a).
2.2.24: Lyndon (1959a).
Exercise 2.2.10: MacDowell and Specker (1961).
Exercise 2.2.19: Henkin (1963).

2.3 This section closely parallels the paper of Vaught (1961), "Denumerable models of complete theories". The notions of countably saturated and atomic models were introduced in that paper. Theorem 2.3.13, the characterization of ω-categorical theories, was proved independently by Engeler (1959), Ryll-Nardzewski (1959), and Svenonius (1959a).

2.3.2–4, 7, 9: Vaught (1961).
2.3.13: Engeler (1959), Ryll-Nardzewski (1959), Svenonius (1959a), Vaught (1961).
2.3.15: Vaught (1961).
Exercises 2.3.15, 16: Ehrenfeucht (see Vaught, 1961).
Exercise 2.3.17: Morley (1970).

2.4 Recursively saturated models were introduced by Barwise and Schlipf (1975), (1976), Ressayre (1973), (1977), and Schlipf (1975). The basic results presented in this section are due to Barwise and Schlipf. They made extensive use of recursively saturated models in generalized recursion theory over admissible sets. For more about this subject see the book Barwise (1975). Partial isomorphisms were introduced by Karp (1965), and are important in the model theory of infinitary logics; see, for example, Barwise (1973). They are related to the earlier characterization of elementary equivalence of Fraisse and Ehrenfeucht outlined in Exercises 1.3.15–1.3.20. Many applications of recursively saturated models in model theory of the type described in this section, and a history of the subject, can be found in Barwise and Schlipf (1976) and in Schlipf (1978).

2.4.1–2.4.3 and 2.4.5: Barwise and Schlipf (1976).
2.44: Karp (1965).
2.4.6: Schlipf (1978).
2.4.7: Pressburger (1930), by elimination of quantifiers.
2.4.8: Barwise and Schlipf (1976).
2.4.9: Craig and Vaught (1958).
2.4.10: Barwise and Schlipf (1976). This is a form of the result that countable recursively saturated models are resplendent.
Exercises 2.4.21–2.4.22: Schlipf (1978).
Exercise 2.4.30: Kunen. The set of axioms is from Clark (1978).
Exercise 2.4.31: Millar (1981).

Exercises 2.4.32–2.4.33: Barwise and Schlipf (1975).
Exercise 2.4.34: Lipschitz and Nadel (1978).

2.5 Lindström's Characterization Theorem was proved in Lindström (1966). It did not attract attention until it was rediscovered and publicized by Friedman (1970). Many variants of the definition of an abstract logic have been proposed, for example Lindström (1969), Friedman (1970), Barwise (1974), and Ebbinghaus (1985). The definition given here is similar to the notion of a regular logic in Ebbinghaus (1985), except that the Finite Occurrence Property is included here as part of the definition, and the substitution property from Ebbinghaus is replaced by a stronger form of the Relativization Property.

2.5.2: Hanf (1964).
2.5.3: Barwise (1974).
Exercise 2.5.1: Hanf (1964).
Exercise 2.5.2: Lindström (1966).
Exercise 2.5.4: Barwise (1974).
Exercise 2.5.6: Lindström (1969).
Exercise 2.5.7: Lindström (1973), (1974), Flum (1985).

Notes for Chapter 3

3.1 The basic notions of elementary chains (Tarski and Vaught, 1957) is of fundamental importance. The notions and basic results on elementary extensions and elementary chains were introduced by Tarski and Vaught (1957). The closely related notion of model completeness is due to Robinson, who proved the basic results and gave several examples of model complete theories in 1956. The last theorem in this section is a nontrivial application of the method of elementary chains due to Shoenfield.

3.3.3–6, 9: Tarski and Vaught (1957).
3.1.7: Łoś (1954), Vaught (1954a).
3.1.11: Shoenfield (unpublished).
Exercises 3.1.1, 3.1.3: Tarski and Vaught (1957).
Exercise 3.1.10: Cohn (1968, p. 33, Proposition 5.9).
Exercise 3.1.15: Chang (1967a).
Exercise 3.1.16: Levy (1960) and Montague (1961).

3.2 Preservation theorems formed the backbone of model theory until 1962. More recently, they were prominent in the early development of model theory for infinitary logic. The various preservation theorems were originally proved by several different methods. The more uniform method presented here was developed in Keisler (1960). Another approach to preservation theorems using saturated models is discussed in Section 5.2.

We have already encountered the notion of countably saturated models in Chapter 2. The notion of countably homogeneous models originates with Craig, Jónsson, Morley, and Vaught; for a detailed discussion, see Morley and Vaught (1962). The notions of T admitting a pair (α, β) of cardinals and the general problem of Löwenheim–Skolem theorems for two (or more) cardinals are due to Vaught; see Morley and Vaught (1962), Vaught (1965a, b).

3.2.2: Łoś (1955b), Tarski (1954).

3.2.3: Łoś and Suszko (1957), Chang (1959).

3.2.4: Lyndon (1959c).

3.2.7: (iii) is due to R.M. Robinson, and (iv) is due to Morley (both unpublished).

3.2.8: (i) and (iii) are due to Vaught, and (ii) is due to Craig.

3.2.9: Vaught (see Morley and Vaught, 1962).

3.2.11: Keisler (1966b).

Exercise 3.2.1: Robinson (1956c).

Exercise 3.2.2: Keisler (1960).

Exercise 3.2.4: Robinson (1956c).

Exercise 3.2.10: Craig (1961).

Exercise 3.2.13: Morley and Vaught (1962).

Exercise 3.2.14: Keisler and Morley (1968).

Exercise 3.2.18: Kueker (unpublished).

Exercise 3.2.19: Keisler (1970).

3.3 Skolem functions date back to Skolem (1920). The important notion of indiscernibles is due to Ehrenfeucht and Mostowski (1956). The ultrafilter theorem, Proposition 3.3.6, was first proved by Tarski; for more discussion on this, see notes for Section 4.1. The remarkable theorem of Ramsey (1930) was discovered by him to settle the following: It is decidable whether a universal sentence has an infinite model. The consequences of his combinatorial theorem far transcend the original purpose for which it was intended.

3.3.6: Tarski (1930b).

3.3.7: Ramsey (1930).

3.3.8–13: Ehrenfeucht and Mostowski (1956).

3.3.14: Morley (1965a).

Exercises 3.3.7, 3.3.10–11:	Ehrenfeucht and Mostowski (1956).
Exercise 3.3.14:	Keisler (unpublished).
Exercise 3.3.15:	Hodges (1969).
Exercise 3.3.16:	Park (M.I.T. thesis).
Exercise 3.3.18:	Keisler (unpublished).
Exercises 3.3.19–20:	Rabin (1965).

3.5 The notions of model completeness and model completion are due to A. Robinson (1956b). Most of the examples and basic results in this section are also due to him. The work was motivated to a large extent by applications to algebraically closed and real closed fields, particularly the nullstellensatz and Hilbert's seventeenth problem. Algebraically prime models were introduced by Robinson (1956b), and are studied, for example, in Baldwin and Kueker (1981). Model companions were introduced by Barwise, Eklof, Robinson, and Sabbagh in 1969. Additional material on model completeness can be found in the expositions of Cherlin (1976), Hirschfeld and Wheeler (1975), Macintyre (1977), Robinson (1973), and Simmons (1975).

3.5.1–3.5.2:	Robinson (1956b).
3.5.4:	Schlipf (1978).
3.5.7–3.5.8:	Lindström (1964).
3.5.11:	Robinson (1956b).
3.5.13–3.5.14:	Robinson (1963), Barwise, Eklof, Robinson, Sabbagh.
3.5.15–3.5.16:	Eklof and Sabbagh (1971).
3.5.17:	Millar (19??).
3.5.18:	Eklof and Sabbagh (1971).
3.5.19:	Shoenfield (1971).
3.5.20:	Millar (19??); see also Hodges (1981), Problem 13, p. 57.
3.5.21:	Simmons (1975).

Exercise 3.5.12:	Millar.
Exercise 3.5.14:	Eklof and Sabbagh (1971).
Exercise 3.5.15:	(a) Cherlin (1973).
	(b) Wheeler (Hirschfeld and Wheeler (1975)).
	(c) Robinson (1971).
Exercise 3.5.16–3.5.18:	Hirschfeld and Wheeler (1975).
Exercise 3.5.19:	Saracino (1973).
Exercise 3.5.20:	Baldwin and Kueker (1981).
Exercise 3.5.21:	Kaiser (1969).

Exercise 3.5.22: Robinson (1971).
Exercise 3.5.23: Robinson (1971a), Barwise and Robinson (1970).
Exercise 3.5.24: Keisler (1973), Simmons (1973).

Notes for Chapter 4

4.1 The ultraproduct construction goes back to the work of Skolem (1934). He invented a 'restricted' ultrapower to construct a nonstandard model of complete arithmetic. More recently, Hewitt (1948) studied what we would now call the ultraproduct of real closed fields. The general reduced product construction was introduced by Loś (1955a), and the fundamental theorem was stated in that paper. Frayne, Morel, Scott, and Tarski discovered the ultraproduct version of the compactness theorem, the natural embedding, and other basic facts which make the ultraproduct an important construction in model theory; see Frayne, Morel and Scott (1962). A review of the role of ultraproducts in model theory can be found in the survey articles of Chang (1967b) and Keisler (1965a). The book by Bell and Slomson (1969) is an exposition of that part of model theory which can be reached using only the ultraproduct construction. Ultraproducts have proved to be useful in other branches of mathematics, particularly algebra (Amitsur, Ax, Kochen) and nonstandard analysis (Luxemburg, Robinson).

4.1.1–4: Tarski (1930b), Stone (1936).
4.1.9: Loś (1955a).
4.1.12–13: Frayne, Morel, and Scott (1962), Kochen (1961).
Exercise 4.1.30: Daigneault (see Kochen, 1961).

4.2 Ulam (1930) proposed the problem of whether measurable cardinals exist, and proved that every measurable cardinal is inaccessible. Subsequently, many other mathematical conditions were found to be equivalent to measurability; see Keisler and Tarski (1964). Recent work on measurable cardinals began with the discovery of Hanf (1964) and Tarski (1962) that the first inaccessible cardinal is not weakly compact and hence not measurable. Keisler (1962) applied the ultraproduct construction to give another proof of the fact that the first inaccessible cardinal is not measurable, and then Scott (1961) used ultraproducts to show that the existence of a measurable cardinal contradicts the axiom of constructibility. The work of Hanf and Scott (1961) and Keisler and Tarski (1964) showed that weakly compact cardinals are very large and measurable cardinals are larger still.

Since then measurable cardinals and ultraproducts have been one of the major themes of research in set theory, particularly in the works of Gaifman, Kunen, Rowbottom, Silver, and Solovay. Some of this work is expounded in Section 7.4 of this book.

4.2.4: Frayne, Morel, and Scott (1962).
4.2.11: Keisler (1962).
4.2.12: Tarski (1962).
4.2.14: First proof: Hanf (1964) and Tarski (1962); this proof: Keisler (1962).
4.2.18: Scott (1961).
4.2.19–23: Scott (1961), Keisler and Tarski (1964).
Exercise 4.2.6: Keisler (1962).
Exercise 4.2.9: Hanf (1964).
Exercises 4.2.10–12: Lévy (1960).
Exercise 4.2.13: Hanf (1964).
Exercises 4.2.14–16: Keisler and Tarski (1964).
Exercise 4.2.19: Hanf and Scott (1961).
Exercise 4.2.20: Vaught (1963a).
Exercise 4.2.21: Keisler (1962), Monk and Scott (1964), Tarski (1962), Hanf and Scott (1961), Erdös and Tarski (1961).

4.3 Certain regular ultrafilters were used by Frayne, Morel, Scott, and Tarski to prove the compactness theorem and Frayne's theorem. They also proposed the cardinality problem and essentially proved Proposition 4.3.7 on cardinalities of ultrapowers. The notion of an α-regular ultrafilter in general was introduced by Keisler (1964c). The two-cardinal theorem 4.3.10 and its corollary were proved by Chang and Keisler. It is an example of a theorem whose statement does not mention ultraproducts, but whose proof requires ultrapowers in an essential way.

4.3.7: Frayne, Morel, and Scott (1962), Keisler (1964c).
4.3.9: Keisler (1964c).
4.3.10–11: Chang and Keisler (1962).
4.3.12: Frayne, Morel, and Scott (1962), Keisler (1967a).
4.3.14: Morley and Vaught (1962).
Exercise 4.3.11: Keisler (unpublished).
Exercise 4.3.12: Chang (1967c) with GCH, Kunen and Prikry (1970) without GCH.
Exercises 4.3.13, 14: Frayne, Morel, and Scott (1962).
Exercises 4.3.15, 16: Keisler (1964c).

Exercise 4.3.28: Hodges (unpublished)
Exercises 4.3.29, 30: Keisler (1963), Kochen (1961).
Exercise 4.3.32: Keisler (1967a).
Exercise 4.3.34: Keisler (see Chang, 1967c).
Exercise 4.3.35: Walkoe (see Keisler, 1968b).
Exercise 4.3.36: Benda (1969), Keisler (1968b).

4.4 Robinson introduced nonstandard analysis in (1961a) and in the book (1966). His original approach used the theory of types. The superstructure approach is a variant of the theory of types which fits better into ordinary set theory, and is due to Robinson and Zakon (1969). Internal set theory was introduced by Nelson (1977). Expositions of the superstructure approach to Robinsonian analysis can be found in the books of Albeverio et al. (1986), Cutland (1988), Davis (1976), Hurd and Loeb (1985), and Stroyan and Bayod (1986). Expositions of the internal set theory approach can be found in the books of Lutz and Goze (1971), Robert (1988) and van den Berg (1987). The article Diener and Stroyan (1988) discusses the relationship between the two approaches. The existence of nonstandard universes and enlargements are proved using the compactness theorem in Robinson (1966) and using ultraproducts in Robinson (1967). Bounded ultrapowers appear in Zakon (1972) and in Cherlin and Hirschfeld (1972). The importance of saturation in nonstandard analysis was emphasized by Luxenburg (1969a). ω_1-saturation is essential in much of current nonstandard analysis. The treatment of superstructures in Robinson and Zakon (1969) is reworked with particular attention to the requirements for a base set in Schmid and Schmidt (1987).

4.4.1–4.4.4: Robinson and Zakon (1969).
4.4.5: Robinson (1966).
4.4.9: Mostowski (1949).
4.4.10–4.4.11: Robinson and Zakon (1969).
4.4.12: Henson (1974).
4.4.16: Crisma and Holzer (1983).
4.4.17: Zakon (1972).
4.4.19–4.4.20: Keisler (1963). Another proof of 4.4.20 is in Zakon (1972).
4.4.23: Robinson (1967).
4.4.25–4.4.26: Nelson (1972).
Exercise 4.4.36: Puritz (1971).

Notes for Chapter 5

5.1 The notions of α-saturated and saturated models go back to the η_α-sets of Hausdorff (1914). Their importance for model theory was not realized and exploited until the late 1950's. Jónsson (1956, 1960) studied the notions of universal (α-universal) and homogeneous (α-homogeneous) models. In the present setting for model theory, these notions are also given in Morley and Vaught (1962). Many results relating saturated and special models to universal and homogeneous models were first proved in that paper. The notion of special models is due to Morley and Vaught (1962). The definition of special models given here is due to Chang and Keisler (1966) and is simpler than the original definition.

5.1.4, 5: Vaught (see Morley and Vaught, 1962).
5.1.8, 9: Morley and Vaught (1962).
5.1.12–14: Vaught (see Morley and Vaught, 1962).
5.1.16, 17: Morley and Vaught (1962).
5.1.19–22: Keisler and Morley (1967).

Most of the easier exercises in this section are either from Morley and Vaught (1962) or from Chang and Keisler (1966).

Exercises 5.1.9–13, 15, 16: Keisler and Morley (1967).
Exercises 5.1.24–5.1.26: Henson (1974).

5.2 Most of the preservation theorems were proved before saturated models were invented. However, saturated models, let to a uniform approach to the subject.

5.2.3, 4: Łoś (1955b), Robinson (1956c), Tarski (1954).
5.2.6: Equivalence of (i) and (ii) is due to Łoś and Suszko (1957) and Chang (1959); (iii) is due to Keisler (1960).
5.2.7, 8. Keisler (1960).
5.2.11: Morley and Vaught (1962).
5.2.13: Lyndon (1959c).
5.2.14, 15: Chang (1959), improving earlier results of Robinson (1951).
5.2.16: Vaught (1963b).

Exercise 5.2.1: Tarski (1955).
Exercise 5.2.2: Vaught (1954b).
Exercise 5.2.3: Tarski (1954), Łoś (1955b).
Exercise 5.2.4: An example was first found by Fraïssé (unpublished).
Exercises 5.2.6, 7: Keisler (1960).
Exercise 5.2.8: Kochen (1961).
Exercise 5.2.10: Chang (1959).
Exercise 5.2.12: Rabin (1960, 1962).

Exercise 5.2.13: Park, M.I.T. thesis.
Exercises 5.2.15, 17: Chang (unpublished).
Exercises 5.2.19–25: Keisler (1965c).
Exercise 5.2.26: Feferman (1968).

5.3 The starting point of Section 5.3 is Beth's theorem and Craig's theorem, both of which were proved in Section 2.2. Definability and interpolation theorems give interesting connections between model-theoretical statements in $\mathscr{L}$ and syntactical statements in expansions of $\mathscr{L}$.

5.3.1: This is a very much simplified version of a result of Chang and Moschovakis (1968). The proof here is given in Kueker (1970).
5.3.2: Chang (1968c).
5.3.3: Svenonius (1959a).
5.3.4, 5: Kueker (1970).
5.3.6: Chang (1964b), Makkai (1964). Vaught first pointed out that GCH is not necessary for the proof; the proof here is due to Chang.

Exercise 5.3.2: Chang (unpublished).
Exercises 5.3.5, 6, 8: Kueker (1970).
Exercise 5.3.9: Robinson (1965).
Exercises 5.3.10, 11: Chang (1964b).
Exercise 5.3.12: Kueker (1970).
Exercise 5.3.15: Chang (1964b).
Exercises 5.3.16, 17: Kueker (1970).
Exercise 5.3.18: Reyes (1970); this proof is due to Chang.
Exercise 5.3.19: This is a negative solution, by Chang, to a problem posed by Mostowski.
Exercises 5.3.20–5.3.23: Schlipf (1978).

5.4 The η_α-sets of Hansdorff are forerunners of ω_α-saturated models.

Tarski's original proof of the completeness of the theory of real closed fields used the method of elimination of quantifiers and gave a decision procedure for the theory. Later Erdös, Gillman and Henrickson (1955) proved that any two real closed fields of power ω_1 whose orders are η_1-sets are isomorphic. Kochen observed that this result plus the existence of ω_1-saturated models of power ω_1 gives a new proof of Tarski's theorem. The present proof is adapted from Kochen's proof.

Theorem 5.4.6 was first proved in the special cases where U is either real closed, algebraically closed, or equivalent to the field of rational numbers by Robinson (1959, 1961b). Robinson asked whether the theorem was true in general. This was settled affirmatively by Keisler (1964a).

Theorem 5.4.12 was proved independently by Ax and Kochen (1965a, b, 1966) and Ershov (1965). The present treatment is largely due to Rowbottom. The applications, including Artin's conjecture, are also given in those papers. Artin's conjecture is a striking example of a problem in algebra which was intractible to purely algebraic methods but which was solved elegantly using methods from model theory. Another treatment based on the elimination of quantifiers has been worked out by P. Cohen (unpublished).

5.4.4: Tarski and McKinsey (1948), Kochen (1961).
5.4.6: Keisler (1964a).
5.4.12–19: Ax and Kochen (1965a, b, 1966), Ershov (1965).
Exercise 5.4.1: Hausdorff (1914).
Exercise 5.4.2: Erdös, Gillman and Henrickson (1955).
Exercise 5.4.4: Robinson (1956b).
Exercises 5.4.8, 9: Keisler (1964a).
Exercise 5.4.10: Robinson (1956b).
Exercise 5.4.12: Presburger (1930).
Exercises 5.4.19–26: Ax and Kochen (1965a, b, 1966) and Ershov (1965).

5.5 The main theorem, Theorem 5.5.10, as well as the definition of invariants are due to Tarski. The result was later rediscovered by Ershov (1964). The present proof is due to Keisler.

5.5.10: Tarski (1949).
Exercises 5.5.8, 10–14: Keisler (unpublished).
Exercise 5.5.15: Mostowski and Tarski (1949). Parts (ii) and (iii) require a key observation of Ehrenfeucht.

Notes for Chapter 6

6.1 The notion of a good ultrafilter, the result that good ultrafilters exist (assuming the GCH) and good ultraproducts are saturated, and the isomorphism theorem for ultrapowers (assuming the GCH) are due to Keisler. The existence of good ultrafilters without the GCH, namely Theorem 6.1.4, is due to Kunen. The isomorphism theorem was proved without the GCH by Shelah (1972b).

6.1.1: Keisler (1960, 1964d).
6.1.4: Keisler (1964b) with GCH; Kunen (1972) without GCH.
6.1.6: Keisler (1964d).
6.1.7: (i) is due to Ketonen (unpublished); (ii), (iii) are due to Kunen (1972).

6.1.9: Keisler (1961, 1964d).
6.1.10–15: Shelah (1972b).
6.1.16, 17: Keisler (1961); GCH eliminated by Shelah (1972b).
Exercise 6.1.1: Keisler (1964d).
Exercise 6.1.3: Keisler (1967a).
Exercises 6.1.4, 5: Keisler (1964b) with GCH; Kunen (1973) without GCH.
Exercise 6.1.6: Keisler (1965b).
Exercise 6.1.10: Chang (1960).
Exercises 6.1.11–13: Shelah (1972b).
Exercise 6.1.14: Keisler (1961).
Exercises 6.1.15–18: Keisler (1967a).
Exercise 6.1.20: Keisler (1967b).
Exercise 6.1.21: Keisler (see Benda, 1970).
Exercise 6.1.22: Benda (1970).
Exercise 6.1.23: Keisler (1967b).
Exercises 6.1.24, 25: Shelah (unpublished).

6.2 Direct products, reduced products and ultraproducts all play important roles in model theory. Reduced products were studied in Frayne, Morel, and Scott (1962); some of the underlying ideas also go back to Chang, Łoś (1955a), and Tarski. The characterization of universal direct product sentences given by Proposition 6.2.8 is a very early result of McKinsey (1943). The fact that Horn sentences are preserved under direct products was proved by Horn (1951), and preservation under reduced products by Chang. The converse result, that reduced product sentences are Horn sentences, was proved by Keisler (1965d), assuming the continuum hypothesis. The continuum hypothesis was eliminated by Galvin (1965).

6.2.1: Frayne, Morel, and Scott (1962).
6.2.2: Chang (see the above paper).
6.2.3: Chang and Morel (1958).
6.2.4–6: Keisler (1965b).
6.2.7: Galvin (1965).
6.2.8: McKinsey (1943).
6.2.9: Lyndon (1959b).
Exercises 6.2.7, 8: Keisler (1965b).
Exercise 6.2.10: Lyndon (1959d).
Exercise 6.2.11: Galvin (1965).
Exercise 6.2.12: Tarski (see Frayne, Morel, and Scott, 1962).

Exercise 6.2.13: Birkhoff (1935). Proof here due to Chang.
Exercise 6.2.14: Keisler (1965d).
Exercise 6.2.15: Benda (1970).

6.3 The notion of a determining sequence is due to Feferman and Vaught (1959), and is descended from earlier ideas of Vaught (1954c) and Mostowski (1952). The definition given here contains refinements added by Weinstein (1965). The general exposition in this section owes much to Weinstein (1965) and Galvin (1965). The results of Galvin and Weinstein were both obtained in 1965 doctoral theses, about the same time as Ershov's (1964) theorem.

6.3.2: Feferman and Vaught (1959).
6.3.3: Weinstein (1965).
6.3.4: Feferman and Vaught (1959).
6.3.6, 8: Weinstein (1965).
6.3.9: Weinstein (1965), Galvin (1965).
6.3.13: Galvin (1965).
6.3.14: Vaught (1954c).
6.3.18, 19: Galvin (1965).
6.3.20: Ershov (1964).
Exercise 6.3.1: Oberschelp (1958).
Exercises 6.3.2–4: Weinstein (1965).
Exercise 6.3.5: Appel (1959). The simple proof here is due to Weinstein.
Exercise 6.3.10: Makkai (1965).
Exercise 6.3.11: Galvin (unpublished).
Exercise 6.3.12: Galvin (1965).
Exercise 6.3.13: Chang, Galvin.
Exercise 6.3.14: Galvin (1965).

6.4 The completion $\mathfrak{A}^{\#}$ of a model $\mathfrak{A}$ was introduced by Rabin (1959). In that paper Rabin proved Theorem 6.4.5 in the special case where α is accessible and the GCH holds. In Keisler (1963), the concept of a limit altrapower was introduced and the three theorems is this section were proved, Theorem 6.4.5 in the present general form. The presentation here is based upon simpler proofs of Theorem 6.4.4 by Lindström and of Theorem 6.4.5 by Chang. Blass (unpublished) has given another proof of Theorem 6.4.10. For a more general construction, the limit ultraproduct, see Keisler (1965a) and Kopperman (1972). The construction outlined in Exercise 6.4.30 is similar to Skolem's original 'restricted' ultrapower construction.

6.4.4: Keisler (1963), Lindström (1968).
6.4.5: Rabin (1959), Keisler (1963), Chang (1965b).
6.4.6–11: Keisler (1963).
6.4.12–14: Keisler (1965a).
Exercise 6.4.12: Keisler (1963).
Exercise 6.4.13: Keisler (1965a).
Exercises 6.4.16–26: Keisler (1963).

6.5 Finite iterations of ultrapowers were developed by Frayne, Morel, and Scott (1962). The infinite iterations were introduced by Gaifman (1967). Our presentation is a simplification of Gaifman's work. Gaifman used a category-theoretic approach instead of the notion of a function which lives on a finite set. Independently, Kunen (1970) developed iterated ultrapowers in essentially the same way as in this section, and generalized the construction even further in order to study models of set theory and measurable cardinals.
6.5.1, 2: Frayne, Morel, and Scott (1962).
6.5.6–10: Gaifman (1967).
6.5.11, 12: Keisler (unpublished).
6.5.14: Gaifman (1967).
Exercises 6.5.1–4: Frayne, Morel, and Scott (1962).
Exercise 6.5.16: Keisler (1963), Kochen (1961).
Exercises 6.5.22–29: Gaifman (1967).
Exercise 6.5.32: Keisler (1965a).
Exercise 6.5.34: Keisler (1967a).

Notes for Chapter 7

7.1 Łoś made his conjecture in Łoś (1954). Before Morley proved the conjecture in general, Vaught had shown the special case that if $\alpha > \omega$ is a limit cardinal and T is categorical in all powers less than α, then T is categorical in power α. The original proof of Morley's theorem used the notion of transcendence rank. This notion provides a powerful way of classifying types of elements. The present simpler proof was devised by Baldwin and Lachlan, using methods of Keisler and Marsh. The notion of a stable theory was introduced by Morley, who used the term totally transcendental theory. The two most exciting recent results in categoricity are 7.1.27, due to Baldwin and Lachlan, and 7.1.25, due to Shelah.
7.1.1: Morley (1965a). (This proof due to Keisler.)
7.1.3: Morley (1965a). (This proof due to Silver (unpublished).)

7.1.4–12: Morley (1965a).
7.1.13: Baldwin and Lachlan (1971).
7.1.14–23: Morley (1965a).
7.1.24: Shelah (1972a).
7.1.24′: Shelah (1978).
7.1.25: Shelah (1970d), (1978).
7.1.26: Baldwin (1973).
7.1.27–7.1.28: Baldwin and Lachlan (1971).
7.1.29–7.1.33: Shelah (1971d).
7.1.34: Shelah (1978).
7.1.35–7.1.36: Shelah (1985), (19??).
Exercise 7.1.3: Silver (unpublished).
Exercises 7.1.4–6: Keisler (1971a).
Exercise 7.1.13: Ehrenfeucht (1957), Morley (1965a).
Exercises 7.1.14, 15: Keisler (1971a), (1967b).
Exercise 7.1.16: Silver (unpublished), Ressayre (1969), Rowbottom (unpublished).
Exercises 7.1.17, 18: Morley (1965c).
Exercise 7.1.19: Keisler (1970).

7.2 The main topics covered in this section are the Morley theorem on omitting types and some theorems of Chang and Vaught on the two-cardinal problem. Apart from the methods of indiscernibles and saturated models, we introduce an idea due to Morley which uses the partition theorem of Erdös and Rado to construct models with special properties. The section ends with a discussion of the gap n conjecture and related problems, which leads naturally to the topics covered in the next section.

7.2.1: Erdös and Rado. The elegant model-theoretic proof given here is due to S. Simpson.
7.2.2–4: Morley (1965b).
7.2.5: Morley and Morley (1967) (with $V = L$); Silver (unpublished, without $V = L$); the present proof is due to Chang (1968b).
7.2.6: Vaught (1965b); the present proof is essentially due to Morley (1965b).
7.2.7: Chang (1965a).
7.2.8: Jensen, see Devlin (1984).
7.2.10: Specker (1949).
7.2.11: Rowbottom and Silver, independently (unpublished).
7.2.12: Mitchell (1972), Silver (in Mitchell 1972), Shelah (1979) (see also Ben-David and Shelah 1986).

7.2.13: Vaught (1965a).
7.2.14: Solovay; see Devlin (1984).
Results preceding 7.2.15: Silver (1971) and Solovay, cf. Kanamori and Magidor (1978).
7.2.15: Silver (1971).

Exercises 7.2.1–4:	Morley (1965b).
Exercise 7.2.5:	Chang (1968b); the counterexample was discovered independently by Shelah (1971a) and Schmerl (unpublished).
Exercise 7.2.6:	Helling (1964).
Exercise 7.2.7:	Morley (1965b), Morley and Morley (1967); see note on 7.2.5.
Exercise 7.2.8:	Howard (unpublished), Morley (unpublished), Shelah (1971a).
Exercises 7.2.10, 12:	Vaught (1965b).
Exercise 7.2.13:	Due to a large number of people; among them, Morley (1965b), Lopez-Escobar (1965b), Helling (1964), Chang (1968b).
Exercise 7.2.15:	Chang (1965a).
Exercise 7.2.16:	Vaught (1965b).
Exercise 7.2.17:	Chang.
Exercise 7.2.18:	Vaught (1965a).
Exercise 7.2.19:	Chang.
Exercises 7.2.20, 21:	Vaught (1965b).
Exercise 7.2.25:	Keisler (1971a).
Exercise 7.2.28:	Specker (1949).
Exercise 7.2.30:	Kunen (unpublished).
Exercise 7.2.31:	Morley (see Keisler and Morley, 1968).
Exercise 7.2.33:	Keisler (1966a).

7.3 Ramsey cardinals were introduced by Erdös and his school, and several of the combinatorial results are due to them; for example, see Erdös, Hajnal, and Rado (1965). The main model-theoretic results in this section are Rowbottom's theorem and Silver's theorem.

7.3.5:	Vaught (1965b).
7.3.6:	Silver (unpublished).
7.3.7:	Erdös and Hajnal (1966).
7.3.8:	Fuhrken (1964).
7.3.9:	Erdös (1942).

7.3.10:	Silver, Shoenfield (both unpublished).
7.3.11:	Erdös and Tarski (1961).
7.3.12:	Hanf (1964).
7.3.13:	Rowbottom (1964).
7.3.14:	Erdös and Hajnal (1958, 1962).
7.3.15, 16:	Rowbottom (1964).
7.3.16 (ii):	Also Erdös and Hajnal (1966).
7.3.17(a), (b):	Prikry (1968); (c) Kunen (1970).
7.3.18:	Silver (1966).
Exercises 7.3.9, 10:	Rowbottom (1964).
Exercises 7.3.12, 13:	Erdös and Hajnal.
Exercise 7.3.15.	For $n = 1$, Galvin (unpublished); in general, Rowbottom (unpublished), Erdös and Hajnal (1966), Chang (unpublished).
Exercise 7.3.16:	Shoenfield (unpublished).
Exercise 7.3.18:	Rowbottom (1964).
Exercise 7.3.19:	Silver (1966).
Exercise 7.3.21:	Hanf (1964).
Exercise 7.3.22:	Helling (1965); proof indicated due to Keisler.
Exercises 7.3.23–26:	Silver (1966).
Exercise 7.3.27:	Rowbottom (1964).
Exercise 7.3.28:	Keisler and Scott (unpublished).
Exercise 7.3.29:	Rowbottom (1964).
Exercise 7.3.30:	Prikry (1968).
Exercises 7.3.31–35:	Silver (1966).
Exercise 7.3.37:	Silver (1966).
Exercise 7.3.40:	Hanf (1964).
Exercise 7.3.41:	Silver (1966).
Exercises 7.3.42–43:	Fuhrken (1964).
Exercise 7.3.44:	MacDowell and Specker (1961).
Exercise 7.3.45:	Fuhrken (1964).
Exercise 7.3.46:	Keisler (1968a).
Exercise 7.3.48:	Schmerl and Shelah (1969).
Exercise 7.3.49:	Schmerl (1969).
Exercise 7.3.50:	McKenzie (1971).

7.4 The axiom of constructibility and the basic properties of constructible sets were introduced by Gödel (1939, 1940) in order to prove that if ZF is consistent, then so is ZFC+GCH. This section is based mainly on the work of Gaifman and Rowbottom, and later Silver. Using iterated ultrapowers, Gaifman proved that if there is a measurable cardinal $\alpha > \omega$, then all the conclusions of Theorem 7.4.7 hold. Independently, Rowbottom proved that if there exists a Ramsey cardinal, then part (i) of Theorem 7.4.7 holds. Using the methods explained in the last section, Silver proved Theorem 7.4.7 in its present form. It is a common improvement of both Gaifman and Rowbottom's result. The other two main results of this section, Theorems 7.4.10 and 7.4.12, are due to Keisler and Rowbottom. Rowbottom first proved that any nontrivial case of Chang's conjecture contradicts the axiom of constructibility

7.4.1–6: Gödel (1939, 1940).
7.4.7, 8: Silver (1966).
7.4.10: Keisler and Rowbottom (1965).
7.4.11: Kunen (1970).
7.4.12: Keisler and Rowbottom (1965).

Exercises 7.4.2, 3:	Silver (1966).
Exercise 7.4.4:	Solovay (1967).
Exercise 7.4.5:	Silver (1966).
Exercises 7.4.6–7:	Rowbottom (1964).
Exercises 7.4.8, 10–12:	Silver (1966).
Exercise 7.4.13:	Kunen (1970).
Exercises 7.4.15–16:	Solovay (1967).
Exercise 7.4.17:	Gaifman (1974).
Exercise 7.4.18:	Chang (1968a).

REFERENCES

J. W. ADDISON, L. HENKIN and A. TARSKI, eds.

(1965) *The Theory of Models* (North-Holland, Amsterdam).

K. I. APPEL

(1959) Horn sentences in identity theory, *J. Symb. Logic* **24**, 306–310.

J. AX and S. KOCHEN

(1965a) Diophantine problems over local fields I, *Am. J. Math.* **87**, 605–630.

(1965b) Diophantine problems over local fields II: A complete set of axioms for p-adic number theory, *Am. J. Math.* **87**, 631–648.

(1966) Diophantine problems over local fields III: Decidable fields, *Ann. Math.* **83**, 437–456.

J. T. BALDWIN and A. H. LACHLAN

(1971) On strongly minimal sets, *J. Symb. Logic* **36**, 79–96.

J. BARWISE

(1975) Some Eastern two-cardinal theorems. *Int. Cong. of Logic, Methodology, and Phil. of Sci. at London, Ont. 1975.*

H. BEHMANN

(1922) Beiträge zur Algebra der Logik, insbesondere zur Entscheidungsproblem, *Math. Ann.* **86**, 163–229.

J. L. BELL and A. B. SLOMSON

(1969) *Models and Ultraproducts* (North-Holland, Amsterdam) ix+322 pp.

M. BENDA

(1969) Reduced products and nonstandard logics, *J. Symb. Logic* **34**, 424–436.

(1970) *Reduced products, filters and Boolean ultrapowers*, Ph. D. Thesis, Univ. of Wisconsin. Also *Ann. Math. Logic* **4** (1972), 1–31.

M. BENDA and J. KETONEN

(1974) Regularity of ultrafilters, *Israel J. Math.* **17**, 231–240.

P. BERNAYS

(1937) A system of axiomatic set theory, Parts I–V, *J. Symb. Logic* **2** (1937) 65–77; **6** (1941) 1–17; **7** (1942) 65–89; **7** (1942) 133–145; **8** (1943) 89–106.

E. W. BETH

(1953) On Padoa's method in the theory of definition, *Koninkl. Ned. Akad. Wetensch. Proc. Ser.* A **56** (= *Indag. Math.* **15**) 330–339.

G. BIRKHOFF

(1935) On the structure of abstract algebras, *Proc. Cambridge Phil. Soc.* **31**, 433–454.

(1961) *Lattice Theory*, A. M. S. Colloq. Publ. Vol. **25** (Am. Math. Soc., Providence, R.I.) xix+283 pp.

G. CANTOR
(1895) Beiträge zur Begründung der transfiniten Mengenlehre, *Math. Ann.* **96**, 481–512. [English transl.: *Contributions to the Founding of the Theory of Transfinite Numbers* (Dover Publications, New York, 1915).]

C. C. CHANG
(1954) Some general theorems on direct products and their applications in the theory of models, *Koninkl. Ned. Akad. Wetensch. Proc. Ser.* A **57** (= *Indag. Math.* **16**) 592–598.
(1959) On unions of chains of models, *Proc. Am. Math. Soc.* **10**, 120–127.
(1960) A lemma on ultraproducts and some applications, Prelim. Rept., *Notices Am. Math. Soc.* **7**, 635.
(1964a) On the formula 'there exists x such that $f(x)$ for all $f \in F$', *Notices Am. Math. Soc.* **11**, 587.
(1964b) Some new results in definability, *Bull. Am. Math. Soc.* **70**, 808–813.
(1965a) A note on the two-cardinal problem, *Proc. Am. Math. Soc.* **16**, 1148–1155.
(1965b) A simple proof of the Rabin–Keisler theorem, *Bull. Am. Math. Soc.* **71**, 642–643.
(1967a) Omitting types of prenex formulas, *J. Symb. Logic* **32**, 61–74.
(1967b) Ultraproducts and other methods of constructing models, *Sets, Models and Recursion Theory*, J. N. Crossley, ed. (North-Holland, Amsterdam) 85–121.
(1967c) Descendingly incomplete ultrafilters, *Trans. Am. Math. Soc.* **126**, 108–118.
(1968a) Infinitary properties of models generated from indiscernibles, *Logic, Methodology, and the Philosophy of Science*, B. van Rootselaar and J. F. Staal, eds. (North-Holland, Amsterdam) 9–21.
(1968b) Some remarks on the model theory of infinitary languages, *The Syntax and Semantics of Infinitary Languages*, Lecture Notes in Math. **72**, J. Barwise, ed. (Springer, Berlin) 36–63.
(1968c) A generalization of the Craig interpolation theorem, *Notices Am. Math. Soc.* **15**, 934, also see *Symposia Mathematica V* (1971) 1–19 (Academic Press).

C. C. CHANG and H. J. KEISLER
(1962) Applications of ultraproducts of pairs of cardinals to the theory of models, *Pacific J. Math.* **12**, 835–845.
(1966) *Continuous Model Theory* (Princeton Univ. Press, Princeton, N.J.) xii+165 pp.

C. C. CHANG and A. C. MOREL
(1958) On closure under direct product, *J. Symb. Logic* **23**, 149–154.

C. C. CHANG and Y. N. MOSCHOVAKIS
(1968) On Σ_1^1-relations on special models, *Notices Am. Math. Soc.* **15**, 934.

D. V. CHOODNOVSKY and G. V. CHOODNOVSKY
(1974) Descendingly incomplete ultrafilters, *Soviet Math. Doklady* **15**, 1472–1476.

A. CHURCH
(1956) *Introduction to Mathematical Logic*, Vol. **1** (Princeton Univ. Press, Princeton, N.J.) x+378 pp.

P. J. COHEN
(1963) The independence of the continuum hypothesis, *Proc. Natl. Acad. Sci. U.S.A.* **50**, 1143–1148; **51** (1964) 105–110.
(1966) *Set Theory and the Continuum Hypothesis* (Benjamin, New York) 154 pp.
(1969) Decision procedures for real and p-adic fields. *Comm. Pure and Applied Math.* **22**, 131–151.

P. M. COHN
(1965) *Universal Algebra* (Harper and Row, London) xv+333 pp.

W. CRAIG
(1957) Three uses of the Herbrand-Gentzen Theorem in relating model theory and

proof theory, *J. Symb. Logic* **22**, 269–285.
(1961) $\aleph_0$-homogeneous relatively universal systems, *Notices Am. Math. Soc.* **8**, 265.

A. EHRENFEUCHT
(1957) On theories categorical in power, *Fund. Math.* **44**, 241–248.
(1961) An application of games to the completeness problem for formalized theories, *Fund. Math.* **49**, 129–141.

A. EHRENFEUCHT and A. MOSTOWSKI
(1956) Models of axiomatic theories admitting automorphisms, *Fund. Math.* **43**, 50–68.

E. ENGELER
(1959) A characterization of theories with isomorphic denumerable models, *Notices Am. Math. Soc.* **6**, 161.

P. ERDÖS
(1942) Some set-theoretical properties of graphs, *Rev. Univ. Nacl. Tucumán Ser.* A **3**, 363–367.

P. ERDÖS, L. GILLMAN and M. HENRIKSEN
(1955) An isomorphism theorem for real closed fields, *Ann. Math. Ser.* 2, **61**, 542–554.

P. ERDÖS and A. HAJNAL
(1958) On the structure of set mappings, *Acta Math. Acad. Sci. Hung.* **9**, 111–131.
(1962) Some remarks concerning our paper 'On the structure of set mappings', *Acta Math. Acad. Sci. Hung.* **13**, 223–226.
(1966) On a problem of B. Jónsson, *Bull. Acad. Polon. Sci. Sér. Sci. Math. Astron. Phys.* **14**, 61–99.

P. ERDÖS, A. HAJNAL and P. RADO
(1965) Partition relations for cardinal numbers, *Acta Math.* **16**, 93–196.

P. ERDÖS and R. RADO
(1956) A partition calculus in set theory, *Bull. Am. Math. Soc.* **62**, 427–489.

P. ERDÖS and A. TARSKI
(1961) On some problems involving inaccessible cardinals, *Essays on the Foundations of Mathematics* (Magnus Press, The Hebrew Univ., Jerusalem) 50–82.

YU. L. ERSHOV
(1964) Decidability of the elementary theory of relatively complemented distributive lattices and the theory of filters, *Algebra i Logika Sem.* **3** (3), 17 (in Russian).
(1965) On the elementary theory of maximal normed fields, *Dokl. Akad. Nauk SSSR* 165 (in Russian; English Transl.: *Sov. Math.* 1390–1393).

S. FEFERMAN
(1968) Persistent and invariant formulas for outer extension, *Comp. Math.* **20**, 29–52.

S. FEFERMAN and R. L. VAUGHT
(1959) The first order properties of algebraic systems, *Fund. Math.* **47**, 57–103.

A. A. FRAENKEL and Y. BAR-HILLEL
(1958) *Foundations of Set Theory* (North-Holland, Amsterdam) x+415 pp.

R. FRAÏSSÉ
(1954) Sur quelques classifications des systèmes de relations, *Publ. Sci. Univ. Alger Sér.* A **1**, 35–182.

T. E. FRAYNE, A. C. MOREL and D. S. SCOTT
(1962) Reduced direct products, *Fund. Math.* **51**, 195–228 (Abstract: *Notices Am. Math. Soc.* **5** (1958) 674).

H. FRIEDMAN
(1975) One hundred and two problems in mathematical logic. *J. Symbolic Logic* **40**, 113–129.

G. Fuhrken
(1962) Bemerkung zu einer Arbeit E. Engelers, *Z. Math. Logik u. Grundl. Math.* **8**, 277–279.
(1964) Skolem-type normal forms for first-order languages with a generalized quantifier, *Fund. Math.* **54**, 291–302.
(1968) A model with exactly one undefinable element, *Coll. Math.* **19**, 183–185.

H. Gaifman
(1967) Uniform extension operators for models, *Sets, Models and Recursion Theory*, J. N. Crossley, ed. (North-Holland, Amsterdam) 122–155.

F. Galvin
(1965) *Horn sentences*, Thesis, Univ. of Minnesota.
(1967) Reduced products, Horn sentences, and decision problems, *Bull. Am. Math. Soc.* **73**, 59–64.
(1968) A generalization of Ramsey's Theorem, *Notices Am. Math. Soc.* **15**, 548.
(1970) Horn sentences, *Ann. Math. Logic* **1**, 389–422.

K. Gödel
(1930) Die Vollständigkeit der Axiome des logischen Funktionenkalküls, *Monatsh. Math. Phys.* **37**, 349–360.
(1931) Über formal unentscheidbare Sätze der Principia Mathematica und verwandter Systeme I, *Monatsh. Math. Phys.* **38**, 173–198.
(1939) Consistency proof for the generalized continuum hypothesis, *Proc. Natl. Acad. Sci. U.S.A.* **25**, 220–224.
(1940) *The Consistency of the Axiom of Choice and of the Generalized Continuum Hypothesis with the Axioms of Set Theory*, Ann. Math. Studies 3 (Princeton Univ. Press, Princeton, N.J.).

G. Grätzer
(1968) *Universal Algebra* (Van Nostrand, New York) xvi + 368 pp.

A. Grzegorczyk, A. Mostowski and C. Ryll-Nardzewski
(1961) Definability of sets in models of axiomatic theories, *Bull. Acad. Polon. Sci. Sér. Sci. Math. Astronom. Phys.* **9**, 163–167.

W. Hanf
(1964) Incompactness in languages with infinitely long expressions, *Fund. Math.* **13**, 309–324.

W. Hanf and D. Scott
(1961) Classifying inaccessible cardinals, *Notices Am. Math. Soc.* **8**, 445.

F. Hausdorff
(1914) *Grundzüge der Mengenlehre* (Leipzig) 467 pp.

M. Helling
(1964) Hanf numbers for some generalizations of first-order languages, *Notices Am. Math. Soc.* **11**, 679.
(1965) Ph.D. Thesis, Univ. of California, Berkeley, Calif.

L. A. Henkin
(1949) The completeness of the first-order functional calculus, *J. Symb. Logic* **14**, 159–166.
(1953) Some interconnections between modern algebra and mathematical logic, *Trans. Am. Math. Soc.* **74**, 410–427.
(1954a) Metamathematical theorems equivalent to the prime ideal theorems for Boolean algebras, Prelim. Rept., *Bull. Am. Math. Soc.* **60**, 387–388.
(1954b) A generalization of the concept of ω-consistency, *J. Symb. Logic* **19**, 183–196.

(1963) An extension of the Craig–Lyndon interpolation theorem, *J. Symb. Logic* **28**, 201–216.

E. HEWITT

(1948) Rings of real-valued continuous functions, *Trans. Am. Math. Soc.* **64**, 45–99.

W. A. HODGES

(1969) *The Ehrenfeucht–Mostowski method of constructing models*, Doctoral Dissertation, Univ. of Oxford, 139 pp.

A. HORN

(1951) On sentences which are true of direct unions of algebras, *J. Symb. Logic* **16**, 14–21.

T. JECH

(1971a) *Lectures in Set Theory*, Lecture Notes in Math. 27 (Springer, Berlin) 137 pp.

(1971b) Trees, *J. Symb. Logic* **36**, 1–14.

B. JÓNSSON

(1956) Universal relational systems, *Math. Scand.* **4**, 193–208.

(1960) Homogeneous universal relational systems, *Math. Scand.* **8**, 137–142.

E. KAMKE

(1950) *The Theory of Sets*, transl. by F. Bagemihl (Dover).

A. KANAMORI, W. REINHARDT, and R. SOLOVAY

(1978) Strong axioms of infinity and elementary embeddings. *Ann. Math. Logic* **13**, 73–116.

H. J. KEISLER

(1960) Theory of models with generalized atomic formulas, *J. Symb. Logic* **25**, 1–26.

(1961) Ultraproducts and elementary classes, *Koninkl. Ned. Akad. Wetensch. Proc. Ser.* A **64** (= *Indag. Math.* **23**) 477–495.

(1962) Some applications of the theory of models to set theory, *Logic, Methodology and Philosophy of Science*, E. Nagel, P. Suppes and A. Tarski, eds. (Stanford Univ. Press, Stanford, Calif.) 80–86.

(1963) Limit ultrapowers, *Trans. Am. Math. Soc.* **107**, 383–408.

(1964a) Complete theories of algebraically closed fields with distinguished subfields, *Michigan Math. J.* **11**, 71–81.

(1964b) Good ideals in fields of sets, *Ann. Math.* **79**, 338–359.

(1964c) On cardinalities of ultraproducts, *Bull. Am. Math. Soc.* **70**, 644–647.

(1964d) Ultraproducts and saturated models, *Koninkl. Ned. Akad. Wetensch. Proc. Ser.* A **67** (= *Indag. Math.* **26**) 178–186.

(1965a) A survey of ultraproducts, *Logic, Methodology and Philosophy of Science*, Y. Bar-Hillel, ed. (North-Holland, Amsterdam) 112–126.

(1965b) Limit ultraproducts, *J. Symb. Logic* **30**, 212–234.

(1965c) Some applications of infinitely long formulas, *J. Symb. Logic* **30**, 339–349.

(1965d) Reduced products and Horn classes, *Trans. Am. Math. Soc.* **117**, 307–328.

(1966a) First order properties of pairs of cardinals, *Bull. Am. Math. Soc.* **72**, 141–144.

(1966b) Some model-theoretic results for ω-logic, *Israel J. Math.* **4**, 249–261.

(1967a) Ultraproducts of finite sets, *J. Symb. Logic* **32**, 47–57.

(1967b) Ultraproducts which are not saturated, *J. Symb. Logic* **32**, 23–46.

(1968a) Models with orderings, *Logic, Methodology and Philosophy of Science*, B. van Rootselaar and J. F. Staal, eds. (North-Holland, Amsterdam) 35–62.

(1968b) Formulas with linearity ordered quantifiers, *The Syntax and Semantics of Infinitary Languages*, Lecture Notes in Math. **72**, J. Barwise, ed. (Springer, Berlin) 96–130.

(1970) Logic with the quantifier 'there exist uncountably many', *Ann. Math. Logic* **1**, 1–93.

(1971a) *Model Theory for Infinitary Logic* (North-Holland, Amsterdam) x+208 pp.
(1971b) On theories categorical in their own power, *J. Symb. Logic* **36**, 240–244.

H. J. KEISLER and M. MORLEY
(1967) On the number of homogeneous models of a given power, *Israel J. Math.* **5**, 73–78.
(1968) Elementary extensions of models of set theory, *Israel J. Math.* **6**, 49–65.

H. J. KEISLER and F. ROWBOTTOM
(1965) Constructible sets and weakly compact cardinals, *Notices Am. Math. Soc.* **12**, 373.

H. J. KEISLER and A. TARSKI
(1964) From accessible to inaccessible cardinals, *Fund. Math.* **53**, 225–308.

J. L. KELLEY
(1955) *General Topology* (Van Nostrand).

J. KETONEN
(1976) Some remarks on ultrafilters, *Notices Am. Math. Soc.* **75**, A-325 (abstract).

E. KLEINBERG
(1972) The equiconsistency of two large cardinal axioms (abstract), *Notices Am. Math. Soc.* **16**, A-329.

S. B. KOCHEN
(1961) Ultraproducts in the theory of models, *Ann. Math. Ser.* 2, **74**, 221–261.

R. KOPPERMAN
(1972) *Model Theory and its Applications* (Allyn and Bacon).

G. KREISEL
(1962) *Additions au Cours Polycopié* (Paris)

G. KREISEL and J. L. KRIVINE
(1967) *Elements of Mathematical Logic: Model Theory* (North-Holland, Amsterdam) 222 pp.

D. W. KUEKER
(1970) Generalized interpolation and definability, *Ann. Math. Logic* **1**, 423–468.

K. KUNEN
(1970) Some applications of iterated ultrapowers in set theory, *Ann. Math. Logic* **1**, 179–227.
(1972) Ultrafilters and independent sets, *Trans. Amer. Math. Soc.*, **172**, 199–206.

K. KUNEN and K. PRIKRY
(1971) On descendingly incomplete ultrafilters, *J. Symb. Logic* **36** (1971) 650–652.

A. LACHLAN
(1971) The transcendence rank of a theory, *Pacific J. Math.* **27**, 119–122.

C. H. LANGFORD
(1927) Some theorems on deducibility, *Ann. Math. Ser.* 2, **28**, 16–40.

A. LÉVY
(1960) Axiom schemata of strong infinity, *Pacific J. Math.* **10**, 223–238.

P. LINDSTRÖM
(1964) On model completeness, *Theoria (Lund)* **30**, 183–196.
(1968) Remarks on some theorems of Keisler, *J. Symb. Logic.* **33**, 571–576.

A. LITMAN and S. SHELAH
(1976) Independence of the gap one conjecture from G. C. H., *Notices Am. Math. Soc.* **23**, A-495 (abstract).

E. G. K. LOPEZ-ESCOBAR
(1965a) An interpolation theorem for denumerably long sentences, *Fund. Math.* **57**, 253–272.

(1965b) Universal formulas in the infinitary languages $L_{\alpha\beta}$, *Bull. Acad. Polon. Sci. Sér. Sci. Math. Astron. Phys.* **13**, 383–388.
(1966) On definable well orderings, *Fund. Math.* **55**, 13–21.

J. Łoś
(1954) On the categoricity in power of elementary deductive systems and some related problems, *Colloq. Math.* **3**, 58–62.
(1955a) Quelques remarques, théorèmes et problèmes sur les classes définissables d'algèbres, *Mathematical Interpretation of Formal Systems* (North-Holland, Amsterdam) 98–113.
(1955b) On the extending of models I, *Fund. Math.* **42**, 38–54.

J. Łoś and R. Suszko
(1957) On the extending of models IV: Infinite sums of models, *Fund. Math.* **44**, 52–60.

L. Löwenheim
(1915) Über Möglichkeiten im Relativkalkül, *Math. Ann.* **76**, 447–470.

R. C. Lyndon
(1959a) An interpolation theorem in the predicate calculus, *Pacific J. Math.* **9**, 155–164.
(1959b) Existential Horn sentences, *Proc. Am. Math. Soc.* **10**, 994–998.
(1959c) Properties preserved under homomorphism, *Pacific J. Math.* **9**, 143–154.
(1959d) Properties preserved in subdirect products, *Pacific J. Math.* **9**, 155.

R. MacDowell and E. Specker
(1961) Modelle der Arithmetik, *Infinitistic Methods* (Pergamon Press, London) 257–263.

M. Magidor
(1976) On existence of nonregular ultrafilters, or variations on a theme by Kunen, *Logic Symposium, Oxford 1976* (abstract).

M. Makkai
(1964) On a generalization of a theorem of E. W. Beth, *Acta Math. Acad. Sci. Hung.* **15**, 227–235.
(1965) A compactness result concerning direct products of models, *Fund. Math.* **57**, 313–325.

J. A. Makowsky
(1974) On some conjectures connected with complete sentences, *Fund. Math.* **81**, 193–202.

A. Malcev
(1936) Untersuchungen aus dem Gebiete der mathematischen Logik, *Rec. Math. N.S.* **1**, 323–336.

E. Marczewski-Szpilrajn
(1930) Sur l'extension de l'ordre partiel, *Fund. Math.* **16**, 386–389.

W. Marsh
(1966) *On ω_1 but not ω-categorical theories*, Ph.D. Thesis, Univ. of Dartmouth.

R. McKenzie
(1971) On semigroups whose proper subsemigroups have lesser power, *Algebra Universalis* **1** (1971) 21–25.

J. C. C. McKinsey
(1943) The decision problem for some classes of sentences without quantifiers, *J. Symb. Logic* **8**, 61–76.

D. J. Monk and D. Scott
(1964) Additions to some theorems of Erdös and Tarski, *Fund. Math.* **53**, 335–343.

M. Morley
(1965a) Categoricity in power, *Trans. Am. Math. Soc.* **114**, 514–538.
(1965b) Omitting classes of elements, *The Theory of Models*, J. W. Addison, L. Henkin and A. Tarski, eds. (North-Holland, Amsterdam) 265–273.

(1965c) Countable models of $\aleph_1$-categorical theories, *Israel J. Math.* **5**, 65–72.
(1970) The number of countable models, *J. Symb. Logic* **35**, 14–18.

M. Morley and V. Morley
(1967) The Hanf number for κ-logic, *Notices Am. Math. Soc.* **14**, 556.

M. Morley and R. Vaught
(1962) Homogeneous universal models, *Math. Scand.* **11**, 37–57.

A. Mostowski
(1949) An undecidable arithmetical statement, *Fund. Math.* **36**, 143–164.
(1952) On direct products of theories, *J. Symb. Logic* **7**, 1–31.

A. Mostowski and A. Tarski
(1949) Arithmetical classes and types of well-ordered systems, Prelim. Rept., *Bull. Am. Math. Soc.* **55**, 65.

A. Oberschelp
(1958) Über die Axiome produkt-abgeschlossener arithmetischer Klassen, *Arch. Math. Logik u. Grundl. Math.* **4**, 95–123.

S. Orey
(1956) On ω-consistency and related properties, *J. Symb. Logic* **21**, 246–252.

L. Pacholski and C. Ryll-Nardzewski
(1970) On countably compact reduced products I, *Fund. Math.* **67**, 155–161.

A. Padoa
(1901) *Essai d'une Théorie Algébrique des Nombres Entiers, précédé d'une Introduction Logique à une Théorie Déductive Quelconque*, Bibliothèque du Congr. Intern. de Philos. 3, Logique et Histoire des Sciences (Paris) 309–365.

Post
(1921) Introduction to a general theory of elementary propositons. *Amer. J. Math.* **43**, 163–185.

M. Presburger
(1930) Über die Völlstandigkeit eines gewissen Systems der Arithmetik ganzer Zahlen in welchem die Addition als einzige Operation hervortritt, *C. R. 1er Congr. des Mathematiciens des Pays Slaves* (Warsaw) 92–101, 395.

K. Prikry
(1968) *Changing measurable into accessible cardinals*, Doctoral Dissertation, Univ. of California, Berkeley, Calif. Also Diss. Math. **68** (1970), 1–55.
(1971) On a problem of Gillman and Keisler, *Ann. Math. Logic* **2**, 179–188.
(1973) On descendingly complete ultrafilters. Pp. 459–488 in *Cambridge Summer School in Math. Logic*, Lecture Notes in Math. **337** (Springer, Berlin).

M. O. Rabin
(1959) Arithmetical extensions with prescribed cardinality, *Koninkl. Ned. Akad. Wetensch. Proc. Ser.* A **62** (= *Indag. Math.* **21**) 439–446.
(1960) Characterization of convex systems of axioms, *Notices Am. Math. Soc.* **7**, 503–504.
(1962) Classes of models and sets of sentences with the intersection property, *Proc. B. Pascal Colloq., Ann. de la Faculté des Sci. de l'Université de Clermont* **1**, 39–53.
(1965) Universal groups of automorphisms of models, *The Theory of Models*, J. W. Addison, L. Henkin and A. Tarski, eds. (North-Holland, Amsterdam) 74–84.

F. P. Ramsey
(1930) On a problem in formal logic, *Proc. London Math. Soc. Ser.* 2, **30**, 264–286.

H. Rasiowa and R. Sikorski
(1950) A proof of the completeness theorem of Gödel, *Fund. Math.* **37**, 193–200.

J. P. RESSAYRE
(1969) Sur les théories du premier ordre catégorique en un cardinal, *Trans. Am. Math. Soc.* **142**, 481–505.

G. E. REYES
(1970) Local definability theory, *Ann. Math. Logic* **1**, 95–137.

I. RESNIKOFF
(1965) Tout ensemble de formules de la logique classique est équivalent à un ensemble indépendant, *C. R. Acad. Sci. Paris* **260**, 2385–2388.

P. RIBENBOIM
(1967) *Théorie des Valuations*, Publ. du Dépt. de Math., Univ. de Montréal, 314 pp.

A. ROBINSON
(1951) *On the Metamathematics of Algebra* (North-Holland, Amsterdam) x+195 pp.
(1956a) A result on consistency and its application to the theory of definition, *Koninkl. Ned. Akad. Wetensch. Proc. Ser.* A **59** (= *Indag. Math.* **18**) 47–58.
(1956b) *Complete Theories* (North-Holland, Amsterdam) 129 pp.
(1956c) On a problem of L. Henkin, *J. Symb. Logic* **21**, 33–35.
(1959) On the concept of a differentially closed field, *Bull. Res. Council Israel Sect.* F **8**, 113–128.
(1961a) Non-standard analysis, *Koninkl. Ned. Akad. Wetensch. Proc. Ser.* A **64** (= *Indag. Math.* **23**) 432–440.
(1961b) Model theory and non-standard arithmetic, *Infinitistic Methods* (Pergamon, London) 265–302.
(1963) *On the Metamathematics of Algebra* (North-Holland, Amsterdam) x+284 pp.
(1965) *Introduction to Model Theory and to the Metamathematics of Algebra*, 2nd ed. (North-Holland, Amsterdam) ix+284 pp. (1st ed., 1963).
(1966) *Non-Standard Analysis* (North-Holland, Amsterdam) xi+293 pp.
(1973) Metamathematical problems. *J. Symbolic Logic* **38**, 500–516.

F. ROWBOTTOM
(1964) *Large cardinals and small constructible sets*, Doctoral Dissertation, Univ. of Wisconsin; *Ann. Math. Logic* **3** (1971) 1–44, also *Ann. Math. Logic* **3** (1971) 1–44.

C. RYLL-NARDZEWSKI
(1952) The role of the axiom of induction in elementary arithmetic, *Fund. Math.* **39**, 239–263.
(1959) On the categoricity in power $\aleph_0$, *Bull. Acad. Polon. Sci. Sér. Sci. Math. Astron. Phys.* **7**, 545–548.

G. SACKS
(1972) *Saturated Model Theory* (W. A. Benjamin).

J. SCHMERL
(1969) On hyperinaccessible-like models (abstract), *Notices Am. Math. Soc.* **19**, 843, also see *J. Symb. Logic* **37**, (1972) 521–530.
(1977) An axiomatization for a class of two-cardinal models. *J. Symb. Logic* **42**, 174–178.

J. SCHMERL and S. SHELAH
(1969) On models with orderings (abstract), *Notices Am. Math. Soc.* **19**, 840, also see *J. Symb. Logic* **37** (1972) 531–537.

D. SCOTT
(1961) Measurable cardinals and constructive sets, *Bull. Acad. Polon. Sci. Sér. Sci. Math. Astron. Phys.* **9**, 521–524.

S. SHELAH
(1969) Stable theories, *Israel J. Math.* **7**, 187–202.

(1970a) On theories T categorical in $|T|$, *J. Symb. Logic* **35**, 73–82.
(1970b) Finite diagrams stable in power, *Ann. Math. Logic* **2**, 69–119.
(1970c) When every reduced product is saturated (abstract), *Notices Am. Math. Soc.* **17**, 453.
(1970d) Solution of Łoś' conjecture for uncountable languages (abstract), *Notices Am. Math. Soc.* **17**, 1968.
(1971a) A note on Hanf numbers, *Pacific J. Math.* **34**, 539–544.
(1971b) A combinatorial problem; stability and order for models and theories in infinitary languages, *Pacific J. Math.* **41**, 247–261.
(1971c) The number of non-isomorphic models of an unstable first-order theory, *Israel J. Math.*, **9**, 473–487.
(1971d) Stability, the f.c.p., and superstability; model-theoretic properties of formulas in first-order theories, *Ann. Math. Logic* **3** (1971) 271–362.
(1972a) Uniqueness and characterization of prime models over sets for totally transcendental first-order theories, J. Symb. Logic **37**, 107–113.
(1972b) Every two elementarily equivalent models have isomorphic ultrapowers, *Israel J. Math.*, **10**, 224–233.
(1972c) Remark to 'Local definability theory' of Reyes, *Ann. Math. Logic* **2**(1971) 441–448.
(1976) Refuting Ehrenfeucht conjecture on rigid models, *Israel J. Math.* **25**, 273–286.
(1977) A two-cardinal theorem and a combinatorial problem, *Proc. Amer. Math. Soc.* **62**, 134–136.
(1980) On a problem of Kurosh, Jónsson groups, and applications, *Word Problems* II (North-Holland, Amsterdam).
(1976c) Ultraproducts of finite cardinals and Keisler order, *Notices Am. Math. Soc.* **23**, A-494 (abstract).

J. R. Shoenfield
(1967) *Mathematical Logic* (Addison-Wesley, Reading, Mass.) vii+344 pp.

J. Silver
(1966) *Some applications of model theory in set theory*, Doctoral Dissertation, Univ. of California, Berkeley, Calif., *Ann. Math. Logic* **3** (1971) 45–110.
(1970) Every analytic set is Ramsey, *J. Symb. Logic* **35**, 60–64.
(1971) The independence of Kurepa's conjecture and the two-cardinal conjectures in model theory, *Axiomatic Set Theory*, D. S. Scott, ed., Proc. Sympos. Pure Math. **13(1)** (Am. Math. Soc.) 383–390.

R. Sikorski
(1964) *Boolean Algebras*, 2nd ed. (Springer, Berlin).

T. Skolem
(1920) Logisch-kombinatorische Untersuchungen über die Erfüllbarkeit oder Beweisbarkeit mathematischer Sätze nebst einem Theorem über dichte Mengen, *Skrifter utgitt av Videnskapsselskapet i Kristiania, I, Mat. Naturv. Kl.* **4**, 36 pp.
(1934) Über die Nicht-Charakterisierbarkeit der Zahlenreihe mittels endlich oder abzählbar unendlicht vieler Aussagen mit ausschliesslich Zahlenvariablen, *Fund. Math.* **23**, 150–161.

R. Solovay
(1967) A nonconstructible Δ_3^1 set of integers, *Trans. Am. Math. Soc.* **127**, 50-75.

E. Specker
(1949) Sur un problème de Sikorski, *Colloq. Math.* **2**, 9–12.

E. Steinitz
(1910) Algebraische Theorie der Körper, *J. Reine Angew. Math.* **137**, 167–309.

M. H. Stone
(1936) The representation theorem for Boolean algebra, *Trans. Am. Math. Soc.* **40**, 37–111.

L. Svenonius
(1959a) $\aleph_0$-categoricity in first-order predicate calculus, *Theoria* (Lund) **25**, 82–94.
(1959b) A theorem on permutations in models, *Theoria* (Lund) **25**, 173–178.

W. Szmielew
(1955) Elementary properties of Abelian groups, *Fund. Math.* **41**, 203–271. (Abstract: *Bull. Am. Math. Soc.* **55** (1949) 65.)

A. Tarski
(1930a) Über einige fundamentale Begriffe der Metamathematik, *C. R. Séances Soc. Sci. Lettres Varsovie Cl.* III **23**, 22–29. [English Transl.: *Logic, Semantics and Metamathematics* (Oxford, 1959) 30–37.]
(1930b) Une contribution à la théorie de la mesure, *Fund. Math.* **15**, 42–50.
(1931) Sur les ensembles définissables de nombres réels I, *Fund. Math.* **17**, 210–239. [English Transl.: *Logic, Semantics and Metamathematics* (Oxford, 1956) 110–142.]
(1935a) Der Wahrheitsbegriff in den formalisierten Sprachen, *Studia Philos.* (Warsaw) **1**, 261–405. [English Transl.: *Logic, Semantics and Metamathematics* (Oxford, 1956) 152–278.]
(1935b) Grundzüge des Systemenkalküls I, *Fund. Math.* **25**, 503–526. [English Transl.: *Logic, Semantics and Metamathematics* (Oxford, 1956) 343–383.]
(1936) Grundzüge des Systemenkalküls II, *Fund. Math.* **26**, 283–301.
(1938) Über unerreichbare Kardinalzahlen, *Fund. Math.* **30**, 68–89.
(1949) Arithmetical classes and types of Boolean algebras, Prelim. Rept., *Bull. Am. Math. Soc.* **55**, 64, 1192.
(1952) Some notions and methods on the borderline of algebra and metamathematics, *Proc. Intern. Congr. of Mathematicians, Cambridge, Mass.* 1950, **1** (Am. Math. Soc., Providence, R. I.) 705–720.
(1954) Contributions to the theory of models I, II, *Koninkl. Ned. Akad. Wetensch. Proc. Ser.* A **57** (= *Indag. Math.* **16**) 572–588.
(1955) Contributions to the theory of models III, *Koninkl. Ned. Akad. Wetensch. Proc. Ser.* A **58** (= *Indag. Math.* **17**) 56–64.
(1962) Some problems and results relevant to the foundations of set theory, *Logic, Methodology and Philosophy of Science*, E. Nagel, P. Suppes and A. Tarski, eds. (University Press, Stanford, Calif.) 125–135.

A. Tarski, A. Mostowski and R. M. Robinson
(1953) *Undecidable Theories* (North-Holland, Amsterdam).

A. Tarski and J. C. C. McKinsey
(1948) *A Decision Method for Elementary Algebra and Geometry*, 2nd ed. (Berkeley, Los Angeles).

A. Tarski and R. Vaught
(1957) Arithmetical extensions of relational systems, *Compositio Math.* **13**, 81–102.

S. Ulam
(1930) Zur Masstheorie in der allgemeinen Mengenlehre, *Fund. Math.* **16**, 140–150.

R. Vaught
(1954a) Applications of the Löwenheim–Skolem–Tarski theorem to problems of completeness and decidability, *Koninkl. Ned. Akad. Wetensch. Proc. Ser.* A **57** (= *Indag. Math.* **16**) 467–472.
(1954b) Remarks on universal classes of relational systems, *Koninkl. Ned. Akad. Wetensch. Proc. Ser.* A **57** (= *Indag. Math.* **16**) 589–591.
(1954c) On sentences holding in direct products of relational system, *Proc. Intern. Congr. of Mathematicians, Amsterdam*, 1954, **2** (Noordhoff, Groningen) 409.

(1958a) Prime models and saturated models, *Notices Am. Math. Soc.* **5**, 780.

(1958b) Homogeneous universal models of complete theories, *Notices Am. Math. Soc.* **5**, 775.

(1961) Denumerable models of complete theories, *Infinitistic Methods* (Pergamon, London) 303–321.

(1963a) Indescribable cardinals, *Notices Am. Math. Soc.* **10**, 126.

(1963b) Elementary classes of models closed under descending intersection, *Notices Am. Math. Soc.* **10**, 126.

(1965a) The Löwenheim-Skolem Theorem, *Logic, Methodology and Philosophy of Science*, Y. Bar-Hillel, ed. (North-Holland, Amsterdam) 81–89.

(1965b) A Löwenheim–Skolem Theorem for cardinals far apart, *The Theory of Models*, J. W. Addison, L. Henkin and A. Tarski, eds. (North-Holland, Amsterdam) 390–401.

J. M. Weinstein

(1965) *First-order properties preserved by direct product*, Ph. D. Thesis, Univ. of Wisconsin, Madison, Wisc.

A. N. Whitehead and B. Russell

(1913) *Principia Mathematica*, **1, 2, 3** (Cambridge Univ. Press, London; 2nd ed., 1925; 2nd ed. of Vols. 1, 2, 1927).

ADDITIONAL REFERENCES

The following reference list contains papers explicitly referred to in the text, and a selection of books and expository articles. Many of these contain more extensive reference lists of papers in particular areas of model theory.

A. Adler and M. Jorgensen

(1972) Descendingly incomplete ultrafilters and the cardinality of ultrapowers, *Canad. J. Math.* **24**, 830–834.

S. Albeverio, J. E. Fenstad, R. Høegh-Krohn and T. Lindstrøm

(1986) *Nonstandard Methods in Stochastic Analysis and Mathematical Physics* (Academic Press) xi + 514 pp.

A. Apter

(1983) On a problem of Silver, *Fund. Math.* **116**, 33–38.

J. T. Baldwin

(1973) α_T is finite for $\aleph_1$-categorical T, *Trans. Amer. Math. Soc.* **181**, 37–51.

(1987) *Classification Theory: Chicago 1985*, Lecture Notes in Math. **1292** (Springer, Berlin) 500 pp.

(1988) *Fundamentals of Stability Theory* (Springer-Verlag) xiii + 447 pp.

J. T. Baldwin and D. Kueker

(1981) Algebraically prime models, *Ann. Math. Logic* **20**, 289–330.

K. J. Barwise

(1973) Back and forth through infinitary logic, *Studies in Model Theory*, M. Morely, ed., MAA Studies **8**, 5–34.

(1974) Axioms for abstract model theory, *Annals of Math. Logic* **7**, 221–265.

(1975) *Admissible Sets and Structures* (Springer-Verlag) xiii + 394 pp.
(1977) *Handbook of Mathematical Logic*, ed. (North-Holland) xi + 1165 pp.

K. J. Barwise and S. Feferman
(1985) *Model-Theoretic Logics* (Springer-Verlag) xviii + 893 pp.

K. J. Barwise and A. Robinsons
(1971) Completing theories by forcing, *Ann. Math. Logic* **2**, 119–142.

K. J. Barwise and J. Schlipf
(1975) On recursively saturated models of arithmetic, *Model Theory and Algebra*, Lecture Notes in Math. **498**, D. H. Saracino and V. B. Weispfenning, eds. (Springer, Berlin) 42–55.
(1976) An introduction to recursively saturated and resplendent models, *J. Symb. Logic* **41**, 531–536.

J. L. Bell and M. Machover
(1977) *A Course in Mathematical Logic* (North-Holland) xviii + 599 pp.

S. Ben-David
(1988) A Laver-type indestructability for accessible cardinals, *Logic Colloquium '86* (North-Holland) 9–19.

I. van den Berg
(1987) *Nonstandard Asymptotic Analysis*, Lecture Notes in Math. **1249** (Springer, Berlin) ix + 187 pp.

S. Buechler
(1984) Kueker's conjecture for superstable theories, *J. Symb. Logic* **49**, 930–934.

M. Canjar
(1988) Countable ultraproducts without CH, *Ann. Pure and Appl. Logic* **37**, 1–80.

G. Cherlin
(1973) Algebraically closed division rings, *J. Symb. Logic* **38**, 493–499.
(1976) *Model Theoretic Algebra, Selected Topics*, Lecture Notes in Math. **521** (Springer, Berlin) 234 pp.
(1984) Totally categorical structures, *Proc. Int. Cong. Math.* (Warsaw 1983) 301–306.

G. Cherlin, L. Harrington, and A. Lachlan
(1985) $\aleph_0$-categorical $\aleph_0$-stable structures, *Ann. Pure and Appl. Logic* **28**, 103–135.

G. Cherlin and J. Hirschfeld
(1972) Ultrafilters and ultraproducts in non-standard analysis, *Contributions to Non-standard Analysis*, W. A. J. Luxemburg and A. Robinson, eds. (North-Holland) 261–280.

K. L. Clark
(1978) Negation as failure, *Logic and Databases*, H. Gallaire and J. Minker, eds. (Plenum Press) 293–322.

W. W. Comfort and S. Negrepontis
(1974) *The Theory of Ultrafilters* (Springer-Verlag) x + 482 pp.

W. Craig and R. L. Vaught
(1958) Finite axiomatizability using additional predicates, *J. Symb. Logic* **23**, 289–308.

L. Crisma and S. Holzer
(1983) Starconcepts, *Boll. Unione Matematica Ital. Serie VI*, **2**, 175–192.

N. Cutland
(1988) *Nonstandard Analysis and its Applications*, ed. (Cambridge Univ. Press) xiii + 346 pp.

M. Davis
(1977) *Applied Nonstandard Analysis* (Wiley) xii + 181 pp.

K. Devlin
(1973) Some weak versions of large-cardinal axioms, *Ann. Math. Logic* **5**, 291–325.
(1984) *Constructibility* (Springer-Verlag) xi + 425 pp.

A. J. Dodd and R. B. Jensen
(1981) The core model, *Ann. Math. Logic* **20**, 43–75.

H.-D. Donder
(1988) Regularity of ultrafilters and the core model, *Israel J. Math.* **63**, 289–322.

H.-D. Donder, R. Jensen, and B. Koppelberg
(1981) Some applications of the core model, *Set Theory and Model Theory*, Lecture Notes in Math. **872** (Springer, Berlin) 55–97.

H.-D. Donder and P. Koepke
(1983) On the consistency strength of 'accessible' Jónsson cardinals and of the weak Chang conjecture, *Ann. Pure and Appl. Logic* **25**, 233–361.

F. R. Drake
(1974) *Set Theory: An Introduction to Large cardinals* (North-Holland) xii + 354 pp.

F. Diener and K. D. Stroyan
(1988) Syntactical methods in infinitesimal analysis, *Cutland* (1988) 258–281.

M. Dickmann
(1975) *Large Infinitary Languages* (North-Holland) xv + 464 pp.

H.-D. Ebbinghaus
(1985) *Extended Logic: The General Framework*, Barwise and Feferman (1985) 25–76.

P. Eklof and G. Sabbagh
(1971) Model-completions and modules, *Ann. Math. Logic* **2**, 251–295.

J. Flum
(1985) *Characterizing Logics*, Barwise and Feferman (1985) 77–120.

J. Flum and M. Ziegler
(1980) *Topological Model Theory*, Lecture Notes in Math. **769** (Springer, Berlin) x + 151 pp.

M. Foreman
(1982) Large cardinals and strong model-theoretic transfer properties, *Trans. Amer. Math. Soc.* **272**, 427–463.

R. Fraïssé
(1986) *Theory of Relations* (North-Holland) xii + 397 pp.

H. Friedman
(1970) Why first-order logic? Manuscript (Stanford Univ. and the Univ. of Wisconsin) 19 pp.

H. Gaifman
(1974) Elementary embeddings of set theory, *Axiomatic Set Theory*, Proc. Sympos. Pure Math. **13**(2), T. J. Jech, ed. (Am. Math. Soc.) 33–101.

K. Gödel
(1986) *Collected Works*, Volume 1, S. Feferman, ed. (Oxford Univ. Press) xvi + 474 pp.

V. Harnik and L. Harrington
(1984) Fundamentals of forking, *Ann. Pure and Appl. Logic* **26**, 245–286.

L. Harrington, M. D. Morley, A. Scedrov and S. G. Simpson
(1985) *Harvey Friedman's Research on the Foundations of Mathematics* (North-Holland) xvi + 408 pp.

L. HARRINGTON, M. MAKKAI, and S. SHELAH
(1984) A proof of Vaught's conjecture for ω-stable theories, *Israel J. Math.* **49**, 259–280.

L. HENKIN
(1974) Proceedings of the Tarski Symposium, L. Henkin et al., eds., *Amer. Math. Soc. Proc. of Symp. in Pure Math.*, xx + 498 pp.

L. HENKIN, J. D. MONK and A. TARSKI
(1971) *Cylindric Algebras*, Part I (North-Holland) vi + 508 pp.
(1985) *Cylindric Algebras*, Part II (North-Holland) vii + 302 pp.

W. HENSON
(1974) The isomorphism property in nonstandard analysis and its use in the theory of Banach spaces, *J. Symb. Logic* **39**, 717–736.

J. HIRSCHFELD and W. WHEELER
(1975) *Forcing, Arithmetic, Division Rings*, Lecture Notes in Math. **454** (Springer, Berlin) 266 pp.

W. HODGES
(1985) *Building Models by Games* (Cambridge Univ. Press) 311 pp.

E. HRUSHOVSKI
(19??) Kueker's conjecture for stable theories, Preprint.

A. HURD and P. LOEB
(1985) *An Introduction to Nonstandard Real Analysis* (Academic Press) xii + 232 pp.

T. JECH
(1978) *Set Theory* (Academic Press) xi + 621 pp.

K. KAISER
(1969) Über eine verallgemeinerung der Robinsonschen modellvervollständigung I, *Z. Math. Logik Grund. Math.* **15**, 37–48.

A. KANAMORI and M. MAGIDOR
(1978) The evolution of large cardinal axioms in set theory, *Higher Set Theory*, Lecture Notes in Math. **669** (Springer, Berlin) 99–275.

C. KARP
(1965) Finite quantifier equivalence, *Addison, Henkin, and Tarski* (1965) 407–412.

H.J. KEISLER
(1973) Forcing and the omitting types theorem, *Studies in Model Theory*, M. Morley, ed., M.A.A. Studies, **8**, 96–133.
(1976) *Foundations of Infinitesimal Calculus* (Prindle, Weber and Schmidt) ix + 214 pp.
(1977) Fundamentals of model theory, *Barwise* (1977) 47–104.
(1984) *An Infinitesimal Approach to Stochastic Analysis*, Mem. Amer. Math. Soc. **48**, no. 297, x + 184 pp.
(1986) *Elementary Calculus* (Prindle, Weber and Schmidt) xi + 913 pp.

E. M. KLEINBERG
(1973) Rowbottom and Jónsson cardinals are almost the same, *J. Symb. Logic* **38**, 423–427.

P. KOEPKE
(1988) Some applications of short core models, *Ann. Pure Appl. Logic* **37**, 179–204.

H. K. KUNEN
(1978) Saturated ideals, *J. Symb. Logic* **43**, 65–76.
(1980) *Set Theory: An Introduction to Independence Proofs* (North-Holland) xvi + 313 pp.

A. Lachlan

(1973) On the number of countable models of a countable superstable theory, *Logic, Methodology and Philosophy of Science* IV (North-Holland) 45–56.

D. Lascar

(1986) Stabilité en théorie des modèles (French) Univ. Cath. de Louvain (Cabay Liabraire-Editeur S.a.) ii + 231 pp.

(1987) *Stability in Model Theory* (English translation of preceding reference) (Wiley) viii + 193 pp.

D. Lascar and B. Poizat

(1979) An introduction to forking, *J. Symb. Logic* **44**, 330–350.

J.-P. Levinski

(1984) Instances of the conjecture of Chang, *Israel J. Math.* **48**, 225–243.

P. Lindström

(1966) First order predicate logic with generalized quantifiers, *Theoria* **32**, 186–195.

(1969) On extensions of elementary logic, *Theoria* **35**, 1–11.

(1973) A characterization of elementary logic, *Modality, Morality, and other Problems of Sense and Nonsense*, B. Hallden, ed. (CWK Gleerup Bokolförlag) 189–191.

(1974) On characterizing elementary logic, *Logical Theory and Semantic Analysis*, S. Stenlund, ed. (D. Reidel Publ. Co.) 129–146.

T. Lindstrøm

(1988) An invitation to nonstandard analysis, *Cutland* (1988) 1–105.

L. Lipschitz and M. Nadel

(1978) The additive structure of models of arithmetic, *Proc. Amer. Math. Soc.* **68**, 331–336.

Lo, Li Bo

(1980) Union and product of models, and homogeneous models (Chinese) Beijing Shifan Daxue Xuebao, 31–39.

(1985) Vaught conjecture for homogeneous models, *J. Symb. Logic* **48**, 539–541.

R. Lutz and M. Goze

(1981) *Nonstandard Analysis. A Practical Guide with Applications*, Lecture Notes in Math. **881** (Springer, Berlin).

W. A. J. Luxemburg

(1969) *Applications of Model Theory of Algebra, Analysis, and Probability*, ed. (Holt) i + 307 pp.

(1969a) A general theory of monads, *Luxemburg* (1969) 18–86.

A. Macintyre

(1977) Model completeness, *Barwise* (1977) 139–180.

A. Macintyre, K. McKenna and L. van den Dries

(1983) Elimination of quantifiers in algebraic structures, *Advances in Math.* **47**, 74–87.

M. Magidor

(1979) On the existence of nonregular ultrafilters and the cardinality of ultrapowers, *Trans. Amer. Math. Soc.* **249**, 97–111.

A. I. Malcev

(1971) *The Metamathematics of Algebraic Systems*, Collected papers, 1936–1967 (North-Holland) xviii + 494 pp.

T. Millar

(19??) To appear.

J. D. Monk

(1976) *Mathematical Logic* (Springer-Verlag), ix + 531 pp.

R. MONTAGUE

(1961) Fraenkel's addition to the axioms of Zermelo, *Essays on the Foundations of Mathematics*, Y. Bar-Hillel, ed. (Magnes Press) 91–114.

E. NELSON

(1977) Internal set theory, *Bull. Amer. Math. Soc.* **83**, 1165–1198.

(1988) The syntax of nonstandard analysis, *Ann. Pure Appl. Logic* **38**, 123–134

J. F. PABION

(1982) Saturated models of Peano Arithmetic, *J. Symb. Logic* **47**, 625–637.

M. G. PERETYATKIN

(1980) An example of an ω_1-categorical complete finitely axiomatizable theory (English translation) *Algebra i Logika* **19**, 224–251.

A. PILLAY

(1983) *An Introduction to Stability Theory* (Clarendon Press, Oxford) xi + 146 pp.

B. POIZAT

(1985) Cours de Théorie des Modèles, Nur al-Mantiq wal-Ma'rifah, 82 Rue Racine, 69100 Villeurbanne, France, 584 pp.

C. PURITZ

(1971) Ultrafilters and standard functions in non-standard arithmetic, *Proc. London Math. Soc.* **22**, 705–733.

J. RESSAYRE

(1973) Boolean models and infinitary first-order languages, *Ann. Math. Logic* **6**, 41–92.

(1977) Models with compactness properties relative to an admissible language, *Annals of Math. Logic* **11**, 31–55.

A. ROBERT

(1988) *Nonstandard Analysis* (Wiley) xx + 156 pp.

A. ROBINSON

(1967) Nonstandard theory of Dedekind rings, *Nederl. Akad. Wetensch. Proc. Ser. A70 and Indag. Math.* **29**, 444–452.

(1971a) Forcing in model theory, *Symposia Math.* **5**, 69–82.

(1971b) Infinite forcing in model theory, *Proc. 2nd Scandinavian Symposium in Logic* (North-Holland) 317–340.

(1971c) Forcing in model theory, *Les Actes du Congrès International des Mathématiciens* **1** (Gautheir-Villars, Paris) 245–250.

(1973) Model theory as a framework for algebra, *Studies in Model Theory*, M. Morley, ed., MAA Studies **8**, 134–157.

(1979) Selected Papers, Vol. I, *Model Theory and Algebra* (Yale Univ. Press) xxxvii + 694 pp.

(1979a) Selected Papers, Vol. II, *Nonstandard Analysis and Philosophy* (Yale Univ. Press) xiv + 582 pp.

A. ROBINSON and E. ZAKON

(1969) A set-theoretical characterization of enlargements, *Luxemburg* (1969) 109–122.

J. ROITMAN

(1982) Non-isomorphic hyper-real fields from non-isomorphic ultrapowers, *Math. Z.* **181**, 93–96.

G. SACKS

(1983) On the number of countable models, *Southeast Asian Conference on Logic* (North-Holland) 185–195.

D. SARACINO

(1973) Model companions for $\aleph_0$-categorical theories, *Proc. Amer. Math. Soc.* **39**, 591–598.

J. SCHLIPF

(1975) Some hyperelementary aspects of model theory, *Ph. D. Thesis, Univ. of Wisconsin, Madison*, 165 pp.

(1978) Toward model theory through recursive saturation, *J. Symb. Logic* **43**, 183–206.

JÜRG SCHMIT and JÜRGEN SCHMIDT

(1987) Enlargements without urelements, *Colloq. Math.* **52**, 1–22.

S. SHELAH

(1978) *Classification Theory and the Number of Nonisomorphic Models* (North-Holland) xvi + 544 pp.

(1979) On successors of singular cardinals, *Logic Colloquium '78* (North-Holland) 357–380.

(1982) *Proper Forcing*, Lecture Notes in Math. **940** (Springer, Berlin) xxix + 496 pp.

(1985) Classification of first order theories which have a structure theory, *Bull. Amer. Math. Soc.* **12**, 227–232.

(1988a) On spread of B.A. and the singular cardinal problem, isometric ultrapowers, *Abstracts Amer. Math. Soc.* **9**, 266.

(1988b) On successors of singulars, *Abstracts Amer. Math. Soc.* **9**, 500.

(19??) *Classification Theory and the Number of Nonisomorphic Models, Completed.* (North-Holland).

(19??a) On isomorphic countable ultraproducts, To appear.

S. SHELAH, L. HARRINGTON, and M. MAKKAI

(1984) A proof of Vaught's conjecture for totally transcendental theories, *Israel J. Math.* **49**, 259–278.

J. SHOENFIELD

(1971) A theorem on quantifier elimination, *Symposia Math.* **5**, 173–176.

J. STEEL

(1978) On Vaught's Conjecture, *Cabal Seminar 76–77*, Lecture Notes in Math. **689** (Springer, Berlin) 193–208.

K. D. STROYAN and J. M. BAYOD

(1986) *Foundations of Infinitesimal Stochastic Analysis* (North-Holland) xii + 478 pp.

H. SIMMONS

(1973) An omitting types theorem with applications to the construction of generic structures, *Math. Scand.* **33**, 46–54.

(1975) Companion Theories (Forcing in Model Theory). Seminaires de Mathemetique pure Mensuel, *Inst. Math. Pure et Appl.* (Univ. Catholique de Louvain) 61 pp.

J. TRYBA

(1984) On Jónsson cardinals with uncountable confinality, *Israel J. Math.* **49**, 315–324.

E. ZAKON

(1974) A new variant of nonstandard analysis, *Victoria Symposium on Nonstandard Analysis*, A. Hurd, ed., Lecture Notes in Math. **369** (Springer, Berlin) 313–339.

B. I. ZILBER

(1980) Totally categorical theories, *Model Theory and Arithmetic*, Lecture Notes in Math. **834** (Springer, Berlin).

(1984) The structure of models of uncountably categorical theories, *Proc. Int. Cong. Math.*, Warsaw 1983, 359–368.

INDEX OF DEFINITIONS

Abelian group 40
absolute formula 562
absorption laws 38
abstract logic 128
additive function 386
additive number theory 43
admit elimination of quantifiers 202
admit a pair of cardinals 154
admit a pair of cardinals omitting a type 532
admit a triple of cardinals 532
algebraic in a theory 496, 529
algebraically closed field 41
algebraically independent 348
algebraically prime model 195
amalgamate 196
amalgamation property 195
antisymmetric 580
antisymmetric property 37
Archimedean ordered field 70
arithmetical predicate 415
arithmetical set 125
arithmetical statement 415
arithmetically saturated model 125
Aronsajn tree 525
arrangement 51, 55
Artin's Conjecture 368
assignment 7
associativity 39
atom of Boolean algebra 39
atomic Boolean algebra 39
atomic element of a Boolean algebra 373
atomic formula 22
atomic model 97
atomic theory 97
atomless Boolean algebra 39
atomless element of a Boolean algebra 373
Automorphism Theorem 172
automorphism type 176
autonomous set of formulas 426
Axiom of Choice 584, 593
Axiom of Constructibility 238
Axiom of Regularity 588
axioms of a theory 12, 37
α-categorical 2, 482, 507
α-complete filter 227
α-enlargement 283
α-good ultrafilter 386
α-homogeneous model 297
α-isomorphism property 305
α-like model 557
α-omit 87
α-realize 87
α-regular filter 248
α-saturated model 256, 283, 292
α-saturated over $V(X)$ 283
α-standard model 529
α-universal model 255, 297
α-universal with respect to K 297

back and forth condition 114
back and forth construction 98
Back and Forth Lemma 298
base set 263
basic elementary class 220
basic formula 50
basic Horn formula 407
Bernays set theory 594
Bernays–Morse set theory 595
Beth's Theorem 90
bit of a tuple in a Boolean algebra 379
Boolean algebra 38
Boolean combination 50
bound variable 22
bounded elementary chain theorem 287

bounded elementary embedding 266
bounded elementary submodel 266
bounded limit ultrapower 457
bounded quantifier 265
bounded quantifier formula 265, 562
bounded ultrapower 269
branch of a tree 244
built-in Skolem functions 165
built up from constants 66

Cantor diagonal method 592
cardinal 586
cardinal exponent 586
cardinal of a language 19
cardinal of a model 21
cardinal power 586
cardinality of a set 586
Cartesian product 579, 580
categorical in power α 2, 482, 507
Cauchy sequence 351
CH 586
chain 580
chain, elementary 140
chain of models 140
Chang's Conjecture 156, 534
Chang's Two Cardinal Theorem 520
Chang–Makkai Theorem 333
characteristic of a field 41
Choice Axiom 593
Church's Thesis 110
classification theory 502
closed set of ordinals 246
closed theory 12, 36
closed unbounded set 246
closed under directed unions 146
closed under elementary equivalence 220
closed under reduced roots 446
closed under roots 354
closed under ultraproducts 220
closed under union of well ordered chains 146
Closure Property 128
cofinal 589
cofinality 589
collection scheme 593
commutative ring 40
compact class of models 446
Compactness Theorem 11, 33
companion operator 205
completable formula 97
complete arithmetic 42
complete, α-(filter) 227
complete embedding 448
complete extension 448
complete formula 97
complete number theory 42
complete, ω- (theory) 82
complete theory 17, 37
Completeness Theorem 8, 32
Completeness Theorem (Extended) 11, 33
Completeness Theorem for ω-logic 82
completion of a model 448
completion, valued field 351
comprehension scheme 595
Comprehensiveness Theorem 284
concatenation 584
concurrent relation 290
condition, forcing 209
conditional sentence 14
congruence relation 64
connected partial ordering 580
consequence 11, 17, 32, 50
consistent 9, 25
consistent with T 79
constant 20
constant symbol 18
Constructibility, Axiom of 238
constructible set 559
constructible universe 559
continuum hypothesis 586
core model 253
corresponding relation 20
cotheory 192
countably compact logic 130
countably homogeneous model 113
countably incomplete filter 226
countably prime model 99
countably saturated model 100
countably universal model 103
cover of a model 404
Craig Interpolant 87
Craig Interpolation Theorem 87
cross section axiom 351

De Morgan laws 38
deducible 9, 25
deduction 9
Deduction Theorem 10, 25
definable element 35, 185
definable function 461
definable subset 259, 559
define explicitly 90, 329
define implicitly 90
definitional extension 185

degree of a field extension 348
degree of a formula 496, 497
dense order 38
descendingly α-complete ultrafilter 258
Detachment, Rule of 9
determined formula 421
diagonal intersection 246
diagram 68
direct factor 321
direct limit 322
direct product 224, 321
direct product of formulas 428
direct system 322
directed relation 146
discrete simple ordering 182
distributive laws 38, 40
divisible torsion-free Abelian group 40
Downward Löwenheim–Skolem–Tarski Theorem 66
Δ_0 formula 110

effective process 423
elementarily equivalent 32
elementarily equivalent, abstract 129
elementary chain 140
Elementary Chain Theorem 140
elementary class 199, 220
elementary diagram 137
elementary embedding 84
Elementary Embedding Theorem 172
elementary extension 84, 136
elementary submodel 84, 136
elimination of quantifiers 50, 202
end extension 85, 163
enlargement 283
enumeration 584
Enumeration Principle 585
equivalence relation 59, 580
equivalent 17
equivalent formulas 50
equivalent logics 129
equivalent over Y 174
Erdös cardinal 555
Erdös notation 554
Erdös–Rado Theorem 510
Ershov's Theorem 440
essentially existential 125, 322
essentially finite direct power 431
existence of nonstandard universes 268
Existence Theorem for Countably Saturated Models 102
Existence Theorem for Recursively Saturated Models 112
Existence Theorem for Special Models 296
existential sentence 34
existential-positive formula 319
existentially closed 192
expansion of a language 19
expansion of a model 20
Expansion Property 128
Expansion Theorem 216, 279
Extended Completeness Theorem 11, 33
Extended Omitting Types Theorem 84
extension of a model 22
extension of a theory 37
external set 276

false in a model 7, 32
field 41
field of characteristic zero 41
field of prime characteristic 41
field of sets 39
filter 46, 211
filter generated by E 212
finite cover property 404
finite direct power sentence 412
finite direct product 406
finite direct product of formulas 428
finite forcing companion 210
finite forcing relation 209
finite intersection property 46, 166, 212
Finite Iteration Theorem 463
finite ordinal 582
finite spectrum problem 597
finitely axiomatizable 12, 37
finitely generated model 73
finitely generic model 210
finitely generic set 209
finitely satisfiable 11
first-order predicate language 18
first-order theory 34
forking 505
formal power series 352
formula 22
Frayne's Theorem 256
Fréchet filter 211
free variable 22
fully compact logic 129
function 20, 579
function symbol 18
Fundamental Theorem of Ultraproducts 217

gap n conjecture 524
GCH 586

generalized continuum hypothesis 586
generalized ω-logic 83
generated from a set of indiscernibles 171
germ of a meromorphic function 371
Gödel Completeness Theorem 33
good ultrafilter 257, 386
group 40

Hanf number 134
Hausdorff Maximal Principle 585
Hensel field 353
Hensel's lemma 353
henselization 354
hereditary set 340
homogeneous model 113, 297
homomorphic 70
homomorphic image 71
homomorphically embedded 71
homomorphism 71
Horn formula 407
huge cardinal 537
hull, Skolem 171
hypothesis of inaccessibility 590

ideal in a Boolean algebra 373
Idealization Scheme 285
idempotent laws 38
identity axioms 25
identity formula 24
identity symbol 22
implicitly define 90
improper filter 211
inaccessible cardinal 590
incompletable formula 97
inconsistent 9, 25
increasing sentence 13
independent set of sentences 18
independent unary relations 179
indescribable cardinal 247
indiscernible 169
individual relative to $V(X)$ 263
induction scheme 42
inductive order, theory of 382
infimum of a set of ordinals 582
infinite forcing companion 209
infinite forcing relation 209
infinite order in a group 78
infinitely generic 209
infinity axiom 592
initial ordinal 586
initial segment 583
inner model 537

inseparable theories 88, 93
integral domain 41
Internal Definition Principle 277
internal formula 285
internal set 276
Internal Set Theory 285
Internalization Scheme 285
Interpolation Theorem 17
interpretation function 20
invariants of a Boolean algebra 375
inverse limit 322
inverse system 322
irreducible formula 496, 498
isomorphic chains 580
isomorphic embedding 22
isomorphic models 21
isomorphism 21
isomorphism of well orderings 583
isomorphism, partial 114
Isomorphism Property 128
Isomorphism Theorem for Ultrapowers 398
IST 285
iterated ultrapower 467, 468

joint embedding property 195
Jónsson algebra 554
Jónsson cardinal 538
Jónsson model 534

Keisler Sandwich Theorem 311
Keisler–Rabin Theorem 450
Kueker Finite Definability Theorem 330
Kueker's conjecture 506
Kurepa family 527
Kurepa Hypothesis 527
Kurepa tree 259

language 18
limit cardinal 586
limit ordinal 582
limit ultrapower 451, 452
limited formula 265, 562
Lindenbaum algebra 47, 74
Lindenbaum's Theorem 10, 26
Lindström's Characterization 132
Lindström's Second Theorem 135
linear order 38
Local Ultrapower Theorem 281
locally omit 80
locally realize 79
logic, ω 82
logical symbols 22

Łoś–Vaught Test 139
Löwenheim number 129
Löwenheim–Skolem–Tarski Theorem 67
Lyndon Interpolation Theorem 92

Mahlo number 246
Mahlo's operation 247
main gap 507
maximal consistent 9, 25
measurable cardinal 229
medium compactness 245
method of diagrams 68
minimal nonstandard universe 291
model companion 197
model complete 186
model completion 201
model, first-order logic 20
model of a sentence 32
model pair 115
model, sentential logic 4, 10
Modus Ponens 9, 25
monotonic function 386
Morley Categoricity Theorem 494
Morley Omitting Types Theorem 512
Morley rank 497, 498
Mostowski collapse 272
Mostowski Collapsing Theorem 271

narrowing a gap 254
n-ary operation 580
n-ary relation 579
natural embedding 221
natural model of set theory 45
natural number 582
neat elementary extension 156
negation normal form (nnf) 93
negative formula 313
negative occurrence 92
nonprincipal filter 166
nonstandard analysis 70
nonstandard element, Peano arithmetic 78
nonstandard model of number theory 42
nonstandard universe 266
normal ultrafilter 239
n-tuple 579
null set axiom 592
number theory 42

O sharp 566
O sharp exists 566
occur negatively 92
occur positively 92
Occurrence Property 128
omit a set of formulas 78
Omitting Types Theorem 80
onto function 580
open formula 51
operation over X 580
order of a group element 40
order of a tree 244
order type 583
order type of a model of set theory 238
ordered field 41
ordered pair 579
ordered tuple 579
ordermorphic embedding 169
ordermorphism 169
ordinal 580
ordinal sum 584
outer extension 125, 322
Overspill Principle 289
ω-categorical theory 104
ω-complete 82
ω-consistent 81
ω-homogeneous 113
ω-logic 82
ω-model 81, 515
ω-rule 82
ω-saturated 100

p-adic numbers 366
p-adic valuation 366
pair, ordered 579
pairs axiom 592
partial isomorphism 114
partial order, theory of 37
partial ordering 580
Peano arithmetic 42
pin down 134
positive diagram 70
positive occurrence 82
positive sentence 13, 72
power of a language 19
power of a model 21
power of a set 586
power set 588
power set axiom 592
preservation theorem 147
preserved by reduced factors 419
preserved under finite intersections 14
preserved under homomorphisms 147
preserved under intersections 15
preserved under submodels 147
preserved under unions of chains 147

Pressburger arithmetic 43
prime model 99
principal filter 46
principal ideal in a Boolean algebra 374
product of ultrafilters 463
proof 25
proper filter 212
pseudo-elementary class 224

Quantifier Axioms 24
Quantifier Property 128
quotient Boolean algebra 374

r.e. 110
Ramsey cardinal 539
Ramsey's Theorem 168
rank function 588
rank, Morley 497, 498
rank of a set 588
rank 1 valued field 371
real closed field 41
real closure 344
realize a set of formulas 77
Realizing and Omitting Types Theorem 172
recursive language 110
recursive set 110
recursively enumerable 110
recursively saturated model 111
reduced power 215
reduced power sentence 412
reduced product 215, 216
reduced product of formulas 428
reduced product sentence 412
reduct of a model 20
reduction of a language 19
reflexive property 37
reflexive relation 580
regular cardinal 589
regular filter 248
regularity axiom 592
relation 20, 579
relation symbol 118
relative algebraic closure 354
relatively compact class of models 260
relativization of a formula 125, 321
Relativization Property 129
relativization to V 594
Renaming Property 128
replacement scheme 593, 595
Representation Theorem for Boolean Algebras 39
Representation Theorem for Nonstandard Universes 457
residue class field 352
Restricted Omitting Types Theorem 94
restriction 580
restriction of a closed theory 37
rigid model 598
ring 40
Robinson Consistency Theorem 91, 119
Robinsonian analysis 262
Rowbottom cardinal 546
Rowbottom's Theorem 545
Rule of Detachment 25
Rule of Generalization 25
rule of inference 25

satisfaction relation, abstract 128
satisfiable 10, 16, 32
satisfy a formula 29
satisfy a set of formulas 77
saturated model 100, 292
saturated over $V(X)$ 283
saturated relative to A 511
Scott's Theorem 238
selective filter 481
selective ultrafilter 262
self-determining set of formulas 425
semantically equivalent 17
semantics 3
semigroup 558
sentence, first-order logic 24
sentence, sentential logic 5
sentenial axioms 24
sequence 584
set relative to $V(X)$ 263
set-theoretic arithmetic 594
Silver's Theorem 547
simple expansion 19
simple model 320
simple order 38
simple ordering 580
singular cardinal 589
Skolem expansion 164
Skolem function 164
Skolem hull 165
Skolem Normal Form Theorem 175
Skolem theory 164
special Aronsajn tree 525
special Horn sentence 419
special model 295
specializing chain 295
spectrum problem 505

stability theory 502
stable in power α 485
stable theory 485
stand representation 172
standard model of number theory 42
standard set 276
Standardization Scheme 285
star of A 267
Stretching Theorem 172
strict basic Horn formula 418
strict Horn formula 418
strict well ordering 580
strong homomorphism 321
strong limit cardinal 586
strongly compact cardinal 245
subdirect product of models 419
subformula 23
submodel 21
submodel complete theory 302
submodel generated by X 34
Subset Theorem 172
subsets scheme 592
substitution 22
subtheory 37
successor cardinal 586
successor ordinal 582
sum, ordinal 584
supercompact cardinal 526
superstable theory 505
superstructure 263
supremum of a set of ordinals 582
Svenonius' Theorem 329
symmetric relation 580
syntax 3
Σ_1-formula 110

Tarski's Theorem 346
tautology 8, 24
term 22
theorem 25
theory 12, 36
theory of a model 37
theory of a type 100
three-cardinal problem 598
torsion group 78
torsion-free group 40
totally indiscernible set 495
totally transcendental theory 500
transcendence basis 181
transcendence rank of a field extension 181
transcendence rank of a formula 497
transcendence rank of a model 497
transcendence rank zero 496
transcendental set in a field extension 348
Transfer Scheme 285
Transfinite Induction 582
transitive model 45
transitive property 37
transitive realization 49
transitive relation 580
transitive submodel 125, 270
tree 244
tree property 244
trivial Boolean algebra 369
trivial filter 211
true in a model 7, 32
truncation 271
truth definition 1
truth table 8
tuple 579
two-cardinal model 533
type 78
type of a model 100
type of a tuple 100

ultrafilter 46, 166, 213
Ultrafilter Theorem 214
ultrapower chain 260
ultrapower extension 260, 448
ultraproduct 215
unbounded set 246
uniform filter 252
union of a chain of models 140
union of a set of models 145, 321
unions axiom 592
Uniqueness Theorem for Countably Saturated Models 103
Uniqueness Theorem for Special Models 300
universal model 103, 297
universal sentence 34
universal-existential sentence 146
universe 20
universe class axiom 595
unstable theory 505
Upward Löwenheim–Skolem–Tarski Theorem 67
Upward Morley Theorem 488

valid sentence 7, 32
valuation ring 352
value group 351
value of a sentence 8
value of a term 27
valued field 351

valued subfield 354
Vaught's conjecture 506, 597

Weak Compactness Theorem 232
weakly compact cardinal 233
weakly forces 209
weakly homogeneous model 303
weakly Horn formula 418
well-founded partial order 78
well-founded relation 237
well-ordered chain 580
well-ordered model 147, 535
well-ordering 580
Well-Ordering Principle 585
witness 61

Zermelo set theory 593
Zermelo–Fraenkel set theory 592
zero sharp 566
zero sharp exists 566
ZF 592
ZF – P 594
ZFC 593
Z-group 370
Zorn's Lemma 585
Z-valued field 371

INDEX OF SYMBOLS

$A \models \varphi$	7
$\models \varphi$	7, 32
$\vdash \varphi$	8, 25
$\Sigma \models \varphi$	11, 32, 50
$\|\mathscr{L}\|$	19
$\mathfrak{A} \cong \mathfrak{B}$	21
$\mathfrak{A} \subset \mathfrak{B}$	22
$t(v_0 \ldots v_n)$	24
$\varphi(v_0 \ldots v_n)$	24
$\equiv$	25
$\Sigma \vdash \varphi$	25
$t[x_0, \ldots, x_q]$	28
$\mathfrak{A} \models \varphi[x_0, \ldots, x_q]$	28
$\mathfrak{A} \models \varphi$	29
$\mathfrak{A} \equiv \mathfrak{B}$	32
$\mathfrak{A} \models \Sigma$	32
$\Sigma \models \varphi$	32, 50
$\mathfrak{A} \overset{\sim}{\subset} \mathfrak{B}$	33
$S_\omega(X)$	36
$T\|\mathscr{L}$	36
ZF	44
$R(\alpha)$	45, 588
$\mathscr{L}_A$	68
$\mathfrak{A}_X$	68
$\Delta_{\mathfrak{A}}$	68
$\mathfrak{A} \simeq \mathfrak{B}$	70
$\Sigma(x_1 \ldots x_n)$	77
$\mathfrak{A} \models \Sigma(x_1 \ldots x_n)$	77
$\mathfrak{A} \prec \mathfrak{B}$	83, 136
$f : \mathfrak{A} \prec \mathfrak{B}$	99
$\Sigma(c_1 \ldots c_n)$	101
Δ_0 formula	110
Σ_1 formula	110
$\mathfrak{A} \cong_p \mathfrak{B}$	114
$(\mathfrak{A}, \mathfrak{B})$	115
$\mathscr{L}^{\mathfrak{A}}$	115
$c^{\mathfrak{A}}$	115
$R^{\mathfrak{A}}$	115
$\varphi^{\mathfrak{A}}$	116
φ^U	125, 314
$(l, \models_l)$	128
$\mathscr{L}_\varphi$	128
$\models_l$	128
$\rho\mathscr{L}$	128
$\rho\mathfrak{A}$	128
$l_{\omega_1,\omega}$	129
Th$(\mathfrak{A})$	137
$\mathfrak{A} \precsim \mathfrak{B}$	137
$\bigcup_{\beta<\alpha} \mathfrak{A}_\beta$	139
Σ_n^0, Π_n^0	143
$D(\mathfrak{A})$	150
$\aleph_n(\alpha)$	154
$\beth_n(\alpha)$	154
$H(X)$	165

$\mathfrak{H}(X)$	165	$\prod_D \mathbf{1}$	417
$[X]^n$	168	$\mathscr{L}^{\#}$	448
ACF	188	$\mathfrak{A}^{\#}$	448
$T_{\forall}$	192	$\mathscr{L}_{\alpha\omega}$	530
$T_{\forall\exists}$	192	$\alpha \rightarrow (\beta)^n_\gamma$	554
$\mathfrak{A} \Vdash \varphi$	209	$\kappa(\beta)$	555
$p \Vdash \varphi$	209	h_{wo}	557
$p \Vdash^w \varphi$	209	ρ_n	558
$S(I)$	211, 588	$L(\alpha)$	559
$f \equiv_D g$	214	L	559
$\prod_D \mathfrak{A}_i$	215	$\varphi^{(y)}$	561
Σ^m_n, Π^m_n	221	$O^{\#}$	566
$\mathscr{L}_\alpha$	228	$\langle x, y \rangle$	579
$H(\alpha)$	235	$\langle x_1, \ldots, x_n \rangle$	579
$V_n(X)$	263	$f \mid Y$	580
$V(X)$	263	$\prod X_i$	580
$\prec_b$	266	${}^X Y$	580
$\langle V(X), V(Y), * \rangle$	266	CH	586
$^*A$	269	GCH	586
BASE(x)	270	$\aleph_\xi$	586
$^\sigma A$	273	$\beth_\xi$	586
$\mathscr{L}_{V(X)}$	282	ω_ξ	586
$\mathscr{L}_{\mathrm{INT}}$	282	ω	586
IST	285	α^+	586
St	285	$S_\alpha(X)$	588
$\forall^{\mathrm{St}} x$	285	$S^\alpha(X)$	588
$\forall^{\mathrm{StFin}} x$	285	$S(X)$	588
$\exists^{\mathrm{St}}$	285	cf(α)	589
η_α	342	ZF	592
$D \times E$	406	ZFC	593
ZFL	415	ZF − P	594